Antimicrobial Resistance in the 21st Century

Understanding and Combating the Rising Threat

Antimicrobial Resistance in the 21st Century

Understanding and Combating the Rising Threat

Edited by

Dipayan Das
Department of Microbiology, Royal School of Biosciences,
The Assam Royal Global University, Guwahati, Assam, India

Birson Ingti
Department of Microbiology, Royal School of Biosciences,
The Assam Royal Global University, Guwahati, Assam, India

Amitabha Bhattacharjee
Department of Microbiology, Assam University, Silchar, Assam, India

Payel Paul
Department of Botany, North Bengal St. Xavier's College,
Jalpaiguri, West Bengal, India

ELSEVIER

ACADEMIC PRESS
An imprint of Elsevier

Academic Press is an imprint of Elsevier
125 London Wall, London EC2Y 5AS, United Kingdom
50 Hampshire Street, 5th Floor, Cambridge, MA 02139, United States

ISBN: 978-0-443-44748-8

For Information on all Academic Press publications
visit our website at https://www.elsevier.com/books-and-journals

Publisher: Megan Ball
Acquisitions Editor: Elizabeth Brown
Editorial Project Manager: Vikrant Dixit
Production Project Manager: R. Aishwarya
Cover Designer: Gopalakrishnan Venkatraman

Typeset by MPS Limited, Chennai, India

Contents

List of contributors

Noor Afsha Department of Biochemistry, Oriental Institute of Science and Technology, Bardhaman, West Bengal, India

Iqbal Ahmad Department of Agricultural Microbiology, Aligarh Muslim University, Aligarh, Uttar Pradesh, India

Ramzan Ahmed Department of Transfusion Medicine and Clinical Microbiology, Faculty of Allied Health Sciences, Chulalongkorn University, Bangkok, Thailand

Tuhina Banerjee Department of Microbiology, Institute of Medical Sciences, Banaras Hindu University, Varanasi, Uttar Pradesh, India

Pakiza Begum Department of Chemistry, Gargoan College, Gargaon, Simaluguri, Sivasagar, Assam, India

Amitabha Bhattacharjee Department of Microbiology, Assam University, Silchar, Assam, India

Deep Bhowmik Department of Microbiology, Assam University, Silchar, Assam, India

Deepshikha Bhowmik MRU, ESIC Medical College and Hospital, Faridabad, Haryana, India

Nuntaree Chaichanawongsaroj Department of Transfusion Medicine and Clinical Microbiology, Faculty of Allied Health Sciences, Chulalongkorn University, Bangkok, Thailand

Ranadhir Chakraborty OMICS Laboratory, Department of Biotechnology, University of North Bengal, Siliguri, West Bengal, India

Pankaj Chetia Molecular Plant Taxonomy and Bioinformatics Research Laboratory, Department of Life Sciences, Dibrugarh University, Dibrugarh, Assam, India

Shiela Chetri ICMR National Institute of Epidemiology, Chennai, Tamil Nadu, India

Vishek Choudhary CSIR-Institute of Himalayan Bioresource Technology, Palampur, Himachal Pradesh, India; Academy of Scientific and Innovative Research (AcSIR), Ghaziabad, Uttar Pradesh, India

Manabendra Dutta Choudhury Rabindranath Tagore University, Hojai, Assam, India

Dipayan Das Department of Microbiology, Royal School of Biosciences, The Assam Royal Global University, Guwahati, Assam, India

Nivedita Dasgupta Multidisciplinary Research Unit, Silchar Medical College and Hospital, Silchar, Assam, India

Sourish Dasgupta Department of Botany, North Bengal St. Xavier's College, Jalpaiguri, West Bengal, India

Bornali Sarmah Dutta Department of Microbiology, Gauhati Medical College & Hospital, Guwahati, Assam, India

Rajkumari Elizabeth Advanced Level Institutional Biotech Hub, Assam University, Silchar, Assam, India

Kinsley Erhons Enerijiofi Department of Biological Sciences, Glorious Vision University, Ogwa, Edo State, Nigeria

Indrani Gogoi Molecular Plant Taxonomy and Bioinformatics Research Laboratory, Department of Life Sciences, Dibrugarh University, Dibrugarh, Assam, India; Department of Botany, Dergaon Kamal Dowerah College (Autonomous), Golaghat, Assam, India

Monalisha Hazarika DBT-APSCS&T Centre of Excellence for Bio-resources and Sustainable Development, Kimin, Arunachal Pradesh, India

Suranjana Chaliha Hazarika Department of Microbiology, Gauhati Medical College & Hospital, Guwahati, Assam, India

Birson Ingti Department of Microbiology, Royal School of Biosciences, The Assam Royal Global University, Guwahati, Assam, India

Vijoy Sankar Kairi Department of Pharmacology, Silchar Medical College and Hospital, Silchar, Assam, India

Sameer Uttamrao Khasbage Department of Pharmacology, All India Institute of Medical Sciences (AIIMS), Raipur, Chhattisgarh, India

Pratik Kolhe Regional Ayurveda Research Institute, Gwalior, Madhya Pradesh, India

Sapna A. Kondalkar Regional Ayurveda Research Institute, Gwalior, Madhya Pradesh, India

Rhitu Kotoky Department of Biology, University of Southern Denmark, Odense, Denmark

Pardeep Kumar Department of Botany, Mata Gujri College, Fatehgarh Sahib, Punjab, India

Sunil Kumar Department of Mechanical Engineering, Amrita Vishwa Vidyapeetham, Chennai Campus, Chennai, Tamil Nadu, India

Pimily Langthasa Department of Zoology, Gargaon College, Gargaon, Simaluguri, Sivasagar, Assam, India

Anand Prakash Maurya Department of Microbiology, Assam University Silchar, Silchar, Assam, India

Tapan Kumar Mohanta Department of Genetics Research, Institute for Medical Research and Consultation, Imam Abdulrahman Bin Faisal University, Dammam, Saudi Arabia

Anupama Moirangthem Department of Microbiology, Royal School of Biosciences, The Assam Royal Global University, Guwahati, Assam, India; School of Health Science and Technology, Medhavi Skills University, Singtam, Sikkim, India

Lekshmi Narendrakumar Centre for Microbial Research, BRIC-Translational Health Science and Technology Institute (THSTI), Faridabad, Haryana, India

Deepshikha Nath Department of Biochemistry, Ruhr University Bochum, Bochum, Germany

Soumitra Nath Department of Biotechnology, Gurucharan University, Silchar, Assam, India

Itishree Pandey Department of Biochemistry, Raipur Institute of Medical Sciences, Raipur, Chhattisgarh, India

Saraswati Patel Saveetha College of Pharmacy, Chennai, Tamil Nadu, India

Kaustuvmani Patowary Department of Microbiology, Faculty of Science, Assam down town University, Guwahati, Assam, India

Anukul Paul OMICS Laboratory, Department of Biotechnology, University of North Bengal, Siliguri, West Bengal, India

Biswajit Paul Central Ayurveda Research Institute, Guwahati, Assam, India

Deepjyoti Paul Centre for Microbial Research, BRIC-Translational Health Science and Technology Institute (THSTI), Faridabad, Haryana, India

Payel Paul Department of Botany, North Bengal St. Xavier's College, Jalpaiguri, West Bengal, India

Gargee Dhar Purkayastha Sri Sri Academy, Siliguri, West Bengal, India

Sreeparna Purkayastha Genes N Life Healthcare Private Limited, Hyderabad, Telangana, India

Pue Rakshit Department of Microbiology, Institute of Medical Sciences, Banaras Hindu University, Varanasi, Uttar Pradesh, India

Dolly Roy Department of Pharmacology, Silchar Medical College and Hospital, Silchar, Assam, India

Ramkrishna Roy Department of Microbiology, Royal School of Biosciences, The Assam Royal Global University, Guwahati, Assam, India; Department of Microbiology, Suri Vidyasagar College, Birbhum, West Bengal, India

Himadri Saikia Pragjyotish College, Guwahati, Assam, India

Shyamalima Saikia Molecular Plant Taxonomy and Bioinformatics Research Laboratory, Department of Life Sciences, Dibrugarh University, Dibrugarh, Assam, India

Samreen Department of Agricultural Microbiology, Aligarh Muslim University, Aligarh, Uttar Pradesh, India

Pratima Sarkar Department of Microbiology, University of North Bengal, Siliguri, West Bengal, India

Saurav Sarkar Department of Pharmaceutical Technology, University of North Bengal, Siliguri, West Bengal, India

Maneesh Paul Satyaseela Lead Co-Inventor Enmetazobactam, College of Pharmaceutical Sciences, Dayananda Sagar University, Bengaluru South, Karnataka, India

Suhita Senapati Accumedicare Laboratories Private Limited, Hyderabad, Telangana, India

Swarnab Sengupta Department of Biomedical Sciences, College of Medicine, Florida State University, Tallahassee, FL, United States

Ajanta Sharma Department of Microbiology, Gauhati Medical College & Hospital, Guwahati, Assam, India

Alok Sharma Regional Ayurveda Research Institute, Gwalior, Madhya Pradesh, India

Varsha Shridhar Deparment of Microbiology and Molecular Diagnostics, Molecular Solutions Care Health LLP, Bengaluru, Karnataka, India

Laishram Shantikumar Singh Department of Microbiology, Faculty of Science, Assam down town University, Guwahati, Assam, India

H. Ashalata Singha University of Science and Technology, Baridua, Meghalaya, India

Brijesh Singh Sisodia Regional Ayurveda Research Institute, Gwalior, Madhya Pradesh, India

Diksha Tarafder Department of Microbiology, North Bengal Medical College and Hospital, Siliguri, West Bengal, India

Divya Varshney Department of Agricultural Microbiology, Aligarh Muslim University, Aligarh, Uttar Pradesh, India

Jayalaxmi Wangkheimayum Department of Microbiology, Assam University, Silchar, Assam, India

About the editors

Dr. Dipayan Das, PhD, serves as an assistant professor in the Department of Microbiology and as an assistant dean (academics) at the Royal School of Biosciences, The Assam Royal Global University. He has over 5 years of teaching experience at undergraduate and postgraduate levels, complemented by 8 years of research experience. Prior to his current position, he was associated with the Centre for Innovative Studies at the University of North Bengal. Dr. Das completed his doctoral research under the supervision of Dr. Palash Mandal in the Department of Botany at the University of North Bengal. His research focuses on the application of nanoparticles for crop improvement, with broader interests spanning nanobiology, microbiology, plant physiology and biochemistry. He is a life member of the Indian Botanical Society and the Microbiological Society of India. Dr. Das has authored over 50 research papers in international journals and 5 academic books. His academic achievements include receiving the CSIR-UGC-NET Fellowship, as well as the InSc Young Scientist Award and InSc Research Excellence Award. He secured first rank (gold medal) in MSc Botany in 2014 and second rank (silver medal) in BSc Botany in 2012 from the University of North Bengal. He has qualified the CSIR-UGC NET five times, along with clearing the West Bengal SET and GATE once each. He was also selected for the DST-INSPIRE Doctoral Fellowship. In addition to his academic and research contributions, Dr. Das has five years of experience mentoring aspirants for the CSIR-UGC NET examination, with many of his students currently pursuing doctoral research across leading institutions in India or serving as faculty members in universities and colleges.

Dr. Birson Ingti, PhD, serves as an assistant professor and coordinator in the Department of Microbiology at The Assam Royal Global University. Dr. Ingti possesses over 11 years of research experience along with 6 years of teaching experience. His research specialisation includes antimicrobial resistance, plasmid-mediated β-lactamase genes and microbial genomics. Dr. Ingti has contributed more than 22 research papers to peer-reviewed journals and authored several book chapters. Dr. Ingti has supervised one PhD scholar and is a recipient of the European Society of Clinical Microbiology and Infectious Diseases (ESCMID) Grant. He has also qualified ICAR-NET. His current research is centred on the inhibition of β-lactamases using medicinal plant extracts.

Dr. Amitabha Bhattacharjee, PhD, serves as a professor in the Department of Microbiology at Assam University, Silchar. He brings over 18 years of teaching experience and more than two decades of research expertise, with a primary focus on antimicrobial resistance, clinical microbiology and molecular diagnostics. He earned his doctoral degree from Banaras Hindu University and has undertaken advanced postdoctoral training in the United Kingdom and France through prestigious DBT and ICMR fellowships. Dr. Bhattacharjee has an extensive publication record, with over 135 peer-reviewed research articles and several book chapters to his credit. He is also the holder of four Indian patents related to antibiotic resistance detection. Throughout his career, he has successfully coordinated 14 funded research projects at both national and international levels, supported by agencies such as DBT, CSIR and DST. His contributions have been recognised through several notable awards, including the ICMR International Fellowship, the DBT-CREST Award and the Asia Pacific Young Investigator Award. In addition to serving as a reviewer for prominent scientific journals, he has actively presented his research at international forums. His ongoing work is centred on identifying novel diagnostic markers for antimicrobial resistance and elucidating the genomic basis of resistance in bacterial pathogens.

Dr. Payel Paul, PhD, is currently serving as an assistant professor in the Department of Botany at North Bengal St. Xavier's College. She has 5 years of teaching experience at the undergraduate level and over 10 years of research experience. Dr. Paul was previously associated with the Taxonomy of Angiosperms and Biosystematics Laboratory, Department of Botany, University of North Bengal, where she worked as a senior research scholar. Her research expertise lies in plant systematics, palynology and molecular systematics. She has a strong research profile, with publications in reputed national and international journals, and has authored 30 research articles and 5 books. Dr. Paul secured First Class

Second Position (Silver Medal) in Master of Science in Botany from the University of North Bengal in 2014. She was awarded the Post-Graduate Indira Gandhi Scholarship during 2012–14 and received the UGC-BSR Fellowship as junior research fellow (JRF) and senior research fellow (SRF) from 2015 to 2020. In addition, she has experience in conducting botany practical classes and examinations for IGNOU and has served as an academic counsellor in life sciences at IGNOU Learning Support Centre 45015 since 2021. Dr. Paul received the Young Scientist Award at the International Conference on Advances in Plant, Microbes and Agricultural Sciences held at the University of North Bengal. She was also honoured with the Research Excellence Award in 2025 by the North Bengal St. Xavier's College.

Preface

Antimicrobial resistance (AMR) has emerged as one of the most formidable challenges to global public health in the 21st century. Once hailed as miracle drugs, antibiotics are increasingly losing their efficacy, rendering common infections difficult, and sometimes impossible, to treat. The accelerating pace of resistance threatens to undermine decades of medical progress, affecting not only clinical outcomes but also food security, economic stability and sustainable development worldwide. This book, *Antimicrobial Resistance in the 21st Century: Understanding and Combating the Rising Threat*, is conceived as a comprehensive and interdisciplinary effort to understand this crisis in depth.

This volume brings together contributions that span fundamental science, clinical perspectives, environmental insights and policy frameworks. It begins by establishing a strong conceptual foundation, outlining the mechanisms, challenges and potential solutions associated with antibiotic resistance. Subsequent chapters delve into the genetic, biochemical and evolutionary dynamics that enable microorganisms to evade antimicrobial action, indicating the remarkable adaptability of bacterial systems under selective pressure.

Recognising that AMR is not confined to laboratories or hospitals, this book expands its scope to examine real-world drivers, including antibiotic misuse, environmental reservoirs and agricultural practices. Dedicated sections address the complexities of resistance in the healthcare system and communities, as well as the diagnostic challenges that hinder timely and effective intervention. These discussions are particularly relevant in an era where rapid, accurate detection is critical for containment and treatment.

Importantly, this book does not limit itself to describing the problem, it actively explores solutions. From next-generation antibiotics and alternative therapeutic strategies to vaccines, infection control and antimicrobial stewardship, the chapters collectively emphasise innovation and responsibility. The inclusion of policy-oriented discussions and global governance strategies reflects the urgent need for coordinated international action, particularly in low- and middle-income countries where the burden of AMR is often greatest.

A distinctive feature of this volume is its attention to sociocultural, behavioural and economic dimensions. Antibiotic use is deeply embedded in human practices, and revealing resistance requires not only scientific advancement but also changes in behaviour, awareness and regulatory frameworks. This book also indicates emerging strategies such as target-centric drug repurposing, offering promising avenues for accelerating therapeutic development.

This work is the result of collaborative efforts by researchers and academicians committed to advancing our understanding of AMR and identifying sustainable solutions. It is intended for microbiologists, healthcare professionals, researchers, policymakers and students who seek a nuanced and integrated perspective on antimicrobial resistance.

We extend our sincere gratitude to all contributors for their scholarly insights and dedication. We also acknowledge the institutions that supported this endeavour and provided an environment conducive to academic excellence.

As editors, we hope this book will serve as both a valuable reference and a catalyst for further research, dialogue and action in combating antimicrobial resistance. The challenge is immense, but with informed, collective effort, it is not insurmountable.

Editors
Dipayan Das
Birson Ingti
Amitabha Bhattacharjee
Payel Paul

Preface

Acknowledgements

The editors of *Antimicrobial Resistance in the 21st Century: Understanding and Combating the Rising Threat* wish to express their sincere gratitude to all those who contributed to the successful completion of this volume.

The editors extend their deepest appreciation to the chapter authors for their scholarly contributions, critical insights and dedication to advancing knowledge in the field of antimicrobial resistance. Their expertise has enabled the development of a comprehensive resource that spans fundamental mechanisms, clinical challenges, environmental dimensions and emerging solutions.

The editors are especially thankful to their respective institutions, The Assam Royal Global University, Guwahati, Assam, India; Assam University, Silchar, Assam, India; and North Bengal St. Xavier's College, Jalpaiguri, West Bengal, India, for providing the academic environment and support necessary for this work. The editors also acknowledge the invaluable contributions of peer reviewers and academic colleagues whose constructive feedback helped refine the content and maintain scientific rigour throughout this book.

The editors would like to express their sincere appreciation to the editorial project manager and the entire publishing team at Elsevier for their professional guidance, coordination and continuous support throughout the development of this book. We are particularly grateful to ELSA for their assistance in enhancing the clarity, consistency and overall quality of the manuscript.

We also recognise the broader scientific community working tirelessly to address antimicrobial resistance, a global challenge that demands collaborative and sustained efforts across disciplines, sectors and nations.

Finally, we express our sincere gratitude to our families and well-wishers for their encouragement, patience and unwavering support throughout this endeavour.

Editors

Dipayan Das

Birson Ingti

Amitabha Bhattacharjee

Payel Paul

Introduction

Antimicrobial resistance (AMR) is an urgent and complex issue for global health systems in the 21st century. AMR is the ability of microorganisms to resist the actions of antimicrobial agents, and it may undo decades of medical progress made possible by antibiotics during the past century (Salam et al., 2023). Antibiotics were once hailed as underpinning modern medicine, but today are losing effectiveness against a wide range of pathogens (germs), leading to longer periods of illness, increased death and higher healthcare costs. AMR has been termed one of the top three global health threats by WHO, and common infections and minor injuries could be fatal without effective interventions (Salam et al., 2023).

AMR is primarily an evolutionary process fuelled by genetic variation and natural selection. Even the tiniest microorganisms can adapt through spontaneous mutations or acquisition of resistance genes by horizontal gene transfer processes (transformation, transduction and conjugation). Such genetic modifications commonly lead to mechanisms that counteract the effects of antimicrobial agents, including enzyme hydrolysis (e.g. β-lactamases), target scope modification, efflux pump activation or diminished membrane permeability (Blair et al., 2015). Although these mechanisms are natural phenomena in nature, their accelerated proliferation is mostly a result of anthropogenic forces.

One of the major causes of AMR is the misuse and overuse of antibiotics in human medicine (Ahmed et al., 2024). Antibiotics are overprescribed in many areas (especially, e.g., to treat infections or without diagnosis confirmation). The situation is worsened by partial treatment courses and self-medication, as these enable partially resistant strains to survive and grow. In addition to clinical usage, extensive application of antibiotics in agriculture and animal husbandry has played a major role in the selection and dissemination of resistant bacteria. Antibiotics are used not just for therapeutic purposes but also as growth promoters or prophylaxis, creating reservoirs of resistance that can be transmitted to humans through the food chain and environment (Van Boeckel et al., 2015).

The environment also contributes to the spread of AMR. The effluents of hospitals, pharmaceutical industries and agricultural runoffs have residual antibiotics and resistant microorganisms, creating more resistance-promoting 'hotspots'. These environments promote the transmission of resistance genes between different microbiota, which also worsens the dissemination of AMR (Endale et al., 2023). The One Health approach is therefore important because it recognises that human health, animal health and environmental health are all closely linked, hence advocating integrated strategies to effectively promote public health since AMR cuts across multiple sectors (Robinson et al., 2016).

Healthcare settings, especially hospitals, are one of the critical battlefields against AMR. Multidrug-resistant organisms like *Staphylococcus aureus*, *Klebsiella pneumoniae* and *Pseudomonas aeruginosa* lead to significant complexity in managing nosocomial infections. The increased use of antibiotics in addition to invasive procedures and patients with impaired immune defences has created a unique environment for the emergence and dissemination of resistant pathogens (Piscitelli et al., 2025). At the same time, however, community-associated resistance is also on the rise in the form of these resistant strains becoming more common outside the hospital and making it that much more difficult to develop treatment regimens based on appropriate dictates for public health (Shinde et al., 2026).

Timely detection and accurate diagnosis are required for the effective management of AMR. Nevertheless, classical diagnostic means are time-consuming and could trigger the delay of suitable treatment. One exciting development in this area is the recent advances in molecular diagnostic tools, including polymerase chain reaction, next-generation sequencing (NGS) and combined biosensor technologies that can provide rapid detection of resistance determinants. These innovations are essential to direct precise treatment and limit the inappropriate use of broad-spectrum antibiotics (Köser et al., 2014).

Due to the AMR crisis, there is an increased need for new antibiotic development and alternative treatment strategies (Brüssow, 2024). Recent advances in drug discovery, particularly high-throughput screening, structure-based design and AI-driven approaches, have renewed interest in discovering new antibiotics (Sanapalli et al., 2026). Alternative therapeutic modalities, including bacteriophage therapy, antimicrobial peptides, immunomodulators and microbiome-modifications, are also at the forefront of research (Das et al., 2026; Rahimian et al., 2024). Another strategy that has been developed is the hit-centred drug repurposing, being able to take advantage of existing drugs already experimented in humans, which are safe against specific pathogens, so they can more effectively face resistant pathogens (Recino et al., 2025).

Preventive approaches also play an equally crucial role in dealing with AMR. Vaccination programmes lower the incidence of infectious diseases and hence lower the need for antibiotic treatment. Strategic infection control methods such as hand hygiene, sterilisation and surveillance systems are crucial in curtailing the dissemination of resistant organisms. Antimicrobial stewardship programmes, through evidence-based guidelines, education and monitoring of prescribing practices, are important for promoting the rational use of antibiotics (Dyar et al., 2017).

Strong policy frameworks and global governance are also needed to tackle AMR. Global effort and regulation require focused international collaboration to secure the right antimicrobials for use while avoiding their misuse. However, AMR is not unique to the more developed nations. Low- and Middle-Income Countries are also faced with challenges that differ from those in High-Income Countries, and they are lacking healthcare infrastructure, diagnostic capacity and regulatory oversight that collectively contribute towards AMR. The resistance dynamics are also shaped by sociocultural factors, including public perceptions and behavioural patterns of antibiotic use, which need to be taken into account in intervention strategies.

AMR has major economic and societal effects. Higher healthcare costs, loss of productivity and failure of essential surgical procedures could spell disaster for economies globally. The pharmaceutical industry has also been challenged, as the high cost and low reward of antibiotic development have led to a decline in new drug approvals. Novel incentive models, based on public–private partnerships, are thus urgently needed to spur research and development in this vital area.

This book offers a comprehensive, multidisciplinary overview. Integrative and cross-disciplinary health-focused reviews bring perspectives from microbiology, clinical medicine, environmental science, pharmacology and public health to facilitate an understanding of the issue. The chapters collectively examine the causes, drivers, impacts and possible solutions to AMR as a planetary threat that calls for sustained and coordinated global action.

AMR is more than a medical problem; it is a societal responsibility — and you are part of the solution! Solving this part requires a multipronged approach, including scientific research, motivation in the misuse of antibiotics, policy diversity and public awareness. This volume contributes to the global challenge of preserving the effectiveness of antimicrobial agents and thereby securing the future of effective healthcare by increasing our understanding of antimicrobial resistance and forming concrete solutions.

References

Ahmed, S.K., Hussein, S., Qurbani, K., Ibrahim, R.H., Fareeq, A., Mahmood, K.A., et al., 2024. Antimicrobial resistance: Impacts, challenges, and future prospects. Journal of Medicine, Surgery, and Public Health 2, 100081. Available at https://doi.org/10.1016/j.glmedi.2024.100081.

Blair, J.M.A., Webber, M.A., Baylay, A.J., Ogbolu, D.O., Piddock, L.J.V., 2015. Molecular mechanisms of antibiotic resistance. Nature Reviews Microbiology 13, 42–51. Available at https://doi.org/10.1038/nrmicro3380.

Van Boeckel, T.P., Brower, C., Gilbert, M., Grenfell, B.T., Levin, S.A., Robinson, T.P., et al., 2015. Global trends in antimicrobial use in food animals. Proceedings of the National Academy of Sciences of the United States of America 112, 5649–5654. Available at https://doi.org/10.1073/pnas.1503141112.

Brüssow, H., 2024. The antibiotic resistance crisis and the development of new antibiotics. Microbial Biotechnology 17. Available at https://doi.org/10.1111/1751-7915.14510 e14510.

Das, D., Khan, T., Kalita, J., Rustagi, S., Panda, S., Singh, N., et al., 2026. Antimicrobial peptides: emerging next-generation strategy for sustainable plant disease management. Frontiers in Antibiotics 5, 1–19. Available at https://doi.org/10.3389/frabi.2026.1766594.

Dyar, O.J., Huttner, B., Schouten, J., Pulcini, C., ESGAP (ESCMID Study Group for Antimicrobial Stewardship), 2017. What is antimicrobial stewardship? Clinical Microbiology and Infection: The Official Publication of the European Society of Clinical Microbiology and Infectious Diseases 23, 793–798. Available at https://doi.org/10.1016/j.cmi.2017.08.026.

Endale, H., Mathewos, M., Abdeta, D., 2023. Potential causes of spread of antimicrobial resistance and preventive measures in One Health perspective: a review. Infection and Drug Resistance 16, 7515. Available at https://doi.org/10.2147/IDR.S428837.

Köser, C.U., Ellington, M.J., Peacock, S.J., 2014. Whole-genome sequencing to control antimicrobial resistance. Trends in Genetics 30, 401–407. Available at https://doi.org/10.1016/j.tig.2014.07.003.

Piscitelli, P., Costigliola, V., Azamfirei, L., 2025. The challenge of antimicrobial resistance in intensive care setting. The Journal of Critical Care Medicine 11, 109. Available at https://doi.org/10.2478/jccm-2025-0020.

Rahimian, M., Jafari-Gharabaghlou, D., Mohammadi, E., Zarghami, N., 2024. A new insight into phage combination therapeutic approaches against drug-resistant mixed bacterial infections. Phage (New Rochelle) 5, 203–222. Available at https://doi.org/10.1089/phage.2024.0011.

Recino, A., Rayner, M.L.D., Rohn, J.L., Della Pasqua, O., Recino, A., Rayner, M.L.D., et al., 2025. Therapeutic innovation in drug repurposing: challenges and opportunities. Drug Discovery Today 30, 104390. Available at https://doi.org/10.1016/j.drudis.2025.104390.

Robinson, T.P., Bu, D.P., Carrique-Mas, J., Fèvre, E.M., Gilbert, M., Grace, D., et al., 2016. Antibiotic resistance is the quintessential One Health issue. Transactions of the Royal Society of Tropical Medicine and Hygiene 110, 377–380. Available at https://doi.org/10.1093/trstmh/trw048.

Salam, M.A., Al-Amin, M.Y., Salam, M.T., Pawar, J.S., Akhter, N., Rabaan, A.A., et al., 2023. Antimicrobial resistance: a growing serious threat for global public health. Healthcare (Basel) 11, 1946. Available at https://doi.org/10.3390/healthcare11131946.

Sanapalli, B.K.R., Palit, S., Deshpande, A., Tokala, R., Sigalapalli, D.K., Sanapalli, V., 2026. Artificial intelligence and the discovery of antibiotics: reinventing with opportunities, challenges and clinical translation. Antibiotics (Basel) 15, 233. Available at https://doi.org/10.3390/antibiotics15020233.

Shinde, A., Mohan, A., Bulusu, V.M., Soni, P., Singh, J., Khadanga, S., et al., 2026. Current status of antimicrobial resistance in Indian healthcare system: combating antimicrobial resistance with precision medicine. Frontiers in Antibiotics 5. Available at https://doi.org/10.3389/frabi.2026.1632790.

Chapter 1

Introduction to antibiotic resistance: mechanisms, challenges, and solutions

Deepjyoti Paul and Lekshmi Narendrakumar
Centre for Microbial Research, BRIC-Translational Health Science and Technology Institute (THSTI), Faridabad, Haryana, India

1.1 Introduction

Antimicrobial resistance (AMR) is recognised as a global problem that results in therapeutic failure and poses a huge threat to human health. Resistant pathogens are associated with higher rates of morbidity and mortality. In the year 2019, the World Health Organisation (WHO) declared AMR to be one of the top 10 global public health threats facing humanity and also forecasted an entry into a postantibiotic era within the current century (Murray et al., 2022). In fact, AMR is estimated to cost the European Union approximately 1.5 billion euros per year, with an associated 25,000 deaths (approximately 23,000 in the United States) each year, and if nothing is done, it is predicted that by 2050, AMR bacteria will account for more than 10 million deaths worldwide annually, with over 9 million deaths occurring in low- and middle-income countries (LMICs) (Dadgostar, 2019; Dhingra et al., 2020; O'Neill, 2014). This major challenge to society is mainly due to a variety of human actions, that is unnecessary prescriptions in healthcare, blanket prescriptions in agriculture, and careless release of antibiotics into the environment, enriching resistance factors in areas like hospitals, communities, wastewater treatment plants, and drinking water (Singer et al., 2016; Manyi-Loh et al., 2018; Durrance-Bagale et al., 2021). As a result, the AMR problem is set to have devastating effects on modern medicine, the food industry, and the environment if left unchecked. The emergence and spread of resistant bacterial pathogens is not only a threat to the health sector but also a challenge in infectious disease management (Baker, 2021; Paul et al., 2022; Lu et al., 2020; Cave et al., 2021). Infections caused by these multidrug-resistant (MDR) pathogens increase the burden of healthcare-associated and community-acquired diseases, undermining the efficacy of antibiotics, which have transformed medicine and saved millions of lives.

1.2 Historical perspective of antibiotics and antibiotic resistance

The introduction of antibiotics into clinical use is indeed one of the most fortunate medical breakthroughs in the history of medicine, alleviating much human misery. Apart from the treatment of many bacterial infections and achieving major advances in medicine, antibiotics are also a critical tool for cancer, surgery, and organ transplants. It was in the late 1890s, when pyocyanase, the first antimicrobial drug used in hospitals, was introduced by two German scientists, Rudolph Emmerich and Oscar Low (Hays et al., 1945). The story of antimicrobial drugs started with this drug, which was isolated from the Gram-negative bacterium *Pseudomonas aeruginosa*, formerly known as *Bacillus pyocyaneus* and found to be active against numerous pathogenic bacteria causing infections like cholera and typhus. However, due to the dubious effectiveness and toxic preparation, in due course of time, the drug was forsaken (Hays et al., 1945).

Paul Ehrlich, who is well known for laying the foundation of the modern antibiotic era and his hypothesis of 'magic bullet', fostered the discovery of arsenic-based drug Salvarsan in 1909, which was found to be effective for the syphilis causing organism *Treponema pallidum* (Gelpi et al., 2015). After a thorough study and screening programme for 5 years from 1904, Salvarsan, the sixth compound in the 600th series tested, showed a promising result for the treatment of patients (Aminov, 2010). This was a remarkable discovery and became the drug of choice for the endemic and incurable disease of that time. Despite its limitations and side effects, Salvarsan, marketed by Hoechst, became a great success and remained as the most commonly used therapeutic option until the advent ofpenicillin in 1940. The accidental discovery of

Antimicrobial Resistance in the 21st Century. DOI: https://doi.org/10.1016/B978-0-443-44748-8.00011-7

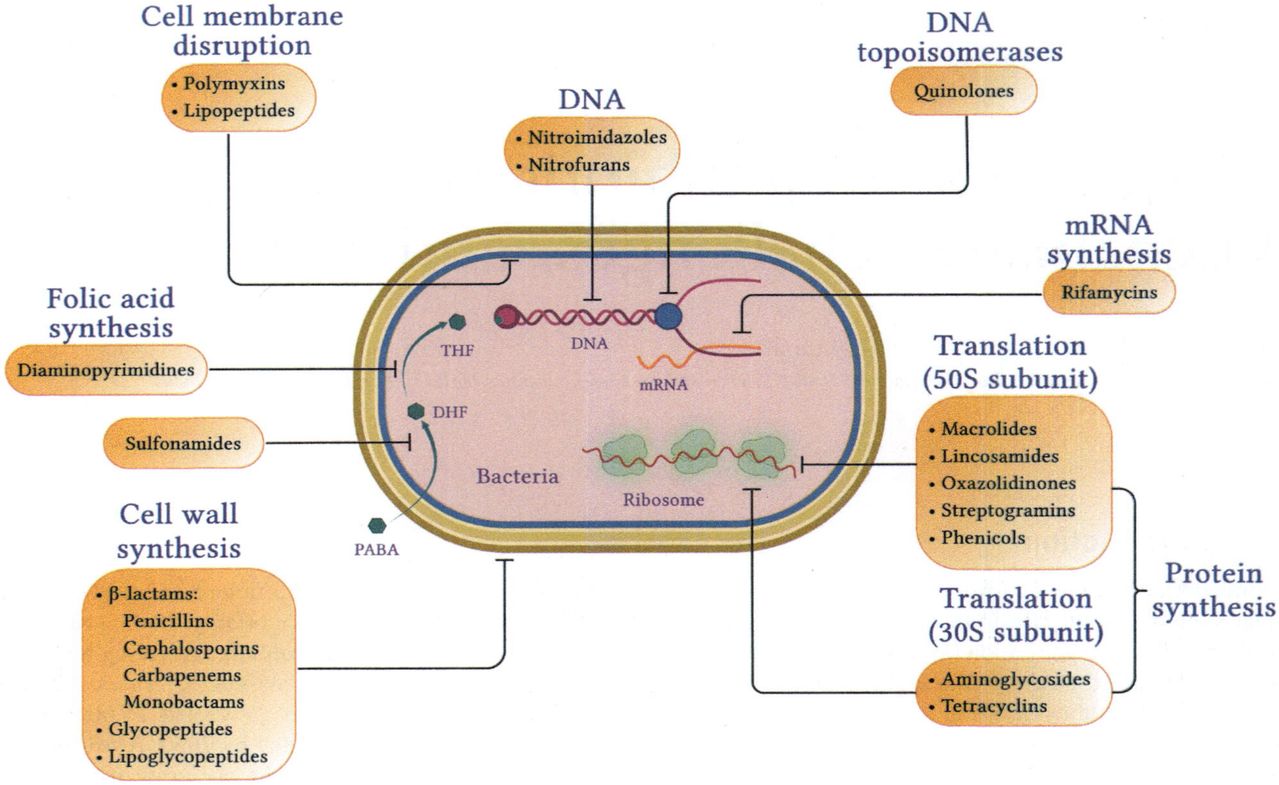

FIGURE 1.1 **Mechanisms of action of major antibacterial agents in bacterial cells.** Overview of the mechanisms of action of various classes of antibiotics.

penicillin by Sir Alexander Fleming in 1928 was known to be a milestone in the history of antimicrobial drug discovery, which is still being used for the treatment of many bacterial infections after its first clinical use in the 1940s. Penicillin, which originated from *Penicillium notatum*, is known to be the first effective and exceptional antibiotic in terms of safety and efficacy, which played a crucial role in saving the lives of many wounded soldiers during World War II (Gaynes, 2017; Boyd et al., 2021). Further, the continuous systematic screening for antibiotics led to the discovery of the first sulpha drugs, Sulfonamidochrysoidine (Prontosil) by Bayer, and later Gerhard Domagk confirmed its efficacy against many bacterial diseases (Yousef et al., 2018). As sulphonamide was very reasonable and easy to resynthesize, various pharmaceutical industries started mass production of sulphonamide derivatives, and as a result, Salvarsan was supplanted by the sulphonamide prodrug Prontosil. Sulphonamides were the first truly effective, broad-spectrum antimicrobials in clinical use, and the legacy of this oldest antibiotic is possibly reflected in its clinical use till today (Christensen, 2021).

Selman Waksman, the Ukrainian scientist known as the father of antibiotics, discovered more than twenty natural inhibitory substances, including actinomycin, streptomycin, and neomycin, by the end of the 1940s and into the early 1950s. Waksman was honoured with the Nobel Prize in medicine in 1952 for his discovery of streptomycin, which was found to be effective against a range of pathogenic bacteria, including tuberculosis. The discovery of Penicillin, Prontosil, and Salvarsan was exemplary, which became the cornerstone for future drug discovery research, and this drew a new era in the antibiotic armamentarium, described as the 'golden age' of antibiotic discovery (1940–62) (Gould, 2016). Antibiotics were celebrated as the ultimate cure for their high efficacy and potency, and were considered the 'miracle drugs' that the medical society had long been longing for. As a result of the initial success of antibiotics, bacterial diseases were naively considered to be permanently defeated (Penesyan et al., 2015) (Fig. 1.1.) But, unfortunately, with the increasing use of antibiotics, the bacteria started developing resistance to their inhibitory effects, which resulted in the lightning rise of AMR.

1.3 Burden and global impact of antibiotic resistance

Bacterial resistance to antibiotics poses a significant threat to human health worldwide, leading to millions of deaths per year. The health burden of AMR is most often described as the number of deaths caused by drug-resistant bacterial infections. This section summarises global, regional, and national patterns of AMR across different infections, their epidemiology, and associated pathogens.

1.3.1 Global epidemiology and disease burden

AMR substantially amplifies the global burden of the most common and critical bacterial infections, such as bloodstream, gastrointestinal, urinary tract, and urogenital gonorrhoeal infections. In a recent study, it was found that approximately 8.9 million deaths were due to bacterial infections in a year and amongst 1.27 million deaths were attributable to AMR and 4.95 million deaths were associated with AMR (Murray et al., 2022). Bloodstream infections are known to be one of the most severe bacterial infections and are the major cause of hospital and ICU admissions worldwide, leading to sepsis, multiorgan failure, and ultimately death. The epidemiology of bloodstream infections differs by several factors, such as regions, healthcare infrastructures, and infection control practices. The rising prevalence and the widespread distribution of certain resistant pathogens increase the complexity of the treatment, prolonging hospital stays, healthcare costs, and associated nosocomial infections. Globally, the resistant pathogens like *Escherichia coli*, *Klebsiella pneumoniae,* and *Staphylococcus aureus* are often responsible for bloodstream infections with regional variation. These pathogens, especially *Acinetobacter* spp. and *K. pneumoniae*, often showed resistance to key 'Watch' antibiotics, including third-generation cephalosporins, carbapenems, fluoroquinolones, and carbapenems, posing major challenges to therapeutic options (Hetsa et al., 2025).

Gastrointestinal infections also pose a major public health challenge worldwide and contribute significantly to the global AMR burden, particularly in low-resource settings with inadequate sanitation and limited access to health care (Ahmed et al., 2024). Although most diarrhoeal gastrointestinal infections are primarily viral, bacterial pathogens such as nontyphoidal *Salmonella* and *Shigella* spp. are the leading causes of bacterial gastroenteritis. The increasing resistance of these pathogens to fluoroquinolones, third-generation cephalosporins, and azithromycin has substantially reduced the available treatment options. Notably, higher levels of resistance in *Salmonella* spp. have been reported in the Southeast Asia region, whereas *Shigella* spp. exhibit elevated AMR levels in both Europe and Southeast Asia (Bastola et al., 2021).

Similarly, urinary tract infections (UTIs) represent one of the most common causes for frequent antibiotic consumption and are a major driver of AMR. Among Gram-negatives, *E. coli* and *K. pneumoniae* are most commonly associated with this infection, with rising resistance towards different classes of antibiotics. The burden of AMR in UTI varies among WHO regions. However, the Southeast regions consistently reported the highest level of resistance in most of the pathogens, while the lowest were reported by the European and Western Pacific regions (Pujades-Rodriguez et al., 2019).

1.3.2 Mortality, morbidity, and healthcare impact

AMR increases the risk of common infections, leading to routine medical procedures being unsafe and infections being untreatable, potentially taking back decades of progress in modern medicine. Recent studies revealed AMR as a leading cause of death worldwide, with the highest mortality in low-resource settings (Salam et al., 2023). AMR continues to exert a profound impact on global mortality, morbidity, and the healthcare system, and only in 2021, bacterial AMR was linked to approximately 1.14 million deaths. The death attributable to AMR increased between 1990 and 2019, and current forecasts estimate this will increase to 1.91 million deaths attributed to AMR, along with over 8 million associated deaths by 2050. AMR-related mortality among children under five years has substantially declined, which is mainly due to vaccinations, improved sanitation, and infection control. In contrast, AMR mortality has sharply risen among adults, particularly those aged above 70 years. This is mainly driven by old age, poor immune system, exposure to health care settings, and the use of many antimicrobials (Tadesse et al., 2023).

The reliance on the last resort of antibiotics, namely, polymyxins, and carbapenems, is associated with increased toxicity and adverse effects, further burdening patients and healthcare providers. Carbapenem-resistant pathogens have shown the most alarming increase in burden, substantially worsening the clinical outcomes. A recent study suggests that although improvements in sepsis management and healthcare quality have averted millions of AMR-associated deaths, these gains have been offset by demographic shifts and the emergence of more resistant pathogens. Strengthening health systems, improving access to effective antibiotics, and enhancing infection prevention could avert far more deaths than drug development alone, underscoring the critical role of healthcare capacity, stewardship, and equitable access in mitigating the global AMR burden.

1.3.3 Economic burden of antibiotic resistance

Apart from the clinical consequences, AMR imposes a significant economic burden on healthcare systems as well as on the national economy. The infections caused by resistant bacteria directly escalate the healthcare costs as it often requires expensive diagnostics, prolonged hospitalisation, and expensive antimicrobial agents, as well as the indirect economic losses arising from the reduced workforce productivity and premature mortality. In a recent study, it has been estimated that the total economic loss caused by AMR will be approximately $300 billion to $1 trillion by 2050 (Pulingam et al., 2022). Another study from the World Bank in the year 2017 reported that AMR would increase the levels of poverty with a huge impact, mainly on the poorest countries. AMR is also affecting the veterinary settings, reducing the output of livestock, as the increased resistance to commonly used antibiotics in the agricultural settings would reduce the efficacy of the treatment on the livestock and spread the infections (Enshaie et al., 2025). Hence, a reduction in livestock production and trade would increase the price of protein sources in the market, which would be of concern to meet the increasing demand of the rising world population (World Bank, 2017; Van Boeckel et al., 2015). As per the World Bank report, the AMR crisis will have a drastic impact on the yield of livestock, especially in LMICs (Lekagul et al., 2019). Earlier studies have reported that the annual Growth Domestic Product may decrease by 1% globally due to the AMR burden and efforts to overcome it. However, for developing countries, it may decrease to 5%–7% by 2050, which will further increase the difference between a developed and a developing nation (World Bank, 2017). AMR greatly influences the economy of a country and impacts the workforce and is estimated to cause a significant decrease in the global trade sector by 2050 (Naylor et al., 2018; Lekagul et al., 2019).

1.3.4 Impact on low- and middle-income countries

AMR has a significant burden in LMICs, particularly sub-Saharan Africa and South Asia. However, the majority of the studies have been conducted in upper-middle-income countries with very few originating from low-income settings. The high burden in LMIC is mainly due to the easy access to over-the-counter antibiotics, weak regulatory systems, poor healthcare infrastructure, less diagnostic capacity and misuse of antibiotics (Otaigbe and Elikwu, 2023). Several environmental factors, like humidity, poor hygiene, and water sanitation, further accelerate the spread of resistance in different settings. The use of antibiotics in agriculture, low health literacy, and limited access to healthcare drive the lack of guidelines, which favours the spread of resistance in LMICs (Manyi-Loh et al., 2018). Another key concern is that insufficient surveillance limits accurate assessment of resistance patterns. Therefore a coordinated One Health approach, addressing antibiotic use across human health, animal agriculture and environmental sectors through strengthened regulation, is essential.

1.4 Mechanisms and genetic basis of antibiotic resistance

The resistance among bacteria is not uniform and may vary greatly within different bacterial groups or species. The two major forms of bacterial resistance are natural or acquired. Natural resistance may be intrinsic or induced. Natural resistance is often expressed in the organism and is defined as a trait that is universally shared within a bacterial species and is not linked with horizontal gene transfer (HGT). The common intrinsic bacterial mechanisms include reduced outer membrane permeability (LPS in Gram-negative) and the natural activity of multidrug efflux pumps. However, in the case of induced resistance, the genes are naturally present in the bacteria, but only expressed under exposure to antibiotics (Reygaert, 2018). On the other hand, the acquired resistance may be due to the acquisition of any genetic material by the bacteria through transformation, transposition or conjugation. The molecular mechanisms driving AMR in bacteria are complex, and it occurs through different mechanisms. The most significant amongst these are enzymatic degradation or modification of antibiotics, alteration of binding targets, and active efflux systems that efflux the drugs from the cell (Vivekanandan et al., 2025). The difference in the resistance mechanisms in Gram-negative and Gram-positive bacteria is mainly due to their distinct cell envelop structures structural differences. .

1.4.1 Enzymatic inactivation of drugs

Enzymatic inactivation remains one of the most effective bacterial defence mechanisms against antimicrobial agents, and this occurs through either direct degradation of the antibiotics or through chemical modification. The bacteria produce enzymes that break down or modify antibacterial agents, mainly including β-lactamase and aminoglycoside-modifying enzymes (AMEs). β-Lactamase degrades the β-lactam antibiotics. However, AMEs chemically modify aminoglycoside antibiotics and make them inactive (WRIGHT, 2005). Among the enzymatic resistance mechanisms, β-lactamases are the most extensively studied enzymes capable of degrading the β-lactam ring of antibiotics, namely, penicillin, cephalosporins, monobactam, and carbapenems. This group of enzymes was categorised by Ambler into four classes (A–D) based on the molecular structure, which includes extended-spectrum β-lactamases (ESBLs), metallo-β-lactamases, AmpC cephalosporinase, and carbapenemases (Bush and Bradford, 2016). AMEs, such as acetyltransferases, phosphotransferases, and aminoglycoside nucleotidyltransferases (ANT), are primarily responsible for the resistance to aminoglycosides, which chemically modify the antibiotic and prevent it from binding to ribosomal targets (Ramirez and Tolmasky, 2010). Additionally, enzymes such as chloramphenicol acetyltransferase and macrolide phosphotransferase contribute to resistance to chloramphenicol and macrolide antibiotics (Paul et al., 2022). The evolution and rapid dissemination of these enzymes are mainly driven by mutations, antibiotic exposures, and HGT facilitated by mobile genetic elements. This enzymatic inactivation of antibiotics underscores the complexity of AMR and raises growing concern about the urgent need for advanced therapeutic strategies to combat it and restore antibiotic efficacy (Narendrakumar et al., 2023).

1.4.2 Modification of targets

Another significant mechanism by which bacteria evade the antimicrobials is through the alteration of the structure of their drug targets, which results in reduced drug binding and effectiveness. Alterations of penicillin-binding proteins, such as mutations or the acquisition of low-affinity PBPs like PBP2a, confer β-lactam resistance, and alterations in ribosomal proteins, particularly 16S rRNA and ribosomal proteins, compromise the binding of aminoglycosides and tetracyclines (Reygaert, 2018). The alterations in the composition of gyrase or topoisomerase IV reduce the drug's ability to bind, which causes resistance to drugs that inhibit nucleic acid synthesis, for example fluoroquinolone (Fig. 1.2).

1.4.3 Reducing drug uptake

In Gram-negative bacteria, the uptake of molecules is primarily mediated by porin proteins across the outer membrane. Bacteria employ outer membrane modifications, particularly porin alterations, to reduce antimicrobial entry, and there are

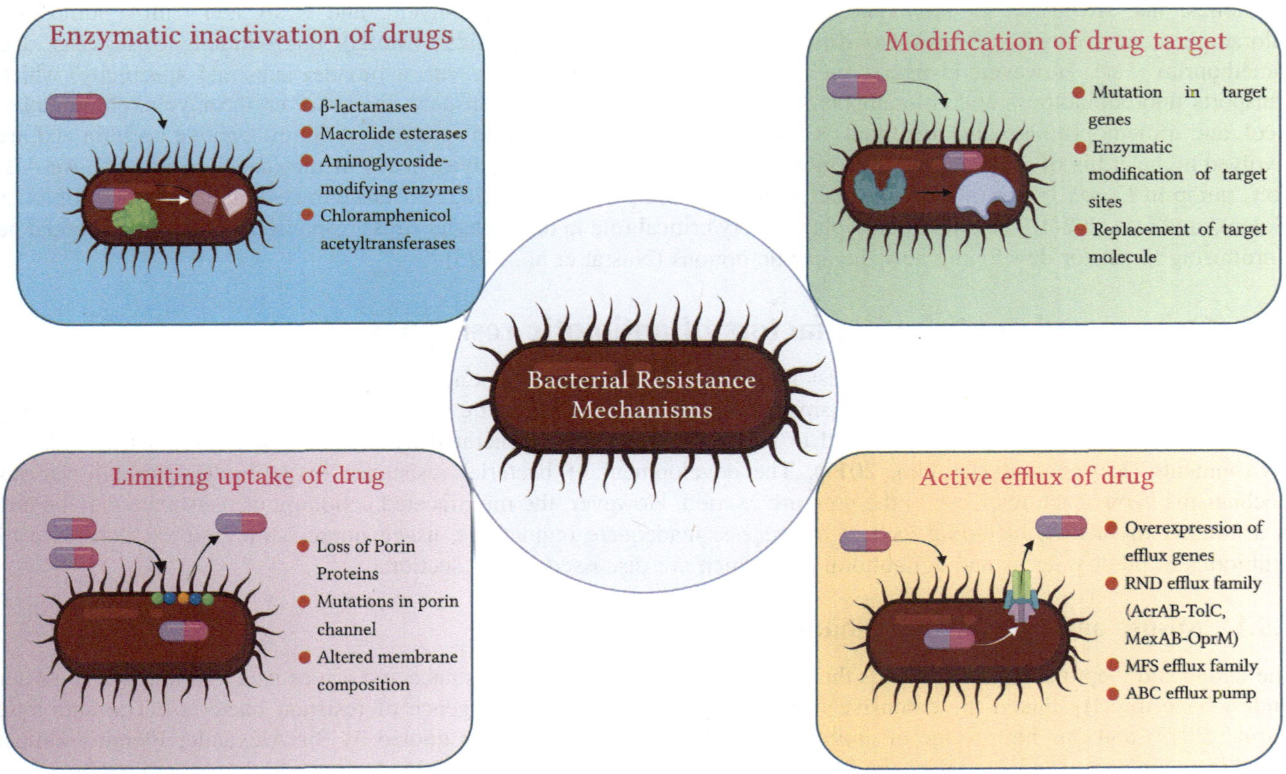

FIGURE 1.2 **Major mechanisms of bacterial antibiotic resistance.** Major mechanisms of antibiotic resistance in bacteria: Enzymatic degradation, target modification, efflux pumps, and reduced permeability.

mainly two ways in which this porin modification can limit drug uptake. A decrease in the number of porins present and mutations that change the selectivity of the porin channel (Fig. 1.2). Several outer membrane proteins, such as OmpF and OmpC in *E. coli*, OprD in *P. aeruginosa,* and OmpK/OmpK36 in *K. pneumoniae*, facilitate the passive influx of hydrophilic antibiotics into the cell (Zhou et al., 2023). Mutations or structural alterations in such porins responsible for drug influx significantly decrease the membrane permeability. For example, the loss or modification of OprD in *P. aeruginosa* decreases carbapenem uptake, resulting in carbapenem resistance. However, alterations in OmpK36 in *K. pneumoniae* confer resistance to cephalosporins and carbapenems. These porin-altered or deficient strains then become MDR with very limited therapeutic options when they cohabit with other additional resistance determinants (Wang et al., 2025).

1.4.4 Efflux out of antimicrobials

Another major bacterial defence mechanism is the efflux pumps, and this is one of the most common resistance mechanisms among a wide range of pathogenic bacteria. Bacteria possess several types of efflux pumps, and the major five families include resistance-nodulation-division (RND), major facilitator superfamily (MFS), small multidrug resistance (SMR), ATP-binding cassette (ABC) transporters, and the multidrug and toxic compound extrusion (MATE). Some of these efflux pumps are expressed constitutively, while others are induced or overexpressed under certain environmental pressure. ABC efflux family made up of six transmembrane segments (TMS) harbours both uptake and efflux transport systems, which transport various molecules such as amino acids, drugs, ions, polysaccharides, proteins and sugars (Gaurav et al., 2023). One notable ABC pump is found in *Vibrio cholerae* (VcaM), which transports fluoroquinolones and tetracycline. The MATE efflux family uses an Na^+ gradient as the energy source and is made up of 12 TMS and is reported to efflux mainly fluoroquinolones and aminoglycosides. NorM pump was the first characterised pump from the family and was found in the chromosomal DNA of *Vibrio parahaemolyticus.* The SMR efflux family members are mainly hydrophobic, and the genes for these pumps are found in different locations, such as chromosomal DNA, plasmids, and transposable elements. These pumps are mainly made up of four TMS and mostly confer resistance to β-lactams and aminoglycosides. The MFS family of efflux pumps is mainly involved in the transport of anionic drugs (e.g., macrolides and tetracyclines), metabolites, and sugars. These pumps are very substrate specific and have different pumps for different antibiotics, like *Acinetobacter baumannii* has the SmvA pump for erythromycin, and CraA and CmlA pumps for chloramphenicol. Likewise, *E. coli* has different efflux pumps for macrolides (MefB), fluoroquinolones (QepA), and trimethoprim (Fsr). However, instances such as NorA pump in *S. aureus* with a broader substrate specificity, which transports fluoroquinolones and chloramphenicol, or LmrS pump, which transports linezolid, erythromycin, chloramphenicol, and trimethoprim are also identified. RND efflux pumps are present in a variety of Gram-negative bacteria and are involved in the efflux of several substrates, including antibiotics, detergents, dyes, heavy metals, and solvents. The AcrAB-TolC pump in *E. coli* is one of the most studied RND efflux pumps conferring resistance to several classes of antibiotics (Solanki and Das, 2024). The efflux pump plays a very critical role in the bacterial stress response mechanism and could be a promising target for developing new therapeutic options (Nusrat et al., 2025).

1.5 Drivers and contributing factors of antibiotic resistance

Microorganisms are known to be the oldest inhabitants of the earth, existing for more than 3.8 billion years, and to survive, they have evolved several mechanisms exhibiting the greatest genetic versatility and adaptability to the changing environment, which allow them to respond to the selective pressure generated by their competitors and from various environments (Kashef and Hamblin, 2017). The development of bacterial resistance to antibiotics through various mechanisms is one such response to the pressure exerted. However, the multifaceted aetiology of resistance may be due to a number of factors, including lack of awareness, inadequate regulations, usage imprecision, and extensive use of antibiotics in poultry farms and agricultural use, which are discussed in this section.

1.5.1 Misuse and overuse of antibiotics

The undue and inept use of antibiotics is the principal cause of bacterial resistance and drives its evolution. Many earlier studies have directly linked the extensive use of antibiotics with the emergence of resistant bacteria ("The antibiotic alarm," 2013) and this high usage of antibiotics perfectly matches the lines quoted by Sir Alexander Fleming almost 70 years ago in early 1945. He raised the fear that 'public will demand' the new miracle drug and thus would begin 'an era of abuses' at that time. The high usage of antibiotics develops the selection pressure in the environment, which in turn induces the resistant genes within a bacterium to transfer to the same or different bacterial species through vertical and

HGT (Narendrakumar et al., 2020). HGT is a driving force in bacterial evolution, mainly facilitated by transposons, mobile genetic elements, and plasmids (Li et al., 2025). The majority of bacterial infections that arise in healthcare settings are due not only to the patient's weak immune system but also to the extensive use of antibiotics, which can transform susceptible bacteria into resistant ones, thereby making them difficult to treat. The over-the-counter sale of antibiotics is also a significant cause for the misuse of antibiotics in countries where antibiotics are unregulated and available online without a prescription, which promotes overuse and makes these lifesaving drugs easily accessible (Kotwani et al., 2021).

1.5.2 Extensive agricultural and veterinary use of antibiotics

A wide range of antibiotics is applied in animal husbandry, aquaculture, poultry farms, and crop production in comparison to humans, and this indiscriminate use of antibiotics is not only to treat infections but also to prevent them and to promote growth, which has threatened the modern antibiotics that are medically important (Ibrahim et al., 2019). In the United States, a bulk amount of streptomycin and oxytetracycline antibiotics are used to spray on orange trees to prevent the citrus greening disease (Vincent et al., 2022). The substantial use of these antibiotics creates antibiotic pollution, resulting in the occurrence of antibiotic residues in animal-derived products, water, and the environment, which can cause toxicity and the development and emergence of resistant pathogens, leading to therapeutic failure. The use of colistin in veterinary sciences has resulted in the emergence of the plasmid-mediated colistin-resistant gene *mcr-1* in 2016 in China (Liu et al., 2016). Later, it was reported in more than 50 countries worldwide, in both hospital and veterinary settings (Wang et al., 2020).

1.5.3 Less availability of new antibiotics

The development of new antibiotics has slowed down dramatically in recent years, leaving very limited treatment options. The reason for this dry pipeline is mainly the cost and complexity of conducting clinical trials, licensing issues, and several other regulatory obstacles (Gargate et al., 2025). Many of the pharmaceutical companies have reduced the number of their research wings as the development of new antibiotics is no longer considered an economically wise investment, and it is mainly because antibiotics are relatively used for a short span of time and are not profitable like other drugs used for chronic treatment (Ayukekbong et al., 2017).

1.5.4 Inappropriate prescription

Lack of an effective prescription system in many countries is also highly responsible for the misuse of antibiotics and consequently the spread of AMR. This problem is fuelled by patient demand for antibiotics and also by the inadequate explanation from the clinicians regarding when these drugs are truly necessary. It is further compounded by diagnostic uncertainty in distinguishing viral from bacterial infections, which often leads to precautionary or inappropriate prescribing (Murray et al., 2022).

1.6 Antibiotic resistance in clinically important pathogens

Antibiotic resistance among clinically significant pathogens has emerged as one of the most significant threats to global public health. The continued use of antimicrobials over the years has led to the development of resistance mechanisms in bacteria. This causes the emergence of MDR, extensively drug-resistant, and pan-drug-resistant phenotypes amongst the bacterial genera and the most significant ones are *P. aeruginosa*, *A. baumannii*, *E. coli*, *K. pneumoniae*, *Enterococcus faecium*, *S. aureus*, *Salmonella enterica*, *Stenotrophomonas* spp., and many others (de Oliveira Santos et al., 2022; Vattanaviboon et al., 2026). These bacteria might have acquired the resistance phenotype as a chromosomal mutation, or through HGT mechanisms such as conjugation, transformation and transduction (Gray-Hammerton et al., 2025). The treatment of infections caused by these resistant pathogens is very difficult, which complicates treatment regimens, prolongs hospital stays, increases healthcare costs, and is linked with higher morbidity and mortality rates worldwide.

1.6.1 Resistance in Gram-positive bacteria

The prevalence of resistance in Gram-positive bacteria and the infections caused by these pathogens are increasing, posing a significant challenge in healthcare settings and the country's economy. It is essential to explore the diverse mechanisms of these bacteria through which they resist the antimicrobial agents. An understanding of these crucial mechanisms will be helpful for the development of novel therapeutic strategies and in combating the risk of AMR. Gram-positive bacteria, particularly *S. aureus*, *Streptococcus pneumoniae*, and *E. faecium*, employ several mechanisms, including alterations to their cell wall structure, efflux pumps, and the production of enzymes, which deactivate the antibiotics (Rajput et al., 2024). These pathogens are also highlighted in the WHO 2024 high-priority pathogen list for their ability to cause serious infections and growing resistance (Sati et al., 2025). Apart from this, the emergence of macrolide and vancomycin-resistant strains further complicated the issues, often in intensive care units and burn wounds (Paul et al., 2023; Karukappadath et al., 2023).

1.6.2 Resistance in Gram-negative bacteria

The continuous evolution of resistant organisms has led to the emergence of pathogens that were once treatable with antibiotics. The Gram-negative bacteria pose a greater therapeutic challenge because of their intrinsic resistance mechanisms and rapid acquisition of resistance determinants (Mehrotra et al., 2023). The critical pathogens like *E. coli*, *K. pneumoniae*, *P. aeruginosa*, and *A. baumannii* are of significant concern and are being reported from both community and hospital settings (Badger-Emeka et al., 2023). The environmental dissemination of resistance genes through wastewater systems, through agricultural and veterinary settings, further contributes to the spread of AMR (Paul et al., 2026). These pathogens have developed resistance against most of the commonly prescribed antibiotics, such as β-lactams, carbapenems, aminoglycosides, and fluoroquinolones, and the key resistance mechanisms include the production of ESBLs, carbapenemase enzymes (bla_{NDM}, bla_{KPC}, bla_{OXA}, $bla_{CTX\text{-}M}$, etc.), alteration of target sites or porin loss, as well as the overexpression of multidrug efflux pumps. Gram-negative bacteria are of special concern due to their resistance to last-resort antibiotics and worldwide dissemination (Breijyeh et al., 2020). The majority of the pathogens responsible for contributing to the burden of AMR in 2019 were GNB, and the infections caused by these pathogens are life-threatening, especially bloodstream infections and associated infections in intensive care units, with a higher mortality (Gauba and Rahman, 2023).

1.6.3 WHO priority pathogens

The WHO has played a pivotal role in combating AMR through different aspects, and among them are policy development, surveillance, and capacity building initiatives. Given the increasing severity of AMR and its being among one of the top 10 global health threats, the WHO has developed the Global Action Plan (GAP) on AMR, in 2015, which provides a framework to improve awareness, strengthen surveillance systems, reduce infection severity, promote proper use of antibiotics, and cutting-edge approaches like therapeutics, diagnostics, and vaccines. The Global Antimicrobial Resistance and Use Surveillance System (GLASS) was established by WHO to standardise and gather data collection across countries. WHO collaborates with different organisations like the Food and Agriculture Organisation (FAO), and the World Organisation for Animal Health (WOAH) to address AMR at the human–animal–environmental interface, which is essential for the one-health approach to support national action plans, especially in LMICs, to mitigate the emergence and spread of resistant pathogens. In the year 2014, the WHO published its first global report on AMR and defined AMR as a serious threat for the future in every region. However, in 2017, WHO published its first-ever 'Bacterial Priority Pathogen List 2017' (BPPL 2017), which is responsible for causing serious health threats to human health. The list was prepared with a goal to prioritise research and development (R&D) investments in AMR, to guide AMR surveillance, and to prevent and control infection strategies worldwide. In the BPPL 2017, the major focus was provided on nosocomial Gram-negative bacterial pathogens such as carbapenem-resistant *A. baumannii*, and carbapenem-resistant Enterobacterales (CRE), which mainly include *E. coli* and *K. pneumoniae*. To address the urgent need for new antibiotics and the severity of AMR, in 2024, the WHO released the second priority pathogen list with advancements from the earlier protocol used for BPPL 2017. The BPPL 2024 categorised the bacterial pathogens into three categories: critical priority, high and medium priority. The critical priority pathogens include carbapenem-resistant *A. baumannii*, third-generation cephalosporin-resistant Enterobacterales, and CRE and the high priority pathogens include fluroquinolone-resistant *Salmonella typhi*, and *Shigella* spp., nontyphoidal *Salmonella*, vancomycin-resistant *E. faecium*, carbapenem-resistant *P. aeruginosa*. The BPPL is a key tool for prioritising R&D, and for global public health policies to combat AMR and for strengthening infection prevention and control (IPC) measures (WHO, 2024).

1.7 Challenges in controlling antibiotic resistance

Rapid population growth and high population density significantly contribute to the emergence and spread of AMR (Endale et al., 2023). In densely populated regions, healthcare facilities are limited, and they also increase the risk of hospital-acquired infections. Larger populations also result in increased use of antibiotics, with self-medication and incomplete treatment courses increasing the selective pressure on the microbes that promote resistance (Endale et al., 2023). Rapid urbanisation, particularly in low and middle-income countries, leads to overcrowding and inadequate sanitation.

These conditions also promote the transmission of infectious diseases, increasing dependence on antibiotics. Most diseases and deaths attributed and associated with AMR are most frequently reported from developing countries. Another major challenge in controlling antibiotic resistance is the increased international and globalised trade, facilitating the rapid cross-border spread of resistant microorganisms (Alenzi et al., 2026). People, animals, and food products move across continents, creating opportunities for resistant bacteria to disseminate beyond their origin. Medical tourism and migration also play a significant role in the global dissemination of antibiotic resistance (Peirano et al., 2014). Other major challenges in controlling antibiotic resistance include the nonavailability of adequate financial resources for addressing AMR and the unavailability of predictive models to forecast outbreaks and resistance spread (Gargate et al., 2025).

1.8 Strategies and solutions to combat antibiotic resistance

Antibiotic resistance demands coordinated, sustained, and evidence-based action to address it effectively. Integrated strategies spanning clinical practices, public health systems, research innovations, and global governance are imperative to combat this growing challenge.

1.8.1 Antibiotic stewardship programmes

Antibiotic stewardship is the careful and responsible management of AMR applied in the context of governance of the health and well-being of the population, which guides health systems at the national and global levels (Hibbard et al., 2024). The major objectives of the antibiotic stewardship programmes (AMS) include:

Improving awareness and understanding of AMR through effective communication, education, and training.
Strengthening the knowledge and evidence base through surveillance and research.
Reducing the incidence of the infection through effective sanitation, hygiene, and infection prevention measures.
Optimising the use of antimicrobial medicines in human and animal health
Developing the economic case for sustainable investment that takes account of the needs of all countries, and increasing investment in new medicines, diagnostic tools, vaccines, and other interventions.

Broadly, the programmes are devised to optimise the use of antimicrobials, improve patient outcomes and reduce AMR (Hibbard et al., 2024). Hence many countries have developed and implemented their National Action Plans on AMR, in which AMS is a key priority (Gunasekara et al., 2025). In most cases, AMS is implemented in both community and critical care units. Effective AMS implementation has been associated with reduced antimicrobial consumption, lower rates of infections, decreased resistance trends, and improved overall quality of care, making it a critical component of modern healthcare systems worldwide (Currie et al., 2020).

1.8.2 Strengthening infection prevention and control

Strengthening IPC is essential to reducing healthcare-associated infections, limiting the spread of AMR, and improving overall patient safety (Ackers et al., 2020). Effective IPC programmes are built on standard precautions such as hand hygiene, appropriate use of personal protective equipment, environmental cleaning, proper sterilisation, and disinfection of medical equipment. Surveillance systems for early detection of outbreaks, adherence to isolation protocols, and implementation of transmission-based precautions further enhance infection control efforts (Meckawy et al., 2022). Continuous training of healthcare workers, strong leadership commitment, adequate staffing, and availability of essential supplies are critical components for sustained success. Integration of IPC with antimicrobial stewardship ensures a coordinated approach to patient care (Ackers et al., 2020). By fostering a culture of safety, accountability, and compliance with evidence-based guidelines, healthcare institutions can significantly reduce infection rates, protect vulnerable populations, and strengthen resilience against emerging infectious threats (Meckawy et al., 2022).

1.8.3 Strengthening antimicrobial resistance surveillance

AMR surveillance includes continuous collection, processing, and analysis of samples to detect resistant pathogens and AMR resistance genes from different sample sources, which is used to inform the AMR burden in clinical practice, public health, and policy decisions. Effective AMR surveillance is multipronged, which integrates microbiology, epidemiology and data science to monitor resistance trends, detect emerging resistance mechanisms and identify outbreaks in a timely manner (Kim et al., 2024). AMR surveillance findings guide IPC strategies. AMR surveillance is also a part of several disease surveillance and monitoring programmes, and effective surveillance systems are in place for this purpose. These surveillance systems can be either active or passive, depending on the monitoring scope of the system. The Active Bacterial Core, which comes under the Emerging Infectious Programme of the Centres for Disease Control and Prevention, was initiated to identify the risk of community-acquired methicillin-resistant *S. aureus* and is an example of a surveillance system (Fridkin et al., 2015). In 2015, the WHO aunched the GLASS, which serves as a collaborative effort to standardise AMR surveillance (WHO, 2015). The GLASS AMR provides a standardised approach to the collection, analysis, and sharing of national AMR data in samples collected routinely for clinical purposes for a set of pathogens that cause common bacterial infections in humans. The 2025 GLASS report was based on over 23 million bacteriologically confirmed infections reported by 104 countries. However, the surveillance data of such efforts might not always be holistic because of differences in regional reporting. GLASS, 2025 AMR surveillance reports that Gram-negative bacterial pathogens, specifically, *E. coli* and *K. pneumoniae* are the most frequently reported drug-resistant bacteria in bloodstream infections globally. Additionally, the WHO along with the FAO, the United Nations, and the WOAH have built a Tripartite Integrated Surveillance System on AMR and Antimicrobial Use, which work towards global web-based repository for resistance, which includes data from different sources such as human, animal, plant and environmental (FAO and WHO, 2019). A list of AMR surveillance programmes are listed in Table 1.1.

1.8.4 Advancing rapid diagnostics

Current rapid diagnostic tests and assays for the detection of resistance genes include both culture-based and nonculture-based molecular-based approaches. These molecular assays are particularly advantageous for their accuracy and multiplex targeting of AMR genes in a single reaction (Kaprou et al., 2021). However, the key limitations of the molecular-based methods are that it often only provides the presence and absence of an AMR gene and not its expression or functionality. Consequently, the detection of a resistance gene does not necessarily equate to phenotypic resistance. This limitation is further compounded in metagenomic samples, where the detected resistance genes may originate from extracellular DNA fragments present in the environmental sample rather than from viable or clinically relevant bacterial cells. Therefore the mere presence of a resistance gene in the environment or complex clinical samples does not always correlate with active resistance in pathogenic bacteria (Hassall et al., 2024). Despite these limitations, the rapid and sensitive detection of resistance genes has made them the cornerstone of modern AMR surveillance and rapid diagnostics (Gan et al., 2024). Apart from conventional polymerase chain reaction (PCR) and quantitative real-time PCR, several next-generation molecular diagnostics have been successfully developed and applied to AMR gene detection. These include loop-mediated isothermal amplification, strand displacement amplification, rolling circle amplification, and helicase-dependent amplification (Gill and Ghaemi, 2008). As compared to the traditional PCRs, the major advantages of the isothermal amplifications include its reduced turnaround time, simplified instrumentation requirements, and operational feasibility in resource-limited, rural, and field settings.

Apart from these molecular techniques, the rapid phenotypic antimicrobial susceptibility tests (AST) method directly measures bacterial growth in the presence of antibiotics, providing a direct sensitivity account. Microfluidic-based growth detection, biosensors, microcalorimetry and automated imaging systems reduce the phenotypic testing time to under 4–6 hours and significantly decrease the practice of empirical therapy. Single-cell confined microchannels have been demonstrated to monitor real-time resistance evolution in bacteria on antibiotic exposures (Lu et al., 2013). Further, single-cell morphological analysis in microfluidic chambers integrating 96-well matrixes and high-throughput analysis provides results comparable to the conventional microdilution assays, able to determine the minimum inhibitory concentrations of the antibiotics used to kill the bacteria (Zhang et al., 2022). Similar microfluidic systems have been utilised to obtain AST results at a minimum time of 30 minutes and was effective in monitoring AST of bacterial pathogens causing UTIs thereby considerably reducing the use of empirical antibiotics (Stamm and Norrby, 2001). Further, biosensors have emerged as an effective tool for the detection of resistant microorganisms and the resistance genes. Several research groups have developed biosensors which detect the *mecA* gene in methicillin-resistant *S. aureus* strains (Jenison et al., 2001). In a similar line, biosensors to detect $bla_{\text{OXA-23}}$ in *A. baumannii*, mutations in *rpoB, gyrA, katG, inhA* in *Mycobacterium tuberculosis*, $bla_{\text{CTX-M}}$ and $bla_{\text{NDM-5}}$ in Gram-negative bacteria were developed, which had a rapid detection time of 30 minutes. Integration of these biosensors and microfluidic systems with artificial intelligence (AI) and digital health systems will further enhance diagnostic capabilities.

1.8.5 Research and development of new therapeutics

The main problem due to the accelerating antibiotic resistance is that the conventional drugs are getting obsolete and the novel drug discovery pathway has become stagnated, with very few newer antimicrobials being identified. Advancing R&D of new therapeutics is therefore essential to sustain modern medicine and global health security (Gargate et al., 2025).

TABLE 1.1 List of major international and national antimicrobial resistance surveillance systems.

Programme	Organisation	Scope	Focus	References
Global Antimicrobial Resistance and Use Surveillance System (GLASS)	World Health Organisation	Global	Standardise global AMR data collection and reporting	Ajulo and Awosile (2024)
Global Antimicrobial Resistance Surveillance System (TrACSS)	Food and Agriculture Organisation (FAO), United Nations, the World Organisation for Animal Health, and the World Health Organisation	Global	Monitor 'One Health' Programme	Drake et al. (2026)
Global Antimicrobial Resistance Monitoring System (GARMIS)	USA Food and Drug Organisation	International	Monitor AMR in imported food and food products	Ajulo and Awosile (2024)
Central Asian and Eastern European Surveillance of Antimicrobial Resistance (CAESAR)	World Health Organisation regional offices, European Society of Clinical Microbiology and Infectious Diseases (ESCMID) and the Dutch National Institute for Public Health and the Environment (RIVM)	Central Asia and Eastern Europe	Monitor the spread of AMR in clinical settings	WHO European Office (2025)
Antimicrobial Resistance Surveillance Programme (AMRSP)	Indian Council of Medical Research (ICMR)	India	Monitor AMR spread and antibiotic usage. Strengthen infection control practices and awareness	Walia et al. (2019)
European Antimicrobial Resistance Surveillance Network (EARS-Net)	European Centre for Disease Control and Prevention	Europe	Monitor resistance trends across Europe	McSorley (2026)
Canadian Antimicrobial Resistance Surveillance System (CARSS)	Canadian Public Health Agency	Canada	Monitor AMR spread in clinical and veterinary settings	Abdesselam et al. (2026)
Japan Nosocomial Infection Surveillance (JANIS)	Ministry of Health and Welfare, Japan	Japan	Monitor nosocomial infections and AMR in clinical settings	Kajihara et al. (2021)
China Antimicrobial Resistance Surveillance System	Chinese Ministry of Health in support with World Health Organisation	China	Multicenter AMR surveillance in clinical pathogens	Qin et al. (2024)
National Antimicrobial Resistance Monitoring System (NARMS)	Centre for Disease Control and Prevention, USA, Food and Drug Administration	United States	Monitor emerging AMR, its spread strategies for AMR de-escalation	Karp et al. (2017)

AMR, Antimicrobial resistance.

Revitalising the antibiotic discovery pipeline to identify novel antibiotics with unique modes of action has become the need of the hour. Despite scientific advances, the antibiotic development pipeline remains stagnant with only 10 antibiotics or antibiotic combinations approved between 2017 and 2023. The UN in 2024 had conducted a high-level meeting on AMR, and the biennial AMR Ministerial Meeting for initiating the identification of two to four new antibiotics for patients by 2030 (Figueras, 2024). Additionally, global initiatives have been working together to fund different phases of antibiotic development. For instance, CARB-X has funded more than 100 projects up to phase 1 clinical trials (Blaskovich and Cooper, 2025). Optimising existing antibiotic scaffolds through structural modification has been a successful method for enhancing the antibiotic efficacy.

Given the evolutionary inevitability of resistance, novel antibiotics alone would not be successful in treating MDR pathogens. Antivirulence agents include molecules that target bacterial pathogenicity factors such as toxins, quorum-sensing molecules, and adhesion molecules that do not directly kill the bacteria but exert lower selective pressure and thereby slow resistance development. Independent studies led by Ping et al. (2017) and Liu et al. (2017) reported Sclareol and Dracorhodin perchlorate which had antitoxin and antiquorum sensing activity against *S. aureus*. Both compounds reduced haemolysis by *S. aureus* by downregulating *hla* and RNAIII expression (Ping et al., 2017; Liu et al., 2017). In the same year, Zhang *et al* had reported Chalcone as a potent antivirulence agent with antitoxin, antiquorum sensing and antibiofilm agent which again acted upon by multiple targets such as *hla* and *agrA* gene (Zhang et al., 2017). A major antivirulence mechanism against *P. aeruginosa* is the reduction of pyocyanin, alginate, elastase, or proteases, which would have detrimental effects on its biofilm formation and pathogenicity (Kim et al., 2018). Apart from new drug discovery, innovative delivery systems such as encapsulations and nanoparticle-based delivery systems play a crucial role in improving therapeutic efficacy, enhancing bioavailability and enabling targeted delivery of drugs (Pradhan et al., 2021). However, one major limitation of antivirulence therapy is that it does not kill the bacterial pathogens and hence cannot be used alone. Most of these antivirulence agents act as antibiotic adjuvants. Further, though there have been several studies on these molecules, additional comprehensive mechanistic studies are important before its clinical application (Totsika, 2016). Bacteriophage therapy is yet another avenue of potential research to treat resistant bacterial pathogens and thus curb AMR. In phage therapy, lytic phages are employed to kill bacterial pathogens (Hibstu et al., 2022). The major advantage of bacteriophage therapy is that it selectively targets and kills bacteria without affecting eukaryotic cells, unlike some antibiotics and small molecules. Further, phages can also be used alongside antibiotics and can be formulated into phage cocktails that can target multiple bacteria. Several studies using a single phage and phage cocktails have been identified to be successful, especially when antibiotics fail. Phage therapy has been considered a last resort to treat infectious diseases. *P. aeruginosa*–specific phages have been tested to understand its ability to reduce cystic fibrosis (Tamma et al., 2022). Additionally, bacteriophages have been tested against *Clostridium difficile*, a Gram-positive spore-forming anaerobic bacteria which is the aetiological agent of *C. difficile* infection (Fujimoto and Uematsu, 2022). However, phage therapy also holds potential drawbacks when it comes to clinical applications. The primary challenge is the safety and efficacy of the selected phage. There have been instances where the bacteriophages themselves have facilitated horizontal transfer of antibiotic resistance genes (Colavecchio et al., 2017). Currently, there is very limited regulatory oversight of phage therapies. Secondly, the bacteria could also evolve resistance to phages like the antibiotics. Despite these limitations, comprehensive research is being carried out in this field. Another major research domain with significant potential to reduce the use of antimicrobials involves the application of beneficial microbes with antimicrobial properties to prevent and treat infections. A wide range of microbes isolated from diverse sources such as human microbiota, fermented food, soil, and environment have demonstrated health-promoting effects. Those that contribute to maintaining general host health are known as 'probiotics' and strains selected and developed for targeted therapeutic intervention against a specific condition are often described as biotherapeutics (Sarita et al., 2024). These beneficial microorganisms exert their protective effects through multiple complementary mechanisms. They modulate the host microbiome by enhancing microbial diversity and promoting the dominance of commensal and symbiotic microflora and suppressing pathogenic microbes through competitive exclusion, nutrient competition and production of antimicrobial molecules such as bacteriocins and organic acids (Latif et al., 2023). Also, they modulate the host immunity and strengthen the epithelial barrier function (Zheng et al., 2023). By preventing pathogen colonisation and reducing infection, these interventions decrease the use of conventional antibiotics. Systematic sourcing, characterisation and clinical evaluation of these beneficial microbes combined with advances in genomics, metabolomics and formulation technologies could enable their safe and effective application across the different sectors.

1.8.6 Regulatory and policy reforms

In response to the global rise in AMR, several regulatory and policy reforms have been implemented at international, national, and regional levels. Intergovernmental organisations such as the FAO, the World Organisation of Animal Health

(OIE) and the WHO, have taken collective efforts globally to minimise the emergence and spread of AMR (Gunasekara et al., 2025). The WHO has recognised AMR as a global priority health issue and the GAP on AMR is in place to monitor and guide the National Action Plan of respective countries (Gunasekara et al., 2025). Additionally, the WHO has adopted a coordinated 'One Health' approach that recognises that AMR is not a problem in human health but is also interconnected with animal and environmental health. One key focus area of the Action Plans is to strengthen government policies related to antimicrobial use. These action plans emphasise improving awareness of AMR, strengthening surveillance and research, optimising antimicrobial use and promoting R&D to identify and develop newer antibiotics and therapeutics (Capuozzo et al., 2024). Several governments have launched information campaigns to educate the public about the negative impact of AMR and strategies for its mitigation. Further, several regulatory and legislative policies have been strengthened in several countries, including bans on the sale of over-the-counter antibiotics, and strict prescription requirements have been implemented to reduce unwanted and overuse of antibiotics (Kotwani et al., 2021). In addition, the integration of electronic data records and national health insurance systems has improved monitoring of patients' treatment histories and patterns of antibiotic use and illness. Despite these efforts, several challenges remain. A major limitation is the absence of a uniformly enforced global regulatory framework, resulting in substantial disparities in implementation, surveillance, and compliance across countries (Endale et al., 2023). Limited resources in LMICs and continued misuse of antibiotics in both human and veterinary sectors are hindering global progress.

1.9 Future perspectives

AMR is evolving faster than the development of new therapeutic options and diagnostic methods, threatening modern medicine, the global health sector, and economy. Addressing this crisis requires a forward-looking, integrated, and systematic approach. The future of AMR control depends on coordinated scientific, clinical, policy, and societal transformations. Rapid, affordable and point-of-care diagnostics will be central for AMR control. Integration of these diagnostic outputs with AI-driven surveillance platforms will facilitate real-time monitoring of resistance patterns and early outbreak detection. Future improvements focusing on the enhancement of genotypic detection with phenotypic resistance could be done, possibly with transcriptomic or proteomic markers that confirm gene expression and functional activity. Strengthening global data-sharing networks to create predictive models capable of predicting the emergence and spread of resistance is of utmost importance. The stagnation of antibiotic and new drug molecule discovery should be lifted through synthetic biology platforms and AI-assisted drug design. AI, machine learning, and systems biology could be effectively harnessed for predicting resistance trends, optimisation of treatment regimens, and identification of novel therapeutic targets. Apart from conventional antibiotic discovery, anti-virulence approaches such as antitoxin and antibiofilm therapies, microbiome modulation, bacteriophage therapy, and host-directed therapeutics, as discussed in the earlier sections of this chapter, need to be explored further to reduce antibiotic use. Research on polyvalent phages that could prevent bacterial resistance should be strengthened. Further, prevention and care remain the major tools to reduce the use of antibiotics and other antimicrobials and thereby prevent the spread of AMR. Investments in identifying newer vaccines and strengthening vaccination programmes remain the cornerstone for prevention strategies. Additionally, improved sanitation and infection control further reduce reliance on antibiotics. AMR is multisectoral and is not restricted by boundaries. Hence strengthening of 'One Health' approaches to restrict AMR in sectors spanning medical health, environmental, veterinary, and plant is important.

1.10 Conclusion

Antibiotic resistance stands as one of the most pressing biological and public health challenges of the 21st century. The complex evolutionary adaptability of microbes has led to the development of different mechanisms to survive the antibiotic pressure. As discussed in this chapter, resistance emerges through a complex and dynamic interplay of genetic mutations and HGT. The different mechanisms of resistance work individually or synergistically, making resilient microbial populations capable of evading the drugs. While microbial evolution cannot be halted, its trajectory can be strategically influenced. Robust antibiotic stewardship, evidence-based policy reforms, and sustained national and global collaboration are imperative to curb the acceleration of antibiotic resistance in microbes. A comprehensive approach integrating mechanistic understanding and practical interventions is imperative in preserving the effectiveness of existing antibiotics and further development of new therapeutic strategies. Further, advances in genomics, AI and rapid molecular diagnostics hold immense potential to reform infection detection and treatment. Additionally, strengthened infection prevention and control strategies at the national and international levels, with incentivised surveillance and vaccination programmes, are essential to reduce transmission of resistant pathogens and minimise antibiotic selective pressure.

Acknowledgement

The authors would like to acknowledge BioRender for providing a platform to design images.

References

Abdesselam, K., Muchaal, P.K., Mishra, A., Ngendabanka, R.J., Amaratunga, K., Mittal, S., et al., 2026. AMR surveillance in Canada: insights from the 2025 priority pathogen list. PLOS ONE 21 (2). Available at https://doi.org/10.1371/journal.pone.0341133, https://journals.plos.org/plosone/article?id=10.1371/journal.pone.0341133.

Ackers, L., Ackers-Johnson, G., Welsh, J., Kibombo, D., Opio, S., 2020. Infection Prevention Control (IPC) and Antimicrobial Resistance (AMR). Springer Science and Business Media LLC, pp. 53–80. Available at http://doi.org/10.1007/978-3-030-62662-4_4.

Ahmed, S.K., Hussein, S., Qurbani, K., Hussein Ibrahim, R., Fareeq, A., Mahmood, K.A., et al., 2024. Antimicrobial resistance: impacts, challenges, and future prospects. Journal of Medicine, Surgery, and Public Health 2, 100081. Available at https://doi.org/10.1016/j.glmedi.2024.100081.

Ajulo, S., Awosile, B., 2024. Balew Arega Negatie, Global antimicrobial resistance and use surveillance system (GLASS 2022): investigating the relationship between antimicrobial resistance and antimicrobial consumption data across the participating countries. PLOS ONE 19 (2), e0297921. Available at https://doi.org/10.1371/journal.pone.0297921.

Alenzi, S.L.O., Alharthi, A.H., Alnefaie, R.A., Alzahrani, A., Alsaadi, O., Albugami, A.N.M., et al., 2026. The Global epidemiology of antimicrobial resistance: trends, determinants, and public health implications. Cureus 18 (1). Available at https://doi.org/10.7759/cureus.100784.

Aminov, R.I., 2010. A Brief history of the antibiotic era: lessons learned and challenges for the future. Frontiers in Microbiology 1. Available at https://doi.org/10.3389/fmicb.2010.00134.

Ayukekbong, J.A., Ntemgwa, M., Atabe, A.N., 2017. The threat of antimicrobial resistance in developing countries: causes and control strategies. Antimicrobial Resistance and Infection Control 6 (1). Available at https://doi.org/10.1186/s13756-017-0208-x, http://www.aricjournal.com/.

Badger-Emeka, L., Al Rashed, A.S., Aljindan, R.Y., Emeka, P.M., Quadri, S.A., Almutairi, H.H., 2023. Incidence of drug-resistant hospital-associated gram-negative bacterial infections, the accompanying risk factors, and clinical outcomes with treatment. Antibiotics 12 (9). Available at https://doi.org/10.3390/antibiotics12091425, http://www.mdpi.com/journal/antibiotics.

Baker, S., 2021. Challenges and opportunities in antimicrobial resistance research. Microbiology 167 (1). Available at https://doi.org/10.1099/mic.0.000999.

Bastola, A., Pyakurel, P., Rayamajhi, R.B., Shrestha, S., Thekkur, P., Pandey, B., et al., 2021. High resistance of *Salmonella* spp. And shigella spp. in blood and stool cultures from the sukraraj tropical and infectious disease hospital, kathmandu, nepal, 2015–2019. Tropical Medicine and Infectious Disease 6 (2). Available at https://doi.org/10.3390/tropicalmed6020059, https://www.mdpi.com/2414-6366/6/2/59/pdf.

Blaskovich, M.A.T., Cooper, M.A., 2025. Antibiotics re-booted—time to kick back against drug resistance. npj Antimicrobials and Resistance 3 (1). Available at https://doi.org/10.1038/s44259-025-00096-1.

Boyd, N.K., Teng, C., Frei, C.R., 2021. Brief overview of approaches and challenges in new antibiotic development: a focus on drug repurposing. Frontiers in Cellular and Infection Microbiology 11. Available at https://doi.org/10.3389/fcimb.2021.684515, http://www.frontiersin.org/Cellular_and_Infection_Microbiology/archive.

Breijyeh, Z., Jubeh, B., Karaman, R., 2020. Resistance of gram-negative bacteria to current antibacterial agents and approaches to resolve it. Molecules 25 (6), 1340. Available at https://doi.org/10.3390/molecules25061340.

Bush, K., Bradford, P.A., 2016. β-Lactams and β-lactamase inhibitors: an overview. Cold Spring Harbor Perspectives in Medicine 6 (8), a025247. Available at https://doi.org/10.1101/cshperspect.a025247.

Capuozzo, M., Zovi, A., Langella, R., Ottaiano, A., Cascella, M., Scognamiglio, M., et al., 2024. Optimizing antibiotic use: addressing resistance through effective strategies and health policies. Antibiotics 13 (12), 1112. Available at https://doi.org/10.3390/antibiotics13121112.

Cave, R., Cole, J., Mkrtchyan, H.V., 2021. Surveillance and prevalence of antimicrobial resistant bacteria from public settings within urban built environments: challenges and opportunities for hygiene and infection control. Environment International 157. Available at https://doi.org/10.1016/j.envint.2021.106836, www.elsevier.com/locate/envint.

Christensen, S.B., 2021. Drugs that changed society: history and current status of the early antibiotics: salvarsan, sulfonamides, and β-lactams. Molecules 26 (19), 6057. Available at https://doi.org/10.3390/molecules26196057.

Colavecchio, A., Cadieux, B., Lo, A., Goodridge, L.D., 2017. Bacteriophages contribute to the spread of antibiotic resistance genes among foodborne pathogens of the Enterobacteriaceae family - a review. Frontiers in Microbiology 8. Available at https://doi.org/10.3389/fmicb.2017.01108, http://journal.frontiersin.org/article/10.3389/fmicb.2017.01108/full.

Currie, K., Laidlaw, R., Ness, V., Gozdzielewska, L., Malcom, W., Sneddon, J., et al., 2020. Mechanisms affecting the implementation of a national antimicrobial stewardship programme; Multi-professional perspectives explained using normalisation process theory. Antimicrobial Resistance and Infection Control 9 (1). Available at https://doi.org/10.1186/s13756-020-00767-w, http://www.aricjournal.com/.

Dadgostar, P., 2019. Antimicrobial resistance: implications and costs. Infection and Drug Resistance 12, 3903–3910. Available at https://doi.org/10.2147/IDR.S234610, https://www.dovepress.com/getfile.php?fileID=54884.

Dhingra, S., Rahman, N.A.A., Peile, E., Rahman, M., Sartelli, M., Hassali, M.A., et al., 2020. Microbial resistance movements: an overview of global public health threats posed by antimicrobial resistance, and how best to counter. Frontiers in Public Health 8. Available at https://doi.org/10.3389/fpubh.2020.535668, http://journal.frontiersin.org/journal/public-health/section/public-health-education-and-promotion#archive.

de Oliveira Santos, J.V., da Costa Júnior, S.D., de Fátima Ramos dos Santos Medeiros, S.M., Cavalcanti, I.D.L., de Souza, J.B., Coriolano, D.L., et al., 2022. Panorama of bacterial infections caused by epidemic resistant strains. Current Microbiology 79 (6). Available at https://doi.org/10.1007/s00284-022-02875-9, link.springer.de/link/service/journals/00284/index.htm.

Drake, A., Sassoon, I., Armitage, J., Abbas, S., Maudling, R., Gupta-Wright, A., et al., 2026. Country governance of antimicrobial resistance (AMR) surveillance: observations on global progress and aid programme effectiveness using data from the tracking AMR country self-assessment survey (TrACSS). Globalization and Health 22 (1). Available at https://doi.org/10.1186/s12992-025-01179-4.

Durrance-Bagale, A., Jung, A.S., Frumence, G., Mboera, L., Mshana, S.E., Sindato, C., et al., 2021. Framing the drivers of antimicrobial resistance in Tanzania. Antibiotics 10 (8). Available at https://doi.org/10.3390/antibiotics10080991, https://www.mdpi.com/2079-6382/10/8/991/pdf.

Endale, H., Mathewos, M., Abdeta, D., 2023. Potential causes of spread of antimicrobial resistance and preventive measures in one health perspective-a review. Infection and Drug Resistance 16, 7515–7545. Available at https://doi.org/10.2147/IDR.S428837, https://www.dovepress.com/getfile.php?fileID=95065.

Enshaie, E., Nigam, S., Patel, S., Rai, V., 2025. Livestock antibiotics use and antimicrobial resistance. Antibiotics 14 (6), 621. Available at https://doi.org/10.3390/antibiotics14060621.

FAO and WHO, 2019. Joint FAO/WHO food standards programme-ad hoc codex intergovernmental task force on antimicrobial resistance. Available at http://www.fao.org/fao-who-codexalimentarius/sh-proxy/en/?lnk=1&url=https%253A%252F%252Fworkspace.fao.org%252Fsites%252Fcodex%252FMeetings%252FCX-804-07%252FWDs%252Famr07_03e.pdf.

Figueras, A., 2024. A global call to action on antimicrobial resistance at the UN general assembly. Antibiotics 13 (10). Available at https://doi.org/10.3390/antibiotics13100915, http://www.mdpi.com/journal/antibiotics.

Fridkin, S.K., Cleveland, A.A., See, I., Lynfield, R., 2015. Emerging infections program as surveillance for antimicrobial drug resistance. Emerging Infectious Diseases 21 (9), 1578–1581. Available at https://doi.org/10.3201/eid2109.150512, http://wwwnc.cdc.gov/eid/article/21/9/pdfs/15-0512.pdf.

Fujimoto, K., Uematsu, S., 2022. Phage therapy for *Clostridioides difficile* infection. Frontiers in Immunology 13. Available at https://doi.org/10.3389/fimmu.2022.1057892.

Gan, M., Zhang, Y., Yan, G., Wang, Y., Lu, G., Wu, B., et al., 2024. Antimicrobial resistance prediction by clinical metagenomics in pediatric severe pneumonia patients. Annals of Clinical Microbiology and Antimicrobials 23 (1). Available at https://doi.org/10.1186/s12941-024-00690-7.

Gargate, N., Laws, M., Rahman, K.M., 2025. Current economic and regulatory challenges in developing antibiotics for Gram-negative bacteria. npj Antimicrobials and Resistance 3 (1). Available at https://doi.org/10.1038/s44259-025-00123-1.

Gauba, A., Rahman, K.M., 2023. Evaluation of antibiotic resistance mechanisms in gram-negative bacteria. Antibiotics 12 (11), 1590. Available at https://doi.org/10.3390/antibiotics12111590.

Gaurav, A., Bakht, P., Saini, M., Pandey, S., Pathania, R., 2023. Role of bacterial efflux pumps in antibiotic resistance, virulence, and strategies to discover novel efflux pump inhibitors. Microbiology 169 (5). Available at https://doi.org/10.1099/mic.0.001333.

Gaynes, R., 2017. The discovery of penicillin—new insights after more than 75 years of clinical use. Emerging Infectious Diseases 23 (5), 849–853. Available at https://doi.org/10.3201/eid2305.161556.

Gelpi, A., Gilbertson, A., Tucker, J.D., 2015. Magic bullet: Paul Ehrlich, Salvarsan and the birth of venereology. Sexually Transmitted Infections 91 (1), 68–69. Available at https://doi.org/10.1136/sextrans-2014-051779, http://sti.bmj.com/content/91/1/68.2.full.pdf+html.

Gill, P., Ghaemi, A., 2008. Nucleic acid isothermal amplification technologies—a review. Nucleosides, Nucleotides and Nucleic Acids 27 (3), 224–243. Available at https://doi.org/10.1080/15257770701845204.

Gould, K., 2016. Antibiotics: from prehistory to the present day. Journal of Antimicrobial Chemotherapy 71 (3), 572–575. Available at https://doi.org/10.1093/jac/dkv484.

Gray-Hammerton, C.J., Hooton, S.P., Sands, K., Walsh, T.R., Orbegozo Rubio, C., Portal, E.A.R., Hudson, C., Stekel, D.J., Dodd, C.E.R., Hobman, J.L., 2025. Sublethal concentrations of antibiotics enhance transmission of antibiotic resistance genes in environmental Escherichia coli. Frontiers in Microbiology 16. Available at https://doi.org/10.3389/fmicb.2025.1675089.

Gunasekara, Y.D., Bailey, K.E., Scarborough, R.O., Sri, A.E., Coppo, M.J.C., Gilkerson, J.R., et al., 2025. National action plan on antimicrobial resistance in selected Asia-Pacific low- and middle-income countries: perspectives of one health stakeholders. One Health 21. Available at https://doi.org/10.1016/j.onehlt.2025.101259, https://www.sciencedirect.com/science/journal/23527714.

Hassall, J., Coxon, C., Patel, V.C., Goldenberg, S.D., Sergaki, C., 2024. Limitations of current techniques in clinical antimicrobial resistance diagnosis: examples and future prospects. npj Antimicrobials and Resistance 2 (1). Available at https://doi.org/10.1038/s44259-024-00033-8.

Hays, E.E., Wells, I.C., Katzman, P.A., Cain, C.K., Jacobs, F.A., Thayer, S.A., et al., 1945. Antibiotic substances produced by pseudomonas aeruginosa. Journal of Biological Chemistry 159 (3), 725–750. Available at https://doi.org/10.1016/s0021-9258(17)41580-8.

Hetsa, B.A., Asante, J., Amoako, D.G., Abia, A.L.K., Mbanga, J., Essack, S.Y., 2025. Multidrug-resistant ESKAPEEc pathogens from bloodstream infections in South Africa: a cross-sectional study assessing resistance to WHO AWaRe antibiotics. Health Science Reports 8 (6). Available at https://doi.org/10.1002/hsr2.70897, onlinelibrary.wiley.com/journal/23988835.

Hibbard, R., Mendelson, M., Page, S.W., Ferreira, J.P., Pulcini, C., Paul, M.C., et al., 2024. Antimicrobial stewardship: a definition with a one health perspective. npj Antimicrobials and Resistance 2 (1). Available at https://doi.org/10.1038/s44259-024-00031-w.

Hibstu, Z., Belew, H., Akelew, Y., Mengist, H.M., 2022. Phage therapy: a different approach to fight bacterial infections. Biologics: Targets and Therapy 16, 173–186. Available at https://doi.org/10.2147/BTT.S381237, https://www.dovepress.com/getfile.php?fileID=84479.

Ibrahim, M., Ahmad, F., Yaqub, B., Ramzan, A., Imran, A., Afzaal, M., et al., 2019. Current trends of antimicrobials used in food animals and aquaculture. Antibiotics and Antimicrobial Resistance Genes in the Environment: Volume 1 in the Advances in Environmental Pollution Research Series. Elsevier, Pakistan, pp. 39–69. Available at http://www.sciencedirect.com/science/book/9780128188828.

Jenison, R., Rihanek, M., Polisky, B., 2001. Use of a thin film biosensor for rapid visual detection of PCR products in a multiplex format. Biosensors and Bioelectronics 16, 757–763. Available at https://doi.org/10.1016/s0956-5663(01)00217-2.

Kajihara, T., Yahara, K., Hirabayashi, A., Shibayama, K., Sugai, M., 2021. Japan nosocomial infections surveillance (JANIS): current status, international collaboration, and future directions for a comprehensive antimicrobial resistance surveillance system. Japanese Journal of Infectious Diseases 74 (2), 87–96. Available at https://doi.org/10.7883/yoken.jjid.2020.499.

Kaprou, G.D., Bergšpica, I., Alexa, E.A., Alvarez-Ordóñez, A., Prieto, M., 2021. Rapid methods for antimicrobial resistance diagnostics. Antibiotics 10 (2), 1–30. Available at https://doi.org/10.3390/antibiotics10020209, https://www.mdpi.com/2079-6382/10/2/209/pdf.

Karp, B.E., Tate, H., Plumblee, J.R., Dessai, U., Whichard, J.M., Thacker, E.L., et al., 2017. National antimicrobial resistance monitoring system: two decades of advancing public health through integrated surveillance of antimicrobial resistance. Foodborne Pathogens and Disease 14 (10), 545–557. Available at https://doi.org/10.1089/fpd.2017.2283, www.liebertonline.com/fpd.

Karukappadath, R.M., Sirbu, D., Zaky, A., 2023. Drug-resistant bacteria in the critically ill: patterns and mechanisms of resistance and potential remedies. Frontiers in Antibiotics 2. Available at https://doi.org/10.3389/frabi.2023.1145190, https://www.frontiersin.org/journals/antibiotics.

Kashef, N., Hamblin, M.R., 2017. Can microbial cells develop resistance to oxidative stress in antimicrobial photodynamic inactivation? Drug Resistance Updates 31, 31–42. Available at https://doi.org/10.1016/j.drup.2017.07.003, http://www.elsevier.com/locate/drup.

Kim, B., Park, J.S., Choi, H.Y., Yoon, S.S., Kim, W.G., 2018. Terrein is an inhibitor of quorum sensing and c-di-GMP in *Pseudomonas aeruginosa*: a connection between quorum sensing and c-di-GMP. Scientific Reports 8 (1). Available at https://doi.org/10.1038/s41598-018-26974-5, www.nature.com/srep/index.html.

Kim, H., Park, J.S., Kim, D., Kim, H.J., Shin, J.H., Kim, Y.A., et al., 2024. Standardization of an antimicrobial resistance surveillance network through data management. Frontiers in Cellular and Infection Microbiology 14. Available at https://doi.org/10.3389/fcimb.2024.1411145, http://www.frontiersin.org/Cellular_and_Infection_Microbiology/archive.

Kotwani, A., Joshi, J., Lamkang, A.S., 2021. Over-the-counter sale of antibiotics in India: a qualitative study of providers' perspectives across two states. Antibiotics 10 (9), 1123. Available at https://doi.org/10.3390/antibiotics10091123.

Latif, A., Shehzad, A., Niazi, S., Zahid, A., Ashraf, W., Iqbal, M.W., et al., 2023. Probiotics: mechanism of action, health benefits and their application in food industries. Frontiers in Microbiology 14. Available at https://doi.org/10.3389/fmicb.2023.1216674, https://www.frontiersin.org/journals/microbiology#.

Lekagul, A., Tangcharoensathien, V., Yeung, S., 2019. Patterns of antibiotic use in global pig production: a systematic review. Veterinary and Animal Science 7, 100058. Available at https://doi.org/10.1016/j.vas.2019.100058.

Li, W., Huang, T., Liu, C., Wushouer, H., Yang, X., Wang, R., Xia, H., Li, X., Qiu, S., Chen, S., Ho, H.C., Huang, C., Shi, L., Guan, X., Tian, G., Liu, G., Ebi, K.L., Yang, L., 2025. Changing climate and socioeconomic factors contribute to global antimicrobial resistance. Nature Medicine 31, 1798–1808. Available at https://doi.org/10.1038/s41591-025-03629-3.

Liu, Y.Y., Wang, Y., Walsh, T.R., Yi, L.X., Zhang, R., Spencer, J., et al., 2016. Emergence of plasmid-mediated colistin resistance mechanism MCR-1 in animals and human beings in China: a microbiological and molecular biological study. The Lancet Infectious Diseases 16 (2), 161–168. Available at https://doi.org/10.1016/S1473-3099(15)00424-7, http://www.journals.elsevier.com/the-lancet-infectious-diseases.

Liu, Y., Shi, D., Guo, Y., Li, M., Zha, Y., Wang, Q., et al., 2017. Dracorhodin *Perochlorate attenuates Staphylococcus aureus* USA300 virulence by decreasing α-toxin expression. World Journal of Microbiology and Biotechnology 33 (1). Available at https://doi.org/10.1007/s11274-016-2129-x.

Lu, Y., Gao, J., Zhang, D.D., Gau, V., Liao, J.C., Wong, P.K., 2013. Single cell antimicrobial susceptibility testing by confined microchannels and electrokinetic loading. Analytical Chemistry 85 (8), 3971–3976. Available at https://doi.org/10.1021/ac4004248.

Lu, J., Sheldenkar, A., Lwin, M.O., 2020. A decade of antimicrobial resistance research in social science fields: a scientometric review. Antimicrobial Resistance and Infection Control 9 (1). Available at https://doi.org/10.1186/s13756-020-00834-2, http://www.aricjournal.com/.

Manyi-Loh, C., Mamphweli, S., Meyer, E., Okoh, A., 2018. Antibiotic use in agriculture and its consequential resistance in environmental sources: potential public health implications. Molecules 23 (4), 795. Available at https://doi.org/10.3390/molecules23040795.

McSorley, J.C., 2026. Analysis of ESAC-Net/EARS-Net Data from 29 EEA Countries for spatiotemporal associations between antimicrobial use and resistance-implications for antimicrobial stewardship? Antibiotics (Basel) 15 (3), 284. Available at https://doi.org/10.3390/antibiotics15030284. Erratum for: Antibiotics (Basel). 2025 14(4),399. https://doi.org/10.3390/antibiotics14040399. (2025).

Meckawy, R., Stuckler, D., Mehta, A., Al-Ahdal, T., Doebbeling, B.N., 2022. Effectiveness of early warning systems in the detection of infectious diseases outbreaks: a systematic review. BMC Public Health 22 (1). Available at https://doi.org/10.1186/s12889-022-14625-4, https://bmcpublichealth.biomedcentral.com/.

Mehrotra, T., Konar, D., Pragasam, A.K., Kumar, S., Jana, P., Babele, P., et al., 2023. Antimicrobial resistance heterogeneity among multidrug-resistant Gram-negative pathogens: phenotypic, genotypic, and proteomic analysis. Proceedings of the National Academy of Sciences of the United States of America 120 (33). Available at https://doi.org/10.1073/pnas.2305465120, www.pnas.org.

Murray, C., Ikuta, K., Sharara, F., et al., 2022. Global burden of bacterial antimicrobial resistance in 2019: a systematic analysis. The Lancet 300, 629–655. Available at https://www.thelancet.com/journals/lancet/article/PIIS0140-6736(21)02724-0/fulltext.

Narendrakumar, L., Chandrika, S.K., Thomas, S., 2020. Adaptive laboratory evolution of Vibrio cholerae to doxycycline associated with spontaneous mutation. International Journal of Antimicrobial Agents 56 (3), 106097. Available at https://doi.org/10.1016/j.ijantimicag.2020.106097.

Narendrakumar, L., Chakraborty, M., Kumari, S., Paul, D., Das, B., 2023. β-Lactam potentiators to re-sensitize resistant pathogens: discovery, development, clinical use and the way forward. Frontiers in Microbiology 13. Available at https://doi.org/10.3389/fmicb.2022.1092556.

Naylor, N.R., Atun, R., Zhu, N., Kulasabanathan, K., Silva, S., Chatterjee, A., et al., 2018. Estimating the burden of antimicrobial resistance: a systematic literature review. Antimicrobial Resistance and Infection Control 7 (1). Available at https://doi.org/10.1186/s13756-018-0336-y, https://aricjournal.biomedcentral.com/.

Nusrat, S., Aliyu, M., Zohora, F.T., 2025. Mechanisms of antimicrobial resistance: from genetic evolution to clinical manifestations. AIMS Microbiology 11 (4), 1007–1034. Available at https://doi.org/10.3934/microbiol.2025045, https://www.aimspress.com/aimspress-data/aimsmicro/2025/4/PDF/microbiol-11-04-045.pdf.

O'Neill, J.I.M., 2014. Antimicrobial resistance: tackling a crisis for the health and wealth of nations. The Review on Antimicrobial Resistance. Available at http://amr-review.org/sites/default/files/160525_Final%20paper_with%20cover.pdf.

Otaigbe, I.I., Elikwu, C.J., 2023. Drivers of inappropriate antibiotic use in low- and middle-income countries. JAC-Antimicrobial Resistance 5 (3). Available at https://doi.org/10.1093/jacamr/dlad062, https://academic.oup.com/jacamr.

Paul, D., Verma, J., Banerjee, A., Konar, D., Das, B., 2022. Antimicrobial resistance traits and resistance mechanisms in bacterial pathogens. In: Kumar, V., Shriram, V., Paul, A., Thakur, M. (Eds.), Antimicrobial Resistance. Springer, Singapore. Available at https://doi.org/10.1007/978-981-16-3120-7_1.

Paul, D., Chawla, M., Ahrodia, T., Narendrakumar, L., Das, B., 2023. Antibiotic potentiation as a promising strategy to combat macrolide resistance in bacterial pathogens. Antibiotics 12 (12), 1715. Available at https://doi.org/10.3390/antibiotics12121715.

Paul, D., Talukdar, D., Kapuganti, R.S., Gupta, V., Narendrakumar, L., Jana, P., et al., 2026. Antibiotic contamination and antimicrobial resistance dynamics in the urban sewage microbiome in India. Nature Communications 17 (1). Available at https://doi.org/10.1038/s41467-025-68034-3, https://www.nature.com/ncomms/.

Peirano, G., Ahmed-Bentley, J., Fuller, J., Rubin, J.E., Pitout, J.D.D., 2014. Travel-related carbapenemase-producing Gram-negative bacteria in Alberta, Canada: The first 3 years. Journal of Clinical Microbiology 52 (5), 1575–1581. Available at https://doi.org/10.1128/JCM.00162-14, http://jcm.asm.org/content/52/5/1575.full.pdf+html.

Penesyan, A., Gillings, M., Paulsen, I., 2015. Antibiotic discovery: combatting bacterial resistance in cells and in biofilm communities. Molecules 20, 5286–5298. Available at https://doi.org/10.3390/molecules20045286.

Ping, O., Mao, S., Xuewen, H., Kaiyu, W., Zhongqiong, Y., Hualin, F., et al., 2017. *Sclareol protects Staphylococcus aureus*-induced lung cell injury via inhibiting alpha-hemolysin expression. Journal of Microbiology and Biotechnology 27 (1), 19–25. Available at https://doi.org/10.4014/jmb.1606.06039, http://www.jmb.or.kr/journal/download.php?Filedir=./submission/Journal/027/&num=8233.

Pradhan, D., Biswasroy, P., Goyal, A., Ghosh, G., Rath, G., 2021. Recent Advancement in nanotechnology-based drug delivery system against viral infections. AAPS PharmSciTech 22 (1). Available at https://doi.org/10.1208/s12249-020-01908-5.

Pujades-Rodriguez, M., West, R.M., Wilcox, M.H., Sandoe, J., 2019. Lower urinary tract infections: management, outcomes and risk factors for antibiotic re-prescription in primary care. EClinicalMedicine 14, 23–31. Available at https://doi.org/10.1016/j.eclinm.2019.07.012, www.thelancet.com/journals/eclinm/issue/current.

Pulingam, T., Parumasivam, T., Gazzali, A.M., Sulaiman, A.M., Chee, J.Y., Lakshmanan, M., Chin, C.F., Sudesh, K., 2022. Antimicrobial resistance: Prevalence, economic burden, mechanisms of resistance and strategies to overcome. European Journal of Pharmaceutical Sciences 170, 106103. Available at https://doi.org/10.1016/j.ejps.2021.106103.

Qin, X., Ding, L., Hao, M., Li, P., Hu, F., Wang, M., 2024. Antimicrobial resistance of clinical bacterial isolates in China: current status and trends. JAC-Antimicrobial Resistance 6 (2). Available at https://doi.org/10.1093/jacamr/dlae052.

Rajput, P., Nahar, K.S., Rahman, K.M., 2024. Evaluation of antibiotic resistance mechanisms in gram-positive bacteria. Antibiotics 13 (12). Available at https://doi.org/10.3390/antibiotics13121197, http://www.mdpi.com/journal/antibiotics.

Ramirez, M.S., Tolmasky, M.E., 2010. Aminoglycoside modifying enzymes. Drug Resistance Updates 13 (6), 151–171. Available at https://doi.org/10.1016/j.drup.2010.08.003, http://www.elsevier.com/locate/drup.

Reygaert, W.C., 2018. An overview of the antimicrobial resistance mechanisms of bacteria. AIMS Microbiology 4 (3), 482–501. Available at https://doi.org/10.3934/microbiol.2018.3.482.

Salam, M.A., Al-Amin, M.Y., Salam, M.T., Pawar, J.S., Akhter, N., Rabaan, A.A., et al., 2023. Antimicrobial resistance: a growing serious threat for global public health. Healthcare (Switzerland) 11 (13). Available at https://doi.org/10.3390/healthcare11131946, www.mdpi.com/journal/healthcare/.

Sarita, B., Samadhan, D., Hassan, M.Z., Kovaleva, E.G., 2024. A comprehensive review of probiotics and human health-current prospective and applications. Frontiers in Microbiology 15. Available at https://doi.org/10.3389/fmicb.2024.1487641, https://www.frontiersin.org/journals/microbiology#.

Sati, H., Carrara, E., Savoldi, A., Hansen, P., Garlasco, J., Campagnaro, E., et al., 2025. The WHO bacterial priority pathogens List 2024: a prioritisation study to guide research, development, and public health strategies against antimicrobial resistance. The Lancet Infectious Diseases 25 (9), 1033–1043. Available at https://doi.org/10.1016/S1473-3099(25)00118-5, https://www.sciencedirect.com/science/journal/14733099.

Singer, A.C., Shaw, H., Rhodes, V., Hart, A., 2016. Review of antimicrobial resistance in the environment and its relevance to environmental regulators. Frontiers in Microbiology 7. Available at https://doi.org/10.3389/fmicb.2016.01728, http://journal.frontiersin.org/article/10.3389/fmicb.2016.01728/full.

Solanki, S., Das, H.K., 2024. Antimicrobial resistance: molecular drivers and underlying mechanisms. JJournal of Medicine, Surgery, and Public Health 3, 100122.

Stamm, W.E., Norrby, S.R., 2001. Urinary tract infections: disease panorama and challenges. Journal of Infectious Diseases 183, S1. Available at https://doi.org/10.1086/318850.

Tadesse, B.T., Keddy, K.H., Rickett, N.Y., Zhusupbekova, A., Poudyal, N., Lawley, T., et al., 2023. Vaccination to reduce antimicrobial resistance burden - data gaps and future research. Clinical Infectious Diseases 77, S597. Available at https://doi.org/10.1093/cid/ciad562, http://cid.oxfordjournals.org/content/by/year.

Tamma, P.D., Souli, M., Billard, M., Campbell, J., Conrad, D., Ellison, D.W., et al., 2022. Safety and microbiological activity of phage therapy in persons with cystic fibrosis colonized with *Pseudomonas aeruginosa*: study protocol for a phase 1b/2, multicenter, randomized, double-blind, placebo-controlled trial. Trials 23 (1). Available at https://doi.org/10.1186/s13063-022-07047-5, https://trialsjournal.biomedcentral.com/.

The antibiotic alarm, 2013. Nature 495 (7440), 141. Available at https://doi.org/10.1038/495141a –141.

Totsika, M., 2016. Benefits and challenges of antivirulence antimicrobials at the dawn of the post-antibiotic Era. Drug Delivery Letters 6 (1), 30–37. Available at https://doi.org/10.2174/2210303106666160506120057.

Van Boeckel, T.P., Brower, C., Gilbert, M., Grenfell, B.T., Levin, S.A., Robinson, T.P., et al., 2015. Global trends in antimicrobial use in food animals. Proceedings of the National Academy of Sciences of the United States of America 112 (18), 5649–5654. Available at https://doi.org/10.1073/pnas.1503141112, http://www.pnas.org/content/112/18/5649.full.pdf.

Vattanaviboon, P., Dulyayangkul, P., Tipanyo, P., Mongkolsuk, S., Charoenlap, N., 2026. Acquired resistance in *Stenotrophomonas maltophilia:* mechanisms underlying the shift from multidrug to pandrug resistance. European Journal of Microbiology and Immunology 16 (1), 40–52. Available at https://doi.org/10.1556/1886.2026.00004.

Vincent, C.I., Hijaz, F., Pierre, M., Killiny, N., 2022. Systemic uptake of oxytetracycline and streptomycin in huanglongbing-affected citrus groves after foliar application and Trunk injection. Antibiotics 11 (8), 1092. Available at https://doi.org/10.3390/antibiotics11081092.

Vivekanandan, K.E., Vinoth Kumar, P., Jaysree, R.C., Rajeshwari, T., 2025. Exploring molecular mechanisms of drug resistance in bacteria and progressions in CRISPR/Cas9-based genome expurgation solutions. Global Medical Genetics 12 (2), 100042. Available at https://doi.org/10.1016/j.gmg.2025.100042.

Walia, K., Madhumathi, J., Veeraraghavan, B., Chakrabarti, A., Kapil, A., Ray, P., et al., 2019. Establishing antimicrobial resistance surveillance & research network in India: journey so far. Indian Journal of Medical Research 149 (2), 164–179. Available at https://doi.org/10.4103/ijmr.IJMR_226_18, http://www.ijmr.org.in/.

Wang, M., Zhang, Y., Pei, F., Liu, Y., Zheng, Y., 2025. Loss of OprD function is sufficient for carbapenem-resistance-only but insufficient for multidrug resistance in *Pseudomonas aeruginosa*. BMC Microbiology 25 (1). Available at https://doi.org/10.1186/s12866-025-03935-3.

Wang, Y., Xu, C., Zhang, R., Chen, Y., Shen, Y., Hu, F., et al., 2020. Changes in colistin resistance and mcr-1 abundance in Escherichia coli of animal and human origins following the ban of colistin-positive additives in China: an epidemiological comparative study. The Lancet Infectious Diseases 20, 1161–1171. Available at https://doi.org/10.1016/s1473-3099(20)30149-3.

WHO, 2015. Global Antimicrobial Resistance Surveillance System-Manual for early Implementation. World Health Organization, Geneva, Switzerland. Available at https://iris.who.int/server/api/core/bitstreams/1d95867b-b505-4f09-bb51-8f3de7a369ab/content.

WHO, 2024. Bacterial Priority Pathogens List, 2024: Bacterial Pathogens of Public Health Importance to Guide Research, Development and Strategies to Prevent and Control Antimicrobial Resistance. World Health Organization, Geneva Licence: CC BY-NC-SA 3.0 IGO. Available at https://www.who.int/publications/i/item/9789240093461.

WHO European Office. AMR dashboard. CAESAR methodology, 2025. Available from: https://whoamrdashboard.shinyapps.io/WHO-AMR-Dashboard/.

World Bank, 2017. Drug-Resistant Infections: A threat to our Economic Future. World Bank, Washington, DC. Available at https://documents1.worldbank.org/curated/en/323311493396993758/pdf/final-report.pdf.

WRIGHT, G., 2005. Bacterial resistance to antibiotics: enzymatic degradation and modification. Advanced Drug Delivery Reviews 57 (10), 1451–1470. Available at https://doi.org/10.1016/j.addr.2005.04.002.

Yousef, F., Mansour, O., Herbali, J., 2018. Sulfonamides: historical discovery development (structure-activity relationship notes). In-Vitro In-Vivo In-Silico Journal 06, 1–15 (2018).

Zhang, B., Teng, Z., Li, X., Lu, G., Deng, X., Niu, X., et al., 2017. Chalcone attenuates *Staphylococcus aureus* virulence by targeting sortase a and alpha-hemolysin. Frontiers in Microbiology 8. Available at https://doi.org/10.3389/fmicb.2017.01715.

Zhang, M., Seleem, M.N., Cheng, J.X., 2022. Rapid antimicrobial susceptibility testing by stimulated raman scattering imaging of deuterium incorporation in a single bacterium. Journal of Visualized Experiments 2022 (180). Available at https://doi.org/10.3791/62398, https://www.jove.com/t/63087/the-microscopy-based-assay-to-study-analyze-recycling-endosomes-using.

Zheng, Y., Zhang, Z., Tang, P., Wu, Y., Zhang, A., Li, D., et al., 2023. Probiotics fortify intestinal barrier function: a systematic review and meta-analysis of randomized trials. Frontiers in Immunology 14. Available at https://doi.org/10.3389/fimmu.2023.1143548, https://www.frontiersin.org/journals/immunology#.

Zhou, G., Wang, Q., Wang, Y., Wen, X., Peng, H., Peng, R., et al., 2023. Outer membrane porins contribute to antimicrobial resistance in gram-negative bacteria. Microorganisms 11 (7), 1690. Available at https://doi.org/10.3390/microorganisms11071690.

Further reading

Anderson, Michael, Schulze, Kai, Cassini, Alessandro, Plachouras, Diamantis, Mossialos, Elias, 2019. A governance framework for development and assessment of national action plans on antimicrobial resistance. The Lancet Infectious Diseases 19 (11), e371. Available at https://doi.org/10.1016/s1473-3099(19)30415-3.

Arcilla, M.S., van Hattem, J.M., Haverkate, M.R., Bootsma, M.C.J., van Genderen, P.J.J., Goorhuis, A., et al., 2017. Import and spread of extended-spectrum β-lactamase-producing Enterobacteriaceae by international travellers (COMBAT study): a prospective, multicentre cohort study. The Lancet Infectious Diseases 17 (1), 78–85. Available at https://doi.org/10.1016/S1473-3099(16)30319-X, http://www.journals.elsevier.com/the-lancet-infectious-diseases.

Diallo, O.O., Baron, S.A., Abat, C., Colson, P., Chaudet, H., Rolain, J.M., 2020. Antibiotic resistance surveillance systems: a review. Journal of Global Antimicrobial Resistance 23, 430–438. Available at https://doi.org/10.1016/j.jgar.2020.10.009, http://www.journals.elsevier.com/journal-of-global-antimicrobial-resistance/.

Drug-Resistant Infections, 2017. A Threat to Our Economic future (March 2017). The World Bank Group. Available at https://documents1.worldbank.org/curated/en/323311493396993758/pdf/final-report.pdf.

Gach, M.W., Lazarus, G., Simadibrata, D.M., Sinto, R., Saharman, Y.R., Limato, R., et al., 2024. Antimicrobial resistance among common bacterial pathogens in Indonesia: a systematic review. The Lancet Regional Health - Southeast Asia 26. Available at https://doi.org/10.1016/j.lansea.2024.100414, https://www.sciencedirect.com/science/journal/27723682.

Ho, C.S., Wong, C.T.H., Aung, T.T., Lakshminarayanan, R., Mehta, J.S., Rauz, S., et al., 2025. Antimicrobial resistance: a concise update. The Lancet Microbe 6 (1). Available at https://doi.org/10.1016/j.lanmic.2024.07.010, https://www.sciencedirect.com/science/journal/26665247.

Jenison, Robert, Yang, Shao, Haeberli, Ayla, Polisky, Barry, 2001. Interference-based detection of nucleic acid targets on optically coated silicon. Nature Biotechnology 19 (1), 62–65. Available at https://doi.org/10.1038/83530.

Naghavi, Mohsen, Vollset, Stein Emil, Ikuta, Kevin S., Swetschinski, Lucien R., Gray, Authia P., Wool, Eve E., et al., 2024. Global burden of bacterial antimicrobial resistance 1990–2021: a systematic analysis with forecasts to 2050. The Lancet 404 (10459), 1199–1226. Available at https://doi.org/10.1016/s0140-6736(24)01867-1.

Schmieder, Robert, Edwards, Robert, 2012. Insights into antibiotic resistance through metagenomic approaches. Future Microbiology 7 (1), 73–89. Available at https://doi.org/10.2217/fmb.11.135.

Sultan, I., Rahman, S., Jan, A.T., Siddiqui, M.T., Mondal, A.H., Haq, Q.M.R., 2018. Antibiotics, resistome and resistance mechanisms: a bacterial perspective. Frontiers in Microbiology 9. Available at https://doi.org/10.3389/fmicb.2018.02066, www.frontiersin.org/Microbiology.

Vong, S., Anciaux, A., Hulth, A., Stelling, J., Thamlikitkul, V., Gupta, S., et al., 2017. Using information technology to improve surveillance of antimicrobial resistance in South East Asia. BMJ (Online) 358, 51–55. Available at https://doi.org/10.1136/bmj.j3781, http://www.bmj.com/.

Wang, Y., Xu, C., Zhang, R., Chen, Y., Shen, Y., Hu, F., et al., 2020. Changes in colistin resistance and mcr-1 abundance in *Escherichia coli* of animal and human origins following the ban of colistin-positive additives in China: an epidemiological comparative study. The Lancet Infectious Diseases 20 (10), 1161–1171. Available at https://doi.org/10.1016/S1473-3099(20)30149-3, http://www.journals.elsevier.com/the-lancet-infectious-diseases.

Chapter 2

The science of resistance: genetic and biochemical mechanisms in antibiotic evasion

Itishree Pandey[1], Deepshikha Bhowmik[2], Deepshikha Nath[3], and Sameer Uttamrao Khasbage[4]
[1]*Department of Biochemistry, Raipur Institute of Medical Sciences, Raipur, Chhattisgarh, India,* [2]*MRU, ESIC Medical College and Hospital, Faridabad, Haryana, India,* [3]*Department of Biochemistry, Ruhr University Bochum, Bochum, Germany,* [4]*Department of Pharmacology, All India Institute of Medical Sciences (AIIMS), Raipur, Chhattisgarh, India*

2.1 Introduction

Antibiotic resistance is a natural evolutionary outcome shaped by the selective pressure imposed by antimicrobial use. When exposed to antibiotics, susceptible bacteria are eliminated while resistant variants proliferate, acquiring resistance through spontaneous genetic mutations or through the horizontal uptake of resistance determinants from neighbouring organisms (Davies and Davies, 2010; Blair et al., 2015; WHO, 2015). This pressure operates across human clinical settings, veterinary environments, and agricultural systems, driving a global convergence of resistance that recognises no geographic boundary.

At the molecular level, resistance is governed by the interplay of genetic and biochemical changes. Genetic alterations, including chromosomal point mutations and horizontal gene transfer (HGT), provide the blueprint, while their biochemical expression, through enzymatic drug inactivation, target modification, reduced membrane permeability, active drug efflux, and biofilm formation, constitutes the functional resistance phenotype. The cumulative burden is substantial: a 2022 global analysis attributed approximately 1.27 million deaths directly to bacterial antimicrobial resistance (AMR) in 2019, with a further 4.95 million deaths in which resistant infection played a contributing role (Murray et al., 2022). The World Health Organisation (WHO) has described this trajectory as a drift toward a post-antibiotic era (WHO, 2015).

India reflects and amplifies these global trends. Resistance among key Gram-negative pathogens such as *Escherichia coli*, *Klebsiella pneumoniae*, and *Acinetobacter baumannii* has reached critical levels, prompting the Indian Council of Medical Research (ICMR) to establish a National Action Plan on AMR. The misuse and over-the-counter availability of antibiotics in many low- and middle-income settings, alongside antibiotic use in livestock production, create conditions where resistance selection and dissemination are continuously reinforced (Arias and Murray, 2009).

This chapter traces the science of resistance from its genetic foundations through its biochemical effectors, examining how these mechanisms evolved, how they spread through microbial communities, and what clinical consequences they produce. It concludes with a critical evaluation of four emerging therapeutic strategies that offer mechanistically distinct alternatives to conventional antibiotics.

2.2 Fundamental concepts in antibiotic action

Antibiotics disrupt essential bacterial physiological processes that are either absent in, or sufficiently distinct from, eukaryotic cells to allow selective toxicity. Understanding the primary molecular targets of antibiotics is a prerequisite to understanding how resistance mechanisms subvert them. The major antibiotic classes and their mechanisms of action are summarised in Table 2.1.

Antimicrobial Resistance in the 21st Century. DOI: https://doi.org/10.1016/B978-0-443-44748-8.00022-1

TABLE 2.1 Classification of antibiotics by target site, mechanism of action, clinical examples, and bactericidal/bacteriostatic outcome.

Antibiotic class	Target site	Mechanism of action	Examples	Outcome	References
Cell wall synthesis	Peptidoglycan (cross-linking)	Beta-lactams bind PBPs and block transpeptidation; vancomycin blocks D-Ala-D-Ala precursors	*Penicillins, Cephalosporins, Carbapenems, Vancomycin*	Osmotic lysis (bactericidal)	Brunton et al. (2023), Katzung (2021), Murray et al. (2020), Rang et al. (2020)
Protein synthesis	30S ribosomal subunit	Block aminoacyl-tRNA entry or cause mRNA misreading	*Aminoglycosides, Tetracyclines*	Inhibited translation	
Protein synthesis	50S ribosomal subunit	Block peptidyl transfer or translocation	*Macrolides, Chloramphenicol, Linezolid*	Growth arrest or cell death	
Nucleic acid synthesis	DNA gyrase/ Topoisomerase IV	Stabilise cleavage complex, induce DNA strand breaks	*Fluoroquinolones*	Bactericidal	
Nucleic acid synthesis	RNA polymerase (beta subunit)	Block initiation of RNA transcription	*Rifampicin*	Bactericidal	
Membrane disruption	Cytoplasmic membrane/LPS	Depolarisation and pore formation	*Polymyxins, Daptomycin*	Rapid cell death (bactericidal)	

PBP, Penicillin-binding protein; *LPS*, lipopolysaccharide.

2.2.1 Cell wall synthesis inhibitors

The bacterial cell wall, composed of cross-linked peptidoglycan, maintains osmotic integrity. Beta-lactam antibiotics bind irreversibly to penicillin-binding proteins (PBPs), transpeptidases essential for cross-linking peptidoglycan strands, leading to structurally weak cell walls and osmotic lysis. Glycopeptides such as vancomycin act earlier in this pathway by binding the D-Ala-D-Ala termini of peptidoglycan precursors, sterically blocking both transglycosylation and transpeptidation. These agents are predominantly bactericidal (Brunton et al., 2023; Katzung, 2021).

2.2.2 Protein synthesis inhibitors

Bacterial ribosomes (70S), assembled from 30S and 50S subunits, differ structurally from eukaryotic ribosomes and are exploited by several antibiotic classes. Aminoglycosides bind the 30S subunit, causing mRNA misreading and production of aberrant proteins; they are bactericidal. Tetracyclines reversibly block aminoacyl-tRNA entry at the 30S subunit, producing bacteriostatic inhibition. At the 50S subunit, macrolides inhibit translocation, chloramphenicol blocks peptidyl transfer, and linezolid prevents formation of the initiation complex. These agents display both bacteriostatic and bactericidal profiles depending on species and drug class (Rang et al., 2020; Murray et al., 2020).

2.2.3 Nucleic acid synthesis inhibitors

Fluoroquinolones act on type II topoisomerases (DNA gyrase and topoisomerase IV), stabilising the enzyme-DNA cleavage complex and inducing double-strand breaks. Rifampicin targets the beta subunit of DNA-dependent RNA polymerase, blocking transcription initiation. Metronidazole undergoes intracellular reductive activation in anaerobes, producing toxic radical intermediates that damage DNA directly. All three classes are bactericidal (Brunton et al., 2023; Katzung, 2021; Rang et al., 2020).

2.2.4 Membrane-disrupting agents

Polymyxins bind the lipid A component of lipopolysaccharide (LPS) in Gram-negative outer membranes, causing membrane disruption and leakage of intracellular contents. Daptomycin inserts into the Gram-positive cytoplasmic membrane in a calcium-dependent manner, causing rapid depolarization. Both classes are bactericidal and are often reserved for resistant organisms due to their toxicity profiles (Brunton et al., 2023; Rang et al., 2020). Fig. 2.1 provides a schematic overview of antibiotic-target sites within the bacterial cell.

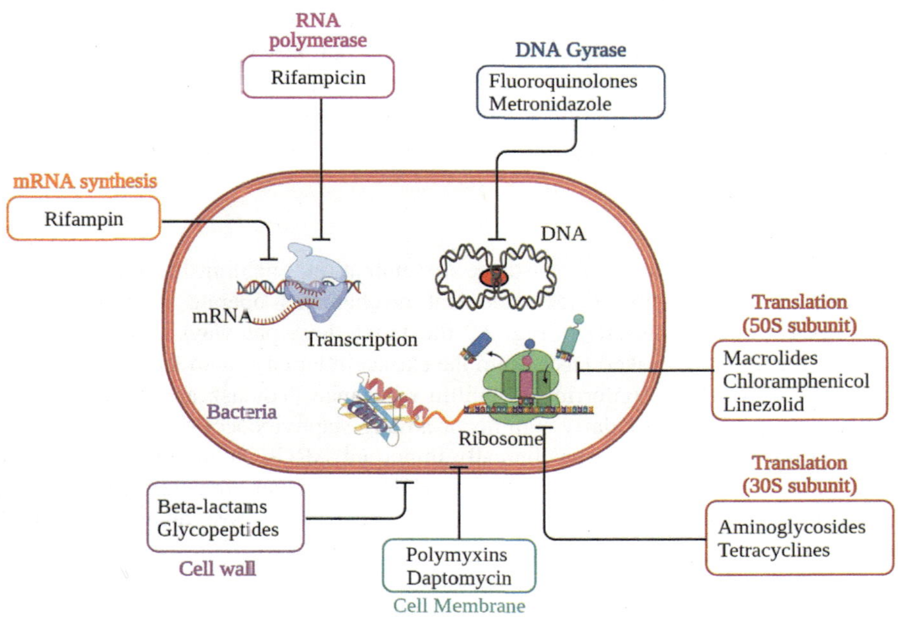

FIGURE 2.1 Schematic representation of the principal antibiotic-target sites within a bacterial cell. Chematic representation of the principal antibiotic target sites within a bacterial cell. Cell wall synthesis inhibitors (beta-lactams and glycopeptides), membrane disruptors (polymyxins and daptomycin), protein synthesis inhibitors (30S and 50S ribosomal subunit agents), and nucleic acid synthesis inhibitors (fluoroquinolones and rifampicin) are indicated at their respective sites of action.

2.3 Mechanisms of antibiotic resistance

Bacteria counteract antibiotic action through an integrated repertoire of genetic and biochemical strategies. These mechanisms are rarely expressed in isolation; their combination, particularly in multidrug-resistant (MDR) and extensively drug-resistant (XDR) organisms, can raise minimum inhibitory concentrations (MICs) far above therapeutically achievable concentrations. Table 2.2 provides a comparative overview of the principal biochemical resistance mechanisms.

2.3.1 Genetic basis of resistance

Antibiotic resistance in bacteria is either intrinsic, arising from inherent structural or physiological features of the organism, or acquired, resulting from de novo mutations or the horizontal uptake of exogenous resistance determinants (Habboush and Guzman, 2018). The collective pool of resistance genes within a bacterial population, whether expressed or silent, is termed the resistome (Kim and Cha, 2021). A recent survey of 25,285 bacterial genomes identified an average of 14.4 antibiotic resistance genes (ARGs) per organism and a total count exceeding 363,000 ARGs, underscoring the vast genetic reservoir available for selection (Liu et al., 2025). Resistance gene categories span all major antibiotic classes, as detailed in Table 2.3 (Lopatkin et al., 2016; Centner et al., 2025).

2.3.1.1 Chromosomal mutation

Spontaneous chromosomal mutations, arising from replication errors, DNA damage, or stress-induced mutagenesis, are the oldest and most fundamental route to resistance. Resistance to streptomycin typically results from amino acid substitutions in ribosomal protein S12 of the 30S subunit (Funatsu and Wittmann, 1972). Rifampicin resistance arises from mutations in the *rpoB* gene encoding the RNA polymerase beta subunit (Tocchini-Valentini et al., 1968). Fluoroquinolone resistance involves mutations in the quinolone resistance-determining regions (QRDRs) of *gyrA* and *parC* (Vivekanandan et al., 2023). Penicillin resistance develops through altered PBPs, elevated beta-lactamase production, or cell envelope changes (Eriksson-Grennberg et al., 1965; Nordström et al., 1970; Suzuki et al., 1978). In organisms such as *Mycobacterium tuberculosis* and *Helicobacter pylori*, chromosomal mutation is effectively the only route to clinical resistance, necessitating multidrug regimens to suppress mutant emergence (Woodford and Ellington, 2007).

Mutation rates in DNA-based microorganisms typically range from 10^{-10} to 10^{-8} per base pair per cell division. Hypermutator strains, characterized by defects in mismatch repair, can amplify this rate by up to 100-fold under antibiotic selection, dramatically accelerating resistance evolution (Baquero et al., 2004; Baquero et al., 2021). Subinhibitory antibiotic concentrations, widespread in both clinical subtherapeutic dosing and in contaminated environments, enrich less-susceptible variants and provide a continuous low-grade selection pressure for resistance emergence (Lambert, 2005).

2.3.1.2 Horizontal gene transfer

HGT is arguably the dominant global driver of resistance dissemination, enabling the interspecies movement of fully formed, functional resistance determinants. Three canonical mechanisms operate transformation, transduction, and conjugation (Jian et al., 2021; Sultan et al., 2018). Fig. 2.2 illustrates these pathways schematically.

Transformation involves the uptake of naked DNA from the extracellular environment; *Streptococcus pneumoniae* uses this route to acquire mosaic PBP genes conferring penicillin resistance (Tomasz and Munoz, 1995). *Transduction* is bacteriophage-mediated gene transfer, particularly significant among staphylococci (Hawkey, 1998), and environmental phage metagenomes have been shown to harbour clinically important ARGs including *blaTEM*, *blaCTX-M*, and *mecA* (Colomer-Lluch et al., 2011). *Conjugation*, the most efficient HGT mechanism, transfers genetic material through a sex pilus and is the primary route for plasmid spread, carrying multiple ARGs across diverse genera (Sultan et al., 2018; Thomas and Nielsen, 2005).

Beyond these classical pathways, outer membrane vesicles (OMVs) can encapsulate and transfer plasmid-encoded ARGs to recipient cells across species boundaries (Rumbo et al., 2011), while bacterial nanotubes, cytoplasmic extensions connecting adjacent cells, facilitate the exchange of both chromosomal and extrachromosomal resistance determinants (Dubey and Ben-Yehuda, 2011; Dubey et al., 2016). Gene transfer agents, phage-derived capsid structures capable of packaging host DNA fragments below 15 kb, represent a further noncanonical transfer mechanism (García-Aljaro et al., 2017).

TABLE 2.2 **Principal biochemical mechanisms of antibiotic resistance: genetic basis, representative examples, effect on antibiotic activity, and clinically relevant pathogens.**

Mechanism	Genetic basis	Examples	Effect on antibiotic activity	Representative pathogens	References
Enzymatic inactivation	Plasmid-borne genes (e.g., bla, aac)	Beta-lactamases (TEM, SHV, NDM-1); aminoglycoside acetyltransferases; CAT	Antibiotic molecule hydrolysed or chemically modified before reaching the target	Staphylococcus *aureus,* Escherichia *coli,* Klebsiella *pneumoniae*	Vivekanandan et al. (2023), Hasan and Al-Harmoosh (2020), Almatroudi (2025)
Target modification	Chromosomal mutation/ acquired genes (e.g., mecA, vanA, erm)	PBP2a (MRSA); D-Ala-D-Lac (VRE); 23 S rRNA methylation	Reduced antibiotic-target binding affinity; cross-resistance to related drug classes	*MRSA, VRE, macrolide-resistant streptococci*	
Reduced membrane permeability	Chromosomal mutation in porin genes (e.g., oprD, ompF)	OprD loss in *Pseudomonas aeruginosa*; OmpF downregulation in *E. coli*	Diminished intracellular antibiotic accumulation	*P. aeruginosa, E. coli,* Mycobacterium *tuberculosis*	
Active efflux	Plasmid or chromosomal overexpression of pump genes	TetA (MFS); AcrAB-TolC (RND); MexAB-OprM (RND); NorA (MFS)	Drug expelled below the lethal threshold, often confers multidrug resistance	*E. coli, P. aeruginosa, S. aureus, M. tuberculosis*	
Biofilm formation	Chromosomal (fimH, fimA, pel, psl, ica operons)	*P. aeruginosa* PAO1; *Staphylococcus epidermidis*; *C. albicans* dual-species biofilms	Physical barrier limits drug diffusion; metabolic dormancy reduces drug efficacy 10–1000x	*P. aeruginosa, S. aureus, S. epidermidis, Candida* spp.	

CAT, Chloramphenicol acetyltransferase; *MFS*, major facilitator superfamily; *PBP*, penicillin-binding protein; *RND*, resistance-nodulation-division.

TABLE 2.3 Major antibiotic resistance gene (ARG) categories, associated genes, resistance mechanisms, dissemination routes, and representative pathogens.

Antibiotic class	Resistance gene(s)	Resistance mechanism	Dissemination route	Pathogens	References
Beta-lactams	*bla (blaTEM, blaSHV, blaCTX-M, blaNDM)*	Beta-lactamase hydrolysis	Plasmid (IncF, IncI), transposon Tn3	*Escherichia coli, K*lebsiella *pneumoniae, P*seudomonas *aeruginosa*	Lopatkin et al. (2016), Galgano et al. (2025), Centner et al. (2025)
Aminoglycosides	*aac, ant, aph*	Enzymatic *N*-acetylation, *O*-adenylation, *O*-phosphorylation	Conjugative plasmids, transposon Tn4001 (IS256)	*Enterobacteriaceae, Staphylococci*	
Tetracyclines	*tet (tetA, tetB, tetM)*	Efflux (MFS); ribosome protection	Plasmid, conjugative transposons (Tn916)	*E. coli, S*taphylococcus *aureus, Campylobacter*	
Glycopeptides	*van (vanA, vanB)*	Target modification (D-Ala-D-Lac substitution)	Tn1546 transposon on conjugative plasmids	*Enterococcus faecalis,* Enterococcus *faecium (VRE)*	
Macrolides	*erm, msr, ere*	23S rRNA methylation; efflux; hydrolysis	Conjugative plasmids, chromosomal integrons	*S. aureus, Streptococcus* spp.	
Quinolones	*qnr, aac(6')-Ib-cr*	Target protection; drug modification	Plasmid (PMQR)	*E. coli, K. pneumoniae*	
Polymyxins	*mcr-1 to mcr-9*	Lipid A modification (phosphoethanolamine addition)	Conjugative plasmid (IncI2, IncHI2)	*E. coli, Salmonella, K. pneumoniae*	

MRSA, Methicillin-resistant *Staphylococcus aureus; PMQR*, plasmid-mediated quinolone resistance; *VRE*, vancomycin-resistant *Enterococcus*.

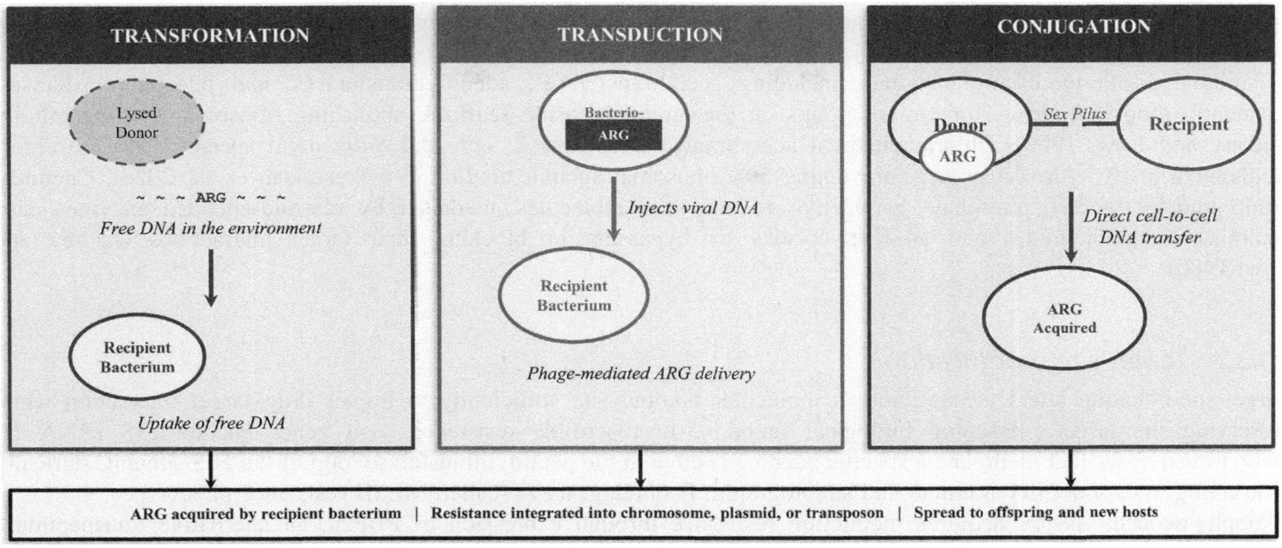

FIGURE 2.2 **Schematic representation of the three canonical horizontal gene transfer pathways in bacteria.** Transformation involves uptake of exogenous DNA from the environment; transduction involves bacteriophage-mediated transfer; conjugation involves direct cell-to-cell transfer through a sex pilus. All three routes facilitate the spread of antibiotic resistance genes (ARGs) between bacterial species.

2.3.1.3 Mobile genetic elements

The movement of ARGs through HGT depends critically on mobile genetic elements (MGEs): plasmids, transposons, and integrons. Plasmids are autonomously replicating extrachromosomal molecules that carry multiple ARGs and spread them through conjugation across taxonomically distant species (Thomas and Nielsen, 2005). Plasmid-borne beta-lactamases, including extended-spectrum beta-lactamases (ESBLs) and carbapenemases, have become defining features of the MDR Gram-negative pathogen landscape (Carcione et al., 2021; Ramirez and Tolmasky, 2010).

Transposons (jumping genes) are DNA segments that relocate within or between DNA molecules via transposase-mediated strand cleavage and resealing. Composite transposons, flanked by insertion sequences (IS), carry ARG cassettes and can move between plasmids and chromosomes, expanding host range. The aminoglycoside resistance transposon Tn4001 (carrying IS256) is an illustrative example (Babakhani and Oloomi, 2018). Noncomposite transposons such as Tn21 harbour integron-associated ARGs and dominate Enterobacteriaceae resistance landscapes (Liebert et al., 1999). Transposable phage Mu, which transmits resistance genes via a virus-like replication mechanism, represents a distinct class of transposable elements (Babakhani and Oloomi, 2018).

Integrons are genetic platforms with site-specific recombination machinery (IntI integrase and attI recombination site) that capture and express gene cassettes in an ordered, promoter-driven array. Class 1 integrons are the most clinically significant, widely distributed in Gram-negative pathogens, including *A. baumannii*, *P*seudomonas *aeruginosa*, *E. coli*, and *K. pneumoniae*, and typically embedded within transposons and conjugative plasmids that extend their horizontal mobility (Cambray et al., 2010; Gillings et al., 2008). Class 2 and 3 integrons are less prevalent but similarly mobile; class 4 chromosomal integrons, first identified in *Vibrio cholerae*, carry large arrays of mostly uncharacterised gene cassettes (Partridge et al., 2009; Clark et al., 2000; Rowe-Magnus et al., 2001). ARG databases, including CARD, AMRFinderPlus, ResFinder, and Pathogenwatch, catalogue these determinants and support clinical and public health surveillance (Centner et al., 2025).

2.3.2 Biochemical mechanisms of resistance

2.3.2.1 Enzymatic drug inactivation

Enzymatic destruction or chemical modification of antibiotics prior to target engagement is one of the most widespread and clinically consequential resistance strategies. Beta-lactamases hydrolyse the beta-lactam ring of penicillins, cephalosporins, and carbapenems. ESBLs, produced predominantly by *K. pneumoniae* and *E. coli*, confer resistance to third-generation cephalosporins and aztreonam. Carbapenemases, including *K. pneumoniae* carbapenemase, NDM-1,

OXA-48, and VIM, hydrolyse carbapenems and represent a last-resort therapy crisis (Ramirez and Tolmasky, 2010; Vakulenko and Mobashery, 2003).

Aminoglycoside-modifying enzymes, including acetyltransferases, adenylyltransferases, and phosphotransferases, covalently modify hydroxyl or amino groups on the aminoglycoside scaffold, abolishing ribosomal binding affinity (Jacoby and Low, 1980). Chloramphenicol acetyltransferases from *E. coli* and *Salmonella* species acetylate chloramphenicol at its hydroxyl groups, preventing 50S ribosomal subunit binding (Vivekanandan et al., 2023). Chemical group transfer (acetyl, phosphate, or adenyl) to antibiotic molecules, mediated by plasmid-encoded enzymes, also neutralises sulphonamides and aminoglycosides by bypassing or blocking their target interactions (Jacoby and Low, 1980).

2.3.2.2 *Target site modification*

Target modification alters the antibiotic's molecular binding site sufficiently to impair drug–target interaction while preserving the target's essential biological function. In macrolide resistance, *erm* genes encode 23S rRNA N-methyltransferases that methylate a specific adenine residue in the peptidyl transferase loop of the 50S subunit, sterically obstructing macrolide, lincosamide, and streptogramin B binding, the so-called MLSB resistance phenotype.

Staphylococcus *aureus* acquires methicillin resistance through expression of PBP2a, an alternative transpeptidase encoded by *mecA* with markedly reduced beta-lactam affinity, allowing cell wall synthesis to continue in the presence of virtually all beta-lactam antibiotics (Kato et al., 2020). Vancomycin resistance in enterococci (VRE) operates through the *vanA*/*vanB* operons, which reprogramme the terminal dipeptide of peptidoglycan precursors from D-Ala-D-Ala to D-Ala-D-Lac or D-Ala-D-Ser, eliminating the vancomycin binding site while maintaining transpeptidation. Fluoroquinolone resistance through chromosomal mutation concentrates in the QRDRs of *gyrA* and *parC*, reducing ternary complex stability between drug, enzyme, and DNA (Vivekanandan et al., 2023; Gellert et al., 1977; Higgins et al., 1978).

2.3.2.3 *Altered membrane permeability*

In Gram-negative organisms, the outer membrane constitutes a selective permeability barrier, and hydrophilic antibiotics, including many beta-lactams, carbapenems, and fluoroquinolones, gain intracellular access predominantly through outer membrane porin channels. Downregulation or structural mutation of specific porins substantially reduces antibiotic influx. Loss or mutation of OprD, the carbapenem-selective porin of *P. aeruginosa*, is a well-characterised mechanism of imipenem resistance (Pagès et al., 2008). In *E. coli*, reduced expression of OmpF and OmpC porins contributes to resistance across multiple antibiotic classes (Hofer and Hofer, 2019; Diene and Rolain, 2014).

M. tuberculosis is intrinsically resistant to many antibiotics due to the exceptional lipid content of its cell wall, particularly mycolic acids, which markedly restrict drug permeability. This intrinsic barrier underpins the requirement for prolonged multidrug regimens in tuberculosis treatment (Vivekanandan et al., 2023). This permeability barrier mechanism frequently operates in concert with active efflux, the combination substantially elevating MICs above clinically achievable concentrations.

2.3.2.4 *Active efflux*

Efflux pumps are membrane-spanning transporter proteins that expel antibiotics from the cytoplasm before they can engage their targets, reducing intracellular concentrations below the minimum inhibitory threshold. Five major superfamilies operate across Gram-positive and Gram-negative bacteria: the major facilitator superfamily (MFS), resistance–nodulation–division (RND), ATP-binding cassette, small multidrug resistance, and multidrug and toxin compound extrusion families (Diene and Rolain, 2014; Hasan and Al-Harmoosh, 2020; Vivekanandan et al., 2023). Fig. 2.3 provides a schematic comparison of these pump families.

RND pumps, which operate as tripartite complexes spanning the inner membrane, periplasm, and outer membrane, are the most clinically significant in Gram-negative bacteria. The AcrAB-TolC system in *E. coli* and the MexAB-OprM system in *P. aeruginosa* confer broad-spectrum resistance covering multiple antibiotic classes simultaneously. The TetA MFS pump specifically expels tetracyclines, while NorA in *S. aureus* handles fluoroquinolones (Hasan and Al-Harmoosh, 2020). Efflux pump expression is frequently inducible through global regulatory networks, enabling bacteria to mount rapid resistance responses without requiring genetic mutation; overexpression through regulatory gene alteration can push MICs above therapeutic concentrations with no change in target gene sequence.

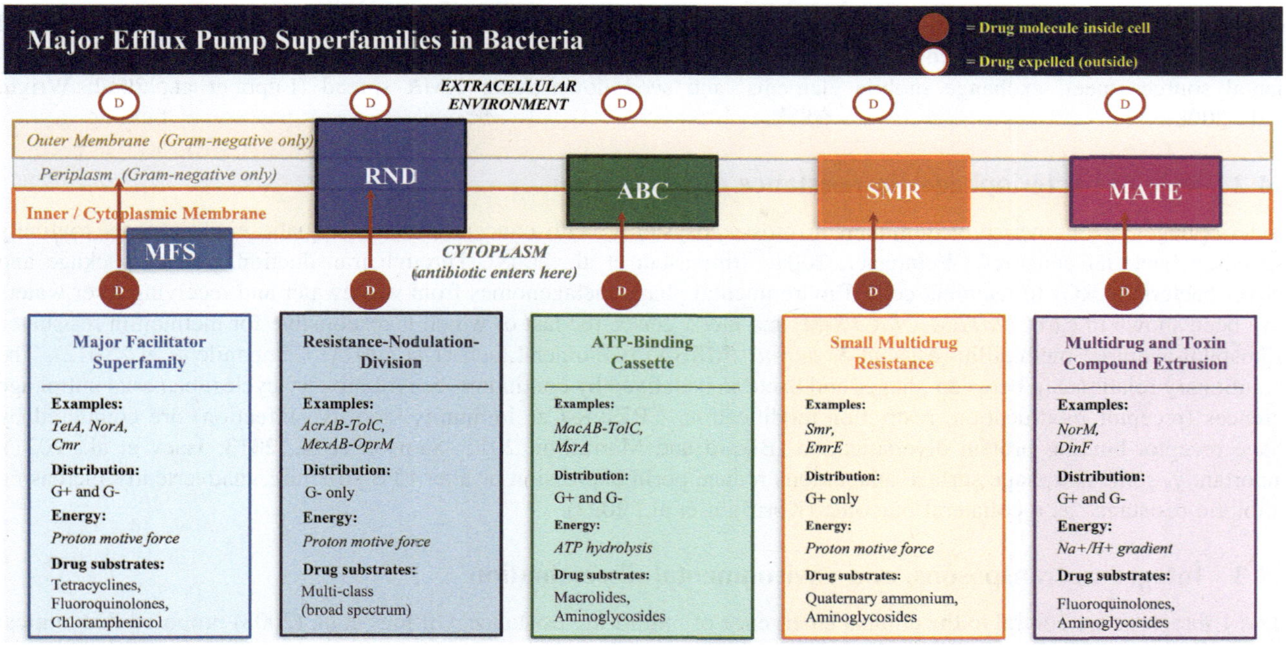

FIGURE 2.3 **Schematic overview of the five major efflux pump superfamilies in bacteria.** MFS (major facilitator superfamily), RND (resistance-nodulation-division), ABC (ATP-binding cassette), SMR (small multidrug resistance), and MATE (multidrug and toxin compound extrusion) pumps differ in energy source, subunit organisation, and distribution across Gram-positive (G+) and Gram-negative (G-) organisms. Clinically important members are listed for each family.

2.3.2.5 *Biofilm formation*

Biofilms are surface-associated bacterial communities encased within a self-produced extracellular matrix of polysaccharides, proteins, and extracellular DNA. This matrix retards antibiotic diffusion, may chemically inactivate certain drugs, and creates steep concentration gradients that protect inner cell layers. Effective biofilm eradication typically requires antibiotic concentrations 10–1000-fold higher than those sufficient to kill planktonic cells, concentrations rarely achievable in tissue (Almatroudi, 2025).

Within biofilms, metabolic dormancy, particularly in nutrient-limited microenvironments, renders bacteria phenotypically tolerant to antibiotics whose activity depends on active growth. A subpopulation of metabolically quiescent persister cells within biofilms tolerates virtually all antibiotic classes through toxin–antitoxin module activation; these persisters survive treatment, resume growth on drug withdrawal, and seed recurrent infection. The high cell density of biofilm communities is also permissive for HGT, potentially accelerating intrabiofilm resistance acquisition and exchange (Almatroudi, 2025; Ridyard and Overhage, 2021).

2.4 Evolution and environmental spread of resistance

Antibiotic resistance does not arise solely within the clinic. It is part of a broader evolutionary and ecological process in which resistance determinants cycle continuously among human, animal, and environmental microbial communities, a framework now encapsulated by the One Health concept.

2.4.1 Selective pressure and environmental reservoirs

Antibiotic use in medicine, livestock production, and aquaculture imposes selective pressure that eliminates susceptible bacteria and enriches resistant variants. Antibiotic residues enter environmental matrices through agricultural runoff, wastewater discharge, and pharmaceutical manufacturing effluents, creating chronic low-level selective pressures in rivers, soils, and sediments (Lupo et al., 2012; Martinez, 2009). Subinhibitory concentrations in these environments do not kill bacteria but enrich less-susceptible variants and promote stepwise mutational resistance accrual (Baquero et al., 2021). Coselection by heavy metals and disinfectants, which colocate on MGEs carrying both metal tolerance and ARGs, sustains resistance selection even in the absence of direct antibiotic pressure (Martinez, 2009). Riverine environments and hospital wastewater represent ecological hotspots where ARG-carrying organisms from human, animal, and environmental sources meet, exchange mobile elements, and seed downstream AMR spread (Lupo et al., 2012; Wright et al., 2008).

2.4.2 Role of bacteriophages in resistance dissemination

Bacteriophages are numerically dominant in most ecosystems, with concentrations in aquatic environments routinely exceeding bacterial densities (Weinbauer, 2004; Srinivasiah et al., 2008). Through transduction, phages package and deliver bacterial ARGs to recipient cells. Environmental phage metagenomes from wastewater and receiving river waters have been shown to carry *blaTEM*, *blaCTX-M*, and *mecA* genes, the last of which is responsible for methicillin resistance in hospital-acquired methicillin-resistant *S. aureus* (MRSA) (Colomer-Lluch et al., 2011; Campanile et al., 2012). The evolutionary relationship between phages and bacteria is defined by continuous coevolutionary cycles: bacterial antiphage defences (receptor modification, restriction-modification, CRISPR-Cas immunity, abortive infection) are countered by phage receptor-binding protein diversification (Bikard and Marraffini, 2012; Samson et al., 2013; Isaev et al., 2021). Importantly, some antiphage surface adaptations reduce porin expression or alter LPS structure, inadvertently increasing antibiotic resistance as a collateral outcome (Kortright et al., 2022).

2.4.3 Integrons, transposons, and environmental dissemination

Class 1 integrons are central to the clinical emergence of multidrug resistance. Gillings et al. (2008) proposed that clinical class 1 integrons evolved from environmental ancestors that originally carried *qac* cassettes encoding quaternary ammonium disinfectant resistance, with antibiotic resistance cassettes subsequently accumulated under clinical selective pressure. Studies have confirmed that quaternary ammonium compound contamination of water bodies selects for bacteria harbouring *qacE* cassettes within class 1 integrons (Gaze et al., 2005), and that heavy metal pollution correlates with elevated integron abundance in aquatic communities (Wright et al., 2008). Elevated class 1 integron and *qac* gene levels have been documented in sewage sludge and pig slurry exposed to detergents and antibiotic residues (Gaze et al., 2011), supporting the conclusion that environmental pollutants are active drivers of integron selection and downstream clinical spread.

2.4.4 Microbial community dynamics and collective resistance

Resistance in natural and clinical settings occurs within complex multispecies communities, where interspecies interactions can profoundly alter individual bacterial selective pressures and produce community-level resistance phenotypes exceeding the capacity of any single organism. Collective antibiotic inactivation, where one species produces a resistance enzyme that detoxifies the shared environment, exemplifies this: beta-lactamase-producing *E. coli* protect coresident *Salmonella enterica* from lethal beta-lactam exposure (Perlin et al., 2009), and similar cooperative inactivation has been documented in polymicrobial clinical infections (Brook, 2009). As a consequence, clinical susceptibility testing on single-species isolates may overestimate therapeutic efficacy in mixed-community infections.

Competition among bacterial species can paradoxically suppress resistance evolution by increasing its fitness cost. Biofilm-resident bacteria tend to express lower-cost but mechanistically weaker resistance compared with planktonic counterparts, consistent with resource allocation trade-offs (Santos-Lopez et al., 2019). The composition of microbial communities also shapes the economics of phage resistance: in mixed-species settings, bacteria preferentially transition from energetically expensive receptor-based phage resistance to less costly CRISPR-mediated immunity (Alseth et al., 2019). These observations collectively argue that AMR must be studied and addressed within realistic multispecies frameworks rather than simple monoculture models (Bottery et al., 2021). The ecological and evolutionary dynamics described above ultimately converge on clinical outcomes: as determinants of resistance cycle between environmental reservoirs and human pathogens, the pool of treatable infections shrinks and the burden on healthcare systems grows, with consequences now measurable at a global scale.

2.5 Clinical impact of antimicrobial resistance

The clinical burden of AMR is both quantifiable and accelerating. The 2022 global burden analysis estimated 1.27 million deaths directly attributable to bacterial AMR and a further 4.95 million deaths in which AMR was a contributing factor in 2019, placing resistance among the leading causes of death worldwide (Murray et al., 2022). The pathogens driving this burden are principally the ESKAPE group: *Enterococcus faecium*, *S. aureus*, *K. pneumoniae*, *A. baumannii*, *P. aeruginosa*, and *Enterobacter* species.

Specific resistance phenotypes have rendered previously treatable infections extremely difficult to manage. MRSA, VRE, ESBL-producing *K. pneumoniae* and *E. coli*, carbapenem-resistant Enterobacteriaceae harbouring NDM-1 and related enzymes, penicillin-resistant pneumococci, and MDR *A. baumannii* dominate healthcare-associated infection landscapes globally. Community-acquired MRSA has further complicated epidemiological control (Dukic et al., 2013).

The direct clinical consequences of resistant infections are well documented. Hospitalisation rates are at least twice those associated with susceptible strains of the same pathogen, with correspondingly prolonged stays and higher mortality (French, 2005; Zhao et al., 2024). Delays in initiating effective therapy, an inevitable consequence of the time required for culture and sensitivity testing, worsen outcomes, particularly in sepsis, where each hour of inappropriate therapy increases mortality risk (Eliopoulos et al., 2003; Edmondson et al., 2024). In resource-limited settings, restricted access to second- and third-line agents can render infections untreatable (Arias and Murray, 2009).

Economic consequences are equally severe. Resistant infections require escalation to more expensive, more toxic second- and third-line agents. A UK economic analysis showed that hospital stays associated with healthcare-acquired infections, the majority caused by MDR organisms, were approximately 2.5 times longer than those of uninfected patients, with costs 2.8 times higher (Plowman et al., 1999). Infection control of MDR outbreaks generates additional costs for isolation, surveillance, and ward closure, and may expose hospitals to litigation even in the absence of negligence (Bryce et al., 2000). Indirect economic burdens, including lost productivity and the downstream costs of prolonged community care, amplify the total impact substantially.

Control of AMR relies on antimicrobial stewardship, vaccination, infection prevention, and international surveillance coordination. However, given the constrained antibiotic development pipeline, increasingly urgent attention is turning toward alternative therapeutic paradigms.

2.6 Emerging therapeutic strategies against resistant pathogens

Four strategies, namely phage therapy, antimicrobial peptides (AMPs), monoclonal antibodies (mAbs), and nanoparticles (NPs), have attracted the most sustained research attention as alternatives or adjuncts to conventional antibiotics. Each is mechanistically distinct and specifically linked to one or more resistance pathways it can circumvent. Table 2.4 provides a comparative summary.

2.6.1 Phage therapy

Bacteriophages infect bacteria through receptor-specific adsorption, inject nucleic acids, replicate using host machinery, assemble progeny virions, and lyse the host cell. Lytic phages accomplish this cycle without genomic integration and are the pharmacological focus of phage therapy (Oromí-Bosch et al., 2023; Wang et al., 2024). Unlike conventional antibiotics, phages kill through entirely different mechanisms, physical membrane disruption at lysis and metabolic hijacking during replication, and retain activity against bacteria that have acquired all established biochemical resistance mechanisms.

Phage therapy offers several properties unavailable to chemical antibiotics: potential co-evolution with bacterial targets under treatment conditions; activity within biofilms through phage-encoded depolymerases that enzymatically degrade biofilm matrix components; no intrinsic mammalian cell toxicity at therapeutic doses; and stability across a broad temperature range (Strathdee et al., 2023; Youssef et al., 2024). Clinical evidence, while still predominantly observational, includes demonstrating the efficacy of the PhagoBurn cocktail against *P. aeruginosa* burn wound infections (Jault et al., 2019) and multiple compassionate-use cases achieving bacteriological clearance in life-threatening MDR infections where antibiotics had failed (Strathdee et al., 2023).

The principal limitation is the narrow host range of individual phages, requiring accurate strain-level pathogen identification before deployment. Bacteria develop phage resistance rapidly through receptor modification, restriction-modification, or CRISPR acquisition (Bikard and Marraffini, 2012; Samson et al., 2013). Phage cocktails combining multiple lytic phages with complementary receptor specificities mitigate this risk and have demonstrated superior outcomes compared with single-phage preparations (Umansky and Fortier, 2023). Regulatory frameworks for phage therapy remain underdeveloped in most jurisdictions, with no standardised approval pathway currently established (Yang et al., 2023). Further randomised controlled trials are required to establish efficacy standards.

TABLE 2.4 Comparative overview of four emerging therapeutic strategies against resistant pathogens: primary target, advantages, limitations, and specific mechanistic linkage to known antibiotic resistance pathways.

Strategy	Primary target	Advantages	Limitations	Linkage to resistance mechanism	References
Phage therapy	Living bacterial cells (lytic phage–receptor interaction)	Active against MDR/XDR strains; bactericidal; co-evolves with target; biofilm-active depolymerases	Narrow host range; phage resistance; regulatory uncertainty; requires pathogen typing	Bypasses all biochemical resistance; kills through a distinct lytic mechanism	Ioannou et al. (2023), Esposito et al. (2023), Mmbando et al. (2025)
Antimicrobial peptides (AMPs)	Bacterial membrane bilayer; intracellular targets (DNA, ribosomes)	Broad spectrum; low resistance risk; active against biofilms; immunomodulatory activity	Proteolytic degradation; cytotoxicity at high doses; high synthesis cost	Membrane disruption circumvents efflux and target-modification resistance	
Monoclonal antibodies (mAbs)	Virulence factors (toxins, adhesins); pathogen surface antigens	High specificity; low toxicity; microbiome-sparing; long half-life	High production cost; narrow antigenic target; unsuitable for empirical therapy	Neutralises virulence without selective pressure for resistance gene acquisition	
Nanoparticles (NPs)	Bacterial membrane; intracellular components; biofilm matrix	ROS-mediated broad-spectrum killing; biofilm penetration; drug-delivery platform	Cytotoxicity; environmental persistence; limited clinical trial data; NP resistance documented	ROS activity overcomes enzymatic inactivation; NP-antibiotic conjugates restore drug activity	

AMPs, Antimicrobial peptides; *mAbs*, monoclonal antibodies; *MDR*, multidrug-resistant; *ROS*, reactive oxygen species; *XDR*, extensively drug resistant.

2.6.2 Antimicrobial peptides

AMPs are short (10–50 amino acids), cationic, amphipathic molecules that are evolutionarily conserved components of innate immunity across virtually all multicellular life forms. Their broad-spectrum activity spans bacteria, fungi, viruses, and parasites, and their simultaneous multiple mechanisms of action substantially reduce the probability that any single mutation confers complete resistance (Ioannou et al., 2023). Membrane-active AMPs interact with the anionic bacterial membrane surface through electrostatic and hydrophobic forces and induce pore formation or membrane dissolution through barrel-stave, toroidal, or carpet mechanisms, directly opposing the impermeability barrier strategies used by resistant bacteria (Xuan et al., 2023). Nonmembrane-disruptive AMPs penetrate cells and inhibit specific intracellular targets, including DNA synthesis, protein translation, and cell wall biosynthesis (Ganesan et al., 2023).

Several AMPs are in clinical use or advanced development. Polymyxins remain active against carbapenem-resistant Gram-negative pathogens at the cost of nephrotoxicity. Cathelicidin LL-37 and its derivatives show preclinical promise against biofilm-forming pathogens, directly addressing a resistance mechanism that antibiotics address poorly (Ridyard and Overhage, 2021). Pexiganan has demonstrated efficacy in diabetic foot infections, and murepavadin (POL7080) exerts antipseudomonal activity through inhibition of lipid A transport rather than direct membrane disruption (Wei et al., 2023). The principal barriers to systemic AMP therapy are proteolytic instability in serum, potential cytotoxicity at therapeutic concentrations, and high production costs for synthetic peptides, challenges being addressed through chemical modification, NP encapsulation, and topical delivery strategies (de Souza et al., 2023).

2.6.3 Monoclonal antibodies

mAbs target virulence factors such as bacterial toxins, adhesins, or immune evasion proteins, rather than essential bacterial survival functions. By neutralising pathogen virulence capacity rather than directly killing the organism, mAbs reduce infection severity while allowing host immunity to clear the pathogen. Because mAb targets are virulence factors rather than proteins under strong purifying selection for metabolic function, the development of resistance is potentially slower than with conventional antibiotics, a strategic advantage (Esposito et al., 2023; Wang et al., 2022).

The hybridoma technology developed by Kohler and Milstein in the 1970s enabled large-scale production of defined mAbs, and subsequent humanisation strategies eliminated immunogenicity barriers to clinical use (Wang et al., 2022). Approved mAbs for bacterial infections include raxibacumab and obiltoxaximab, both targeting *Bacillus anthracis* protective antigen toxin, and bezlotoxumab (Zinplava), which targets *Clostridioides difficile* toxin B and reduces recurrence rates in high-risk patients. MEDI4893, targeting *S. aureus* alpha-toxin, is in clinical evaluation for MRSA pneumonia prevention (Mmbando et al., 2025; Mazur et al., 2023). High per-dose cost from biologic manufacturing and the narrow antigenic specificity of individual mAbs remain the primary barriers to broader adoption; combination mAb strategies targeting multiple nonoverlapping epitopes are under active investigation (Mazur et al., 2023; Pantaleo et al., 2022).

2.6.4 Nanoparticles

NPs engineered at the 1–100 nm scale offer multiple antimicrobial properties that simultaneously address several resistance mechanisms. Their small size enables penetration into biofilm matrices and entry into host cell compartments harbouring intracellular pathogens. Metal and metal oxide NPs (silver, zinc oxide, titanium dioxide) generate reactive oxygen species (ROS) that induce oxidative damage to bacterial membranes, proteins, and nucleic acids through mechanisms independent of specific proteinaceous targets, thereby retaining activity against organisms that have acquired target-modification or enzymatic inactivation resistance (Sahli et al., 2022). Silver NPs have demonstrated activity against MRSA and MDR Gram-negative pathogens in vitro, and their combination with conventional antibiotics can restore drug susceptibility in resistant strains through membrane-sensitising effects.

As drug-delivery platforms, NPs encapsulate antibiotics within polymeric or lipid carriers, thereby protecting them against enzymatic degradation (directly countering enzymatic inactivation), extending pharmacokinetic half-life, and enabling stimuli-responsive release at infection sites (Elumalai et al., 2024). ROS-mediated activity against biofilm matrices also addresses a resistance mechanism poorly reached by conventional antibiotics. Important limitations include the potential cytotoxicity of metal NPs to host cells, environmental persistence and ecotoxicity concerns, the absence of robust randomised controlled trial evidence, and documented in vitro NP resistance through ROS detoxification mechanisms. Rigorous pharmacokinetic-pharmacodynamic characterisation and clinical trial validation remain essential prerequisites for NP-based agents entering the therapeutic mainstream (Mmbando et al., 2025).

2.7 Conclusion

Antibiotic resistance represents the convergence of evolutionary biology, microbial ecology, and clinical medicine. Its mechanisms, spanning chromosomal mutation, HGT, enzymatic drug inactivation, target modification, membrane impermeability, active efflux, and biofilm formation, operate simultaneously and synergistically in MDR pathogens. Their dissemination through plasmids, transposons, integrons, phages, and novel transfer mechanisms, including OMVs and nanotubes, ensures that resistance determinants flow continuously between clinical, environmental, and animal reservoirs in a self-sustaining cycle.

The clinical consequences of this crisis are now firmly quantified, with AMR among the leading global causes of death and an escalating economic burden on health systems. The emerging therapeutic strategies reviewed here, phage therapy, AMPs, mAbs, and NPs, each offer mechanistically distinct avenues that sidestep established resistance pathways, though none has yet achieved the breadth of clinical deployment needed to substitute for conventional antibiotics. Their rational integration into the antimicrobial armamentarium, guided by precision diagnostics, combination strategies, and rigorous clinical evaluation, represents the most plausible near-term pathway to managing resistance at the bedside.

Parallel commitments to antimicrobial stewardship, infection prevention, surveillance, vaccination, and international policy coordination are equally indispensable. Ultimately, addressing AMR requires a shift from viewing resistance as a clinical problem alone toward recognising it as a shared evolutionary challenge that must be met across human medicine, veterinary practice, agriculture, and environmental science together.

References

Almatroudi, A., 2025. Biofilm resilience: molecular mechanisms driving antibiotic resistance in clinical contexts. Biology 14 (2), 165. Available at https://doi.org/10.3390/biology14020165.

Alseth, E.O., Pursey, E., Luján, A.M., McLeod, I., Rollie, C., Westra, E.R., 2019. Bacterial biodiversity drives the evolution of CRISPR-based phage resistance. Nature 574 (7779), 549–552. Available at https://doi.org/10.1038/s41586-019-1662-9, http://www.nature.com/nature/index.html.

Arias, C.A., Murray, B.E., 2009. Antibiotic-resistant bugs in the 21st century - a clinical super-challenge. New England Journal of Medicine 360 (5), 439–443. Available at https://doi.org/10.1056/NEJMp0804651, http://content.nejm.org/cgi/reprint/360/5/439.pdf.

Babakhani, S., Oloomi, M., 2018. Transposons: the agents of antibiotic resistance in bacteria. Journal of Basic Microbiology 58 (11), 905–917. Available at https://doi.org/10.1002/jobm.201800204, http://onlinelibrary.wiley.com/journal/10.1002/(ISSN)1521-4028.

Baquero, F., Martínez, J.L., F. Lanza, V., Rodríguez-Beltrán, J., Galán, J.C., San Millán, A., et al., 2021. Evolutionary pathways and trajectories in antibiotic resistance. Clinical Microbiology Reviews 34 (2) e00050-19. Available from. Available at https://doi.org/10.1128/CMR.00050-19, https://journals.asm.org/doi/10.1128/CMR.00050-19.

Baquero, M.R., Nilsson, A.I., Turrientes, M.D.C., Sandvang, D., Galán, J.C., Martínez, J.L., et al., 2004. Polymorphic mutation frequencies in *Escherichia coli*: emergence of weak mutators in clinical isolates. Journal of Bacteriology 186 (16), 5538–5542. Available at https://doi.org/10.1128/JB.186.16.5538-5542.2004.

Bikard, D., Marraffini, L.A., 2012. Innate and adaptive immunity in bacteria: mechanisms of programmed genetic variation to fight bacteriophages. Current Opinion in Immunology 24 (1), 15–20. Available at https://doi.org/10.1016/j.coi.2011.10.005.

Blair, J.M.A., Webber, M.A., Baylay, A.J., Ogbolu, D.O., Piddock, L.J.V., 2015. Molecular mechanisms of antibiotic resistance. Nature Reviews. Microbiology 13 (1), 42–51. Available at https://doi.org/10.1038/nrmicro3380, http://www.nature.com/nrmicro/index.html.

Bottery, M.J., Pitchford, J.W., Friman, V.P., 2021. Ecology and evolution of antimicrobial resistance in bacterial communities. ISME Journal 15 (4), 939–948. Available at https://doi.org/10.1038/s41396-020-00832-7, http://www.nature.com/ismej/marketing/aims_scope.html.

Brook, I., 2009. The role of beta-lactamase-producing-bacteria in mixed infections. BMC Infectious Diseases 9 (1). Available at https://doi.org/10.1186/1471-2334-9-202.

Brunton, L.L., Hilal-Dandan, R., Knollmann, B.C. (Eds.), 2023. Goodman & Gilman's The Pharmacological Basis of Therapeutics, 14th ed. McGraw Hill, New York.

Bryce, E.A., Tiffin, S.M., Isaac-Renton, J.L., Wright, C.J., 2000. Evidence of delays in transferring patients with methicillin-resistant *Staphylococcus aureus* or vancomycin-resistant enterococcus to long-term–care facilities. Infection Control & Hospital Epidemiology 21 (4), 270–271. Available at https://doi.org/10.1086/501757.

Cambray, G., Guerout, A.M., Mazel, D., 2010. Integrons. Annual Review of Genetics 44, 141–166. Available at https://doi.org/10.1146/annurev-genet-102209-163504.

Campanile, F., Bongiorno, D., Falcone, M., Vailati, F., Pasticci, M.B., Perez, M., et al., 2012. Changing Italian nosocomial-community trends and heteroresistance in *Staphylococcus aureus* from bacteremia and endocarditis. European Journal of Clinical Microbiology & Infectious Diseases 31 (5), 739–745. Available at https://doi.org/10.1007/s10096-011-1367-y.

Carcione, D., Siracusa, C., Sulejmani, A., Leoni, V., Intra, J., Carcione, D., et al., 2021. Old and new beta-lactamase inhibitors: molecular bases and clinical perspectives. Antibiotics 10 (8), 995. Available at https://doi.org/10.3390/antibiotics10080995, https://doi.org/10.3390/antibiotics10080995.

Centner, C.M., Dolby, G., Kgagudi, P., Zadoks, R.N., Shittu, A., Baba, A., et al., 2025. The global landscape of antibiotic resistance gene databases: a comprehensive review. mSystems. Available at https://doi.org/10.1128/msystems.01357-24 e01357-24.

Clark, C.A., Purins, L., Kaewrakon, P., Focareta, T., Manning, P.A., 2000. The Vibrio cholerae 01 chromosomal integron. Microbiology (Reading, England) 146 (10), 2605–2612. Available at https://doi.org/10.1099/00221287-146-10-2605, http://mic.microbiologyresearch.org/content/journal/micro.

Colomer-Lluch, M., Jofre, J., Muniesa, M., Aziz, R., 2011. Antibiotic resistance genes in the bacteriophage DNA fraction of environmental samples. PLoS One 6 (3), e17549. Available at https://doi.org/10.1371/journal.pone.0017549.

Davies, J., Davies, D., 2010. Origins and evolution of antibiotic resistance. Microbiology and Molecular Biology Reviews 74 (3), 417–433. Available at https://doi.org/10.1128/MMBR.00016-10Canada, http://mmbr.asm.org/cgi/reprint/74/3/417.

de Souza, G.Hd.A., Rossato, L., de Oliveira, A.R., Simionatto, S., 2023. Antimicrobial peptides against polymyxin-resistant Klebsiella pneumoniae: a patent review. World Journal of Microbiology and Biotechnology 39 (3). Available at https://doi.org/10.1007/s11274-023-03530-6, https://www.springer.com/journal/11274.

Diene, S.M., Rolain, J.M., 2014. Mechanisms of bacterial resistance: focus on efflux pumps, drug permeability, and target modification. Future Microbiology 9 (5), 575–580. Available at https://doi.org/10.2217/fmb.14.16, https://doi.org/10.2217/fmb.14.16 Future Microbiology. **9** (3) (2014), 277–280.

Dubey, G.P., Ben-Yehuda, S., 2011. Intercellular nanotubes mediate bacterial communication. Cell 144 (4), 590–600. Available at https://doi.org/10.1016/j.cell.2011.01.015, https://www.sciencedirect.com/journal/cell.

Dubey, G.P., Malli Mohan, G.B., Dubrovsky, A., Amen, T., Tsipshtein, S., Rouvinski, A., et al., 2016. Architecture and characteristics of bacterial nanotubes. Developmental Cell 36 (4), 453–461. Available at https://doi.org/10.1016/j.devcel.2016.01.013.

Dukic, V.M., Lauderdale, D.S., Wilder, J., Daum, R.S., David, M.Z., 2013. Epidemics of community-associated methicillin-resistant *Staphylococcus aureus* in the United States: a meta-analysis. PLoS One 8 (1). Available at https://doi.org/10.1371/journal.pone.0052722, http://www.plosone.org/article/fetchObjectAttachment.action?uri=info%3Adoi%2F10.1371%2Fjournal.pone.0052722&representation=PDF.

Edmondson, D.G., Carlsson, A.M., Norris, S.J., Edmondson, D.G., Carlsson, A.M., Norris, S.J., 2024. Delay in diagnosis of Lyme disease: impact on patient outcomes. Frontiers in Medicine 11, 1369779. Available at https://doi.org/10.3389/fmed.2024.1369779 https://doi.org/10.3389/fmed.2024.1369779. Available from.

Eliopoulos, G.M., Cosgrove, S.E., Carmeli, Y., 2003. The impact of antimicrobial resistance on health and economic outcomes. Clinical Infectious Diseases 36 (11), 1433–1437. Available at https://doi.org/10.1086/375081.

Elumalai, K., Srinivasan, S., Shanmugam, A., Elumalai, K., Srinivasan, S., Shanmugam, A., 2024. Nanoparticle-based drug delivery systems in cancer treatment. Biomedical Technology 5, 109–122. Available at https://doi.org/10.1016/j.bmt.2023.09.001, http://www.keaipublishing.com/en/journals/biomedical-technology/.

Eriksson-Grennberg, K.G., Boman, H.G., Jansson, J.A.T., Thorén, S., 1965. Resistance of *Escherichia coli* to penicillins I. Genetic study of some ampicillin-resistant mutants. Journal of Bacteriology 90 (1), 54–62. Available at https://doi.org/10.1128/jb.90.1.54-62.1965.

Esposito, S., Amirthalingam, G., Bassetti, M., Blasi, F., De Rosa, F.G., Halasa, N.B., et al., 2023. Monoclonal antibodies for prophylaxis and therapy of infectious diseases. Frontiers in Immunology 14, 1232246. https://doi.org/10.3389/fimmu.2023.1232246. Frontiers in Immunology. **14** (2023). Available at https://doi.org/10.3389/fimmu.2023.1162342, https://www.frontiersin.org/journals/immunology#.

French, G.L., 2005. Clinical impact and relevance of antibiotic resistance. Advanced Drug Delivery Reviews 57 (10), 1514–1527. Available at https://doi.org/10.1016/j.addr.2005.04.005.

Funatsu, G., Wittmann, H.G., 1972. Ribosomal proteins. XXXIII. Location of amino-acid replacements in protein S12 isolated from Escherichia coli mutants resistant to streptomycin. Journal of Molecular Biology 68 (3), 547–550. Available at https://doi.org/10.1016/0022-2836(72)90108-8.

Galgano, M., Pellegrini, F., Catalano, E., Capozzi, L., Del Sambro, L., Sposato, A., et al., 2025. Acquired bacterial resistance to antibiotics and resistance genes: from past to future. Antibiotics 14 (3), 222. Available at https://doi.org/10.3390/antibiotics14030222.

Ganesan, N., Mishra, B., Felix, L.O., Mylonakis, E., 2023. Antimicrobial peptides and small molecules targeting the cell membrane of *Staphylococcus aureus*. Microbiology and Molecular Biology Reviews 87 (2). Available at https://doi.org/10.1128/mmbr.00037-22.

García-Aljaro, C., Ballesté, E., Muniesa, M., 2017. Beyond the canonical strategies of horizontal gene transfer in prokaryotes. Current Opinion in Microbiology 38, 95–105. Available at https://doi.org/10.1016/j.mib.2017.04.011.

Gaze, W.H., Abdouslam, N., Hawkey, P.M., Wellington, E.M.H., 2005. Incidence of class 1 integrons in a quaternary ammonium compound-polluted environment. Antimicrobial Agents and Chemotherapy 49 (5), 1802–1807. Available at https://doi.org/10.1128/aac.49.5.1802-1807.2005.

Gaze, W.H., Zhang, L., Abdouslam, N.A., Hawkey, P.M., Calvo-Bado, L., Royle, J., et al., 2011. Impacts of anthropogenic activity on the ecology of class 1 integrons and integron-associated genes in the environment. ISME Journal 5 (8), 1253–1261. Available at https://doi.org/10.1038/ismej.2011.15.

Gellert, M., Mizuuchi, K., O'Dea, M.H., Itoh, T., Tomizawa, J.-I., 1977. Nalidixic acid resistance: a second genetic character involved in DNA gyrase activity. Proceedings of the National Academy of Sciences 74 (11), 4772–4776. Available at https://doi.org/10.1073/pnas.74.11.4772.

Gillings, M., Boucher, Y., Labbate, M., Holmes, A., Krishnan, S., Holley, M., et al., 2008. The evolution of class 1 integrons and the rise of antibiotic resistance. Journal of Bacteriology 190 (14), 5095–5100. Available at https://doi.org/10.1128/JB.00152-08.

Habboush, Y., Guzman, N., 2018. Antibiotic Resistance. StatPearls [Internet]. StatPearls Publishing, Treasure Island (FL) Available at https://www.ncbi.nlm.nih.gov/books/NBK513277/.

Hasan, T.H., Al-Harmoosh, R.A., 2020. Mechanisms of antibiotics resistance in bacteria. Systematic Reviews in Pharmacy 11 (6), 817–823. Available at https://doi.org/10.31838/srp.2020.6.118, http://www.sysrevpharm.org//fulltext/196-1593155101.pdf?1598520613.

Hawkey, P.M., 1998. The origins and molecular basis of antibiotic resistance. British Medical Journal 317 (7159), 657–660. Available at https://doi.org/10.1136/bmj.317.7159.657, http://www.bmj.com/.

Higgins, N.P., Peebles, C.L., Sugino, A., Cozzarelli, N.R., 1978. Purification of subunits of *Escherichia coli* DNA gyrase and reconstitution of enzymatic activity. Proceedings of the National Academy of Sciences 75 (4), 1773–1777. Available at https://doi.org/10.1073/pnas.75.4.1773.

Hofer, U., Hofer, U., 2019. Reduced permeability as a mechanism of antimicrobial resistance in Gram-negative bacteria. Nature Reviews. Microbiology 17, 590. Available at https://doi.org/10.1038/s41579-019-0265-x.

Ioannou, P., Baliou, S., Kofteridis, D.P., 2023. Antimicrobial peptides in infectious diseases and beyond—a narrative review. Life (Chicago, Ill. 1978) 13 (8), 1651. Available at https://doi.org/10.3390/life13081651.

Isaev, A.B., Musharova, O.S., Severinov, K.V., 2021. Microbial arsenal of antiviral defenses – part I. Pleiades Journals, Russian Federation Biochemistry (Moscow) 86 (3), 319–337. Available at https://doi.org/10.1134/S0006297921030081, https://link.springer.com/journal/10541.

Jacoby, G.A., Low, K.B., 1980. Genetics of antimicrobial resistance. National Academy of Sciences. The Effects on Human Health of Subtherapeutic Use of Antimicrobials in Animal Feeds. National Academies Press, Washington DC, pp. 175–215.

Jault, P., Leclerc, T., Jennes, S., Pirnay, J.P., Que, Y.A., Resch, G., et al., 2019. Efficacy of a bacteriophage cocktail to treat burn wounds infected by *Pseudomonas aeruginosa* (PhagoBurn): a randomised, controlled, double-blind phase 1/2 trial. Lancet Infectious Diseases 19 (1), 35–45. Available at https://doi.org/10.1016/S1473-3099(18)30482-1, http://www.journals.elsevier.com/the-lancet-infectious-diseases.

Jian, Z., Zeng, L., Xu, T., Sun, S., Yan, S., Yang, L., et al., 2021. Antibiotic resistance genes in bacteria: occurrence, spread, and control. Journal of Basic Microbiology 61 (12), 1049–1070. Available at https://doi.org/10.1002/jobm.202100201.

Kato, H., Yamada, K., Tada, T., Miyoshi-Akiyama, T., Kirikae, T., 2020. Penicillin-binding protein 2a is an important factor for the survival of methicillin-resistant *Staphylococcus aureus* in the presence of beta-lactam antibiotics. Journal of Infection and Chemotherapy 26 (7), 723–728. Available at https://doi.org/10.1016/j.jiac.2019.12.013.

Katzung, B.G. (Ed.), 2021. Basic and Clinical Pharmacology, 15th ed. McGraw Hill, New York.

Kim, D.W., Cha, C.J., 2021. Antibiotic resistome from the one-health perspective: understanding and controlling antimicrobial resistance transmission. Experimental and Molecular Medicine 53 (3), 301–309. Available at https://doi.org/10.1038/s12276-021-00569-z, https://www.nature.com/emm/.

Kortright, K.E., Done, R.E., Chan, B.K., Souza, V., Turner, P.E., 2022. Selection for phage resistance reduces virulence of *Shigella flexneri*. Applied and Environmental Microbiology 88 (2). Available at https://doi.org/10.1128/AEM.01514-21, https://journals.asm.org/doi/10.1128/AEM.01514-21.

Lambert, P.A., 2005. Bacterial resistance to antibiotics: modified target sites. Advanced Drug Delivery Reviews 57 (10), 1471–1485. Available at https://doi.org/10.1016/j.addr.2005.04.003.

Liebert, C.A., Hall, R.M., Summers, A.O., 1999. Transposon Tn21, flagship of the floating genome. Microbiology and Molecular Biology Reviews 63 (3), 507–522. Available at https://doi.org/10.1128/mmbr.63.3.507-522.1999, http://mmbr.asm.org/.

Liu, J., Ma, Y., Sun, Y., Zhao, W., Chen, H., Mao, M., et al., 2025. Epidemiological characteristics of antibiotic resistance genes in various bacteria worldwide. Environment International 198, 109410. Available at https://doi.org/10.1016/j.envint.2025.109410.

Lopatkin, A.J., Huang, S., Smith, R.P., Srimani, J.K., Sysoeva, T.A., Bewick, S., et al., 2016. Antibiotics as a selective driver for conjugation dynamics. Nature Microbiology 1 (6). Available at https://doi.org/10.1038/nmicrobiol.2016.44, www.nature.com/nmicrobiol/.

Lupo, A., Coyne, S., Berendonk, T.U., 2012. Origin and evolution of antibiotic resistance: the common mechanisms of emergence and spread in water bodies. Frontiers in Microbiology 3. Available at https://doi.org/10.3389/fmicb.2012.00018, http://www.frontiersin.org/Journal/DownloadFile.ashx?pdf=1&FileId=3997&articleId=18700&Version=1&ContentTypeId=21&FileName=fmicb-03-00018.pdf.

Martinez, J.L., 2009. Environmental pollution by antibiotics and by antibiotic resistance determinants. Environmental Pollution 157 (11), 2893–2902. Available at https://doi.org/10.1016/j.envpol.2009.05.051.

Mazur, N.I., Terstappen, J., Baral, R., Bardají, A., Beutels, P., Buchholz, U.J., et al., 2023. RSV prevention: vaccine and monoclonal antibody landscape. Lancet Infectious Diseases 23 (1), e2–e21. https://doi.org/10.1016/S1473-3099(22)00291-2. The Lancet Infectious Diseases. **23** (1) (2023), e2, Available from. Available at https://doi.org/10.1016/S1473-3099(22)00291-2, http://www.journals.elsevier.com/the-lancet-infectious-diseases.

Mmbando, G.S., Ally, O., Misinzo, G., 2025. Current approaches and tools for combating antibiotic resistance. Discover Applied Sciences 7 (9). Available at https://doi.org/10.1007/s42452-025-07108-9, https://link.springer.com/journal/42452.

Murray, C.J., Ikuta, K.S., Sharara, F., Swetschinski, L., Robles Aguilar, G., Gray, A., et al., 2022. Global burden of bacterial antimicrobial resistance in 2019: a systematic analysis. The Lancet 399 (10325), 629–655. Available at https://doi.org/10.1016/S0140-6736(21)02724-0, http://www.journals.elsevier.com/the-lancet/.

Murray, P.R., Rosenthal, K.S., Pfaller, M.A., 2020. Medical Microbiology Medical Microbiology. Elsevier, United States, pp. 1–855. Available at https://www.elsevierhealth.com.au/medical-microbiology-9780323673228.html, 10.1016/C2018-0-00092-4.

Nordström, K., Burman, L.G., Eriksson-Grennberg, K.G., 1970. Resistance of *Escherichia coli* to penicillins VIII. Physiology of a class II ampicillin-resistant mutant. Journal of Bacteriology 101 (3), 659–668. Available at https://doi.org/10.1128/jb.101.3.659-668.1970.

Oromí-Bosch, A., Antani, J.D., Turner, P.E., 2023. Developing phage therapy that overcomes the evolution of bacterial resistance. Annual Review of Virology 10 (1), 503–524. Available at https://doi.org/10.1146/annurev-virology-012423-110530, http://www.annualreviews.org/journal/virology.

Pagès, J.M., James, C.E., Winterhalter, M., 2008. The porin and the permeating antibiotic: a selective diffusion barrier in Gram-negative bacteria. Nature Reviews. Microbiology 6 (12), 893–903. Available at https://doi.org/10.1038/nrmicro1994, http://www.nature.com/nrmicro/index.html.

Pantaleo, G., Correia, B., Fenwick, C., Joo, V.S., Perez, L., Pantaleo, G., et al., 2022. Antibodies to combat viral infections: development of vaccines, passive immunotherapy and antiviral compounds. Nature Reviews. Drug Discovery 21 (9), 676–696. Available at https://doi.org/10.1038/s41573-022-00495-3, https://www.nature.com/nrd/.

Partridge, S.R., Tsafnat, G., Coiera, E., Iredell, J.R., 2009. Gene cassettes and cassette arrays in mobile resistance integrons. FEMS Microbiology Reviews 33 (4), 757–784. Available at https://doi.org/10.1111/j.1574-6976.2009.00175.x.

Perlin, M.H., Clark, D.R., McKenzie, C., Patel, H., Jackson, N., Kormanik, C., et al., 2009. Protection of *Salmonella* by ampicillin-resistant *Escherichia coli* in lethal drug concentrations. Proceedings of the Royal Society B 276 (1674), 3759–3768. Available at https://doi.org/10.1098/rspb.2009.0997, http://rspb.royalsocietypublishing.org/content/276/1674/3759.full.pdf.

Plowman, R., Graves, N., Griffin, M., Roberts, J.A., Swan, A.V., Cookson, B., et al., 1999. Socio-Economic Burden of Hospital Acquired Infection. Health Protection Agency, London.

Ramirez, M.S., Tolmasky, M.E., 2010. Aminoglycoside modifying enzymes. Drug Resistance Updates 13 (6), 151–171. Available at https://doi.org/10.1016/j.drup.2010.08.003, http://www.elsevier.com/locate/drup.

Rang, H.P., Dale, M.M., Ritter, J.M., Flower, R.J., Henderson, G., 2020. Rang and Dale's Pharmacology, 9th ed. Elsevier, Amsterdam.

Ridyard, K.E., Overhage, J., 2021. The potential of human peptide LL-37 as an antimicrobial and anti-biofilm agent. Antibiotics 10 (6), 650. Available at https://doi.org/10.3390/antibiotics10060650.

Rowe-Magnus, D.A., Guerout, A.-M., Ploncard, P., Dychinco, B., Davies, J., Mazel, D., 2001. The evolutionary history of chromosomal super-integrons provides an ancestry for multiresistant integrons. Proceedings of the National Academy of Sciences 98 (2), 652–657. Available at https://doi.org/10.1073/pnas.98.2.652.

Rumbo, C., Fernández-Moreira, E., Merino, M., Poza, M., Mendez, J.A., Soares, N.C., et al., 2011. Horizontal transfer of the OXA-24 carbapenemase gene via outer membrane vesicles: a new mechanism of dissemination of carbapenem resistance genes in *Acinetobacter baumannii*. Antimicrobial Agents and Chemotherapy 55 (7), 3084–3090. Available at https://doi.org/10.1128/AAC.00929-10Spain, http://aac.asm.org/cgi/reprint/55/7/3084.

Sahli, C., Moya, S.E., Lomas, J.S., Gravier-Pelletier, C., Briandet, R., Hémadi, M., 2022. Recent advances in nanotechnology for eradicating bacterial biofilm. Theranostics 12 (5), 2383–2405. Available at https://doi.org/10.7150/thno.67296, https://www.thno.org/v12p2383.

Samson, J.E., Magadán, A.H., Sabri, M., Moineau, S., 2013. Revenge of the phages: defeating bacterial defences. Nature Reviews Microbiology 11 (10), 675–687. Available at https://doi.org/10.1038/nrmicro3096.

Santos-Lopez, A., Marshall, C.W., Scribner, M.R., Snyder, D.J., Cooper, V.S., 2019. Evolutionary pathways to antibiotic resistance are dependent upon environmental structure and bacterial lifestyle. eLife 8. Available at https://doi.org/10.7554/elife.47612, https://elifesciences.org/download/aHR0cHM6Ly9jZG4uZWxpZmVzY2llbmNlcy5vcmcvYXJ0aWNsZXMvNTAxNzUvZWxpZmUtNTAxNzUtdjIucGRm/elife-50175-v2.pdf?_hash=y6tEhe0k4BK%2FFnb%2BSPk1ewaVnqPuyoDqsq643o%2ByNoM%3D.

Srinivasiah, S., Bhavsar, J., Thapar, K., Liles, M., Schoenfeld, T., Wommack, K.E., 2008. Phages across the biosphere: contrasts of viruses in soil and aquatic environments. Research in Microbiology 159 (5), 349–357. Available at https://doi.org/10.1016/j.resmic.2008.04.010.

Strathdee, S.A., Hatfull, G.F., Mutalik, V.K., Schooley, R.T., 2023. Phage therapy: from biological mechanisms to future directions. Cell 186 (1), 17–31. Available at https://doi.org/10.1016/j.cell.2022.11.017, https://www.sciencedirect.com/journal/cell.

Sultan, I., Rahman, S., Jan, A.T., Siddiqui, M.T., Mondal, A.H., Haq, Q.M.R., 2018. Antibiotics, resistome and resistance mechanisms: a bacterial perspective. Frontiers in Microbiology 9. Available at https://doi.org/10.3389/fmicb.2018.02066, www.frontiersin.org/Microbiology.

Suzuki, H., Nishimura, Y., Hirota, Y., 1978. On the process of cellular division in *Escherichia coli*: a series of mutants of *E. coli* altered in the penicillin-binding proteins. Proceedings of the National Academy of Sciences 75 (2), 664–668. Available at https://doi.org/10.1073/pnas.75.2.664.

Thomas, C.M., Nielsen, K.M., 2005. Mechanisms of, and barriers to, horizontal gene transfer between bacteria. Nature Reviews. Microbiology 3 (9), 711–721. Available at https://doi.org/10.1038/nrmicro1234.

Tocchini-Valentini, G.P., Marino, P., Colvill, A.J., 1968. Mutant of *E. coli* containing an altered DNA-dependent RNA polymerase. Nature 220 (5164), 275–276. Available at https://doi.org/10.1038/220275a0.

Tomasz, A., Munoz, R., 1995. β-Lactam antibiotic resistance in gram-positive bacterial pathogens of the upper respiratory tract: a brief overview of mechanisms. Microbial Drug Resistance 1 (2), 103–109. Available at https://doi.org/10.1089/mdr.1995.1.103.

Umansky, A.A., Fortier, L.C., 2023. The long and sinuous road to phage-based therapy of *Clostridioides difficile* infections. Frontiers in Medicine 10. Available at https://doi.org/10.3389/fmed.2023 1259427 Available from: journal.frontiersin.org/journal/medicine.

Vakulenko, S.B., Mobashery, S., 2003. Versatility of aminoglycosides and prospects for their future. Clinical Microbiology Reviews 16 (3), 430–450. Available at https://doi.org/10.1128/CMR.16.3.430-450.2003.

Vivekanandan, K.E., Kumar, P.V., Jaysree, R.C., Rajeshwari, T., 2023. Exploring molecular mechanisms of drug resistance in bacteria and progressions in CRISPR/Cas9-based genome expurgation solutions. Biomedicine and Pharmacotherapy 162, 114612. Available at https://doi.org/10.1016/j.biopha.2023.114612.

Wang, B., Du, L., Dong, B., Kou, E., Wang, L., Zhu, Y., 2024. Current knowledge and perspectives of phage therapy for combating refractory wound infections. International Journal of Molecular Sciences 25 (10), 5465. Available at https://doi.org/10.3390/ijms25105465.

Wang, Z., Wang, G., Lu, H., Li, H., Tang, M., Tong, A., 2022. Development of therapeutic antibodies for the treatment of diseases. Molecular Biomedicine 3 (1). Available at https://doi.org/10.1186/s43556-022-00100-4, https://www.springer.com/journal/43556.

Wei, X., Gao, J., Xu, C., Pan, X., Jin, Y., Bai, F., et al., 2023. Murepavadin induces envelope stress and enhances the efficacy of beta-lactams in *Pseudomonas aeruginosa*. Microbiology Spectrum 11 (5) e01257-23. Available from. Available at https://doi.org/10.1128/spectrum.01257-23, https://journals.asm.org/doi/pdf/10.1128/spectrum.01257-23.

Weinbauer, M.G., 2004. Ecology of prokaryotic viruses. FEMS Microbiology Reviews 28 (2), 127–181. Available at https://doi.org/10.1016/j.femsre.2003.08.001.

Woodford, N., Ellington, M.J., 2007. The emergence of antibiotic resistance by mutation. Clinical Microbiology and Infection 13 (1), 5–18. Available at https://doi.org/10.1111/j.1469-0691.2006.01492.x.

World Health Organization (WHO), 2015. Global Action Plan on Antimicrobial Resistance. Geneva: WHO. Available from: https://www.who.int/publications/i/item/9789241509763 (accessed 2025).

Wright, M.S., Baker-Austin, C., Lindell, A.H., Stepanauskas, R., Stokes, H.W., McArthur, J.V., 2008. Influence of industrial contamination on mobile genetic elements: class 1 integron abundance and gene cassette structure in aquatic bacterial communities. ISME Journal 2 (4), 417–428. Available at https://doi.org/10.1038/ismej.2008.8.

Xuan, J., Feng, W., Wang, J., Wang, R., Zhang, B., Bo, L., et al., 2023. Antimicrobial peptides for combating drug-resistant bacterial and fungal infections. Drug Resistance Updates 68, 100954. Available at https://doi.org/10.1016/j.drup.2023.100954, https://doi.org/10.1016/j.drup.2023.100954.

Yang, Q., Le, S., Zhu, T., Wu, N., Yang, Q., Le, S., et al., 2023. Regulations of phage therapy across the world. Frontiers in Microbiology 14, 1113065. https://doi.org/10.3389/fmicb.2023.1113065. Frontiers in Microbiology. 1664302X. **14** (2023), Available from. Available at https://doi.org/10.3389/fmicb.2023.1250848, https://www.frontiersin.org/journals/microbiology#.

Youssef, R.A., Sakr, M.M., Shebl, R.I., Saad, B.T., Aboshanab, K.M., Youssef, R.A., et al., 2024. Genomic characterization and evaluation of two microencapsulated lytic phages against MDR *Salmonella serovars*. Annals of Clinical Microbiology and Antimicrobials 23 (1), 1–17. Available at https://doi.org/10.1186/s12941-024-00680-5.

Zhao, J., Chan, M., Tang, Q., Zhang, A., Zhao, J., Chan, M., et al., 2024. Estimating health and economic burdens of multidrug-resistant hospital-associated infections caused by major gram-negative pathogens in Singapore. Antimicrobial Resistance and Infection Control 13 (1), 11. Available at https://doi.org/10.1186/s13756-024-01366-7.

Chapter 3

Survival of the fittest: the adaptive evolution of antibiotic-resistant bacteria

Anukul Paul and Ranadhir Chakraborty
OMICS Laboratory, Department of Biotechnology, University of North Bengal, Siliguri, West Bengal, India

3.1 Introduction

Resistance to antimicrobial agents has become a major factor in global morbidity and mortality, a finding that contrasts dramatically with early optimism in the mid-20th century, when the discovery of antibiotics was heralded as a final victory over infectious diseases. It soon became evident that pathogenic microorganisms possessed an extraordinary ability to adapt to the pressure of antimicrobials and could become resistant to every agent deployed. Most clinically important bacteria could become resistant through either intrinsic mechanisms or by acquiring characteristics from other species in their ecological networks. Intrinsic resistance refers to inherent, species-specific traits such as outer-membrane permeability in *Pseudomonas aeruginosa* (Nikaido, 2003). In contrast, acquired resistance arises through chromosomal mutations or horizontal gene transfer (HGT), for example, acquisition of $bla_{\text{CTX-M}}$ via plasmids in *Escherichia coli* (Cantón et al., 2012). The fundamental basis of these alterations was the ability to take up fewer antimicrobial compounds, to alter drug targets, to use enzymes to inactivate antibiotics, or actively pump out toxic molecules. Each of these was an evolutionary solution to survive under selective pressure (C Reygaert, 2018).

Antibiotic resistance is a global health problem that arises through the movement of bacteria and genes between humans, animals, and the environment. Countless barriers to transmission reduce the flow of germs and genes, yet pathogens continually acquire new resistance determinants from other species. This makes it more challenging for us to prevent and cure bacterial infections. Evolutionary events generating novel resistance determinants in pathogens are rare, difficult to predict, and can have significant consequences. On the other hand, the spread of resistant strains, which are already widespread, is common, quantifiable, and more predictable, but each event has limited consequences. Determining the routes and drivers of environmental evolution and the transmission of antibiotic resistance, as well as the bottlenecks involved, are integral to understanding and addressing the resistance crisis as a whole (Larsson and Flach, 2022). The rise in antibiotic resistance among bacteria, viewed from an evolutionary biology perspective, can be attributed to the principle of differential fitness, wherein the survival and reproductive success of bacterial populations depend on their fitness in the presence of antimicrobial agents.

Multidrug-resistant (MDR) bacterial strains have increasingly risen as a serious, global health concern, with both Gram-negative and Gram-positive pathogens demonstrating remarkable abilities to adapt to the pressure exerted by antimicrobials by acquiring resistance elements already in circulation within the larger bacterial gene pool. This adaptive capacity was enabled largely by mobile genetic elements (MGEs), which allowed genetic material to move within genomes or transfer between cells, including insertion sequences (ISs), transposons, integrons with their associated gene cassettes, plasmids, and integrative conjugative elements. Acting together, these elements enabled widespread horizontal gene exchange and provided key vehicles through which antibiotic resistance had been acquired and spread across diverse bacterial taxa. These processes were previously observed in clinically significant organisms in the ESKAPEE category, such as *Enterococcus faecium*, *Staphylococcus aureus*, *Klebsiella pneumoniae*, *Acinetobacter baumannii*, *P. aeruginosa*, *Enterobacter* species, and *E. coli*, which had become some of the most challenging organisms to treat in the hospital environment (Partridge et al., 2018). These genomic studies have revealed that the antibiotic resistance effects of these organisms are usually cloned and transferred through the work of MGEs. The resistance to carbapenem and cephalosporin was highly correlated to the presence of the IS elements in *Acinetobacter*, most of which included outward-facing promoters, which had the capacity to

Antimicrobial Resistance in the 21st Century. DOI: https://doi.org/10.1016/B978-0-443-44748-8.00014-2

express β-lactamase at a clinical level. Their functions determine the evolutionary directions of priority bacterial pathogens that were influenced by MGEs (Noel et al., 2022).

It is no secret that prokaryotic MGEs play a predominant role in the evolution of bacteria and enable the dissemination of accessory functions, such as antibiotic resistance, through the presence of transposons, integrons, bacteriophages, and plasmids. Although these elements were traditionally studied in isolation, limiting broader insight into their evolutionary and ecological behaviour, recent advances have begun to reveal their collective influence on genome plasticity and HGT. Results from large-scale comparative analyses indicated that MGEs accounted for a substantial fraction of bacterial genomes and were widely distributed across both habitats and taxonomic groups when identified either by using conserved recombinase markers or through pangenome-based boundary estimation. Transposable elements were the most abundant category and outnumbered the phage-derived elements by a considerable margin, with both making extensive contributions to gene movement within and between species. Several MGE-mediated horizontal transfers were detected across various ecological niches, indicating widespread genetic exchange and frequent comobilisation of elements such as transposons and integrons. These conclusions strongly reiterated that MGEs were the dominant vehicles for the spread of antibiotic resistance determinants and continued to underpin the resistant bacterial populations through underlying evolutionary processes (Khedkar et al., 2022).

Most bacteria in natural environments exist in structured, surface-associated communities called biofilms, in which the constituent cells usually show physiological characteristics that are radically different from those of planktonic populations. Biofilm-associated infections have been implicated in a wide range of clinical conditions and are often chronic, persistent, and recalcitrant to conventional antibiotic therapy. The underlying reasons for this recalcitrance lie in several intrinsic features: a protective extracellular matrix, reduced cellular growth rates, adjustments in antibiotic target expression, and the efficient transfer of resistance genes among neighbours. Consequently, biofilms not only form physical impediments to the external environment but are also dynamic evolutionary hotspots that can generate and disseminate antibiotic resistance. Nevertheless, cooperative, competitive, and metabolically interdependent multispecies biofilm interactions further affected antimicrobial tolerance and altered the behaviour of the entire polymicrobial community. These features underscore the role of biofilm biology in the evolution of resistant bacteria in an adaptive way and the strategy to deal with such infections (Liu et al., 2024).

In this review, we explain what we know at present regarding the roles of the environment, including antibiotic pollution in the evolution of resistance, transmission, and a view of the situation of antibiotic resistance in the clinic. Simultaneously, research on infection management, approaches to antibiotic availability, monitoring systems, and alternative treatment technologies had reshaped the attempt to contain resistant pathogens. New uses of artificial intelligence were also a potential source of assistance in susceptibility testing, real-time surveillance, refinement of diagnostics, and the discovery of antimicrobials. All these developments indicated the urgency of the problem of AMR and the necessity of accelerating technological and scientific work to address its growing influence (Ho et al., 2025). With the conventional antimicrobial strategy, other types of technological advancements are starting to offer alternative methods to solve the issue, including re-engineering the resistance mechanisms of microbial systems, reconstituting the antibiotic sensitivity in the resistant strains, or editing the resistance determinants using any of the gene-editing systems, including the CRISPR–Cas9 system (Ali et al., 2018). We provide an overview of current evidence, describe risk scenarios, establish methods for surveillance and for assessing potential drivers, and finally outline actions to mitigate risks. Understanding the evolutionary and genetic dynamics that enabled these organisms to bypass antimicrobial therapy provided an essential basis for addressing the antibiotic resistance problem explored throughout this chapter.

3.2 Evolutionary drivers of antibiotic resistance

3.2.1 Selective pressure and survival of the fittest

The exceptional ability of bacteria to resist antibiotics emerged through continuous selective pressure, thereby engineering genetic and phenotypic diversity over millions of years and allowing only those mutants capable of countering the action of antimicrobials to survive and propagate. Antibiotic exposure served as a powerful ecological filter that killed the susceptible members of a population while allowing rare mutations or horizontally acquired resistance determinants to become dominant under the pressure of such drugs. This 'survival of the fittest' dynamics had been possible due to the immense genetic plasticity of bacteria, enabling them to acquire adaptive mutations, to reprogramme gene expression networks, or to take up MGEs carrying resistance genes by transformation, transduction, or conjugation, all well-established evolution trajectories for adaptation under antibiotic pressure. Subinhibitory antibiotic concentrations in clinical, environmental, and agricultural settings further increased selection pressure that enriched resistant subpopulations, even at low drug concentrations insufficient to completely inhibit bacterial growth. Hence, resistance traits arising either through chromosomal mutation or acquired through plasmids, transposons, and integrons are quickly amplified in microbial communities whenever exposure to antimicrobials provides an evolutionary advantage and the treatment with antibiotics becomes a potent evolutionary force. This interplay of selective pressure and adaptive genetic variation is at the core of microbial population evolution towards antibiotic resistance, reminding us why modern antibiotic use, even today, shapes the evolutionary profile of pathogenic bacteria (Munita and Arias, 2016).

Resistance pathways were formed by the interaction of deterministic forces (predominantly natural selection, with mutations or horizontally acquired genes) and stochastic forces (drift, random fluctuations, and population fragmentation) that afforded windows of opportunity for adaptive traits to emerge and persist. In the presence of antibiotic stress, bacteria expressing functions that incidentally reduced drug susceptibility, whether through pre-existing physiological mechanisms, newly acquired resistance genes, or spontaneous mutations, showed increased fitness and were selectively enriched as the susceptible cells were eliminated. These determinants of resistance did not originally evolve for antibiotic resistance but became advantageous when exposure to antibiotics provided an ecological sieve rewarding these traits and illustrating the classic principle that 'every gene is everywhere, but the environment selects'. Moreover, antibiotic exposure not only selected resistant phenotypes but also enhanced genetic variation by inducing stress responses, promoting mutagenesis, and favouring gene amplification or horizontal transfer that expanded the pool of selectable variants (Baquero et al., 2021). Resistance genes existed in environmental microorganisms for millennia as part of the huge resistome maintained by antibiotic-producing bacteria like *Streptomyces*, where resistance originally evolved as a self-protection strategy. When human activities, especially the agricultural use of antibiotics and environmental contamination, amplified the selection pressures, resistant variants emerged as the 'fittest' members of microbial communities that flourished where susceptible competitors were eliminated. The same selective forces drove the mobilisation of the resistance genes into pathogenic bacteria from environmental reservoirs through HGT mechanisms such as transformation, transduction, and conjugation, thereby hastening the recruitment of advantageous traits across ecological boundaries. Proof of ancient resistance determinants in permafrost samples, as well as more recent genetic findings such as CTX-M β-lactamases and *qnr*, a quinolone resistance gene traced to environmental organisms, has shown that selection continuously favours resistant variants over evolutionary timescales (Perry and Wright, 2013).

Low doses of antibiotics act as chemical signalling that changes the bacterial gene expression, virulence, biofilm formation, and quorum-sensing pathways, indirectly affecting the susceptibility and ecological behaviour, enhancing the fitness in the case of antibiotic stress. Thus, all these effects illustrate that even nontoxic doses of antibiotics function as strong evolutionary filters consistently acting in favour of the 'fittest' microbial phenotypes, which are best adapted to persist and ultimately to thrive in an antibiotic-rich environment (Andersson and Hughes, 2014). In summary, selective pressure provides the evolutionary driving force that favours the 'survival of the fittest' bacterial lineages, those capable of withstanding the antimicrobial onslaught, leading to the global rise of MDR pathogens (Laxminarayan et al., 2013).

3.2.2 Mutation, fitness landscapes, and adaptive trajectories

Drug-sensitivity mutations arose through various routes of genetic change. However, their evolution was deeply affected by the targeted biochemical pathways imposed by the presence or absence of antibiotics and the linkages between the genes targeted by mutations. Using high-resolution phenotyping, it was found that each deletion mutant has a unique 'phenotypic signature' profile for drug and stress challenges, thus showing how the concerted effects of small genetic changes were coordinated to improve or diminish cell viability according to the targeted pathways affected by antibiotics. These quantitative fitness responses mapped the functional architecture of bacteria, showing that antibiotic exposure reshaped the adaptive landscape by rewarding mutations that enhanced tolerance through changes in membrane integrity, efflux capacity, metabolic flux, or DNA repair pathways. Some examples include mutants unable to express components of the electron transport chain, making them more tolerant of aminoglycosides due to reduced accumulation (Nichols et al., 2011).

The key lesson from studies of experimental evolution concerned the role of sequential mutation in the evolution of antibiotic resistance, which reshaped the fitness landscape and limited the evolutionary trajectories available to bacterial populations. The evolution of resistance to chloramphenicol, doxycycline, and other antibiotics under continued antibiotic pressure in the morbidostat revealed the variety of mutation subsets conferring resistance to each drug, clearly demonstrating multiple mutational routes to equivalent fitness improvements under ribosomal-mediated stress. In contrast, resistance evolution under trimethoprim followed a much more constrained mutational trajectory. Nearly all independently evolving bacterial populations acquired sequential mutations in the gene encoding dihydrofolate reductase (*dhfr*) or its promoter region. Because trimethoprim targets a single essential enzyme, the number of beneficial mutational pathways was limited, resulting in stepwise evolution of resistance driven by strong epistatic interactions. These findings demonstrate that each antibiotic imposes a distinct fitness landscape, which determines both the predictability of evolutionary outcomes and the rate at which resistance evolves (Toprak et al., 2012). Sub-MIC concentrations of streptomycin do not selectively enrich for preexisting high-fitness mutations; rather, they allow the population to follow alternative evolutionary routes involving the accumulation of several small-effect mutations that lead to high-level streptomycin resistance. Adaptive evolution strongly depends on the selection intensity and the type of fitness landscape; weak selective conditions enable the survival and diversification of the population, the emergence of the mutator phenotype, and accelerate adaptive evolution through the accumulation of beneficial mutations by several orders of magnitude (Wistrand-Yuen et al., 2018). These mutations that provided new resistance functions generated a new fitness landscape where the mutational robustness decreased, and the values of epistasis and landscape ruggedness increased. These led to the landscape having steep drops and multiple peaks of local maximum fitness values. Some evolutionary paths were made inaccessible by evolution (Steinberg and Ostermeier, 2016). The evolutionary course of such adaptation was determined by the pathogen transmission mode, its ecological niche, and the genetic basis of resistance, whereby certain bacteria, such as *Mycobacterium tuberculosis*, acquired the same set of resistance mutations during treatment, while others, such as *Klebsiella pneumoniae* and *Salmonella typhi*, accumulated a basketful of mutations or horizontally transferred genes, turning regional clones into global leaders. The development of dominant high-risk clones, such as methicillin-resistant *S. aureus* (MRSA) ST22, *E. coli* ST131, or *K. pneumoniae* ST258, occurred according to an ordered sequence, whereby the 'kingmaker mutations' or MGEs initially provided resistance, favouring the acquisition of secondary mutations increasing their 'fitness bonus' under selective antibiotic pressure (Baker et al., 2018).

3.2.3 Trade-offs and compensatory evolution

The evolutionarily conferred resistance trait is prone to impose measurable physiological or metabolic costs on bacteria, giving rise to a typical trade-off with resistant bacteria exhibiting decreased fitness, virulence, or competitiveness in non-antibiotic environments; nevertheless, abundant evidence supports the rapid evolution of bacteria with compensatory capabilities capable of counteracting such fitness costs associated with resistance. Recent systems-based analyses integrating experimental evolution with quantitative genomic data have demonstrated that the evolution of antimicrobial resistance (AMR) can be partially predictable across biological scales (Charlebois, 2023). Compensatory evolution can occur through multiple processes, such as intragenic or secondary mutations within the same gene or another gene influencing the target or its regulatory network, gene duplication to counteract metabolic defects, metabolic network rearrangement to rebalance resistance via altered metabolic costs, and resistance modifiers interacting with resistance mutations to counteract fitness costs (Hernando-Amado et al., 2017; Durão et al., 2018). Often, there will be an effect on essential biological processes, such as protein biosynthesis, transcription, cell wall biosynthesis, and regulatory signals, leading to substantial selective disadvantages that are, in principle, penetrable by the outgrowth of sensitive lineages when the selective (antibiotic) pressure is removed. Conversely, compensatory mutations in *rpsL* for streptomycin resistance, extragenic substitutions in *rpsD* and *rpsE* genes, and genomic amplifications, such as *metZ/metW*, illustrate the microorganism's ability to mitigate the selective disadvantages imposed by ribosomal, fluoroquinolone, and fusidic acid resistance mechanisms, respectively (Andersson and Hughes, 2010). Many mutations that conferred rifampin resistance to *E. coli* incurred high fitness penalties, mainly reflected in reduced growth rates or impaired transcriptional efficiency, thereby exemplifying the concept of evolutionary trade-offs in resistance (Reynolds, 2000). Resistant strains evolved along fitness peaks by climbing new topographic summits instead of reverting to the ancestral form, thereby allowing 'adaptive valleys' to form, making it very unlikely for these bacteria to reverse their resistance. These results demonstrate that compensatory evolution not only stabilises bacterial resistance but also allows resistant strains to persist for many years after exposure to antibiotics (Levin et al., 2000).

3.2.4 Ecological niches shaped by antibiotic exposure

Antibiotics in soil, sediment, rhizosphere, and aquatic ecosystems are more likely to occur at subinhibitory concentrations. They are involved in functions that are not lethal but are used as signalling molecules in gene expression, quorum sensing, virulence, and metabolic interactions. For example, sub-MIC levels of tetracycline and macrolide antibiotics such as azithromycin have been shown to modulate global gene expression, biofilm formation, and quorum sensing in *P. aeruginosa* and other bacterial species, illustrating signalling-mediated physiological changes at non-lethal concentrations (Nolan and Behrends, 2021; Romero et al., 2011). In low-pressure ecosystems, resistance genes evolved for other functions unrelated to antibiotic detoxification, including housekeeping metabolic functions, structural functions, or intracellular signal trafficking, until high levels of selective pressure from anthropogenic antibiotic use in hospital or farm settings refunctioned them as resistance genes. Ecosystems exposed to high levels of anthropogenic antibiotics, including wastewater, river tributaries receiving farm runoff, and soil receiving manure, are hotspots of HGT, allowing environmental and associated human bacterial strains to come into contact and increase the load of clinically useful resistance genes carried by vectors such as plasmids, integrons, and transposons. These genetic elements are maintained in the environment and are rarely reversible and highly stable under current ecological conditions because of co-selection (e.g. resistance genes on one plasmid), ecological selective advantage provided by traits (e.g. siderophores and microcins), and the plasmid-encoded toxin–antitoxin system (Martinez, 2009; Wein et al., 2019).

Exposure to antibiotics causes drastic shifts in bacterial niches due to altered selective pressure and lateral gene transfer (LGT) rates in human-affected ecosystems. With time, antibiotics, heavy metals, and biocides concentrated in hospital waste, farms, and natural systems downstream of human influence form a long-lasting and heterogeneous selective environment that favours bacteria harbouring MGEs encoding adaptive functions such as AMR mechanisms. These factors induce physiological responses, such as SOS responses, leading to the excision, mobilisation, transfer, and recombination of plasmids, integrons, transposons, and prophages, thereby enhancing the generation and transfer of mobile DNA. Biofilms formed in water supply systems, sewage facilities, and aquatic sediments are primary sites of large numbers of LGT, allowing frequent transfer and fixation of foreign DNA, along with the maintenance of large bacterial numbers that are constantly exposed to subinhibitory concentrations of antibiotics. Human-induced ecological change, mainly due to their huge biomass, has further concentrated bacterial communities in environments such as the human gut, where rates of LGT are 20–50 times higher than in natural environments, thereby enhancing the transfer and fixation of advantageous traits (Gillings, 2017). The richness and abundance of AMR genes (ARGs), as well as mobile elements such as ISs, are also high in untreated influent. The receiving habitats are also influenced by the effluent, which adds nutrients such as nitrogen, phosphorus, and organic carbon. The effluent also contains a wide range of organic materials like antimicrobial agents, herbicides, and pharmaceuticals, which can sustain selective pressures and coselect for resistance (Read et al., 2024). Evidence from a worldwide survey of municipal sewage has shown that environments rich in both antibiotics and biocides can act as selective filters, conferring a selective advantage on bacteria with resistance phenotypes while simultaneously deselecting bacteria with resistance genotypes that incur high costs in low-selection environments. Wastewater communities, which represent a zone of overlap between human and environmental bacteria, can either enhance resistance selection if antibiotic levels surpass the minimal selective threshold or favour the dominance of wastewater-adapted bacteria with reduced susceptibility or resistance to antibiotics when selective pressure is not high, in a selective pressure-dependent process of niche evolution (Yu et al., 2025). The human-induced disturbances, such as climate change, land-use modification, urbanisation, and pollution, further increase the selectivity of these landscapes, promoting the hotspots of wastewater treatment plants (WWTPs), effluent-receiving streams, farmlands amended with farm manure, and cities where ARGs, MGEs, and resistant bacteria congregate and spread (Zambrano, 2023).

3.3 Genetic mechanisms underpinning resistance

3.3.1 Spontaneous chromosomal mutations

The emergence of a resistance-conferring mutation depends not simply on one inherent mutation rate, but on many factors, such as the length of the gene, how many sites allow a permissible mutation that confers resistance without the loss of viability, and the existence of a large number of genes with mutations causing resistance in either an independent or cooperative way. Structural properties of genes, such as unstable nucleotide tracts, chromosome position relative to the origin of replication, and copy number, have an additional impact on mutability, resulting in wide variation in the frequency of resistant mutant emergence. Physiological stress, such as starvation and exposure to antibiotics, can boost mutation rate by stress-induced mutagenesis pathways, particularly through the activation of error-prone polymerases produced as a result of the SOS response, whereas hypermutable subpopulations resulting from flaws in mismatch repair systems can produce exceptionally high mutation rates leading to accelerated adaptation under antibiotic pressure (Martinez and Baquero, 2000). Such mutations often alter genes encoding critical cellular targets, most prominently the targets of rifamycins and fluoroquinolones (i.e. *rpoB* and DNA topoisomerases), which reduce antibiotic binding and hence confer resistance. Mutations can also alter membrane permeability or efflux potential, as illustrated by loss or reduced expression of porins such as OprD in *P. aeruginosa*, or regulatory mutations that constitutively express multidrug efflux pumps like AcrAB and the Mex pumps, which confer broad-spectrum resistance phenotypes. Overproduction of chromosomal β-lactamases (e.g. AmpC) most often arises via promoter or regulatory mutations that drive constitutive expression of the enzyme, which renders most cephalosporins ineffective. In species such as *M. tuberculosis*, where plasmid-mediated resistance mechanisms occur very rarely, spontaneous chromosomal mutations are the only path available for the evolution of resistance, permitting stepwise acquisition of mutations that yield MDR and extensively drug-resistant (XDR) strains (Blázquez and Gómez-Gómez, 2014). β-Lactamases probably originated from ancestral DD-peptidases via divergent mutational pathways that altered key residues, loop conformation, and active-site geometry, thereby allowing enzymes to acylate and deacylate β-lactam antibiotics efficiently while losing their original function in processing cell wall precursors. Spontaneous mutations that altered the ω-loop, the SXN/YXN catalytic motifs, or the positions of residues surrounding the active-site serine caused radical changes in substrate binding and reactivity, increasing β-lactam turnover and providing resistance mechanisms at the chromosomal level (Knox et al., 1996).

3.3.2 Activation of efflux pumps

Activation of efflux pumps is one of the major genetic mechanisms that confer improved bacterial survival during antibiotic selection pressure, most importantly in Gram-negative bacteria (Fig. 3.1). Efflux pumps in the major facilitator superfamily (MFS), multidrug and toxic compound extrusion (MATE), resistance nodulation cell division (RND), small multidrug resistance (SMR), and ATP-binding cassette (ABC) superfamily work by extruding a broad array of toxic compounds, including multiple antimicrobials, through either proton motive force or ATP hydrolysis, thus conferring a natural, broad-spectrum resistance mechanism. Activation of efflux pumps fails to confer high-level resistance in most instances; instead, it improves bacterial survival under antibiotic selection pressure, facilitates the acquisition of new resistance mutations, particularly in target enzymes, and, in instances involving the overexpression of pumps exporting multiple substrates, leads to cross-resistance (Webber and Piddock, 2003).

Multidrug efflux systems, particularly the RND efflux pumps AcrAB–TolC in *E. coli* and MexAB–OprM in *P. aeruginosa*, are chromosomal efflux systems for exporting a broad series of structurally unrelated antibiotics directly from the periplasmic space into the environment, making it difficult for drug accumulation within bacterial cells, hence protecting the organism against inhibitory concentrations of drugs (Li et al., 2015). When wild-type *P. aeruginosa* was first treated with gentamicin, it led to surviving cells adapted to withstand the killing effect, along with reduced

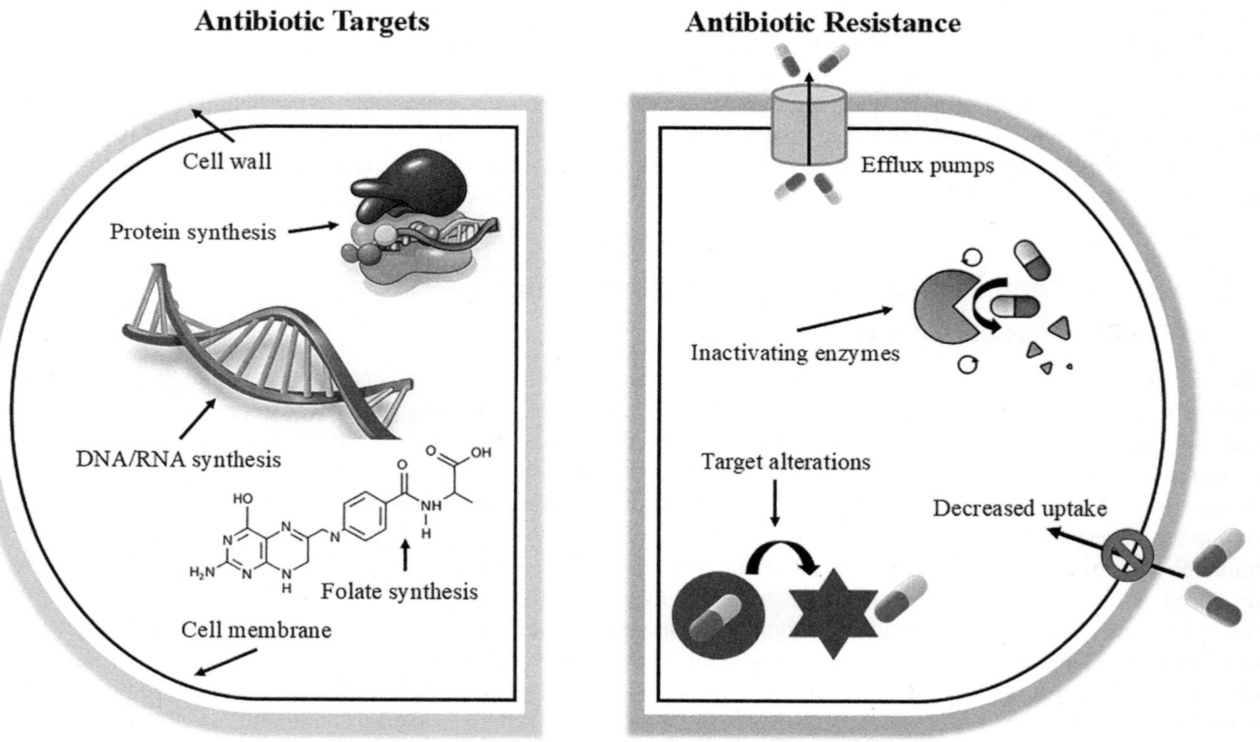

FIGURE 3.1 **Schematic illustration of major antibiotic targets and corresponding mechanisms of antibiotic resistance in bacterial cells.** The left panel depicts principal cellular targets of antimicrobial agents, including inhibition of cell wall and cell membrane integrity, protein synthesis at the ribosome, DNA and RNA synthesis, and folate biosynthesis pathways. The right panel illustrates key resistance strategies evolved by bacteria to counter antibiotic activity, such as active drug efflux mediated by efflux pumps, enzymatic inactivation of antibiotics, modification or protection of antibiotic target sites, and reduced intracellular drug accumulation due to decreased uptake or altered membrane permeability.

susceptibility, a sensitivity not mediated by mutations but rather by the efflux pump, demonstrating its regulated expression against drug concentrations by the bacteria (Hocquet et al., 2003). Efflux pump genes are also readily transmitted via plasmids among pathogenic organisms, thereby promoting the rapid spread of MDR bacteria (Blair et al., 2015).

3.3.3 Target modification and reduced permeability

Bacteria often resist antibiotics by either modifying the target for antibiotics or limiting the entry of antibiotics, and both methods combined make up the most important mechanisms of resistance to antibiotics (Fig. 3.1). Target modification can occur by mutations, enzymatic change, or target substitution, all of which decrease the affinity for antibiotics but retain the original function (Munita and Arias, 2016). Target modification results from mutations or post-translational modifications that decrease the binding affinity between an antibiotic and its target while leaving the target's physiological function intact. Such modifications arise from point mutations in genes encoding topoisomerases conferring fluoroquinolone resistance, in *rpoB* conferring rifampin resistance, or in 23S rRNA that diminishes linezolid binding, thereby preventing inhibition even at therapeutic concentrations. Additionally, methylation of ribosomal RNA by erm or cfr transferases in inactivated cells renders them unresponsive to macrolides, lincosamides, streptogramins, phenicols, and oxazolidinones, thereby preventing inhibition. Finally, apart from chemical resistance, certain organisms alter cell membrane structure. For instance, *S. aureus* and enterococci alter phospholipid composition or alter membrane charge to reduce daptomycin binding; *K. pneumoniae* alters lipid A, reducing colistin binding (Wright, 2007). The most important modification could be those concerning penicillin-binding proteins (PBPs), in which changes in protein structure diminish β-lactam binding and enable the synthesis of the cell wall to proceed despite the presence of β-lactam antibiotics; the most famous example is the appearance of the *mecA* gene in *S. aureus*, which codes for the PBP2a protein showing greatly reduced affinity for β-lactams. Other target modifications are those concerning glycopeptide resistance, in which *van* gene carriage alters the peptidoglycan precursors, reducing vancomycin binding affinity (C Reygaert, 2018).

3.3.4 Horizontal gene transfer

HGT is considered one of the most powerful evolutionary forces shaping bacterial antibiotic resistance. HGT is primarily mediated by three mechanisms: conjugation, transduction, and transformation, thereby enabling rapid transfer between strains and species (Fig. 3.2). Transformation is the process of uptake and chromosomal integration of free DNA fragments released from dead bacterial cells by a competent recipient cell. Conjugation depicts plasmid-mediated DNA transfer from a donor (F^+) cell to a recipient (F^-) cell through direct cell-to-cell contact via a conjugative pilus, resulting in the generation of a new F^+ cell. Transduction shows bacteriophage-mediated transfer of bacterial DNA, in which fragments of host DNA are packaged into viral particles and subsequently delivered to a new recipient cell, leading to genetic recombination and formation of a transduced cell. Conjugative plasmids are major platforms for resistance transfer because they combine modules for replication, maintenance, and mating-pair formation, enabling self-transmission and mobilisation of non-conjugative elements across diverse microbial communities. Transposons and integrons further enhance genetic mobility by capturing gene cassettes, including many clinically important resistance genes, and inserting them into new genomic locations, thus creating combinatorial diversity and enabling rapid adaptation. HGT events are not random; they are strongly influenced by ecological connectivity, population density, and selective pressures such as antibiotic exposure, which enrich for mobile elements carrying advantageous traits (Frost et al., 2005).

The conjugative plasmids appear to be particularly important since they can mobilise not only themselves but also nonconjugative plasmids and transposons, including those from the Tn3-family and Tn21-subfamily, thereby amplifying the spread of β-lactamases, aminoglycoside-modifying enzymes, and last-line resistance genes, such as bla_{NDM} or *mcr-1*. Integrons further speed up the evolution of resistance by capturing a wide variety of gene cassettes and reorganising them under the increasing selective pressure, while the families of transposons, such as Tn402, Tn7-like elements, and composite IS26-mediated structures, repeatedly reorder genomic contexts for optimal expression of the resistance genes (Partridge et al., 2018). Early observations of transferable virulence factors in pneumococci established transformation as a major route of DNA uptake, and subsequent discoveries of plasmid-, phage-, and transposon-mediated transfer revealed the extensive gene flux that shapes bacterial genomes. A considerable proportion of bacterial chromosomes consists of horizontally acquired genes, although only those that confer a selective advantage or are carried by self-transmissible MGEs remain in the population. Barriers include DNA restriction, surface exclusion, and replication incompatibilities, but note that these constraints reduce, but do not prevent, the spread of foreign genes (Thomas and Nielsen, 2005).

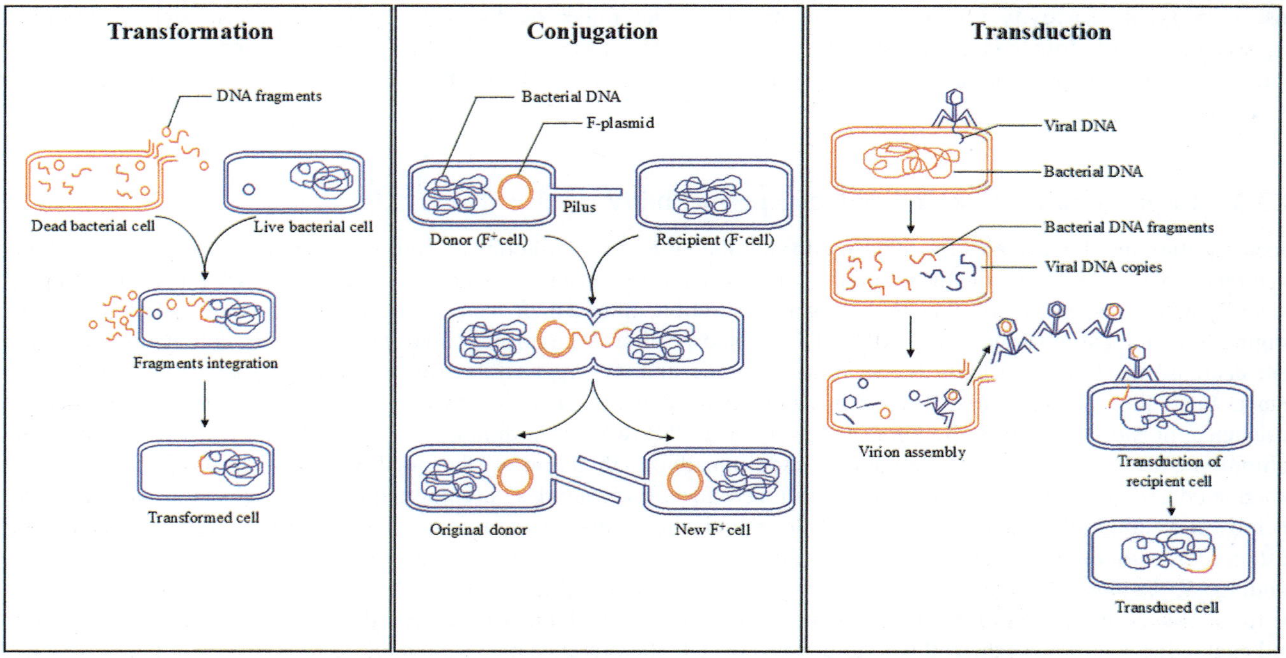

FIGURE 3.2 **Schematic representation of the three principal mechanisms of horizontal gene transfer in bacteria: transformation, conjugation, and transduction.** Transformation is the process of uptake and chromosomal integration of free DNA fragments released from dead bacterial cells by a competent recipient cell. Conjugation depicts plasmid-mediated DNA transfer from a donor (F^+) cell to a recipient (F^-) cell through direct cell-to-cell contact via a conjugative pilus, resulting in the generation of a new F^+ cell. Transduction shows bacteriophage-mediated transfer of bacterial DNA, in which fragments of host DNA are packaged into viral particles and subsequently delivered to a new recipient cell, leading to genetic recombination and formation of a transduced cell.

3.4 Mobile genetic elements and the architecture of the resistome

3.4.1 Types of mobile genetic elements: plasmids, integrons, transposons

MGEs include distinct but interrelated classes of DNA that enable both intracellular and intercellular mobility of genetic material, and the core of this network comprises plasmids, integrons, and transposons. Plasmids are described as self-replicating replicons consisting of a conserved backbone for replication, copy-number control, and partitioning, together with diverse accessory genes encoding antibiotic resistance, virulence, symbiosis, or metabolic traits. Plasmids, especially multiresistance plasmids widely distributed among *Enterobacteriaceae*, *P. aeruginosa*, *A. baumannii*, staphylococci, and enterococci, represent very effective vehicles for intercellular DNA transfer by conjugation or mobilisation while maintaining resistance genes due to replication control, stable inheritance systems, and accessory maintenance modules (Partridge et al., 2018; Smillie et al., 2010; Toussaint and Chandler, 2012). By contrast, transposons are promiscuously recombining DNA segments that mobilise themselves or others within and between replicons using transposases or site-specific recombinases, generating genomic rearrangements and often capturing and spreading resistance genes through hopping onto plasmids or phages. Transposons, including ISs, composite transposons, Tn3-family elements, Tn7-like transposons, and many resistance island-forming units, mediate intracellular recombination and DNA mobility through cut-and-paste or replicative mechanisms, often capturing resistance genes or integrons and relocating them within or between chromosomes and plasmids; notable examples include IS26-driven gene accretion, Tn21-subfamily integron-bearing transposons, Tn1546 carrying *vanA*, and Tn4401 carrying *bla*$_{KPC}$ (Partridge et al., 2018; Toussaint and Chandler, 2012). Integrons are another important class of MGE: while not self-mobile, they encode integrase enzymes which catalyse the site-specific insertion and excision of mobile gene cassettes, forming tandem arrays of resistance determinants that are expressed from integron-associated promoters. The latter systems are often embedded within transposons or conjugative plasmids, massively increasing their mobility across bacterial populations (Frost et al., 2005). Class 1 integrons associated with Tn402-like transposons represent the most clinically relevant examples because they carry various cassette-borne ARGs and are widespread among Gram-negative pathogens (Partridge et al., 2018; Toussaint and Chandler, 2012).

3.4.2 Plasmid-mediated dissemination and host-range expansion

Conjugative plasmids, especially those of the Enterobacteriaceae and Enterococcaceae families, act as important vectors of AMR, transferring ARGs from the invading nosocomial pathogens and the resident members of the human intestinal microbiome, an important biologic niche of plasmid transfer in hospital settings. These plasmid-bacteria relationships can become highly adapted in hospital settings, such as observed global dissemination of *K. pneumoniae* ST11 and ST405 with the *OXA-48* plasmid, and *E. coli* ST131 with the plasmid-borne IncFII encoding the $bla_{\text{CTX-M}}$ gene. Plasmid host-range expansion is driven by plasmid fitness costs, conjugative transfer, and compensatory evolution. Initial fitness burdens imposed by newly acquired plasmids are gradually alleviated by chromosomal or plasmid-borne mutations, promoting stable maintenance in new hosts. Meanwhile, conserved plasmid backbones ensure replication and stability, whereas accessory regions acquire resistance genes through interactions with insertion sequences, transposons, and integrons. Conjugation contributes significantly to host-range expansion, as plasmids may successfully transfer into recipients in which they cannot replicate; in such events, associated mobile elements can mobilise resistance genes onto alternative replicons, enabling stable inheritance. Studies also show that simple mutational changes in replication initiator proteins or promoter elements can rapidly broaden a plasmid's host range under selective pressure, while the ability of plasmids to evade restriction-modification barriers further enhances their dissemination potential across ecological niches (Baharoglu and Mazel, 2014). Epidemic plasmid families such as IncF, IncI, IncA/C, and IncH dominate global AMR spread because they harbour extensive assortments of β-lactamases, AmpC enzymes, carbapenemases, aminoglycoside-modifying enzymes, and plasmid-mediated quinolone resistance genes, and because they circulate widely in human, animal, and environmental isolates. Broad-host-range plasmids such as IncA/C, IncH, IncP, IncL/M, IncN, and IncQ possess replication and maintenance systems capable of functioning across multiple genera, enabling them to colonise *E. coli*, *K. pneumoniae*, *Salmonella enterica*, *Aeromonas*, *Vibrio*, and additional Gram-negative hosts. Additionally, plasmids often form cointegrates or multireplicon plasmids, expanding their compatibility profiles and allowing stable persistence in new hosts, as seen with IncF plasmids carrying multiple replicons (FII, FIA, FIB) or IncA/C plasmids acquiring IncR elements (Rozwandowicz et al., 2018).

3.4.3 Integrons and gene cassette dynamics

Integrons are ancient and highly adaptable gene systems that allow acquisition, rearrangement, and expression of gene cassettes and therefore constitute powerful drivers for the evolution of the bacterial genome and AMR. Integrons serve as powerful genetic platforms for capturing, rearranging, and expressing exogenous open reading frames embedded within gene cassettes, enabling bacteria to acquire new adaptive traits, most notably antibiotic resistance, quickly. Integrons consist of a site-specific recombinase gene (*intI*), a primary recombination site (*attI*), and an outward-facing promoter (*Pc*), which is responsible for transcription of integrated cassettes, which usually contain single genes flanked by distinctive *attC* recombination sites. Mobile integrons (MIs), especially classes 1, 2, and 3, are usually located in transposons and plasmids and contain small arrays that are enriched for ARGs, whereas chromosomal integrons (CIs), especially superintegrons, contain hundreds of cassettes encoding diverse functions and are stabilised by toxin–antitoxin cassettes that inhibit large-scale deletions (Cambray et al., 2010). This dynamic gene capture system not only enhances the spread of AMR but also broadly contributes to bacterial genome plasticity, ecological diversification and long-term evolutionary success (Mazel, 2006). Integrons display staggering levels of evolutionary diversity, with CIs, especially in the bacteria *Vibrio*, *Xanthomonas*, *Treponema*, and *Pseudomonas*, possessing tens to hundreds of cassettes, while the mobile clinical integrons of classes 1, 2, and 3 have appeared more recently through transposons and have spread antibiotic resistance worldwide (Bhat et al., 2023). Intergenic cassettes exhibit a dynamic nature, characterised by frequent insertions, deletions, and rearrangements, resulting in considerable diversity in the genomic makeup. The presence of promoters within some cassettes enables them to express and function independently of their placement within the array. Environmental analysis has illustrated the fact that the majority of the ORFs within the cassettes code for unknown functions and possess potential for adaptation to the surroundings, while the clinical integrons have preferentially selected cassettes carrying antibiotic resistance genes like *qac*, *aadA*, and *dfrA* due to the selective pressures exerted by disinfectants, heavy metals, and antibiotics (Gillings, 2014). Integron activity is tightly coupled to host regulatory systems: the LexA-controlled SOS response sharply increases IntI expression, accelerating cassette excision and rearrangement during environmental stress or HGT events such as conjugation or transformation, thereby enabling 'adaptation on demand' in fluctuating environments. CIs in environmental bacteria often contain dozens to hundreds of cassettes, as seen in *Vibrio cholerae* superintegrons, and serve as long-term evolutionary reservoirs. In contrast, MIs carried on plasmids and transposons have played a central role in disseminating multidrug resistance among clinically important Gram-negative pathogens (Escudero et al., 2015).

3.4.4 Transposons and insertion sequences

Transposons and ISs (IS elements) function as highly mobile intracellular genetic modules that reshape bacterial genomes by catalysing nonhomologous DNA rearrangements and mobilising resistance determinants across replicons (Frost et al., 2005). IS elements such as IS26, IS257, and ISAba1 typically encode a transposase and mobilise adjacent DNA either through replicative or conservative transposition, frequently generating target site duplications that mark insertion events; when two copies flank a gene, they form composite transposons capable of moving resistance determinants as a unit. Other major contributors include Tn21-subfamily transposons, which frequently harbour class 1 integrons, and Tn7-like transposons, which utilise specialised target-selection pathways enabling integration at defined chromosomal sites or conjugative plasmids, thereby facilitating both vertical and horizontal spread of resistance determinants (Partridge et al., 2018). IS-mediated insertions, deletions, and recombination events diversify virulence repertoires, expand host range, and enable rapid ecological adaptation. Importantly, the presence of related Tn3-like elements in *Xanthomonas, Pseudomonas,* and *Stenotrophomonas* underscores their evolutionary importance as drivers of LGT (Ferreira et al., 2015).

3.4.5 Interplay between mobile genetic elements and bacterial evolutionary fitness

MGEs such as plasmids, integrons, transposons, and ISs have a profound impact on bacterial evolutionary fitness by reshaping genomes, modulating physiology, and changing adaptive potential (Fig. 3.3). Although the acquisition of an MGE often has the initial cost of a metabolic burden as cells need to replicate and sustain extra DNA, the possession of many MGEs has strong adaptive benefits, especially if they encode antibiotic resistance, virulence factors, or stress-response genes. Evidence has been shown that some resistance plasmids not only confer minimal fitness costs but also can increase growth or biofilm formation, showing that MGEs can increase ecological competitiveness in the absence of antibiotic pressure (i.e. extended-spectrum β-lactamase [ESBL] plasmids increasing extracellular matrix production and altering virulence-related gene expression). Moreover, host-plasmid interactions are highly genotype-specific, with epistatic effects that either impose a burden due to an MGE or confer a net benefit, as in the case of fitness cost of carbapenemase-encoding plasmids, which varies dramatically across different host genetic backgrounds, allowing them to be stably maintained in certain commensal lineages. Coselection, therefore, further reinforces the maintenance of MGEs, as the co-selection of groups of resistance genes on the same plasmid or integron guarantees that exposure to any one antibiotic will select for the entire element, selecting for long-term stability of bacterial communities. MGEs also exist as toxin–antitoxin systems that kill plasmid-free segregants, imposing a plasmid of stable inheritance during cell division. At a wider, evolutionary level, MGEs are engines of genome plasticity, capable of achieving rapid reorganisation, acquiring new functions, and adapting to new niches.MGEs play a disproportionate role in the emergence of pathogenic strains, metabolic innovation, and the diversification of ecological strategies in prokaryotes (Jordt et al., 2020).

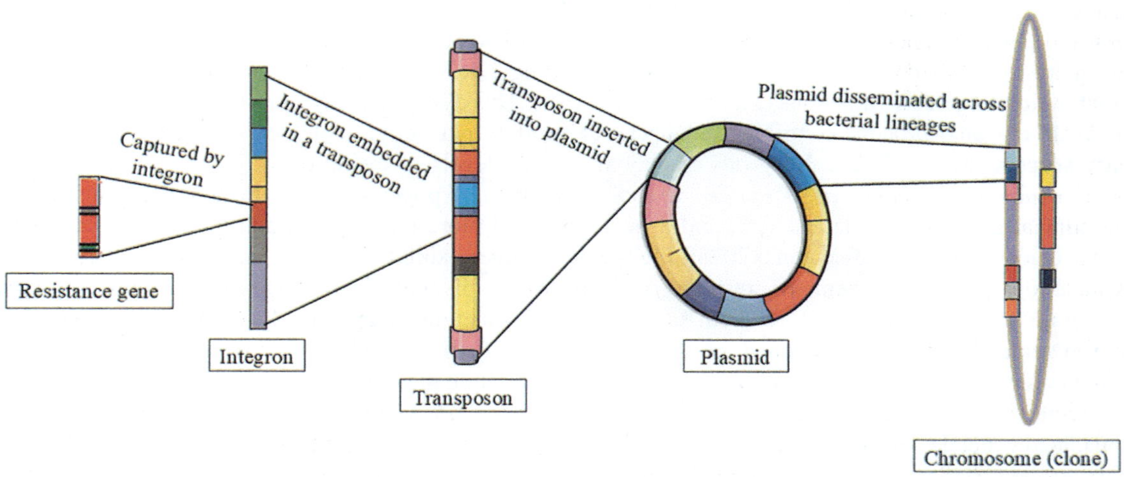

FIGURE 3.3 **Schematic overview of the roles of mobile genetic elements in the acquisition, assembly, and dissemination of antibiotic resistance genes in bacteria.** Resistance genes are mobilised and rearranged through the coordinated actions of plasmids, transposons, insertion sequences, and integrons, enabling their movement between chromosomal DNA and extrachromosomal elements. Conjugation facilitates plasmid-mediated horizontal transfer between bacterial cells, while transduction by bacteriophages contributes to intercellular gene movement. Transposition and recombination events drive the capture, reshuffling, and accumulation of resistance determinants within integrons and plasmids, promoting the formation of multidrug resistance regions and enhancing the evolutionary adaptability and host-range expansion of resistant bacterial populations.

Although, due to adaptive properties such as antibiotic resistance, the acquisition of plasmid DNA often boosts the growth and competitiveness of microbial populations, the acquisition of plasmid DNA frequently leads to decreased growth and competitiveness of those populations due to replication-associated burden, interference with the host chromosomal processes and deleterious protein–protein interactions, especially between plasmid-encoded replication proteins, such as Rep, and essential helicases, which lead to increased stress responses. Bacteria often harbour several MGEs at once, and the interactions among these elements are often additive or subtractive (fertility inhibition, induction of the SOS response, activation of phage, mobilisation of genomic islands, antagonistic competition for cellular resources), resulting in downstream effects on host physiology and evolutionary dynamics. Such interactions not only affect the short-term survival of MGE-bearing cells but also long-term evolutionary outcomes of MGE-bearing cells, such as synergistic or antagonistic epistasis among coexisting MGEs, which can determine plasmid persistence, can affect plasmid distribution across bacterial taxa, and can, in turn, affect the success of particular plasmid-host associations in both natural and clinical environments. Compensatory evolution further alters this interplay, with mutations in either the plasmid or chromosome mitigating initial fitness costs and, in some cases, rendering hosts 'plasmid permissive', allowing bacteria to tolerate and sustain a greater variety of MGEs and consequently driving the stability and spread of multidrug resistance in microbial communities (San Millan and MacLean, 2017). Experimental evolution showed that bacteria can quickly remodel plasmid architecture through IS26-mediated deletions, removing genetically expensive regions, such as conjugation machinery and toxin–antitoxin systems; these deletions significantly increase host growth rate and enhance plasmid stability over hundreds of generations, providing a good example of how MGEs evolve to be more compatible with their new hosts while losing functions that are dispensable in their new environment. Strikingly, such compensatory restructuring occurred easily in *E. coli* but not in *K. pneumoniae*, underscoring that different host species vary in their 'evolvability' and ability to produce MGE-mediated adaptations, a factor that influences which bacterial lineages become effective long-term reservoirs of multidrug resistance plasmids. These findings highlight that MGEs and their bacterial hosts participate in a constant co-evolutionary process whereby MGEs shape bacterial fitness landscapes and bacteria, in turn, influence MGE structure and expression, which ultimately shapes the persistence, dissemination potential and evolutionary course of resistance plasmids in clinical and environmental settings.

3.4.6 Assembly of resistomes in clinical and environmental microbiomes

The emergence of resistomes in the clinical and environmental microbiome is mainly due to continuous HGT, which has introduced many resistance factors into bacterial communities, thereby promoting their persistence when they confer a selective advantage. Significant parts of bacterial genomes consist of genes acquired horizontally, including those conferring antibiotic resistance, which are retained only if they confer a benefit on the host or if the MGEs that contain them provide their own transfer and maintenance mechanisms. Transformation, conjugation, and transduction are modes of interspecific and inter-environmental exchange of DNA, while such processes as homologous recombination and recircularization of plasmids are modes of integration of segments of incoming DNA into the host genome. Significantly, environmental reservoirs like soil, water, biofilms, decomposing biomass, and host-related niches provide a plentiful supply of extracellular DNA (eDNA) that can be taken up by competent bacteria, leading to resistome assembly, including among distantly related taxa. Obstacles such as restrictions, modification systems, surface exclusion, and host-range constraints reduce, but rarely stop, gene acquisition, suggesting that bacterial genomes are in constant flux, with MGEs dynamically changing resistance profiles across ecosystems. As a result, resistomes form dynamic libraries that are affected by the availability of DNA, the mobility of MGEs, and ecological selection, forming complex reservoirs that facilitate the transmission of resistance in both environmental and clinical settings (Wright, 2007).

The environmental resistome, which includes all genes conferring resistance, either directly or indirectly, is ancient, highly diverse, and closely linked to antibiotic-producing organisms such as *Streptomyces*, which harbour many self-immunity factors that are evolutionary precursors of contemporary clinical resistance genes. Despite the limited number of environmental microbes being culturable, metagenomic and functional metagenomic techniques have uncovered vast reservoirs of resistance factors in soil, water, and wastewater, many of which structurally or functionally resemble genes found in pathogens. MGEs promote gene transfer at ecological boundaries where environmental, commensal, and pathogenic groups intersect. This makes resistomes dynamic collections shaped by ecological proximity, selective pressures, and the ongoing dispersal of antibiotic resistance genes via the mobilome across microbial environments (Perry and Wright, 2013). Soil, regarded as one of the oldest and most extensive reservoirs of resistance genes, contains an impressive variety of ARGs, many of which existed before the advent of human antibiotic use and exhibit significant sequence similarity with those identified in clinical pathogens. These environmental reservoirs facilitate the assembly of the resistome via HGT, thereby facilitating the transfer of ARGs among saprophytic soil bacteria, commensal species, and pathogenic strains. Research comparing soil and gut microbiomes reveals that a portion of ARGs is shared between the two environments, suggesting that clinically important resistomes can arise from larger environmental gene reservoirs, even if only a small percentage of environmental genes transfer and adapt in bacteria associated with humans. The makeup of resistomes is additionally influenced by ecological elements like microbial diversity, nutrient gradients, antibiotic-producing species, and selective pressures from human activities (e.g. urban development, antibiotic contamination), which affect both the prevalence and transfer potential of ARGs (Nesme and Simonet, 2015).

3.5 Ecological and environmental reservoirs of resistance

3.5.1 River systems, wastewater, and natural water bodies as reservoirs

River systems, wastewater channels, and natural water bodies are important ecological reservoirs for the survival and spread of antibiotic-resistant bacteria (ARB) and ARGs. Effluent from hospitals, agricultural sites, and urban areas routinely releases resistant pathogens and commensals into water environments, where they mix with native microbial communities and exchange genetic information via MGEs. These aquatic systems are also exposed to large amounts of antimicrobial substances, soaps, antiseptics, industrial contaminants, and heavy metals that create selective environments that favour the development and persistence of resistant populations. River systems that receive various effluent discharges show cumulative effects, with ARGs being carried several kilometres downstream and potentially contaminating previously untouched regions after hydrological events like overflows or floods, emphasising the function of natural water bodies as active reservoirs and channels for the spread of resistance across clinical and environmental divides (Manaia, 2023). Rivers and surface waters located downstream of WWTPs often hold detectable levels of antibiotics like sulphonamides, macrolides, fluoroquinolones, and cephalosporins, typically at concentrations high enough to promote the selection of resistant organisms, including producers of ESBL, while the attachment of antibiotics to sediments and organic matter extends their persistence in the environment. The merging of bacteria from human, animal, and environmental origins within river systems and wastewater treatment facilities turns aquatic environments into 'genetic reactors', where recombination, HGT, and selection happen simultaneously, speeding up the rise of new resistance profiles that could subsequently re-enter clinical and agricultural contexts (Fig. 3.4) (Baquero et al., 2008; Pruden et al., 2006).

3.5.2 Hospital and municipal wastewater as hotspots of adaptation

WWTPs act as 'convergence zones' where bacteria from various ecological backgrounds blend, engage, and experience HGT, facilitated by elevated cell concentrations, biofilm development, and exposure to stressors from pollutants that encourage ARG mobilisation. WWTPs, while intended to minimise microbial and chemical loads, serve as notable point sources of ARGs, since activated sludge and treated effluents hold considerable quantities of resistance determinants and MGEs, including integrons, IS elements, and transposons, which then flow into receiving rivers and biofilm-rich substrates downstream. Urban wastewater treatment plants (UWTPs) frequently discharge effluents with significant amounts of integrons, plasmids, and various resistance factors into nearby surface waters, where they interact with indigenous microbiota and broaden the environmental resistome (Fig. 3.4). Despite the use of biological treatment, effluents may discharge 10^9–10^{12} CFU/day/population, and a minimum of 10^7–10^{10} bacteria with acquired resistance features, thus giving the magnitude of the influx of microbes to aquatic environments (Rizzo et al., 2013). Extensive metagenomic research has shown that untreated and treated sewage contain a range of resistomes that mirror closely human antibiotic use patterns, with these ARG signatures being transferred to adjacent rivers and microbial communities downstream (Hendriksen et al., 2019). Data from wastewater surveillance show that hospital discharges contain significantly higher levels and more clinically significant ARGs, in particular β-lactamases, including carbapenemases like bla_{NDM}, bla_{KPC}, and bla_{VIM}, and multidrug resistance factors compared to adjacent areas, thus highlighting that wastewater serves as a pathway that concentrates high-risk resistance factors before their release into the natural aquatic environment. Metagenomic monitoring also demonstrates how selective culture enrichment reveals the presence of many low-abundance but clinically important ARGs, which cannot otherwise be detected, highlighting the importance of effluent systems being able to harbour a cryptic resistome, which may assist in downstream dissemination into water bodies and ecosystems. Due to their continuous collection and transport of ARGs released by human populations, particularly from hospital hotspots experiencing strong antibiotic selective pressure, these water systems are crucial in the environmental assembly, persistence, and dissemination of resistomes, thus enhancing their importance as reservoirs that can impact AMR dynamics within the One Health framework (Acosta et al., 2025).

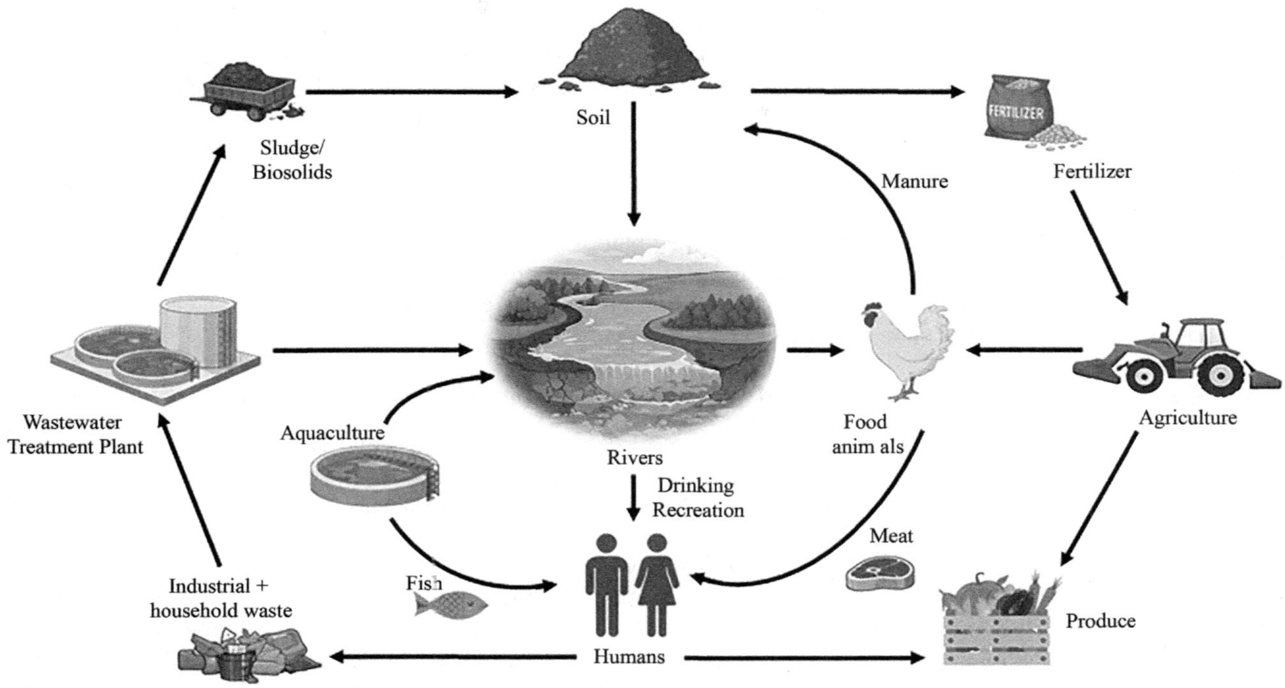

FIGURE 3.4 **Conceptual diagram illustrating the interconnected environmental reservoirs and transmission pathways of antibiotic resistance across human, agricultural, and aquatic systems.** Wastewater treatment plants receive industrial and household waste containing antibiotics, resistant bacteria, and resistance genes, which can be redistributed into aquatic environments and sludge or biosolids applied to soil. Agricultural practices, including the use of manure and fertilisers, facilitate the transfer of resistance determinants between soil, crops, livestock, and food animals. Rivers and other surface waters act as integrative hubs linking wastewater effluents, aquaculture, drinking and recreational water use, and human exposure.

3.5.3 Agricultural runoff and animal-associated resistomes

Agricultural runoff and animal-associated resistomes constitute significant pathways for the environmental emergence and spread of antibiotic resistance, as high-density livestock production systems continuously expose animal gut microbiomes to low-level antimicrobials, metals, and other stressors. Long-standing evidence indicates that administering antibiotics to livestock rapidly increases antibiotic-resistant strains in their gut microbiota, which can then be transmitted to farm workers and nearby communities through direct contact, faecal shedding, groundwater pollution, and tainted meat products (Spellberg et al., 2016). The use of subtherapeutic antimicrobials in livestock increases resistant bacterial populations and MGEs in the intestines, which are then discharged into the environment via untreated animal waste containing leftover antimicrobials, resistant bacteria, and resistance genes. When these wastes are kept in open lagoons or applied to land, they interact extensively with soils, sediments, water bodies, and wildlife, promoting HGT events that accelerate the expansion of the environmental resistome. Soils in agricultural settings that receive manure exhibit increased levels of ARGs, integrons, and resistance plasmids, and these resistome profiles persist for months to years and are often linked to antimicrobial residues. Additionally, runoff containing these waste-derived microbiomes, together with selective agents, can transport resistance traits across landscapes and into aquatic ecosystems, enabling the evolution, recombination, persistence, and eventual reintroduction of resistance genes into human-associated microbiomes (Fig. 3.4) (You and Silbergeld, 2014).

In extensive pig farming operations in China, manure, compost, and treated soils exhibit remarkably diverse, abundant, and highly concentrated ARGs across all major resistance categories. High-throughput qPCR profiling was performed to identify 149 unique ARGs, among which the most enriched genes showed increases of 192-fold to more than 28,000-fold relative to antibiotic-free controls, reflecting extreme selective pressures due to the high use of antimicrobials and heavy metals as feed additives. Significantly, transposases, particularly those of the IS6 family such as IS26, were shown to be enriched by up to 90,000-fold, with the abundances of the ARG-transposases being shown to be almost perfectly correlated with each other (r^2=0.96), reflecting the high potential for HGT and the transfer of resistance factors inside and outside of the agricultural environment (Zhu et al., 2013). Around 80% of all antimicrobials are used in animal farming, creating strong selective pressure that promotes the acquisition of antibiotic resistance genes (ARGs) within the gut microbiota. These resistant bacteria and their ARGs are subsequently excreted in manure, which also contains residual antibiotics, often at concentrations ranging from 1–10 mg/kg and occasionally exceeding 200 mg/kg. Once applied to land, manure serves as a significant source of antibiotic-resistant elements that infiltrate agroecosystems, resulting in a sharp increase in ARG richness in soils, up to 32.5%–70.5% compared to unamended soils within a day of application, influenced by the direct introduction of manure-associated resistomes and selection pressures from remaining antibiotics (Hilaire et al., 2022). Metagenomic analysis of cattle manure, catch-basin water, and soils treated with manure showed that bovine faecal microbiomes contain the highest concentration of ARGs, predominantly featuring tetracycline and macrolide resistance genes that reflect patterns of antimicrobial usage in feedlots (e.g. routine in-feed supplementation with chlortetracycline or tylosin). These ARGs, especially ribosomal protection proteins such as tetQ, tetW, tet40, and tetO, enter nearby catch-basin waters via rainfall-induced runoff, creating resistomes in downstream areas that somewhat mirror those present in feedlot manure (Zaheer et al., 2019).

3.5.4 Gene exchange networks in mixed environmental–clinical habitats

Gene exchange networks in combined environmental-clinical habitats serve as dynamic interfaces where bacteria from various sources, such as soil, water, animals, and humans, engage and share resistance traits, resulting in highly interconnected mobilomes that surpass ecological limits. Mixed environmental-clinical habitats such as wastewater treatment facilities, river sediments receiving hospital effluents, agricultural soils enriched with manure, and human-associated microbiomes create highly interconnected gene-exchange networks driven by various MGEs, especially integrative and conjugative elements (ICEs), which facilitate rapid LGT among phylogenetically distant bacterial groups. ICEs merge with host chromosomes, are excised in response to environmental signals, and are transmitted via conjugation, serving as connections that unite environmental microbiomes with clinical pathogens. The broad-host range of these symbionts, indicated by the presence of elements such as Tn916, ICEclcB13, SXT, and ICEMlSymR7A, allows for gene transfer between soil, water, wastewater, commensal, and pathogenic populations, which are strongly connected with activated sludge or wastewater systems, where interactions between species can occur frequently. These components, besides carrying their own cargo, including genes for antibiotic resistance, virulence, and metabolism, are also capable of transferring plasmids, genomic islands, and integrative mobilisable elements, creating genetic mosaics and spreading adaptive characteristics across ecological boundaries. The regulatory responsiveness of ICEs to stress cues (such as SOS induction, antibiotic exposure, and nutrient availability) further enhances the ability of ICEs to transfer clinically significant genes from environmental reservoirs into human-associated microbiomes. Consequently, HGT mediated by ICE provides a dynamic interconnected network that enables environmental bacteria to directly affect the evolutionary trajectory of pathogens, accelerating the formation and dissemination of antibiotic resistance factors in the natural and clinical environments (Wozniak and Waldor, 2010). Environmental microbiomes are often long-term reservoirs for ARGs, where they persist, accumulate mutations, and are transferred between hosts before infecting pathogenic bacteria, potentially resulting in human treatment failures. Significantly, surveillance focused on the abundance of ARGs ignores the true threat because the genetic background, specifically plasmid linkage, determines whether an ARG detected in the environment will be able to pass into clinical settings. Innovations such as targeted amplicon sequencing, ddPCR linkage analysis, and long-read or Hi-C metagenomics now enable scientists to decipher these networks by determining which ARG variants spread between environments, which hosts they inhabit, and which MGEs facilitate their transfer (Klümper et al., 2025).

3.5.5 Environmental persistence and selection pressures

Antibiotics and resistance genes serve as dual contaminants, with antibiotics applying consistent local selective pressure and resistance genes disseminating worldwide as self-replicating units on MGEs. Even without antibiotics, some resistance genes endure because of low or offset fitness costs, coselection with metals or biocides, and association with advantageous ecological traits, enabling resistant populations to remain established in the environment (Baquero et al., 2009). The persistent presence of antibiotics in the environment creates ongoing, widespread selection pressures that promote the growth, persistence, and evolution of ARB across clinical, agricultural, and natural ecosystems. Significant portions of prescribed antibiotics, varying between 20% and 80% depending on their category, are excreted unchanged by humans and animals into wastewater, manure, sludge, rivers, lakes, soils, and even food items, creating widespread reservoirs of sub-MIC antibiotic levels that persist long enough to affect microbial evolution. These environmentally distributed sublethal levels not only favour preexisting resistant mutants to grow but also enhance mutagenesis, stimulate HGT, and influence bacterial physiology, collectively speeding up resistance evolution well beyond clinical environments (Andersson and Hughes, 2014). Very low antibiotic levels, reaching 230 times below the MIC for susceptible strains, can specifically promote the growth of resistant mutants, challenging the conventional belief that selection occurs only between the MICs of resistant and susceptible populations. Competition experiments reveal that even minimal levels of tetracycline, ciprofloxacin, and streptomycin at nanogram or picogram concentrations reduce the growth rate of susceptible cells, while resistant cells remain unaffected, thereby conferring a significant fitness benefit on resistant mutants. These sub-MIC levels reflect antibiotic concentrations typically observed in aquatic habitats, soil, wastewater discharges, manure-affected areas, and specific sites within human and animal bodies, suggesting that environmental contamination with small antibiotic traces can significantly influence resistance patterns (Gullberg et al., 2011).

3.6 Adaptive survival strategies beyond classical resistance

3.6.1 Biofilm formation and biofilm-associated tolerance

The formation of biofilms provides bacteria with an effective survival strategy, as the extracellular polymeric substance (EPS) matrix establishes a protected, well-structured environment that significantly enhances resistance to antimicrobial pressure, allowing groups to endure antimicrobial pressure much more effectively than planktonic cells. Biofilms form through a controlled shift from independent cells to surface-adhered communities encapsulated in a self-generated extracellular matrix made up of exopolysaccharides, eDNA, proteins, and lipids. This matrix forms sharp gradients of nutrients, oxygen, and waste products that lead to diverse physiological subpopulations, including slow-growing and dormant cells that are naturally resistant to various antibiotics (Hall and Mah, 2017; Stewart and Franklin, 2008). Biofilms exhibit emergent characteristics, such as increased resistance to antibiotics, disinfectants, dehydration, and other environmental stressors, that are not present in free-floating planktonic cells. This defence occurs through different matrix-mediated processes, such as EPS constituents absorbing, neutralising, or interacting with antimicrobial substances; steep physiological gradients form slow-growing, dormant, or viable but nonculturable subpopulations that are naturally more resistant, as well as the matrix helping to retain extracellular enzymes and signalling compounds with enhanced collective survival (Flemming et al., 2016). Bacteria within these biofilms display significant resistance to antibiotics, phagocytic cells, and antibodies, mainly due to the matrix's ability to hinder the access of antimicrobial agents, reactive oxygen species, and immune components, while also promoting slow-growing or nutrient-poor subpopulations that are naturally less vulnerable to antibiotic elimination (Costerton et al., 1999). eDNA found in biofilms increases tolerance by chelating cations, activating stress-response pathways, and promoting HGT. Biofilms trigger specific tolerance mechanisms, including the upregulation of efflux pumps, signalling of stringent responses, defences against oxidative stress, and the formation of persister cells (Hall and Mah, 2017). Bacteria within biofilms exhibit increased mutation rates, regulation of virulence and structural genes driven by quorum sensing, and the ability for prolonged inflammatory stimulation, facilitating chronic infections linked to foreign bodies, cystic fibrosis, and long-lasting wounds (Høiby et al., 2011; Marti et al., 2014).

3.6.2 Persister cells and transient resistance phenotypes

Persister cells constitute a minor, phenotypically unique subgroup that boosts bacterial survival by transitioning into a dormant, nonreplicating state that is naturally tolerant, rather than genetically resistant, to antibiotics, enabling them to endure otherwise fatal drug exposure and to repopulate after treatment pressure decreases. These cells emerge randomly in expanding populations, but their numbers rise during stressful conditions, especially in biofilms and persistent infections, where immune access is restricted, and antibiotics are frequently applied. Persisters evade elimination because bactericidal antibiotics depend on active cellular functions (such as translation, DNA replication, and cell wall synthesis) to inflict lethal harm, whereas dormancy inhibits these interactions. Mechanistically, the formation of persisters includes redundant pathways, such as toxin–antitoxin systems and stress-induced responses like SOS-mediated activation of TisB, which reduces proton motive force and ATP levels, leading cells into dormancy (Harms et al., 2016; Lewis, 2010). Importantly, persisters can resume normal growth upon removal of antibiotics, and their descendants remain fully susceptible to infection, indicating that persistence is a transient, non-heritable tolerance mechanism rather than a mechanism of resistance (Balaban et al., 2004).

Persisters fundamentally differ from resistant cells in that the process of survival is not associated with rising MIC values, nor does persistence enable the cells to grow in the presence of antibiotics, but rather persistence manifests itself as a biphasic killing curve, in which the majority of cells die rapidly, but a small proportion persists for long durations of time, even at drug concentrations well above the MIC. This phenotype for tolerance arises through two general mechanisms: triggered persistence, which is initiated by stress factors such as starvation, high cell density, acidification or immune factors, and spontaneous persistence, in which rare cells switch to a tolerant state randomly during uniform exponential growth. The formation of persisters is closely linked to physiological diversity, reduced concentrations of

ATP, stress responses, and genetic variations such as the *hipA7* allele, which can drastically increase persistence rates without exhibiting traditional resistance (Balaban et al., 2019). Persisters retain significantly reduced intracellular antibiotic levels, around one-fifth of the overall population, as they sustain increased efflux pump activity despite suppressed growth and metabolism. Single-cell analyses revealed that persisters exhibit rapid antibiotic efflux, with efflux rates nearly three times those of standard cells, allowing them to eliminate drugs before lethal interactions occur. This phenotype is supported by the increased expression of multidrug efflux genes, especially *tolC*, *acrA/B*, and *emrA/B*, which show notable transcriptional enhancement in persisters compared to the overall population (Pu et al., 2016).

3.6.3 Metabolic adaptation and stress-response pathways

Metabolic adaptation and stress-response mechanisms are crucial for survival during antibiotic exposure, as they slow or redirect essential physiological processes to reduce antibiotic effectiveness. Numerous tolerant phenotypes result from slowed growth rates, nutrient scarcity, or the triggering of global stress responses, all of which reduce the efficacy of bactericidal antibiotics that mainly attack actively proliferating cells. The rigorous response, resulting from nutrient limitation and facilitated by the accumulation of ppGpp, is highlighted as an important metabolic switch that inhibits biosynthetic pathways and promotes a transient non-growing phase, thereby extending the minimum duration for population killing (MDK). Similarly, environmental pressures, such as starvation, stress originating from hosts, and halting of growth due to antibiotics, can cause the cells to go into low metabolism or dormant stages, which greatly enhance tolerance by slowing down important cellular processes, such as protein production, DNA replication or maintenance of cell walls, which are the primary target of many antibiotics (Brauner et al., 2016).

M. tuberculosis persisters, non-dividing or slow-dividing cells surviving drug treatments, essentially rely on extensive metabolic reprogramming to survive unfavourable conditions such as low oxygen, nutrient deprivation, acidic conditions, and antibiotic stress. These stresses trigger transitions to low-energy physiological states that are mediated by pathways such as the glyoxylate shunt, modified menaquinone synthesis, triacylglycerol accumulation, cytochrome bd-driven respiration, and reduced ATP synthesis, leading to decreased metabolic activity and sensitivity to drugs. At the heart of this adaptation is a stringent response, driven by RelA and the alarmone ppGpp, that inhibits rRNA production, reduces the expression of ribosomal and metabolic genes, and supports survival in energy-restricted environments. Extra stress-response systems, such as toxin–antitoxin mechanisms, universal stress proteins, pathways for proteasomal degradation, and translation machinery, further strengthen a physiology adapted for persistence instead of growth (Martínez and Rojo, 2011; Zhang et al., 2012). Metabolic control influences antibiotic uptake by adjusting porins and transporters; for instance, carbon catabolite repression via the Crc regulator modifies the expression of OprD and GlpT, thereby impacting susceptibility to β-lactams and fosfomycin (Martínez and Rojo, 2011). Metabolic changes such as dormancy, reduced growth, and transitions to stationary-phase states significantly reduce antibiotic efficacy, resulting in phenotypic resistance independent of genetic alterations. For instance, slow-growing or resting cells become resistant to β-lactams, aminoglycosides, and various other medications since essential antibiotic targets or absorption mechanisms are inactive (Dörr et al., 2010).

3.6.4 Evolutionary benefits of persistence strategies

Persisters lack heritable resistance mutations; instead, they result from various physiological 'glitches', random occurrences, and stress-induced reactions that halt growth temporarily, making cells resistant to antibiotic killing while remaining fully sensitive again upon resuming replication. Mathematical modelling together with experimental results suggests that these transiently tolerant subpopulations are a form of a bet-hedging strategy that permits bacterial communities to survive intermittent antibiotic assaults that would otherwise annihilate them. Populations that produce persisters more frequently obtain a survival benefit in repeated antibiotic exposure, whereas those lacking persistence characteristics are more rapidly eliminated, as evidenced by competition trials with *Bacillus subtilis, in* which competence-proficient strains outperformed competence-deficient mutants during penicillin pulses. While persistence may not have evolved specifically in response to antibiotic resistance, its incidental benefits still provide important evolutionary benefits in terms of preserving population survival, assisting in relapse after treatment, and serving as a temporal reservoir for the potential emergence of genetically resistant mutants (Windels et al., 2019; Levin et al., 2014). Tolerance mutations significantly raise the chances that infrequent resistance mutations will persist during antibiotic treatment, as tolerant cells sustain viable population levels throughout the therapy, thus prolonging the mutational window and diminishing the probability that new mutants will be eradicated during drug exposure. Continuous exposure to ampicillin led to the selection of tolerance mutations that prolonged the bacterial lag phase, enabling cells to endure antibiotic treatment without altering the MIC. Mathematical modelling indicated that in the absence of tolerance, partial-resistance mutations would generally be eliminated, needing more than 100 treatment cycles to become established; however, under a tolerant environment, they quickly became fixed within the population (Levin-Reisman et al., 2017).

3.7 Evolutionary trajectories of priority multidrug-resistant pathogens (case studies)

3.7.1 *Mycobacterium tuberculosis*—stepwise evolution of multidrug resistance

The incremental development of multidrug resistance in *M. tuberculosis* represents the ongoing buildup of chromosomal mutations that modify drug targets, activation mechanisms, or metabolic activities in response to therapeutic selection pressure. Resistance primarily develops via random genetic mutations, as *M. tuberculosis* has no traditional HGT mechanisms; therefore, insufficient drug exposure resulting from incorrect prescriptions, patient non-compliance, or variations in pharmacokinetics fosters environments that favour the selection of these mutants and reinforces their prevalence in bacterial populations. Initial resistance frequently develops to isoniazid and rifampicin due to mutations in *katG*, *inhA*, and *rpoB*, with common mutations like *katG* S315T or *rpoB* S531L allowing survival without significantly impacting fitness, thus facilitating transmission. Throughout treatment, whether it succeeds or not, further mutations build up in genes like *pncA*, *rpsL*, *rrs*, *embB*, and *gyrA*/*gyrB*, increasing resistance to pyrazinamide, aminoglycosides, ethambutol, and fluoroquinolones, respectively. While these mutations might initially lead to fitness costs, *M. tuberculosis* frequently acquires compensatory mutations, particularly in *rpoC* or *rpoA*, that restore replicative ability and improve the transmissibility of otherwise disadvantageous resistant strains. Whole-genome sequencing (WGS) has shown that numerous MDR and XDR lineages, especially those of the Beijing genotype, spread via continuous transmission rather than through new resistance acquisition, highlighting how sequential mutations, compensatory adaptations, and selective amplification under antibiotic stress collectively influence the evolutionary path of this globally important pathogen (Gagneux, 2018; Zhang and Yew, 2015; Somoskovi et al., 2001).

These mutational events occur simultaneously at infection sites in respiratory and extrapulmonary tissues, often forming branching evolutionary paths rather than linear sequences. This implies that different clonal variants can coexist, compete, or sequentially replace each other as selective pressures influence the genomic environment of the pathogen, such as exposure to antibiotics. Notably, variants can differ by multiple SNPs even within the same host, which indicates continuous diversification that could then be disseminated, thus contributing to the emergence and spread of resistant strains (Pérez-Lago et al., 2014). Uncommon and low-frequency mutations, such as *the rrs* g878a polymorphism, are involved in this step-by-step process by producing small streptomycin resistance, which is validated by recombineering and MIC tests. In contrast to frequent mutations in *rpsL* or the 1400 region of *rrs*, this rare variant does not induce cross-resistance to amikacin, kanamycin, or capreomycin, supporting the notion that resistance arises from distinct, target-specific mutational occurrences rather than wide-ranging genetic changes (Domenech et al., 2022).

3.7.2 *Staphylococcus aureus*—methicillin-resistant *S. aureus* evolution, staphylococcal cassette chromosome *mec*, and fitness compensation

The evolutionary path of *S. aureus* towards methicillin resistance demonstrates successive phases of adaptation influenced by strong antibiotic selection, HGT, and compensatory evolution. Initial penicillinase-producing strains, including the widespread phage type 80/81 lineage, showcased the remarkable capability of the species to swiftly adapt to therapeutic challenges, a trend that reoccurred with the rise of MRSA after gaining the *mecA* gene located within the mobile staphylococcal cassette chromosome *mec* (SCCmec), which encodes the low-affinity PBP2a, granting resistance to nearly all β-lactams. The acquisition of SCCmec is widely recognised as the first and defining stage in MRSA evolution, after which clones undergo further adaptation through the acquisition or deletion of additional MGEs and single-nucleotide variants that improve survival in specific ecological niches such as hospitals, communities, or livestock settings. Different MRSA lineages, such as HA-MRSA, CA-MRSA, and those associated with livestock, have emerged separately multiple times, each influenced by the stable presence of particular MGEs (e.g. SCCmec II/III in HA-MRSA; SCCmec IV along with Panton–Valentine leukocidin (PVL)-encoding φ2 phage in CA-MRSA). These clones frequently develop compensatory mutations that minimise the fitness burden of resistance; for instance, fluoroquinolone-resistant HA-MRSA CC22 necessitated particular *gyrA*/*parC* mutations for effective dissemination. Even though MGE-encoded resistance is heavily dependent on it, the evolution of MRSA is further shaped by host adaptation elements such as immune-evasion genes from the φ3 prophage, thereby enhancing its colonisation and transmission capabilities. Together, these processes demonstrate the evolution of MRSA via the gradual acquisition of resistance factors, meticulously shaped by ecological selection and compensatory genetic alterations that reinforce resistance among highly successful global clones

(Chambers and DeLeo, 2009; Lindsay, 2019). CA-MRSA strains subsequently acquired additional factors that enhanced both virulence and fitness, including increased expression of alpha-toxin and phenol-soluble modulins, and, in certain clones such as USA300, the gain of the arginine catabolic mobile element, which improves skin colonisation and resistance to host-derived polyamines. The development of virulence in CA-MRSA involved a refined equilibrium between resistance and toxin production, addressing the usual trade-off in MRSA, where heightened resistance often diminishes virulence gene expression. The precise integration of SCCmec with compensatory regulatory modifications, together with the sporadic acquisition of prophage-encoded toxins such as PVL, enabled CA-MRSA to achieve remarkable epidemiological success and to position itself as a significant global MDR pathogen (Deurenberg and Stobberingh, 2009).

3.7.3 *Klebsiella pneumoniae*—New Delhi metallo-β-lactamase-1, *Klebsiella pneumoniae* carbapenemase, and plasmid-driven dissemination

The evolutionary success of MDR *K. pneumoniae* can be largely credited to the swift horizontal acquisition and worldwide spread of plasmid-encoded carbapenemase genes, particularly bla_{NDM} *and* bla_{KPC}, which have turned this species into a prevalent hospital and emerging community pathogen. *K. pneumoniae* carbapenemase (KPC) enzymes (Ambler class A), first discovered in the United States during the mid-1990s, became globally widespread due to the proliferation of high-risk clonal lineages like ST258, with the resistance gene bla_{KPC} consistently located within the transposon Tn4401, facilitating efficient mobilisation across various plasmid types and bacterial hosts. Conversely, New Delhi metallo-β-lactamase-1 (NDM-1) (Ambler class B metallo-β-lactamase) shows a distinctly different evolutionary path, marked by spread among various *K. pneumoniae* lineages and unrelated *Enterobacteriaceae* species through highly diverse conjugative plasmids that frequently coharbour ESBLs, aminoglycoside resistance genes, and other factors, leading to significant multidrug or pandrug resistance (Nordmann et al., 2011). Even though the hypervirulent CG23 lineage traditionally showed no AMR, recent genomic monitoring indicates that occasional human isolates have obtained substantial conjugative plasmids, such as IncA/C2 and IncN types, which contain various resistance factors, facilitating the dissemination of carbapenemase and presenting the risk of convergence between hypervirulence and MDR traits. Although obstacles like dense K1 capsule production and functioning CRISPR–Cas9 systems might restrict plasmid absorption, once they are formed, these plasmids can sustain stability, as shown by the prolonged retention of an IncFII AMR plasmid in the equine sublineage for more than twenty years. These results emphasise how the plasmid-driven acquisition of NDM/KPC-type carbapenemases can convert *K. pneumoniae* lineages from local pathogens into widespread high-risk clones, illustrating the vital role of horizontal plasmid transfer in influencing MDR evolutionary pathways (Munoz-Price et al., 2013).

3.7.4 Comparative evolutionary insights and implications

A comparative examination of priority MDR pathogens shows that despite their different ecology and genetic structure, their evolutionary paths converge on a few basic principles shaped by selection in interrelated environments. Antibiotic resistance did not originate in clinical pathogens but arose from the environmental microbiosphere, where resistance genes evolved over extended evolutionary periods for functions such as metabolism, detoxification, and signalling, well before the advent of human antibiotic use. ARGs create a massive shared resistome that includes pathogenic, commensal, and environmental bacteria, with MGEs like plasmids, integrons, transposons, bacteriophages, and gene transfer agents serving as the main channels for the spread of resistance (Von Wintersdorff et al., 2016). In pathogens like *M. tuberculosis*, the evolution of resistance is primarily driven by gradual chromosomal mutations due to strong within-host selection, while *S. aureus* and *K. pneumoniae* demonstrate the effectiveness of HGT, where MGEs like SCCmec, integrons, transposons, and plasmids facilitate the swift acquisition and spread of resistance among lineages and species. In all systems, the emergence of resistance is limited by fitness costs; however, these are often alleviated by compensatory evolution, co-selection with ecologically beneficial traits, or stabilisation within genetic frameworks featuring toxin–antitoxin systems, enabling resistant clones to survive even without antibiotics. Crucially, the evolution of modern resistance is best considered as a multilandscape phenomenon, integrating natural and anthropogenic ecosystems (e.g. wastewater) and the host being treated, where each environment imposes unique selective pressures that collectively accelerate the emergence of resistance (MacLean and Millan, 2019). Among these pathogens, the evolution of resistance is additionally influenced by convergent and parallel evolution, with similar genes and pathways consistently chosen by selection in separate hosts, emphasising the predictability of adaptive antibiotic responses (Didelot et al., 2016).

3.8 Technological advances in deciphering antimicrobial resistance evolution

3.8.1 Next-generation sequencing and resistome/metagenome profiling

Next-generation sequencing (NGS) technologies have fundamentally transformed the study of AMR evolution by allowing detailed characterisation of bacterial genomes, resistomes, and intricate microbial communities at an unparalleled scale and speed. WGS of clinical isolates allows for the identification of resistance factors, the inference of evolutionary connections, and the tracking of transmission routes with more accuracy than conventional phenotypic or marker-based techniques, thus connecting genotype to AMR and epidemic dissemination in almost real time. WGS has also shown that bacteria can evolve significantly over the course of an infection, demonstrating short-term evolutionary rates much higher than those inferred from long-term phylogenetic studies, thereby facilitating rapid adaptation to antibiotics within individual patients. Simultaneously, shotgun metagenomic sequencing has broadened the examination of the resistome beyond culturable pathogens, uncovering vast reservoirs of ARGs in environmental, animal, and human microbiomes. These methods have shown that resistance genes are often linked to MGEs and can move across taxonomic and ecological divides, often appearing before they emerge in clinical pathogens. Improvements in long-read sequencing and hybrid assembly techniques have enhanced the clarity of plasmids, integrons, and genomic rearrangements influencing resistance evolution, while more advanced bioinformatic pipelines now allow for comparative resistome profiling across spatial, temporal, and ecological gradients (Didelot et al., 2012).

NGS-based resistome and metagenome profiling has offered compelling proof that environmental microbiomes serve as active and current sources of clinically significant ARGs, rather than just remote evolutionary forerunners. Through high-throughput functional metagenomics combined with de novo assembly workflows, it was shown that bacteria from soil contain a wide variety of resistance factors across several antibiotic categories, including β-lactams, aminoglycosides, tetracyclines, sulphonamides, and amphenicols, many demonstrating 100% nucleotide identity with resistance genes present in prominent human pathogens. Significantly, these genes were often found in proximity to integrons, transposases, and other mobilisation components, providing direct genomic evidence of recent HGT events between environmental bacteria and clinical pathogens (Forsberg et al., 2012). Comparative studies of more than a thousand metagenomes demonstrated that relatively small sets of selected resistance genes can be used to rank environments by total resistance gene abundance and diversity, underscoring the scalability of NGS-based monitoring. Metagenomic profiling has shown that ecosystems impacted by anthropogenic antibiotic pollution, such as wastewater, pharmaceutical release, agricultural activities, and contaminated soils, harbour particularly diverse and abundant resistomes, including resistance genes that were not previously observed in known human pathogens, which are of importance as a reservoir for future clinical resistance (Bengtsson-Palme, 2018). Genome-resolved metagenomics and single-cell genomics provide an opportunity to reconstruct metagenome-assembled genomes (MAGs) and single-amplified genomes (SAGs), which allow linking resistance determinants to specific host and environment in the ecosystem, without cultivation bias. Standardised frameworks such as MISAG and MIMAG have improved comparative resistome analyses by defining criteria for genome completeness, contamination, and metadata reporting, making studies more reproducible and enabling cross-study integration. Combined with continuous improvements in sequencing depth, read length, and bioinformatics approaches, these have allowed comprehensive profiling of resistance reservoirs, transmission modes, and the evolutionary dynamics of AMR at an unprecedented scale and resolution, providing key insights into the assembly, maintenance, and spread of resistomes across microbiomes (Bowers et al., 2018; Loman and Pallen, 2015; van Dijk et al., 2014).

3.8.2 CRISPR-based genome editing for functional analysis

The significance of AMR research using the genome-editing technique of CRISPR in experimental medicine was described as a revolutionary tool in functional genomics. CRISPR–Cas systems, which initially developed as adaptive immune defences in bacteria and archaea, have been reengineered to facilitate accurate, RNA-guided genome editing in various organisms, thus permitting direct functional analysis of resistance genes, regulatory networks, and fitness factors under antibiotic stress. It was emphasised that CRISPR–Cas9 and its modifications enable targeted gene knockouts, knock-ins, transcriptional repression (CRISPRi), and activation (CRISPRa), which have been increasingly utilised in high-throughput screenings to discover essential genes, resistance-associated loci, and compensatory pathways related to antimicrobial tolerance (Barrangou and Doudna, 2016). RNA-guided CRISPR–Cas systems, especially Cas9, enable precise deletions, insertions, or point mutations through sequence-specific DNA cleavage combined with homologous recombination, thus allowing direct exploration of resistance factors, regulatory mechanisms, and fitness-related mutations. In bacteria, CRISPR-based editing enhances recombination efficiency while selectively targeting non-edited cells, leading to nearly total recovery of desired mutants and allowing for the concurrent modification of multiple locations. This approach has been applied successfully to many species, such as *Streptococcus pneumoniae* and *E. coli*, in the analysis of gene functionality, validation of mutations linked to resistance and epistatic interactions that lead to adaptive routes (Jiang et al., 2013).

CRISPR-based genome editing has expanded beyond gene modification to become a powerful functional tool for AMR analysis, specifically targeting resistant bacterial populations. Genome-targeting systems were developed to induce lethal double-strand breaks in bacterial chromosomes targeted to endogenous sequences, enabling the removal of bacterial strains using genomic information alone and in a programmable fashion. Employing native type I and type II CRISPR–Cas systems, effective sequence-specific targeting was demonstrated to occur irrespective of genomic location, transcriptional activity or strand orientation and was capable of distinguishing between closely related strains differing by only a few nucleotides. Significantly, the method made it possible to precisely regulate the makeup of bacterial populations in mixed bacterial cultures by controlling the different types of CRISPR RNAs delivered, something that is not possible with standard antibiotics. From an AMR perspective, the proposal to use the capabilities of CRISPR systems as 'intelligent antimicrobials' for the functional validation of resistance factors, the targeted elimination of MDR strains, and the experimental exploration of evolutionary hypotheses related to resistance persistence and population dynamics. These approaches, with the help of the technology of the CRISPR variants, can thus be considered as an important link between genomic monitoring and functional intervention in the research of AMR (Gomaa et al., 2014).

3.8.3 Evolution-aware computational models and phylogenetic reconstructions

Models taking into consideration evolution and phylogenetic reconstructions based on WGS are critical tools for understanding the dynamics of AMR evolution within the host and population scale. Genome-scale analyses have shown that bacterial pathogens can evolve rapidly under antibiotic pressure over short clinical timescales. Phylogenetic models that incorporate recombination, HGT, and within-host diversity help distinguish transmission events from the de novo emergence of resistance and identify convergent evolution across different bacterial populations. According to these models, resistance is frequent and is usually achieved by recognisable genetic targets, including the *rpoB*, *gyrA*, and *katG* genes, where compensatory mutations help to become resistant and assist them to survive and proliferate with time. Importantly, evolution-based studies have found the shortcomings of applying traditional low-resolution typing methods and emphasised the necessity to incorporate population genetics, bottleneck modelling of transmission, and consideration of sampling in time to be able to infer AMR dynamics (Didelot et al., 2016).

The various methodologies, which include population genetics, epidemiological modelling and phylogenetic analysis, assist in reconstituting the resistance trajectories by means of various hierarchical levels that include bacterial clones, resistance genes, and MGEs. Linear, gene-centred models cannot adequately explain the evolution of resistance because AMR is influenced by complex, multilevel selection processes involving bacterial populations, plasmids, integrons, and transposons, all acting simultaneously. Resistance thresholds at which reversal is unlikely can be predicted using computational models, which include ecological factors like antibiotic gradient, population density, connectedness of the genetic exchange network, and the cost of fitness. Phylogenetic studies also suggest the importance of high-risk clones and adaptable MGEs as evolutionary 'intermediaries' in enabling fast horizontal transfer/coselection of resistance factors between clinical and environmental reservoirs. Importantly, these evolution-driven models offer a conceptual and quantitative foundation for eco-evolutionary interventions (to aid the development of strategies) that respond to pathogens and evolutionary processes, including infiltration, promiscuity, adaptability, and endurance, which promote AMR development in the global microbiome (Baquero et al., 2011).

3.8.4 Integrated 'multiomics' approaches

Integrated multiomics approaches have evolved into a powerful framework for investigating the evolutionary dynamics of AMR by simultaneously analysing genomic, transcriptomic, proteomic, metabolomic, and microbiome dimensions to enable a holistic understanding of the emergence and maintenance of resistance traits. Integration of a multiomics approach further extends the associative relationships among the respective omics, enabling reconstitution of the information cascade from genetic variation and MGEs to the modification of gene expression, protein activity, metabolic reorganisation, and community interplay that, in combination, determine the evolution of resistance. These methods have demonstrated the ability to identify causal effects of response behaviours, important points of control and regulation, and the ways in which environmental factors alter resistance by promoting epigenetic, transcriptional, and metabolic changes. These methods will be especially useful in AMR to connect the composition of the resistome with the functional expression, fitness cost, compensatory evolution or interactions between the host and the microbiome, which gives a holistic and mechanistically informed picture of development and sustainability of resistance, which genomics alone cannot (Hasin et al., 2017). WGS combined with computational metabolomics revealed that nutrient availability and metabolic states were significant bottlenecks on the rate and pattern of resistance evolution in terms of which form of resistance and compensatory mutations could be selectively attained. Importantly, the combination of multiomics allowed identifying compensatory metabolic pathways and other weaknesses in resistant populations that could not have been foreseen by using genomic data alone, revealing the usefulness of systems-level methodologies in AMR evolution and developing evolution-guided therapeutic and intervention strategies (Zampieri et al., 2017). The combination of genomics, transcriptomics, proteomics, metabolomics, and epigenomics data, using mathematical and computational modelling approaches, including network-based models, Bayesian inference, multivariate regression, and data fusion, has the potential to reconstruct regulatory circuits, metabolic constraints, and signalling pathways linking resistance genes to functional outcomes. Considered within the AMR evolution, integrating various omics would offer a systems-level view of the relationship between resistome composition, gene expression change, protein function, and metabolism change to give a more profound understanding of how resistance occurs, concentrates, and maintains itself in both clinical and environmental microbiomes (Bersanelli et al., 2016).

3.9 Evolution-informed strategies to mitigate resistance

3.9.1 One Health surveillance

One Health surveillance has been identified as a foundation-based, evolution-informed approach to mitigate AMR, as it explicitly combines human, animal, and environmental health domains into a single analysis. AMR is a quintessential One Health problem because resistance genes and resistant bacteria arise, evolve, and travel across interrelated ecosystems, driven by the shared use of antibiotics, waste streams, and the globalised movement of people, animals, and food products. Some surveillance confined to a clinical setting was revealed to be inadequate because subtherapeutic use of antibiotics in livestock, aquaculture, and agriculture generates powerful selective pressures leading to the fixation and amplification of resistance genes, which can then be transmitted to pathogens in humans through food chains, direct contact, or through the environment, such as water and soil (Robinson et al., 2016). One Health surveillance demands concerted data collection on antimicrobial consumption, AMR gene prevalence and transmission pathways from across sectors and regions, and harmonised methodologies and international data sharing. By embedding surveillance within an evolutionary perspective, One Health frameworks enable early detection of emerging resistance, identification of high-risk transmission interfaces, and development of context-specific interventions that address not only pathogens but also the ecological and socio-economic drivers sustaining AMR on a global scale (Destoumieux-Garzón et al., 2018).

3.9.2 Evolution-aware antibiotic stewardship

Evolution-aware antibiotic stewardship was strengthened by the concept of collateral sensitivity. Resistance evolution to one antibiotic can systematically increase bacterial susceptibility to others through predictable perturbations of cellular networks. Unlike conventional stewardship approaches that rely on reducing antibiotic use or assuming fitness costs of resistance, collateral sensitivity–based cycling actively selects against resistant subpopulations by exploiting their newly acquired vulnerabilities. Experimental validation showed that sequential use of antibiotics with reciprocal collateral sensitivity (e.g. gentamicin and cefuroxime) eliminated resistant lineages more efficiently than standard cycling and reduced the mutant prevention concentration, thereby narrowing the evolutionary window for resistance emergence (Imamovic and Sommer, 2013; Nichol et al., 2019). Resistance, therefore, does not simply arise from among the 'most resistant' mutants, but comes from lineages that can adapt variably from intermediate selective pressures, indicating the importance of drug concentration gradients in enabling stepwise resistance evolution. Inappropriate dosing of antibiotics, including suboptimal gradients generated by uneven distributions of antibiotic concentration in tissues or environments, may inadvertently promote the evolution of resistance through incremental adaptive steps. In terms of stewardship, strategies to reduce prolonged exposure to intermediate concentrations of antibiotics, favouring dosing regimens and treatment designs that limit evolutionary 'paths' to resistance. By considering spatial structure, mutation rates, and evolutionary pathways, evolution-aware stewardship focuses not only on reducing antibiotic use but also on designing treatment strategies that limit opportunities for bacteria to evolve resistance (Baym et al., 2016).

3.9.3 Novel therapeutic strategies: antiresistance adjuvants, phage therapy

New treatment plans that clearly consider the principles of evolution are under development to address the weaknesses of the traditional antibiotics and overcome the problem of multidrug resistance. The genome-targeting capabilities of the system CRISPR–Cas are modifiable to be used as highly specific and programmable antimicrobials, which can selectively kill bacterial strains by simply determining the identity of their DNA sequences. This can be used to kill resistant bacteria by targeting the key chromosomal loci or resistance determinants without placing broad-spectrum selective pressure on the nontarget microbiota using antibiotics, which is a significant evolutionary benefit compared to conventional antibiotics. Notably, this method provides the opportunity to differentiate between closely related strains and quantitatively regulate the population of microbes, which is practically unrealistic with conventional medications or lytic bacteriophages (Gomaa et al., 2014). Some of the evolutionary benefits of phage therapy over conventional antibiotics include high specificity, phage amplification at the site of infection, and minimal disturbance of the commensal microbiota. Notably, even though bacterial resistance to phages is unavoidable, phage resistance is often associated with fitness costs, including reduced virulence or a lack of antibiotic efflux, which can convert bacteria back to antibiotic sensitivity and, therefore, be used therapeutically (Kortright et al., 2019).

3.9.4 Environmental interventions and wastewater management

UWTPs constitute significant intervention sites in the environmental aspect of the AMR problem since they are overwhelmed with large doses of antibiotics, resistant bacteria, clinical, domestic, and industrial resistance gene traits which promote their selection, maintenance, and horizontal transmission. The treatment processes that exist in the conventional wastewater facilities expose the dense and metabolically active microbial communities to the subinhibitory levels of antibiotics, which may enhance their selection by the resistant phenotypes and plasmid-mediated gene transfer. Even though wastewater treatment can indeed cause a total bacteria count decline, various studies have found that it is not selective in the reduction of the ARB or resistance genes, and many studies, too, have indicated that it can, in fact, enhance certain resistance phenotypes by exerting selective forces during wastewater treatment. Treatment technologies and disinfection processes, such as tertiary filtration and oxidation, are therefore considered increasingly relevant in resistance mitigation efforts, but not all methods are effective in the case of MGEs and eDNA. Evolution-inspired wastewater management is not limited to efficiency of removal, but also takes into account the disruption of certain environments and hotspots of gene transfer, and risk-based limits of acceptable resistance load in effluents, and will transform wastewater systems into active control nodes, but not passive passageways, in the global resistome cycle (Rizzo et al., 2013). Extensive surveillance investigations have revealed that treated wastewater effluents always possess varied and highly abundant ARGs, some of which are clinically relevant genes, such as *sul1*, *tetM*, bla_{TEM} and class 1 integron integrase *intI1*, are discharged into the rivers and the surface waters. Significantly, the spread of ARGs is contingent not only on the degree of effluent but also on hydrological dilution, treatment structure, and biological treatment processes, thereby emphasising the importance of evolution-conscious infrastructure design. These findings can inform the One Health paradigm because they demonstrate that the overall management of human wastewater directly influences environmental resistomes, shaping the evolutionary context in which resistance characteristics can be reintroduced into clinical scenarios. Thus WWTP effluents and receiving waters need to be supervised together to quickly identify the formation of resistance genes and a means of countering wastewater treatment so that selective pressures and HGT are minimised in the natural environment (Cacace et al., 2019).

3.10 Conclusions and future perspectives

3.10.1 Synthesis of key insights (evolutionary foundations of antimicrobial resistance)

Under an evolutionary perspective, AMR continues to be the most active homoeostasis between the high levels of positive selective pressure caused by the use of antibiotics and the costs of maintaining resistance due to genetic changes. The majority of chromosomal mutations that confer antibiotic resistance have quantifiable costs to fitness under drug-free conditions due to their interference with fundamental cellular functions or their higher energy requirements. Such costs are not evolutionarily stable and uniform; they vary among species of bacteria and among classes of antibiotics and genetic backgrounds, and are frequently avoided by rapidly compensating evolution, epistatic interactions, and genetic co-selection with other beneficial phenotypes. Notably, resistance mutations, which encode resistance of a high degree (high MICs), are more costly, but the clinical populations are likely to become enriched with low-cost or compensated resistance variants, which explains the presence of AMR despite the elimination of antibiotics (MacLean and San Millan, 2015). Collectively, these lessons underscore the fact that AMR cannot be described as a short-lived phenomenon that is produced on the shelf of drug exposure but rather an adaptive evolutionary result that is determined by the availability of mutations, fitness landscapes, compensatory pathways, and ecological conditions. Any long-range strategy for AMR management must then include the principles of evolution, since it is obvious that resistance can be stabilised, diversified, and maintained unless selective environments and pathways of transmission are addressed concurrently.

3.10.2 Remaining research gaps

The most undecided vital question is disentangling the direct mechanistic effects of antibiotics on HGT from indirect effects through population dynamics and selection. Antibiotics, however, do not directly affect the efficiency of conjugation as previously thought; instead, they affect the diffusion of resistance through the relative fitness and growth of donors, recipients, and transconjugants. This raises a broader issue in current AMR research: the majority of models and surveillance systems are not able to effectively incorporate ecological context, physiological condition, spatial distribution, and population-scale dynamics when calculating the spread of resistance (Lopatkin et al., 2016). These gaps will be overcome by integrative experimental systems, high-resolution time data, and evolution-conscious models not only based on simplified assumptions to understand the action of antibiotics on the emergence of resistance, not in isolation through mutation and selection, but in complex population and ecological interactions. Some of the most significant unresolved issues include the absence of standardised definitions, experimental frameworks, and quantitative measures of reliably distinguishing between persistence, resistance, tolerance, and heteroresistance. Although persister cells are now known to be a distinct, phenotypically apparent subpopulation capable of surviving exposure to lethal antibiotics without conferring hereditary antibiotic resistance, their role in relapse, chronic infection, and subsequent genetic resistance has not been well studied, particularly in vivo. Major gaps also exist in the molecular processes that underlie phenotypic heterogeneity in the environmental and host-based induction stimuli that select for high-persistence phenotypic types during recurrent antibiotic exposures (Balaban et al., 2019). These gaps will need to be addressed by means of standardised methodologies, single-cell-based methods, and integrative in vivo models, which will be useful in reliably establishing the clinical relevance of persistence and its application in the evolutionary stepping stone to stable AMR.

3.10.3 Future directions: evolutionary management of antimicrobial resistance

Future efforts to fight AMR have to go beyond reactive drug development and adopt evolutionary management as a core guiding principle. Resistance evolution is not random but follows partly predictable trajectories that are determined by underlying cellular networks, resulting in the phenomenon of collateral sensitivity, in which resistance to one antibiotic increases susceptibility to another. By systematically determining the collateral sensitivity networks of multiple antibiotics, rationally designed drug-cycling protocols can actively select against resistance rather than just slowing its emergence. A key difference between this approach and the traditional cycling of antibiotics is that the latter depends on resistance fitness costs that are often overcome by compensatory mutations or that persist long after drug withdrawal. Importantly, the effectiveness of collateral sensitivity-based cycling in eradicating resistant subpopulations in an experimental setting was validated to be more efficient, minimise mutant selection windows, and decrease the concentrations needed to suppress resistance (Imamovic and Sommer, 2013). Combining longitudinal susceptibility testing and genomic surveillance might help bridge the gap between in vitro evolution and real-world dynamics, which will enable adaptive, evolution-aware treatment regimens. Collectively, these approaches are pointing toward a future where AMR is controlled using evolutionary control strategies that are predictive, data-driven, and explicitly embrace the unpredictability of evolution, rather than trying to eliminate it (Nichol et al., 2019). Embedding such evolution-aware frameworks into stewardship programs, with genomic surveillance and real-time susceptibility profiling as support, is a promising way forward to sustainable antibiotic use and long-term containment of resistance evolution rather than the inevitable escalation of resistance.

3.10.4 Implications for policy, clinical practice, and environmental security

AMR already constitutes one of the most significant causes of death, with a disproportionate impact on low and middle-income nations, creating a critical requirement for policy frameworks and is concerned with equity of access to surveillance capacity, laboratory capacity, and working antimicrobials. Clinically, resistance to first-line antibiotics, particularly β-lactams and fluoroquinolones, was also causing additional deaths, and there is a dire need to reinforce infection prevention and control programmes and adopt evidence-based antibiotic stewardship programmes, with rapid development of vaccines and novel therapeutics against high-burden pathogen-drug combinations. AMR as an environmental security issue by asserting that poor sanitation, poor wastewater management, and environmental spreading of resistant pathogens and genes contributed to transmission outside the clinical setting. All these knowledge points lead to the conclusion that AMR mitigation needs the formulation of coordinated One Health-oriented policies that combine human health, healthcare systems, and environmental management to decrease the level of selection pressure to break transmission chains and protect global health security (Murray et al., 2022). As a policy consideration, the findings showed the necessity of harmonised global laws, appropriate monitoring of antimicrobial use in humans, animals, and the environment, and a gradual elimination of antimicrobials as growth promoters (Van Boeckel et al., 2015). Urgent need for regulatory frameworks that address antibiotic emissions from agriculture, WWTPs, pharmaceutical manufacturing, and aquaculture, as these environments were identified as major reservoirs and amplification points for ARGs. Consequently, effective AMR mitigation demands integrated policies that combine surveillance, stewardship, improved waste management, and international governance to safeguard public health, maintain the viability of clinical treatment, and ensure environmental security in the face of accelerating global resistance evolution (Pruden et al., 2013).

Acknowledgements

Authors sincerely acknowledge the Department of Biotechnology (DBT), Govt. of India, as Anukul Paul derived financial support through a DBT-sponsored research project (SAN No. 102/IFD/SAN/995/2023-24). The authors gratefully acknowledge Miss Nabanita Roy, former M.Sc. student of the Department of Biotechnology, University of North Bengal, for her valuable assistance in the design and preparation of figures used in this chapter.

Conflict of interest

The authors have no competing interests of any kind to declare concerning the publication of this research work.

AI disclosure

During the preparation of this work the author(s) used ChatGPT (OpenAI) to assist with language refinement, restructuring of text for clarity and coherence, and improvement of academic writing style. The tool was also used to support figure caption drafting and editorial formatting. All scientific content, interpretations, and conclusions were developed, verified, and approved by the author, who takes full responsibility for the accuracy, originality, and integrity of the chapter. After using this tool/service, the author(s) reviewed and edited the content as needed and take(s) full responsibility for the content of the publication.

References

Acosta, N., Lee, J., Bautista, M.A., Bhatnagar, S., Li, C., Waddell, B.J., et al., 2025. Metagenomic analysis after selective culture enrichment of hospital and community wastewater enhances antimicrobial resistance gene detection. mBio 16 (9), 1–21. Available at https://doi.org/10.1128/mbio.01672-25, https://journals.asm.org/journal/mbio.

Ali, J., Rafiq, Q.A., Ratcliffe, E., 2018. Antimicrobial resistance mechanisms and potential synthetic treatments. Future Science OA 4 (4). Available at https://doi.org/10.4155/fsoa-2017-0109, http://www.future-science.com/loi/fso?open=2015.

Andersson, D.I., Hughes, D., 2010. Antibiotic resistance and its cost: Is it possible to reverse resistance? Nature Reviews Microbiology 8 (4), 260–271. Available at https://doi.org/10.1038/nrmicro2319.

Andersson, D.I., Hughes, D., 2014. Microbiological effects of sublethal levels of antibiotics. Nature Reviews Microbiology 12 (7), 465–478. Available at https://doi.org/10.1038/nrmicro3270, http://www.nature.com/nrmicro/index.html.

Baharoglu, Z., Mazel, D., 2014. SOS, the formidable strategy of bacteria against aggressions. FEMS Microbiology Reviews 38 (6), 1126–1145. Available at https://doi.org/10.1111/1574-6976.12077.

Baker, S., Thomson, N., Weill, F.X., Holt, K.E., 2018. Genomic insights into the emergence and spread of antimicrobial-resistant bacterial pathogens. Science 360 (6390), 733–738. Available at https://doi.org/10.1126/science.aar3777, http://science.sciencemag.org/content/360/6390/733/tab-pdf.

Balaban, N.Q., Merrin, J., Chait, R., Kowalik, L., Leibler, S., 2004. Bacterial persistence as a phenotypic switch. Science 305 (5690), 1622–1625. Available at https://doi.org/10.1126/science.1099390.

Balaban, N.Q., Helaine, S., Lewis, K., Ackermann, M., Aldridge, B., Andersson, D.I., et al., 2019. Definitions and guidelines for research on antibiotic persistence. Nature Reviews Microbiology 17 (7), 441–448. Available at https://doi.org/10.1038/s41579-019-0196-3, http://www.nature.com/nrmicro/index.html.

Baquero, F., Martínez, J.L., Cantón, R., 2008. Antibiotics and antibiotic resistance in water environments. Current Opinion in Biotechnology 19 (3), 260–265. Available at https://doi.org/10.1016/j.copbio.2008.05.006.

Baquero, F., Alvarez-Ortega, C., Martinez, J.L., 2009. Ecology and evolution of antibiotic resistance. Environmental Microbiology Reports 1 (6), 469–476. Available at https://doi.org/10.1111/j.1758-2229.2009.00053.x.

Baquero, F., Coque, T.M., De La Cruz, F., 2011. Ecology and evolution as targets: the need for novel eco-evo drugs and strategies to fight antibiotic resistance. Antimicrobial Agents and Chemotherapy 55 (8), 3649–3660. Available at https://doi.org/10.1128/AAC.00013-11Spain, http://aac.asm.org/cgi/reprint/55/8/3649.

Baquero, F., Martínez, J.L., Lanza, V.F., Rodríguez-Beltrán, J., Galán, J.C., San Millán, A., et al., 2021. Evolutionary pathways and trajectories in antibiotic resistance. Clinical Microbiology Reviews 34 (4). Available at https://doi.org/10.1128/CMR.00050-19, https://journals.asm.org/doi/10.1128/CMR.00050-19.

Barrangou, R., Doudna, J.A., 2016. Applications of CRISPR technologies in research and beyond. Nature Biotechnology 34 (9), 933–941. Available at https://doi.org/10.1038/nbt.3659, ://www.nature.com/nbt/index.html.

Baym, M., Lieberman, T.D., Kelsic, E.D., Chait, R., Gross, R., Yelin, I., et al., 2016. Spatiotemporal microbial evolution on antibiotic landscapes. Science 353 (6304), 1147–1151. Available at https://doi.org/10.1126/science.aag0822, http://science.sciencemag.org/content/sci/353/6304/1147.full.pdf.

Bengtsson-Palme, J., 2018. The diversity of uncharacterized antibiotic resistance genes can be predicted from known gene variants—but not always. Microbiome 6 (1). Available at https://doi.org/10.1186/s40168-018-0508-2.

Bersanelli, M., Mosca, E., Remondini, D., Giampieri, E., Sala, C., Castellani, G., et al., 2016. Methods for the integration of multi-omics data: mathematical aspects. BMC Bioinformatics 17 (S2). Available at https://doi.org/10.1186/s12859-015-0857-9.

Bhat, B.A., Mir, R.A., Qadri, H., Dhiman, R., Almilaibary, A., Alkhanani, M., et al., 2023. Integrons in the development of antimicrobial resistance: critical review and perspectives. Frontiers in Microbiology 14. Available at https://doi.org/10.3389/fmicb.2023.1231938, ://www.frontiersin.org/journals/microbiology#.

Blair, J.M.A., Webber, M.A., Baylay, A.J., Ogbolu, D.O., Piddock, L.J.V., 2015. Molecular mechanisms of antibiotic resistance. Nature Reviews Microbiology 13 (1), 42–51. Available at https://doi.org/10.1038/nrmicro3380, http://www.nature.com/nrmicro/index.html.

Blázquez, J., Gómez-Gómez, J.M., 2014. Evolution of antibiotic resistance by hypermutation. Evolutionary Biology of Bacterial and Fungal Pathogens. wiley, Spain, pp. 319–331. Available at https://doi.org/10.1128/9781555815639.ch27, https://onlinelibrary.wiley.com/doi/book/10.1128/9781555815639.

Bowers, R.M., Kyrpides, N.C., Stepanauskas, R., Harmon-Smith, M., Doud, D., Reddy, T.B.K., et al., 2018. Tanja woyke, erratum: corrigendum: minimum information about a single amplified genome (MISAG) and a metagenome-assembled genome (MIMAG) of bacteria and archaea. Nature Biotechnology 36 (2), 196. Available at https://doi.org/10.1038/nbt0218-196a 196.

Brauner, A., Fridman, O., Gefen, O., Balaban, N.Q., 2016. Distinguishing between resistance, tolerance and persistence to antibiotic treatment. Nature Reviews Microbiology 14 (5), 320–330. Available at https://doi.org/10.1038/nrmicro.2016.34, http://www.nature.com/nrmicro/index.html.

C Reygaert, W., 2018. An overview of the antimicrobial resistance mechanisms of bacteria. AIMS Microbiology 4 (3), 482–501. Available at https://doi.org/10.3934/microbiol.2018.3.482.

Cacace, D., Fatta-Kassinos, D., Manaia, C.M., Cytryn, E., Kreuzinger, N., Rizzo, L., et al., 2019. Antibiotic resistance genes in treated wastewater and in the receiving water bodies: a pan-European survey of urban settings. Water Research 162, 320–330. Available at https://doi.org/10.1016/j.watres.2019.06.039, www.elsevier.com/locate/watres.

Cambray, G., Guerout, A.M., Mazel, D., 2010. Integrons. Annual Review of Genetics 44, 141–166. Available at https://doi.org/10.1146/annurev-genet-102209-163504.

Cantón, R., González-Alba, J.M., Galán, J.C., 2012. CTX-M enzymes: origin and diffusion. Frontiers in Microbiology 3. Available at https://doi.org/10.3389/fmicb.2012.00110, http://www.frontiersin.org/Journal/DownloadFile.ashx?pdf=1&FileId=7243&articleId=20597&Version=1&ContentTypeId=21&FileName=fmicb-03-00110.pdf.

Chambers, H.F., DeLeo, F.R., 2009. Waves of resistance: *Staphylococcus aureus* in the antibiotic era. Nature Reviews Microbiology 7 (9), 629–641. Available at https://doi.org/10.1038/nrmicro2200.

Charlebois, D.A., 2023. Quantitative systems-based prediction of antimicrobial resistance evolution. npj Systems Biology and Applications 9 (1). Available at https://doi.org/10.1038/s41540-023-00304-6, https://www.nature.com/npjsba/.

Costerton, J.W., Stewart, P.S., Greenberg, E.P., 1999. Bacterial biofilms: a common cause of persistent infections. Science 284 (5418), 1318–1322. Available at https://doi.org/10.1126/science.284.5418.1318.

Destoumieux-Garzón, D., Mavingui, P., Boetsch, G , Boissier, J., Darriet, F., Duboz, P., et al., 2018. The one health concept: 10 years old and a long road ahead. Frontiers Media S.A., France. Frontiers in Veterinary Science 5. Available at https://doi.org/10.3389/fvets.2018.00014, https://www.frontiersin.org/articles/10.3389/fvets.2018.00014/full.

Deurenberg, R.H., Stobberingh, E.E., 2009. The molecular evolution of hospital- and community-associated methicillin-resistant *Staphylococcus aureus*. Current Molecular Medicine 9 (2), 100–115. Available at https://doi.org/10.2174/156652409787581637Netherlands, ://www.ingentaconnect.com/content/ben/cmm/2009/00000009/00000002/art00002.

Didelot, X., Bowden, R., Wilson, D.J., Peto, T.E.A., Crook, D.W., 2012. Transforming clinical microbiology with bacterial genome sequencing. Nature Reviews Genetics 13 (9), 601–612. Available at https://doi.org/10.1038/nrg3226.

Didelot, X., Walker, A.S., Peto, T.E., Crook, D.W., Wilson, D.J., 2016. Within-host evolution of bacterial pathogens. Nature Reviews Microbiology 14 (3), 150–162. Available at https://doi.org/10.1038/nrmicro.2015.13, http://www.nature.com/nrmicro/index.html.

van Dijk, E.L., Auger, H., Jaszczyszyn, Y., Thermes, C., 2014. Ten years of next-generation sequencing technology. Trends in Genetics 30 (9), 418–426. Available at https://doi.org/10.1016/j.tig.2014.07.001.

Domenech, P., Mouhoub, E., Reed, M.B., 2022. Experimental confirmation that an uncommon rrs gene mutation (g878a) of *Mycobacterium tuberculosis* confers resistance to streptomycin. Antimicrobial Agents and Chemotherapy 66 (3). Available at https://doi.org/10.1128/aac.01915-21, https://journals.asm.org/doi/10.1128/aac.01915-21.

Dörr, T., Vulić, M., Lewis, K., Waldor, M.K., 2010. Ciprofloxacin causes persister formation by inducing the TisB toxin in *Escherichia coli*. PLoS Biology 8 (2), e1000317. Available at https://doi.org/10.1371/journal.pbio.1000317.

Durão, P., Balbontín, R., Gordo, I., 2018. Evolutionary mechanisms shaping the maintenance of antibiotic resistance. Trends in Microbiology 26 (8), 677–691. Available at https://doi.org/10.1016/j.tim.2018.01.005.

Escudero, J.A., Loot, C., Nivina, A., Mazel, D., 2015. The integron: adaptation on demand. Microbiology Spectrum 3 (2). Available at https://doi.org/10.1128/microbiolspec.MDNA3-0019-2014, http://www.asmscience.org/docserver/fulltext/microbiolspec/3/2/MDNA3-0019-2014.pdf?expires=1456326499&id=id&accname=guest&checksum=48AE7F3A480B157BBB848B4BF5BE909F.

Ferreira, R.M., de Oliveira, A.C.P., Moreira, L.M., Belasque, J., Gourbeyre, E., Siguier, P., et al., 2015. A TALE of transposition: Tn3-like transposons play a major role in the spread of pathogenicity determinants of xanthomonas citri and other xanthomonads. mBio 6 (1). Available at https://doi.org/10.1128/mBio.02505-14, http://mbio.asm.org/content/6/1/e02505-14.full.pdf.

Flemming, H.C., Wingender, J., Szewzyk, U., Steinberg, P., Rice, S.A., Kjelleberg, S., 2016. Biofilms: an emergent form of bacterial life. Nature Reviews Microbiology 14 (9), 563–575. Available at https://doi.org/10.1038/nrmicro.2016.94, http://www.nature.com/nrmicro/index.html.

Forsberg, K.J., Reyes, A., Wang, B., Selleck, E.M., Sommer, M.O.A., Dantas, G., 2012. The shared antibiotic resistome of soil bacteria and human pathogens. Science 337 (6098), 1107–1111. Available at https://doi.org/10.1126/science.1220761, http://www.sciencemag.org/content/337/6098/1107.full.pdf.

Frost, L.S., Leplae, R., Summers, A.O., Toussaint, A., 2005. Mobile genetic elements: the agents of open source evolution. Nature Reviews Microbiology 3 (9), 722–732. Available at https://doi.org/10.1038/nrmicro1235.

Gagneux, S., 2018. Ecology and evolution of *Mycobacterium tuberculosis*. Nature Reviews Microbiology 16 (4), 202–213. Available at https://doi.org/10.1038/nrmicro.2018.8.

Gillings, M.R., 2014. Integrons: past, present, and future. Microbiology and Molecular Biology Reviews 78 (2), 257–277. Available at https://doi.org/10.1128/MMBR.00056-13, http://mmbr.asm.org/content/78/2/257.full.pdf+html.

Gillings, M.R., 2017. Lateral gene transfer, bacterial genome evolution, and the *Anthropocene*. Annals of the New York Academy of Sciences 1389 (1), 20–36. Available at https://doi.org/10.1111/nyas.13213, http://onlinelibrary.wiley.com/journal/10.1111/(ISSN)1749-6632.

Gomaa, A.A., Klumpe, H.E., Luo, M.L., Selle, K., Barrangou, R., Beisel, C.L., 2014. Programmable removal of bacterial strains by use of genome-targeting CRISPR-Cas systems. mBio 5 (1). Available at https://doi.org/10.1128/mBio.00928-13, http://mbio.asm.org/content/5/1/e00928-13.full.pdf.

Gullberg, E., Cao, S., Berg, O.G., Ilbäck, C., Sandegren, L., Hughes, D., et al., 2011. Selection of resistant bacteria at very low antibiotic concentrations. PLoS Pathogens 7 (7). Available at https://doi.org/10.1371/journal.ppat.1002158Sweden, http://www.plospathogens.org/article/fetchObjectAttachment.action?uri=info%3Adoi%2F10.1371%2Fjournal.ppat.1002158&representation=PDF.

Hall, C.W., Mah, T.F., 2017. Molecular mechanisms of biofilm-based antibiotic resistance and tolerance in pathogenic bacteria. FEMS Microbiology Reviews 41 (3), 276–301. Available at https://doi.org/10.1093/femsre/fux010, http://femsre.oxfordjournals.org/.

Harms, A., Maisonneuve, E., Gerdes, K., 2016. Mechanisms of bacterial persistence during stress and antibiotic exposure. Science 354 (6318). Available at https://doi.org/10.1126/science.aaf4268.

Hasin, Y., Seldin, M., Lusis, A., 2017. Multi-omics approaches to disease. Genome Biology 18 (1). Available at https://doi.org/10.1186/s13059-017-1215-1.

Hendriksen, R.S., Munk, P., Njage, P., van Bunnik, B., McNally, L., Lukjancenko, O., et al., 2019. Global monitoring of antimicrobial resistance based on metagenomics analyses of urban sewage. Nature Communications 10 (1). Available at https://doi.org/10.1038/s41467-019-08853-3, http://www.nature.com/ncomms/index.html.

Hernando-Amado, S., Sanz-García, F., Blanco, P., Martínez, J.L., 2017. Fitness costs associated with the acquisition of antibiotic resistance. Essays in Biochemistry 61 (1), 37–48. Available at https://doi.org/10.1042/EBC20160057, http://essays.biochemistry.org/content/61/1/37.full-text.pdf.

Hilaire, S.S., Chen, C., Pan, Z., Radolinski, J., Stewart, R.D., Maguire, R.O., et al., 2022. Subsurface manure injection reduces surface transport of antibiotic resistance genes but may create antibiotic resistance hotspots in soils. Environmental Science and Technology 56 (21), 14972–14981. Available at https://doi.org/10.1021/acs.est.2c00981, http://pubs.acs.org/journal/esthag.

Ho, C.S., Wong, C.T.H., Aung, T.T., Lakshminarayanan, R., Mehta, J.S., Rauz, S., et al., 2025. Antimicrobial resistance: a concise update. The Lancet Microbe 6 (1). Available at https://doi.org/10.1016/j.lanmic.2024.07.010, ://www.sciencedirect.com/science/journal/26665247.

Hocquet, D., Vogne, C., El Garch, F., Vejux, A., Gotoh, N., Lee, A., et al., 2003. Patrick plésiat, MexXY-OprM efflux pump is necessary for adaptive resistance of *Pseudomonas aeruginosa* to aminoglycosides. Antimicrobial Agents and Chemotherapy 47 (4), 1371–1375. Available at https://doi.org/10.1128/aac.47.4.1371-1375.2003.

Høiby, N., Ciofu, O., Johansen, H.K., Song, Z.J., Moser, C., Jensen, P.Ø., et al., 2011. The clinical impact of bacterial biofilms. International Journal of Oral Science 3 (2), 55–65. Available at https://doi.org/10.4248/IJOS11026.

Imamovic, L., Sommer, M.O.A., 2013. Use of collateral sensitivity networks to design drug cycling protocols that avoid resistance development. Science Translational Medicine 5 (204). Available at https://doi.org/10.1126/scitranslmed.3006609Denmark, http://stm.sciencemag.org/content/5/204/204ra131.full.pdf.

Jiang, W., Bikard, D., Cox, D., Zhang, F., Marraffini, L.A., 2013. RNA-guided editing of bacterial genomes using CRISPR-Cas systems. Nature Biotechnology 31 (3), 233–239. Available at https://doi.org/10.1038/nbt.2508.

Jordt, H., Stalder, T., Kosterlitz, O., Ponciano, J.M., Top, E.M., Kerr, B., 2020. Coevolution of host–plasmid pairs facilitates the emergence of novel multidrug resistance. Nature Ecology and Evolution 4 (6), 863–869. Available at https://doi.org/10.1038/s41559-020-1170-1, www.nature.com/natecolevol/.

Khedkar, S., Smyshlyaev, G., Letunic, I., Maistrenko, O.M., Coelho, L.P., Orakov, A., et al., 2022. Landscape of mobile genetic elements and their antibiotic resistance cargo in prokaryotic genomes. Nucleic Acids Research 50 (6), 3155–3168. Available at https://doi.org/10.1093/nar/gkac163, https://academic.oup.com/nar/issue.

Klümper, U., Fang, P., Li, B., Xia, Y., Frigon, D., Hamilton, K.A., et al., 2025. Towards the integration of antibiotic resistance gene mobility into environmental surveillance and risk assessment. npj Antimicrobials and Resistance 3 (1). Available at https://doi.org/10.1038/s44259-025-00154-8.

Knox, J.R., Moews, P.C., Frere, J.M., 1996. Molecular evolution of bacterial β-lactam resistance. Chemistry and Biology 3 (11), 937–947. Available at https://doi.org/10.1016/S1074-5521(96)90182-9, http://www.cell.com.

Kortright, K.E., Chan, B.K., Koff, J.L., Turner, P.E., 2019. Phage therapy: a renewed approach to combat antibiotic-resistant bacteria. Cell Host and Microbe 25 (2), 219–232. Available at https://doi.org/10.1016/j.chom.2019.01.014, http://www.elsevier.com.

Larsson, D.G.J., Flach, C.F., 2022. Antibiotic resistance in the environment. Nature Reviews Microbiology 20 (5), 257–269. Available at https://doi.org/10.1038/s41579-021-00649-x, http://www.nature.com/nrmicro/index.html.

Laxminarayan, R., Duse, A., Wattal, C., Zaidi, A.K.M., Wertheim, H.F.L., Sumpradit, N., et al., 2013. Antibiotic resistance-the need for global solutions. The Lancet Infectious Diseases 13 (12), 1057–1098. Available at https://doi.org/10.1016/S1473-3099(13)70318-9.

Levin, B.R., Concepción-Acevedo, J., Udekwu, K.I., 2014. Persistence: a copacetic and parsimonious hypothesis for the existence of non-inherited resistance to antibiotics. Current Opinion in Microbiology 21, 18–21. Available at https://doi.org/10.1016/j.mib.2014.06.016, http://www.elsevier.com/locate/mib.

Levin, B.R., Perrot, V., Walker, N., 2000. Compensatory mutations, antibiotic resistance and the population genetics of adaptive evolution in bacteria. Genetics 154 (3), 985–997.

Levin-Reisman, I., Ronin, I., Gefen, O., Braniss, I. Shoresh, N., Balaban, N.Q., 2017. Antibiotic tolerance facilitates the evolution of resistance. Science 355 (6327), 826–830. Available at https://doi.org/10.1126/science.aaj2191, http://science.sciencemag.org/content/sci/355/6327/826.full.pdf.

Lewis, K., 2010. Persister cells. Annual Review of Microbiology 64 (1), 357–372. Available at https://doi.org/10.1146/annurev.micro.112408.134306.

Li, X.Z., Plésiat, P., Nikaido, H., 2015. The challenge of efflux-mediated antibiotic resistance in Gram-negative bacteria. American Society for Microbiology, Canada Clinical Microbiology Reviews 28 (2), 337–418. Available at https://doi.org/10.1128/CMR.00117-14, http://cmr.asm.org/content/28/2/337.full.pdf+html.

Lindsay, J.A., 2019. Staphylococci: evolving genomes. Microbiology Spectrum 7 (6). Available at https://doi.org/10.1128/microbiolspec.GPP3-0071-2019, https://journals.asm.org/doi/pdf/10.1128/microbiolspec.gpp3-0071-2019.

Liu, H.Y., Prentice, E.L., Webber, M.A., 2024. Mechanisms of antimicrobial resistance in biofilms. npj Antimicrobials and Resistance 2 (1). Available at https://doi.org/10.1038/s44259-024-00046-3.

Loman, N.J., Pallen, M.J., 2015. Twenty years of bacterial genome sequencing. Nature Reviews Microbiology 13 (12), 787–794. Available at https://doi.org/10.1038/nrmicro3565, http://www.nature.com/nrmicro/index.html.

Lopatkin, A.J., Huang, S., Smith, R.P., Srimani, J.K., Sysoeva, T.A., Bewick, S., et al., 2016. Antibiotics as a selective driver for conjugation dynamics. Nature Microbiology 1 (6). Available at https://doi.org/10.1038/nmicrobiol.2016.44, www.nature.com/nmicrobiol/.

MacLean, R.C., Millan, A.S., 2019. The evolution of antibiotic resistance. Science 365 (6458), 1082–1083. Available at https://doi.org/10.1126/science.aax3879, https://science.sciencemag.org/content/365/6458/1082/tab-pdf.

MacLean, R.C., San Millan, A., 2015. Microbial evolution: towards resolving the plasmid paradox. Current Biology 25 (17), R764. Available at https://doi.org/10.1016/j.cub.2015.07.006, ://www.elsevier.com/journals/current-biology/0960-9822.

Manaia, C.M., 2023. Framework for establishing regulatory guidelines to control antibiotic resistance in treated effluents. Critical Reviews in Environmental Science and Technology 53 (6), 754–779. Available at https://doi.org/10.1080/10643389.2022.2085956, www.tandf.co.uk/journals/titles/10643389.asp.

Marti, E., Variatza, E., Balcazar, J.L., 2014. The role of aquatic ecosystems as reservoirs of antibiotic resistance. Trends in Microbiology 22 (1), 36–41. Available at https://doi.org/10.1016/j.tim.2013.11.001.

Martinez, J.L., 2009. The role of natural environments in the evolution of resistance traits in pathogenic bacteria. Proceedings of the Royal Society B: Biological Sciences 276 (1667), 2521–2530. Available at https://doi.org/10.1098/rspb.2009.0320, http://rspb.royalsocietypublishing.org/content/276/1667/2521.full.pdf.

Martinez, J.L., Baquero, F., 2000. Mutation frequencies and antibiotic resistance. Antimicrobial Agents and Chemotherapy 44 (7), 1771–1777. Available at https://doi.org/10.1128/aac.44.7.1771-1777.2000.

Martínez, J.L., Rojo, F., 2011. Metabolic regulation of antibiotic resistance. FEMS Microbiology Reviews 35 (5), 768–789. Available at https://doi.org/10.1111/j.1574-6976.2011.00282.x.

Mazel, D., 2006. Integrons: agents of bacterial evolution. Nature Reviews Microbiology 4 (8), 608–620. Available at https://doi.org/10.1038/nrmicro1462.

Munita, J.M., Arias, C.A., 2016. Mechanisms of antibiotic resistance. Microbiology Spectrum 4 (2). Available at https://doi.org/10.1128/microbiolspec.VMBF-0016-2015, ://www.asmscience.org/docserver/fulltext/microbiolspec/4/2/VMBF-0016-2015.pdf?expires=1486108180&id=id&accname=45052&checksum=43A031364F99D87142C9C34E0F0E2DC9.

Munoz-Price, L.S., Poirel, L., Bonomo, R.A., Schwaber, M.J., Daikos, G.L., Cormican, M., et al., 2013. Clinical epidemiology of the global expansion of *Klebsiella pneumoniae* carbapenemases. The Lancet Infectious Diseases 13 (9), 785–796. Available at https://doi.org/10.1016/S1473-3099(13)70190-7.

Murray, C.J., Ikuta, K.S., Sharara, F., Swetschinski, L., Robles Aguilar, G., Gray, A., et al., 2022. Global burden of bacterial antimicrobial resistance in 2019: a systematic analysis. The Lancet 399 (10325), 629–655. Available at https://doi.org/10.1016/S0140-6736(21)02724-0, http://www.journals.elsevier.com/the-lancet/.

Nesme, J., Simonet, P., 2015. The soil resistome: a critical review on antibiotic resistance origins, ecology and dissemination potential in telluric bacteria. Environmental Microbiology 17 (4), 913–930. Available at https://doi.org/10.1111/1462-2920.12631.

Nichols, R.J., Sen, S., Choo, Y.J., Beltrao, P., Zietek, M., Chaba, R., et al., 2011. Phenotypic landscape of a bacterial cell. Cell 144 (1), 143–156. Available at https://doi.org/10.1016/j.cell.2010.11.052, ://www.sciencedirect.com/journal/cell.

Nichol, D., Rutter, J., Bryant, C., Hujer, A.M., Lek, S., Adams, M.D., et al., 2019. Antibiotic collateral sensitivity is contingent on the repeatability of evolution. Nature Communications 10 (1). Available at https://doi.org/10.1038/s41467-018-08098-6, http://www.nature.com/ncomms/index.html.

Nikaido, H., 2003. Molecular basis of bacterial outer membrane permeability revisited. Microbiology and Molecular Biology Reviews 67 (4), 593–656. Available at https://doi.org/10.1128/mmbr.67.4.593-656.2003.

Noel, H.R., Petrey, J.R., Palmer, L.D., 2022. Mobile genetic elements in Acinetobacter antibiotic-resistance acquisition and dissemination. Annals of the New York Academy of Sciences 1518 (1), 166–182. Available at https://doi.org/10.1111/nyas.14918, http://onlinelibrary.wiley.com/journal/10.1111/(ISSN)1749-6632.

Nolan, C., Behrends, V., 2021. Sub-inhibitory antibiotic exposure and virulence in *Pseudomonas aeruginosa*. Antibiotics 10 (11), 1393. Available at https://doi.org/10.3390/antibiotics10111393.

Nordmann, P., Naas, T., Poirel, L., 2011. Global spread of carbapenemase-producing *Enterobacteriaceae*. Emerging Infectious Diseases 17 (10), 1791–1798. Available at https://doi.org/10.3201/eid1710.110655.

Partridge, S.R., Kwong, S.M., Firth, N., Jensen, S.O., 2018. Mobile genetic elements associated with antimicrobial resistance. Clinical Microbiology Reviews 31 (4). Available at https://doi.org/10.1128/CMR.00088-17, http://cmr.asm.org/.

Pérez-Lago, L., Comas, I., Navarro, Y., González-Candelas, F., Herranz, M., Bouza, E., et al., 2014. Whole genome sequencing analysis of intrapatient microevolution in *Mycobacterium tuberculosis*: potential impact on the inference of tuberculosis transmission. Journal of Infectious Diseases 209 (1), 98–108. Available at https://doi.org/10.1093/infdis/jit439.

Perry, J.A., Wright, G.D., 2013. The antibiotic resistance "mobilome": searching for the link between environment and clinic. Frontiers in Microbiology 4 (MAY). Available at https://doi.org/10.3389/fmicb.2013.00138, http://www.frontiersin.org/Journal/FullText.aspx?ART_DOI=10.3389/fmicb.2013.00138&name=Antimicrobials,_Resistance_and_Chemotherapy&x=y#h9.

Pruden, A., Pei, R., Storteboom, H., Carlson, K.H., 2006. Antibiotic resistance genes as emerging contaminants: studies in northern Colorado. Environmental Science and Technology 40 (23), 7445–7450. Available at https://doi.org/10.1021/es060413l.

Pruden, A., Joakim Larsson, D.G., Amézquita, A., Collignon, P., Brandt, K.K., Graham, D.W., et al., 2013. Management options for reducing the release of antibiotics and antibiotic resistance genes to the environment. Environmental Health Perspectives 121 (8), 878–885. Available at https://doi.org/10.1289/ehp.1206446, http://ehp.niehs.nih.gov/wp-content/uploads/121/8/ehp.1206446.pdf.

Pu, Y., Zhao, Z., Li, Y., Zou, J., Ma, Q., Zhao, Y., et al., 2016. Enhanced efflux activity facilitates drug tolerance in dormant bacterial cells. Molecular Cell 62 (2), 284–294. Available at https://doi.org/10.1016/j.molcel.2016.03.035, www.molecule.org.

Read, D.S., Gweon, H.S., Bowes, M.J., Anjum, M.F., Crook, D.W., Chau, K.K., et al., 2024. Dissemination and persistence of antimicrobial resistance (AMR) along the wastewater-river continuum. Water Research 264. Available at https://doi.org/10.1016/j.watres.2024.122204, https://www.sciencedirect.com/science/journal/00431354.

Reynolds, M.G., 2000. Compensatory evolution in rifampin-resistant *Escherichia coli*. Genetics 156 (4), 1471–1481.

Rizzo, L., Manaia, C., Merlin, C., Schwartz, T., Dagot, C., Ploy, M.C., et al., 2013. Urban wastewater treatment plants as hotspots for antibiotic resistant bacteria and genes spread into the environment: a review. Science of the Total Environment 447, 345–360. Available at https://doi.org/10.1016/j.scitotenv.2013.01.032.

Robinson, T.P., Bu, D.P., Carrique-Mas, J., Fèvre, E.M., Gilbert, M., Grace, D., et al., 2016. Antibiotic resistance is the quintessential one health issue. Transactions of the Royal Society of Tropical Medicine and Hygiene 110 (7), 377–380. Available at https://doi.org/10.1093/trstmh/trw048, http://trstmh.oxfordjournals.org/content/by/year/2013.

Romero, D., Traxler, M.F., López, D., Kolter, R., 2011. Antibiotics as signal molecules. Chemical Reviews 111 (9), 5492–5505. Available at https://doi.org/10.1021/cr2000509.

Rozwandowicz, M., Brouwer, M.S.M., Fischer, J., Wagenaar, J.A., Gonzalez-Zorn, B., Guerra, B., et al., 2018. Plasmids carrying antimicrobial resistance genes in Enterobacteriaceae. Journal of Antimicrobial Chemotherapy 73 (5), 1121–1137. Available at https://doi.org/10.1093/jac/dkx488.

San Millan, A., MacLean, R.C., 2017. Fitness costs of plasmids: a limit to plasmid transmission. Microbiology Spectrum 5 (5). Available at https://doi.org/10.1128/microbiolspec.MTBP-0016-2017.

Smillie, C., Garcillán-Barcia, M.P., Francia, M.V., Rocha, E.P.C., De La Cruz, F., 2010. Mobility of plasmids. Microbiology and Molecular Biology Reviews 74 (3), 434–452. Available at https://doi.org/10.1128/MMBR.00020-10France, http://mmbr.asm.org/cgi/reprint/74/3/434.

Somoskovi, A., Parsons, L.M., Salfinger, M., 2001. The molecular basis of resistance to isoniazid, rifampin, and pyrazinamide in *Mycobacterium tuberculosis*. Respiratory Research 2 (3), 164–168. Available at https://doi.org/10.1186/rr54.

Spellberg, B., Hansen, G.R., Kar, A., Cordova, C.D., Price, L.B., Johnson, J.R., 2016. Antibiotic resistance in humans and animals. NAM Perspectives 6 (6). Available at https://doi.org/10.31478/201606d.

Steinberg, B., Ostermeier, M., 2016. Shifting fitness and epistatic landscapes reflect trade-offs along an evolutionary pathway. Journal of Molecular Biology 428 (13), 2730–2743. Available at https://doi.org/10.1016/j.jmb.2016.04.033.

Stewart, P.S., Franklin, M.J., 2008. Physiological heterogeneity in biofilms. Nature Reviews Microbiology 6 (3), 199–210. Available at https://doi.org/10.1038/nrmicro1838.

Thomas, C.M., Nielsen, K.M., 2005. Mechanisms of, and barriers to, horizontal gene transfer between bacteria. Nature Reviews Microbiology 3 (9), 711–721. Available at https://doi.org/10.1038/nrmicro1234.

Toprak, E., Veres, A., Michel, J.B., Chait, R., Hartl, D.L., Kishony, R., 2012. Evolutionary paths to antibiotic resistance under dynamically sustained drug selection. Nature Genetics 44 (1), 101–105. Available at https://doi.org/10.1038/ng.1034.

Toussaint, A., Chandler, M., 2012. Prokaryote genome fluidity: toward a system approach of the mobilome. Methods in Molecular Biology 804, 57–80. Available at https://doi.org/10.1007/978-1-61779-361-5_4.

Van Boeckel, T.P., Brower, C., Gilbert, M., Grenfell, B.T., Levin, S.A., Robinson, T.P., et al., 2015. Global trends in antimicrobial use in food animals. Proceedings of the National Academy of Sciences of the United States of America 112 (18), 5649–5654. Available at https://doi.org/10.1073/pnas.1503141112, http://www.pnas.org/content/112/18/5649.full.pdf.

Von Wintersdorff, C.J.H., Penders, J., Van Niekerk, J.M., Mills, N.D., Majumder, S., Van Alphen, L.B., et al., 2016. Dissemination of antimicrobial resistance in microbial ecosystems through horizontal gene transfer. Frontiers in Microbiology 7. Available at https://doi.org/10.3389/fmicb.2016.00173, http://journal.frontiersin.org/article/10.3389/fmicb.2016.00173/full.

Webber, M.A., Piddock, L.J.V., 2003. The importance of efflux pumps in bacterial antibiotic resistance. Journal of Antimicrobial Chemotherapy 51 (1), 9–11. Available at https://doi.org/10.1093/jac/dkg050.

Wein, T., Hülter, N.F., Mizrahi, I., Dagan, T., 2019. Emergence of plasmid stability under non-selective conditions maintains antibiotic resistance. Nature Communications 10 (1). Available at https://doi.org/10.1038/s41467-019-10600-7, http://www.nature.com/ncomms/index.html.

Windels, E.M., Michiels, J.E., Fauvart, M., Wenseleers, T., Van den Bergh, B., Michiels, J., 2019. Bacterial persistence promotes the evolution of antibiotic resistance by increasing survival and mutation rates. ISME Journal 13 (5), 1239–1251. Available at https://doi.org/10.1038/s41396-019-0344-9, http://www.nature.com/ismej/marketing/aims_scope.html.

Wistrand-Yuen, E., Knopp, M., Hjort, K., Koskiniemi, S., Berg, O.G., Andersson, D.I., 2018. Evolution of high-level resistance during low-level antibiotic exposure. Nature Communications 9 (1). Available at https://doi.org/10.1038/s41467-018-04059-1, http://www.nature.com/ncomms/index.html.

Wozniak, R.A.F., Waldor, M.K., 2010. Integrative and conjugative elements: mosaic mobile genetic elements enabling dynamic lateral gene flow. Nature Reviews Microbiology 8 (8), 552–563. Available at https://doi.org/10.1038/nrmicro2382.

Wright, G.D., 2007. The antibiotic resistome: the nexus of chemical and genetic diversity. Nature Reviews Microbiology 5 (3), 175–186. Available at https://doi.org/10.1038/nrmicro1614.

You, Y., Silbergeld, E.K., 2014. Learning from agriculture: understanding low-dose antimicrobials as drivers of resistome expansion. Frontiers in Microbiology 5. Available at https://doi.org/10.3389/fmicb.2014.00284, http://journal.frontiersin.org/Journal/10.3389/fmicb.2014.00284/full.

Yu, Z., Gray, D.A., Fick, J., Waters, N., Lindberg, R., Grabic, R., et al., 2025. Antibiotic resistance selection and deselection in municipal wastewater from 47 countries. Nature Communications 16 (1). Available at https://doi.org/10.1038/s41467-025-65670-7, ://www.nature.com/ncomms/.

Zaheer, R., Lakin, S.M., Polo, R.O., Cook, S.R., Larney, F.J., Morley, P.S., et al., 2019. Comparative diversity of microbiomes and Resistomes in beef feedlots, downstream environments and urban sewage influent. BMC Microbiology 19 (1). Available at https://doi.org/10.1186/s12866-019-1548-x, ://www.biomedcentral.com/bmcmicrobiol/.

Zambrano, M.M., 2023. Interplay between antimicrobial resistance and global environmental change. Annual Review of Genetics 57, 275–296. Available at https://doi.org/10.1146/annurev-genet-022123-113904, ://www.annualreviews.org/journal/genet.

Zampieri, M., Enke, T., Chubukov, V., Ricci, V., Piddock, L., Sauer, U., 2017. Metabolic constraints on the evolution of antibiotic resistance. Molecular Systems Biology 13 (3). Available at https://doi.org/10.15252/msb.20167028.

Zhang, Y., Yew, W.W., 2015. Mechanisms of drug resistance in *Mycobacterium tuberculosis*: update 2015. International Journal of Tuberculosis and Lung Disease 19 (11), 1276–1289. Available at https://doi.org/10.5588/ijtld.15.0389, http://docstore.ingenta.com/cgi-bin/ds_deliver/1/u/d/ISIS/83247304.1/iuatld/ijtld/2015/00000019/00000011/art00004/E0A6F4460C8F491814461094295267934A50B7C455.pdf?link=http://www.ingentaconnect.com/error/delivery&format=pdf.

Zhang, Y., Yew, W.W., Barer, M.R., 2012. Targeting persisters for tuberculosis control. Antimicrobial Agents and Chemotherapy 56 (5), 2223–2230. Available at https://doi.org/10.1128/AAC.06288-11, http://aac.asm.org/content/56/5/2223.full.pdf+html.

Zhu, Y.G., Johnson, T.A., Su, J.Q., Qiao, M., Guo, G.X., Stedtfeld, R.D., et al., 2013. Diverse and abundant antibiotic resistance genes in Chinese swine farms. Proceedings of the National Academy of Sciences of the United States of America 110 (9), 3435–3440. Available at https://doi.org/10.1073/pnas.1222743110China, http://www.pnas.org/content/110/9/3435.full.pdf+html.

Chapter 4

Fuelling the resistance: the impact of antibiotic misuse and overuse

Deepshikha Bhowmik[1], Saraswati Patel[2], Himadri Saikia[3], H. Ashalata Singha[4], and Itishree Pandey[5]

[1]*MRU, ESIC Medical College and Hospital, Faridabad, Haryana, India,* [2]*Saveetha College of Pharmacy, Chennai, Tamil Nadu, India,* [3]*Pragjyotish College, Guwahati, Assam, India,* [4]*University of Science and Technology, Baridua, Meghalaya, India,* [5]*Raipur Institute of Medical Sciences, Raipur, Chhattisgarh, India*

4.1 An introduction to antimicrobial resistance: the silent pandemic

Antimicrobial resistance (AMR) is a major public health concern in the 21st century, representing a significant threat to human, animal, and environmental health within the One Health framework (WHO, 2022a,b). This is because once microbes (such as bacteria or viruses) mutate and become resistant to the medications designed to kill them, treatments that had previously been successful can fail after decades of use (WHO, 2021). Excessive usage of antimicrobials in human medicine, agriculture, and veterinary practice has placed substantial selective pressure on the development and proliferation of microbial strains with elevated levels of resistance. Although resistance to antimicrobial agents is a natural biological process, human activities such as inappropriate antibiotic use, over-prescription, and environmental contamination have greatly accelerated the emergence of resistant pathogens (Davies and Davies, 2010; Van Boeckel et al., 2019).

The consequences of AMR are substantial, including an increase in morbidity and mortality from what was once considered minor infections, increased healthcare costs associated with treatment failures, and an inability to treat severe infections. Several international initiatives have recently been undertaken through the World Health Organisation (WHO), the Food and Agriculture Organisation (FAO), and the World Organisation for Animal Health (OIE) to combat AMR, using a coordinated One Health approach to address the interrelationship between human, animal, and environmental health. Misuse of antibiotics in clinical, agricultural, or environmental areas will influence resistance patterns across these different domains (Muteeb et al., 2023; FAO/OIE/WHO, 2019).

In 2019 drug-resistant infections accounted for more than 1.27 million fatalities globally. Without intervention, we may find ourselves back in a scenario where even a minor cut or infection could lead to death (WHO, 2022a,b). The causes of AMR differ across regions and involve factors such as insufficient regulation of antimicrobial use (AMU), lack of proper diagnostics, inappropriate prescribing practices, and the unrestricted availability of antimicrobial medications. Moreover, in numerous low- and middle-income countries (LMICs), antimicrobials are often sold without a prescription and are commonly utilised as a workaround for poor infection control, biosecurity measures, and hygiene standards in both healthcare and agricultural environments (Okeke et al., 2021).

Hospital waste, emissions from pharmaceutical manufacturing, and runoff from agricultural activities lead to regions with high concentrations of resistant genes and bacteria, transforming the environment into a persistent reservoir for AMR. These reservoirs facilitate the ongoing transfer of resistance across various ecosystems, emphasising the urgent need for a comprehensive One Health approach that links human, animal, and environmental health to improve management and surveillance (Larsson et al., 2023; Bengtsson-Palme and Larsson, 2016).

This chapter examines the consequences of improper and excessive antibiotic use, with a focus on environmental aspects, including pharmaceutical contamination and wastewater discharge. It evaluates existing strategies for mitigation, identifies regulatory deficiencies, and emphasises methods to enhance stewardship and monitoring, particularly in LMICs.

Antimicrobial Resistance in the 21st Century. DOI: https://doi.org/10.1016/B978-0-443-44748-8.00010-5

4.2 A century of struggle: historical insights into the antimicrobial resistance era

The narrative of AMR has evolved following the first discovery of antibiotics. The first antibiotic, penicillin, was discovered by Alexander Fleming in 1928, marking the onset of a medical revolution that saved countless lives which would have otherwise succumbed to infections thought to be fatal (Laxminarayan et al., 2013). Fleming also cautioned that improper use of penicillin could lead to an increase in resistant microorganisms. This warning served as a warning, and AMR has developed into a significant public health issue affecting people around the world and has become a source of global concern (WHO, 2015a,b).

Antibiotics have been commercially available since the 1940s and have been utilised in both human and animal medicine to improve health and extend life expectancy. In the early 1940s and 1950s, antibiotics were being used extensively in both human and veterinary medicine, and this extensive use led to the development of resistant bacterial strains. Within a couple of years of penicillin's introduction, penicillin-resistant *Staphylococcus aureus* emerged, demonstrating that bacterial adaptations to drugs can occur rapidly once drugs are introduced (Davies and Davies, 2010). Over the next several decades, the emergence of resistance to multiple drugs occurred with streptomycin, tetracyclines, and sulphonamides, among others; and in hospitals, multidrug-resistant organisms, including Pseudomonas aeruginosa and Enterobacteriaceae, were increasingly found (Chambers and DeLeo, 2009).

The emergence of methicillin-resistant *S. aureus* (MRSA) by the 1980s was a major setback after the introduction of methicillin in 1959, which had seemed to offer a solution to the problem of penicillin-resistant staphylococci (Chambers and DeLeo, 2009). The overuse of antibiotics, especially during the 1990s and 2000s, in all three areas of medicine, veterinary medicine, and agriculture, caused the rapid development of antibiotic-resistant genes among different species and habitats. The emergence of extensively drug-resistant tuberculosis (XDR-TB), carbapenem-resistant Enterobacteriaceae (CRE), and colistin-resistant bacteria is an example of the dangers of a possible postantibiotic world (O'Neil, 2016).

Between 2012 and 2021, the use of broad-spectrum antibiotics in the hospital system increased by approximately 15% overall, and the use of carbapenems rose by 34%, while the use of last-resort drugs for multidrug-resistant infections almost doubled (WHO, 2021). The above data illustrate that the human component of AMR is significant and has developed into a long-standing aspect of medicine and agriculture.

For many years, antibiotics have been overprescribed and improperly prescribed in humans, often prescribed for viral infections or at less-than-optimal dosages. For many years, antibiotics were used not only to treat diseases in animals but also to promote growth in domesticated animals. The emergence of antibiotic resistance within food production systems was influenced by this factor (Van Boeckel et al., 2019).

Consequently, tackling AMR is now recognised as a pressing global health concern that necessitates the collaboration of governmental bodies at both national and international levels, healthcare professionals, researchers, and the public. Global initiatives include the WHO's Global Action Plan on Antimicrobial Resistance (AMR) established in 2015 (WHO, 2015a,b) and the 'One Health' approach, which seeks to integrate the human, animal, and environmental domains to mitigate antibiotic resistance and maintain the effectiveness of current antimicrobial drugs (FAO/OIE/WHO, 2019). The narrative of AMR is not just about the evolution and adaptation of bacteria; it reflects the impacts of human actions and offers important insights into the management, preservation, and responsible application of antimicrobial therapies for the future.

4.3 Current global scenario of antimicrobial resistance: One Health perspective

The Lancet Global Research on Antimicrobial Resistance (GRAM) 2019 showcased a significant advancement in the understanding of the global impact of bacterial AMR. The study indicated that around 8.9 million deaths globally were caused by bacterial infections in 2019; of these, 1.27 million were linked to AMR, while roughly 4.95 million deaths resulted from infections with antibiotic-resistant bacteria (Murray et al., 2022). According to recent WHO reports (2024), AMR is expected to lead to between 36 and 39 million deaths worldwide in the next 25 years, unless we have some form of intervention. The WHO additionally cautions that the global antibiotic pipeline is in a dual crisis, characterised by insufficient development of new medications and a lack of innovation in the discovery of new antibiotics. According to Dr Yukiko Nakatani, WHO Assistant Director General: 'Unless there is much greater investment in R&D and also efforts to get both old and new treatments to the people who need them most, drug-resistant diseases will continue to be a spreading public health threat' (Murray et al., 2022).

In addition, a 2023 Lancet study estimated that the death rate due to drug-resistant disease could be 2 million deaths per year by 2050 (up from an average of 1 million deaths per year from 1990 to 2021). If deaths associated with drug-resistant disease are included, AMR may be responsible for approximately 10 million deaths each year by mid-century – a staggering figure that emphasises why many experts refer to AMR as a 'silent pandemic' (Venter et al., 2023) (Fig. 4.1).

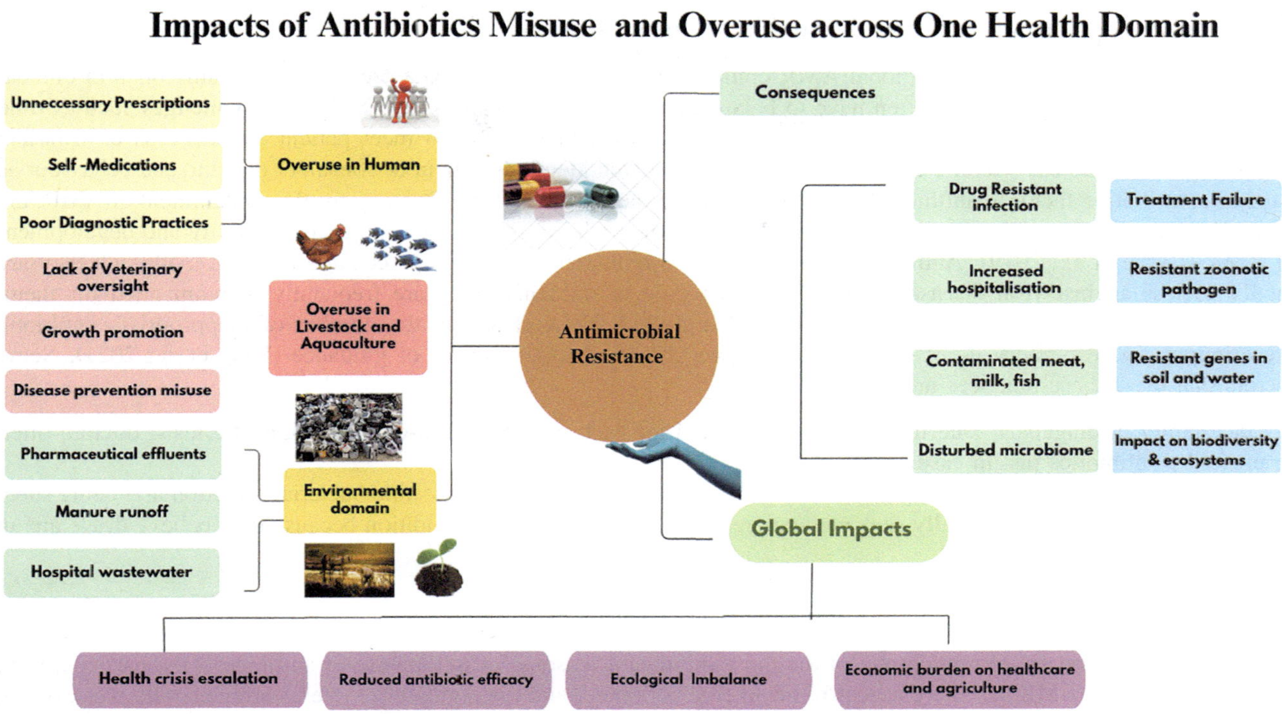

FIGURE 4.1 **Impacts of antibiotic overuse and misuse across the One Health domain.** A diagram titled "Impacts of Antibiotics Misuse and Overuse across One Health Domain" illustrates antimicrobial resistance as the central theme, with contributing factors on the left and resulting effects on the right. The left side shows three categories: overuse in humans, including unnecessary prescriptions, self-medication, and poor diagnostic practices; overuse in livestock and aquaculture, including lack of veterinary oversight, growth promotion, and misuse for disease prevention; and environmental sources, including pharmaceutical effluents, manure runoff, and hospital wastewater. On the right, the consequences are presented in paired outcomes: drug-resistant infections alongside treatment failure, increased hospitalisation alongside resistant zoonotic pathogens, contaminated meat, milk, and fish alongside resistant genes in soil and water, and disturbed microbiome alongside impacts on biodiversity and ecosystems. At the bottom, broader global impacts are highlighted, including health crisis escalation, reduced antibiotic efficacy, ecological imbalance, and economic burden on healthcare and agriculture, with visual icons such as pills, animals, and environmental elements reinforcing the interconnected One Health perspective.

4.4 Drivers and impacts of misuse and overuse of antibiotics

4.4.1 Human healthcare: the frontline of misuse

Antibiotics are used therapeutically, prophylactically, and empirically in human healthcare. They are considered among the greatest medical breakthroughs, dramatically reducing deaths and illnesses from bacterial diseases in both humans and animals. Antibiotics can be used successfully to prevent severe infections and make risky surgical operations much safer when they are appropriately used for patients who undergo cancer treatment (chemotherapy), organ transplantations, and other high-risk surgeries.

In addition to their clinical applications, antibiotics are prevalent in agriculture, aquaculture, and livestock production, where they play a crucial role in maintaining animal health and ensuring food security. However, the misuse and overuse of antibiotics in all sectors have become significant contributors to antibiotic resistance, which poses a considerable threat to global health (Mittal et al., 2020). As bacteria develop resistance, infections tend to last longer, become more challenging to treat, and could lead to an increase in mortality rates. Diseases that were once easily managed—like pneumonia, sepsis, and urinary tract infections—are progressively proving less responsive to conventional antibiotics (WHO, 2020). This escalating issue is evident in the rising instances of drug-resistant infections, such as XDR-TB, CRE, and MRSA (Laxminarayan et al., 2016).

As effective antibiotics become limited, routine medical care also becomes riskier. Without reliable treatments for infections, procedures such as organ transplants, cancer therapy, and caesarean deliveries carry greater danger. The WHO warns that the world may be approaching a 'post-antibiotic era', where even minor injuries or common infections could once again become life-threatening (WHO, 2020).

4.4.2 Inappropriate prescribing by healthcare practitioners

A major contributing factor to the global AMR epidemic is still the inappropriate prescription of antibiotics in clinical settings. Healthcare professionals often have to balance their clinical responsibilities with daily challenges such as time constraints, high patient loads, a lack of diagnostic resources, and pressure to meet patient expectations as frontline stewards of antibiotic use (Nammi et al., 2025). Previous research identifies patient or consumer demand and behaviour as a major contributing factor to antibiotic abuse. According to the WHO's Global Strategy for the containment of AMR, the cost-effective use of antimicrobials enhances the clinical effect while suppressing drug-related toxicity and development of AMR (Byrne et al., 2019). Antimicrobial medication is redundant in a huge percentage of cases. Although they are totally useless against viral diseases like the common cold or influenza, they are frequently recommended for them. According to a 2021 systematic analysis published in JAMA Network Open, the number of inappropriate antibiotic prescriptions for respiratory illnesses remains high worldwide, despite decades of guidance (King et al., 2021). Some factors that drive these practices are:

1. Uncertainty in diagnosis: Due to time constraints and the lack of rapid point-of-care diagnostic tests, doctors often prescribe antibiotics just in case.
2. Defensive medicine: Overprescription is caused by a fear of overlooking a possible bacterial infection and being sued.
3. Patient pressure: Patients usually seek and anticipate prescriptions for any condition because they believe drugs are an effective treatment.

4.4.3 Self-medication and access without prescription

Antibiotic misuse and abuse remain one of the most significant reasons for the misuse of antimicrobials, particularly in developing nations (LMICs), where access to healthcare is limited, and the regulatory systems are generally weak (Ocan et al., 2015). Antibiotics are readily available in many developing countries for purchase at local pharmacies without the need for a physician's prescription. As a result, residents of developing countries will commonly treat themselves for acute respiratory infections (common cold, influenza) with antibiotics, since they are easily accessible and the patients are unaware that these conditions are caused by viruses rather than bacteria. These factors contribute to the misuse of antimicrobials by promoting the use of the wrong type of antimicrobial agent, incorrect dose of antimicrobial agent, and the completion of an inadequate course of antimicrobial treatment—all of which provide a favourable environment for the emergence and proliferation of antimicrobial-resistant pathogens. A 2023 article published in BMJ Global Health stated that a large number of countries still permit the sale of antibiotics over the counter without a prescription, primarily due to poorly regulated pharmaceutical markets and a lack of public knowledge regarding appropriate AMU (Auta et al., 2023). Reports from research studies conducted in India, Lebanon, and other economically disadvantaged countries have demonstrated a high prevalence of self-medication with antibiotics, as people can obtain these agents without prescriptions from pharmacies. AMR increased quickly because the pandemic raised the demand for drugs to treat viral infections. Many people in a study from Lebanon believed that antibiotics could effectively treat viral diseases, such as the common cold. This shows there are still misunderstandings about how to use antibiotics properly. Outpatient data from India indicate a preference for broad-spectrum antibiotics, often used before the specific microbe is identified. This puts unnecessary pressure on bacterial populations and worsens the problem of AMR (Zheng et al., 2025).

In addition to the aforementioned problems, socioeconomic obstacles to accessing healthcare, such as high medical service costs, remote locations, and a lack of knowledge about antibiotic use, continue to worsen the problem. In general, people who cannot afford to visit formal healthcare providers will seek out alternative forms of health care, such as informal healthcare providers, based on their own personal experience with antibiotics as a rapid solution to their illness. Therefore, more than just restricting the distribution of antibiotics through pharmacies will be necessary to address this issue. A multifaceted approach, including public education campaigns to raise awareness of the risks of self-medication, improving access to affordable health care, and improving the delivery of healthcare services within communities, will be required to address antibiotic misuse. Promotion of rational antibiotic use through culturally sensitive public health education programmes and improved dispensing practices is necessary to prevent self-medication and limit the growth of AMR (WHO, 2022a,b) (Fig. 4.1).

4.4.4 Hospital-acquired infections and empiric therapy

Hospitals represent the major hotbeds of resistance among pathogenic organisms. MRSA and CRE are two examples of the so-called superbugs, which thrive in patients suffering from serious illnesses, largely due to the widespread use of broad-spectrum antibacterial agents as empiric therapy or medications administered prior to the specific pathogen being identified. Antifungal treatments in hospital settings are likewise associated with the increasing frequency and spreading of pan-resistant fungi, including *Candida auris* (Murray et al., 2022).

4.5 Agriculture and livestock: the hidden driver

Antibiotics used in agriculture, particularly in the livestock and aquaculture industries, represent a significant source of the current global crisis related to AMR. The use of antibiotics in modern animal husbandry is common for several reasons, including treatment of infections, disease prevention (prophylactic use), and enhancement of animal growth rate (subtherapeutic use). These uses of antibiotics lead to increased productivity and reduced loss in the short term; however, they provide an environment that supports the development and dissemination of antimicrobial-resistant bacteria.

Enshaie et al. (2025) performed a comprehensive review of AMU in the livestock industry and concluded that antimicrobial exposure in livestock was a key contributor to the global AMR crisis. Antimicrobial-resistant bacteria developed in animals may be transmitted to humans via the food chain, contact with animals, or contaminated environments. Although there have been some policy efforts to limit the nontherapeutic use of antibiotics, enforcement has been inconsistent; therefore, misuse of antibiotics continues due to a lack of monitoring and weak regulation (Lai et al., 2025). Many small-holding dairy and poultry producers in India administer antibiotics to their animals without the oversight of veterinarians; instead, they rely on either informal recommendations from others or past experiences when determining dosages for administration. Due to this lack of regulation, uncontrolled AMU in animals has led to zoonotic transmission of resistant bacteria and antibiotic residue in products obtained from animals (Enshaie et al., 2025).

In many LMICs where there is still widespread availability of nonprescribed antibiotics, this will continue to add fuel to the fire. The 2023 BMJ Global Health study demonstrated that unregulated pharmaceutical markets and a lack of knowledge amongst farmers create an environment which allows for self-medication and incorrect use of drugs (Auta et al., 2023). Often, farmers choose to use antibiotics rather than good hygiene, adequate space for their animals, and/or good biosecurity as a way to immediately improve their production levels while accepting the longer-term negative impacts on their ecosystem and health.

There are several ways that resistant bacteria found in livestock can be transmitted to humans. Direct contact with animals, eating meat, milk, or eggs from animals treated with antibiotics, or exposure to contaminated soil and water are all potential routes for these resistant microorganisms to enter the human body (Ardakani et al., 2023). Once these resistant organisms have entered the human body, they can become part of the individual's microbiome, thus making it more difficult for antibiotics to work effectively in treating infections (WHO, 2022a,b). Additionally, when antibiotics are used in animal agriculture and then excreted in urine and faeces, they contaminate manure and wastewater, creating a constant flow of pollutants and a continued process of selecting for antibiotic-resistant bacteria. The presence of these pollutants not only diminishes soil quality but also contaminates nearby water bodies, adversely affecting aquatic life and ultimately reentering the human food chain (Nammi et al., 2025).

A comprehensive global strategy is essential to avert the inappropriate use of antibiotics in agriculture. Governments and policymakers must implement laws and regulations that ban the use of antibiotics for nonmedical purposes, require veterinarians to prescribe antibiotics, and strengthen monitoring and surveillance. Informing farmers and consumers about alternative approaches such as vaccinations, probiotics, proper hygiene, and animal husbandry can help lessen dependency on antibiotics. Educational initiatives are equally vital as the pursuit of new alternatives to antibiotics. Programmes must be created to educate farmers and consumers on the significance of sustainable agricultural practices and their contribution to the 'One Health' initiative aimed at safeguarding human, animal, and environmental health (Nammi et al., 2025).

4.6 Aquaculture: the emerging concern

AMR has become an urgent global health threat as a result of the excessive and improper use of antibiotics in aquaculture around the world. Antibiotic resistance occurs when bacteria evolve mechanisms to resist antibiotics, making them less effective. Bacteria develop resistance to antibiotics by sharing genetic material that gives them an advantage. This is a problem that is being seen in many countries with growing aquaculture industries (Adenaya et al., 2023).

While the increased density of fish and shrimp farming has greatly contributed to the world's economy, it has also created an ideal environment for antibiotic resistance to emerge and spread. A practice that is commonly used in the aquaculture industry is the direct disposal of antibiotic-laced water from fish farms into surrounding ecosystems (Kuznetsova and Lee, 2001). When this happens, it changes sediment, coastal waters, and the natural microbial community and promotes the survival and proliferation of antibiotic-resistant bacteria. The continued accumulation of antibiotics in coastal environments can be hazardous because it disrupts the balance of native microbial populations and their roles in driving important biogeochemical cycles and organic matter decomposition (Desbois, 2025).

Antibiotics are used both therapeutically and prophylactically in high-density aquaculture systems to reduce the incidence of disease. Aquatic species, like terrestrial livestock, are often raised in close proximity to one another and experience increased physiological stress and disease susceptibility. To prevent loss of fish due to disease, antibiotics are applied to water or feed; however, global data on the use of antibiotics in aquaculture is limited and probably underestimates the extent of their application (Schar et al., 2020). These practices accelerate the selection of drug-resistant microorganisms, which can spread through the environment and ultimately harm human health.

The transfer of resistance genes via mobile genetic elements links environmental reservoirs to human pathogens, thereby posing a direct threat to clinical medicine. Antibiotics in aquatic systems also negatively affect nontarget organisms and critical ecological processes, potentially causing long-term effects on biodiversity and ecosystem stability (Deng et al., 2025).

There are several concerns regarding antibiotic residues in farmed fish that can affect consumers, including the potential for antibiotic-resistant bacteria to develop, hypersensitivity reactions, bone marrow suppression, mutagenesis, teratogenesis, and carcinogenic effects (Limbu et al., 2021). In addition to harming the environment, the use of antibiotics in aquaculture leads to the development of antibiotic-resistant organisms, which may have long-term detrimental effects on human and animal health and well-being. As a result, the overuse and abuse of antibiotics in aquaculture pose major worldwide health risks. According to the current study, microorganisms from aquaculture systems frequently exhibit antibiotic resistance patterns that are similar to those of clinically significant human infections. This suggests that antibiotic resistance in aquatic pathogens and antibiotic resistance in human pathogens are clearly related (Cabello et al., 2016; Watts et al., 2017; Miranda et al., 2018; Done et al., 2015), supporting the contention that there is a direct link between aquatic and human AMR. Additionally, patients may experience allergic and toxic reactions if they consume low concentrations of antibiotics, such as nitrofurans or chloramphenicol, through contaminated seafood. Long-term exposure to minute amounts of antibiotic residues can also lead to the creation of novel antibiotic-resistant organisms and may pose unidentified health problems in the future. All things considered, these findings show how urgently better antibiotic stewardship, stronger biosecurity procedures on aquaculture farms, and greater regulation of aquaculture operations are needed. To safeguard aquatic ecosystems and human health, the aquaculture sector will also need to implement environmentally friendly production methods and create alternatives to conventional disease management techniques, such as probiotics, vaccines, and improved water quality control (Zhou et al., 2021; Gao et al., 2022; Schar et al., 2020).

4.7 Environmental dimension of antimicrobial resistance: the connecting reservoir

The environment strongly influences the emergence and dissemination of AMR. Evidence suggests that soil, water bodies, and garbage serve as reservoirs for resistant organisms, in addition to the abuse of antibiotics in humans and animals (Liguori et al., 2022). Antibiotics are released into the environment through pharmaceutical waste and improperly handled sewage, leading to subtherapeutic levels of resistance. This emphasises the importance of treating AMR as an ecological problem requiring all-encompassing solutions (Samreen et al., 2021).

4.7.1 Pharmaceutical manufacturing pollution

Discharge of antibiotics from manufacturing facilities into the environment is one of the largest sources of high antibiotic concentrations and the largest source of resistance genes in the environment (Larsson, 2014; Pal et al., 2016). The regulatory framework governing antibiotic production and waste discharges varies significantly by country; some countries have very strict requirements for production and waste discharge, whereas others have very little, if any, regulation. Since licensing of environmental discharges is typically governed by the federal government, state governments, or municipalities, enforcement of the regulatory framework also can differ significantly.

As noted above, the manufacturing process for antibiotics involves multiple steps and is part of a large, complex global supply chain. Therefore both consumers and manufacturers of antibiotics have difficulty identifying potential environmental risks in the supply chain and developing strategies to mitigate them. While several organisations have recommended best practices, including limits on antibiotic discharge from their operations, most are voluntary, and few have been implemented as enforceable regulations (Bengtsson-Palme and Larsson, 2016).

In response to the gap between recommendations and the implementation of regulatory practices, industry associations have begun developing and implementing new practices and tools to encourage responsible production. One example of such an association is the Pharmaceutical Supply Chain Initiative (PSCI), developed in 2006, which encourages its member companies to follow ethical and responsible business practices throughout the pharmaceutical supply chain. As of 2018, PSCI has 26 member companies and uses collaborative programmes and training to promote shared learning among members, demonstrating that collective action is more likely to achieve greater success than individual company actions (Meena et al., 2022).

In addition, thirteen major pharmaceutical companies have signed on to the Industry Roadmap for Progress on Combating Antimicrobial Resistance, which includes commitments to review their supply chains for antimicrobials, reduce antimicrobial waste, and establish goals to reduce the amount of antimicrobials discharged into the environment. A significant hurdle facing the roadmap is increasing membership in the initiative, particularly companies that serve local markets in LMICs, where regulatory oversight may be limited (Meena et al., 2022).

4.7.2 Wastewater treatment challenges

The primary pathway by which antibiotics, resistant microorganisms, and resistance genes enter the environment is wastewater. Wastewater from both residential and hospital settings has residual, unmetabolized drugs, resistant microorganisms, and mobile genetic elements shed by patients. If this wastewater is discharged into receiving bodies of water either untreated or only partially treated, it creates an ideal environment for the development and dissemination of resistance (Yakubu, 2017; Liu et al., 2021). While traditional wastewater treatment plants (WWTPs) are capable of effectively removing organic pollutants and pathogens, they are not designed to remove antibiotics or resistance genes. Therefore drugs such as fluoroquinolones, macrolides, and beta-lactams, along with multidrug-resistant bacteria, are present in WWTP-treated effluent. Resistant organisms and their resistance genes have been demonstrated to survive secondary and/or tertiary treatment and be discharged into rivers, lakes, and irrigation systems. Additionally, sludges produced by WWTPs are often used as fertiliser and can transport resistance genes into soil (Chu et al., 2018). This problem is most severe in LMICs where wastewater infrastructure is typically inadequate. In numerous locations around the world, raw sewage is discharged into surface waters without any form of treatment. Some hospitals do not possess pretreatment systems for their effluent, resulting in the discharge of disinfectants, antimicrobial agents, and resistant bacteria into surrounding waterways and public sewer systems (Hounmanou et al., 2025).

To mitigate the risk of AMR associated with wastewater, improved infrastructure and novel treatment methods are necessary. Several technologies, including ozonation, membrane bioreactor, activated carbon filtration, and advanced oxidation process, are effective in eliminating antibiotic residues and resistance genes; however, due to the high cost of these technologies, they cannot currently be widely applied (Guerra-Rodríguez et al., 2023; Le et al., 2018; Spit et al., 2022; Azuma et al., 2022). Setting discharge regulations, enhancing AMR monitoring in wastewater systems, and overseeing hospital effluent discharges are essential initial measures. If we fail to tackle these challenges, wastewater will persist as a major contributor to environmental AMR and hinder global mitigation initiatives (Chau et al., 2022).

4.7.3 Hospital effluents as reservoirs of resistance

While carrying out tests in the lab, providing clinical services, and conducting research, hospitals produce large volumes of liquid waste. Healthcare facilities generate wastewater that is high in organic debris and new contaminants such as bacteria, parasites, medications, and disinfectants. Public health and aquatic habitats may suffer if this untreated effluent is released into the environment. One of the main sources of antibiotic residues that encourage the growth and survival of antibiotic-resistant bacteria is hospital wastewater. Antibiotic resistance and hospital wastewater have been linked in several settings. For example, high quantities of ciprofloxacin (up to 236.6μg/L) were found in wastewater drainages in Ujjain, India, and *E. coli* isolated from those samples were resistant to both ciprofloxacin and other antibiotics such as amikacin (Diwan et al., 2010). Similar trends were seen in South Africa, where hospital effluent levels of cefazolin and sulfamethoxazole were connected with their frequent use in those establishments (Deedat et al., 2025). Hospital wastewater had much greater levels of resistance genes, including sul1, blaNDM-1, and intI1, than municipal wastewater, according to metagenomic investigations conducted in the Kathmandu Valley (Nepal) (Thakali et al., 2021).

Hospital wastewater contributes to antibiotic resistance through multiple interconnected processes. First, they contain high concentrations of antibiotic residues, which impose a strong selective pressure favouring the growth of resistant strains while suppressing the susceptible ones. Second, the diverse microbial populations present in the effluents create an ideal environment for horizontal gene transfer (HGT) to occur. Third, many resistance genes are encoded on mobile genetic elements such as plasmids and integrons (e.g., intI1). Fourth, once resistant bacteria and their genes enter sewer systems or natural waterways, they may remain viable in the environment and travel throughout ecosystems, affecting humans, animals, and the ecosystem at large (Schuster et al., 2022).

There is an urgent need to establish guidelines for the discharge of hospital effluent, including limits for antibiotic concentrations, and to align with current discussions in the European Union on environmental policy for antibiotic residue limitations. In addition to developing guidelines for hospital effluent discharge, there is also a need to strengthen infection control practices in hospitals to decrease the number of multidrug-resistant organisms shed into the effluent (Abosse et al., 2024). Improving antibiotic stewardship through improved prescribing practices will reduce the amount of antibiotics prescribed unnecessarily, thus reducing selective pressure for resistance (Tacconelli et al., 2014; Bansal, 2019). Finally, integrating routine monitoring of antibiotic residues, resistant bacteria, and resistance genes into national AMR action plans will enable more targeted and efficient interventions and enhance One Health strategies, as recommended by the WHO and FAO (Barlam et al., 2016).

4.7.4 Role of environmental surveillance

Tracking AMR in the environment is vital for reducing the presence and impact of AMR within a One Health framework. Environmental media (water, soil, and waste) collect and store resistant bacteria, antibiotic-resistance genes, and antibiotics from human, animal, and industrial sources; clinical sites do not provide such large reservoirs of resistant bacteria and antibiotic-resistance genes (Hart et al., 2023). The tracking of these components in environmental media gives us significant information on how the AMR spreads out of the confines of the traditional healthcare setting (Anjum et al., 2021).

It also serves as an early warning system. If AMR is identified in the environment, it will enable policymakers to respond to the potential public health issue before it develops into a full-blown epidemic (Ashwini et al., 2023). Examples include identifying the presence of carbapenamase genes (blaNDM-1, blaKPC) in hospital and municipal wastewater, indicating the presence of unknown resistance reservoirs before the occurrence of a clinical outbreak (Hendriksen et al., 2019). In addition to clinical surveillance systems, the epidemiological data generated by the collection of environmental data is useful and complementary to clinical data. Wastewater-based surveillance can give a country/region a total estimate of its resistance burden and therefore capture symptomatic and asymptomatic carriers (Mao et al., 2020). Finally, the analysis of WWTP influents and effluents can help determine how well current technologies are removing antimicrobials and antibiotic-resistance genes (Hultman et al., 2018). As such, integrating environmental surveillance into national and international AMR plans is vital for a comprehensive One Health response. Organisations such as the World Health Organisation (WHO) have recommended that environmental data be used to fill knowledge gaps and assist in directing intervention efforts (WHO, 2015a,b). Of particular urgency for developing this capability are lower income countries that have inadequate sanitation and weak regulatory environments that increase the likelihood and rate of AMR emergence and spread (Singer et al., 2016).

4.8 Consequences of antibiotic misuse and overuse

4.8.1 Economic consequences

The economic burden of antibiotic resistance is substantial. Antibiotic resistance may increase the number of people in extreme poverty by approximately 28 million by 2050, as estimated by the World Bank (2019), while it would reduce the global GDP by 3.8%. In addition to the aforementioned factors of cost, the increased expenses for longer hospitalisations and additional care from second-line treatments, along with enhanced infection control measures, have greatly raised the costs of healthcare. The CDC (2023) reports that the United States will pay approximately $21 billion more per year for the treatment of drug-resistant infections. Furthermore, AMR represents another important impact on the economy, due to the loss of productivity. This is especially true for LMICs, which are often plagued by infectious disease issues that pose serious public health challenges, resulting in larger lost productivity than other regions. Therefore, AMR has impacts on not only the health sector, but also societal stability, food security, and employment (CDC, 2023).

4.8.2 Environmental and ecological consequences

Overuse of antibiotics has equally alarming effects on the environment. Microbial populations that are essential to soil fertility, nutrient cycling, and ecosystem balance are disrupted by residual antibiotics in the environment (Gaze et al., 2013). Antibiotic-contaminated aquatic environments may include resistance genes that proliferate across microbial communities, resulting in long-term AMR reservoirs (Berendonk et al., 2015). Because ecosystems are interrelated, environmental resistance can readily affect the health of humans and animals, which emphasises the necessity for integrated treatments.

4.8.3 Cultural and social aspects of antimicrobial resistance

Antibiotic usage is significantly shaped by social norms and cultural values, influencing overuse, abuse, and misuse behaviour (Melaku and Assegid, 2025). As a result of these, microorganisms have evolved to withstand therapies that once suppressed them. This increases the risk of infection and makes it more difficult to cure (Ferraz, 2024; Uddin et al., 2021). Bacteria lose their susceptibility to antibiotics not only due to overuse and abuse but also due to the use of antibiotics in agriculture, a lack of new antibiotics, travel abroad, unhygienic conditions that promote bacterial growth, and HGT mechanisms. Furthermore, the development of resistant infections is made worse by insufficient infection control practices and poor sanitation in communities and healthcare facilities (Ferraz, 2024; Ventola, 2015).

Many people hold cultural beliefs about antibiotics, such as the belief that they are essential for future medical needs and the practice of self-medication (Imanpour and McMaughan, 2024). Antibiotic usage can be decreased in some cultures by relying on traditional medicine, highlighting the need for culturally appropriate healthcare options (González-Villoria et al., 2024). The development and transmission of resistant diseases are impacted by these behaviours. Poverty, societal conditions, and cultural customs further encourage people to purchase drugs from unauthorised drug outlets, self-medicate for common diseases, visit traditional healers, and ask the neighbourhood for medications (Huttner et al., 2019). Traditional practitioners' medications are frequently unidentified pharmacological compounds combined with antimicrobials in inadequate dosages, which also encourage AMR and prolong immediate treatment at allopathic medical facilities. Patients frequently seek treatment regardless of the type of infection and skip required examinations during consultations because they are motivated by a desire to recover quickly and at low cost (Carmeli et al., 2016).

The absence of infrastructure, driven by weak economies, corruption, and inadequate competence in many low- and middle-income nations, has resulted in a high burden of infectious diseases. Due to insufficient attention to preventive measures, including water, sanitation, and hygiene, it is extremely difficult to provide universal access to basic healthcare services for large populations, especially in rural regions (Abalkhail and Alslamah, 2022; Maillard et al., 2020; Melaku and Assegid, 2025). Antimicrobials are frequently misused due to inadequate regulatory oversight (Pokharel et al., 2019). The availability of antimicrobials has been mistakenly linked to universal health coverage in rural and underresourced areas of many low and middle-income nations, where access to qualified healthcare providers is severely limited (Willis and Chandler, 2019). People buy over-the-counter antibiotics for mild to moderate ailments, many of which are viral infections that resolve on their own (Huttner et al., 2019). Furthermore, inadequate medication surveillance and medication regulation expose populations to inferior and counterfeit drugs. Implementing a strict legislation to restrict the unregulated over-the-counter dispensing of antimicrobials is challenging due to insufficient administration intermingled with the various socioeconomic, cultural, and behavioural factors that drive healthcare seeking (Pokharel et al., 2019).

4.9 Regulatory and policy landscape

4.9.1 Policy and regulation

Global, ongoing governmental policies addressing the development of AMR will be needed to ensure achievement of numerous important SDGs concerning health, responsible use of products, and international cooperation (Alhassan et al., 2025). The most effective methods for the government to prevent the misuse of antibiotics will include enforcement of prescriptions, prohibition of antibiotic use for animal growth promotion, and establishment of programmes to monitor both antibiotic use and levels of drug resistance (WHO, 2022a,b).

Although Denmark and the Netherlands have demonstrated that a significant reduction in antibiotic use in agriculture can occur without a decrease in farm productivity, replicating these results in other geographic locations is difficult. The primary reasons for this difficulty are cost-related issues, the need for greater technical capabilities, and reliance on antibiotics as a low-cost method for managing disease (Lhermie et al., 2022). Therefore increasing collaboration among policymakers and developing institutional capacity in resource-constrained environments are two key priorities for achieving such successes (FAO, 2024; Rahman et al., 2022).

As researchers become increasingly interested in alternative uses of antibiotics, such as natural antimicrobials, several studies indicate that these may also be useful in poultry and aquaculture (Callaway et al., 2021; Vladkova et al., 2024). However, there has been slow progress toward the commercialisation of these alternatives primarily due to inadequate financing and complex regulatory processes (FAO-OIE-WHO, 2021). Greater financial investments will be required to increase production of these products, ensure quality control, and establish long-term efficacy in practical applications (Franklin et al., 2024).

Technology has vast potential to make a positive impact. In terms of how antibiotics are used in medicine, artificial intelligence (AI) and other technologies can assist with rapid diagnosis of illnesses and avoid overprescribing antibiotics, which would result in better outcomes in patients, particularly those in resource-poor countries (WHO, 2021; Franklin et al., 2024). Veterinary telemedicine may also support responsible antibiotic use through remote consultations and the ability to rapidly track the development of disease outbreaks (Sihombing et al., 2023).

Resilient financial arrangements will be necessary for long-term efforts against AMR. In lower-income areas, collaborative platforms like the Antimicrobial Resistance Multi-Partner Trust Fund support important research, build capacity, and implement policy. To close the financial gap hindering the adoption of complete AMR solutions, it is vital to expand these methods. Ultimately, removing institutional barriers, promoting cross-sector cooperation, and ensuring fair resource distribution are essential to an effective global approach (Noyes et al., 2021; Ponzo et al., 2024).

4.9.2 Global antimicrobial resistance action plans

The global fight against AMR underwent an ocean of change in 2015 when the World Health Assembly approved the Global Action Plan (GAP-AMR) (WHO, 2015). This framework provides countries with a model for developing their own National Action Plans, with a focus on infection prevention, awareness, surveillance, and optimising medicine utilisation (WHO, 2019).

To track advancements, the WHO, the FAO, and the World Organisation for Animal Health (OIE) developed a Monitoring and Evaluation framework (WHO, 2019, Tripartite, 2024). When the United Nations Environment Programme (UNEP) joined this coalition in 2022, it experienced a significant transformation and became a 'Quadripartite' endeavour (Quadripartite, 2022). This modification promotes a unified 'One Health' approach across human, animal, and ecosystem health and formally recognises the crucial role environmental contamination plays in the propagation of AMR (WHO, 2022a,b).

The WHO's first global guideline on solid waste management and wastewater treatment with reference to the manufacture of antibiotics was the most significant event of 2024. To help fill a regulatory gap in the control of antibiotic pollution, this advice provides predicted no-effect concentration (PNEC)-based environmental/ecosystem and human health targets, covering all phases of active pharmaceutical ingredient manufacture to packaging (WHO, 2024).

Despite this progress, implementation remains inconsistent, particularly in LMICs. While most nations have developed action plans, translating them into enforced laws, robust surveillance, and sustained funding continues to be a major challenge, underscoring the gap between global strategy and local reality (Quadripartite Joint Secretariat, 2024; The World Bank, 2023).

4.9.3 National antimicrobial resistance action plans

Most high-income countries and some LMICs have developed NAPs, although their scope and implementation vary widely (van Heuvel et al., 2022). Six strategic targets, including awareness, surveillance, infection prevention, and optimised antibiotic use, were outlined in India's First National Action Plan on AMR (2017–21) (GOI, 2017). A national programme for laboratory surveillance, stewardship, and infection control has been the focus of activities coordinated by the National Centre for Disease Control (NCDC) (Mutua et al., 2020). Promoting State Action Plans to localise interventions was a crucial move, although state-by-state progress has been inconsistent. The National Medical Commission's 2024 release of stewardship modules, which aim to convert NAP objectives to the practitioner level, is one recent development. But there are still regulatory gaps, especially regarding the control of antibiotic residue emissions from pharmaceutical production into the environment (Barik et al., 2023).

In addition to encouraging increased cooperation among member states, the European Union's first AMR Action Plan (2011–16) placed a strong emphasis on responsible antibiotic use, surveillance, and research (European Commission, 2011). This was followed by the creation of the EU One Health Action Plan in 2017, which was founded on three pillars: supporting worldwide AMR action, promoting research and innovation, and making the EU a best-practice region (European Commission, 2017). In 2023, the EU Council reiterated these commitments and specifically included environmental monitoring and emission management in the AMR strategy (Council of the European Union, 2023). The Farm to Fork Strategy also advocates a 50% reduction in antibiotic use in animals by 2030 to link AMR policy to broader sustainability goals (European Commission, 2020).

Building on a national plan from 2014, the United States' National Action Plan for Combating Antibiotic-Resistant Bacteria (CARB) was introduced in 2015. Its first five goals were to speed up R&D, improve surveillance, and decrease resistant infections (The White House, 2014). These fundamental aims were kept in the revised CARB Plan for 2020–25, which improved its objectives in light of experience by emphasising environmental aspects, stewardship, and measurable results (The White House, 2020). Scaling up environmental monitoring and the general implementation of quick diagnostics continue to be obstacles, despite notable advancements in areas such as hospital stewardship and national surveillance (CDC, 2019).

4.9.4 Regulations on antibiotic manufacturing and environmental safety

The production of antibiotics is now recognised as a key environmental driver of AMR, and pharmaceutical effluents (with high levels of APIs) can be used to select for resistant bacteria and resistance genes. Nevertheless, most countries do not have any enforceable antibiotic-specific discharge standards (Kotwani et al., 2021). In 2024, the WHO published the initial global technical guidance on wastewater and solid waste management for the manufacture of antibiotics. This report establishes PNECs for antibiotics to reduce the selection of resistance, suggests risk-based strategies for dealing with effluents, and provides gradual implementation strategies. The advice focuses on the integration of human health and ecological endpoints and suggests transparency, audit systems, and verification by a third party (WHO, 2024). To supplement the WHO advice, AMR Industry Alliance, with the help of British Standards Institution (BSI), has established a voluntary standard of manufacture of antibiotics, the Antibiotic Manufacturing Standard. This criterion compels companies to implement wastewater treatment technologies that can achieve concentrations below the boundaries of resistance-selecting concentrations and to develop certification plans to do so. Even though it is voluntary, it indicates growing pressure on drug manufacturing corporations worldwide to act in a way that shows environmental responsibility (AMR Industry Alliance, 2022). To consider pharmaceuticals, especially antibiotics, as emerging contaminants, the European Union is updating its Water Framework Directive and associated environmental quality laws. The new Watch List mechanism assists in tracking priority antibiotics in surface waters, and continuous revisions are expected to incorporate effect-oriented approaches to measure ecological and AMR-related risks (Gomez Cortes et al., 2025). Although the National Action Plan on AMR (2017–21) in India pointed to the necessity of containing the pharmaceutical pollution, there are still no enforceable limits of antibiotic residues in effluents. In a recent evaluation of Indian pharmaceutical units, it was found that most claimed to have effluent treatment plants (ETPs), but monitoring was insufficient, and most smaller units lacked complex treatment systems such as zero liquid discharge (ZLD). This is a loophole that has been caused by poor regulatory watchdogs and enforcement in the majority of LMICs (Narain et al., 2022; Hibbard et al.,2024).

4.10 Strategies to mitigate antibiotic misuse and overuse

According to the WHO, AMR is one of the top 10 global public health concerns the world is currently facing (WHO, 2021). The misuse and overuse of antibiotics in human health, animal agriculture, and the environment at large is the main cause of this rapidly worsening epidemic. Reducing this hazard is a multifaceted, ongoing process that calls for a concerted 'One Health' strategy that integrates environmental, animal, and human health.

4.10.1 Antibiotic stewardship and management across human health, veterinary, and agricultural sectors

Antimicrobial stewardship (AMS) is a collective term that refers to the utilisation of antimicrobials in a way that maintains their effectiveness while ensuring their continued accessibility to individuals who seek them. It integrates two distinct concepts of the ideal use of antimicrobials: conservation, which involves selecting when to use and when not to use antimicrobials, and optimised usage, which includes considerations for how to utilise antimicrobials properly. AMS thereby signifies an exhaustive and contextualised strategy for AMU, taking into account not solely the quantity of antimicrobials employed, but also the various methods by which and purposes for which a variety of stakeholders (health care providers, patients, and other individuals) use antimicrobials, as well as the circumstances in which these selections are made. The notion of AMS is becoming an integral component of global initiatives to manage AMR, and various locally targeted stewardship ventures initiated and carried out during the last 20 years have attempted to mitigate or augment AMU in both the public health and animal welfare sectors. The first explicit reference to 'antimicrobial stewardship' dates back to the 1990s, in recognition of increasing alarm over rapidly developing antibiotic resistance in hospitals; however, actions that ought to be acknowledged as AMS programmes today have already been put to use since the 1970s. In the course of the 1990s, AMS emerged as a term that applied to programmes or interventions intended to mitigate AMU and hence the selection burden for AMR in healthcare facilities. AMU's growing recognition in community-based and animal health contexts has broadened its applications. The importance of AMS in veterinary care became evident in the wider context of human health in the 1990s, with recommendations on AMU in animals, as well as principles of AMS for animal health, emerging subsequently (Dyar et al., 2017).

4.10.2 Antimicrobial stewardship programs

Institutional initiatives called antimicrobial stewardship programmes (ASPs) aim to maximise the usage of antibiotics. According to Dyar et al. (2017), their goal is to ensure that patients receive the appropriate antibiotic at the appropriate dose for the appropriate duration. Research demonstrates that ASPs can reduce antibiotic use by up to 30% without compromising patient outcomes. Diagnostic tests, formulary restrictions, prescriber feedback, and healthcare professional education are important elements.

International organisations have created classification systems that rank antimicrobials according to their significance for human medicine to improve global AMS. The regulatory decisions and veterinary use of critically important medications are limited by the European Medicines Agency's (EMA) AMEG A-B-C-D classification and the WHO's Critically Important Antimicrobials (CIA) list. Notably, to maintain their effectiveness for use in human healthcare, third- and fourth-generation cephalosporins, fluoroquinolones, and colistin are restricted by stringent veterinary usage regulations (EFSA, 2019; McCullough et al., 2016).

4.10.3 Public education and awareness

Antibiotic resistance is still not well understood by the general public. By outlining when antibiotics are suitable and stressing adherence to recommended regimens, educational efforts can dramatically minimise overuse (Cartelet, 2021). Social media campaigns and school-based programmes have been successful in altering behaviour, especially in younger demographics. Proper regulations and legislation should be formulated for the manufacture and over-the-counter sale of antibiotics, including monitoring antibiotic use in both the human and animal sectors. To guarantee compliance with international best practices, veterinary professionals are essential in teaching farmers, monitoring antibiotic usage, and implementing AMS policies (Czaplewski et al., 2016; Jensen, 1998).

4.10.4 Research and innovation

Innovation is essential to outpacing the evolution of germs. The development of new antibiotics has been significantly delayed by financial and scientific challenges, necessitating alternative approaches. These include treatments that change the microbiome, antimicrobial peptides, and bacteriophage therapy (Jensen, 1998). Purchasing rapid diagnostic equipment can also help doctors distinguish between bacterial and viral illnesses, reducing the need for unnecessary prescription drugs.

To effectively combat AMR, veterinary antibiotic stewardship programmes must include sustainable alternatives. Research indicates that nonantibiotic treatments such as propolis, essential oils, plant extracts, antibacterial peptides, and medium-chain fatty acids exhibit antimicrobial activity and may reduce reliance on traditional antibiotics (Alhassan et al., 2025). To reduce infection risks, nations with effective AMR-reduction programmes have already begun implementing these substitutes alongside improved husbandry techniques, probiotics, and immunisation protocols (McCullough et al., 2016; WHO, 2022a,b).

Natural product-based treatments have emerged as potential alternatives to traditional antibiotics in recent developments in veterinary care. According to recent studies, probiotics, essential oils, bacteriophages, and plant-based antimicrobials all have broad-spectrum antimicrobial efficacy while lowering the competition for resistance (Callaway et al., 2021; Vladkova et al., 2024). Target specificity, minimal alteration of the gut microbiota, and a lower likelihood of developing resistance are just a few benefits of these alternative therapies. Antibacterial peptides, flavonoids, and polyphenols are examples of plant-derived substances that have shown strong antimicrobial activity against harmful bacteria while boosting host protection (WHO, 2017; Vladkova et al., 2024). Propolis and medium-chain fatty acids, which are frequently used in animal fodder, have been found to promote gut health, minimise bacterial infections, and reduce dependence on existing antibiotics (Smits et al., 2021; FAO, 2024). Furthermore, bacteriophage therapy, which utilises the administration of viruses that target pathogenic bacteria, offers a highly specialised, self-replicating substitute to antibiotics, with potential uses in poultry, aquaculture, and cattle management (Callaway et al., 2021). Countries that have used natural product-based initiatives have reported favourable outcomes in terms of lowering AMU. In France, replacing in-feed antibiotics with essential oils and organic acids resulted in a substantial decrease in colistin-resistant *Escherichia coli* in poultry production facilities (EMA and EFSA, 2017). Germany and Switzerland have investigated phytogenic fodder supplements, such as garlic and oregano extracts, to reduce pathogenic bacteria while promoting animal growth. In China, probiotic supplementation in swine production effectively reduced the use of tetracyclines and macrolides, thereby restricting the transmission of resistance genes (Ahmad et al., 2023).

4.10.5 Surveillance system: progress and persistent gaps

In recent years, due to prolonged antibiotic use, AMR has emerged among pathogens. AMR is acknowledged as a significant public health concern of the 21st century, threatening the efficient prevention and treatment of an expanding range of diseases brought on by different pathogenic microflora that are no longer treatable with conventional drugs. AMR's progression is a normal occurrence. It was first reported in the 1940s, soon after the discovery of penicillin. However, in recent decades, human influence has significantly accelerated and intensified its evolution (Martínez and Baquero, 2014).

Genetic alterations that enable bacteria to withstand exposure to antimicrobial agents are the main cause of antibiotic resistance. HGT and mutation are the two primary methods. Drug efficacy may be decreased by spontaneous changes in bacterial DNA that change target locations or metabolic pathways. However, through conjugation, transformation, or transduction, HGT allows bacteria to obtain resistance genes from other microorganisms (Tesema and Birhanu, 2024). Various processes enable multidrug resistance, often combined to make bacteria resistant to a combination of antibiotic classes. The rapid adaptation of microorganisms demonstrates the necessity of judicious antibiotic use.

4.10.6 Barriers to effective antibiotic surveillance

AMR is still typically triggered by misuse of antibiotics in the production of animal food, especially in LMICs with less established veterinary services and weaker regulatory frameworks (ElSayed et al., 2025; Toghroli et al., 2024). Global efforts to enhance regulatory oversight and increase public knowledge of the risks associated with AMR face significant challenges in regions where financial and technological limitations make it difficult to implement suitable regulations (Rahman et al., 2022; WHO, 2021; FAO, 2024).

A serious flaw in surveillance and stewardship initiatives is the absence of efficient diagnostic techniques in LMICs, which impedes the prudent use of antibiotics (Bandyopadhyay and Samanta, 2020). Inadequate diagnostic skills often result in empirical treatment, which raises the possibility of antibiotic abuse (Sihombing et al., 2023; WHO, 2013). The economic constraints and limited availability of veterinary care resources also limit access to diagnostics, leading patients, veterinarians, and livestock producers to rely on nonevidence-based methods (Martínez and Baquero, 2014; Ponzo et al., 2024). Limitations also exist with data collection and access to data within surveillance systems that track antibiotic use and resistance patterns (FAO-OIE-WHO, 2021; Velazquez-Meza et al., 2022). Investments in cost-effective, high-efficiency diagnostic technologies and the establishment of national and regional-level surveillance systems would help bridge this gap (Singer et al., 2016; Samreen et al., 2021). To improve AMR surveillance, it is important to develop surveillance programmes in healthcare and agriculture to track the movement of resistance genes through soil and water systems (WHO, 2017). Also, implementing environmentally responsible practices such as establishing robust wastewater treatment plants, developing regulations for manure disposal, and utilising an ecosystem-wide approach to track AMR will greatly contribute to protecting environmental health and limiting the spread of resistant bacteria (Ponzo et al., 2024; Robinson et al., 2016).

4.10.7 Global collaboration and the 'One Health' approach

'One Health' illustrates that human health is linked to the health of animals and the environment; as a result of this relationship, global coordination is required among medical, veterinary, agricultural, and environmental professionals to address AMR effectively (FAO, 2024), and the GLASS programme provides data for such coordination to occur. Ultimately, coordinated action is needed to stop the spread of drug-resistant microorganisms.

Because AMR has drivers and potential effects on the well-being of humans, animals, the health of plants, and the environment, it has increasingly been addressed through a One Health approach. A One Health strategy recognises that the wellness of humans, wild and domesticated animals, plants, and the entire ecosystem is closely linked and interdependent, and significantly includes stewardship across all fields as one of its main underlying concepts (Hibbard et al., 2024). International programmes like the FAO Action Plan on Antimicrobial Resistance and the WHO Global Action Plan on AMR highlight resource equity, cooperative innovation, and coordinated regulations as essential to long-term AMR mitigation (Robinson et al., 2016; Cameron et al., 2025). It is essential to incorporate these strategies within the One Health paradigm, especially since environmental contamination and transmission of zoonotic infections are still significant drivers of resistance (Dyar et al., 2017).

4.11 Challenges and gaps

Although there is increased awareness of the environmental aspects of AMR, there are still significant problems in translating global guidance into practical action. To start with, regulatory loopholes are notable: although the WHO has recently suggested PNECs for antibiotic residues as part of its technical guidance on wastewater management in antibiotic production, few countries have established legally binding discharge standards, and the practice remains mostly voluntary (WHO, 2015a,b). Second, there is still an uneven and fragmented coverage of surveillance. The increasing incorporation of AMR surveillance in HICs across human, animal, and environmental spheres is straining the limited capacity of LMICs, where the bulk of antibiotic production is concentrated. Poor infrastructure and incomplete laboratory chains, as well as insufficient investment, imply that the ETPs are frequently not functioning at their full capacity, and the systematic checking of the antibiotic remainder, the bacteria exhibiting resistance, or the resistance genes is uncommon (Mudenda et al., 2023; Suleman et al., 2025). Lack of alignment between the good manufacturing practice (GMP) systems and the environmental safety standards. Product safety is guaranteed by GMP, which does not target effluent discharges, whereas most national environmental regulations target chemical pollutants in general and do not account for AMR selection pressure from antibiotic residues (Pruden et al., 2013). Weak enforcement capacity is another long-standing barrier. If regulatory frameworks are in place, the inspection mechanisms are not well-intended, and the fines for non-compliance are minimal. The lack of full implementation is also attributed to smaller pharmaceutical producers, especially in LMICs, citing the economic cost of implementing more advanced treatment systems such as ZLD (Hejna et al., 2022). Lastly, there are still significant evidence gaps. In Europe and North America, there is better documentation of environmental levels of antibiotics, ARGs, and ARBs than in Asia and Africa. Further, ecological resistance selection thresholds remain in development, and longitudinal research relating manufacturing discharges to clinical AMR effects is limited (Bengtsson-Palme and Larsson, 2016; Vestel et al., 2022).

4.12 Future directions

The fact that antibiotic residues, ARB, and ARGs persist in hospital and industrial effluents underscores the need for a multisectoral intervention addressing the environmental aspect of AMR. The latest modifications to the Standard of the AMR Industry Alliance (2025) of the Antibiotic Manufacturing reflect the trend toward more stringent discharge levels and more advanced wastewater treatment conditions, although globally, the implementation is not even, especially in those countries where the production capacities are high (Aslam and Aljasir, 2025). Outside the production process, new data indicate that industrialisation and climate change are increasing resistance by establishing ecological hotspots, where high temperatures and uncontrolled discharges enhance ARG survival and transmission (Oyelayo et al., 2025; Bhatia, 2024). This is a major concern in areas such as South Asia, where the lack of well-developed wastewater systems is combined with intensive pharmaceutical production (Bhatia, 2024). At the technological scale, ozonation, activated carbon, and anaerobic membrane bioreactors have been proposed as effective for treating ARB and ARGs in effluents from the healthcare and livestock industries. Nevertheless, their scalability is constrained by economic realities, and incentive-based models and global financial mechanisms to support LMICs are necessary (Pham et al., 2024). The integration of surveillance remains a significant gap. Even though the role of the environment is recognised in the majority of national action plans, the systematic monitoring of antibiotic residues, ARB, and ARGs remains weak, particularly in LMICs. By including environmental indicators in AMR National Action Plans, early alerts on resistance hotspots would be established, and the effectiveness of regulatory interventions would be benchmarked.

4.13 Conclusion

Global health faces an unprecedented threat from AMR. For decades, the unrelenting misuse and overuse of antibiotics in human medicine, animal agriculture, and the broader environment has contributed to this developing silent epidemic. The issue, which was once confined to hospitals, has grown into a much larger and far more complicated problem with insufficient regulation and enforcement, no limits on the amount of pharmaceutical waste that can be released into the environment, and excessive use of antibiotics throughout all sectors of society.

A cohesive 'One Health' strategy is necessary to fight AMR. However, due to limited resources and surveillance, low- and middle-income nations face the greatest obstacles to the consistent implementation of global efforts. Environmental contamination from hospital and industrial waste is a serious yet neglected problem that requires more stringent regulations and sophisticated remediation. Three commitments are essential going forward: shared accountability, creativity, and cooperation. These requirements include creating novel treatments and diagnostics, utilising technology such as AI, and guaranteeing fair access through strict laws and international cooperation.

References

Abalkhail, A., Alslamah, T., 2022. Institutional factors associated with infection prevention and control practices globally during the infectious pandemics in resource-limited settings. Vaccines 10 (11), 1811. Available at https://doi.org/10.3390/vaccines10111811.

Abosse, J.S., Megersa, B., Zewge, F., Eregno, F.E., 2024. Healthcare waste management and antimicrobial resistance: a critical review. Journal of Water and Health 22 (11), 2076–2093. Available at https://doi.org/10.2166/wh.2024.232, https://iwaponline.com/jwh/article-pdf/22/11/2076/1510465/jwh2024232.pdf.

Adenaya, A., Berger, M., Brinkhoff, T., Ribas-Ribas, M., Wurl, O., 2023. Usage of antibiotics in aquaculture and the impact on coastal waters. Marine Pollution Bulletin 188, 114645. Available at https://doi.org/10.1016/j.marpolbul.2023.114645.

Ahmad, N., Joji, R.M., Shahid, M., 2023. Evolution and implementation of One Health to control the dissemination of antibiotic-resistant bacteria and resistance genes: a review. Frontiers in Cellular and Infection Microbiology 12. Available at https://doi.org/10.3389/fcimb.2022.1065796, http://www.frontiersin.org/Cellular_and_Infection_Microbiology/archive.

Alhassan, M.Y., Kabara, M.K., Ahmad, A.A., Abdulsalam, J., Habib, H.I., 2025. Revisiting antibiotic stewardship: veterinary contributions to combating antimicrobial resistance globally. Bulletin of the National Research Centre 49 (1). Available at https://doi.org/10.1186/s42269-025-01317-3.

AMR Industry Alliance, 2022. Antibiotic manufacturing standard.

Anjum, M.F., Schmitt, H., Börjesson, S., Berendonk, T.U., Donner, E., Stehling, E.G., et al., 2021. The potential of using *E. coli* as an indicator for the surveillance of antimicrobial resistance (AMR) in the environment. Current Opinion in Microbiology 64, 152–158. Available at https://doi.org/10.1016/j.mib.2021.09.011, http://www.elsevier.com/locate/mib.

Ardakani, Z., Canali, M., Aragrande, M., Tomassone, L., Simoes, M., Balzani, A., et al., 2023. Evaluating the contribution of antimicrobial use in farmed animals to global antimicrobial resistance in humans. One Health 17. Available at https://doi.org/10.1016/j.onehlt.2023.100647, http://www.journals.elsevier.com/one-health/.

Ashwini, A., Jamwal, P., Vanak, A.T., 2023. Environmental surveillance of antimicrobial resistance in a rapidly developing catchment. Environmental Monitoring and Assessment 195 (1). Available at https://doi.org/10.1007/s10661-022-10630-7, https://www.springer.com/journal/10661.

Aslam, B., Aljasir, S.F., 2025. Climate change and AMR: interconnected threats and one health solutions. Antibiotics 14 (9). Available at https://doi.org/10.3390/antibiotics14090946, http://www.mdpi.com/journal/antibiotics.

Auta, A., Hadi, M.A., Oga, E., Adewuyi, E.O., Abdu-Aguye, S.N., Adeloye, D., et al., 2023. Global access to non-prescription antibiotics: a systematic review and meta-analysis. BMJ Global Health 8 (2), e010464 (2023).

Azuma, T., Katagiri, M., Sekizuka, T., Kuroda, M., Watanabe, M., 2022. Inactivation of bacteria and residual antimicrobials in hospital wastewater by ozone treatment. Antibiotics 11 (7), 862. Available at https://doi.org/10.3390/antibiotics11070862.

Bandyopadhyay, S., Samanta, I., 2020. Antimicrobial resistance in the agri-food chain and companion animals as a re-emerging menace in the post-COVID epoch. Frontiers in Veterinary Science 7, 620 2020.

Bansal, O.P., 2019. Antibiotics in hospital effluents and their impact on the antibiotic-resistant bacteria and remediation of the antibiotics: a review. Network Pharmacology 4 (3-4), 6–30 (2019).

Barlam, T.F., Cosgrove, S.E., Abbo, L.M., Macdougall, C., Schuetz, A.N., Septimus, E.J., et al., 2016. Implementing an antibiotic stewardship program: guidelines by the infectious diseases society of America and the society for healthcare epidemiology of America. Clinical Infectious Diseases 62 (10), e51. Available at https://doi.org/10.1093/cid/ciw118, http://cid.oxfordjournals.org/content/by/year.

Barik, S.R., Pradhan, H.S., Prakash, J, Satpathy, S.K., 2023. A Comparative Analysis of National and State Level Action Plans on Antimicrobial Resistance in India. Journal of Multidisciplinary Research in Healthcare 9, 37–50. Available at https://doi.org/10.15415/jmrh.2023.92004.

Bengtsson-Palme, J., Larsson, D.G.J., 2016. Concentrations of antibiotics predicted to select for resistant bacteria: proposed limits for environmental regulation. Environment International 86, 140–149. Available at https://doi.org/10.1016/j.envint.2015.10.015, www.elsevier.com/locate/envint.

Berendonk, T.U., Manaia, C.M., Merlin, C., Fatta-Kassinos, D., Cytryn, E., Walsh, F., et al., 2015. Tackling antibiotic resistance: the environmental framework. Nature Reviews Microbiology 13 (5), 310–317. Available at https://doi.org/10.1038/nrmicro3439, http://www.nature.com/nrmicro/index.html.

Bhatia, R., 2024. Environmental aspects of antimicrobial resistance in India: current progress & way forward. Indian Journal of Medical Research 159 (1), 10–15. Available at https://doi.org/10.4103/ijmr.ijmr_2104_23.

Byrne, M.K., Miellet, S., McGlinn, A., Fish, J., Meedya, S., Reynolds, N., et al., 2019. The drivers of antibiotic use and misuse: the development and investigation of a theory driven community measure. BMC Public Health 19 (1). Available at https://doi.org/10.1186/s12889-019-7796-8, http://www.biomedcentral.com/bmcpublichealth.

Cabello, F.C., Godfrey, H.P., Buschmann, A.H., Dölz, H.J., 2016. Aquaculture as yet another environmental gateway to the development and globalisation of antimicrobial resistance. The Lancet Infectious Diseases 16 (7), e127. Available at https://doi.org/10.1016/S1473-3099(16)00100-6, http://www.journals.elsevier.com/the-lancet-infectious-diseases.

Callaway, T.R., Lillehoj, H., Chuanchuen, R., Gay, C.G., 2021. Alternatives to antibiotics: a symposium on the challenges and solutions for animal health and production. Antibiotics 10 (5). Available at https://doi.org/10.3390/antibiotics10050471, https://www.mdpi.com/2079-6382/10/5/471/pdf.

Cameron, A., Connolly, J., Esiovwa, R., Henriquez, F.L., Hursthouse, A., Mukherji, S., et al., 2025. Mind the gaps': stakeholder perspectives on addressing antimicrobial resistance in the environment in the Indian context. Global Health Action 18 (1). Available at https://doi.org/10.1080/16549716.2025.2491200, https://www.tandfonline.com/loi/zgha20.

Carmeli, Y., Armstrong, J., Laud, P.J., Newell, P., Stone, G., Wardman, A., et al., 2016. Ceftazidime-avibactam or best available therapy in patients with ceftazidime-resistant Enterobacteriaceae and *Pseudomonas aeruginosa* complicated urinary tract infections or complicated intra-abdominal infections (REPRISE): a randomised, pathogen-directed, phase 3 study. The Lancet Infectious Diseases 16 (6), 661–673. Available at https://doi.org/10.1016/S1473-3099(16)30004-4, http://www.journals.elsevier.com/the-lancet-infectious-diseases.

Cartelet, C., 2021. Exploring the ethical tensions surrounding antimicrobial stewardship in UK companion animal veterinary practice [Doctoral dissertation, University of Nottingham]. University of Nottingham Repository.

Centres for Disease Control and Prevention, 2019. Antibiotic resistance threats in the United States. U.S. Department of Health and Human Services.

Centres for Disease Control and Prevention, 2023. Antibiotic resistance threats in the United States. U.S. Department of Health and Human Services.

Chambers, H.F., DeLeo, F.R., 2009. Waves of resistance: *Staphylococcus aureus* in the antibiotic era. Nature Reviews. Microbiology 7 (9), 629–641. Available at https://doi.org/10.1038/nrmicro2200.

Chau, K.K., Barker, L., Budgell, E.P., Vihta, K.D., Sims, N., Kasprzyk-Hordern, B., et al., 2022. Systematic review of wastewater surveillance of antimicrobial resistance in human populations. Environment International 162, 107171. Available at https://doi.org/10.1016/j.envint.2022.107171.

Chu, B.T.T., Petrovich, M.L., Chaudhary, A., Wright, D., Murphy, B., Wells, G., et al., 2018. Metagenomics reveals the impact of wastewater treatment plants on the dispersal of microorganisms and genes in aquatic sediments. Applied and Environmental Microbiology 84 (5). Available at https://doi.org/10.1128/AEM.02168-17, http://aem.asm.org/content/84/5/e02168-17.full.pdf.

Council of the European Union, 2023. Recommendation on stepping up EU action to combat antimicrobial resistance in a One Health approach.

Czaplewski, L., Bax, R., Clokie, M., Dawson, M., Fairhead, H., Fischetti, V.A., et al., 2016. Alternatives to antibiotics-a pipeline portfolio review. The Lancet Infectious Diseases 16 (2), 239–251. Available at https://doi.org/10.1016/S1473-3099(15)00466-1, http://www.journals.elsevier.com/the-lancet-infectious-diseases.

Davies, J., Davies, D., 2010. Origins and evolution of antibiotic resistance. Microbiology and Molecular Biology Reviews 74 (3), 417–433. Available at https://doi.org/10.1128/MMBR.00016-10Canada, http://mmbr.asm.org/cgi/reprint/74/3/417.

Deedat, F.Z., Faya, A.M., Gumbi, B.P., Johnston, D.M.G., Karpoormath, R., Essack, S.Y., 2025. The association between antibiotic use in hospitals and residual antibiotic concentrations in hospital effluents: a pilot study. JAC-Antimicrobial Resistance 7 (3). Available at https://doi.org/10.1093/jacamr/dlaf070, https://academic.oup.com/jacamr.

Deng, Y., Tan, A., Zhao, F., Wang, F., Gong, H., Lai, Y., et al., 2025. Global distribution of antimicrobial resistance genes in aquaculture. One Health Advances 3 (1). Available at https://doi.org/10.1186/s44280-025-00071-5.

Desbois, A.P., 2025. Aquaculture requires special consideration in national antibiotic use strategies. Science of the Total Environment 845, 157115 (2025).

Diwan, V., Tamhankar, A.J., Khandal, R.K., Sen, S., Aggarwal, M., Marothi, Y., et al., 2010. Antibiotics and antibiotic-resistant bacteria in waters associated with a hospital in Ujjain, India. BMC Public Health 10. Available at https://doi.org/10.1186/1471-2458-10-414.

Done, H.Y., Venkatesan, A.K., Halden, R.U., 2015. Does the recent growth of aquaculture create antibiotic resistance threats different from those associated with land animal production in agriculture? AAPS Journal 17 (3), 513–524. Available at https://doi.org/10.1208/s12248-015-9722-z, http://www.springerlink.com/content/1550-7416/.

Dyar, O.J., Huttner, B., Schouten, J., Pulcini, C., 2017. What is antimicrobial stewardship? Clinical Microbiology and Infection 23 (11), 793–798. Available at https://doi.org/10.1016/j.cmi.2017.08.026.

EFSA, 2019. Technical specifications on harmonised monitoring of antimicrobial resistance in zoonotic and indicator bacteria from food-producing animals and food. EFSA Journal 17 (6), e05709.

ElSayed, N., Amine, A., El-Attar, L., Amin, M.E.K., 2025. Antimicrobial use in animal farms in Egypt: rates, patterns, and determinants. Journal of the Egyptian Public Health Association 100 (1). Available at https://doi.org/10.1186/s42506-024-00180-w, https://www.springer.com/journal/42506.

EMA and EFSA, 2017. Joint scientific opinion on measures to reduce the need to use antimicrobial agents in animal husbandry in the European Union, and the resulting impacts on food safety (RONAFA). EFSA Journal 15 (1) 2017.

Enshaie, A., Nigam, A., Patel, P., Rai, A., 2025. Livestock antibiotics use and antimicrobial resistance. Journal of Global Health and Veterinary Medicine 14 (2), 55–68 (2025).

European Commission, 2011. Action plan against the rising threats from antimicrobial resistance 2011–2016.

European Commission, 2017. A European One Health Action Plan against antimicrobial resistance.

European Commission, 2020. Farm to fork strategy: for a fair, healthy and environmentally friendly food system.

FAO, 2024. FAO Action Plan on Antimicrobial Resistance 2021–2025.

FAO/OIE/WHO, 2019. Antimicrobial Resistance: A Manual for Developing National Action Plans. Tripartite Collaboration (2019).

FAO-OIE-WHO, 2021. One Health Approach to AMR Mitigation in the Asia-Pacific Region. World Health Organisation (2021).

Ferraz, M.P., 2024. Antimicrobial resistance: the impact on and from society according to the one health approach. Societies 14 (9), 187 (2024).

Franklin, A.M., Weller, D.L., Durso, L.M., McLain, J.E., 2024. A one health approach for monitoring antimicrobial resistance. Frontiers in Water 6, 1359109 (2024).

Gao, Y., Li, X., Zhang, M., Hu, J., 2022. Chronic low-level antibiotic exposure alters gut microbiota and metabolic health. Science of the Total Environment 806, 150646 (2022).

Gaze, W.H., Krone, S.M., Joakim Larsson, D.G., Li, X.Z., Robinson, J.A., Simonet, P., et al., 2013. Influence of humans on evolution and mobilization of environmental antibiotic resistome. Emerging Infectious Diseases 19 (7). Available at https://doi.org/10.3201/eid1907.120871, http://wwwnc.cdc.gov/eid/article/19/7/12-0871_article.htm.

Gomez Cortes, L., Porcel Rodriguez, E., Marinov. D., Sanseverino, I., Lettieri, T., 2025. Selection of substances for the 5th watch list under the water framework directive. PublicatioAl Meslaman., A. Z. (2023). antibiotic resistance in low- and middle-income countries: current practices and its global implications. * Expert Review of Anti-Infective Therapy 21 (10), 1075–1086 (2023).

González-Villoria, A.M., García Quiroz, A.D., Osorio Guzmán, E.U., Suarez-Herrera, J.C., Abeldaño Zuñiga, R.A., 2024. Knowledge, attitudes, and practices of parents in the use of antibiotics: a case study in a Mexican Indigenous community. Healthcare (Switzerland) 12 (3). Available at https://doi.org/10.3390/healthcare12030294, www.mdpi.com/journal/healthcare/.

Government of India, 2017. *National Action Plan on Antimicrobial Resistance (NAP-AMR) 2017–2021* Ministry of Health and Family Welfare (2017).

Guerra-Rodríguez, S., Abeledo-Lameiro, M.J., Polo-López, M.I., Plaza-Bolaños, P., Agüera, A., Rodríguez, E., et al., 2023. Pilot-scale sulfate radical-based advanced oxidation for wastewater reuse: simultaneous disinfection, removal of contaminants of emerging concern, and antibiotic resistance genes. Chemical Engineering Journal 477, 146916. Available at https://doi.org/10.1016/j.cej.2023.146916.

Hart, A., Warren, J., Wilkinson, H., Schmidt, W., 2023. Environmental surveillance of antimicrobial resistance (AMR), perspectives from a national environmental regulator in 2023. Eurosurveillance 28 (11). Available at https://doi.org/10.2807/1560-7917.ES.2023.28.11.2200367, https://www.eurosurveillance.org/docserver/fulltext/eurosurveillance/28/11/eurosurv-28-11-6.pdf?expires=1696654126&id=id&accname=guest&checksum=C4028F75D7637F4F33A3EBFAE2D62D8A.

Hejna, M., Kapuścińska, D., Aksmann, A., 2022. Pharmaceuticals in the aquatic environment: a review on eco-toxicology and the remediation potential of algae. International Journal of Environmental Research and Public Health 19 (13), 7717. Available at https://doi.org/10.3390/ijerph19137717.

Hendriksen, R.S., Munk, P., Njage, P., van Bunnik, B., McNally, L., Lukjancenko, O., et al., 2019. Global monitoring of antimicrobial resistance based on metagenomics analyses of urban sewage. Nature Communications 10 (1). Available at https://doi.org/10.1038/s41467-019-08853-3, http://www.nature.com/ncomms/index.html.

van Heuvel, L., Caini, S., Dückers, M.L.A., Paget, J., 2022. Assessment of the inclusion of vaccination as an intervention to reduce antimicrobial resistance in AMR national action plans: a global review. Globalization and Health 18 (1). Available at https://doi.org/10.1186/s12992-022-00878-6, https://globalizationandhealth.biomedcentral.com/.

Hibbard, R., Mendelson, M., Page, S.W., Ferreira, J.P., Pulcini, C., Paul, M.C., et al., 2024. Antimicrobial stewardship: a definition with a One Health perspective. npj Antimicrobials and Resistance 2 (1). Available at https://doi.org/10.1038/s44259-024-00031-w.

Hounmanou, Y.M.G., Houefonde, A., Nguyen, T.T., Dalsgaard, A., 2025. Mitigating antimicrobial resistance through effective hospital wastewater management in low- and middle-income countries. Frontiers in Public Health 12. Available at https://doi.org/10.3389/fpubh.2024.1525873.

Hultman, J., Tamminen, M., Pärnänen, K., Cairns, J., Karkman, A., Virta, M., 2018. Host range of antibiotic resistance genes in wastewater treatment plant influent and effluent. FEMS Microbiology Ecology 94 (4). Available at https://doi.org/10.1093/FEMSEC/FIY038, http://femsec.oxfordjournals.org/.

Huttner, B., Saam, M., Moja, L., Mah, K., Sprenger, M., Harbarth, S., et al., 2019. How to improve antibiotic awareness campaigns: findings of a WHO global survey. BMJ Global Health 4 (3), e001239. Available at https://doi.org/10.1136/bmjgh-2018-001239.

Imanpour, S., McMaughan, D.J., 2024. You never know when you will need an antibiotic': a qualitative study of structural barriers and cultural assumptions in antibiotic misuse among immigrants in the United States. Journal of Health Services Research and Policy 29 (4), 266–274. Available at https://doi.org/10.1177/13558196241252748, https://journals.sagepub.com/home/HSR.

Jensen, B., 1998. The impact of feed additives on the microbial ecology of the gut in young pigs. Journal of Animal and Feed Sciences 7 (Suppl. 1), 45–64. Available at https://doi.org/10.22358/jafs/69955/1998.

King, L.M., Bartoces, M., Fleming-Dutra, K.E., Roberts, R.M., Hicks, L.A., 2021. Changes in US outpatient antibiotic prescriptions from 2011 to 2016. JAMA Network Open 4 (11), e2137514 (2021).

Kotwani, A., Joshi, J., Kaloni, D., 2021. Pharmaceutical effluent: a critical link in the interconnected ecosystem promoting antimicrobial resistance. Environmental Science and Pollution Research 28 (25), 32111–32124. Available at https://doi.org/10.1007/s11356-021-14178-w, https://link.springer.com/journal/11356.

Kuznetsova, M., Lee, C., 2001. Enhanced extracellular enzymatic peptide hydrolysis in the sea-surface microlayer. Marine Chemistry 73 (3-4), 319–332. Available at https://doi.org/10.1016/S0304-4203(00)00116-X.

Lai, C., Wang, Z., Gu, T., Jian, L., Meng, X., Meng, Q., Gao, D., et al., 2025. Overall Evaluation of Antibiotics Occurrence from Large-Scale Livestock Farms in Sichuan Basin, China: Spatial Distribution, Source Apportionment, and Risk Assessment. Toxics 13, 154. Available at https://doi.org/10.3390/toxics13030154.

Larsson, D.G.J., 2014. Pollution from drug manufacturing: review and perspectives. Philosophical Transactions of the Royal Society B: Biological Sciences 369 (1656), 20130571. Available at https://doi.org/10.1098/rstb.2013.0571.

Larsson, D.G.J., Gaze, W.H., Laxminarayan, R., Topp, E., 2023. AMR, one health and the environment. Nature Microbiology 8 (5), 754–755. Available at https://doi.org/10.1038/s41564-023-01351-9.

Laxminarayan, R., Duse, A., Wattal, C., Zaidi, A.K.M., Wertheim, H.F.L., Sumpradit, N., et al., 2013. Antibiotic resistance-the need for global solutions. The Lancet Infectious Diseases 13 (12), 1057–1098. Available at https://doi.org/10.1016/S1473-3099(13)70318-9.

Laxminarayan, R., Matsoso, P., Pant, S., Brower, C., Røttingen, J.A., Klugman, K., et al., 2016. Access to effective antimicrobials: a worldwide challenge. The Lancet 387 (10014), 168–175. Available at https://doi.org/10.1016/S0140-6736(15)00474-2, http://www.journals.elsevier.com/the-lancet/.

Le, T.H., Ng, C., Tran, N.H., Chen, H., Gin, K.Y.H., 2018. Removal of antibiotic residues, antibiotic resistant bacteria and antibiotic resistance genes in municipal wastewater by membrane bioreactor systems. Water Research 145, 498–508. Available at https://doi.org/10.1016/j.watres.2018.08.060, www.elsevier.com/locate/watres.

Lhermie, G., Ndiaye, Y., Rushton, J., Raboisson, D., 2022. Economic evaluation of antimicrobial use practices in animal agriculture: a case of poultry farming. JAC-Antimicrobial Resistance 4 (6). Available at https://doi.org/10.1093/jacamr/dlac119, https://academic.oup.com/jacamr.

Liguori, K., Keenum, I., Davis, B.C., Calarco, J., Milligan, E., Harwood, V.J., et al., 2022. Antimicrobial resistance monitoring of water environments: a framework for standardized methods and quality control. Environmental Science and Technology 56 (13), 9149–9160. Available at https://doi.org/10.1021/acs.est.1c08918, http://pubs.acs.org/journal/esthag.

Limbu, S.M., Chen, L.Q., Zhang, M.L., Du, Z.Y., 2021. A global analysis on the systemic effects of antibiotics in cultured fish and their potential human health risk: a review. Reviews in Aquaculture 13 (2), 1015–1059. Available at https://doi.org/10.1111/raq.12511, http://onlinelibrary.wiley.com/journal/10.1111/(ISSN)1753-5131.

Liu, C., Tan, L., Zhang, L., Tian, W., Ma, L., 2021. A review of the distribution of antibiotics in water in different regions of china and current antibiotic degradation pathways. Frontiers in Environmental Science 9. Available at https://doi.org/10.3389/fenvs.2021.692298, journal.frontiersin.org/journal/environmental-science.

Maillard, J.Y., Bloomfield, S.F., Courvalin, P., Essack, S.Y., Gandra, S., Gerba, C.P., et al., 2020. Reducing antibiotic prescribing and addressing the global problem of antibiotic resistance by targeted hygiene in the home and everyday life settings: a position paper. American Journal of Infection Control 48 (9), 1090–1099. Available at https://doi.org/10.1016/j.ajic.2020.04.011, http://www.journals.elsevier.com/ajic-american-journal-of-infection-control/.

Mao, K., Zhang, H., Yang, Z., 2020. Can a paper-based device trace COVID-19 sources with wastewater-based epidemiology? Environmental Science and Technology 54 (7), 3733–3735. Available at https://doi.org/10.1021/acs.est.0c01174, http://pubs.acs.org/journal/esthag.

Martínez, J.L., Baquero, F., 2014. Emergence and spread of antibiotic resistance: setting a parameter space. Upsala Journal of Medical Sciences 119 (2), 68–77. Available at https://doi.org/10.3109/03009734.2014.901444, http://informahealthcare.com/toc/ups/114/2?cookieSe=1.

McCullough, A.R., Parekh, S., Rathbone, J., Del Mar, C.B., Hoffmann, T.C., 2016. A systematic review of the public's knowledge and beliefs about antibiotic resistance. Journal of Antimicrobial Chemotherapy 71 (1), 27–33. Available at https://doi.org/10.1093/jac/dkv310, http://jac.oxfordjournals.org/.

Meena, D.K., Mathaiyan, J., Gauthaman, G., 2022. WHO AWaRe strategy and antimicrobial stewardship to combat antimicrobial resistance: an Indian perspective. International Journal of Health and Allied Sciences 11 (1). Available at https://doi.org/10.55691/2278-344x.1008.

Melaku, T., Assegid, L., 2025. Prescription in peril: the sociology of antibiotics and antimicrobial resistance in low resource settings. Discover Social Science and Health 5 (1). Available at https://doi.org/10.1007/s44155-025-00225-1.

Miranda, C.D., Tello, A., Keen, P.L., 2018. Mechanisms of antimicrobial resistance in finfish aquaculture environments. Frontiers in Microbiology 9, 1281 (2018).

Mittal, A.K., Bhardwaj, R., Mishra, P., Rajput, S.K., 2020. Antimicrobial misuse/overuse: adverse effect, mechanism, challenges and strategies to combat resistance. The Open Biotechnology Journal 14 (1) (2020).

Mudenda, S., Chabalenge, B., Daka, V., Mfune, R.L., Salachi, K.I., Mohamed, S., et al., 2023. Global strategies to combat antimicrobial resistance: a one health perspective. Pharmacology & Pharmacy 14 (08), 271–328. Available at https://doi.org/10.4236/pp.2023.148020.

Murray, C.J., Ikuta, K.S., Sharara, F., Swetschinski, L., Robles Aguilar, G., Gray, A., et al., 2022. Global burden of bacterial antimicrobial resistance in 2019: a systematic analysis. The Lancet 399 (10325), 629–655. Available at https://doi.org/10.1016/S0140-6736(21)02724-0, http://www.journals.elsevier.com/the-lancet/.

Muteeb, G., Khan, S.A., Rahman, R., Bhat, M.A., 2023. A one health perspective on antimicrobial resistance: Integrating human, animal, and environmental health for sustainable solutions. Journal of Global Infectious Diseases 15 (2), 45–53 (2023).

Mutua, F., Sharma, G., Grace, D., Bandyopadhyay, S., Shome, B., Lindahl, J., 2020. A review of animal health and drug use practices in India, and their possible link to antimicrobial resistance. Antimicrobial Resistance and Infection Control 9 (1). Available at https://doi.org/10.1186/s13756-020-00760-3, http://www.aricjournal.com/.

Nammi, J., Pasala, R., Andhe, N., Vasam, R., Poruri, A.D., Sherikar, R.R., 2025. Antibiotic misuse: an in-depth examination of its global consequences and public health challenges. Cureus 17 (6). Available at https://doi.org/10.7759/cureus.85941.

Narain, J.P., Kassim, A., Bhatia, R., 2022. Pharmaceutical effluents in low- and middle-income countries: regulatory gaps and AMR risks. *WHO South-East Asia Journal of Public Health 11 (2), 89–97 (2022).

Noyes, N.R., Slizovskiy, I.B., Singer, R.S., 2021. Beyond antimicrobial use: a framework for prioritizing antimicrobial resistance interventions. Annual Review of Animal Biosciences 9, 313–332. Available at https://doi.org/10.1146/annurev-animal-072020-080638, http://www.annualreviews.org/journal/animal.

Ocan, M., Obuku, E.A., Bwanga, F., Akena, D., Richard, S., Ogwal-Okeng, J., et al., 2015. Household antimicrobial self-medication: a systematic review and meta-analysis of the burden, risk factors and outcomes in developing countries. BMC Public Health 15 (1). Available at https://doi.org/10.1186/s12889-015-2109-3, http://www.biomedcentral.com/bmcpublichealth.

Okeke, I.N., Ihekweazu, C., Wessels, M.R., Rotimi, V.O., Shittu, A.O., 2021. Antimicrobial resistance in Africa: challenges and opportunities. The Lancet Microbe 2 (8), e415–e416 2021).

O'Neill, J., 2016. Tackling drug-resistant infections globally: final report and recommendations. Review on Antimicrobial Resistance 2016.

Oyelayo, E.A., Taiwo, T.J., Oyelude, S.O., Alao, J.O., 2025. The global impact of industrialisation and climate change on antimicrobial resistance: assessing the role of Eco-AMR Zones. Environmental Monitoring and Assessment 197 (6). Available at https://doi.org/10.1007/s10661-025-14086-3, https://www.springer.com/journal/10661.

Pal, C., Bengtsson-Palme, J., Kristiansson, E., Larsson, D.G.J., 2016. The structure and diversity of human, animal and environmental resistomes. Microbiome 4. Available at https://doi.org/10.1186/s40168-016-0199-5, http://www.microbiomejournal.com/.

Pham, M.N., Nishimura, F., Lan, J.C.W., Khoo, K.S., 2024. Recent advancement of eliminating antibiotic resistance bacteria and antibiotic resistance genes in livestock waste: a review. Environmental Technology and Innovation 36. Available at https://doi.org/10.1016/j.eti.2024.103751, https://www.sciencedirect.com/science/journal/23521864.

Pokharel, S., Raut, S., Adhikari, B., 2019. Tackling antimicrobial resistance in low-income and middle-income countries. BMJ Global Health 4 (6), e002104. Available at https://doi.org/10.1136/bmjgh-2019-002104.

Ponzo, E., De Gaetano, S., Midiri, A., Mancuso, G., Giovanna, P., Giuliana, D., et al., 2024. The antimicrobial resistance pandemic is here: implementation challenges and the need for the one health approach. Hygiene 4 (3), 297–316. Available at https://doi.org/10.3390/hygiene4030024, www.mdpi.com/journal/hygiene.

Pruden, A., Joakim Larsson, D.G., Amézquita, A., Collignon, P., Brandt, K.K., Graham, D.W., et al., 2013. Management options for reducing the release of antibiotics and antibiotic resistance genes to the environment. Environmental Health Perspectives 121 (8), 878–885. Available at https://doi.org/10.1289/ehp.1206446, http://ehp.niehs.nih.gov/wp-content/uploads/121/8/ehp.1206446.pdf.

Quadripartite Joint Secretariat, 2024. Global progress report on antimicrobial resistance. World Health Organisation.

Quadripartite Joint Secretariat on Antimicrobial Resistance, 2022. The Quadripartite organisations.

Rahman, M.M., Alam Tumpa, M.A., Zehravi, M., Sarker, M.T., Yamin, M., Islam, M.R., et al., 2022. An overview of antimicrobial stewardship optimization. Antibiotics 11 (5), 667 (2022).

Robinson, T.P., Bu, D.P., Carrique-Mas, J., Fèvre, E.M., Gilbert, M., Grace, D., et al., 2016. Antibiotic resistance is the quintessential one health issue. Transactions of the Royal Society of Tropical Medicine and Hygiene 110 (7), 377–380. Available at https://doi.org/10.1093/trstmh/trw048, http://trstmh.oxfordjournals.org/content/by/year/2013.

Samreen, Ahmad, I., Malak, H.A., Abulreesh, H.H., 2021. Environmental antimicrobial resistance and its drivers: a potential threat to public health. J Glob Antimicrob Resist 27, 101–111. Available at https://doi.org/10.1016/j.jgar.2021.08.001.

Schar, D., Klein, E.Y., Laxminarayan, R., Gilbert, M., Van Boeckel, T.P., 2020. Global trends in antimicrobial use in aquaculture. Scientific Reports 10 (1). Available at https://doi.org/10.1038/s41598-020-78849-3, www.nature.com/srep/index.html.

Schuster, D., Axtmann, K., Holstein, N., Felder, C., Voigt, A., Färber, H., et al., 2022. Antibiotic concentrations in raw hospital wastewater surpass minimal selective and minimum inhibitory concentrations of resistant *Acinetobacter baylyi* strains. Environmental Microbiology 24 (12), 5721–5733. Available at https://doi.org/10.1111/1462-2920.16206, http://onlinelibrary.wiley.com/journal/10.1111/(ISSN)1462-2920.

Sihombing, B., Bhatia, R., Srivastava, R., Aditama, T.Y., Laxminarayan, R., Rijal, S., 2023. Response to antimicrobial resistance in South-East Asia Region. The Lancet Regional Health - Southeast Asia 18. Available at https://doi.org/10.1016/j.lansea.2023.100306, https://www.thelancet.com/lansea/about.

Singer, A.C., Shaw, H., Rhodes, V., Hart, A., 2016. Review of antimicrobial resistance in the environment and its relevance to environmental regulators. Frontiers in Microbiology 7. Available at https://doi.org/10.3389/fmicb.2016.01728, http://journal.frontiersin.org/article/10.3389/fmicb.2016.01728/full.

Smits, C.H. M., Li, D., Patience, J.F., den Hartog, L.A., 2021. Animal nutrition strategies and options to reduce the use of antimicrobials in animal production. FAO Animal Production and Health Paper No. 184. Food and Agriculture Organisation of the United Nations.

Spit, T., van der Hoek, J.P., de Jong, C., van Halem, D., de Kreuk, M., Perez, B.B., 2022. Removal of antibiotic resistance from municipal secondary effluents by ozone-activated carbon filtration. Frontiers in Environmental Science 10. Available at https://doi.org/10.3389/fenvs.2022.834577, journal.frontiersin.org/journal/environmental-science.

Suleman, M., Clark, J.R., Bull, S., Jones, J.D., 2025. Ethical argument for establishing good manufacturing practice for phage therapy in the UK. Journal of Medical Ethics 51 (5). Available at https://doi.org/10.1136/jme-2023-109423, http://jme.bmj.com/.

Tacconelli, E., Cataldo, M.A., Dancer, S.J., De Angelis, G., Falcone, M., Frank, U., et al., 2014. ESCMID guidelines for the management of the infection control measures to reduce transmission of multidrug-resistant Gram-negative bacteria in hospitalized patients. Clinical Microbiology and Infection 20 (1), 1–55. Available at https://doi.org/10.1111/1469-0691.12427.

Tesema, M.Y., Birhanu, A.G., 2024. One health initiative to mitigate the challenge of antimicrobial resistance in the perspectives of developing countries. Bulletin of the National Research Centre 48 (1). Available at https://doi.org/10.1186/s42269-024-01176-4.

Thakali, O., Malla, B., Tandukar, S., Sthapit, N., Raya, S., Furukawa, T., et al., 2021. Release of antibiotic-resistance genes from hospitals and a wastewater treatment plant in the Kathmandu Valley, Nepal. Water 13 (19), 2733. Available at https://doi.org/10.3390/w13192733.

The White House, 2014. National Strategy for Combating Antibiotic-resistant Bacteria. Executive Office of the President (2014).

The White House, 2020. *National Action Plan for Combating Antibiotic-resistant Bacteria 2020–2025* Executive Office of the President (2020).

The World Bank, 2023. Pulling together to beat superbugs: Knowledge and implementation gaps in addressing antimicrobial resistance.

Toghroli, R., Hassani, L., Aghamolaei, T., Rajabi, A., Zare, S., Dabiran, S., 2024. Barriers faced by veterinarians against antimicrobial resistance. BMC Infectious Diseases 24 (1), 455 (2024).

Tripartite AMR Monitoring and Evaluation (MandE) Framework, 2024 One Health Global. Retrieved 2024.

Uddin, T.M., Chakraborty, A.J., Khusro, A., Zidan, B.R.M., Mitra, S., Emran, T.B., et al., 2021. Antibiotic resistance in microbes: history, mechanisms, therapeutic strategies and future prospects. Journal of Infection and Public Health 14 (12), 1750–1766. Available at https://doi.org/10.1016/j.jiph.2021.10.020, http://www.elsevier.com/wps/find/journaldescription.cws_home/716388/descriptio.

Van Boeckel, T.P., Brower, C., Gilbert, M., Grenfell, B.T., Levin, S.A., Robinson, T.P., et al., 2019. Global trends in antimicrobial use in animals. Science (New York, N.Y.) 347 (6229), 564–567 (2019).

Velazquez-Meza, M.E., Galarde-López, M., Carrillo-Quiróz, B., Alpuche-Aranda, C.M., 2022. Antimicrobial resistance: one health approach. Veterinary World 15 (3), 743–749. Available at https://doi.org/10.14202/vetworld.2022.743-749.

Venter, N., Ashley, E.A., Dittrich, S., Fernandes, M., 2023. Global burden of bacterial antimicrobial resistance and projections to 2050. The Lancet 402 (10397), 123–135 (2023),.

Ventola, C.L., 2015. The antibiotic resistance crisis: Part 1: causes and threats. P and T 40 (4), 277–283. Available at http://www.ptcommunity.com/system/files/pdf/ptj4004277.pdf.

Vestel, J., Caldwell, D.J., Tell, J., Constantine, L., Häner, A., Hellstern, J., et al., 2022. Default predicted no-effect target concentrations for antibiotics in the absence of data for the protection against antibiotic resistance and environmental toxicity. Integrated Environmental Assessment and Management 18 (4), 863–867. Available at https://doi.org/10.1002/ieam.4560, http://www.interscience.wiley.com/jpages/1551-3777.

Vladkova, T.G., Smani, Y., Martinov, B.L., Gospodinova, D.N., 2024. Recent progress in terrestrial biota-derived anti-biofilm agents for medical applications. Applied Microbiology (Switzerland) 4 (3), 1362–1383. Available at https://doi.org/10.3390/applmicrobiol4030094, https://www.mdpi.com/journal/applmicrobiol.

Watts, J.E.M., Schreier, H.J., Lanska, L., Hale, M.S., 2017. The rising tide of antimicrobial resistance in aquaculture: sources, sinks and solutions. Marine Drugs 15 (6). Available at https://doi.org/10.3390/md15060158, http://www.mdpi.com/1660-3397/15/6/158/pdf.

WHO, 2015. Global Action Plan on Antimicrobial Resistance. World Health Organisation.

WHO, 2015b. Library cataloguing-in-publication data global action plan on antimicrobial resistance; 2015; ISBN 9789241509763.

WHO, 2021. Global action plan on antimicrobial resistance. Available from: https://www.emro.who.int/health-topics/drug-resistance/global-action-plan.html. (accessed 20.01.25).

Willis, L.D., Chandler, C., 2019. Quick fix for care, productivity, hygiene and inequality: reframing the entrenched problem of antibiotic overuse. BMJ Global Health 4, e001590. Available at https://doi.org/10.1136/bmjgh-2019-001590.

World Health Organisation, 2013. Integrated surveillance of antimicrobial resistance in foodborne bacteria.

World Health Organisation, 2015. Global action plan on antimicrobial resistance.

World Health Organisation, 2017. Guidelines on the use of Medically Important Antimicrobials in Food-producing Animals. 2017.

World Health Organisation, 2019. Monitoring and evaluation of the global action plan on antimicrobial resistance: a framework for countries. 2019.

World Health Organisation, 2022. Antimicrobial Resistance: Global Report on Surveillance. WHO Press, Geneva, Switzerland 2022.

World Health Organisation, 2020. Antimicrobial Resistance: Global Report on Surveillance. 2020.

World Health Organisation, 2024. Technical briefing on the management of manufacturing effluents to minimise antimicrobial resistance.

World Health Organisation, Food and Agriculture Organisation of the United Nations, World Organisation for Animal Health, and United Nations Environment Programme. (2022b). *One Health Joint Plan of Action (2022-2026)*. Food and Agriculture Organisation of the United Nations.

Yakubu, O., 2017. Addressing environmental health problems in ogoniland through implementation of United Nations environment program recommendations: environmental management strategies. Environments 4 (2), 28. Available at https://doi.org/10.3390/environments4020028.

Zheng, S., Li, Y., Chen, C., Wang, N., Yang, F., 2025. Solutions to the dilemma of antibiotics use in livestock and poultry farming: regulation policy and alternatives. Toxics 13 (5), 348. Available at https://doi.org/10.3390/toxics13050348.

Zhou, L., Limbu, S.M., Shen, M., Zhai, W., Qiao, F., He, A., et al., 2021. Environmental concentrations of antibiotics impair zebrafish gut health. Environmental Pollution 268, 115706 (2021).

Chapter 5

Hidden hotspots: environmental drivers of antibiotic resistance

Anand Prakash Maurya
Department of Microbiology, Assam University Silchar, Silchar, Assam, India

5.1 Introduction

Environmental pollution remains one of the greatest problems faced by mankind, as it leads to increased mortality and has become one of the most serious global challenges. In the last few years, worldwide awareness of the effects of pollution on the environment and human health has increased considerably. Environmental pollution is mainly caused by chemical substances known as pollutants. Due to an increase in the human population globally, the use of these substances (pollutants) in the form of pesticides, detergents, dyes, hydrocarbons, etc., has increased along with urbanisation, industrialisation, and excessive exploitation of natural resources. Therefore the accumulation of chemical pollutants, also known as anthropogenic pollutants, in the environment has increased, gradually affecting ecosystems (Häder et al., 2020). It has been observed that the discharge of pollutants from industries, hospitals, urban runoff, water treatment plants, sewage treatment plants, and agricultural wastes into soil and water affects the overall distribution of microorganisms and exerts selective pressure that may lead to the emergence of antimicrobial-resistant bacteria (ARBs). However, one of the major reasons behind the emergence of ARBs is the increasing use and misuse of antibiotics.

A study documented that; 67% global consumption of antibiotics has risen from 63,151±1560 to 105,596±3605 t between 2010 and 2030 (Boeckel et al., 2015). Another study reported that antibiotic usage across the globe surpasses 100,000 t/year, which is of increasing concern. Antibiotics are pervasive in the atmosphere, and substantial amounts of the antibiotics and resistant genes have been found in the fresh waters, soils (Chen et al., 2017), surface and groundwater (Guo et al., 2017), waste water (Yadav and Kapley, 2019), coastal areas, and even deep ocean sediments (Su et al., 2014; Ma et al., 2017; Jang et al., 2018).

Contaminated soils are considered the primary reservoir of ARBs, and a wide variety of antibiotic-resistance genes (ARGs) are frequently identified in these ARBs (Table 5.1). Global spread of ARBs currently poses a risk to human health as it is linked with the failure of antibiotic treatments and results in increased mortality (Ashbolt et al., 2013). Therefore it is presumed that a lack of knowledge about antibiotic-resistant pathogens and inadequate monitoring will lead to increased problems with antimicrobial resistance (AMR). Therefore more focus has to be given to the role of ecological factors that contribute to antibiotic resistance.

Earlier reviews elaborated the influence of anthropogenic contaminants (i.e., heavy metals, poly aromatic hydrocarbons (PAHs), biocides, detergents etc.) on growth of ARBs and the persistence of ARGs (Singer et al., 2016; Bengtsson-palme et al., 2018) in soil (Gorovtsov et al., 2018; Cycoń et al., 2019), waste water (Karkman et al., 2018), estuarine sediments (Rodgers et al., 2019), urban runoff (Almakki et al., 2019), livestock waste (He et al., 2020), aquatic ecosystems (Häder et al., 2020), drinking water systems (Sanganyado and Gwenzi, 2019; López-Pacheco et al., 2019), and in natural environment with their adverse effect on the aquatic lives (Nguyen et al., 2019). In addition, some authors have highlighted the mechanisms and physiological roles of efflux pumps in the development of AMR (Piddock, 2006; Poole, 2007; Sun et al., 2014; Blanco et al., 2016; Du et al., 2018).

Thus, taking into account the importance of AMR and the role of environmental pollutants on bacteria, this review underlined the occurrence and distribution of ARBs and ARGs in the environmental contaminants, the contribution of toxic chemicals, hydrocarbons, antibiotics, and other anthropogenic pollutants in the expansion of ARBs/ARGs in the soil, water, sediments or other environmental resources like seawater. This review also elaborated on the responses of

Antimicrobial Resistance in the 21st Century. DOI: https://doi.org/10.1016/B978-0-443-44748-8.00021-X

ARBs/ARGs to anthropogenic pollutants and illustrated possible mechanisms underlying the expansion of these ARGs in different environments. Also, greater emphasis is placed on the knowledge gap and on promising recommendations to diminish the threat of AMR to the environment.

5.2 Occurrence and distribution of antimicrobial-resistant bacteria in farming, agriculture, plants, activated sludge, and soils contaminated environments

AMR is a result of growth and progression through natural selection. ARBs present in the environment can be spread due to overuse or misuse of antibiotics (Table 5.1). Some authors documented that there is a direct association between the consumption of antibiotics and the development of antibiotic resistance in bacteria (Olesen et al., 2018). These antibiotics at sub-inhibitory and subtherapeutic concentrations may induce the development of ARBs in the environment through modifications in gene regulation (Zhang et al., 2018). However, horizontal gene transmission (HGT) may be deliberated as a significant driver for capturing and transmission of ARGs in the pathogenic and environmental bacteria (Martínez et al., 2015). Treatment procedures were found incompetent in eliminating the ARBs from the water, as discharges of treatment plants usually contain the ARBs, and these are coming directly to the normal environment, where they spread the resistance to other bacteria present in soil and water through HGT (Ben et al., 2017). It has been documented that individuals can be directly exposed to ARBs by using contaminated water, bathing, coming in contact with soil and eating food that was produced using contaminated water (Leonard et al., 2018).

5.2.1 In agricultural settings

Due to increased use of antibiotics in agricultural settings, it also disturbs the soil microbiome, as about 90% of the given antibiotics to livestock are usually excreted into the environment through urine or stool, and these are normally spread as fertiliser in agricultural fields, and sometimes spread into the water (Bartlett et al., 2013). There were reports that showed the positive association prevails between the usage of manure and ARBs in environmental soil and water (Peak et al., 2007; Bibbal et al., 2007). It has been demonstrated that oxytetracycline was very toxic for soil bacteria, and among all the bacteria, the resistance rate was highest in Gram-negative bacteria (Chen et al., 2013). In another report, it has been noticed that the presence of sulfadiazine in the soil affects the soil bacteria, which increases with the concentration of sulfadiazine. However, another report documented that there were changes in the diversity of soil bacteria when tetracycline, ciprofloxacin, and sulfadimethoxine (100 mg/kg) in the soil plant system were used for agriculture. The bacterial diversity was found more in the proteobacteria, but resistance rates were increased with the time in case of ciprofloxacin and sulfadimethoxine, while in case of tetracycline, it was decreased with the time (Li et al., 2016). It was recorded that penicillin playing negative role in soil enzyme actions and interrupts homoeostasis in the soil (Telesinski et al., 2018), while it was reported that there were no imperative influences of oxytetracycline on the alkaline phosphate action of soil (Li et al., 2015).

Valcarcel et al. (2011) reported that nearly 50%–80% of the antibiotics are excreted through the urine and due to mixing of sewage water with the environmental soil or water, causing the antibiotic contamination. In Spain, norfloxacin (10 ng/L), tetracycline (23 ng/L), ofloxacin (179 ng/L), trimethoprim (424 ng/L), ciprofloxacin (3 ng/L), metronidazole (1195.5 ng/L), and erythromycin (320.5 ng/L) antibiotics were found in the Tagus, Guadarrama, Jarma, and Henares rivers (Valcárcel et al., 2011). However, concentrations of 0.10–16.6 ng/L of trimethoprim, erythromycin, and sulfamethoxazole were found in the Yellow and Bohai Seas, posing a threat to marine animals and other organisms (Zhang et al., 2013).

5.2.2 In the farming sector

It was reported that higher levels of ARBs and ARGs were attributable to the routine use of antibiotics in the farming sector (Thompson et al., 2020). Yang et al. (2014) reported that, in farms, chicken compost fertiliser contains gentamicin, amoxicillin, kanamycin, and cephalexin, which enhances 21%–66% of the growth of ARBs, and among them, 8%–20% were resistant to many antibiotics (Yang et al., 2014). However, in China, in seawater, tetracycline (2.11–9.23 ng/L) and sulphonamides (1.42– 71.32 and 2.18–63.87 mg/kg) were recorded in the aquatic and sedimentary organisms (Na et al., 2013). Further, it has been estimated that macrolide concentrations in the manures varied from 0.07–0.14, 1.05–2.1, and 0.62–1.24 mg/kg; however, the concentration of tetracyclines and sulphonamides varied from 0.49, 8.44, and 1.39 mg/kg and 1.65, 16.56, and 15.62 mg/kg separately (Kim et al., 2011). While in another report, it was confirmed that manure was the main cause for contamination in the groundwater, and it was found that the concentration of sulphonamides was 0.01–3460.57 ng/L and ciprofloxacin was 323.75 ng/L (Cabeza et al., 2012; Jurado et al., 2012).

TABLE 5.1 Occurrence and distribution of antibiotic resistance attributes in the anthropogenic contaminants.

Sl. no.	Type of contaminants	Location	Source	Antibiotic resistance attributes			References
				ARBs	ARGs	MGEs	
1	Activated sludge samples	Ankleshwar, Gujarat, India	Effluent treatment plants	*Pseudomonas aeruginosa, Klebsiella pneumoniae strain M743, Escherichia coli aacA4, Acinetobacter baumannii strain, Vibrio cholerae strain MO10*	*pKJK5, IncP-1beta, pB8, CmeABC* operon	Efflux pumps, class 1 integrons, plasmids	Yadav and Kapley (2019)
2	Drinking water, river water, faeces, sewage, sludge, swine and wastewater	China	Environmental samples from multiple ecosystems	*E. coli, K. pneumoniae, P. aeruginosa, Salmonella enterica, A. baumannii*	*intI1, sul1, tetA, tetB, tetC, VEB-6, VEB 1,c, dfrA17-aadA5, aadA1, aadA2, aadA4, aadA5, aacA29b, qacE*	Efflux pump, class 1 integrons	Ma et al. (2017)
3	Urban river water samples	Wenzhou, China	Wen-Rui Tang River watershed of eastern China	*Bacteroides, Burkholderia, Acidovorax, Rhodobacter, Flavobacterium,*	*aadA, aacA_aphD, aacC2, qacE,* bla_{GES}, bla_{OXA1}, bla_{OXA30}, $bla_{OXA10\text{-}01}$, $bla_{OXA10\text{-}02}$, bla_{PER}, bla_{TEM}, bla_{VEB}, and *cfxA*	Efflux pumps Class 1 integrons	Zhou et al. (2017)
4	Coastal aquaculture	South Jeolla province and Jeju Island, South Korea	Effluents of coastal aquaculture	*Vibrio, Proteobacteria, Marinomonas, Bacteroidetes, Gammaproteobacteria, Pseudoalteromonas, Flavobacteriia, Fusobacteria, Acinobacteria, Cyanobacteria*	*qnrD, qnrS, aac (60)-Ib-cr, tetA, tetB, tetD, tetE, tetG, tetH, tetM, tetQ, tetX, tetZ, tetBP, floR, ermC, oqxA, TEM, CTX, SHV, sul1*	Efflux pump transporters, Class 1 integrons	Jang et al. (2018)
5	Clay loam and paddy soil	Shaoxing City, Zhejiang Province, P.R. China	Agriculture soil	–	*tetA, tetC, tetW, sul1, sul2, IntI1*	Class 1 integrons	Li et al. (2016)b)
6	Residues in the river basin of the Dongjiang	South China	Estuary and sediments of the Dongjiang River basin	–	*sul1, sul2, sul3, ermB, ermC, ermF, tetA, intI1, intI2*	Class 1 integrons, class 2 integrons	Su et al. (2014)

(*Continued*)

TABLE 5.1 (Continued)

Sl. no.	Type of contaminants	Location	Source	Antibiotic resistance attributes			References
				ARBs	ARGs	MGEs	
7	Soil samples of residential areas	Western Australia	Sandy and residential areas soil	–	*bla* $_{CTX}$, *bla* $_{OXA}$, *bla* $_{TEM}$, *tetB, tetC, tetD, tetA, tetE, tetG, tetR, tetM, tetW, tetO, tetQ, tetS, Tn917, sul1, sul2,*	Class 1 integrons, Transposons	Knapp et al. (2017)
8	Soil samples	Guangzhou, China	Soil from the petrochemical plant	*Proteobacteria, Bacteroidetes, Acidobacteria, Chlamydiae, Chloroflexi, Firmicutes, Gemmatimonadetes, Nitrospirae*	*acrA, acrB, amrB, bacA, ceoB, macB, mexB, tolC, amrB, bacA, ceoB*	Efflux Pumps	Chen et al. (2017)
9	Wastewater treatment plants	Beijing, China	Aerobic activated sludge and the anaerobically digested sludge	*Proteobacteria, Nitrospirae, Bacteroidetes, Actinobacteria, Firmicutes, Cloacimonetes*	*intI1, ISSsp4, ISMsa21, ISMba16, sul1*	Class 1 integrons, Insertion sequences	Guo et al. (2017)
10	Forest soil samples	Rotenburg, Germany	Forest-polluted soil contaminated with herbicide DALAPON (2,2-dichloropropionate)	*Xanthomonas maltophilia strain RS6, Comamonas acidovorans strain RS7, and Alcaligenes xylosoxidans ssp. denitrificans strain RS9*	*AmpR, TetR, Sul, KanR*	–	Schwarze et al. (1997)
11	Waste water discharges	Spain	Activated sludge of an urban wastewater treatment plant	Burkholderiaceae, Comamonadaceae, Microbacteriaceae, Cytophagaceae, Flavobacteriaceae, Cryomorphagaceae	*intI1, intI2, intI4, intI5, intI7, intI8, intI9 and intI10, ISCR1, ISCR4, ISCR7, Tn3, Tn21, Tn1721, Tn2501, Tn3926*	Integrases, integrin integrases, transposases, and ISCR elements	Lekunberri et al. (2018)
12	Wastewater	China	Effluents, activated sludge, anaerobic digestion sludge	*Streptococcus, Acinetobacter, Mycobacterium, Vibrio, Flavobacterium, Bifidobacterium, Moraxella, Francisella, Syntrophomonas, Kosmotoga, Thermovirga*	*acrA, aacA, aphD, mefE, mdtO, TolC, tetA, mdfA, MdtL, ermB, mexD, mexB, VEB-1, VEB-3, OXA-10, OXA-46*	ABC-type efflux pump, transporter protein	Yang et al. (2014)
13	Faecal samples	Alberta Canada	Beef feedlots, downstream environments, and urban sewage influent	*Actinobacteria, Acidobacteria, Bacteriodetes, Proteobacteria, Fusobacteria, Firmicutes*	*TetQ, MefA, MefB, MSR, OXA, CTX-M, SMR, Sul, qacE*	Efflux pump, Class 1 integron	Zaheer et al. (2019)

14	Multi-Contaminated Tinto River Estuary (surface water and sediments)	Spain	Phosphogypsum deposits, copper metallurgy plant, fertilisers plants, chemicals production plant, hydrocarbons storage, pharmaceuticals and food supplements plant, pigments plant	*Bacillus, Altererythrobacter, Brevundimonas, Ponticaulis, Erythrobacter, Ochrobactrum, Mesonia,*	*TetR, TmR, EmR/KmR, RpR/TcR*	Efflux pump	Eduardo-Correia et al. (2020)
15	Natural Hot Spots (Water and sediment samples)	Lucky Strike hydrothermal vent area, North Atlantic	Sea floor sediments	*Dokdonia donghaensis, Sulfitobacter dubius, Oceanibulbus indolifex, Mesoflavibacter zeaxanthinifaciens, Maribacter dokdonensis, Erythrobacter citreus, Aurantimonas coralicida*	*blaA, blaC, ParE, gyrB, MATE MDR pump family gene, macA, macB, RND, MFP* subunit gene	Efflux pump, efflux pump membrane protein, plasmids, transporter protein	Farias et al. (2015)
16	Copper Tailing dam	Northern China	Soil samples	–	*tetA, tetS, tetX, arsB, arsC, pbrT, czcA, ereA,* $bla_{CTX\text{-}M}$, bla_{SHV}, *tetC, tetG, tetA/P, tetM, tetO, tetQ, tetW, tetB/P, mphA, sulII*	Efflux pump, class 1 integron	Chen et al. (2019)

ARGs, Antibiotic-resistance genes; *MATE*, multidrug and toxic compound extrusion; *MDR*, multidrug resistance; *MFP*, membrane fusion protein; *MGEs*, mobile genetic elements; *RND*, resistance nodulation division.

5.2.3 In the environmentally contaminated soil

A report assessed that ARBs were found in the environmental soil polluted with heavy metals, where lands were used for different agricultural purposes (Safari Sinegani and Younessi, 2017). A degree of resistance was observed for amoxicillin and ampicillin in soil isolates, and agricultural soil showed a high prevalence of resistance to vancomycin, streptomycin, and tetracycline. Additionally, multiple antibiotic resistances have been noticed in the agricultural soils of amoxicillin, vancomycin, doxycycline, ampicillin, and streptomycin (Safari Sinegani and Younessi, 2017). However, 77% of the isolates were resistant to beta-lactam antibiotics, 40% to vancomycin, 31.43% to tetracyclines, and 28.57% to streptomycin (Younessi et al., 2019). These numbers indicate that the occurrence of an enormous number of ARBs in soils suggests the presence of an environmental reservoir of soil ARBs.

5.2.4 In plants

Further, antibiotic concentration in plants was also recorded, and it was found that the concentration of quinolone was 8.37–6532 mg/kg in the aquatic plants, while in the animals and birds it was 17.8–167 mg/kg (Li et al., 2012). While, the concentration of tetracycline, oxytetracycline, and chlortetracycline was found in between 1.9–5.6, 78–330, and 92–481 mg/kg in the coriander plants, however surprisingly in the leaves of radish there were various concentration of chloramphenicol (8–30 mg/kg), sulfadoxine (0.2–0.6 mg/kg), sulfamethoxazole (0.9–2.7 mg/kg) and sulfachloropyridazine (0.1–0.5 mg/kg) were found to be present. While in leaves of celery plants, antibiotics lincomycin (5–20 mg/kg), ofloxacin (1.7–3.6 mg/kg), and pefloxacin (1.1 mg/kg) were found to be present in the winter season (Hu et al., 2016).

5.2.5 In activated sludge

The most abundant ARBs were proteobacteria, which carried aminoglycosides, beta-lactams, colistin, fluoroquinolones, trimethoprim, rifampicin, tetracyclines, and sulphonamides (Yadav and Kapley, 2019). The higher abundance of mobile genetic elements (MGEs) like IncP-1 plasmid pKJK5, IncP-1beta, and pB8 was also found, and these are involved in playing a role in HGT. Authors further reported that, in the sludge samples, there were multidrug efflux pumps (e.g. *Campylobacter jejuni, CmeABC* operon) that played the main role in the release of toxic metabolites (Yadav and Kapley, 2019). Therefore it can be speculated that efflux pumps also play an important role in the expansion of ARBs and ARGs in the environment.

5.3 Rate and existence of antimicrobial-resistant bacteria and antibiotic-resistance genes in the environments and treatment plants

Sewage, treatment plant, hospitals, farms, house wastes, etc., contain many ARBs, ARGs and pharmaceutical effluents such as antibiotics. The presence of antibiotics creates the selective pressure which enhances the progression of ARGs in the environment. Reports suggested that, when tylosin was used in the sheep, it was observed that tylosin was present in the urine and faeces of sheep for up to four days, and about 11% of the used tylosin was excreted in the urine and faeces (Ishikawa et al., 2018). In a study from China, the authors quantify the presence of class 1 integrons in 64 environmental samples collected from 8 different ecosystems. Among them, the richness of class 1 integrons varied from 3.83×10^{-4} to $4.26\times10^{\circ}$*intI1*/cell, and the most abundant common resistance was observed in the integrons, which were aminoglycosides (57%) followed by trimethoprim and beta-lactams (Ma et al., 2017).

5.3.1 In treatment plants

In sewage and wastewater treatment plants, the presence of antibiotics, heavy metals, and chemical substances creates the selective pressure that enhances the HGT (Karkman et al., 2018). Hendriksen et al. (2019) performed metagenomic analysis in the 79 treatment plants situated in 60 different countries. They found that there were differences in the occurrence of ARGs in South America/ North America/Europe/Asia/Oceania/Africa (Hendriksen et al., 2019). Among them, the cluster of Europe/North America/ and Oceania restricted a very small number of ARGs which encoding for high resistance to macrolides; however cluster of South America/Asia/Africa has the ARGs, which were resistant to sulphonamides and phenicols. While Brazil, Vietnam and India had different ARG patterns, it was proposed that they were the hotspots of the emergence of new ARBs and ARGs (Hendriksen et al., 2019). In another report, 42 ARG subtypes were detected in activated sludge samples and 51 ARGs in the digested sludge. In these sludge samples, integrons (*intI1*), transposons, insertion sequences (*ISSsp4, ISMsa21, ISMba16*) and plasmids were also found that transfer ARGs into the bacteria present in the environment (Guo et al., 2017).

5.3.2 In aquatic environments

The presence of ARBs and ARGs in the aquatic environments has multiple mechanisms. It was documented that 363.4 tons of antibiotics were used by Chilean Marine salmonid farms in the year 2016, which may have played a role in the development of resistance in aquatic environments (Sandner-Miranda et al., 2018). However, it may be considered that the chief sources of ARBs and ARGs could be humans and veterinary settings, where gut bacteria are exposed to high doses of many antibiotics, but reports suggest that antibiotic pollution in the aquatic environments encourages the development of ARBs in the normal environment (Hatosy and Martiny, 2015; Fekadu et al., 2019). In another report, tetracycline, macrolide, sulphonamide resistance genes and integrons were found to be present in the residues in the river basin of Dongjiang, South China. The most predominant genes were sulphonamide resistance genes (*sul1, sul2*, and *sul3*) with 5.14×1010 and 7.36×109 copies/g in to the different time of year, followed by macrolide resistance genes (*ermB, ermC*, and*ermF*) with 1.57×1010 and 2.46×109 copies/g, integrons (*intI1* and *intI2*) with 6.64×109 and 8.68×108 copies/g, and *tetA* was the most predominant among the tetracycline resistance genes, with 5.71×109 and 3.33×108 copies/g in arid and rainy times (Su et al., 2014).

5.3.3 In coastal aquaculture

In the coastal aquaculture of South Korea, there were 24 different types of resistance carrying determinants were present, that encoding quinolone resistance (*qnrD, qnrS, aac(60)-Ib-cr*), macrolide resistance (*ermC*), multidrug resistance (MDR) (*oqxA*), tetracycline resistance (*tetA, tetB, tetD, tetE, tetG, tetH, tetM, tetQ, tetX, tetZ, tetBP*), florfenicol resistance (floR), beta-lactams resistance (bla_{TEM}, bla_{CTX}, bla_{SHV}), and a class 1 integrons-integrase gene (*intI1*). Among these resistant types, the most predominant were tetB and tetD (74%–98%) and *int1* genes showed the progressive association with *tetB, tetD, tetE, tetH, tetX, tetZ, tetQ,* and *sul1* (Jang et al., 2018). However, in a study from India, around 7% of soil samples, 42% of well water, and 100% of river sediment, were positive for qnr genes (fluoroquinolone resistant). However, this study was unsuccessful in demonstrating any close association between environmental exposure to qnr genes and the presence of qnr genes in humans, as the prevalence of qnr genes in humans was 91% (Rutgersson et al., 2014).

5.3.4 In soils and sediments

It has been documented that 70% of the ARGs were composed of the efflux pumps that encode the ARGs (resistance to acriflavine and fluroquinolones) in the contaminated soils (Chen et al., 2017). Reports suggested that ARGs in the PAH contaminated soils were associated with beta-lactams, fluoroquinolones, aminoglycosides, macrolides, tetracyclines, etc. Among these, ARGs are associated with almost 80% of resistance to aminoglycosides, which is the most commonly used medicine (Chen et al., 2017). Further, it has been reported that the *tetM* and *tetQ* genes were dominant in the animal compost that was treated with tetracycline (Schwaiger et al., 2010). The presence of sulphamethoxazole (*sul 1* and *sul 2*) and bla_{TEM} were high in farms where high antibiotic doses were used most frequently (Heuer et al., 2011). Reports also indicated that ARGs were quite high in pig manure compared with cow and sheep manure, and it has also been recommended that the presence of ARGs might be directly linked to antibiotics used in animal farming (McKinney et al., 2010).

Reports suggested that, ARGs are commonly found in the manure, sediments, wastewater, and lagoon slurry at very high levels, that is up to 28,000 times in comparison to soil/sediments and/or running water (Zhu et al., 2013). The absolute abundance of ARGs in livestock waste varies from 10^6 to 10^{12} copies/mL (Sim et al., 2011), in estuaries 1.48×10^6 copies/g (Zhu et al., 2017), in source water 7.78×103±7.13×103 copies/mL(Lu et al., 2018), in river water 1.51×104±1.49×104 copies/mL (Lu et al., 2018), in surface water 3.61×10^{10} to 3.02×10^{11} copies/L (Zhou et al., 2017), in soil 4.27×10^8 to 2.64×10^{10} copies/g, in subtropical urban reservoir 4.03×10^6 to 3.72×10^9 copies/L (Fang et al., 2019), in sediments 2.62×10^8 to 8.67×10^{10} copies/g (He et al., 2020), and in sea water 1.06×10^7 to 8.61×10^7 copies/L (Su et al., 2020). Therefore, the high level of absolute abundance in the livestock waste and estuaries, sea, sediments or in water suggests that antibiotics or wastewater from medical facilities/Industrial areas might be the chief source of expansion of these ARBs and ARGs.

5.4 Influence of pharmaceutical effluents, toxic chemicals, heavy metals, antibiotics, herbicides, and biocides on the rate and progression of antimicrobial-resistant bacteria and antibiotic-resistance genes

The anthropogenic contaminants act as a selective pressure to promote the expansion of ARBs/ARGs, which mainly happens through the coselection mechanisms. Coselection mechanisms (coresistance or cross-resistance) are conferred when the concentration of complexes present is much less than the lethal dose (Webber et al., 2015; Maurya et al., 2020). The dissemination and evolution of resistance enrich the ARGs through mutations and MGEs (Ashbolt et al., 2013). Specifically, the transmission dynamics of ARGs may be reflected as a significant factor for the growth and expansion of ARBs in the normal environments (Martínez et al., 2015).

Antibiotic pressure created due to overuse and misuse of antibiotics results in an increase in abundance of ARBs and ARGs in clinical and nonclinical environments, the farming sector, and marine environments (Ashbolt et al., 2013; Hu et al., 2016). Research indicates that chemicals also encourage the antibiotic pressure, which includes the antibiotics, nano-materials, and disinfectants (Seiler and Berendonk, 2012; Li et al., 2016; Ding et al., 2016; Zhang et al., 2017). It was found that heavy metals also exist in the normal environments, and some of the anthropogenic activities expressively speed up the discharge and increase of metals in the environment(Rodríguez Martín et al., 2015). Also, animal manures pollute the microflora of soil with many ARGs, and 55% of swine farms (out of 80%) and 75% of cattle farms (out of 95%) were found progressive for the occurrence of ARBs in the animal farms in the Netherlands (Berendsen et al., 2015).

Reports documented that coselection through genomic relationships may usually occur throughout the genes that contribute to susceptibility to biocides, antibiotics and heavy metals. However, colocation of these genes on MGE increased the probability of transmission of co-resistance to other bacteria (Ianeva, 2009; Seiler and Berendonk, 2012). A report documented that *Acinetobacter baumannii* in agricultural soils, sewage water and petroleum contaminated areas showed resistance to multiple heavy metals such as mercury (Hg), arsenic (As), and silver (Ag) (Kim et al., 2017; Huang et al., 2017). Further, it has been found that efflux pumps are the most widespread bacterial mechanism conferring resistance to heavy metals, which may be mediated either by ATP hydrolysis or by proton electrochemical gradients (Ianeva, 2009). However, basal intensities of the efflux pump are usually not enough to carry the heavy metal resistance genes in the ARBs, but due to the changes in the expression through mutations in the promoter region or regulators of the efflux pump, ARBs cause the overexpression of an efflux pump and therefore cause the antibiotic resistance (Blanco et al., 2016; Li et al., 2016).

Antibiotics are also present in agricultural soil, acting as a pool for the spread of ARBs that may be released in surface water compartments through overflow, draining activities and filtration (Lee et al., 2017). It has been seen that, 30%–90% of a substantial amount of used antibiotics are discharged through the urine or faecal excretion (Shao et al., 2019). These can also spill into the aquatic environment, where they give direct influence on the fish and therefore ARBs were continuously increasing in the marine environments (Cantas et al., 2013). Fick et al. (2009) reported the presence of fluoroquinolones in the pharmaceutical effluents in wastewater treatment plants. It was found that there was a presence of 14 mg/L of ciprofloxacin in the discharges in treatment plants, and there was 6.5 mg/L in two indigenous lagoons (Fick et al., 2009). However, in a study in China, a progressive relationship was seen among the organic carbon and dissolved oxygen with the presence of 285 ARGs in this polluted town river (Zhou et al., 2017); while it has been established through the research that the release of treated and unprocessed sewage water into the rivers increases the bacterial diversity, ARBs and ARGs with MGEs (Lekunberri et al., 2018). On the other hand, like antibiotics, heavy metals also enter the environment, and they are stored in the water, soil, sediments from agricultural activities, and industry effluents. Studies supported that copper, cadmium, nickel, and zinc were found to be present in the storm water, which could also be the promoter for expansion of ARBs and ARGs (Lundy et al., 2012).

It has been found that chlorohexidine digluconate exposure raises the risks of resistance in the bacteria (Cieplik et al., 2019). When 310 gram-positive bacteria were exposed to chlorhexidine, there was an adequate positive association of streptococci between reduced susceptibility to ampicillin, chlorhexidine, tetracycline and aminoglycoside antibiotics was observed (Cheng et al., 2019). However, among 87 isolates of seafoods, an adequate positive association has been found for cetrimide, triclosan, and hexadecylpyridinium chloride with cefotaxime, and chloramphenicol with triclosan, as well as co-trimoxazole with thymol (Romero et al., 2017). There were eightfold increases in the MIC values for tetracyclines, chloramphenicol, and nalidixic acid when *Salmonella enteritidis* was exposed to chlorine for a short period of time (Potenski et al., 2003). It was documented that triclosan usually causes overexpression of the *MexAB-OprM* efflux pump, which indicates multiple antibiotic resistance in *Pseudomonas aeruginosa*, but there was a substantial association between the sulfamerizine and tetracycline resistance genes *tet*(B) and *tet*(C) found in the sewage water plant (Li et al., 2015; Chuanchuen et al., 2001). Reports also show that even lower concentrations of heavy metals, biocides, antibiotics and disinfectants in groundwater and sewage water also affect the competence of HGT (Karkman et al., 2018). However, in another report, it was found that less exposure to chlorine did not affect the competency of their gene transfer in plasmid RP4, but high exposure resulted in a decrease in the competency of HGT (Li et al., 2016)b).

Therefore, it is a fact that ARBs evolve through mutations due to the low concentration of antibiotics in aquatic environments (Kohanski et al., 2010), and even if a part of the resistant mutation is in a very small portion, still, selection of ARBs will happen for many generations. Therefore, further attention is needed for aquatic and natural environments where the origin of ARBs and ARGs is high.

5.5 Response mechanisms against anthropogenic pollutants by antimicrobial-resistant bacteria and antibiotic-resistance genes

5.5.1 Interaction mechanisms: poly aromatic hydrocarbons versus antimicrobial-resistant bacteria/antibiotic-resistance genes

Exposure to pollutants can play a life-threatening role in public health by altering therapeutic effects through potential modulation of nucleotide bases in living beings. Therefore, investigation of response mechanisms to environmental pollutants by ARBs and ARGs has become an emerging research target in environmental science and technology. Some research groups have predicted a few possible mechanisms; however, more vivid evidence is still needed. Mostly, several studies have been carried out on substituted and non-substituted hydrophobic PAHs. The PAHs groups of chemicals, such as naphthalene and anthracene, are rich in π-electrons and hydrophobic in nature, which facilitate their interaction with bacterial phospholipid cell wall, plasmid, and DNA bases (Kang et al., 2015; Shou et al., 2019; Hu et al., 2015). It has been reported that PAHs mostly interact with the plasmid through noncovalent bonds(Boyland and Green, 1962; Johnson et al., 2010; Wei et al., 2010). However, computational studies have revealed that small PAHs interact with free extracellular DNA through (1) van der Waals' forces, (2) π–π stacking and (3) hydrogen bonding (H-bond) (A-B) (Li et al., 2011). While experimentally, it has also been anticipated that in the presence of ions such as calcium (Ca^{2+}), the non-covalent interaction of PAHs can be reversed, which endorses that Ca^{2+} can form the strong electrovalent bonds with the phosphate (PO_4^{3-}) groups of DNA and weaken the PAHs–DNA interactions (Hu et al., 2015). This experiment concluded that DNA has a stronger interaction towards polar ions, such as the Ca^{2+} ion, than with hydrophobic molecules. The ionic interactions of the Ca^{2+} ion with the phosphate group have been depicted in Fig. 5.1.

Recently, with limitations, an antihydrolysis mechanism has been established using substituted (both polar and nonpolar chemical groups) aromatic compounds such as 1,3-dichlorobenzene, *m*-phenylenediamine, aniline, and 2-naphthol. These mechanisms revealed that the substituted aromatic pollutants initiate to cross the ARGs through the extracellular polymeric substance permeable barrier into the interior of cells, and enabling the spreading of AMR (Shou et al., 2019). Upon comparing the evidence with PAHs and substituted aromatic compounds, it can be suggested that the substitution in aromatic rings (such as NH_2, OH, and Cl) can alter the physicochemical properties (e.g. molecular polarity, hydrophobicity, electron-donating character, etc.) (Shou et al., 2019). Hence, depending on the types of interactions between the aromatics and nucleotide bases, hydrocarbons can alter the positions or sequences of the bases, which could be harmful to living beings (Figs.5.2 and 5.3).

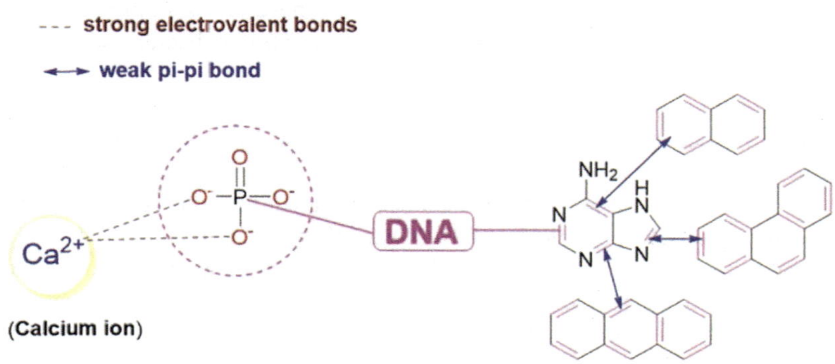

FIGURE 5.1 Ionic electrovalent interactions of calcium ion with deoxyribonucleic acid. Ionic electrovalent interactions of Ca^{2+} with the phosphate group of DNA.

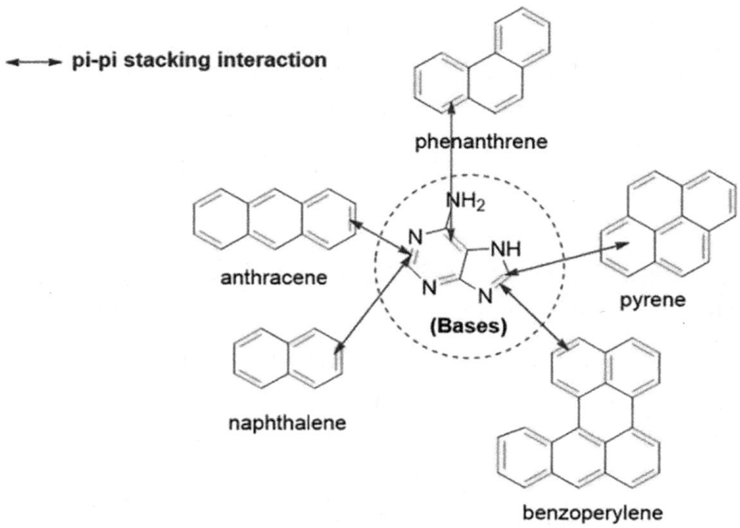

FIGURE 5.2 **π–π stacking interactions between polycyclic aromatic hydrocarbons and nucleoside bases.** π–π stacking interaction: interactions of PAHs with nucleoside bases.

5.5.2 Role of efflux pump in conferring resistance to anthropogenic pollutants

Efflux pumps are also called transporters that are actively involved in the transport of heavy metals, antibiotics, hydrocarbons, etc., which in turn help bacteria to become resistant (Soto, 2013; Ma et al., 2017; Maurya et al., 2020). Antimicrobial efflux pump in bacteria specifies the strategic mechanisms to develop the antibiotic/biocides/hydrocarbon resistance in the environment. About 6%–18% of the membrane proteins of the bacterial genomes represent the bacterial efflux pump (Piddock, 2006). The genetic attributes encoding the efflux pumps are commonly found to be present on a chromosome or in a plasmid (Piddock, 2006; Blanco et al., 2016), for example, RND (resistance nodulation division) pumps were found to be present on the IncH1 plasmid in *Citrobacter freundii* (Blair et al., 2015). RND efflux pump, commonly reported in the gram-negative organisms, comprises a transporter protein known as AcrB (commonly found to be located in the cytoplasmic membrane), an accessory protein AcrA (found to be located in the periplasmic space), and an outer membrane protein channel/outer membrane protein TolC (Nikaido, 2011; Routh et al., 2010). Therefore, to make the bacteria resistant, transporters export the pollutants such as antibiotics, biocides, heavy metals, or hydrocarbons from the cytoplasm and also export from the outer membrane of bacteria (Delmar et al., 2014; Salam, 2020). There are some other efflux pumps which were reported in the bacteria Fig. 5.4, such as MFS (major facilitator super family), MATE (multidrug and toxic compound extrusion), SMR (small multidrug resistance), and ABC (ATP-binding cassette) and are commonly found in gram-positive and gram-negative bacteria (Li et al., 2015). These efflux pumps use ATP or the proton motive force to export the contaminants from the bacterial cells. Besides efflux pumps, bacteria also develop AMR by acquiring resistance genes through different gene transfer mechanisms.

It has been experimentally proven that ARBs can exchange their genes in the soil with other bacteria through gene transfer mechanisms like transformation, conjugation, etc. Also, there were many MGEs, such as plasmids, integrons, insertion sequences, transposons, etc., present in the bacteria that carry the ARGs. Due to the presence of these MGE, acquired resistance is happening through the HGT. It has also been noticed that soils act as the chief reservoir of the ARBs and from where bacteria acquire the ARGs (Table 5.1) (Wintersdorff et al., 2016; Lau et al., 2017; Maurya et al., 2021). Further, it was well noted that integrons can capture ARGs from the environment and exchange the gene cassettes through site-specific recombination in other bacteria, therefore they play a major role in the dissemination of ARGs into different environments (Ling et al., 2013). Among all the integrons, class 1 integrons were the most common and frequent in the environment and clinical settings. There were 5' and 3' conserved segments in the integrons; the 5' segment involved in the integration and excision of gene cassette and carrying *intI* gene, the 3' segment encoding the sulphonamide (*sulI*) and quaternary ammonium compounds (*qacE*) (Partridge et al., 2009; Ling et al., 2013). Thus, it becomes necessary to comprehend the contagion profiles of different antibiotic resistance types with special reference to land uses and other human activities.

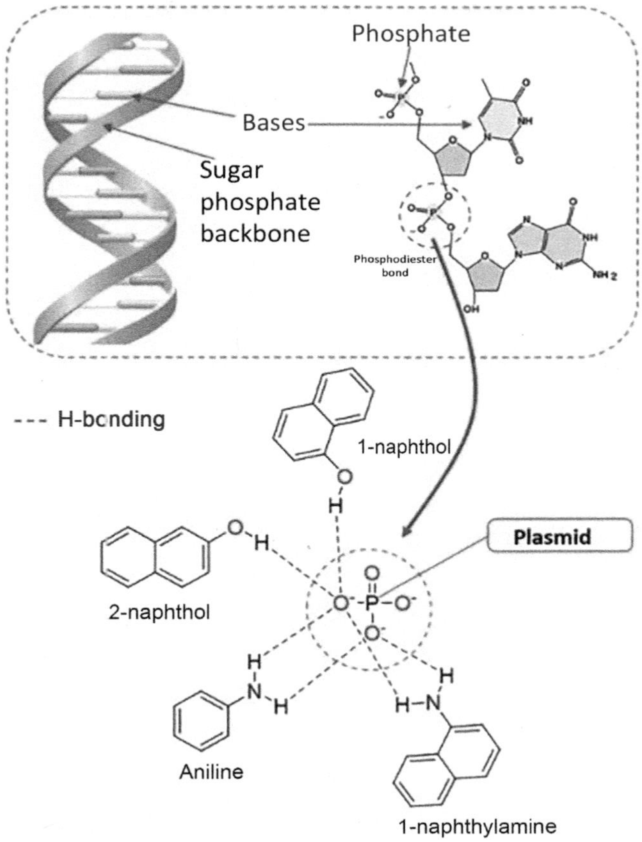

FIGURE 5.3 **Role of hydrogen bonding in binding of polar PAHs to plasmid phosphate backbone.Hydrogen bonding (H-bond) interaction: interactions of polar group-substituted PAHs with the phosphate group of the plasmid.**

5.5.3 Development of resistance while exposure to hydrocarbons

It has been well established that hydrocarbon efflux pumps belong to the RND efflux pump family, which is involved in the transport of hydrocarbons and antibiotics as well. However, the complete role of RND efflux pumps is not well-known but it has been presumed that, these are involved in the transport of the toxic compounds from the bacteria, and also provide the defence to the chemical compounds such as heavy metals, or hydrocarbons or detergents, or dyes, which are present as an environmental contaminants (Alvarez-Ortega et al., 2013; Routh et al., 2010). The RND efflux pump is specific to Gram-negative bacteria and consists of three major components: two are membrane transport proteins located in the outer and inner membranes, while the third is usually present in the periplasm, linking the first two components (Martinez et al., 2009). Previous study indicated that substrate binding may take place in the periplasm of the inner membrane protein (Bina et al., 2008). However, the toxicity of hydrocarbons depends on their preferential dissolution in hydrophobic rather than hydrophilic environments. After bacteria are exposed to hydrocarbons, these are readily dissolved in the cell membrane, leading to ineffectiveness and damage to membrane-associated processes (Sikkema et al., 1995; Yang and Hinner, 2015).

Basically, there are three main resistance strategies against the action of hydrocarbons: to reduce the entry, to degrade the internalised hydrocarbons, and to enhance their transport through the RND efflux pump (Fernández and Hancock, 2012; Soto, 2013). However, the physical properties of the bacterial membrane can be modified by changes in the amount of the fatty acids, cyclopropanation of the membrane lipids, or *cis/trans* isomerisation of unsaturated fatty acids (Sikkema et al., 1995; Martinez et al., 2009; Fernández and Hancock, 2012; Soto, 2013). Further, there is no doubt that transport of hydrocarbons is the mechanism that considerably makes a high impact on the hydrocarbon resistance.

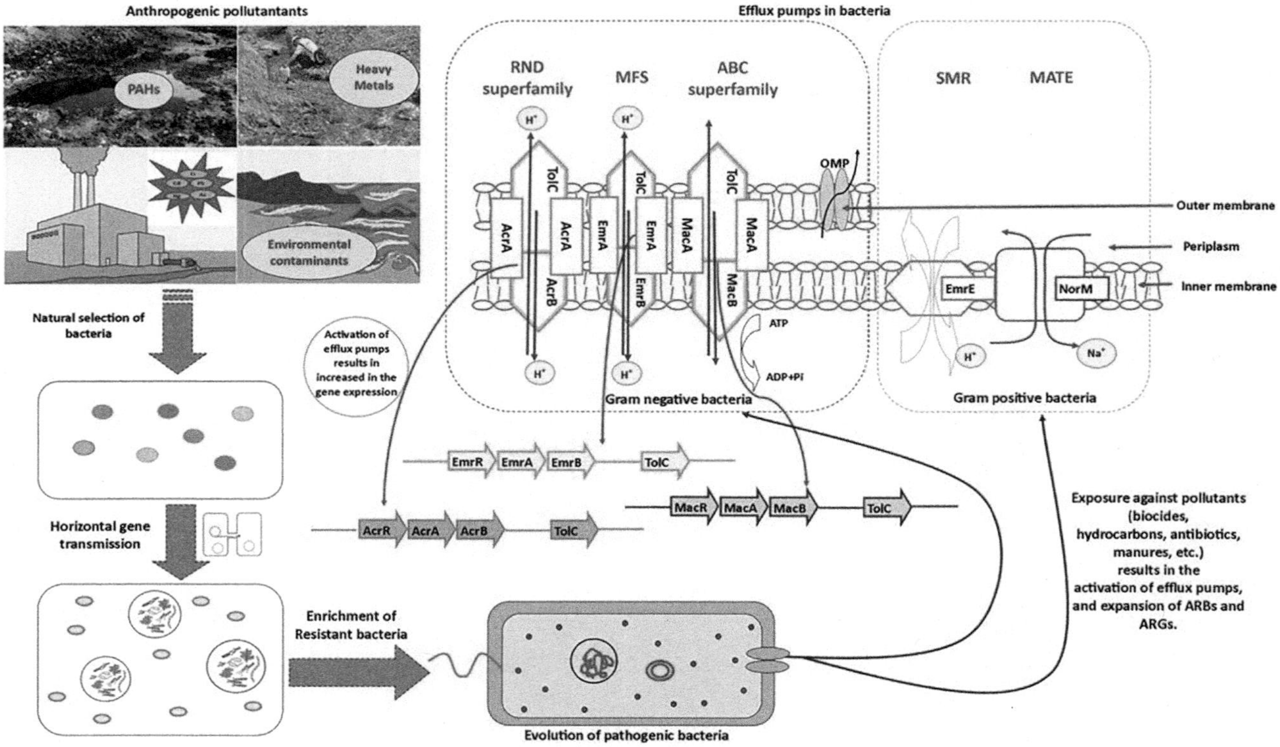

FIGURE 5.4 **Efflux pump systems in bacteria, showing pollutant-induced overexpression and ARG dissemination with active extrusion of toxic compounds.** Resistance nodulation division (RND), major facilitator superfamily (MFS), and ATP-binding cassette (ABC) efflux pumps are present in Gram-negative bacteria, and small multidrug resistance (SMR) and multidrug and toxic compound extrusion (MATE) efflux pumps are present in Gram-positive bacteria. RND, MFS, and ABC efflux pumps of gram-negative bacteria comprise a transporter protein (*AcrB, EmrB, MacB*); an accessory protein (*AcrA, EmrA, MacA*); and an outer membrane protein channel (*TolC*). Exposure to anthropogenic pollutants results in the natural selection and enrichment of antibiotic-resistant bacteria (via horizontal gene transmission). Anthropogenic activity causes mutations or alterations in the efflux pumps, porin proteins, or transporter proteins of the pathogenic bacteria, which results in the over-expression of efflux pumps. Activation of the efflux pumps leads to the increased gene expression/over-expression that enhances the fate of ARGs; and transport of the pollutants (hydrocarbons, biocides, antibiotics, manures, etc.) from the cytoplasm to the outer membrane; and from the periplasm to the outer membrane by efflux pumps.

Reports documented that *Pseudomonas putida* DOT-T1E and S12 have the ability to survive in the toluene-saturated medium. It has also been seen that the hydrocarbon efflux pump SrpABC of S12, and TtgGHI of DOT-T1E were involved in the solvent tolerance (Kieboom et al., 1998; Rojas et al., 2001) and MexEF/OprN of *P. putida* KT2440 were involved in the trinitrotoluene tolerance (Fernández et al., 2009). However, it was experimentally observed that overexpression of TtgGHI of *P. putida* DOT-T1E resulted in the discharge of tetracycline, ampicillin, erythromycin, or norfloxacin (Molina-Santiago et al., 2014). While in the case of *Pseudomonas fluorescens* LP6a, the EmhABC efflux pump had been characterised to transport the hydrocarbons (such as fluoranthrene or phenanthrene) and antibiotics (such as chloramphenicol). Further, chloramphenicol and fluoranthrene or phenanthrene act as substrates for the EmhABC efflux system (Hearn et al., 2003; Adebusuyi and Foght, 2011).

Metagenomic profiling indicates that PAHs act as substrates for efflux transporters such as *acrA*, *acrB*, *amrB* and *tolC* (Chen et al., 2017). However, in *P. putida* B6-2, SrpABC, and TtgABC had been involved in the expulsion of hydrocarbons. Among them, TtgABC had shown increased expression and contributed to antibiotic resistance, while SrpABC was involved in the release of the toxic compounds produced during the metabolism, and also contributed to antibiotic resistance (Yao et al., 2017). Hearn et al. (2003) documented that the interruption of the emhB gene of RND efflux pump showed decreased resistance to nalidixic acid and chloramphenicol; however, disruption did not affect the resistance to tetracycline, erythromycin, or streptomycin (Hearn et al., 2003).

5.5.4 Development of resistance while exposure to biocides

In day-to-day life, the most commonly used biocides are triclosan, chlorohexidine, acids, anilides, halogenated compounds, organic acids, phenolics, oxidising agents, aldehydes, diamides, alcohols, quaternary ammonium compounds (QACs) or benzalkonium chloride (Gadea et al., 2017; Elekhnawy et al., 2020). These biocides alter the bacterial cell envelope, phospholipids, outer membrane proteins or the ionic charges of the cell surface, which finally leads to a decrease in bacterial cell wall membrane permeability (Denyer and Maillard, 2002; Halder et al., 2015). Report documented that, due to the presence of efflux pumps AcrEF-TolC, AcrABTolC, EmrE in *E. coli*, adaptation of QACs leads to bacterial resistance to phenicolic complexes (Pérez et al., 2012; Slipski et al., 2018). Other reports also documented that, in *P. aeruginosa*: MexAB-OprM, MexCD-OprJ, MexEF-OprN and MexJK efflux pumps, in *Salmonella*: AcrABTolC, in *Staphylococcus aureus*: QacA-D, QacG, QacH and Smr efflux pumps are involved in the expansion of biocide resistance (Kaatz et al., 1993; Chuanchuen et al., 2002; Poole, 2005; Alcalde-Rico et al., 2018). It was found that integron-carrying efflux transporters (*qacE*) were also associated with resistance to biocides (Li and Nikaido, 2009), which indicates that the efflux pumps also play important role in the development of intrinsic resistance to biocides.

It has been reported that assembly of the AcrB efflux pump, marA (transcriptional activator) and upregulation of biofilm were the essential mechanisms to promote the MDR in the biocide-tolerant bacteria (Li and Nikaido, 2009). While overexpression of AcrAB–TolC with broad substrate specificity confers reduced resistance to detergents, dyes, or antibiotics (Piddock, 2006). However, in *S. aureus*, overexpression of norC, mdeA, mepA, and norA efflux pump was increasing resistance to dyes and biocides (DeMarco et al., 2007). In another report, adeABC, amvA, abeS efflux pump was found to be involved in the inactivation of aminoglycoside antibiotics and biocides tolerant in the *Acinetobacter* spp. (Eze et al., 2018). Some other studies also documented that the outer membrane porin *kpnO* in *K*lebsiella *pneumoniae*; *adeABC,* and *adeIJK* in *A.baumannii* were not only involved in transporting the biocides but also involved in the efflux of disinfectants and detergents (Fernández and Hancock, 2012; Eze et al., 2018).

5.5.5 Development of resistance while exposure to heavy metals

The most well-known heavy metals are copper (Cu), mercury (Hg), lead (Pb), zinc (Zn), and cadmium (Cd), which are mostly present in the soil in contaminated areas. The main reservoirs of the heavy metals are: urban areas with huge population density and faster anthropogenic activities. The toxicity of metals is commonly determined by the pH of soil, texture, metal concentration, or by estimation of organic carbon (Müller et al., 2020). It has been established that metals participate in the formation of strong coordination bonds with the thiol groups (—SH) and oxygen receptors in biomolecules, and thereafter they enter into the protein structure, changing their functions through the mutations in the DNA (Limo et al., 2018). There are several reports documented on the heavy metal resistance mechanisms: (1) decreasing the metal ions intensification into the bacterial cells through efflux pumps; (2) insertion of cations inside the molecules having the sulfhydryl group; (3) transformation of the heavy metals into the less toxic species; and (4) alteration of cell wall permeability to metals (Nies, 2003; Seiler and Berendonk, 2012).

Removal of heavy metals from the cell wall of bacteria is dependent on energy-dependent ion efflux that acts as proton antiporters/efflux transporters. Basically, there are three main types of efflux transporters that are involved in the transport of heavy metals from the bacterial cell: (1) CBA transporters, (2) P-type ATPase, and (3) cation diffusion facilitator (CDF) Fig. 5.5 (Prabhakaran et al., 2016). These transporters export the metal ion from the cytoplasm to the periplasm, and from the periplasm to the outer membrane. Among all three efflux transporters, CBA transporters play the most important role and are also required in nodulation, cell division and resistance to metal (Nies and Silver, 1995). These transporters are forming three-component protein complexes, such as (1) membrane fusion protein, (2) RND protein, and (3) outer membrane factor (Nies, 2003; Prabhakaran et al., 2016). Further, CDF and P-type ATPase transporters help in the transport of metal ions from the cytoplasm to the periplasm, while CBA transporters detoxify the metals entering into the periplasmic space Fig. 5.5. CBA transporters also remove ions from the periplasm by the ATPase and CDF transporters Fig. 5.5 (Prabhakaran et al., 2016). Additionally, P-type ATPase exports the metal ions from cytoplasm to periplasm using ATP as an energy source; CBA transporters bridge the whole cell wall and export the metals from periplasm to cytoplasm using chemoosmotic gradient; while CDF transporters expel the metal ions from the cytoplasm to periplasm through the proton motive force (Ma et al., 2009).

It has been revealed that bacterial heavy metal resistance structures are usually located on plasmids, chromosomes, or in transposons. Among them, plasmid encoding systems mostly include the pumping of toxic metals; however, in several circumstances, these are found to be located in the chromosome by forming very complex structures (Nies, 2003). It was documented that, in the soil concentration of heavy metal resistance is more than the antibiotics and hence, plasmid encoding the metal resistance easily transmitted from environmental bacteria to pathogenic bacteria, where they also

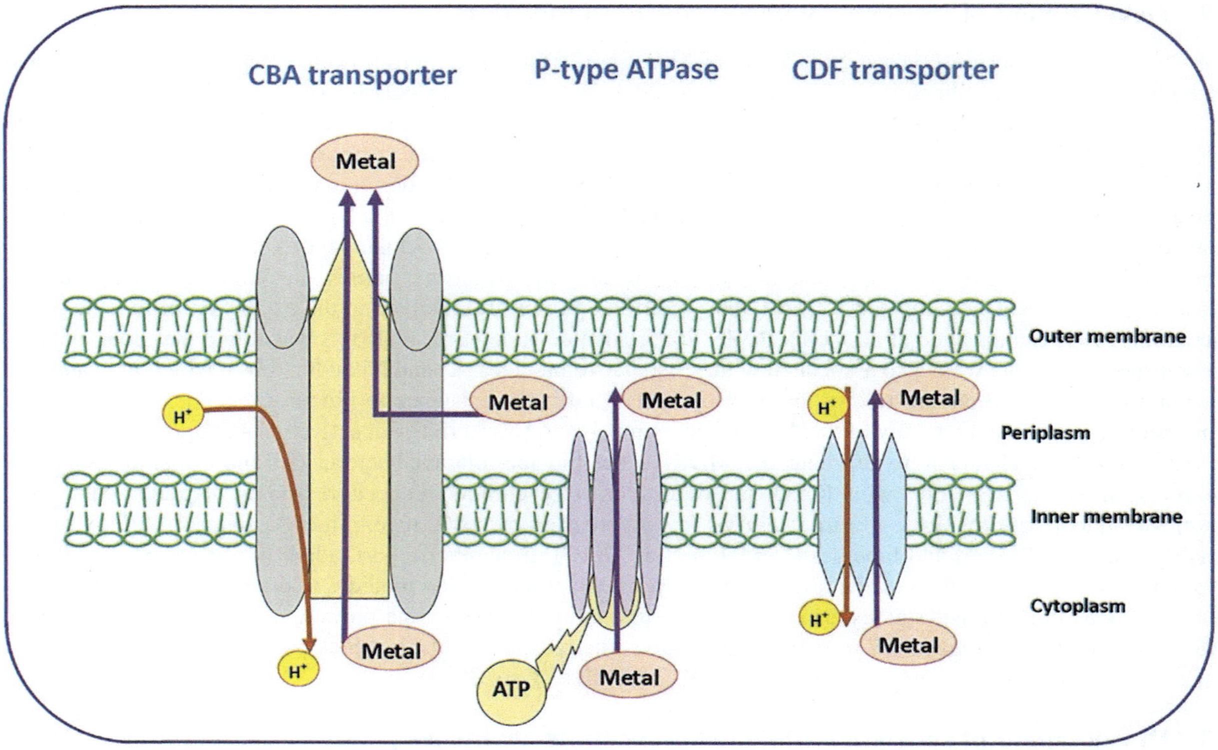

FIGURE 5.5 **Efflux pump transporters mediate the export of heavy metals from bacterial cells, contributing to metal tolerance and resistance.** Efflux pump transporters involved in transportation of heavy metals from the bacterial cell (Prabhakaran et al., 2016). These export the metal ion from cytoplasm to periplasm, and periplasm to outer membrane. Cation diffusion facilitator (CDF) and P-type ATPase helps in the export of heavy metals from cytoplasm to periplasm, whereas CBA detoxify the heavy metals entering into the periplasmic space. CBA remove heavy metals from the periplasm by the ATPase and CDF transporters. P-type ATPase exports the metal ions from cytoplasm to periplasm using ATP as an energy source; CBA transporters bridge the whole cell wall and export the heavy metals from periplasm to cytoplasm using chemoosmotic gradient; however, CDF transporters transport the metal ions from the cytoplasm to periplasm through the proton motive force.

proliferate the heavy metal resistance genes in combination with antibiotic resistance (Das et al., 2016). Researchers found that Cd resistance was present in the *cad* operon (i.e. plasmid-borne operon) in the *Staphylococcus* spp. via the *CadA* pump present in the pI258 plasmid and Cd exporting P-type ATPase (Nucifora et al., 1989), while in *P. aeruginosa* it was present on the czc operon (Chakraborty and Das, 2014). Further, in the heavy metal contaminated soil, 44.33% of the bacteria were resistant to tetracycline antibiotics, 31.21% were resistant to amoxicillin in the mining wastes (Safari Sinegani and Younessi, 2017). While 9.75% of the bacteria were resistant to streptomycin, and 15.48% were resistant to vancomycin, in the pasture soils. Authors also observed the high level of lead (Pb) contamination in mining waste soil (65.8%), pasture (31.57%), and agricultural (7.55%) soil. In addition to this, 50% of the gram-negative isolates were showing coresistance to nickel (Ni), amoxicillin, amikacin, and vancomycin (Safari Sinegani and Younessi, 2017), which indicates that the main reservoir for antibiotic-resistant bacteria is heavy metal-contaminated soil.

5.5.6 Development of resistance while exposure to fertilisers and manures

The microbial population present in the animal manure are highly diverse and usually depends on the animal's species. The main manure-borne pathogens are *E. coli*, *Clostridium*, Listeria monocytogenes, *Bacillus*, *Salmonella* spp., *Enterococcus*, *Campylobacter* spp., and *Yersinia enterocolitica* (Lima et al., 2020). Fertilisers used to enhance the soil productivity have shown a great impact on soil microbial diversity, and using the antibiotics in the livestock changes the microbial diversity, resulting in shaping the antibiotic resistome (Wang et al., 2020).

Based on the chemical structure of antibiotics (such as sulphamethoxazole, tetracycline, chlorotetracycline, oxytetracycline) used in livestock, huge proportions of antibiotics enter the soils when using animal manure

(He et al., 2020). It has been seen that routine use of tetracycline and amoxicillin in the livestock and farming sectors increases the abundance of ARBs and ARGs (Cheng et al., 2019). Zhu et al. (2013) reported that the *tet* gene was the most highly diverse in the animal manure, and is actively involved in the efflux of tetracycline antibiotics, and also plays a role in the enzyme modification of the antibiotics (Zhu et al., 2013). Further, abundance of *sul1, sul2, tetW, tetQ, tetM* and *tetO* genes were found in greater concentrations in animal manure because of their high use in animals (Li et al., 2016; Kyselková et al., 2015a; Guo et al., 2018), while in the case of swine manure, high concentration of *erm* (erythromycin ribosome methylation) genes (*ermA, ermB, ermC, ermF, ermG, ermT, ermQ* and *ermX*) were found in the swine waste ponds (Chen et al., 2007; Van den Meersche et al., 2019; Macedo et al., 2020).

5.6 Future prospects

Antibiotics used in the clinical setting, livestock, and agriculture are entering the environmental soil and water through several routes, disturbing the biodiversity of soil bacteria, and increasing AMR. Through the various chemical bonds and physical forces, the soil particle engrossed the antibiotics. Hence, the first immediate measure should be to evaluate the AMR in the groundwater, aquatic environments, seawater, rivers, coastal areas, manures and other contaminated soil areas, which will help in reducing the burden of resistance in the environment.

In several cases, antibiotics are out of use due to the prevalence of their resistance and treatment choices for certain pathogens have become very limited, whereas bacterial infections still annually claim a hundred thousand lives worldwide. Hence, the significance of antibiotics cannot be overestimated as we are reliant upon them for treating the diseases. To date, many recommendations have been made and proposed, and several reports have been documented, but the expansion of AMR is relentless. Therefore, a proper risk management plan for the disposal of anthropogenic pollutants should be enforced instantly, which will help to control ARBs and ARGs in the environment.

5.7 Conclusion

The presence of anthropogenic pollutants in the environment acts as a cofactor for the expansion of ARGs. Presently, a very limited number of factors are known that help in the expansion of ARGs, and hence, broad proof is needed to understand the factors that help in the expansion of ARGs in the environment. Since pharmaceutical effluents, hydrocarbons, antibiotics, biocides, pesticides, heavy metals, in the natural environments have shown the direct deleterious effect on human health and aquatic life, and therefore are of growing concern. If this worldwide ecological contamination continues in future, the occurrence and expansion of these ARGs in the environment will also probably remain the same. Therefore, reducing the industrial emissions, improving management of animal manures, improving and guaranteeing the good governance, public health expenditure, access to clean and drinking water, refining the sanitation practices in all sectors have become essential to slow down or stop the global burden of AMR.

References

Adebusuyi, A.A., Foght, J.M., 2011. An alternative physiological role for the EmhABC efflux pump in *Pseudomonas fluorescens* cLP6a. BMC Microbiology 11. Available at https://doi.org/10.1186/1471-2180-11-252.

Alcalde-Rico, M., Olivares-Pacheco, J., Alvarez-Ortega, C., Cámara, M., Martínez, J.L., 2018. Role of the Multidrug resistance efflux pump MexCD-OprJ in the *Pseudomonas aeruginosa* quorum sensing response. Frontiers in Microbiology 9. Available at https://doi.org/10.3389/fmicb.2018.02752.

Almakki, A., Jumas-Bilak, E., Marchandin, H., Licznar-Fajardo, P., 2019. Antibiotic resistance in urban runoff. The Science of the Total Environment 667, 64–76. Available at https://doi.org/10.1016/j.scitotenv.2019.02.183.

Alvarez-Ortega, C., Olivares, J., Martínez, J.L., 2013. RND multidrug efflux pumps: what are they good for? Frontiers in Microbiology 4. Available at https://doi.org/10.3389/fmicb.2013.00007, http://www.frontiersin.org/Journal/DownloadFile.ashx?pdf=1&FileId=15454&articleId=40475&Version=1&ContentTypeId=21&FileName=fmicb-04-00007.pdf.

Ashbolt, N.J., Amézquita, A., Backhaus, T., Borriello, P., Brandt, K.K., Collignon, P., et al., 2013. Human health risk assessment (HHRA) for environmental development and transfer of antibiotic resistance. Environmental Health Perspectives 121 (9), 993–1001. Available at https://doi.org/10.1289/ehp.1206316.

Bartlett, J.G., Gilbert, D.N., Spellberg, B., 2013. Seven ways to preserve the miracle of antibiotics. Clinical Infectious Diseases 56 (10), 1445–1450. Available at https://doi.org/10.1093/cid/cit070.

Ben, W., Wang, J., Cao, R., Yang, M., Zhang, Y., Qiang, Z., 2017. Distribution of antibiotic resistance in the effluents of ten municipal wastewater treatment plants in China and the effect of treatment processes. Chemosphere 172, 392–398. Available at https://doi.org/10.1016/j.chemosphere.2017.01.041.

Bengtsson-palme, J., Kristiansson, E., Larsson, D.G.J., 2018. Environmental factors influencing the development and spread of antibiotic resistance. FEMS Microbiology Reviews 68–80. Available at https://doi.org/10.1093/femsre/fux053.

Berendsen, B.J.A., Wegh, R.S., Memelink, J., Zuidema, T., Stolker, L.A.M., 2015. The analysis of animal faeces as a tool to monitor antibiotic usage. Talanta 132, 258–268. Available at https://doi.org/10.1016/j.talanta.2014.09.022, https://www.journals.elsevier.com/talanta.

Bibbal, D., Dupouy, V., Ferré, J.P., Toutain, P.L., Fayet, O., Prère, M.F., 2007. A. Bousquet-mélou, impact of three ampicillin dosage regimens on selection of ampicillin resistance in enterobacteriaceae and excretion of bla TEM genes in swine feces. Applied and Environmental Microbiology 73 (15), 4785–4790. Available at https://doi.org/10.1128/aem.00252-07.

Bina, X.R., Provenzano, D., Nguyen, N., Bina, J.E., 2008. Vibrio cholerae RND family efflux systems are required for antimicrobial resistance, optimal virulence factor production, and colonization of the infant mouse small intestine. Infection and Immunity 76 (8), 3595–3605. Available at https://doi.org/10.1128/IAI.01620-07.

Blair, J.M.A., Webber, M.A., Baylay, A.J., Ogbolu, D.O., Piddock, L.J.V., 2015. Molecular mechanisms of antibiotic resistance. Nature Reviews Microbiology 13 (1), 42–51. Available at https://doi.org/10.1038/nrmicro3380, http://www.nature.com/nrmicro/index.html.

Blanco, P., Hernando-Amado, S., Reales-Calderon, J., Corona, F., Lira, F., Alcalde-Rico, M., et al., 2016. Bacterial multidrug efflux pumps: much more than antibiotic resistance determinants. Microorganisms 4 (1), 14. Available at https://doi.org/10.3390/microorganisms4010014.

Boyland, E., Green, B., 1962. The interaction of polycyclic hydrocarbons and purines. British Journal of Cancer 16 (2), 347–360. Available at https://doi.org/10.1038/bjc.1962.40.

Cabeza, Y., Candela, L., Ronen, D., Teijon, G., 2012. Monitoring the occurrence of emerging contaminants in treated wastewater and groundwater between 2008 and 2010. The Baix Llobregat (Barcelona, Spain). Journal of Hazardous Materials 239-240, 32–39. Available at https://doi.org/10.1016/j.jhazmat.2012.07.032.

Cantas, L., Shah, S.Q.A., Cavaco, L.M., Manaia, C.M., Walsh, F., Popowska, M., et al., 2013. A brief multi-disciplinary review on antimicrobial resistance in medicine and its linkage to the global environmental microbiota. Frontiers in Microbiology 4 (MAY). Available at https://doi.org/10.3389/fmicb.2013.00096, http://www.frontiersin.org/Journal/FullText.aspx?ART_DOI=10.3389/fmicb.2013.00096&name=Antimicrobials,_Resistance_and_Chemotherapy&x=y#h10.

Chakraborty, J., Das, S., 2014. Characterization and cadmium-resistant gene expression of biofilm-forming marine bacterium *Pseudomonas aeruginosa* JP-11. Environmental Science and Pollution Research 21 (24), 14188–14201. Available at https://doi.org/10.1007/s11356-014-3308-7.

Chen, J., Yu, Z., Michel, F.C., Wittum, T., Morrison, M., 2007. Development and application of real-time PCR assays for quantification of erm genes conferring resistance to macrolides-lincosamides-streptogramin B in livestock manure and manure management systems. Applied and Environmental Microbiology 73 (14), 4407–4416. Available at https://doi.org/10.1128/AEM.02799-06.

Chen, W., Liu, W., Pan, N., Jiao, W., Wang, M., 2013. Oxytetracycline on functions and structure of soil microbial community. Journal of Soil Science and Plant Nutrition 13 (4), 967–975. Available at https://doi.org/10.4067/S0718-95162013005000076China, http://www.scielo.cl/pdf/jsspn/v13n4/aop7613.pdf.

Chen, B., He, R., Yuan, K., et al., 2017. Polycyclic aromatic hydrocarbons (PAHs) enriching antibiotic resistance genes (ARGs) in the soils. Environmental Pollution (Barking, Essex : 1987) 220, 1005–1013. Available at https://doi.org/10.1016/j.envpol.2016.11.047.

Cheng, G., Ning, J., Ahmed, S., et al., 2019. Selection and dissemination of antimicrobial resistance in Agri-food production. Antimicrob Resist Infect Control 8, 158. Available at https://doi.org/10.1186/s13756-019-0623-2.

Chuanchuen, R., Beinlich, K., Hoang, T.T., et al., 2001. Cross-resistance between triclosan and antibiotics in Pseudomonas aeruginosa is mediated by multidrug efflux pumps: Exposure of a susceptible mutant strain to triclosan selects nfxB mutants overexpressing MexCD-OprJ. Antimicrobial Agents and Chemotherapy 45, 428–432. Available at https://doi.org/10.1128/AAC.45.2.428-432.2001.

Chuanchuen, R., Narasaki, C.T., Schweizer, H.P., 2002. The MexJK efflux pump of *Pseudomonas aeruginosa* requires OprM for antibiotic efflux but not for efflux of triclosan. Journal of Bacteriology 184 (18), 5036–5044. Available at https://doi.org/10.1128/JB.184.18.5036-5044.2002.

Cieplik, F., Jakubovics, N.S., Buchalla, W., Maisch, T., Hellwig, E., Al-Ahmad, A., 2019. Resistance toward chlorhexidine in oral bacteria-is there cause for concern? Frontiers in Microbiology 10. Available at https://doi.org/10.3389/fmicb.2019.00587, https://www.frontiersin.org/journals/microbiology#.

Cycoń, M., Mrozik, A., Piotrowska-Seget, Z., 2019. Antibiotics in the soil environment—degradation and their impact on microbial activity and diversity. Front Microbiol 10. Available at https://doi.org/10.3389/fmicb.2019.00338.

Das, S., Dash, H.R., Chakraborty, J., 2016. Genetic basis and importance of metal resistant genes in bacteria for bioremediation of contaminated environments with toxic metal pollutants. Applied Microbiology and Biotechnology 100 (7), 2967–2984. Available at https://doi.org/10.1007/s00253-016-7364-4, link.springer.de/link/service/journals/00253/index.htm.

Delmar, J.A., Su, C.C., Yu, E.W., 2014. Bacterial multidrug efflux transporters. Annual Review of Biophysics 43 (1), 93–117. Available at https://doi.org/10.1146/annurev-biophys-051013-022855, http://arjournals.annualreviews.org/loi/biophys.

DeMarco, C.E., Cushing, L.A., Frempong-Manso, E., Seo, S.M., Jaravaza, T.A.A., Kaatz, G.W., 2007. Efflux-related resistance to norfloxacin, dyes, and biocides in bloodstream isolates of *Staphylococcus aureus*. Antimicrobial Agents and Chemotherapy 51 (9), 3235–3239. Available at https://doi.org/10.1128/AAC.00430-07.

Denyer, S.P., Maillard, J.-Y., 2002. Cellular impermeability and uptake of biocides and antibiotics in Gram-negative bacteria. Journal of Applied Microbiology 92 (1), 35S. Available at https://doi.org/10.1046/j.1365-2672.92.5s1.19.x.

Ding, C., Pan, J., Jin, M., Yang, D., Shen, Z., Wang, J., et al., 2016. Enhanced uptake of antibiotic resistance genes in the presence of nanoalumina. Nanotoxicology 10 (8), 1051–1060. Available at https://doi.org/10.3109/17435390.2016.1161856.

Du, D., Wang-Kan, X., Neuberger, A., et al., 2018. Multidrug efflux pumps: structure, function and regulation. Nature Reviews. Microbiology 16, 523–539.

Eduardo-Correia, B., Morales-Filloy, H., Abad, J.P., 2020. Bacteria from the multi-contaminated Tinto River estuary (SW, Spain) show high multi-resistance to antibiotics and point to *Paenibacillus* spp. as antibiotic-resistance-dissemination players. Frontiers in Microbiology 10. Available at https://doi.org/10.3389/fmicb.2019.03071, https://www.frontiersin.org/journals/microbiology#.

Elekhnawy, E., Sonbol, F., Abdelaziz, A., Elbanna, T., 2020. Potential impact of biocide adaptation on selection of antibiotic resistance in bacterial isolates. Future Journal of Pharmaceutical Sciences 6 (1). Available at https://doi.org/10.1186/s43094-020-00119-w.

Eze, E.C., Chenia, H.Y., El Zowalaty, M.E., 2018. Acinetobacter baumannii biofilms: effects of physicochemical factors, virulence, antibiotic resistance determinants, gene regulation, and future antimicrobial treatments. Infection and Drug Resistance 11, 2277–2299. Available at https://doi.org/10.2147/IDR.S169894, https://www.dovepress.com/getfile.php?fileID=46237.

Fang, P., Peng, F., Gao, X., Xiao, P., Yang, J., 2019. Decoupling the dynamics of bacterial taxonomy and antibiotic resistance function in a subtropical urban reservoir as revealed by high-frequency sampling. Frontiers in Microbiology 10. Available at https://doi.org/10.3389/fmicb.2019.01448.

Farias, P., Santo, C.E., Branco, R., Francisco, R., Santos, S., Hansen, L., et al., 2015. Natural hot spots for gain of multiple resistances: arsenic and antibiotic resistances in heterotrophic, aerobic bacteria from marine hydrothermal vent fields. Applied and Environmental Microbiology 81 (7), 2534–2543. Available at https://doi.org/10.1128/aem.03240-14.

Fekadu, S., Alemayehu, E., Dewil, R., Van der Bruggen, B., 2019. Pharmaceuticals in freshwater aquatic environments: a comparison of the African and European challenge. Science of The Total Environment 654, 324–337. Available at https://doi.org/10.1016/j.scitotenv.2018.11.072.

Fernández, M., Duque, E., Pizarro-Tobías, P., et al., 2009. Microbial responses to xenobiotic compounds. Identification of genes that allow Pseudomonas putida KT2440 to cope with 2,4,6-trinitrotoluene. Microb Biotechnol 2, 287–294. Available at https://doi.org/10.1111/j.1751-7915.2009.00085.x.

Fernández, L., Hancock, R.E.W., 2012. Adaptive and mutational resistance: role of porins and efflux pumps in drug resistance. Clinical Microbiology Reviews 25 (4), 661–681. Available at https://doi.org/10.1128/CMR.00043-12Canada, http://cmr.asm.org/content/25/4/661.full.pdf+html.

Fick, J., Söderström, H., Lindberg, R.H., Phan, C., Tysklind, M., Larsson, D.G.J., 2009. Contamination of surface, ground, and drinking water from pharmaceutical production. Environmental Toxicology and Chemistry 28 (12), 2522–2527. Available at https://doi.org/10.1897/09-073.1.

Gadea, R., Fernández Fuentes, M.Á., Pérez Pulido, R., Gálvez, A., Ortega, E., 2017. Effects of exposure to quaternary-ammonium-based biocides on antimicrobial susceptibility and tolerance to physical stresses in bacteria from organic foods. Food Microbiology 63, 58–71. Available at https://doi.org/10.1016/j.fm.2016.10.037, http://www.elsevier.com/inca/publications/store/6/2/2/8/3/3/index.htt.

Gorovtsov, A.V., Sazykin, I.S., Sazykina, M.A., 2018. The influence of heavy metals, polyaromatic hydrocarbons, and polychlorinated biphenyls pollution on the development of antibiotic resistance in soils. Environ. Sci. Pollut. Res. 25, 9283–9292.

Guo, J., Li, J., Chen, H., et al., 2017. Metagenomic analysis reveals wastewater treatment plants as hotspots of antibiotic resistance genes and mobile genetic elements. Water Research 123, 468–478. Available at https://doi.org/10.1016/j.watres.2017.07.002.

Häder, D.P., Banaszak, A.T., Villafañe, V.E., Narvarte, M.A., González, R.A., Helbling, E.W., 2020. Anthropogenic pollution of aquatic ecosystems: emerging problems with global implications. Science of the Total Environment 713. Available at https://doi.org/10.1016/j.scitotenv.2020.136586, www.elsevier.com/locate/scitotenv.

Halder, S., Yadav, K.K., Sarkar, R., Mukherjee, S., Saha, P., Haldar, S., et al., 2015. Alteration of Zeta potential and membrane permeability in bacteria: a study with cationic agents. SpringerPlus 4 (1), 1–14. Available at https://doi.org/10.1186/s40064-015-1476-7, http://www.springerplus.com/archive.

Hatosy, S.M., Martiny, A.C., 2015. The ocean as a global reservoir of antibiotic resistance genes. Applied and Environmental Microbiology 81 (21), 7593–7599. Available at https://doi.org/10.1128/AEM.00736-15, http://aem.asm.org/content/81/21/7593.full.pdf.

He, Y., Yuan, Q., Mathieu, J., Stadler, L., Senehi, N., Sun, R., et al., 2020. Antibiotic resistance genes from livestock waste: occurrence, dissemination, and treatment. npj Clean Water 3 (1). Available at https://doi.org/10.1038/s41545-020-0051-0.

Hearn, E.M., Dennis, J.J., Gray, M.R., Foght, J.M., 2003. Identification and characterization of the emhABC efflux system for polycyclic aromatic hydrocarbons in *Pseudomonas fluorescens* cLP6a. Journal of Bacteriology 185 (21), 6233–6240. Available at https://doi.org/10.1128/JB.185.21.6233-6240.2003.

Hendriksen, R.S., Munk, P., Njage, P., van Bunnik, B., McNally, L., Lukjancenko, O., et al., 2019. Global monitoring of antimicrobial resistance based on metagenomics analyses of urban sewage. Nature Communications 10 (1). Available at https://doi.org/10.1038/s41467-019-08853-3, http://www.nature.com/ncomms/index.html.

Heuer, H., Solehati, Q., Zimmerling, U., Kleineidam, K., Schloter, M., Müller, T., et al., 2011. Accumulation of sulfonamide resistance genes in arable soils due to repeated application of manure containing sulfadiazine. Applied and Environmental Microbiology 77 (7), 2527–2530. Available at https://doi.org/10.1128/aem.02577-10.

Hu, X., Kang, F., Gao, Y., Zhou, Q., 2015. Bacterial diversity losses: A potential extracellular driving mechanism involving the molecular ecological function of hydrophobic polycyclic aromatic hydrocarbons. Biotechnol. Reports 5, 27–30.

Hu, Y., Yang, X., Li, J., et al., 2016. The bacterial mobile resistome transfer network connecting the animal and human microbiomes. Applied and Environmental Microbiology 82, 6672–6681. Available at https://doi.org/10.1128/AEM.01802-16.

Huang, Z., Lu, Q., Wang, J., Chen, X., Mao, X., He, Z., et al., 2017. Inhibition of the bioavailability of heavy metals in sewage sludge biochar by adding two stabilizers. PLOS ONE 12 (8), e0183617. Available at https://doi.org/10.1371/journal.pone.0183617.

Ianeva, O.D., 2009. Mechanisms of bacteria resistance to heavy metals. Mikrobiolohichnyĭ zhurnal (Kiev, Ukraine: 1993) 71 (6), 54–65.

Ishikawa, N.K., Touno, E., Higashiyama, Y., Sasamoto, M., Soma, M., Yoshida, N., et al., 2018. Determination of tylosin excretion from sheep to assess tylosin spread to agricultural fields by manure application. Science of the Total Environment 633, 399–404. Available at https://doi.org/10.1016/j.scitotenv.2018.03.216, www.elsevier.com/locate/scitotenv.

Jang, H.M., Kim, Y.B., Choi, S., Lee, Y., Shin, S.G., Unno, T., et al., 2018. Prevalence of antibiotic resistance genes from effluent of coastal aquaculture, South Korea. Environmental Pollution 233, 1049–1057. Available at https://doi.org/10.1016/j.envpol.2017.10.006, www.elsevier.com/inca/publications/store/4/0/5/8/5/6.

Johnson, E.R., Keinan, S., Mori-Sánchez, P., Contreras-García, J., Cohen, A.J., Yang, W., 2010. Revealing noncovalent interactions. Journal of the American Chemical Society 132 (18), 6498–6506. Available at https://doi.org/10.1021/ja100936w.

Jurado, A., Vàzquez-Suñé, E., Carrera, J., López de Alda, M., Pujades, E., Barceló, D., 2012. Emerging organic contaminants in groundwater in Spain: a review of sources, recent occurrence and fate in a European context. Science of The Total Environment 440, 82–94. Available at https://doi.org/10.1016/j.scitotenv.2012.08.029.

Kaatz, G.W., Seo, S.M., Ruble, C.A., 1993. Efflux-mediated fluoroquinolone resistance in *Staphylococcus aureus*. Antimicrobial Agents and Chemotherapy 37 (5), 1086–1094. Available at https://doi.org/10.1128/AAC.37.5.1086.

Kang, F., Hu, X., Liu, J., Gao, Y., 2015. Noncovalent binding of polycyclic aromatic hydrocarbons with genetic bases reducing the in vitro lateral transfer of antibiotic resistant genes. Environmental Science & Technology 49 (17), 10340–10348. Available at https://doi.org/10.1021/acs.est.5b02293.

Karkman, A., Do, T.T., Walsh, F., Virta, M.P.J., 2018. Antibiotic-resistance genes in waste water. Trends in Microbiology 26 (3), 220–228. Available at https://doi.org/10.1016/j.tim.2017.09.005, www.elsevier.com/locate/tim.

Kieboom, J., Dennis, J.J., Zylstra, G.J., De Bont, J.A.M., 1998. Active efflux of organic solvents by *Pseudomonas putida* S12 is induced by solvents. Journal of Bacteriology 180 (24), 6769–6772. Available at https://doi.org/10.1128/jb.180.24.6769-6772.1998, http://jb.asm.org/.

Kim, K.R., Owens, G., Kwon, S.I., So, K.H., Lee, D.B., Ok, Y.S., 2011. Occurrence and environmental fate of veterinary antibiotics in the terrestrial environment. Water, Air, and Soil Pollution 214 (1-4), 163–174. Available at https://doi.org/10.1007/s11270-010-0412-2, http://www.kluweronline.com/issn/0049-6979/.

Kim, M., Jung, J.H., Ha, S.Y., An, J.G., Shim, W.J., Yim, U.H., 2017. Long-term monitoring of PAH contamination in sediment and recovery after the hebei spirit oil spill. Archives of Environmental Contamination and Toxicology 73 (1), 93–102. Available at https://doi.org/10.1007/s00244-017-0365-1, link.springer.de/link/service/journals/00244/index.htm.

Knapp, C.W., Callan, A.C., Aitken, B., Shearn, R., Koenders, A., Hinwood, A., 2017. Relationship between antibiotic resistance genes and metals in residential soil samples from Western Australia. Environmental Science and Pollution Research 24 (3), 2484–2494. Available at https://doi.org/10.1007/s11356-016-7997-y, http://www.springerlink.com/content/0944-1344.

Kohanski, M.A., DePristo, M.A., Collins, J.J., 2010. Sublethal antibiotic treatment leads to multidrug resistance via radical-induced mutagenesis. Molecular Cell 37 (3), 311–320. Available at https://doi.org/10.1016/j.molcel.2010.01.003.

Kyselková, M., Jirout, J., Vrchotová, N., et al., 2015. Spread of tetracycline resistance genes at a conventional dairy farm. Front Microbiol 6. Available at https://doi.org/10.3389/fmicb.2015.00536.

Lau, Engelen, C.H., van, K., Gordon, S., et al., 2017. Novel antibiotic resistance determinants from agricultural soil exposed to antibiotics widely used in human medicine and animal farming. Applied and Environmental Microbiology 83, e00989-17.

Lee, C.R., Lee, J.H., Park, M., Park, K.S., Bae, I.K., Kim, Y.B., et al., 2017. Biology of *Acinetobacter baumannii:* pathogenesis, antibiotic resistance mechanisms, and prospective treatment options. Frontiers in Cellular and Infection Microbiology 7. Available at https://doi.org/10.3389/fcimb.2017.00055, https://fjfsdata01prod.blob.core.windows.net/articles/files/249706/pubmed-zip/.versions/1/.package-entries/fcimb-07-00055/fcimb-07-00055.pdf.

Lekunberri, I., Balcázar, J.L., Borrego, C.M., 2018. Metagenomic exploration reveals a marked change in the river resistome and mobilome after treated wastewater discharges. Environmental Pollution 234, 538–542. Available at https://doi.org/10.1016/j.envpol.2017.12.001, www.elsevier.com/inca/publications/store/4/0/5/8/5/6.

Leonard, A.F.C., Yin, X.L., Zhang, T., Hui, M., Gaze, W.H., 2018. A coliform-targeted metagenomic method facilitating human exposure estimates to *Escherichia coli*-borne antibiotic resistance genes. FEMS Microbiology Ecology 94 (3). Available at https://doi.org/10.1093/femsec/fiy024.

Li, X.Z., Nikaido, H., 2009. Efflux-mediated drug resistance in bacteria: an update. Drugs 69 (12), 1555–1623. Available at https://doi.org/10.2165/11317030-000000000-00000.

Li, F., Li, X., Liu, X., et al., 2011. Noncovalent interactions between hydroxylated polycyclic aromatic hydrocarbon and DNA: Molecular docking and QSAR study. Environmental Toxicology and Pharmacology 32, 373–381. Available at https://doi.org/10.1016/j.etap.2011.08.001.

Li, W., Shi, Y., Gao, L., et al., 2012. Occurrence of antibiotics in water, sediments, aquatic plants, and animals from Baiyangdian Lake in North China. Chemosphere 89, 1307–1315. Available at https://doi.org/10.1016/j.chemosphere.2012.05.079.

Li, X.Z., Plésiat, P., Nikaido, H., 2015. The challenge of efflux-mediated antibiotic resistance in Gram-negative bacteria. Clinical Microbiology Reviews 28 (2), 337–418. Available at https://doi.org/10.1128/CMR.00117-14, http://cmr.asm.org/content/28/2/337.full.pdf+html.

Li, D., Zeng, S., He, M., Gu, A.Z., 2016. Water disinfection byproducts induce antibiotic resistance-role of environmental pollutants in resistance phenomena. Environmental Science and Technology 50 (6), 3193–3201. Available at https://doi.org/10.1021/acs.est.5b05113, http://pubs.acs.org/journal/esthag.

Ling, A.L., Pace, N.R., Hernandez, M.T., Lapara, T.M., 2013. Tetracycline resistance and class 1 integron genes associated with indoor and outdoor aerosols. Environmental Science & Technology 47, 4046–4052. Available at https://doi.org/10.1021/es400238g.

López-Pacheco, I.Y., Silva-Núñez, A., Salinas-Salazar, C., et al., 2019. Anthropogenic contaminants of high concern: Existence in water resources and their adverse effects. The Science of the Total Environment 690, 1068–1088.

Lima, T., Domingues, S., Silva, G.J.D., 2020. Manure as a potential hotspot for antibiotic resistance dissemination by horizontal gene transfer events. Veterinary Sciences 7 (3). Available at https://doi.org/10.3390/VETSCI7030110, https://res.mdpi.com/d_attachment/vetsci/vetsci-07-00110/article_deploy/vetsci-07-00110.pdf.

Limo, M.J., Sola-Rabada, A., Boix, E., Thota, V., Westcott, Z.C., Puddu, V., et al., 2018. Interactions between metal oxides and biomolecules: from fundamental understanding to applications. Chemical Reviews 118 (22), 11118–11193. Available at https://doi.org/10.1021/acs.chemrev.7b00660.

Lu, J., Tian, Z., Yu, J., Yang, M., Zhang, Y., 2018. Distribution and abundance of antibiotic resistance genes in sand settling reservoirs and drinking water treatment plants across the Yellow River, China. Water 10 (3), 246. Available at https://doi.org/10.3390/w10030246.

Lundy, L., Ellis, J.B., Revitt, D.M., 2012. Risk prioritisation of stormwater pollutant sources. Water Research 46 (20), 6589–6600. Available at https://doi.org/10.1016/j.watres.2011.10.039.

Ma, Z., Jacobsen, F.E., Giedroc, D.P., 2009. Coordination chemistry of bacterial metal transport and sensing. Chemical Reviews 109 (10), 4644–4681. Available at https://doi.org/10.1021/cr900077w, http://pubs.acs.org/doi/pdfplus/10.1021/cr900077w.

Ma, L., Li, B., Jiang, X.T., Wang, Y.L., Xia, Y., Li, A.D., et al., 2017. Catalogue of antibiotic resistome and host-tracking in drinking water deciphered by a large scale survey. Microbiome 5. Available at https://doi.org/10.1186/S40168-017-0369-0, https://microbiomejournal.biomedcentral.com/articles/10.1186/s40168-017-0369-0.

Macedo, G., Hernandez-Leal, L., van der Maas, P., Heederik, D., Mevius, D., Schmitt, H., 2020. The impact of manure and soil texture on antimicrobial resistance gene levels in farmlands and adjacent ditches. Science of The Total Environment 737, 139563. Available at https://doi.org/10.1016/j.scitotenv.2020.139563.

Martinez, J.L., Sánchez, M.B., Martínez-Solano, L., Hernandez, A., Garmendia, L., Fajardo, A., et al., 2009. Functional role of bacterial multidrug efflux pumps in microbial natural ecosystems. FEMS Microbiology Reviews 33 (2), 430–449. Available at https://doi.org/10.1111/j.1574-6976.2008.00157.x.

Martínez, J.L., Coque, T.M., Baquero, F., 2015. What is a resistance gene? Ranking risk in resistomes. Nature Reviews Microbiology 13 (2), 116–123. Available at https://doi.org/10.1038/nrmicro3399.

Maurya, A.P., Rajkumari, J., Bhattacharjee, A., Pandey, P., 2020. Development, spread and persistence of antibiotic resistance genes (ARGs) in the soil microbiomes through co-selection. Reviews on Environmental Health 35 (4), 371–378. Available at https://doi.org/10.1515/reveh-2020-0035, http://www.degruyter.com/view/j/reveh.

Maurya, A.P., Rajkumari, J., Pandey, P., 2021. Enrichment of antibiotic resistance genes (ARGs) in polyaromatic hydrocarbon–contaminated soils: a major challenge for environmental health. Environ Sci Pollut Res. Available at https://doi.org/10.1007/s11356-020-12171-3.

McKinney, C.W., Loftin, K.A., Meyer, M.T., Davis, J.G., Pruden, A., 2010. Tet and sul antibiotic resistance genes in livestock lagoons of various operation type, configuration, and antibiotic occurrence. Environmental Science and Technology 44 (16), 6102–6109. Available at https://doi.org/10.1021/es9038165.

Molina-Santiago, C., Daddaoua, A., Fillet, S., et al., 2014. Interspecies signalling: Pseudomonas putida efflux pump TtgGHI is activated by indole to increase antibiotic resistance. Environmental Microbiology 16, 1267–1281. Available at https://doi.org/10.1111/1462-2920.12368.

Müller, A., Österlund, H., Marsalek, J., Viklander, M., 2020. The pollution conveyed by urban runoff: a review of sources. Science of The Total Environment 709, 136125. Available at https://doi.org/10.1016/j.scitotenv.2019.136125.

Na, G., Fang, X., Cai, Y., Ge, L., Zong, H., Yuan, X., et al., 2013. Occurrence, distribution, and bioaccumulation of antibiotics in coastal environment of Dalian, China. Marine Pollution Bulletin 69 (1-2), 233–237. Available at https://doi.org/10.1016/j.marpolbul.2012.12.028.

Nies, D.H., 2003. Efflux-mediated heavy metal resistance in prokaryotes. FEMS Microbiology Reviews 27 (2-3), 313–339. Available at https://doi.org/10.1016/S0168-6445(03)00048-2, http://femsre.oxfordjournals.org/.

Nikaido, H., 2011. Structure and mechanism of RND-type multidrug efflux pumps. Advances in enzymology and Related Areas of Molecular Biology 77, 1–60.

Nucifora, G., Chu, L., Misra, T.K., Silver, S., 1989. Cadmium resistance from *Staphylococcus aureus* plasmid pI258 cadA gene results from a cadmium-efflux ATPase. Proceedings of the National Academy of Sciences 86 (10), 3544–3548. Available at https://doi.org/10.1073/pnas.86.10.3544.

Nguyen, C.C., Hugie, C.N., Kile, M.L., Navab-Daneshmand, T., 2019. Association between heavy metals and antibiotic-resistant human pathogens in environmental reservoirs: A review. Front. Environ. Sci. Eng. 13 (3), 46.

Nies, D.H., Silver, S., 1995. Ion efflux systems involved in bacterial metal resistances. Journal of Industrial Microbiology 14, 186–199. Available at https://doi.org/10.1007/BF01569902.

Olesen, S.W., Barnett, M.L., Macfadden, D.R., Brownstein, J.S., Hernández-Díaz, S., Lipsitch, M., et al., 2018. The distribution of antibiotic use and its association with antibiotic resistance. eLife 7. Available at https://doi.org/10.7554/eLife.39435, https://elifesciences.org/download/aHR0cHM6Ly9jZG4uZWxpZmVzY2llbmNlcy5vcmcvYXJ0aWNsZXMvMzk0MzUvZWxpZmUtMzk0MzUtdjIucGRm/elife-39435-v2.pdf?_hash=dtAtMK4pOQh1DkIuBNDK%2BLZKhA20MSzeIzpuKovlCDM%3D.

Partridge, S.R., Tsafnat, G., Coiera, E., Iredell, J.R., 2009. Gene cassettes and cassette arrays in mobile resistance integrons: Review article. FEMS Microbiology Reviews 33, 757–784.

Peak, N., Knapp, C.W., Yang, R.K., Hanfelt, M.M., Smith, M.S., Aga, D.S., et al., 2007. Abundance of six tetracycline resistance genes in wastewater lagoons at cattle feedlots with different antibiotic use strategies. Environmental Microbiology 9 (1), 143–151. Available at https://doi.org/10.1111/j.1462-2920.2006.01123.x.

Pérez, A., Poza, M., Fernández, A., del Carmen Fernández, M., Mallo, S., Merino, M., et al., 2012. Involvement of the AcrAB-TolC efflux pump in the resistance, fitness, and virulence of enterobacter cloacae. Antimicrobial Agents and Chemotherapy 56 (4), 2084–2090. Available at https://doi.org/10.1128/aac.05509-11.

Piddock, L.J.V., 2006. Clinically relevant chromosomally encoded multidrug resistance efflux pumps in bacteria. Clinical Microbiology Reviews 19 (2), 382–402. Available at https://doi.org/10.1128/cmr.19.2.382-402.2006.

Poole, K., 2005. Efflux-mediated antimicrobial resistance. Journal of Antimicrobial Chemotherapy 56 (1), 20–51. Available at https://doi.org/10.1093/jac/dki171.

Poole, K., 2007. Efflux pumps as antimicrobial resistance mechanisms. Annals of Medicine 39, 162–176.

Potenski, C.J., Gandhi, M., Matthews, K.R., 2003. Exposure of *Salmonella* enteritidis to chlorine or food preservatives increases susceptibility to antibiotics. FEMS Microbiology Letters 220 (2), 181–186. Available at https://doi.org/10.1016/S0378-1097(03)00099-5, http://femsle.oxfordjournals.org/.

Prabhakaran, P., Ashraf, M.A., Aqma, W.S., 2016. Microbial stress response to heavy metals in the environment. RSC Advances 6 (111), 109862–109877. Available at https://doi.org/10.1039/c6ra10966g, http://pubs.rsc.org/en/journals/journalissues.

Rodríguez Martín, J.A., De Arana, C., Ramos-Miras, J.J., Gil, C., Boluda, R., 2015. Impact of 70 years urban growth associated with heavy metal pollution. Environmental Pollution 196, 156–163. Available at https://doi.org/10.1016/j.envpol.2014.10.014.

RodgersDanner, M.C., Robertson, A., Behrends, V., Reiss, J., 2019. Antibiotic pollution in surface fresh waters: Occurrence and effects. The Science of the Total Environment 664, 793–804.

Rojas, A., Duque, E., Mosqueda, G., Golden, G., Hurtado, A., Ramos, J.L., et al., 2001. Three efflux pumps are required to provide efficient tolerance to toluene in *Pseudomonas putida* DOT-T1E. Journal of Bacteriology 183 (13), 3967–3973. Available at https://doi.org/10.1128/jb.183.13.3967-3973.2001.

Romero, J.L., Grande Burgos, M.J., Pérez-Pulido, R., Gálvez, A., Lucas, R., 2017. Resistance to antibiotics, biocides, preservatives and metals in bacteria isolated from seafoods: co-selection of strains resistant or tolerant to different classes of compounds. Frontiers in Microbiology 8. Available at https://doi.org/10.3389/fmicb.2017.01650, https://www.frontiersin.org/journals/microbiology#.

Routh, M.D., Zalucki, Y., Su, C.C., et al., 2010. Efflux pumps of the Resistance-nodulation-division family: A perspective of their structure, function, and regulation in gram-negative bacteria. Advances in Enzymology and Related Areas of Molecular Biology 77 (1), 109–146. Available at https://doi.org/10.1002/9780470920541.ch3.

Rutgersson, C., Fick, J., Marathe, N., Kristiansson, E., Janzon, A., Angelin, M., et al., 2014. Fluoroquinolones and qnr genes in sediment, water, soil, and human fecal flora in an environment polluted by manufacturing discharges. Environmental Science & Technology 48 (14), 7825–7832. Available at https://doi.org/10.1021/es501452a.

Safari Sinegani, A.A., Younessi, N., 2017. Antibiotic resistance of bacteria isolated from heavy metal-polluted soils with different land uses. Journal of Global Antimicrobial Resistance 10, 247–255. Available at https://doi.org/10.1016/j.jgar.2017.05.012, http://www.journals.elsevier.com/journal-of-global-antimicrobial-resistance/.

Salam, L.B., 2020. Unravelling the antibiotic and heavy metal resistome of a chronically polluted soil. 3 Biotech 10 (6). Available at https://doi.org/10.1007/s13205-020-02219-z, http://www.springerlink.com/content/2190-572x/.

Sandner-Miranda, L., Vinuesa, P., Cravioto, A., Morales-Espinosa, R., 2018. The genomic basis of intrinsic and acquired antibiotic resistance in the genus serratia. Frontiers in Microbiology 9 (MAY). Available at https://doi.org/10.3389/fmicb.2018.00828.

Sanganyado, E., Gwenzi, W., 2019. Antibiotic resistance in drinking water systems: Occurrence, removal, and human health risks. Science of the Total Environment 669, 785–797. Available at https://doi.org/10.1016/j.scitotenv.2019.03.162.

Schwaiger, K., Hölzel, C., Bauer, J., 2010. Resistance gene patterns of tetracycline resistant *Escherichia coli* of human and porcine origin. Veterinary Microbiology 142 (3-4), 329–336. Available at https://doi.org/10.1016/j.vetmic.2009.09.066.

Schwarze, R., Brokamp, A., Schmidt, F.R.J., 1997. Isolation and characterization of dehalogenases from 2,2- dichloropropionate-degrading soil bacteria. Current Microbiology 34 (2), 103–109. Available at https://doi.org/10.1007/s002849900152.

Seiler, C., Berendonk, T.U., 2012. Heavy metal driven co-selection of antibiotic resistance in soil and water bodies impacted by agriculture and aquaculture. Frontiers in Microbiology 3. Available at https://doi.org/10.3389/fmicb.2012.00399, http://www.frontiersin.org/Journal/DownloadFile.ashx?pdf=1&FileId=14564&articleId=18889&Version=1&ContentTypeId=21&FileName=fmicb-03-00399.pdf.

Shao, J.L., Lai, B., Jiang, W., Wang, J.T., Hong, Y.H., Chen, F.B., et al., 2019. Diversity and co-occurrence patterns of soil bacterial and fungal communities of chinese cordyceps habitats at shergyla mountain, tibet: Implications for the occurrence. Microorganisms 7 (9). Available at https://doi.org/10.3390/microorganisms7090284, https://www.mdpi.com/2076-2607/7/9/284/pdf.

Shou, W., Kang, F., Huang, S., Yan, C., Zhou, J., Wang, Y., 2019. Substituted aromatic-facilitated dissemination of mobile antibiotic resistance genes via an antihydrolysis mechanism across an extracellular polymeric substance permeable barrier. Environmental Science & Technology 53 (2), 604–613. Available at https://doi.org/10.1021/acs.est.8b05750.

Sikkema, J., de Bont, J.A., Poolman, B., 1995. Mechanisms of membrane toxicity of hydrocarbons. Microbiological Reviews 59 (2), 201–222. Available at https://doi.org/10.1128/mmbr.59.2.201-222.1995.

Singer, A.C., Shaw, H., Rhodes, V., Hart, A., 2016. Review of antimicrobial resistance in the environment and its relevance to environmental regulators. Front Microbiol 7, 1–22. Available at https://doi.org/10.3389/fmicb.2016.01728.

Sim, W.J., Lee, J.W., Lee, E.S., Shin, S.K., Hwang, S.R., Oh, J.E., 2011. Occurrence and distribution of pharmaceuticals in wastewater from households, livestock farms, hospitals and pharmaceutical manufactures. Chemosphere 82 (2), 179–186. Available at https://doi.org/10.1016/j.chemosphere.2010.10.026, www.elsevier.com/locate/chemosphere.

Slipski, C.J., Zhanel, G.G., Bay, D.C., 2018. Biocide selective TolC-independent efflux pumps in enterobacteriaceae. Journal of Membrane Biology 251 (1), 15–33. Available at https://doi.org/10.1007/s00232-017-9992-8, link.springer.de/link/service/journals/00232/index.htm.

Soto, S.M., 2013. Role of efflux pumps in the antibiotic resistance of bacteria embedded in a biofilm. Virulence 4 (3), 223–229. Available at https://doi.org/10.4161/viru.23724, http://www.landesbioscience.com/journals/virulence/2012VIRULENCE0085R.pdf?nocache=1859583232.

Su, H.C., Pan, C.G., Ying, G.G., Zhao, J.L., Zhou, L.J., Liu, Y.S., et al., 2014. Contamination profiles of antibiotic resistance genes in the sediments at a catchment scale. Science of the Total Environment 490, 708–714. Available at https://doi.org/10.1016/j.scitotenv.2014.05.060, www.elsevier.com/locate/scitotenv.

Su, S., Li, C., Yang, J., et al., 2020. Distribution of antibiotic resistance genes in three different natural water bodies-a lake, river and sea. International Journal of Environmental Research and Public Health 17. Available at https://doi.org/10.3390/ijerph17020552.

Sun, J., Deng, Z., Yan, A., 2014. Bacterial multidrug efflux pumps: Mechanisms, physiology and pharmacological exploitations. Biochemical and Biophysical Research Communications 453, 254–267. Available at https://doi.org/10.1016/j.bbrc.2014.05.090.

Telesinski, A., Platkowski, M., Cybulska, K., Telesinska, N., Wrobel, J., Pawlowska, B., 2018. Response of soil enzymes to two antibiotics: polymyxin B and penicillin G. Parlar scientific publications. Poland Fresenius Environmental Bulletin 27 (5), 3837–3845. Available at http://www.psp-parlar.de/feb_auswahl.asp.

Thompson, F., de Oliveira, B.C., Cordeiro, M.C., Masi, B.P., Rangel, T.P., Paz, P., et al., 2020. Severe impacts of the Brumadinho dam failure (Minas Gerais, Brazil) on the water quality of the Paraopeba River. Science of the Total Environment 705. Available at https://doi.org/10.1016/j.scitotenv.2019.135914, www.elsevier.com/locate/scitotenv.

Valcárcel, Y., González Alonso, S., Rodríguez-Gil, J.L., et al., 2011. Detection of pharmaceutically active compounds in the rivers and tap water of the Madrid Region (Spain) and potential ecotoxicological risk. Chemosphere 84, 1336–1348. Available at https://doi.org/10.1016/j.chemosphere.2011.05.014.

Van den Meersche, T., Rasschaert, G., Haesebrouck, F., Van Coillie, E., Herman, L., Van Weyenberg, S., et al., 2019. Presence and fate of antibiotic residues, antibiotic resistance genes and zoonotic bacteria during biological swine manure treatment. Ecotoxicology and Environmental Safety 175, 29–38. Available at https://doi.org/10.1016/j.ecoenv.2019.01.127.

Wang, F., Han, W., Chen, S., Dong, W., Qiao, M., Hu, C., et al., 2020. Fifteen-year application of manure and chemical fertilizers differently impacts soil ARGs and microbial community structure. Frontiers in Microbiology 11. Available at https://doi.org/10.3389/fmicb.2020.00062.

Webber, M.A., Whitehead, R.N., Mount, M., Loman, N.J., Pallen, M.J., Piddock, L.J.V., 2015. Parallel evolutionary pathways to antibiotic resistance selected by biocide exposure. Journal of Antimicrobial Chemotherapy 70 (8), 2241–2248. Available at https://doi.org/10.1093/jac/dkv109, http://jac.oxfordjournals.org/.

Wei, Y., Lin, Y., Zhang, A.Q., Guo, L.H., Cao, J., 2010. Evaluation of the noncovalent binding interactions between polycyclic aromatic hydrocarbon metabolites and human p53 cDNA. Science of the Total Environment 408 (24), 6285–6290. Available at https://doi.org/10.1016/j.scitotenv.2010.09.024.

Wintersdorff, C.J.von, Penders, J., Niekerk, J.M.van, et al., 2016. Dissemination of antimicrobial resistance in microbial ecosystems through horizontal gene transfer. Front Microbiol 7, 173.

Yadav, S., Kapley, A., 2019. Exploration of activated sludge resistome using metagenomics. Science of the Total Environment 692, 1155–1164. Available at https://doi.org/10.1016/j.scitotenv.2019.07.267.

Yang, Y., Li, B., Zou, S., et al., 2014. Fate of antibiotic resistance genes in sewage treatment plant revealed by metagenomic approach. Water Research 62, 97–106. Available at https://doi.org/10.1016/j.watres.2014.05.019.

Yang, N.J., Hinner, M.J., 2015. Getting across the cell membrane: an overview for small molecules, peptides, and proteins. Methods in Molecular Biology 1266, 29–53. Available at https://doi.org/10.1007/978-1-4939-2272-7_3, http://www.springer.com/series/7651.

Yao, X., Tao, F., Zhang, K., Tang, H., Xu, P., Rebecca, E., 2017. Parales, multiple roles for two efflux pumps in the polycyclic aromatic hydrocarbon-degrading *Pseudomonas putida* strain B6-2 (DSM 28064). Applied and Environmental Microbiology 83 (24). Available at https://doi.org/10.1128/aem.01882-17.

Younessi, N., Safari Sinegani, A.A., Khodakaramian, G., 2019. Detection of antibiotic resistance genes in culturable bacteria isolated from soils around mines in Hamedan, Iran. International Journal of Environmental Science and Technology 16 (12), 7643–7652. Available at https://doi.org/10.1007/s13762-018-02178-2.

Zaheer, R., Lakin, S.M., Polo, R.O., Cook, S.R., Larney, F.J., Morley, P.S., et al., 2019. Comparative diversity of microbiomes and Resistomes in beef feedlots, downstream environments and urban sewage influent. BMC Microbiology 19 (1). Available at https://doi.org/10.1186/s12866-019-1548-x, http://www.biomedcentral.com/bmcmicrobiol/.

Zhang, R., Tang, J., Li, J., et al., 2013. Antibiotics in the offshore waters of the Bohai Sea and the Yellow Sea in China: Occurrence, distribution and ecological risks. Environmental Pollution (Barking, Essex : 1987) 174, 71–77. Available at https://doi.org/10.1016/j.envpol.2012.11.008.

Zhang, Y., Gu, A.Z., He, M., Li, D., Chen, J., 2017. Subinhibitory concentrations of disinfectants promote the horizontal transfer of multidrug resistance genes within and across genera. Environmental Science and Technology 51 (1), 570–580. Available at https://doi.org/10.1021/acs.est.6b03132, http://pubs.acs.org/journal/esthag.

Zhang, Y., Gu, A.Z., Cen, T., Li, X., He, M., Li, D., et al., 2018. Sub-inhibitory concentrations of heavy metals facilitate the horizontal transfer of plasmid-mediated antibiotic resistance genes in water environment. Environmental Pollution 237, 74–82. Available at https://doi.org/10.1016/j.envpol.2018.01.032, www.elsevier.com/inca/publications/store/4/0/5/8/5/6.

Zhou, Z.C., Zheng, J., Wei, Y.Y., Chen, T., Dahlgren, R.A., Shang, X., et al., 2017. Antibiotic resistance genes in an urban river as impacted by bacterial community and physicochemical parameters. Environmental Science and Pollution Research 24 (30), 23753–23762. Available at https://doi.org/10.1007/s11356-017-0032-0, http://www.springerlink.com/content/0944-1344.

Zhu, Y.G., Johnson, T.A., Su, J.Q., Qiao, M., Guo, G.X., Stedtfeld, R.D., et al., 2013. Diverse and abundant antibiotic resistance genes in Chinese swine farms. Proceedings of the National Academy of Sciences of the United States of America 110 (9), 3435–3440. Available at https://doi.org/10.1073/pnas.1222743110China, http://www.pnas.org/content/110/9/3435.full.pdf+html.

Zhu, Y.G., Zhao, Y., Li, B., Huang, C.L., Zhang, S.Y., Yu, S., et al., 2017. Continental-scale pollution of estuaries with antibiotic resistance genes. Nature Microbiology 2. Available at https://doi.org/10.1038/nmicrobiol.2016.270, www.nature.com/nmicrobiol/.

Chapter 6

Hospital battleground: addressing antimicrobial resistance in healthcare settings

Nivedita Dasgupta[1], Dolly Roy[2], and Vijoy Sankar Kairi[2]
[1]*Multidisciplinary Research Unit, Silchar Medical College and Hospital, Silchar, Assam, India,* [2]*Department of Pharmacology, Silchar Medical College and Hospital, Silchar, Assam, India*

6.1 Introduction

Antimicrobial resistance (AMR), in recent years, has come up as one of the most significant threats to contemporary medicine and global public health. Antibiotics transformed the treatment of infectious diseases during the twentieth century; however, their effectiveness is increasingly compromised by the adaptive capacity of microorganisms. The growing prevalence of resistant pathogens undermines therapeutic success, complicates routine clinical management, and threatens the safety of medical advances that depend on effective antimicrobial therapy. As resistance expands, infections that were easily treatable in the past now require more complex and costly interventions, and are often associated with increased morbidity and mortality (World Health Organisation, 2014). Resistance arises through natural evolutionary processes driven by genetic mutation and horizontal gene transfer among microbial populations. Human practices have accelerated these mechanisms substantially, particularly through inappropriate antibiotic prescribing, excessive use of broad-spectrum agents, suboptimal dosing strategies, and prolonged therapy durations. These practices exert selective pressure that eliminates susceptible organisms while allowing resistant strains to survive and proliferate, facilitating their dissemination across healthcare environments and communities (Ventola, 2015). Beyond individual treatment failure, AMR has broad clinical and economic implications. Resistant infections lead to prolonged hospitalisation, increased healthcare costs, and a greater risk of complications. Moreover, the effectiveness of modern interventions such as chemotherapy, organ transplantation, and complex surgical procedures relies heavily on antimicrobial prophylaxis and treatment. As resistance increases, the safety of these procedures is compromised, placing additional strain on healthcare systems (Friedman et al., 2016). AMR poses a complex health challenge in which human, animal, and environmental systems are interconnected. Antibiotic use across various sectors of agriculture and veterinary practice plays a pivotal role in the emergence and dissemination of resistant organisms, which may infect humans and animals through direct contact, food chains, or environmental exposure via water, soil, and air. Recognising these interconnections highlights the importance of integrated approaches that address antimicrobial use across sectors (Robinson et al., 2016).

6.2 Mechanisms of antimicrobial resistance in healthcare settings

Antibiotic resistance in hospitals emerges from interactions between microbial genetic adaptation and environmental selection pressures characteristic of healthcare environments. High antibiotic exposure, vulnerable patient populations, and close contact between individuals create ideal conditions for resistant organisms to evolve and spread. At the microbial level, resistance commonly arises through enzymatic drug inactivation, alteration of antibiotic targets, reduced intracellular drug accumulation, and metabolic bypass mechanisms. Enzymatic degradation, such as the production of β-lactamases, allows bacteria to neutralise antibiotics before they can penetrate their targets effectively. Structural changes in drug targets can reduce binding affinity, rendering therapy ineffective, while efflux pumps and decreased membrane permeability limit intracellular drug concentrations. Together, these strategies enable microorganisms to survive

Antimicrobial Resistance in the 21st Century. DOI: https://doi.org/10.1016/B978-0-443-44748-8.00004-X

antimicrobial exposure and establish persistent infections (Nasrollahian et al., 2024). Resistance genes frequently spread from bacteria to other bacteria through various mechanisms such as horizontal gene transfer, including conjugation, transformation, and transduction. Mobile genetic elements such as plasmids and integrons facilitate the rapid dissemination of multidrug resistance traits among diverse bacterial species. These processes are amplified within biofilms that form on medical devices or environmental surfaces, where high bacterial density and reduced antimicrobial penetration promote gene exchange and persistence (Liu et al., 2024). Hospital environments, especially intensive care units, generate intense selective pressure due to frequent use of broad-spectrum antibiotics and prolonged treatment courses. Once resistant strains emerge, clonal expansion allows them to spread efficiently among patients, healthcare workers, and surfaces. Biofilms further protect the microorganisms from host defence mechanisms, immune responses, and antimicrobial therapy, by creating reservoirs that sustain transmission within healthcare facilities (Khan et al., 2018). Although bacterial resistance remains the most prominent concern, resistance also occurs among viruses, fungi, and parasites. Viral resistance often arises through mutations affecting antiviral drug targets, as seen in HIV treatment, where high mutation rates necessitate combination therapy strategies. Antifungal resistance is increasingly observed in species such as Candida and Aspergillus, particularly among immunocompromised populations exposed to prolonged antifungal therapy. Similarly, drug-resistant parasites such as Plasmodium species continue to challenge malaria treatment efforts, highlighting the broader scope of AMR beyond bacterial pathogens (Visser et al., 2014).

6.3 Drivers of antibiotic resistance in healthcare systems

Antibiotic resistance in hospitals is driven by a complex interaction of clinical practices, system-level factors, and microbial dynamics. Among these, antibiotic overuse and misuse remain the most influential contributors. Prescribing antibiotics for viral infections or continuing empirical therapy without reassessment exposes microbial populations to unnecessary selective pressure, accelerating the emergence of resistant strains (CDC, 2024). Patients admitted to hospitals, particularly in critical care or oncology settings, often receive multiple antibiotic courses, increasing cumulative exposure and promoting resistance within their microbiota. Inadequate infection prevention and control (IPC) measures, including poor hand hygiene and insufficient environmental cleaning, facilitate transmission of resistant organisms between patients. Invasive devices such as catheters and ventilators further increase risk by disrupting natural barriers and promoting biofilm formation. Delayed or inaccurate diagnostics also contribute by encouraging prolonged empirical use of broad-spectrum agents. Combined with high patient turnover and environmental reservoirs, these factors create healthcare settings where resistant organisms thrive and spread efficiently. Horizontal gene transfer between bacteria further accelerates this process, particularly under sustained antibiotic pressure (Tenover, 2006).

6.4 Antibiotic overuse and selection pressure

Antibiotic exposure does not create resistance directly; rather, it selects for resistant organisms already present within microbial populations. Repeated or unnecessary exposure eliminates susceptible organisms, allowing resistant variants to dominate. This effect is amplified by the use of broad-spectrum antibiotics, inappropriate dosing strategies, and unnecessarily prolonged therapy durations. Such practices disrupt the normal microbiota, enabling expansion of multidrug-resistant organisms and increasing risk of opportunistic infections. Patients may become reservoirs of resistant bacteria even after clinical recovery, facilitating onward transmission. At the system level, escalating resistance often drives clinicians to initiate broader empiric therapy, creating a self-reinforcing cycle that further accelerates the development of resistance (OECD, Stemming the Superbug Tide).

6.4.1 Strategies to combat antimicrobial resistance

Addressing AMR requires coordinated action across clinical, institutional, and policy levels. Key strategies include antimicrobial stewardship (AMS), strengthened infection prevention measures, improved surveillance, and development of new diagnostics and therapeutics. Effective stewardship programmes aim to optimise antimicrobial use by providing a protocol for appropriate drug selection, dosing, route, and duration while minimising unnecessary exposure. Recent studies have suggested that antimicrobial stewardship programmes (ASPs) reduce mortality, decrease healthcare costs, shorten hospital stays, and lower rates of resistant infections without compromising patient outcomes. By promoting rational prescribing and ongoing review of therapy, stewardship interventions help preserve antimicrobial effectiveness while maintaining high-quality patient care (Timbrook et al., 2016).

6.5 Antimicrobial stewardship: principles and core elements

AMS refers to coordinated strategies designed to improve antimicrobial use and optimise clinical outcomes, which directly or indirectly minimise the unintended consequences such as resistance and toxicity. Stewardship programmes emphasise evidence-based prescribing, integration of microbiological data, and continuous reassessment of therapy to ensure that antibiotics are used only when clinically indicated (Dyar et al., 2017). The Centres for Disease Control and Prevention outlines core elements essential for effective hospital stewardship programmes, including leadership commitment, accountability, pharmacy expertise, implementation of targeted interventions, monitoring of antibiotic use and resistance patterns, reporting of outcomes, and ongoing education. These elements support structured and sustainable approaches adaptable to diverse healthcare settings (CDC, Core Elements of Hospital Antibiotic Stewardship Programs, 2019). Multidisciplinary collaboration is fundamental to successful stewardship implementation. Infectious disease physicians, pharmacists, microbiologists, nurses, and hospital administrators contribute complementary expertise that improves decision-making, enhances adherence to guidelines, and promotes shared responsibility for antimicrobial use (Barlam et al., 2016).

6.5.1 Implementation strategies in clinical practice

Evidence-based prescribing guidelines provide clinicians with structured recommendations for antibiotic selection, dosing, and duration, reducing variability in care and limiting unnecessary use of broad-spectrum agents. Incorporating local resistance patterns ensures relevance to regional epidemiology and enhances treatment effectiveness. Prospective audit and feedback represents a key stewardship intervention in which antimicrobial prescriptions are reviewed after initiation, allowing timely recommendations such as deescalation, discontinuation, or optimisation of therapy. This collaborative approach improves prescribing behaviour while maintaining clinician autonomy and patient safety (Davey et al., 2017). Formulary restriction and preauthorisation strategies complement audit processes by limiting access to selected high-risk antibiotics unless specific clinical criteria are met. These measures help prevent inappropriate initiation of broad-spectrum agents and encourage adherence to established guidelines (Barlam et al., 2016).

6.5.2 Role of healthcare professionals in stewardship

Successful stewardship requires active participation from all members of the healthcare team. Physicians play a central role by integrating clinical evidence, microbiological data, and patient-specific factors into prescribing decisions. Pharmacists contribute by reviewing prescriptions for accuracy and safety, monitoring drug interactions, and supporting dose optimisation. Nurses provide essential patient monitoring and education, reinforcing adherence to therapy. Regular reassessment of antimicrobial therapy, typically within 48–72 hours, allows clinicians to refine treatment based on laboratory results and clinical response. Engaging patients in responsible antibiotic use, including adherence to prescribed regimens and avoidance of self-medication, further strengthens stewardship efforts.

6.6 Multidisciplinary approaches to strengthen hospital response to antimicrobial resistance

Despite extensive global and national efforts to combat AMR, the challenge continues to intensify (Tigges et al., 2024). Many global and national initiatives operate in isolation and often lack coordination and the comprehensive engagement of all stakeholders necessary to achieve meaningful and substantial impact . To address this critical gap, the Indian Medical Association, in collaboration with NITI Aayog, MOHFW, Directorate General of Health Services, World Health Organisation - India, NCDC, and 52 medical associations/organisations, has established the National Alliance of Medical Professionals on Antimicrobial Resistance (NAMP-AMR) (Singh et al., 2024). This alliance aims to bridge these gaps by uniting medical professionals across disciplines and specialities, enabling a unified approach that amplifies impact, drives cohesive policy changes, and expands educational outreach to maximise the impact. Through this integrated and collaborative approach involving multistakeholder strategy, NAMP-AMR is poised to drive significant advancements in addressing AMR (Sharma et al., 2023; WHO, 2016).

However, AMR is far from being exclusively a healthcare-related problem. The WHO's list of main drivers for AMR include the misuse and overuse of antimicrobials in both the clinical and farm context, the lack of access to clear water, bad practises in the sanitation and hygiene of both humans and animals, poor access to early diagnostics, a lack of legislation on the matter and last, but definitely not least, a generalised lack of awareness and knowledge on the subject. The AMR crisis is therefore a great example of a problem that should be approached under the holistic perspective of "One Health" (Schneider et al., 2018).

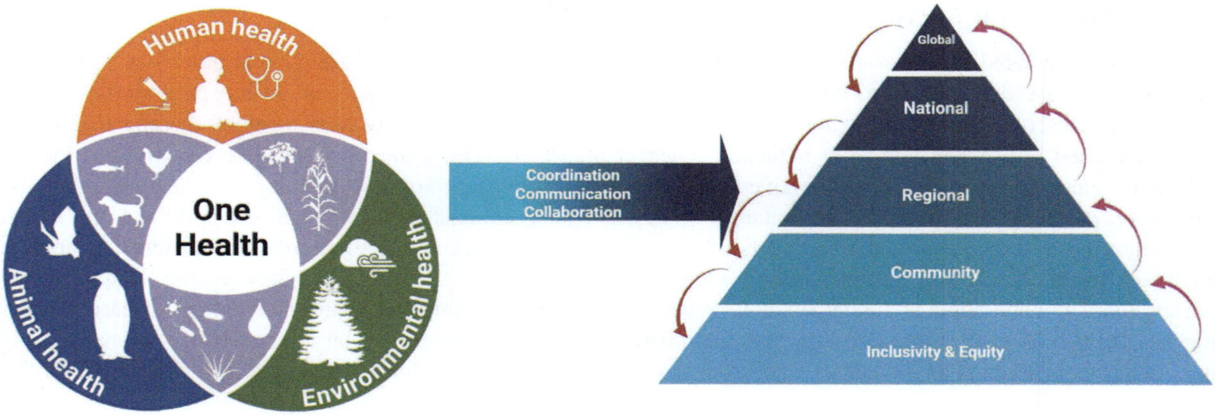

FIGURE 6.1 **Framework to tackle antimicrobial resistance by One Health perspective.** Conceptual framework of the One Health approach illustrating the interconnectedness of human, animal, and environmental health with multilevel governance model.

One Health is the understanding that AMR extends the boundaries of healthcare alone Fig. 6.1. According to the World Health Organisation, the primary drivers of AMR include the inappropriate and excessive use of antimicrobials in both medical and agricultural settings; inadequate access to safe drinking water; poor sanitation and hygiene practices among human and animals; limited availability of timely diagnostic tools; insufficient regulatory frameworks; and, critically, widespread gaps in awareness and awareness and understanding of AMR (Caudell et al.,2020). Consequently, AMR represents a complex, multifaceted challenge that necessitates a holistic "One Health" One Health concept as provided by the One Health Initiative Task Force of the American Veterinary Medical Association, which would be "the collaborative efforts of multiple disciplines working locally, nationally, and globally, to achieve optimal health for people, animals and our environment." (Santos-Lopez et al., 2019).

6.7 Bridging critical care and public health in the fight against resistance

Bridging critical care and public health in the fight against resistance requires an integrated, systems-level approach. It involves aligning hospital-based AMS, surveillance, and infection control practices. These efforts must be connected with community-driven prevention strategies, environmental monitoring, and policy interventions. It can be achieved by adopting a framework described below:

Step 1: select and adapt evidence-based practices

To implement evidence based practices (EBPs) within the real world of AMR, the first step should be chosen based on effectiveness but adapted to the local context. A true One Health approach demands the involvement of policymakers, regulatory authorities, professionals and communities across human, animal and environmental sectors. Also, this approach should ensure that interventions are feasible, equitable and sustainable across all sectors of AMR governance (Adsul et al., 2024). In many low- and middle-income countries (LMICs) where there is either overuse or inappropriate use of antibiotics in both human and veterinary, along with the coexistence of environmental contamination, few interventions have been developed and tested with One Health approach (Ogyu et al., 2020). Effective EBPs require the involvement of national and regional authorities, along with frontline implementers, who can formulate and adapt policies tailored to local epidemiological contexts and specific contexts using local epidemiological data, rather than relying on a structured framework.

Step 2: Assess barriers and enablers to implementing EBPs

After selecting evidence-based practices, the next step is to identify the barriers and facilitators that will help implement the process and establish the strategies. The consolidated framework and the theoretical domain framework are two identified multilevel determinants that decide the success and failure of the strategies. However, current analysis showed that this framework is focused only on isolated factors, thereby ignoring One Health synergies and missing opportunities for interdisciplinary collaboration. In order to assess implementation barriers and enablers, it is suggested to integrate both the theoretical domain framework and the consolidated framework (Damschroder et al., 2022). This comprehensive assessment offers a strong foundation for the development of evidence-based implementation strategies.

Step 3: Select, use, and adapt implementation strategies

After identifying barriers and enablers, the next step is to identify the techniques or methods which can be implemented for a sustainable clinical programme or practice. The best approach is to accurately identify and rectify strategies with stakeholders across different settings. However, in practical scenarios, the coordination is often lacking in their applications. A recent systematic review of AMS interventions reported that implementation efforts are commonly confined to educational activities or audit-and-feedback mechanisms, with less consideration of how these strategies interact with local governance structures, healthcare infrastructure, or prevailing professional norms (Conlin et al., 2024). Moreover, the majority of AMS interventions lacked a systematic approach to strategy design or contextual adaptation, underscoring a recurring limitation in global AMR initiatives (Conlin et al., 2024). Although implementation science has developed and evaluated a wide range of strategies to support the uptake and integration of evidence-based practices, and frameworks such as the Expert Recommendations for Implementing Change (ERIC) offer valuable starting points, their effective use depends on participatory adaptation and refinement (Waltz et al., 2015). Within a One Health framework, particular emphasis should be given to strategies that are related to local contexts, involve sustained stakeholder engagement, and incorporate frequent monitoring and evaluation (Powell et al., 2015).

Step 4: Evaluate and sustain implementation

Evaluation of AMS interventions should be grounded in mixed-methods, theory-driven frameworks to capture not only measurable outcomes but also the complex contextual factors that shape implementation and impact. Frameworks such as RE-AIM (Reach, Effectiveness, Adoption, Implementation, and Maintenance) enable systematic assessment of the population coverage of interventions, their clinical and behavioural effects, levels of uptake across settings and professional groups, precision of implementation, and long-term sustainability (Winkler et al., 2025). Complementing this, the Consolidated Framework for Implementation Research (CFIR) provides a structured lens to examine how intervention characteristics, inner and outer settings, individual actors, and implementation processes interact to influence success or failure (Birken et al., 2017). Together, these frameworks support a comprehensive understanding of both "what works" and "why it works" in specific contexts.

The EQUIPS-ICU study is frequently considered the best approach in LMIC hospital settings (Wagstaff et al., 2025). By combining quantitative indicators—such as antimicrobial consumption, appropriate prescribing, and patient outcomes—with qualitative insights from healthcare workers, the study demonstrated how AMS interventions can be evaluated in a way that is both context-sensitive and methodologically rigorous. Importantly, EQUIPS-ICU highlighted the value of continuous feedback loops, allowing findings to inform ongoing refinement of stewardship strategies rather than serving solely as end-point evaluations.

The sustained impact of AMS interventions ultimately depends on local stakeholder ownership, ensuring that clinicians, administrators, and policymakers perceive stewardship activities as integral to routine practice rather than externally imposed. Cost-effectiveness is also critical, particularly in resource-constrained settings, where interventions must demonstrate value relative to competing health priorities. Finally, long-term sustainability requires continuous surveillance within a One Health framework, integrating data from human health, animal health, and environmental sectors to monitor antimicrobial use and resistance trends over time. Such integrated surveillance supports adaptive stewardship strategies that remain responsive to evolving epidemiological and system-level challenges (Ogyu et al.,2020).

6.8 Strengthening governance, practice, and research to combat antimicrobial resistance

6.8.1 Policy: strengthening governance and regulatory frameworks

Effective governance is central to addressing AMR, yet many countries remain at the stage of policy formulation rather than implementation. India, along with many countries, has created National Action Plans in coordination with the World Health Organisation's Global Action Plan on Antimicrobial Resistance, but the implementation of these plans is not very effective. To put the words into action, governments must ensure constant financial support, impose rules strictly, and ensure that there is clear accountability among different government departments. Successful governance requires teamwork among human health, animal health, and environmental sectors, followed by periodic evaluation.

To effectively respond to AMR, we need to embrace a One Health approach that recognises the coordinated interactions among human, animal, and environmental health. This requires the judicious use of antibiotics in livestock, poultry, and aquaculture, strengthening veterinary oversight, and promoting preventive measures such as vaccination and improved farm biosecurity. It is equally crucial to regulate pharmaceutical manufacturing effluents, hospital wastewater, and agricultural runoff to limit environmental dissemination of antimicrobial residues. Cross-sectoral data sharing among public health, veterinary, and environmental agencies is essential to identify resistance patterns and transmission hotspots (Laxminarayan et al., 2013).

Rules and regulations play an important role in antibiotic overuse. In India, it is important to strictly enforce Schedule H1, which restricts over-the-counter sales of antibiotics without a prescription. Routine monitoring of pharmacies, associated with penalties for noncompliance, can further reduce inappropriate dispensing. Additionally, prohibiting the use of antibiotics for growth promotion in food animals is a key strategy to reduce the development of antibiotic resistance in agricultural settings.

Investment in water, sanitation, and hygiene infrastructure is fundamental to reducing infection burden and, consequently, antibiotic consumption. Strengthening hospital infection control facilities, expanding access to clean water, and promoting community-level hygiene education are essential components of a preventive public health strategy that complements AMS efforts (WHO, 2019).

6.8.2 Practice: implementing stewardship in healthcare settings

Within healthcare systems, ASPs play a vital role in limiting irrational antibiotic use and promoting its proper use. Effective ASPs incorporate prospective audit and feedback mechanisms, formulary restrictions on last-resort antibiotics, and monitoring by the healthcare team after changes to therapy resulting from recommendations made by the AMS team and microbiological results. Continuous professional training for clinicians, pharmacists, and nurses is necessary to maintain good prescribing practices (Barlam et al., 2016).

IPC measures are equally vital in reducing healthcare-associated infections (HAIs), which are major drivers of antibiotic use. Strict adherence to hand hygiene protocols, routine environmental decontamination, proper sterilisation of medical equipment, and systematic surveillance of HAIs are indispensable components of hospital safety and quality improvement.

Diagnostic stewardship further strengthens antimicrobial management by minimising empirical and unnecessary prescriptions. The integration of rapid diagnostic tests, routine culture and sensitivity testing, and molecular diagnostics enables clinicians to distinguish bacterial from viral infections and tailor therapy accordingly. Expanding diagnostic capacity in district and rural hospitals is particularly important in resource-constrained settings.

Evidence-based prescribing should be guided by local antibiogram data, encouraging the use of narrow-spectrum agents and timely deescalation of therapy. This approach not only preserves antibiotic efficacy but also reduces adverse drug effects and healthcare costs (CDC, 2019).

6.8.3 Research: strengthening evidence and innovation

Robust surveillance systems are foundational to understanding AMR dynamics. National networks such as the Indian Council of Medical Research's AMR Surveillance Network (AMRSN) must be expanded and standardised to generate high-quality, real-time resistance data. These data should inform clinical guidelines, policy decisions, and resource allocation at national and regional levels.

Implementation science is increasingly recognised as a critical field in AMR research, as it bridges the gap between theoretical efficacy and real-world effectiveness. Studies should focus on identifying structural and behavioural barriers to stewardship adoption, evaluating scalable intervention models, and developing context-specific strategies for low-resource settings. Such research is essential for ensuring that AMR interventions are feasible, sustainable, and equitable (ICMR, 2022).

Given the global rise of multidrug-resistant pathogens, investment in novel therapeutics is imperative. Research priorities include the development of new antibiotics, bacteriophage therapy, vaccines, and alternative treatment modalities such as microbiome-based interventions and immunotherapies. Preventive strategies through vaccination have the potential to substantially reduce antibiotic demand.

Understanding the sociocultural drivers of antibiotic misuse is equally important. Research should examine patient perceptions, health-seeking behaviours, and economic factors influencing antibiotic consumption. Engaging communities in participatory approaches can foster trust, improve adherence to medical guidance, and enhance the effectiveness of public health campaigns (Charani and Holmes, 2019).

6.8.4 Bridging the gaps between policy, practice, and research

The integration of surveillance data into clinical practice is essential for evidence-based decision-making. Facility-specific antibiograms should guide empirical treatment protocols, particularly in critical care settings, thereby reducing reliance on broad-spectrum antibiotics.

Digital health innovations offer promising pathways to translate policy into practice. Mobile applications for antibiotic prescribing, electronic medical records with stewardship alerts, and clinical decision-support tools can facilitate guideline adherence and reduce inappropriate prescribing (Baalen et al., 2021).

Bridging research and policy requires active engagement between scientists and decision-makers. The role of "policy entrepreneurs"—individuals who translate scientific evidence into actionable policy—has become increasingly significant in shaping AMR governance.

Sustainable capacity building can be achieved through a train-the-trainer model, whereby selected healthcare professionals receive specialised AMR training and subsequently disseminate knowledge within their institutions. This approach fosters local ownership, strengthens institutional capacity, and ensures long-term impact beyond externally driven initiatives (Rowe et al., 2005).

6.9 Conclusion

AMR represents a critical challenge to modern healthcare, driven by complex interactions between microbial evolution, clinical practice, and healthcare systems. Hospitals provide environments where selective pressure, vulnerable patient populations, and opportunities for transmission converge, accelerating the emergence and spread of resistant organisms. Comprehensive strategies centred on AMS, infection prevention, and multidisciplinary collaboration are essential to mitigate this threat. By promoting responsible antimicrobial use and integrating evidence-based practices into routine care, healthcare systems can preserve the effectiveness of existing therapies and safeguard future treatment options.

Acknowledgement

The author gratefully acknowledges the financial support provided by the MRU, SMCH funded by Department of Health Research (DHR)/ Indian Council of Medical Research (ICMR). The authors would also like to acknowledge Biorender for designing the graphical illustration.

AI disclosure

During the preparation of this work, the author(s) used ChatGPT to For References. After using this tool/service, the author(s) reviewed and edited the content as needed and take(s) full responsibility for the content of the publication.

References

Adsul, P., Shelton, R.C., Oh, A., Moise, N., Iwelunmor, J., Griffith, D.M., 2024. Challenges and opportunities for paving the road to global health equity through implementation science. Annual Review of Public Health 45 (1), 27–45. Available at https://doi.org/10.1146/annurev-publhealth-060922-034822.

Baalen, S.V., Boon, M., Verhoef, P., 2021. From clinical decision support to clinical reasoning support systems. Journal of Evaluation in Clinical Practice 27 (3), 520–528. Available at https://doi.org/10.1111/jep.13541.

Barlam, T.F., Cosgrove, S.E., Abbo, L.M., Macdougall, C., Schuetz, A.N., Septimus, E.J., et al., 2016. Implementing an antibiotic stewardship program: guidelines by the Infectious Diseases Society of America and the Society for Healthcare Epidemiology of America. Clinical Infectious Diseases 62 (10), e51. Available at https://doi.org/10.1093/cid/ciw118, http://cid.oxfordjournals.org/content/by/year.

Birken, S.A., Powell, B.J., Presseau, J., Kirk, M.A., Lorencatto, F., Gould, N.J., et al., 2017. Combined use of the consolidated framework for implementation research (CFIR) and the theoretical domains framework (TDF): a systematic review. Implementation Science 12 (1). Available at https://doi.org/10.1186/s13012-016-0534-z, http://www.implementationscience.com/.

Caudell, M.A., Dorado-Garcia, A., Eckford, S., Creese, C., Byarugaba, D.K., Afakye, K., et al., 2020. Towards a bottom-up understanding of antimicrobial use and resistance on the farm: a knowledge, attitudes, and practices survey across livestock systems in five African countries. Public Library of Science, Kenya PLoS ONE 15 (1). Available at https://doi.org/10.1371/journal.pone.0220274, https://journals.plos.org/plosone/article/file?id=10.1371/journal.pone.0220274&type=printable.

Centers for Disease Control and Prevention (CDC), 2024. Antimicrobial Resistance in Health Care: Causes and How It Spreads. CDC. Available at https://www.cdc.gov/antimicrobial-resistance/index.html.

Centers for Disease Control and Prevention, 2019. Core elements of hospital antibiotic stewardship program. U.S. Department of Health and Human Services, CDC, Atlanta (GA). Available at https://www.cdc.gov/antibiotic-use/hcp/core-elements/hospital.html.

Charani, E., Holmes, A., 2019. Antibiotic stewardship—twenty years in the making. Antibiotics 8 (1). Available at https://doi.org/10.3390/antibiotics8010007.

Conlin, M., Hamard, M., Agrinier, N., Birgand, G., 2024. Assessment of implementation strategies adopted for antimicrobial stewardship interventions in long-term care facilities: a systematic review. Clinical microbiology and Infection: The Official Publication of the European Society of Clinical Microbiology and Infectious Diseases 30 (4), 431–444.

Damschroder, L.J., Reardon, C.M., Widerquist, M.A.O., Lowery, J., 2022. The updated consolidated framework for implementation research based on user feedback. Implementation Science 17 (1). Available at https://doi.org/10.1186/s13012-022-01245-0.

Davey, P., Marwick, C.A., Scott, C.L., Charani, E., Mcneil, K., Brown, E., et al., 2017. Interventions to improve antibiotic prescribing practices for hospital inpatients. Cochrane Database of Systematic Reviews 2017 (2). Available at https://doi.org/10.1002/14651858.CD003543.pub4, http://as.wiley.com/WileyCDA/Brand/id-6.html.

Dyar, O.J., Huttner, B., Schouten, J., Pulcini, C., 2017. What is antimicrobial stewardship? Clinical Microbiology and Infection 23 (11), 793–798. Available at https://doi.org/10.1016/j.cmi.2017.08.026.

Friedman, N.D., Temkin, E., Carmeli, Y., 2016. The negative impact of antibiotic resistance. Clinical Microbiology and Infection 22 (5), 416–422. Available at https://doi.org/10.1016/j.cmi.2015.12.002.

Indian Council of Medical Research, 2022. Annual report: antimicrobial resistance surveillance network (AMRSN). New Delhi: ICMR. Available at https://www.icmr.gov.in/reports.

Khan, A., Miller, W.R., Arias, C.A., 2018. Mechanisms of antimicrobial resistance among hospital-associated pathogens. Expert Review of Anti-Infective Therapy 16 (4), 269–287. Available at https://doi.org/10.1080/14787210.2018.1456919, http://www.tandfonline.com/loi/ierz20.

Laxminarayan, R., Duse, A., Wattal, C., Zaidi, A.K.M., Wertheim, H.F.L., Sumpradit, N., et al., 2013. Antibiotic resistance-the need for global solutions. The Lancet Infectious Diseases 13 (12), 1057–1098. Available at https://doi.org/10.1016/S1473-3099(13)70318-9.

Liu, H.Y., Prentice, E.L., Webber, M.A., 2024. Mechanisms of antimicrobial resistance in biofilms. npj Antimicrobials and Resistance 2 (1). Available at https://doi.org/10.1038/s44259-024-00046-3.

Nasrollahian, S., Graham, J.P., Halaji, M., 2024. A review of the mechanisms that confer antibiotic resistance in pathotypes of *E. coli*. Frontiers in Cellular and Infection Microbiology 14. Available at https://doi.org/10.3389/fcimb.2024.1387497, http://www.frontiersin.org/Cellular_and_Infection_Microbiology/archive.

Ogyu, A., Chan, O., Littmann, J., Pang, H.H., Lining, X., Liu, P., et al., 2020. National action to combat AMR: a one-health approach to assess policy priorities in action plans. BMJ Global Health 5 (7), e002427. Available at https://doi.org/10.1136/bmjgh-2020-002427.

Powell, B.J., Waltz, T.J., Chinman, M.J., Damschroder, L.J., Smith, J.L., Matthieu, M.M., et al., 2015. A refined compilation of implementation strategies: results from the expert recommendations for implementing change (ERIC) project. Implementation Science 10 (1). Available at https://doi.org/10.1186/s13012-015-0209-1, http://www.implementationscience.com/.

Robinson, T.P., Bu, D.P., Carrique-Mas, J., Fèvre, E.M., Gilbert, M., Grace, D., et al., 2016. Antibiotic resistance is the quintessential one health issue. Transactions of the Royal Society of Tropical Medicine and Hygiene 110 (7), 377–380. Available at https://doi.org/10.1093/trstmh/trw048.

Rowe, A.K., De Savigny, D., Lanata, C.F., Victora, C.G., 2005. How can we achieve and maintain high-quality performance of health workers in low-resource settings? Lancet 366 (9490), 1026–1035. Available at https://doi.org/10.1016/S0140-6736(05)67028-6.

Santos-Lopez, A., Marshall, C.W., Scribner, M.R., Snyder, D.J., Cooper, V.S., 2019. Evolutionary pathways to antibiotic resistance are dependent upon environmental structure and bacterial lifestyle. eLife 8. Available at https://doi.org/10.7554/eLife.47612.

Schneider, S., Salm, F., Vincze, S., Moeser, A., Petruschke, I., Schmücker, K., et al., 2018. Perceptions and attitudes regarding antibiotic resistance in Germany: a cross-sectoral survey amongst physicians, veterinarians, farmers and the general public. Journal of Antimicrobial Chemotherapy 73 (7), 1984–1988. Available at https://doi.org/10.1093/jac/dky100, http://jac.oxfordjournals.org/.

Sharma, A., Thakur, N., Thakur, A., Chauhan, A., Babrah, H., 2023. The challenge of antimicrobial resistance in the indian healthcare system. Cureus 15 (7). Available at https://doi.org/10.7759/cureus.42231.

Singh, V.P., Jha, D., Rehman, B.U., Dhayal, V.S., Dhar, M.S., Sharma, N., 2024. A mini-review on the burden of antimicrobial resistance and its regulation across one health sectors in India. Journal of Agriculture and Food Research 15. Available at https://doi.org/10.1016/j.jafr.2024.100973, https://www.sciencedirect.com/science/journal/26661543.

Tenover, F.C., 2006. Mechanisms of antimicrobial resistance in bacteria. American Journal of Medicine 119 (6), S3. Available at https://doi.org/10.1016/j.amjmed.2006.03.011.

Tigges, P., Greser, A., Gágyor, I., Kraft, J., Maun, A., Schmiemann, G., et al., 2024. Addressing AMR and planetary health in primary care: the potential of general practitioners as change agents. Frontiers in Public Health 12. Available at https://doi.org/10.3389/fpubh.2024.1383423, http://journal.frontiersin.org/journal/public-health/section/public-health-education-and-promotion#archive.

Timbrook, T.T., Hurst, J.M., Bosso, J.A., 2016. Impact of an antimicrobial stewardship program on antimicrobial utilization, bacterial susceptibilities, and financial expenditures at an academic medical center. Hospital Pharmacy 51 (9), 703–711. Available at https://doi.org/10.1310/hpj5109-703, http://archive.hospital-pharmacy.com/doi/pdf/10.1310/hpj5109-703.

Ventola, C.L., 2015. The antibiotic resistance crisis: Part 1: causes and threats. Medi Media USA Inc, undefined P and T 40 (4), 277–283. Available at http://www.ptcommunity.com/system/files/pdf/ptj4004277.pdf.

Visser, S.N., Danielson, M.L., Bitsko, R.H., Holbrook, J.R., Kogan, M.D., Ghandour, R.M., et al., 2014. Trends in the parent-report of health care provider-diagnosed and medicated attention-deficit/hyperactivity disorder: United States, 2003–2011. Journal of the American Academy of Child & Adolescent Psychiatry 53, 34–46.e2. Available at 10.1016/j.jaac.2013.09.001.

Wagstaff, D., Amuasi, J., Arfin, S., Aryal, D., Nor, M.B.M., Bonney, J., et al., 2025. Evidence based QUality Improvement for Prescribing Stewardship in ICU (EQUIPS-ICU): protocol for type III hybrid implementation-effectiveness study. Implementation Science 20 (1). Available at https://doi.org/10.1186/s13012-024-01413-4.

Waltz, T.J., Powell, B.J., Matthieu, M.M., Damschroder, L.J., Chinman, M.J., Smith, J.L., et al., 2015. Use of concept mapping to characterize relationships among implementation strategies and assess their feasibility and importance: Results from the Expert Recommendations for Implementing Change (ERIC) study. Implementation Science 10 (1). Available at https://doi.org/10.1186/s13012-015-0295-0, http://www.implementationscience.com/.

Winkler, A.S., Brux, C.M., Carabin, H., das Neves, C.G., Häsler, B., Zinsstag, J., et al., 2025. The lancet one health commission: harnessing our interconnectedness for equitable, sustainable, and healthy socioecological systems. The Lancet 406 (10502), 501–570. Available at https://doi.org/10.1016/S0140-6736(25)00627-0, https://www.sciencedirect.com/science/journal/01406736.

World Health Organization, 2016. Global Action Plan on Antimicrobial Resistance. WHO, Geneva (2016). Available at https://www.who.int/publications/i/item/9789241509763.

World Health Organization and UNICEF, 2019. Water, Sanitation and Hygiene in Health Care Facilities. WHO, Geneva (2019).

World Health Organization. Antimicrobial resistance: global report on surveillance 2014. Geneva: World Health Organization. (2014), Available at https://www.who.int/publications/i/item/9789241564748.

Chapter 7

The rise of antibiotic resistance: Impacts and Solutions in community health

Shiela Chetri[1], Suhita Senapati[2], Sreeparna Purkayastha[3], Deep Bhowmik[4], Monalisha Hazarika[5], Kinsley Erhons Enerijiofi[6], and Sunil Kumar[7]

[1]*ICMR National Institute of Epidemiology, Chennai, Tamil Nadu, India,* [2]*Accumedicare Laboratories Private Limited, Hyderabad, Telangana, India,* [3]*Genes N Life Healthcare Private Limited, Hyderabad, Telangana, India,* [4]*Department of Microbiology, Assam University, Silchar, Assam, India,* [5]*DBT-APSCS&T Centre of Excellence for Bio-resources and Sustainable Development, Kimin, Arunachal Pradesh, India,* [6]*Department of Biological Sciences, Glorious Vision University, Ogwa, Edo State, Nigeria,* [7]*Department of Mechanical Engineering, Amrita Vishwa Vidyapeetham, Chennai Campus, Chennai, Tamil Nadu, India*

7.1 Introduction

Since their introduction in the 1940s, antibiotics have been the primary therapeutic representatives for combating bacterial infections. Initially celebrated as ‘miracle drugs’, their irrational and extreme use over time has played a pivotal role in the emergence and global spread of multidrug-resistant bacterial strains. Historically, troubles regarding AMR were largely restricted to hospital environments, while community-based transmission received limited attention. However, resistance has now accelerated alarmingly within outpatient and community settings as well. Today, AMR is massive across all levels of healthcare, from hospitals to clinics, physicians’ offices, and even home care environments. This escalating public health crisis is further aggravated by the slow pace of research and development of new antimicrobial agents intended to counter future resistance threats (Hooton and Levy, 2001).

In recent times, demographic shifts and alterations in healthcare service structures have resulted in a notable rise in the population residing in the community, which necessitates specialised care due to an elevated risk of infection (Maillard et al., 2020). In these environments, the major demographic is the elderly population, categorised by diminished immune capacity to combat infections, which is exacerbated by the prevalence of comorbid conditions such as diabetes, etc. A reduction in immunity normally begins at the age of 50. Another group of the population with diminished immunity encompasses young individuals, freshly discharged patients, and their relatives, alongside those with compromised immune systems due to chronic or progressive conditions such as HIV/AIDS, or as a consequence of ongoing treatment with immunosuppressant or other pharmaceuticals (Maillard et al., 2020). Immunocompromised persons often rely on several medications, primarily antibiotics, to combat infections, which may subsequently heighten resistance to other infections, such as those caused by Clostridium difficile. In everyday surroundings such as shopping complexes, public transport, and fitness centres, hygiene standards are often far less rigorous than those maintained in clinical settings. This lack of adequate hygiene routines creates favourable conditions for the spread of infectious diseases within the community (Fig. 7.1. Insufficient sanitation and hygiene habits substantially contribute to the transmission of various illnesses, including respiratory tract infections, gastrointestinal disorders, and *Staphylococcus aureus*-related skin infections. Including the elderly, communal living arrangements extremely intensify the risk of infection spread. Factors such as faecal incontinence and shared facilities facilitate the quick transmission of enteric and foodborne pathogens. Subsequently, these conditions increase the occurrence of *Campylobacter*-induced diarrhoea and salmonellosis among older adults living in such environments. A growing population of vulnerable 'at-risk' persons is now obtaining care beyond conventional hospital environments (Scott, 2013). Furthermore, immunocompromised patients within the community should be aware of opportunistic organisms such as *Klebsiella* spp., *Escherichia coli*, and *P. aeruginosa*, which are frequently associated with hospital-acquired illnesses (Bloomfield and Ackerley, 2024). Pathogens can infiltrate the home via domestic pets or animals, hospitalised, infected, or colonised persons, and contaminated food or drinking

Antimicrobial Resistance in the 21st Century. DOI: https://doi.org/10.1016/B978-0-443-44748-8.00005-1

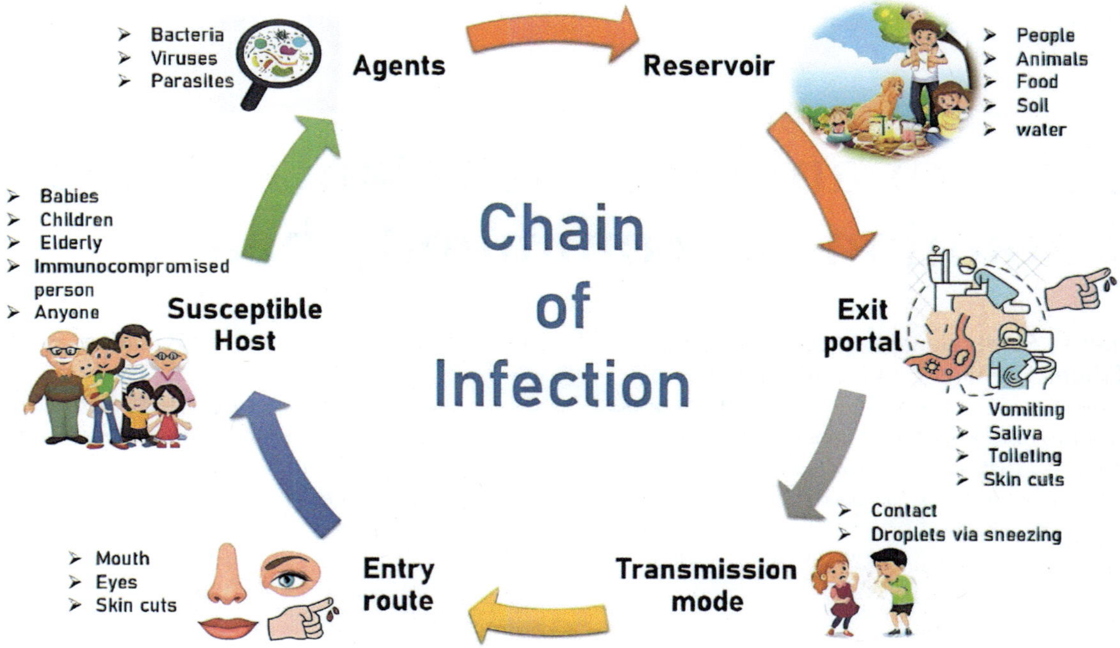

FIGURE 7.1 **The chain of infection within home and in our daily life**, chain of infection shows the sequential links in disease transmission. The cycle consists of: (1) infectious agents (eg, bacteria, viruses, parasites); (2) reservoirs in which pathogens survive (e.g. people, animals, foodstuffs, soil, and water); (3) portals of exit such as saliva and vomit, excreta and skin lesions, etc.; (4) modes of transmission, including direct contact and respiratory droplets; (5) portals of entry through the mouth/eyes/breaches in integrity of the skin; (6) susceptible hosts-infants/young children/the elderly/immunocompromised individual/general population. Disrupting any link in this chain is fundamental to infectious prevention and control.

water. The key to preventing the dissemination of infection – regardless of the setting – is to break the chain of infection (Liu et al., 2018; Bloomfield and Ackerley, 2024). These microbes are perpetually expelled from the mentioned sources, and they rapidly disseminate throughout the household environment. This spread occurs primarily through various paths, containing direct physical connection, the handling and ingesting of polluted food, contact with frequently touched surfaces, the use of shared utensils, and even through the air, thereby increasing the risk of transmission within the home (Bloomfield and Ackerley, 2024).

Furthermore, multiple prior investigations have repeatedly shown that these pathogens can be transmitted from infected to healthy individuals via a variety of channels that are prevalent in everyday life and regular household surroundings. These modes of transmission include, but are not limited to, direct hand-to-hand contact, touching contaminated surfaces, exposure to faeces, inhalation of respiratory droplets, and handling or consuming infected food. A study carried out in Culiacán, Mexico, identified *Salmonella* spp. in 173 out of 180 kitchen cleaning cloths (96%) collected from household kitchens (Chaidez et al., 2014). Although reports indicate that pathogens such as Salmonella can proliferate in fabric, creating a permanent reservoir, additional analysis is needed to substantiate this assertion. The magnitude of exposure to harmful pathogens determines the probability of contamination. Though distinct organisms and circumstances have varying 'infectious doses'. that is the quantity of virus particles or bacterial cells needed to significantly increase the probability of contamination (Bloomfield and Ackerley, 2024; Yezli and Otter, 2011). The infective amount for frequently faced bacteria, including *Shigella*, *Campylobacter*, enterohemorrhagic *Escherichia coli* (EHEC), and *Clostridium difficile*, as well as viruses like norovirus, etc., can be minimal (1–500 cells or particles) (Kothary and Babu, 2001; Larson et al., 1978; Robilotti, Deresinski and Pinsky, 2015; Winther, 2011; Atmar et al., 2014; Haas et al., 2014). In contrast, pathogens such as *Salmonella* frequently need a far greater infectious dosage (usually more than 10^6 cells) (Kothary and Babu, 2001). This threshold, however, may change based on variables, including the mode of exposure and the host's susceptibility. Notably, infections can occur at much lower dosages in high-risk populations, including the elderly, young children, and immunocompromised individuals (Kothary and Babu, 2001). Although dealing with an increasing number of affected roles in the convergence, especially at the household level, might relieve the stress on overworked healthcare systems, this advantage may be jeopardised by insufficient contamination regulator procedures in the household. As a result, immediate focus has to be placed on preventing infections in medical institutions as well as limiting their spread in households and public spaces.

7.2 Evolutionary pathways and routes towards antibiotic resistance

Antibiotic resistance may develop via mutation within the bacterial genome or through the incorporation of exogenous DNA. Mutations occur spontaneously in response to antibiotic exposure and can be integrated into the genome of the host organism, whether human or animal. However, such occurrences are exceedingly uncommon in nature and occur independently of the genetic reservoirs found in other species. But, when it comes to acquiring novel resistance determinants from various environments – such as soil, water, or other ecosystems with highly diverse ecological niches – these sources provide a significantly larger genetic pool than that found in human and domesticated animal microbiotas (Schulz et al., 2017; Morar and Wright, 2010). Unquestionably, the most remarkable feature of the environmental microbiome is its vast genetic diversity, serving as a wide reservoir of genes that pathogenic microorganisms can hypothetically acquire or exploit to enhance their survival under antibiotic stress. It is now well determined that every antibiotic developed to date, be it synthetic or natural, has encountered resistance from at least specified target pathogens. This implies that nature may already have resistance mechanisms towards any future antibiotics, underscoring the urgent need for a fundamentally different approach for antibiotic design and their proper implementation. Throughout the ages, most antibiotic resistance genes (ARGs) are assumed to have gradually evolved from ancestral genes that once gratified different biological roles (Morar and Wright, 2010; Andersson et al., 2015). Moreover, the selection of unusual genotypes accompanied by acquired resistance resulting from mobilisation and/or horizontal gene transfer, which would otherwise vanish from the population in the absence of selective pressure, is probably a crucial bottleneck (Kimura and Ohta, 1969). Throughout these phases, compensatory mutations may emerge somewhere in the bacterium's genome harbouring the antibiotic resistance gene, possibly alleviating related fitness costs. This can occur either by reducing niche overlap with competing organisms or by enhancing the bacterium's competitive ability (Letten et al., 2021). In clinical settings, the emergence of ARGs necessitates a precise temporal and spatial alignment of events to arise (Waglechner and Wright, 2017). Fundamentally, the external environment may influence part, all, or none of the evolutionary processes that result in the development of ARGs (Fig. 7.2. An extensive and imperious knowledge gap exists regarding our insufficient comprehension of the specific sites and settings that facilitate the advent of novel resistance mechanisms in clinically pertinent bacteria.

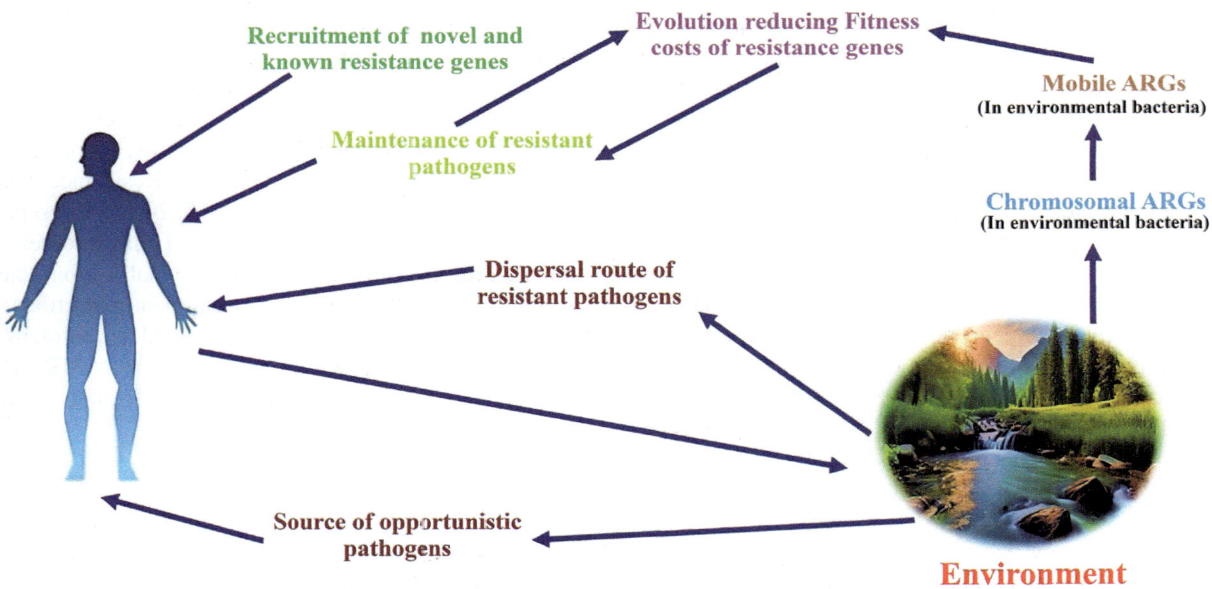

FIGURE 7.2 **Overview of the central role of environment in antibiotic resistance evolution and propagation**, a conceptual diagram of bidirectional human–environment interactions in antimicrobial resistance. The left features a human silhouette, and the right contains a natural ecosystem (river and forest), both connected (by arrows). In environmental bacteria, mobile ARGs are derived from chromosomal antibiotic resistance genes (ARGs) found in the environment. Arrows indicate dispersal pathways of resistant pathogens from the environment and to humans, the environmental niche of opportunistic pathogens, recruitment of novel and existing resistance genes into human hosts, maintenance of resistant pathogens, and evolutionary mechanisms reducing the fitness cost associated with resistance genes. *ARGs*, Antibiotic resistance genes.

7.3 International status of antibiotic resistance

7.3.1 African region

Regardless of development levels, the challenge of AMR affects all continents and countries. In Africa, AMR is a growing and profound menace that is exceeding TB, HIV, and malaria in annual deaths. According to some reports, about 700,000 lives are lost annually, and these statistics could rise steeply to 10 million by 2050 if left unattended (Antimicrobial Resistance: New Report Warns of Growing Threat-growing-threat/, 2024). This results from various factors, including insufficient sanitation and hygiene, inadequate medical management, a lack of capacity-building initiatives, subpar antibiotics, poor stewardship, and others (Jayaprakasan et al., 2025). Despite the AMR crisis raging on the continent, African countries have insufficient stewardship and implementation of AMR. A lack of microbial laboratories and trained personnel in AMR contributes to the challenges. Between 2016 and 2019, data were collected from 205 laboratories across 14 African countries and analysed under the Mapping AMR and Antimicrobial Use Project (MAAP), which uncovered that only 1.3% of laboratories performed bacteriological testing. It was also found that 80% of labs conducted fewer than 1000 AST tests, whereas in countries like Nigeria, Burkina Faso, and Chile, less than 50% of the population is covered for AMR testing (Antimicrobial Resistance in Africa – Findings from the Mapping AMR & AMU Partnership MAAP Project, 2022). Only a small portion of microbiological labs are equipped for AMR testing. National programmes by the WHO, the African CDC, and the Fleming Fund play an important role in building up potential laboratories, skilled people, and quality surveillance. According to the WHO, although 43 African nations are enroled in the Global Antimicrobial Resistance and Use Surveillance System (GLASS), only 15 have submitted data from their national surveillance programme. Algeria, Angola, Burkina, Togo, Burundi, Cameroon, and Namibia joined GLASS in 2020, and the Democratic Republic of Congo joined in 2021, strengthening the surveillance system. The WHO, through its Regional Strategy for AMR 2023–30, has endorsed 47 member States in the African Region to create and improve NAPs on AMR (World Health Organization, 2022; Global Antimicrobial Resistance and Use Surveillance System GLASS report, 2025. World Health Organization., 2022).

7.3.2 The United States

AMR is one of the most pressing threats of the 21st century. The Centres for Disease Control and Prevention (CDC), in 2013, first announced its Antimicrobial Resistance Threat Report in the United States (US) about AMR as an urgent national priority and since then has consistently been ranking AMR among the topmost health concerns. As per the AR Report 2019, 2.8 million AMR infections are recorded nationally in the United States each year, leading to 35,000 deaths. This report also categorised carbapenem-resistant *Acinetobacter, Clostridioides difficile*, carbapenem-resistant Enterobacteriaceae (CRE) as 'Urgent Threat', and ESBL-producing Enterobacteriaceae, MRSA, drug-resistant *Salmonella typhi, and Shigella. Sp.*, and six other microbes under 'Serious Threat' (CDC. Antibiotic Resistance Threats in the United States, 2019; Atlanta, GA: U.S. Department of Health and Human Services, CDC; 2019, 2019). The reported cases of multidrug-resistant *Candida auris* had increased five times between 2019 and 2022, and 4000 clinical cases were reported in 2023 (CDC 2023 Candida auris: Information for Healthcare Professionals, 2023). The National Action Plan for Combating Antibiotic-Resistant Bacteria (CARB) has set goals to minimise AMR infections, strengthen inspections, and promote innovation (Cocorullo et al., 2024). While certain AMR infections are controlled under national action plans and stewardship plans, the COVID-19 pandemic uncovered the vulnerability and surges in hospital-onset AMR (CDC. Antibiotic Resistance Threats in the United States, 2019; Atlanta, GA: U.S. Department of Health and Human Services, CDC; 2019, 2019).

7.3.3 Europe

AMR is increasingly accepted as one of the most critical health challenges in Europe as it impairs the efficiency of antibiotics and imperils advanced modern healthcare. Northern and Western Europe showcase a smaller number of resistance levels compared to Southern and Eastern Europe, which signifies more constructive containment and stewardship (European Centre for Disease Prevention and Control & World Health Organization. Regional Office for Europe. 2022; World Health Organization, 2022). Within the European Union (EU), growing evidence indicates widespread AMR among Gram-negative bacteria, including third-generation cephalosporin and carbapenem-resistant *Klebsiella pneumoniae, Escherichia coli, and Pseudomonas aeruginosa* (WHO 2023 WHO–ECDC Report: Antimicrobial resistance threatens patient safety in the European Region, 2023). The carbapenem resistance is lower in *K. pneumoniae* compared to *E. coli and P. aeruginosa* (European Centre for Disease Prevention and Control, 2017). According to the

European Antimicrobial Resistance Surveillance Network (EARS-Net) report, the proportion of carbapenem resistance in *Acinetobacter sp.* is high in Hungary at 53% and Italy at 79%, while it is low at 4% in Germany. Norway, Sweden, and Denmark are lower than 1% (Ayobami et al., 2020). In 2020 670,000 cases of AMR infections were reported by the WHO, among which 33,000 deaths were caused by direct causes. The European Centre for Disease Prevention and Control (ECDC) cautioned against the rising number of infections and consequential deaths for all antibiotic resistance combinations. Experts predict that AMR may cause around 39 million deaths across the globe from 2025 to 2050 (European Centre for Disease Prevention and Control & World Health Organization. Regional Office for Europe. 2022; Antimicrobial resistance surveillance in Europe 2022 – 2020 data. World Health Organization. Regional Office for Europe., 2022; European Commission 2023 EU action on antimicrobial resistance, 2023). To address this chaos, the EU adopted the One Health Action Plan, and the European Medicines Agency (EMA) has supported antibiotic pioneering by research scholars and pharmaceutical companies (Antibiotic-resistance deaths to surge from 2025-2050, 2024). Globally, 193 UN member states declared to reduce AMR deaths by 10% by 2030 with an extended global action plan.

7.3.4 Australia

Australia has made a modest development to decrease antibiotic consumption. The Antimicrobial Use and Resistance in Australia (AURA) reported that the community's antimicrobial prescriptions fell from 26.6 million in 2017 to 21.8 million in 2022. Despite this, antibiotic utilisation remains high, with one in three Australians receiving at least one prescription (Australian Commission on Safety and Quality in Health Care, 2023). Alarmingly, antimicrobial use in aged care increased by 11.1% between 2022 and 2023, which raises concern about the vulnerability of the elderly population. Vancomycin-resistant Enterococci (VRE) have escalated from 5.3% in 2006 to 42.5% in 2023 (Department of Health and Aged Care 2024 Australia's National Antimicrobial Resistance Strategy – 2020 and beyond, 2024. The WHO in 2023 predicted 39–40 million mortalities between 2025 and 2050 if effective precautions are not taken. These statistics emphasise the urgency of addressing AMR as a 'silent pandemic' (Global Antimicrobial Resistance and Use Surveillance System GLASS report, 2025). The northern and remote western parts of Australia have persistently reported higher resistance levels. This is due to limited healthcare access, overuse of broad-spectrum antibiotics, and unique community health dynamics (Australian Commission on Safety and Quality in Health Care, 2024). The Australian government recently assigned $22.5 million to enhance surveillance and stewardship across human and animal health from 2024 to 2027 (Department of Health and Aged Care 2024 Australia's National Antimicrobial Resistance Strategy – 2020 and beyond, 2024).

7.3.5 South America

The resistance rates in South America are alarmingly high. The occurrence of carbapenem-resistant *Pseudomonas aeruginosa* (CRPA), as per a metaanalysis, is 40.9%, which is above the global average of 34.7% (Ramatla et al., 2025). The CRE accounted for 6.4% of all the isolates with the bla_{NDM} gene, which increased from 4.1% in 2015 to 39.4% in 2022 (de Araujo et al., 2025). Another study revealed that 29.8% of enterobacterales are colistin-resistant (Ramatla et al., 2025), 27.3% of *E. coli* are ESBL-resistant, and 52.4% of *Klebsiella sp.* are also ESBL-resistant. Country-specific CRE prevalence is reported as Argentina (10.7%), Brazil (17.3%), Colombia (18.2%), Ecuador (10%), and Venezuela (15%) (Jones et al., 2013).

7.3.6 Asia

Asia is the host at the centre of the AMR calamity (Fig. 7.3, facing an escalated threat across healthcare, agriculture, and environmental sectors. The WHO has recognised South-East Asia and the Western Pacific as hotbeds for AMR, with China, India, Pakistan, Bangladesh, and Vietnam being the epicentre of it. It is estimated that by 2050, over 4.3 million deaths will take place in the Asian-Pacific region and globally, a loss of 2 trillion USD annually (The Guardian 2025 Superbugs could kill millions more and cost $2tn a year by 2050, 2025). A recent metaanalysis indicated that in South-East Asia, the incidence of extended-spectrum β-lactamase (ESBL) and multidrug-resistant (MDR) producing *Klebsiella pneumoniae* is estimated at 55% (confidence interval (CI): 9–97) and 27% (CL: 5–57), respectively, with Thailand recording the highest occurrence (Salawudeen et al., 2023). Another study detected the prevalence of carbapenem resistance (CR) to be 31.3% (confidence interval 21.6–41), with Southeast Asian countries being the highest, followed by Southeast Asian countries (SEA) (Jayathilaka et al., 2025). The WHO GLASS incorporated 17 Asian countries, providing valuable insights. The GLASS 2023 report published that CR in *K. pneumoniae* surpassed 50% in India, 40% in Pakistan, and 35% in Vietnam.

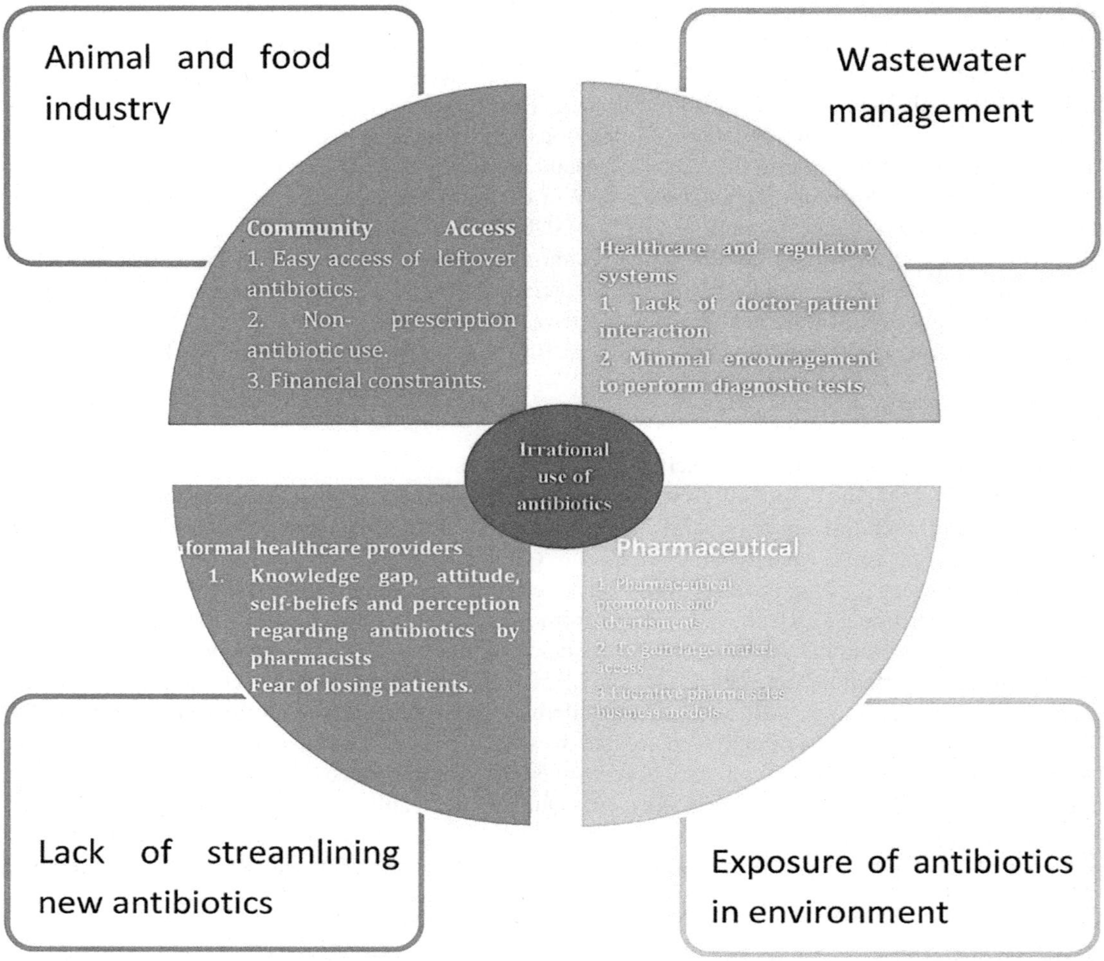

FIGURE 7.3 **Factors involved in irrational consumption of antibiotics**, drivers of irrational use of antibiotics at multisectoral level. The schema points to several interlocking factors across community practices (availability of leftover antibiotics, use without prescription, and financial barriers), healthcare and regulatory systems (limited doctor–patient interface and inadequate incentives for diagnostic testing), informal subsectors of healthcare providers (knowledge deficits, attitudes, self-beliefs, and fear of losing patients) and influences from translation areas of the pharmaceutical industry (promotion, advertising strategies in markets that they are looking to expand into as well as incentives-based sales models). Wider systemic factors include practices in the animal and food industry, gaps in wastewater management, a lack of streamlined development of new antibiotics, and exposures to antibiotic residues in environments.

7.4 Status of antibiotic resistance in India

India has one of the world's highest antibiotic consumption rates, which is over 10.7 units per person per year. Factors like over-the-counter antibiotic sales and irrational prescriptions drive misuse. As per 'The Centre for Disease Dynamics, Economics & Policy' (CDDEP) 2021 report, the consumption rates have increased by 103% between 2000 and 2015 (Knipe et al., 2021). The Indian Council of Medical Research's AMR Surveillance Network (ICMR AMRSN) reports that *E. coli* is nearly 72% resistant to fluoroquinolones and 60% to cephalosporin. In addition, 65% of *K. pneumoniae* and more than 80% are resistant to carbapenem, and 14%–15% of *Enterococcus faecium* are resistant to vancomycin (ICMR 2022, 2022). Moreover, India accounts for 27% of global multidrug-resistant tuberculosis (MDR-TB) (World Health Organization, 2022; Global Tuberculosis Report 2022. World Health Organization, Geneva., 2022). The economic consequences of AMR in India are noteworthy. By 2050, the annual GDP would be at a loss of $20 billion due to AMR, as estimated by the World Bank (Jonas et al., 2017. Drug-resistant infections: a threat to our economic future. World Bank Rep, 2, pp.1–132, 2017). Individual patients with resistant infections often have to pay two to three times more due to extended hospital stays and the use of last-resort antibiotics. AMR is now not only limited to healthcare, but its presence has extended to poultry, agricultural practices, and environmental reservoirs. A study from Punjab has reported that over

60% of *E. coli* isolates from broiler chickens were resistant to multiple antibiotic classes (World Health Organization, 2022; Global Tuberculosis Report 2022. World Health Organization, Geneva., 2022). An investigation in Hyderabad's pharmaceutical industrial area found ciprofloxacin residues as high as 28 mg/L in water bodies (Larsson, de Pedro and Paxeus, 2007).

A strengthened One Health Approach, integrating healthcare, veterinary, and environmental sectors, is essential to curb the rising threat. Under the Antimicrobial Resistance Strategic Action Plan (KARSAP), Kerala has suspended licences of 450 pharmacies and cancelled 2 for selling antibiotics without prescriptions. To make all the hospitals and healthcare centres antibiotic-smart, a colour coding system, such as blue covers to identify antibiotics, is proposed in Kerala (State intensifies measures to curb antibiotic misuse, 2025). The Centre for Cellular and Molecular Platforms (C-CAMP), Bengaluru, in collaboration with the Global AMR Innovation Fund (GAMRIF), aims to detect resistant microbes, antibiotic residues, and prevention strategies. The programme would encourage innovations suitable for low- and middle-income countries.

7.5 Irrational use of antibiotics by unskilled healthcare service providers

This group of healthcare providers contributes significantly to the global public health epidemic. However, their lack of professional skills occasioned by their level of education, training, and exposure contributes to improper use of antibiotics. This section will through light into the causes, consequences, and recommendations for possible solutions to the irrational use of antibiotics by unskilled healthcare service providers for the overall benefit of man, the healthcare system, and the world at large. According to the research conducted by Costelloe and colleagues (Costelloe et al., 2010) and Morgan and coworkers (Morgan et al., 2011), there is a direct link between the inappropriate or improper consumption of antibiotics and the development and transmission of antibiotic resistance. In this context, both healthcare providers and patients bear the substantial responsibility. This section will delve deeper into the causes that lead to irrational use of antibiotics by unskilled healthcare service providers:

7.5.1 Lack of knowledge and training

The unskilled healthcare service providers lack the prerequisite knowledge and training about antibiotics, their appropriate use, and the consequences of misuse. This contributes to inappropriate and wrong prescription, dispensing, and advice on antibiotics to unsuspecting patients (Larsson and Flach, 2022).

7.5.2 Overprescription and misuse

Overprescription and misuse are inevitable among these groups of healthcare workers. It is even worrisome that antibiotics are overprescribed for viral, protozoa, and other nonbacterial infectious agents. There are cases where antibiotics have been used in the treatment of the common cold and nausea. These inadvertently lead to increased antibiotic resistance and risks of adverse drug reactions, thereby increasing public health epidemic of the teeming population (Ahmed et al., 2024; Murray et al., 2022).

7.5.3 Contribution to antibiotic resistance

The irrational use of antibiotics aggravates the development and spread of antibiotic-resistance bacteria within the environment. This makes infections difficult to treat, thereby leading to the eventual spread of antibiotic-resistance bacteria and even resulting in treatment failures (Van Boeckel et al., 2019; Mulchandani et al., 2025).

7.5.4 Patient demand and expectations

Some patients may request antibiotics from unskilled healthcare service providers even when there is no need for them. These categories of staff may feel pressured to prescribe antibiotics just to satisfy the lustful desire of the patients arising from the inability of these categories of health care workers to take a professional stand. Undoubtedly, this unethical practise leads to an increase antibiotic resistance spread, rendering treatment difficult to handle (Mulchandani et al., 2025).

7.5.5 Inadequate diagnosis

The lack of training and continuous re-training of the unskilled healthcare providers may be responsible for inadequate or poor diagnosis of infections and infectious agents. This can result in the wrong use of antibiotics. For instance, broad-spectrum antibiotics may be the drug of choice for a particular infection in place of targeted treatment, and the reverse is the case for broad-spectrum response (Mulchandani et al., 2025).

7.5.6 Self-medication

In most circumstances, the unskilled healthcare providers intentionally or unintentionally encourage self-medication among patients and nonpatients alike. This can lead to wrong intake beyond the required dosage, increase cases of antibiotic resistance and, in extreme cases, may lead to death of the patients (Ahmed et al., 2024).

7.6 Consequences of irrational use of antibiotics by unskilled healthcare service providers

The section discusses the concerns that arise when antibiotics are misused by the unskilled healthcare service providers (Fig. 7.3).

7.6.1 Increased antimicrobial resistance

Once antibiotics are being prescribed by unskilled healthcare service providers who are not trained, there are bound to be errors in administration and dosage. These scenarios lead to increased and continuous cases of antibiotic resistance and compromise of public health (Handa et al., 2024).

7.6.2 Treatment failures

This can lead to additional healthcare costs arising from long stays at health facilities. These failures in treatment options arising from antibiotics misuse in humans and animals continuously increase the development and spread of antibiotic resistance. In extreme cases, there are possibilities of death which can deplete the population (Mulchandani et al., 2025; Salam et al., 2023).

7.6.3 Adverse reactions

The intake of drugs prescribed by unskilled healthcare service providers can result in lots of adverse drug reactions. This could be due to limited access to diagnostic tests, which leads to prophylaxis treatment, thereby increasing the risks of irrational antibiotic use. Adverse drug reactions may also result from improper adherence to the protocols in drug administration (Murray et al., 2022; Handa et al., 2024).

7.6.4 Increased healthcare costs

The indiscriminate prescription of antibiotics by unprofessional health care providers and their use by unsuspecting patients can lead to increased health care costs. This is because the drugs may not be able to handle the pathogens responsible for the ailments, thereby leading to the purchase of more drugs (Handa et al., 2024).

7.6.5 Morbidity and mortality

In extreme cases where there has been unwholesome consumption of antibiotics, especially specific target antibiotics in place of broad-based or vice versa, there is the increased probability of organ and system destruction. Arising from these disruptions in body organs, ultimately, deaths of the patients are inevitable gradually leading to the reduction in humans or animals' population (Mulchandani et al., 2025; Salam et al., 2023).

7.7 Major factors inducing the spread of antibiotic resistance in community

The primary purpose of antibiotics was to treat severe bacterial infections, but excessive usage has led to their hoarding in the environment. The vast presence of antibiotics has made it evidently resistant to microorganisms making it even more difficult to treat growing infections. Emerging issues include multidrug resistance, biofilm-associated tolerance, and global dissemination of genes like NDM-1. These trends pose a major threat to public health and highlight the need for global surveillance and responsible antibiotic use (Walsh et al., 2011; Uruén et al., 2021).

7.7.1 Overuse and misuse of antibiotics

There has been a nationwide growth of 103% in the usage of antibiotics from 2000 to 2015 because of inadequate information and supply without any proper prescription in countries like India and a few other middle-income countries (Klein et al., 2018). Medication by oneself without proper diagnosis by the concerned expert and over-the-counter accessibility has been very prominent in some countries of Asia, as well as Africa and Europe, which has been one of the major factors of AMR in the public (Table 7.1. There has been a common practice of people in such countries where they consume antibiotics for even viral infections, which usually do not require any antibiotic medication, which eventually leads to resistance of antibiotics for therapeutic usage. The socio-economic condition of a country shows an influential part in the utilisation of antibiotics or any other antimicrobial agents (World Health Organization 2013 Integrated surveillance of antimicrobial resistance, 2013). Fake drug supply is another reason for AMR. It has been reported that in 2019 an estimation of over 4.95 million deaths globally was related to AMR, emphasising the grave threat it poses to global health. A raising body of demonstrate shows that individual knowledge, attitudes, and beliefs about antimicrobials significantly influence their use. These behavioural factors manifest notably in consultation practices, where individuals may request antibiotics unnecessarily from general physicians or engage in self-medication practices without proper diagnosis or prescription (Murray et al., 2022).

7.7.1.1 Limited access to diagnostics

The inadequate knowledge of diagnostic leads to wrong identification of proper microbial infection, like distinguishing a viral with a bacterial infection, and hence the unnecessary use of antibiotics. Lack of proper antibiotic susceptibility labs all over is one of the primary reasons for the occurrence of antibiotic resistance in the nation and other developing countries (Ayukekbong, Ntemgwa and Atabe, 2017). Deficient data from community surveillance impedes the efficacy of operational activities by obscuring local resistance patterns.

7.7.1.2 Application of antibiotics in agriculture and veterinary

For the growth and development of disease control in livestock, antibiotics are being used in their feed. AMR also might cause the zoonotic spread of resistant bacteria between animals and humans by means of dairy products, meat, or even direct contact. The bacteria that are resistant to multiple drugs have been found in both meat and fresh produce, leading to a severe public health concern (Mezali and Hamdi, 2012). The rural areas worldwide are often seen to be in close contact

TABLE 7.1 Major sources and mechanisms of antibiotic resistance in India.

Source	Mechanism of resistance spread	Common resistant pathogens
Human Clinical Settings	Selection pressure due to overuse and misuse; nosocomial transmission	*Klebsiella pneumoniae*, *Escherichia coli*, *Acinetobacter* spp.
Pharmaceutical Industry Waste	Horizontal gene transfer in contaminated rivers and soil	Environmental reservoirs of resistance genes
Livestock and Poultry Farming	Spread via faecal-oral route, contaminated meat, and water	*Salmonella* spp., *E. coli*, *Enterococcus* spp.
Aquaculture	Antibiotics mixed with feed and water; gene transfer in aquatic bacteria	*Vibrio* spp., *Aeromonas* spp.
Community Settings	Self-medication, incomplete dosage, OTC availability	Community-acquired *E. coli* and *Streptococcus* spp.
Environmental Sources	Runoff from hospitals, agriculture, open defecation sites	Mixed environmental microbiota with resistant genes
Travel and Medical Tourism	Global movement of resistant strains (e.g. NDM-1 gene)	*E. coli*, *Pseudomonas aeruginosa*, *Klebsiella* spp.

with animals and other livestock, which primarily leads to antimicrobial infections that often are left untreated or with wrong treatment.

7.7.1.3 Poor infection control practices

Improper hygiene maintenance and lack of sanitation in public places such as hospitals and offices lead to the development of primary as well as secondary infections in a community, which includes humans and animals as well. Overcrowding of people in different public sectors leads to the widespread of resistant microbes. Dissemination of resistant pathogenic microbes like bacteria and viruses is accelerated by personal interaction, waterborne, and foodborne diseases. Enhancing personal hygiene and sanitation will mitigate the proliferation of drug-resistant organisms (Zhang et al., 2024).

7.7.1.4 Environmental factors for transmission causing resistance

Ecological sections such as soil, water bodies, air, and wildlife are increasingly recognised as significant pools and show paths for antibiotic residues and AMR genes. Untreated sewage, agricultural runoff, industrial discharge, and hospital effluents introduce antibiotics and resistant bacteria into rivers, lakes, and groundwater. Additionally, wildlife acts as a vector, acquiring and disseminating resistance through contact with contaminated environments. Contaminants like heavy metals and microplastics further aggravate resistance selection by coselecting for ARGs, fostering horizontal gene transfer among microbes (Smith et al., 2004). Antibiotic resistance is not driven only by the existence of antimicrobial agents. Pollutants of the environment, such as heavy metals, disinfectants, biocides, and other xenobiotics, play a significant role in reassuring the coselection mechanisms (Table 7.2). These mixtures impose nonantibiotic discerning pressures that promote the coselection of AMR genes (ARGs) by favouring bacterial populations expressing both metal resistance genes and ARGs typically located on mobile genetic elements such as jumping genes and transposons. This interface facilitates horizontal gene transfer (HGT), an initial mechanism driving the spread of resistance traits between environmental and pathogenic bacteria, even in the absence of direct antibiotic exposure. For example, bacteria subjected to metals such as zinc, copper, or mercury often develop multidrug resistance due to the genetic relation between metal and antibiotic resistance factors. Such coselection mechanisms are commonly examined in agricultural soils, wastewater treatment plants, and industrial effluent sites, representing a subtle yet influential environmental driver of AMR (Berendonk et al., 2015).

7.7.2 Policy, regulation and stewardship failures

One of the major features driving the global acceleration of AMR is the absence of robust regulatory frameworks governing the distribution, sale, and disposal of antibiotics. In many developing economies, antibiotics can be easily obtained without a prescription, often sold by unlicensed individuals. Furthermore, the improper disposal of expired

TABLE 7.2 Pollutants involved in environmental coselection of antimicrobial resistance (AMR).

Pollutant	Source	Mechanism of coselection	Impact on AMR
Heavy metals (e.g. Cu, Zn, Hg, Cd)	Mining, agriculture, industry	Coresistance via plasmids carrying metal and antibiotic resistance genes	Selection of multidrug-resistant strains
Disinfectants (e.g. QACs)	Hospitals, household cleaning products	Cross-resistance due to efflux pumps and membrane adaptation	Promotes resistance to both biocides and antibiotics
Pesticides and herbicides	Agricultural runoff	Alter microbial metabolism and gene expression	Indirect selection of resistant phenotypes
Pharmaceutical residues	Improper disposal, hospital waste	Maintains low-level selective pressure	Sustains and amplifies existing resistance traits
Industrial solvents and surfactants	Industrial waste streams	Induce stress responses and gene transfer	Encourage persistence of resistance elements in microbial DNA

antibiotics through household waste has led to the widespread disgrace of the environment with antibiotic residues, leading to resistance through environmental exposure and gene selection (Wales and Davies, 2015). Though significant progress has been made worldwide with the formulation of National Action Plans (NAPs) under the World Health Organisation's Global AMR Strategy, execution and enforcement at the community and primary healthcare levels remain ineffective. Weak institutional ability, inadequate funding, and poor coordination among sectors often result in fragmented oversight and unsuccessful policy execution. Subsequently, even where national policies exist, their practical impact remains limited, allowing the irrational use of antibiotics to resume largely unchecked (Chokshi et al., 2019). To curb the community-level spread of resistance, it is imperative to strengthen regulatory systems, ensure difficult enforcement of existing laws, and establish comprehensive antibiotic stewardship programmes that endorse responsible and sustainable antimicrobial use.

7.7.2.1 Patient compliance as a driver of antimicrobial resistance

When antibiotics are taken irregularly or prematurely discontinued, bacteria are exposed to subtherapeutic concentrations, which fail to eliminate the pathogen completely. This selective pressure allows partially resistant organisms to survive and mutate or acquire resistance genes, contributing to the evolution of resistant strains (Ventola, 2015). Moreover, self-medication, sharing leftover antibiotics, or stopping treatment once symptoms improve are widespread behaviours, especially in regions with low health literacy or weak healthcare access.

7.8 Actions required to curb antibiotic misuse and prevent resistance

Antibiotic resistance (ABR), an increasing worldwide issue, occurs when antibiotics that were created particularly to treat infectious is no longer effective (Zhu, Huang and Yang, 2022). The term AMR is broader and includes resistance to medications used to treat diseases caused by bacteria as well as other microorganisms like viruses (like HIV), fungi (like candida), and parasites (like malaria). Our desire to reduce the use of antibiotics counteracts incentives for the pharmaceutical company to create new therapies; therefore business objectives are frequently focused on investing in medications to treat chronic conditions that demand long-term use. Developing and underdeveloped countries are especially vulnerable to increased AMR due to a range of contributing factors. These include the widespread overuse of antibiotics, their use in agriculture, substandard drug quality, weak regulatory oversight, and limited access to proper healthcare. Additional challenges, such as malnutrition, frequent or prolonged infections, and the inability to afford newer, more effective treatments, further compound the problem (Md Haroon et al., 2020). Even if new effective antibiotics are developed, international cooperation and commitment are desperately needed to create a 'one health' plan that addresses regulations, independent monitoring, and policies (Harbarth et al., 2015; Asokan and Asokan, 2016).

7.9 Strategies to control antimicrobial resistance

7.9.1 Good hygiene and proper sanitation

Along with the high-level use of antibiotics, some environmental factors, such as crowdedness and inadequate sanitation, also assist the emergence and transmission of resistant microbes. Resistance can be spread through contact among people, contaminated food or water, or vectors. Prevention of the transmission of resistant microbes can be accomplished by enhancing the basic sanitation and hygiene (Smith et al., 2004). Improvements in hospital infection inhibition and restraint will decrease nosocomial spread of ARB, such as *S. aureus*.

7.9.2 Vaccination

AMR has become one of the biggest risks to public health globally in the 21st century, posing serious challenges to the treatment of infections and the effectiveness of modern medicine. Maximising the potential of vaccines and monoclonal antibodies, infections and diseases caused by AMR microbes can be prevented, thereby reducing the need for antibiotics and the selective pressure that supports the emergence of resistance (La Guidara et al., 2024). Conversely, the likelihood of generating resistance mechanisms against vaccines is incredibly low (Micoli et al., 2021), and even in the few cases where resistance has been documented (Cherry, 2012; Romanò et al., 2014), a decrease in the burden of disease has been attained. This is a significant distinction from antibiotics, as the development of resistance might totally eliminate the impact on a patient. Vaccines shield the patient against potential infections by being given before bacteria begin to multiply. In contrast, antibiotics treat persistent infectious disorders and come into contact with more bacteria that are likely to become resistant strains. Vaccines trigger immune responses against many targets (referred to as antigens) and/or multiple epitopes of the same antigen (polyclonal antibodies), in contrast to antibiotics that target a particular bacterial target. Subsequently, the probability of vaccine-revealed mutants occurring is extensively decreased, since multiple simultaneous mutations within frequent epitopes would be necessary (Micoli et al., 2021).

Monoclonal antibodies (mAbs) have the competence to predicament to an extensive range of antigens, whether they are energetic for the microbe's existence or not, comprising components like virulence factors, dissimilar antibiotics, which frequently dislocate dynamic bacterial functions. Additionally, because mAbs are bioproducts, they are inherently safe, and clinical trials have shown little to no toxicity thus far (Lanini et al., 2022), while small molecules, despite being less costly than mAbs, are linked to an increased risk of adverse effects.

7.9.3 Bacteriophage-based therapies

Bacteriophage therapy has appeared as a potential alternative to conventional antibiotics (Olawade et al., 2024) and is effective in treating multidrug-resistant (MDR) bacterial infections (Romero-Calle et al., 2019). Due to a constant rise in antibiotic resistance globally, infectious diseases caused by bacteria, like *S. aureus*, *Pseudomonas aeruginosa*, and *Escherichia coli,* have become more challenging to treat (Trend et al., 2018; Cocorullo et al., 2024). Phage therapy has proven effective in clearing infections that did not respond to standard antibiotics in several clinical cases. One notable example involves the successful use of phage to treat persistent wound infections caused by methicillin-resistant *S. aureus* (MRSA), particularly when all other treatment options had failed (Peng et al., 2019). Phages specifically engineered to target *Pseudomonas aeruginosa*, a major pathogen in cystic fibrosis infections, have established encouraging results by dropping bacterial loads and improving patient outcomes, particularly in cases where the bacteria have developed resistance to multiple antibiotics (Chang et al., 2018; Rossitto et al., 2018). These successful applications emphasise the emerging agreement of phage therapy as a powerful alternative when conventional antibiotic treatments fail.

7.9.4 Phage therapy combined with antibiotics

Phage therapy alone has shown significant potential; however, when combined with antibiotics, its therapeutic efficacy can be substantially improved, specifically in the treatment of chronic bacterial infections (Kraus et al., 2024). This combined phenomenon, known as phage–antibiotic synergy (PAS) (Subramanian, 2024), appears when the joint action of antibiotics and bacteriophages yields a greater antibacterial effect than either treatment could achieve separately (Chaudhry et al., 2017; Simon et al., 2021). Interestingly, certain antibiotics have been reported to accelerate phage production within bacterial hosts, resulting in the formation of larger phage plaques (Kamal and Dennis, 2015). A major challenge in healing bacterial infections lies in the presence of biofilms, dense microbial communities covered in a protective extracellular matrix that shields bacteria from antibiotics. Phages suggest a distinct advantage in this context, as they can penetrate and degrade biofilms, lowering bacterial density and making the remaining microbes more susceptible to antibiotic action. Several studies have validated this synergistic interaction, pointing out that combining phage with antibiotics leads to faster and more effective bacterial eradication compared to either treatment used alone (Chatterjee et al., 2019; Azam et al., 2024; Luo et al., 2024; Al-Anany et al., 2024). For instance, phage–antibiotic sequences have substantially improved outcomes in persistent respiratory infections caused by *Pseudomonas aeruginosa*, a pathogen notorious for forming resilient biofilms (Chaudhry et al., 2017; Oechslin et al., 2017; Cafora et al., 2019; Law et al., 2019). Furthermore, phages can reduce bacterial polysaccharides, thereby improving antibiotic penetration into bacterial colonies. These cooperative mechanisms collectively underscore the promise of phage–antibiotic combination therapy in fighting multidrug-resistant infections. Subsequently, there is increasing advocacy for using phage in tandem with antibiotics to accomplish superior bacterial clearance than phage therapy alone can deliver (Torres-Barceló and Hochberg, 2016). This integrated methodology shows efficacy against infections caused by methicillin-resistant *S. aureus* (MRSA) and multidrug-resistant *Enterococcus* strains.

7.9.5 Probiotics

Probiotics are live microorganisms with positive effects when administered under proper circumstances to a host (Hill et al., 2014). Probiotics possess antagonistic qualities preventing the development of gut pathogens by (1) generation of biologically active metabolites such as bacteriocins, organic acids, antimicrobial peptides, and antioxidants (Vieco-Saiz et al., 2019); (2) competition for binding sites for the nutrients (Knipe et al., 2021); and (3) modifying functions of immune system (Azad et al., 2018). The second and third mechanisms are inherent to probiotic cells – stemming from their ability to bind and colonise, as well as their indirect interaction with immune cells. In contrast, the first mechanism resembles the direct molecular action of antibiotics against pathogens. Beyond the transmission of resistance genes – which is usually assessed before microbes are classified as probiotics – these beneficial organisms help lower the risk of disease resistance by employing multiple antimicrobial strategies. Probiotics can modulate the gut microbiota and strengthen the immune system of the host to reduce bacterial infections and dysbiosis, respectively, or they can directly act through antibiosis through the generation of metabolites like bacteriocins, antioxidant compounds, organic acids, and by competing for nutrients and space (Helmy et al., 2023).

7.9.6 Immuno-antibiotics: an emerging strategy for tackling antibiotic resistance

Traditionally, antibiotics have been estimated solely for their strict bacteriostatic or bactericidal effects. On the other hand, growing evidence shows that antimicrobials can work alongside the host's intrinsic, unaffected approach, delivering decisive impacts supporting bacterial exclusion. This could result in more prompt and extensive effects, which would reduce the likelihood that resistance will emerge among remaining bacteria (Berti et al., 2020). Reports indicate that naturally occurring defensive host peptides have evolved to defend the host from infections by preventing pathogens. Their antimicrobial actions closely resemble those of therapeutic peptide antibiotics like colistin and daptomycin (Rima et al., 2021). Beyond their primary in vitro mechanisms, antibiotics are associated with enhancing and optimising immune responses of the host (Watson et al., 2020). This incorporates their competence to influence bacterial virulence factors and modulate immune pathways that shape the host's response to infection (Berti et al., 2020). Experimental research by Volk et al. revealed that patients with MRSA bacteraemia who established β-lactam adjunctive therapy alongside standard antibiotics exhibited improved levels of interleukin-1 (IL-1) and declined levels of interleukin-10 (IL-10), conversing that antibiotics can influence the host's immune response (Volk et al., 2020).

Dual-acting immuno-antibiotics (DAIAs) represent a promising class of antimicrobials targeting two key bacterial pathways: riboflavin biosynthesis and methyl-D-erythritol phosphate (MEP) pathway for production of isoprenoid (Singh et al., 2021). This two-pronged approach is designed not only to kill MDR bacteria but also to enhance the body's natural immune response, offering a powerful synergy between antimicrobial activity and immune system activation.

7.9.7 The characteristics of stakeholders and government supervisory agencies

Challenging AMR is not just the job of scientists and medical authorities. Everyone from the public to numerous stakeholders has an important role to play (Blakely et al., 2006). At the same time, continued education is essential to help healthcare providers grasp the importance of responsible prescribing and making treatment decisions based on solid evidence (Bhutta and Balchin, 1996). AMR should be considered a major public health issue by the government of every nation. Regulations and policies should be established to ensure that antimicrobials are accessed and used responsibly. Media professionals must receive sufficient training on how to communicate scientific and medical information in simple terms through a variety of channels to educate the public about the potential risks of AMR and the behaviours that encourage the emergence of AMR organisms (Blakely et al., 2006). In a number of developing countries, the positive aspects of an educational programme aimed at journalists have been effective in combating HIV/AIDS, and a similar strategy for AMR is probably going to have additional benefits (Martinez-Cajas et al., 2008). Regulatory agencies and governments have introduced several initiatives to combat AMR, but the majority of these efforts have taken place primarily in developed countries and are lacking in the underdeveloped nations (World Health Organization, 2014; World Health Organization, 2015). Some African countries are part of the 2009 Global Antibiotic Resistance Partnership (GARP) launched to help develop practical policy recommendations on antibiotic resistance in low- and middle-income nations. However, only India, Nepal, South Africa, Kenya, Mozambique, Tanzania, Vietnam, and Uganda currently have active national GARP groups.

7.9.8 Antibiotic stewardship: a necessity for future

Tackling antibiotic misuse and encouraging ethical use requires a collective approach engaging different healthcare experts. Pharmacists play a crucial role in informing patients about the proper use of antibiotics, confirming their appropriate dispensing, and monitoring for difficult drug reactions. Likewise, nurses contribute considerably by keeping patients informed about their treatment, maintaining cautious monitoring for infection symptoms, and applying infection control practices (Kilari and Oroszi, 2024). The predominant goal of antimicrobial stewardship is to endorse the judicious and safe use of antimicrobials, optimise patient results, decrease the emergence of AMR, and curb the spread of infections triggered by multidrug-resistant (MDR) organisms (Aryee and Price, 2015).

Using antibiotics responsibly means finding the right balance between a person's right to receive proper treatment and society's need to preserve the long-term effectiveness of these vital medicines (Dyar et al., 2017). The approach to accomplish this involves the implementing policies that include evidence-based prescription guidelines (Bertollo, Lutkemeyer and Levin, 2018). To optimise patient outcomes and reduce the evolution of antibiotic resistance, a multidisciplinary approach to antibiotic stewardship is essential in the future.

7.10 Conclusion

Antibiotic resistance is the ultimate outcome of an assemblage of factors, mainly including the acquirement of foreign genetic elements, antibiotic selective pressure or exposure, and clonal propagation. Several methodologies were developed over time to understand these causes, such as network tracing, ecological studies, and molecular and traditional epidemiological studies. However, the minimization of antimicrobial usage in humans as well as in animals should be focused. However, certain research indicates that cutting back on antimicrobial usage does not consistently result in a proportional decline in resistance (Enne et al., 2001), while other studies have revealed more encouraging outcomes (Guillemot et al., 1998). Moreover, instead of targeting to eradicate antibiotic consumption in humans and animals entirely, it would be more beneficial to focus on minimising the practice of narrow-spectrum antibiotics and those prescribed for extended duration. Evidences indicate that some antibiotic use patterns may raise the risk of harbouring penicillin-resistant *Streptococcus pneumoniae* (PRSP) (Guillemot et al., 1998). In the end, antibiotic resistance within the community is a complicated and increasingly serious issue, and to address this, a thorough and multifaceted strategy is required, one that surpasses simply cutting back on antibiotic usage whenever and wherever possible. This chapter seeks to deliver a thorough overview that encompasses the evolutionary pathways of antibiotic resistance and extends to the national and international status of the problem, the misuse and overuse of antibiotics especially by untrained healthcare providers, the factors that contribute to this issue, and the current approaches used to address the emergence and transmission of antibiotic resistance in the community settings. Consequently, the existence of resistance factors that actually contribute to making the microorganism pathogenic is likely to be involved to cause slight or no damage to their bacterial host, thus making it difficult to eliminate from the pathogen population. The variety of resistance genes existing within environment indicates the abundance of several more resistance genes accessible for the pathogens to get involved in developing resistance. The resistance genes common among the bacteria associated with human microbes probably will not be eradicated even under normal environment where antibiotics are not present. Therefore the mobile resistance genes circulating within human pathogens may readily reappear under antibiotic exposure or during treatment. The introduction of novel resistance genes in pathogens, conversely, could have crucial consequences for human health as these mobile resistance genes aim at the newly established antibiotics, or more efficacious resistance mechanisms against the existing drugs, would further confine the available therapeutic options.

References

Ahmed, S.K., Hussein, S., Qurbani, K., Hussein Ibrahim, R., Fareeq, A., Mahmood, K.A., et al., 2024. Antimicrobial resistance: Impacts, challenges, and future prospects. Journal of Medicine, Surgery, Public. Health 2, 100081. Available at https://doi.org/10.1016/j.glmedi.2024.100081.

Al-Anany, A.M., Fatima, R., Nair, G., Mayol, J.T., Hynes, A.P., 2024. Temperate phage-antibiotic synergy across antibiotic classes reveals new mechanism for preventing lysogeny. American Society for Microbiology, mBio 15 (6) Available from. Available at https://doi.org/10.1128/mbio.00504-24, https://journals.asm.org/doi/10.1128/mbio.00504-24.

Andersson, D.I., Jerlström-Hultqvist, J., Näsvall, J., 2015. Evolution of new functions de novo and from preexisting genes. Cold Spring Harbor Perspectives in Biology 7 (6), a017996. Available at https://doi.org/10.1101/cshperspect.a017996.

Antibiotic-resistance deaths to surge from 2025-2050, study says, (2024). Available from: https://www.reuters.com/business/healthcare-pharmaceuticals/antibiotic-resistance-deaths-surge-2025-2050-study-says-2024-09-16/.

Antimicrobial Resistance in Africa – Findings from the Mapping AMR & AMU Partnership (MAAP) Project, (2022). Available from: https://onehealthtrust.org/news-media/webinars/antimicrobial-resistance-in-africa-findings-from-the-mapping-amr-amu-partnership-maap-project.

Antimicrobial Resistance: New Report Warns of Growing Threat-growing-threat/, (2024). Available from: https://africacdc.org/news-item/antimicrobial-resistance-new-report-warns-of-growing-threat/.

Aryee, A., Price, N., 2015. Antimicrobial stewardship – can we afford to do without it? British Journal of Clinical Pharmacology 79 (2), 173–181. Available at https://doi.org/10.1111/bcp.12417.

Asokan, G.V., Asokan, V., 2016. Bradford Hill's criteria, emerging zoonoses, and one health. Journal of Epidemiology and Global Health 6 (3), 125–129. Available from. Available at https://doi.org/10.1016/j.jegh.2015.10.002, https://www.atlantis-press.com/journals/jegh.

Atmar, R.L., Opekun, A.R., Gilger, M.A., Estes, M.K., Crawford, S.E., Neill, F.H., et al., 2014. Determination of the 50% human infectious dose for norwalk virus. The Journal of Infectious Diseases 209 (7), 1016–1022. Available from. Available at https://doi.org/10.1093/infdis/jit620, http://jid.oxfordjournals.org/content/current.

Australian Commission on Safety and Quality in Health Care, 2023. AURA 2023: fifth Australian report on antimicrobial use and resistance in human health. Sydney: ACSQHC. Available from: https://www.safetyandquality.gov.au/our-work/antimicrobial-resistance/antimicrobial-use-and-resistance-australia-aura/aura-2023-fifth-australian-report-antimicrobial-use-and-resistance-human-health.

Australian Commission on Safety and Quality in Health Care, 2024. Australian Passive AMR Surveillance: An update of resistance trends in multidrug-resistant organisms – 2006 to 2023. Sydney; ACSQHC. Available from: https://www.amr.gov.au/australias-response/objective-5-integrated-surveillance-and-response-resistance-and-usage/surveillance-antimicrobial-use-and-resistance-human-health.

Ayobami, O., Willrich, N., Suwono, B., Eckmanns, T., Markwart, R., 2020. The epidemiology of carbapenem-non-susceptible Acinetobacter species in Europe: analysis of EARS-Net data from 2013 to 2017. Antimicrobial Resistance Infect. Control 9 (1) Available from. Available at https://doi.org/10.1186/s13756-020-00750-5, http://www.aricjournal.com/.

Ayukekbong, J.A., Ntemgwa, M., Atabe, A.N., 2017. The threat of antimicrobial resistance in developing countries: causes and control strategies. Cameroon Antimicrobial Resistance Infect. Control. 6 (1) Available from. Available at https://doi.org/10.1186/s13756-017-0208-x, http://www.aricjournal.com/.

Azad, M.A.K., Sarker, M., Wan, D., 2018. Immunomodulatory effects of probiotics on cytokine profiles. BioMed Research International 2018 Available from. Available at https://doi.org/10.1155/2018/8063647, http://www.hindawi.com/journals/biomed/.

Azam, A.H., Sato, K., Miyanaga, K., Nakamura, T., Ojima, S., Kondo, K., et al., 2024. Selective bacteriophages reduce the emergence of resistant bacteria in bacteriophage-antibiotic combination therapy. Microbiology Spectrum 12 (6) Available from. Available at https://doi.org/10.1128/spectrum.00427-23, https://journals.asm.org/doi/10.1128/spectrum.00427-23.

Berendonk, T.U., Manaia, C.M., Merlin, C., Fatta-Kassinos, D., Cytryn, E., Walsh, F., et al., 2015. Tackling antibiotic resistance: the environmental framework. Nature Reviews Microbiology 13 (5), 310–317. Available from. Available at https://doi.org/10.1038/nrmicro3439, http://www.nature.com/nrmicro/index.html.

Berti, A., Rose, W., Nizet, V., Sakoulas, G., 2020. Antibiotics and innate immunity: a cooperative effort toward the successful treatment of infections. Open Forum Infectious Diseases 7 (8) Available from. Available at https://doi.org/10.1093/ofid/ofaa302, https://academic.oup.com/ofid.

Bertollo, L.G., Lutkemeyer, D.S., Levin, A.S., 2018. Are antimicrobial stewardship programs effective strategies for preventing antibiotic resistance? A systematic review. Journal of Infection Control 46 (7), 824–836. Available from. Available at https://doi.org/10.1016/j.ajic.2018.01.002, http://www.journals.elsevier.com/ajic-american-journal-of-infection-control/.

Bhutta, T.I., Balchin, C., 1996. Assessing the impact of a regulatory intervention in Pakistan. Social Science & Medicine 42 (8), 1195–1202. Available at https://doi.org/10.1016/0277-9536(95)00392-4.

Blakely, J.T.M., Sinkowitz-Cochran, R.L., Jarvis, W.R., 2006. Infectious diseases physicians' preferences for continuing medical education on antimicrobial resistance and other general topics. Infection Control & Hospital Epidemiology 27 (8), 873–875. Available at https://doi.org/10.1086/505922.

Bloomfield, S., Ackerley, L., 2024. Developing better understanding of hygiene is key to developing hygiene behaviour change in home and everyday life settings. Perspectives in Public Health 144 (6), 354–362. Available at https://doi.org/10.1177/17579139231163734.

Cafora, M., Deflorian, G., Forti, F., Ferrari, L., Binelli, G., Briani, F., et al., 2019. Phage therapy against *Pseudomonas aeruginosa* infections in a cystic fibrosis zebrafish model. Scientific Reports 9 (1). Available at https://doi.org/10.1038/s41598-018-37636-x.

CDC, 2023. *Candida auris*: Information for Healthcare Professionals. Centers for Disease Control and Prevention. Available from: http://https://www.cdc.gov/fungal/candida-auris/index.html.

CDC, 2019. Antibiotic Resistance Threats in the United States, 2019. Atlanta, GA: U.S. Department of Health and Human Services, CDC. Available from: http://doi.org/10.15620/cdc:82532.

Chaidez, C., Soto-Beltran, M., Gerba, C.P., Tamimi, A.H., 2014. Reduction of risk of Salmonella infection from kitchen cleaning clothes by use of sodium hypochlorite disinfectant cleaner. Letters in Applied Microbiology 59 (5), 487–492. Available at https://doi.org/10.1111/lam.12321.

Chang, R.Y.K., Chen, K., Wang, J., Wallin, M., Britton, W., Morales, S., et al., 2018. Proof-of-principle study in a murine lung infection model of antipseudomonal activity of phage PEV20 in a dry-powder formulation. Antimicrobial Agents Chemotherapy 62 (2) Available from. Available at https://doi.org/10.1128/AAC.01714-17, http://aac.asm.org/content/62/2/e01714-17.full.pdf+html.

Chatterjee, A., Johnson, C.N., Luong, P., Hullahalli, K., McBride, S.W., Schubert, A.M., et al., 2019. Bacteriophage resistance alters antibiotic-mediated intestinal expansion of enterococci. Infection and Immunity 87 (6) Available from. Available at https://doi.org/10.1128/IAI.00085-19, http://iai.asm.org/content/by/year.

Chaudhry, W.N., Concepción-Acevedo, J., Park, T., Andleeb, S., Bull, J.J., Levin, B.R., et al., 2017. Synergy and order effects of antibiotics and phages in killing *Pseudomonas aeruginosa* biofilms. PLoS One 12 (1), e0168615. Available at https://doi.org/10.1371/journal.pone.0168615.

Cherry, J.D., 2012. Epidemic pertussis in 2012 - the resurgence of a vaccine-preventable disease. Journal of Medicine 367 (9), 785–787. Available from. Available at https://doi.org/10.1056/NEJMp1209051, http://www.nejm.org/doi/pdf/10.1056/NEJMp1209051.

Chokshi, A., Sifri, Z., Cennimo, D., Horng, H., 2019. Global contributors to antibiotic resistance. Journal of Global Infectious Diseases 11 (1), 36–42. Available from. Available at https://doi.org/10.4103/jgid.jgid_110_18, http://www.jgid.org/.

Cocorullo, M., Stelitano, G., Chiarelli, L.R., 2024. Phage therapy: an alternative approach to combating multidrug-resistant bacterial infections in cystic fibrosis. International Journal of Molecular Sciences 25 (15), 8321. Available at https://doi.org/10.3390/ijms25158321.

Costelloe, C., Metcalfe, C., Lovering, A., Mant, D., Hay, A.D., 2010. Effect of antibiotic prescribing in primary care on antimicrobial resistance in individual patients: systematic review and meta-analysis. BMJ (Clinical Research ed.) 340 (may18 2), c2096. Available at https://doi.org/10.1136/bmj.c2096.

de Araujo, L.G., Cedeño, K., Bomfim, A.P., de Oliveira Silva, M., Mendes, A.V., Barberino, M.G., et al., 2025. Carbapenem-resistant Enterobacterales bloodstream infections related to death in two Brazilian tertiary hospitals. BMC Infectious Diseases 25 (1) Available from. Available at https://doi.org/10.1186/s12879-025-11115-x, https://bmcinfectdis.biomedcentral.com/.

Department of Health and Aged Care, 2024. Australia's National Antimicrobial Resistance Strategy – 2020 and beyond. Canberra: Australian Government. Available from: https://www.amr.gov.au/resources/australias-national-antimicrobial-resistance-strategy-2020-and-beyond.

Dyar, Huttner, B., Schouten, J., Pulcini, C., 2017. What is antimicrobial stewardship? Clinical Microbiology and Infection: Official Publication of the European Society of Clinical Microbiology and Infectious Diseases. 23 (11), 793–798. Available at https://doi.org/10.1016/j.cmi.2017.08.026.

Enne, V.I., Livermore, D.M., Stephens, P., Hall, L.M.C., 2001. Persistence of sulphonamide resistance in *Escherichia coli* in the UK despite national prescribing restriction. Lancet 357 (9265), 1325–1328. Available from. Available at https://doi.org/10.1016/S0140-6736(00)04519-0, http://www.journals.elsevier.com/the-lancet/.

European Centre for Disease Prevention and Control & World Health Organization, 2022. Regional Office for Europe. (2022) . Antimicrobial resistance surveillance in Europe 2022 – 2020 data. World Health Organization. Regional Office for Europe. Available from: https://iris.who.int/handle/10665/351141.

European Centre for Disease Prevention and Control, 2017. Surveillance of antimicrobial resistance in Europe – Annual report of the European Antimicrobial Resistance Surveillance Network (EARS-Net) 2017. Stockholm: ECDC. Available from: https://www.ecdc.europa.eu/en/publications-data/surveillance-antimicrobial-resistance-europe-2017.

European Commission, 2023. EU action on antimicrobial resistance. Directorate-General for Health and Food Safety. Available from: https://health.ec.europa.eu/antimicrobial-resistance.

Global Antimicrobial Resistance and Use Surveillance System (GLASS) report, 2025. Antibiotic use data for 2022. Geneva: World Health Organization. Available from: https://www.who.int/publications/i/item/9789240108127.

Guillemot, D., Carbon, C., Balkau, B., Geslin, P., Lecoeur, H., Vauzelle- Kervroedan, F., et al., 1998. Low dosage and long treatment duration of β-lactam: risk factors for carriage of penicillin-resistant *Streptococcus pneumoniae*. American Medical Association, Fr. JAMA 279 (5), 365–370. Available from. Available at https://doi.org/10.1001/jama.279.5.365, http://jama.jamanetwork.com/journal.aspx.

Haas, C.N., Rose, J.B., Gerba, C.P., 2014. Quantitative Microbial Risk Assessment. Wiley. Available at http://doi.org/10.1002/9781118910030.

Handa, V.L., Patel, B.N., Bhattacharya, D.A., Kothari, R.K., Kavathia, D.G., Vyas, B.R.M., 2024. A study of antibiotic resistance pattern of clinical bacterial pathogens isolated from patients in a tertiary care hospital. Frontiers in Microbiology 15 Available from. Available at https://doi.org/10.3389/fmicb.2024.1383989, https://www.frontiersin.org/journals/microbiology#.

Harbarth, S., Balkhy, H.H., Goossens, H., Jarlier, V., Kluytmans, J., Laxminarayan, R., et al., 2015. Antimicrobial resistance: one world, one fight!. 4 (1) Available from. Available at https://doi.org/10.1186/s13756-015-0091-2, https://www.aricjournal.com/.

Helmy, Y.A., Taha-Abdelaziz, K., Hawwas, H.A.E.H., Ghosh, S., AlKafaas, S.S., Moawad, M.M.M., et al., 2023. Antimicrobial resistance and recent alternatives to antibiotics for the control of bacterial pathogens with an emphasis on foodborne pathogens. Antibiotics 12 (2) Available from. Available at https://doi.org/10.3390/antibiotics12020274, http://www.mdpi.com/journal/antibiotics.

Hill, C., Guarner, F., Reid, G., Gibson, G.R., Merenstein, D.J., Pot, B., et al., 2014. Expert consensus document: the international scientific association for probiotics and prebiotics consensus statement on the scope and appropriate use of the term probiotic. Nat. Rev. Gastroenterology Hepatology 11 (8), 506–514. Available from. Available at https://doi.org/10.1038/nrgastro.2014.66, https://www.nature.com/nrgastro/index.html.

Hooton, T.M., Levy, S.B., 2001. Antimicrobial resistance: a plan of action for community practice. American Family Physician 63 (6), 1087–1096.

ICMR, 2022. Annual Report of the Antimicrobial Resistance Surveillance Network. Indian Council of Medical Research, New Delhi. Available from: https://www.icmr.gov.in/icmrobject/custom_data/pdf/resource-guidelines/AMRSN_Annual_Report_2022.pdf.

Jayaprakasan, A.K., Ghosh, A., Banerjee, B., Singh, M.M., Solanki, H.K., 2025. Growing burden of antimicrobial resistance and the need for effective stewardship: an Indian perspective. One Health Bulletin 5 (4), 158–164. Available at https://doi.org/10.4103/ohbl.ohbl_10_25.

Jayathilaka, N., Shehana, S., Nakkawita, D., Senaratne, T., 2025. Prevalence and molecular epidemiology of carbapenem resistance in Asia: a systematic review and meta-analysis. Systematic Reviews 14 (1) Available from. Available at https://doi.org/10.1186/s13643-025-02776-5, https://systematicreviewsjournal.biomedcentral.com/.

Jonas, O.B., Irwin, A., Berthe, F.C.J., Le Gall, F.G., Marquez, P.V., 2017. 2017. Drug-resistant infections: a threat to our economic future. World Bank Reports 2, 1–132. Available from. Available at https://www.worldbank.org/en/topic/health/publication/drug-resistant-infections-a-threat-to-our-economic-future.

Jones, R.N., Guzman-Blanco, M., Gales, A.C., Gallegos, B., Castro, A.L.L., Martino, M.D.V., et al., 2013. Susceptibility rates in Latin American nations: report from a regional resistance surveillance program (2011). The Journal of Infectious Diseases 17 (6), 672–681. Available at https://doi.org/10.1016/j.bjid.2013.07.002.

Kamal, F., Dennis, J.J., 2015. Burkholderia cepacia complex phage-antibiotic synergy (PAS): antibiotics stimulate lytic phage activity. Applied and Environmental Microbiology 81 (3), 1132–1138. Available from. Available at https://doi.org/10.1128/AEM.02850-14, http://aem.asm.org/content/81/3/1132.full.pdf.

Kilari, V.B., Oroszi, T., 2024. The misuse of antibiotics and the rise of bacterial resistance: a global concern. Pharmacology & Pharmacy. 15 (12), 508–523. Available at https://doi.org/10.4236/pp.2024.1512028.

Kimura, M., Ohta, T., 1969. The average number of generations until fixation of a mutant gene in a finite population. Genetics 61 (3), 763–771. Available at https://doi.org/10.1093/genetics/61.3.763.

Klein, E.Y., Van Boeckel, T.P., Martinez, E.M., Pant, S., Gandra, S., Levin, S.A., et al., 2018. Global increase and geographic convergence in antibiotic consumption between 2000 and 2015. Proceedings of the National Academy of Sciences of the United States of America. 115 (15) E3463, Available from. https://doi.org/10.1073/pnas.1717295115, http://www.pnas.org/content/pnas/115/15/E3463.full.pdf.

Knipe, H., Temperton, B., Lange, A., Bass, D., Tyler, C.R., 2021. Probiotics and competitive exclusion of pathogens in shrimp aquaculture. Reviews in Aquaculture 13 (1), 324–352. Available from. Available at https://doi.org/10.1111/raq.12477, http://onlinelibrary.wiley.com/journal/10.1111/(ISSN)1753-5131.

Kothary, M.H., Babu, U.S., 2001. Infective dose of foodborne pathogens in volunteers: a review. 21 (1), 49–68. Available from. Available at https://doi.org/10.1111/j.1745-4565.2001.tb00307.x, https://onlinelibrary.wiley.com/journal/10.1111/(ISSN)1745-4565.

Kraus, S., Fletcher, M.L., Łapińska, U., Chawla, K., Baker, E., Attrill, E.L., et al., 2024. Phage-induced efflux down-regulation boosts antibiotic efficacy. PLoS Pathogens 20 (6) Available from. Available at https://doi.org/10.1371/journal.ppat.1012361, https://journals.plos.org//plospathogens/article?id=10.1371/journal.ppat.1012361.

La Guidara, C., Adamo, R., Sala, C., Micoli, F., 2024. Vaccines and monoclonal antibodies as alternative strategies to antibiotics to fight antimicrobial resistance. International Journal of Molecular Sciences 25 (10), 5487. Available at https://doi.org/10.3390/ijms25105487.

Lanini, S., Milleri, S., Andreano, E., Nosari, S., Paciello, I., Piccini, G., et al., 2022. Safety and serum distribution of anti-SARS-CoV-2 monoclonal antibody MAD0004J08 after intramuscular injection. Nature Communications 13 (1). Available at https://doi.org/10.1038/s41467-022-29909-x.

Larson, H.E., Price, A.B., Honour, P., Borriello, S.P., 1978. Clostridium difficile and the ætiology of pseudomembranous colitis. Lancet 311 (8073), 1063–1066. Available at https://doi.org/10.1016/s0140-6736(78)90912-1.

Larsson, D.G.J., Flach, C.F., 2022. Antibiotic resistance in the environment. Nature Reviews Microbiology 20 (5), 257–269. Available from. Available at https://doi.org/10.1038/s41579-021-00649-x, http://www.nature.com/nrmicro/index.html.

Larsson, D.G.J., de Pedro, C., Paxeus, N., 2007. Effluent from drug manufactures contains extremely high levels of pharmaceuticals. Journal of Hazardous Materials 148 (3), 751–755. Available at https://doi.org/10.1016/j.jhazmat.2007.07.008.

Law, N., Logan, C., Yung, G., Furr, C.L.L., Lehman, S.M., Morales, S., et al., 2019. Successful adjunctive use of bacteriophage therapy for treatment of multidrug-resistant Pseudomonas aeruginosa infection in a cystic fibrosis patient. Infection 47 (4), 665–668. Available from. Available at https://doi.org/10.1007/s15010-019-01319-0, link.springer.de/link/service/journals/15010/index.htm.

Letten, A.D., Hall, A.R., Levine, J.M., 2021. Using ecological coexistence theory to understand antibiotic resistance and microbial competition. Nature Ecology & Evolution 5 (4), 431–441. Available at https://doi.org/10.1038/s41559-020-01385-w.

Liu, C.M., Stegger, M., Aziz, M., Johnson, T.J., Waits, K., Nordstrom, L., et al., 2018. Escherichia coli ST131-H22 as a foodborne uropathogen. mBio 9 (4). Available at https://doi.org/10.1128/mBio.00470-18.

Luo, J., Liu, M., Ai, W., Zheng, X., Liu, S., Huang, K., et al., 2024. Synergy of lytic phage pB23 and meropenem combination against carbapenem-resistant *Acinetobacter baumannii*. Antimicrobial Agents Chemotherapy 68 (6) Available from. Available at https://doi.org/10.1128/aac.00448-24, https://journals.asm.org/doi/10.1128/aac.00448-24.

Maillard, J.Y., Bloomfield, S.F., Courvalin, P., Essack, S.Y., Gandra, S., Gerba, C.P., et al., 2020. Reducing antibiotic prescribing and addressing the global problem of antibiotic resistance by targeted hygiene in the home and everyday life settings: a position paper. Journal of Infection Control 48 (9), 1090–1099. Available from. Available at https://doi.org/10.1016/j.ajic.2020.04.011, http://www.journals.elsevier.com/ajic-american-journal-of-infection-control/.

Martinez-Cajas, J.L., Invernizzi, C.F., Ntemgwa, M., Schader, S.M., Wainberg, M.A., 2008. Benefits of an educational program for journalists on media coverage of HIV/AIDS in developing countries. Journal of the International AIDS Society 11 (1), 2. Available from. Available at https://doi.org/10.1186/1758-2652-11-2, http://onlinelibrary.wiley.com/journal/10.7448/(ISSN)1758-2652.

Md Haroon, R., Rahman, M., Sultana, H., Islam, K., Al Rakib, M.M.N., Kalam, M.A., et al., 2020. Antibacterial resistance patterns of bacteria isolated from clinical specimens at Uttara IbnSina Diagnostic Centre. Journal of Microbiology Research 14 (5), 175–181. Available at https://doi.org/10.5897/AJMR2020.9324 Available from.

Mezali, L., Hamdi, T.M., 2012. Prevalence and antimicrobial resistance of *Salmonella* isolated from meat and meat products in Algiers (Algeria). Foodborne Pathogens and Disease 9 (6), 522–529. Available at https://doi.org/10.1089/fpd.2011.1032.

Micoli, F., Bagnoli, F., Rappuoli, R., Serruto, D., 2021. The role of vaccines in combatting antimicrobial resistance. Nature Reviews Microbiology 19 (5), 287–302. Available at https://doi.org/10.1038/s41579-020-00506-3.

Morar, M., Wright, G.D., 2010. The genomic enzymology of antibiotic resistance. Annual Review of Genetics 44, 25–51. Available at https://doi.org/10.1146/annurev-genet-102209-163517.

Morgan, D.J., Okeke, I.N., Laxminarayan, R., Perencevich, E.N., Weisenberg, S., 2011. Non-prescription antimicrobial use worldwide: a systematic review. The Lancet Infectious Diseases 11 (9), 692–701. Available from. Available at https://doi.org/10.1016/S1473-3099(11)70054-8, https://www.sciencedirect.com/science/journal/14733099.

Mulchandani, R., Tiseo, K., Nandi, A., Klein, E., Gandra, S., Laxminarayan, R., et al., 2025. Global trends in inappropriate use of antibiotics, 2000–2021: scoping review and prevalence estimates. BMJ Public Health 3 (1), e002411. Available at https://doi.org/10.1136/bmjph-2024-002411.

Murray, C.J., Ikuta, K.S., Sharara, F., Swetschinski, L., Robles Aguilar, G., Gray, A., et al., 2022. Global burden of bacterial antimicrobial resistance in 2019: a systematic analysis. Lancet 399 (10325), 629–655. Available from. Available at https://doi.org/10.1016/S0140-6736(21)02724-0, http://www.journals.elsevier.com/the-lancet/.

Oechslin, F., Piccardi, P., Mancini, S., Gabard, J., Moreillon, P., Entenza, J.M., et al., 2017. Synergistic interaction between phage therapy and antibiotics clears *Pseudomonas Aeruginosa* infection in endocarditis and reduces virulence. The Journal of Infectious Diseases 215 (5), 703–712. Available from. Available at https://doi.org/10.1093/infdis/jiw632, http://jid.oxfordjournals.org/content/current.

Olawade, D.B., Fapohunda, O., Egbon, E., Ebiesuwa, O.A., Usman, S.O., Faronbi, A.O., et al., 2024. Phage therapy: a targeted approach to overcoming antibiotic resistance. Microbial Pathogenesis 197 Available from. Available at https://doi.org/10.1016/j.micpath.2024.107088, https://www.sciencedirect.com/science/journal/08824010.

Peng, C., Hanawa, T., Azam, A.H., LeBlanc, C., Ung, P., Matsuda, T., et al., 2019. Silviavirus phage ϕMR003 displays a broad host range against methicillin-resistant Staphylococcus aureus of human origin. Applied Microbiology and Biotechnology 103 (18), 7751–7765. Available from. Available at https://doi.org/10.1007/s00253-019-10039-2, link.springer.de/link/service/journals/00253/index.htm.

Ramatla, T., Nkhebenyane, J., Lekota, K.E., Thekisoe, O., Monyama, M., Achilonu, C.C., et al., 2025. Global prevalence and antibiotic resistance profiles of carbapenem-resistant Pseudomonas aeruginosa reported from 2014 to 2024: a systematic review and meta-analysis. Frontiers in Microbiology 16 Available from. Available at https://doi.org/10.3389/fmicb.2025.1599070, https://www.frontiersin.org/journals/microbiology#.

Rima, M., Rima, M., Fajloun, Z., Sabatier, J.-M., Bechinger, B., Naas, T., 2021. Antimicrobial peptides: a potent alternative to antibiotics. Antibiotics 10 (9), 1095. Available at https://doi.org/10.3390/antibiotics10091095.

Robilotti, E., Deresinski, S., Pinsky, B.A., 2015. Norovirus. Clinical Microbiology Reviews 28 (1), 134–164. Available from. Available at https://doi.org/10.1128/CMR.00075-14, http://cmr.asm.org/content/28/1/134.full.pdf.

Romanò, L., Paladini, S., Galli, C., Raimondo, G., Pollicino, T., Zanetti, A.R., 2014. Hepatitis B vaccination. Human Vaccines & Immunotherapeutics 11 (1), 53–57. Available at https://doi.org/10.4161/hv.34306.

Romero-Calle, D., Guimarães Benevides, R., Góes-Neto, A., Billington, C., 2019. Bacteriophages as alternatives to antibiotics in clinical care. Antibiotics 8 (3), 138. Available at https://doi.org/10.3390/antibiotics8030138.

Rossitto, M., Fiscarelli, E.V., Rosati, P., 2018. Challenges and promises for planning future clinical research into bacteriophage therapy against *Pseudomonas aeruginosa* in cystic fibrosis. An argumentative review. Frontiers in Microbiology 9 (MAY) Available from. Available at https://doi.org/10.3389/fmicb.2018.00775, https://www.frontiersin.org/articles/10.3389/fmicb.2018.00775/full.

Salam, M.A., Al-Amin, M.Y., Salam, M.T., Pawar, J.S., Akhter, N., Rabaan, A.A., et al., 2023. Antimicrobial resistance: a growing serious threat for global public health. Healthcare (Switz.) 11 (13) Available from. Available at https://doi.org/10.3390/healthcare11131946, www.mdpi.com/journal/healthcare/.

Salawudeen, A., Raji, Y.E., Jibo, G.G., Desa, M.N.M., Neoh, H.M., Masri, S.N., et al., 2023. Epidemiology of multidrug-resistant *Klebsiella pneumoniae* infection in clinical setting in South-Eastern Asia: a systematic review and meta-analysis. Antimicrobial Resistance & Infection Control 12 (1) Available from. Available at https://doi.org/10.1186/s13756-023-01346-5, https://aricjournal.biomedcentral.com/.

Schulz, F., Eloe-Fadrosh, E.A., Bowers, R.M., Jarett, J., Nielsen, T., Ivanova, N.N., et al., 2017. Towards a balanced view of the bacterial tree of life. Microbiome 5 (1), 140. Available at https://doi.org/10.1186/s40168-017-0360-9.

Scott, E., 2013. Community-based infections and the potential role of common touch surfaces as vectors for the transmission of infectious agents in home and community settings. American Journal of Infection Control 41 (11), 1087–1092. Available at https://doi.org/10.1016/j.ajic.2013.05.014.

Simon, K., Pier, W., Krüttgen, A., Horz, H.-P., 2021. Synergy between Phage Sb-1 and Oxacillin against Methicillin-Resistant *Staphylococcus aureus*. Antibiotics 10 (7), 849. Available at https://doi.org/10.3390/antibiotics10070849.

Singh, K.S., Sharma, R., Reddy, P.A.N., Vonteddu, P., Good, M., Sundarrajan, A., et al., 2021. Retraction Note: IspH inhibitors kill Gram-negative bacteria and mobilize immune clearance. Nature 599 (7885). Available at https://doi.org/10.1038/s41586-021-03961-x 518–518.

Smith, M.A., Garbharran, H., Edwards, M.J.O., O'Hara-Murdock, P., 2004. Health promotion and disease prevention through sanitation education in South African Zulu and Xhosa Women. Journal of Transcultural Nursing 15 (1), 62–68. Available at https://doi.org/10.1177/1043659603259972.

State intensifies measures to curb antibiotic misuse, 2025. Available from: https://timesofindia.indiatimes.com/city/thiruvananthapuram/state-intensifies-measures-to-curb-antibioticmisuse/articleshow/121005989.cms.

Subramanian, A., 2024. Emerging roles of bacteriophage-based therapeutics in combating antibiotic resistance. Frontiers in Microbiology 15. Available at https://doi.org/10.3389/fmicb.2024.1384164.

The Guardian, 2025, July 20. Superbugs could kill millions more and cost $2tn a year by 2050, models show. Available from: https://www.theguardian.com/society/2025/jul/20/superbugs-could-kill-millions-more-and-cost-2tn-a-year-by-2050-models-show.

Torres-Barceló, C., Hochberg, M.E., 2016. Evolutionary rationale for phages as complements of antibiotics. Trends in Microbiology 24 (4), 249–256. Available from. Available at https://doi.org/10.1016/j.tim.2015.12.011, www.elsevier.com/locate/tim.

Trend, S., Chang, B.J., O'Dea, M., Stick, S.M., Kicic, A., 2018. Use of a primary epithelial cell screening tool to investigate phage therapy in cystic fibrosis. Frontiers in Pharmacology 9 Available from. Available at https://doi.org/10.3389/fphar.2018.01330, http://www.frontiersin.org/Pharmacology.

Uruén, C., Chopo-Escuin, G., Tommassen, J., Mainar-Jaime, R.C., Arenas, J., 2021. Biofilms as promoters of bacterial antibiotic resistance and tolerance. Antibiotics 10 (1), 1–36. Available from. Available at https://doi.org/10.3390/antibiotics10010003, https://www.mdpi.com/2079-6382/10/1/3/pdf.

Van Boeckel, T.P., Pires, J., Silvester, R., Zhao, C., Song, J., Criscuolo, N.G., et al., 2019. Global trends in antimicrobial resistance in animals in low- and middle-income countries. Science (New York, N.Y.) 365 (6459) Available from. Available at https://doi.org/10.1126/science.aaw1944, https://science.sciencemag.org/content/365/6459/eaaw1944/tab-pdf.

Ventola, C.L., 2015. The antibiotic resistance crisis: Part 1: causes and threats. undefined P T 40 (4), 277–283. Available from. Available at http://www.ptcommunity.com/system/files/pdf/ptj4004277.pdf.

Vieco-Saiz, N., Belguesmia, Y., Raspoet, R., Auclair, E., Gancel, F., Kempf, I., et al., 2019. Benefits and inputs from lactic acid bacteria and their bacteriocins as alternatives to antibiotic growth promoters during food-animal production. Frontiers in Microbiology 10. Available at https://doi.org/10.3389/fmicb.2019.00057.

Volk, C.F., Burgdorf, S., Edwardson, G., Nizet, V., Sakoulas, G., Rose, W.E., 2020. Interleukin (IL)-1β and IL-10 host responses in patients with staphylococcus aureus bacteremia determined by antimicrobial therapy. Clinical Infectious Diseases:An Official Publication of the Infectious Diseases Society of America 70 (12), 2634–2640. Available from. Available at https://doi.org/10.1093/cid/ciz686, http://cid.oxfordjournals.org/content/by/year.

Waglechner, N., Wright, G.D., 2017. Antibiotic resistance: it's bad, but why isn't it worse? BMC Biology 15 (1). Available at https://doi.org/10.1186/s12915-017-0423-1.

Wales, A.D., Davies, R.H., 2015. Co-selection of resistance to antibiotics, biocides and heavy metals, and its relevance to foodborne pathogens. Antibiotics 4 (4), 567–604. Available from. Available at https://doi.org/10.3390/antibiotics4040567, http://www.mdpi.com/2079-6382/4/4/567/pdf.

Walsh, T.R., Weeks, J., Livermore, D.M., Toleman, M.A., 2011. Dissemination of NDM-1 positive bacteria in the New Delhi environment and its implications for human health: An environmental point prevalence study. The Lancet Infectious Diseases 11 (5), 355–362. Available at https://doi.org/10.1016/S1473-3099(11)70059-7.

Watson, K., Russell, C.D., Baillie, J.K., Dhaliwal, K., Fitzgerald, J.R., Mitchell, T.J., et al., 2020. Developing novel host-based therapies targeting microbicidal responses in macrophages and neutrophils to combat bacterial antimicrobial resistance. Frontiers in Immunology 11 Available from. Available at https://doi.org/10.3389/fimmu.2020.00786, https://www.frontiersin.org/journals/immunology#.

WHO, 2023. WHO–ECDC Report: Antimicrobial resistance threatens patient safety in the European Region. World Health Organization Regional Office for Europe. Available from: https://www.who.int/europe/news/item/14-04-2023-who-ecdc-report–antimicrobial-resistance-threatens-patient-safety-in-european-region.

Winther, B., 2011. Rhinovirus infections in the upper airway. Proceedings of the American Thoracic Society 8 (1), 79–89. Available at https://doi.org/10.1513/pats.201006-039rn.

World Health Organization, 2013. Integrated surveillance of antimicrobial resistance. Available from: http://apps.who.int/iris/bitstream/10665/91778/1/9789241506311_eng.pdf.

World Health Organization, 2022. Global antimicrobial resistance and use surveillance system (GLASS) report 2022. World Health Organization. Available from: https://www.who.int/publications/i/item/9789240108127.

World Health Organization, 2022. Global Tuberculosis Report 2022. World Health Organization, Geneva. Available from: https://www.who.int/publications/i/item/9789240061729.

World Health Organization, 2014. Additional global, regional and national strategies and plans to address antimicrobial resistance. Available from: http://www.cddep.org/garp/home.

World Health Organization, 2015. WHO report finds systems to combat antibiotic resistance lacking. Available from: http://www.who.int/mediacentre/news/releases/2015/antibiotic-resistance-lacking/en/.

Yezli, S., Otter, J.A., 2011. Minimum infective dose of the major human respiratory and enteric viruses transmitted through food and the environment. Food and Environmental Virology 3 (1), 1–30. Available at https://doi.org/10.1007/s12560-011-9056-7.

Zhang, T., Nickerson, R., Zhang, W., Peng, X., Shang, Y., Zhou, Y., et al., 2024. The impacts of animal agriculture on One Health—Bacterial zoonosis, antimicrobial resistance, and beyond. One Health 18, 100748. Available at https://doi.org/10.1016/j.onehlt.2024.100748.

Zhu, Y., Huang, W.E., Yang, Q., 2022. Clinical perspective of antimicrobial resistance in bacteria. Infection and Drug Resistance 15, 735–746. Available from. Available at https://doi.org/10.2147/IDR S345574, https://www.dovepress.com/getfile.php?fileID=78731.

Chapter 8

Tackling antibiotic resistance: challenges in everyday environments

Dipayan Das
Department of Microbiology, Royal School of Biosciences, The Assam Royal Global University, Guwahati, Assam, India

8.1 Introduction

Antibiotic resistance has emerged from a clinical problem to a complex environmental crisis that extends beyond the traditional healthcare system (Larsson and Flach, 2022). The discovery and widespread use of antibiotics transformed modern medicine during the twentieth century (Muteeb et al., 2023). The extensive and regular indiscriminate application of antibiotics has caused strong selective pressure on bacterial populations. This pressure has accelerated the evolution and spreading of diverse resistance mechanisms (Galgano et al., 2025). The misuse of antibiotics in both human and veterinary medicine is widely recognised and continues to contribute to the process of antimicrobial resistance (AMR) (Caneschi et al., 2023). Due to the alarming situation, the environmental factors contributing to AMR in the global population have recently received considerable scientific attention (Alwisi et al., 2025; Schages et al., 2020).

Everyday environments (where people live, work, and interact with natural systems) function as a central crossing point for the exchange of resistant bacteria and resistance genes (Cave et al., 2021). These environments include households, municipal water distribution systems, agricultural soils, food production units, and community spaces. Their interconnected nature forms complex transmission networks. Such networks allow the movement of antibiotic resistance genes (ARGs) and antibiotic-resistant bacteria (ARB) across ecological compartments and geographical regions (Diala et al., 2025; Schages et al., 2020). Different ecosystems exhibit substantially different mechanisms in the persistence and amplification of AMR. Aquatic systems, soils, and atmospheric particulates carry resistant microorganisms. These compartments also provide favourable conditions for horizontal gene transfer (HGT), thereby allowing the emergence of resistant bacteria (Liu et al., 2022). The subinhibitory concentrations of antibiotics, heavy metals, and other anthropogenic contaminants also generate selective pressure. This coselection process supports the maintenance and spread of resistance traits within microbial communities (Swarthout et al., 2022).

This review evaluates the current understanding of antibiotic resistance in day-to-day environments. The study is focused on existing surveillance gaps, challenges in risk assessment, and the development of effective mitigation strategies. Evidence published is examined to characterise environmental reservoirs, define transmission pathways, evaluate prevalence patterns, and identify the principal barriers to effectively controlling this growing global health concern.

8.2 Background and theoretical foundations

The concept of antibiotic resistance as an environmental phenomenon is based on the recognition that resistance genes are ancient (Taneja and Sharma, 2019). In recent years, anthropogenic activities have played a significant role in accelerating the selection, amplification, and dissemination of these resistance determinants (Czatzkowska et al., 2022). The environmental resistome encompasses the total pool of antibiotic resistance genes present within environmental microbiomes. This reservoir includes both intrinsic and acquired resistance determinants. Importantly, these genes can be transferred to clinically relevant pathogens through HGT mechanisms such as conjugation, transformation, and transduction (Saima et al., 2019). Thus the environmental bacteria act as a genetic pool for resistance traits that may emerge in human pathogens.

The One Health concept provides the theoretical concept for understanding antibiotic resistance (Velazquez-Meza et al., 2022). It deals with the interconnected problem across human, animal, and environmental health domains (Aslam et al., 2021).

Antimicrobial Resistance in the 21st Century. DOI: https://doi.org/10.1016/B978-0-443-44748-8.00017-8

This approach recognises that targeting only one sector is insufficient to control resistance effectively (Cella et al., 2023). Instead, integrated strategies are required to deal with interactions among these domains (Puvača et al., 2022; Sarkar et al., 2023). The nature connects human and animal populations through shared water bodies, agricultural systems, food chains, and contaminated surfaces or materials (Tayyaba and Ahmed, 2022). These linkages facilitate the circulation of resistant bacteria and resistance genes across ecological compartments.

Several concepts help to understand the environmental dimension of antibiotic resistance. First, the principle of coselection explains how exposure to nonantibiotic substances (such as heavy metals, biocides, and pesticides) can maintain resistance genes in bacterial populations even in the absence of antibiotic pressure (Maldonado et al., 2025). Second, the concept of environmental hotspots identifies specific locations where conditions favour the emergence and amplification of resistance, including wastewater treatment plants (WWTPs), agricultural facilities, and areas with high anthropogenic impact (Nji et al., 2021; Zhao et al., 2023). Third, the recognition of mobile genetic elements (such as plasmids, integrons, and transposons) as vehicles for the dissemination of resistance genes indicates the dynamic nature of the environmental resistome (Nji et al., 2021).

8.3 Environmental reservoirs of antibiotic resistance

8.3.1 Household environments

Domestic households act as a reservoir of antibiotic-resistant microbes (Fig. 8.1). They function as interfaces that connect human microbiota, food matrices, water systems, and environmental bacterial communities. A comprehensive investigation by Schages et al. (2020) showed that private households in Germany have substantial levels of β-lactamase genes and antibiotic-resistant bacteria, with shower drains identified as the principal reservoir. Quantitative PCR analysis of 54 households revealed a significantly higher relative abundance of ARGs in shower drains (0.2367 ARG copies per 16 S rRNA gene copy) compared to dishwashers (0.1329 copies) and washing machines (0.0006 copies) (Schages et al., 2020). Class 1 integrons (intI1) are an important genetic element facilitating HGT and were detected in 96.3% of households, indicating extensive potential for resistance gene mobilisation (Schages et al., 2020).

The household resistome comprises a broad spectrum of AMR factors. Among these, blaCMY-2, blaACT/MIR, and blaOXA-48 have been reported as the predominant β-lactamase genes in domestic environments. The detection of

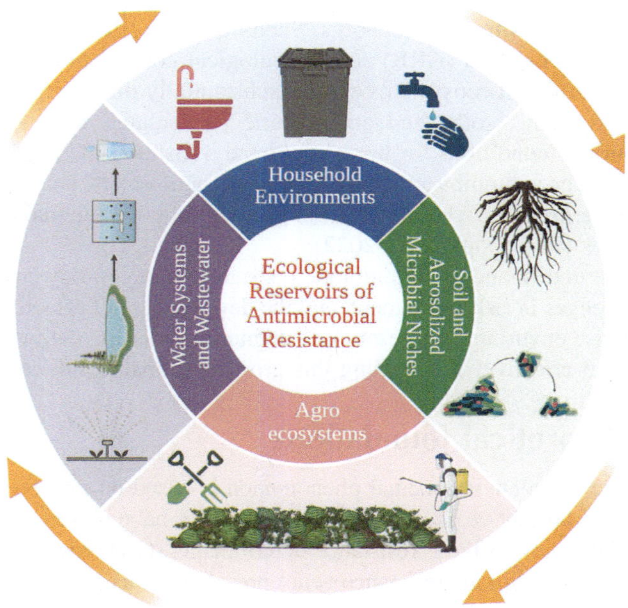

FIGURE 8.1 Ecological compartments contributing to the environmental persistence and spread of antimicrobial resistance. Illustration displaying the main environmental sources that support and spread AMR. It includes household settings (sanitation and waste), water and wastewater systems, agricultural fields, and soil and air environments. These areas are connected and allow antibiotic-resistant bacteria and resistance genes to survive and move between them. The circular arrows indicate ongoing exchange among these sectors, largely influenced by human activities, emphasising the need for a One Health approach. *AMR*, Antimicrobial resistance.

carbapenemase-encoding genes is particularly alarming. Carbapenems are regarded as last-resort antibiotics for the treatment of multidrug-resistant bacterial infections. Their compromised efficacy will have significant threat to clinical therapeutics. A positive correlation between genotypic profiles and phenotypic resistance patterns has been observed in bacterial isolates recovered from households (Schages et al., 2020).

Domestic hygiene practices are important for modulating the household burden of antibiotic resistance. Automated dishwashing and laundering have been shown to significantly reduce antibiotic-resistant bacteria. The efficacy of these processes increases with higher washing temperatures and extended cycle durations (Schages et al., 2020). The persistence of low bacteria counts after laundering remains a concern. Residual contamination on textiles may facilitate indirect transmission within the household environment. These findings showed the necessity for optimised and evidence-based hygiene protocols to minimise the risk of resistance dissemination (Schages et al., 2020).

8.3.2 Water systems and wastewater

Water systems are a major route for the distribution of antibiotic resistance across the environment and into human populations (Table 8.1). Municipal WWTPs are recognised as hotspots for antibiotic resistance genes and antibiotic-resistant bacteria. These facilities receive heterogeneous inputs from domestic households, hospitals, and industrial sources (Li et al., 2022).

It was reported that ARGs in the aquatic ecosystem have reached alarming levels (Li et al., 2021). Numerous investigations have reported the extensive occurrence of resistance determinants in surface waters, groundwater, drinking water distribution systems, and recreational water bodies (Mutuku et al., 2022; Niegowska and Wögerbauer, 2022). Water systems have developed as a source for the spreading of AMR, facilitating the transmission of antibiotic-resistant bacteria and ARGs among environmental reservoirs, animals, and human populations (Nava et al., 2022). Singh et al. reported the role of water bodies in this transmission system and stated that conventional water treatment processes are insufficient for the complete removal of these resistance determinants (Singh et al., 2022). The detection of antibiotic-resistant bacteria in drinking water systems is another concern from the aspect of public health (Xi et al., 2009). Wastes from hospitals and animal farming processes act as major hotspots of resistance, discharging contaminants into the nearest river system. These rivers are frequently used for drinking, sanitation, and recreational purposes, thereby increasing the risk of human exposure (Mohamad et al., 2025).

Research in South Africa has documented the presence of mobile colistin resistance genes (mcr-3 and mcr-5) in river and storm water, indicating the environmental diffusion of resistance to critically important antibiotics (Snyman et al., 2021). Similarly, investigations conducted in agricultural systems have reported that interconnected surface water bodies located near pastures and greenhouse facilities carry a diverse pool of antibiotic resistance genes. Among these, sulphonamide resistance genes are frequently predominant. The widespread occurrence of the class 1 integron-integrase gene (intI1) indicates that HGT is involved in the distribution and persistence of ARGs within these aquatic environments (Zhao et al., 2023).

The naturalisation of *Escherichia coli* in wastewater environments has emerged as a significant concern. Yu et al. reported the presence of wastewater-adapted *E. coli* strains that have stress-resistance factors that can survive under conventional treatment conditions, including chlorination, thermal exposure, and ultraviolet irradiation (Yu et al., 2022). These strains also have a diverse type of antibiotic resistance genes. The cooccurrence of treatment-resistance traits and AMR indicates a process of coselection and adaptive evolution within wastewater systems (Yu et al., 2022). Such adaptation may compromise the effectiveness of two basic public health interventions (water treatment and antibiotic therapy), thereby causing a high risk for the persistence and spreading of resistant pathogens (Abebe et al., 2024; Duarte et al., 2022).

8.3.3 Farming environments and practices

The agricultural sector functions as an important reservoir and amplification site for antibiotic resistance (Manyi-Loh et al., 2018). It is primarily mediated by the extensive use of antibiotics in livestock production (Mann et al., 2021). The application of manure as a fertiliser contributes to the persistence and dissemination of antibiotic resistance determinants (Karwowska, 2024). Animal waste acts as a primary contributor to establishing an interconnection between agricultural activities and environmental AMR. Agricultural runoff and contaminated irrigation water act as major sources of antibiotic resistance genes and antibiotic-resistant bacteria (Zhao et al., 2023).

Zhao et al. (2023) reported a widespread occurrence of antibiotic resistance genes in surface waters located near pastures and greenhouse cultivation areas. ARGs belonging to multiple antibiotic classes were frequently detected together, indicating the concurrent use of different antibiotics in agricultural practices. The study further identified HGT as a major mechanism behind the dissemination and persistence of ARGs in these environments. This was particularly evident for sulphonamide resistance genes, which showed strong signatures of mobility and distribution across sites (Zhao et al., 2023).

TABLE 8.1 Environmental reservoirs of antibiotic resistance.

Environmental condition	Sources	Resistance determinants	Abundance	References
Household	Shower drains, dishwashers, and washing machines	β-lactamases (blaCMY-2, blaACT/MIR, blaOXA-48), class 1 integrons (intI1)	Shower drains: 0.2367 ARG copies/16 S rRNA; intI1 in 96.3% of households	Schages et al. (2020)
Water systems	Wastewater treatment plants, surface waters, drinking water, rivers, storm water	Diverse ARGs, including sulphonamide, tetracycline, β-lactamase, and mobile colistin resistance (mcr-3, mcr-5)	Widespread detection; WWTPs identified as major hotspots	Li et al. (2022); Mohamad et al. (2025); Mutuku et al. (2022); Singh et al. (2022); Zhao et al. (2023)
Agriculture	Pastures, greenhouses, livestock facilities, agricultural runoff, animal waste	Sulphonamide ARGs (dominant), tetracycline, β-lactamase genes	ARGs are prevalent in all surface waters near agricultural areas	Malagón-Rojas et al. (2020); Swarthout et al. (2022); Zhao et al. (2023)
Community contexts	Household water, sewage drains, clinic sources (LMICs)	blaTEM (47%), tetM (38%), ermB (31%)	86% of samples resistant to ≥1 antibiotic; 58% multidrug resistant	Diala et al. (2025); Nji et al. (2021)
Soil compartments	Agricultural soils, biosolid-amended soils, contaminated sites	Diverse ARGs influenced by manure application and wastewater irrigation	Increased abundance in anthropogenically impacted areas	Li et al. (2022); Malagón-Rojas et al. (2020); Ndukwe and Ibrahim (2025); Sassi et al. (2025)
Hospital/Clinical waste	Hospital wastewater, clinical effluents	Extended-spectrum β-lactamases (ESBLs), carbapenemases	High concentrations of ARGs and ARB	Mutuku et al. (2022); Sarkar et al. (2023); Sassi et al. (2025)
Solid waste	Municipal solid waste landfills, composting facilities	Diverse ARGs and ARB	Favourable conditions for AMR growth and gene transfer	Sarkar et al. (2023)

In low- and middle-income countries, the human–animal interface in agricultural systems creates significant challenges. Closeness between livestock and household members, limited biosecurity, and inadequate sanitation infrastructure facilitate the circulation of antibiotic-resistant bacteria and resistance genes. Swarthout et al. (2022) reported considerable evidence of antibiotic resistance exchange between domesticated animals and humans in direct contact within LMIC settings. Their findings indicate transmission of whole bacterial strains as well as HGT occurring within households and backyard farming environments. Environmental matrices (including water sources, soil, and animal-derived food products) were identified as important pathways for dissemination. However, the directionality and relative contribution of these transmission routes remain insufficiently determined (Swarthout et al., 2022). Pesticide application represents another indirect source of antibiotic resistance within agricultural systems. Exposure to pesticides and other agrochemicals can increase the prevalence of AMR in soil environments. As reported by Malagón-Rojas et al., the natural environment functions as a reservoir and diffusion pathway for ARGs (Malagón-Rojas et al., 2020). Such observations indicate that resistance development is not exclusively linked to antibiotic misuse but also to coselective pressures exerted by nonantibiotic chemicals. This coselection phenomenon indicates the requirement for integrated management strategies that take into consideration multiple anthropogenic stress factors influencing environmental microbiomes (Malagón-Rojas et al., 2020).

8.3.4 Soil and air compartments

Soil ecosystems that carry out agricultural activities, wastewater irrigation, or biosolid application contain diverse communities of antibiotic-resistant bacteria and resistance genes (Zeng et al., 2025). The persistence of antibiotics and resistance genes in soil can extend for months to years, creating long-term reservoirs that can contaminate crops, groundwater, and surface waters (Sassi et al., 2025). Airborne dissemination of antibiotic resistance is comparatively underexplored but is increasingly recognised as a significant environmental concern. Bioaerosols generated from WWTPs, intensive livestock production units, and composting facilities serve as important reservoirs of antibiotic-resistant bacteria and antibiotic resistance genes (Bruni et al., 2020). These bioaerosol particles can remain suspended in the atmosphere and be transported over considerable distances, facilitating the spatial spread of resistance determinants beyond their sources (Maldonado et al., 2025). Puvača et al. (2022) observed that resistant microorganisms and their resistance genes can enter air compartments through various pathways, including aerosolisation from wastewater treatment processes and agricultural activities.

8.4 Mechanisms of resistance emergence and dissemination

8.4.1 Horizontal gene transfer

HGT is a major mechanism through which antibiotic resistance genes are transferred among bacterial populations in environmental compartments (Fig. 8.2) (Michaelis and Grohmann, 2023). In contrast to vertical gene transfer, which occurs during cellular replication, HGT leads to resistance development from phylogenetically unrelated organisms (Aminov, 2011). This exchange occurs through well-established mechanisms, including conjugation, transformation, and transduction (Saima et al., 2019). Environmental matrices provide favourable conditions for HGT. High bacterial densities, metabolically diverse microbial associations, and the widespread occurrence of mobile genetic elements collectively enhance the probability of genetic exchange and persistence of resistance traits (Vinayamohan et al., 2022).

Integrons, particularly class 1 integrons, play a fundamental role in the acquisition, assembly, and dissemination of antibiotic resistance genes (Ghaly et al., 2017). These mobile genetic elements function as genetic scaffolds for cassette insertion, promoting the mobilisation and spread of adaptive traits through HGT (Tokuda and Shintani, 2024). The widespread detection of the intI1 gene in households (96.3% of households) causes environmental distribution (Schages et al., 2020). The strong positive correlation between intI1 and β-lactamase–encoding genes indicates active plasmid-mediated mobilisation and coselection of resistance factors within domestic microbial communities (Ur Rahman and Mohsin, 2019; Schages et al., 2020). Mobile genetic elements such as plasmids, transposons, and integrons allow the movement of resistance genes between bacterial species and across environmental compartments (Nji et al., 2021).

The aquatic environment provides particularly favourable conditions for genetic recombination and the emergence of resistance (Saima et al., 2019). WWTPs represent convergence zones for diverse anthropogenic and environmental microbiota, thereby creating conditions that intensify horizontal gene exchange (Yan et al., 2025). The selective pressure exerted by residual antibiotics, heavy metals, and biocides in these environments helps in the maintenance and spread of colocated resistance genes (Mutuku et al., 2022).

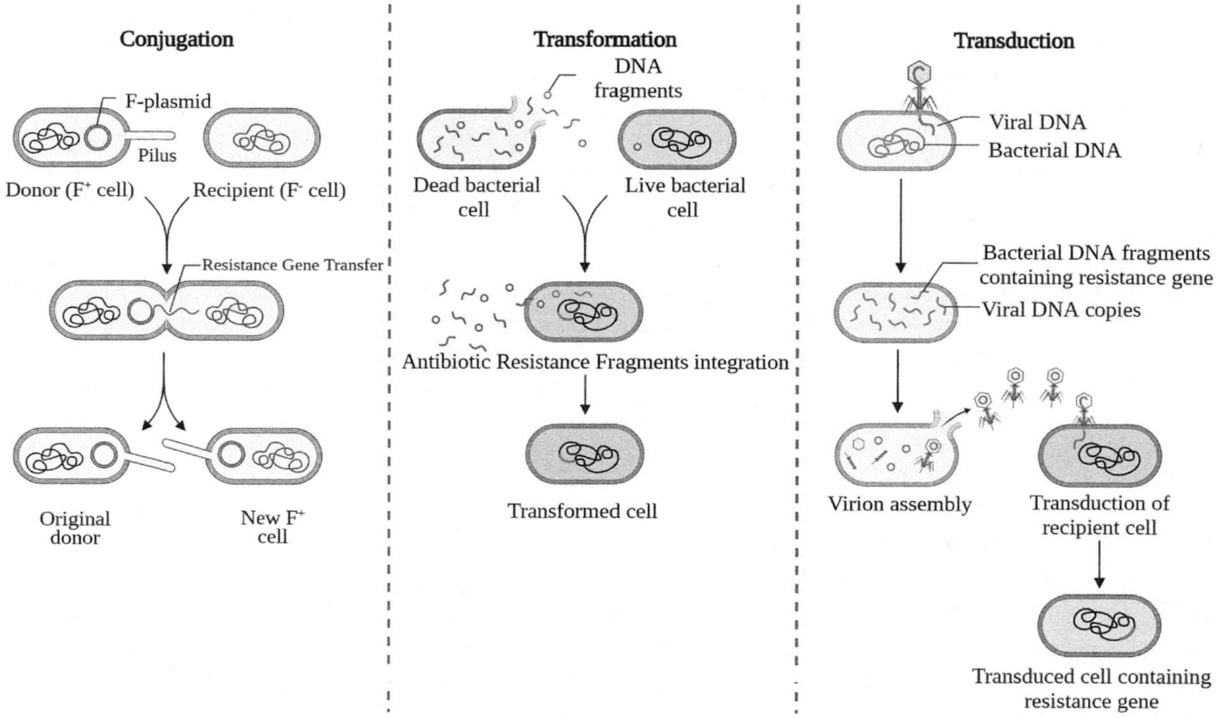

FIGURE 8.2 **Mechanism of transmission of antibiotic resistance genes through horizontal gene transfer.** Schematic representation of the three principal mechanisms of HGT involved in the spread of antibiotic resistance genes among bacteria. (Left) Conjugation: Direct cell-to-cell transfer of plasmid DNA from a donor (F^+) to a recipient (F^-) cell through conjugative pilus, resulting in acquisition of resistance determinants. (Middle) Transformation: Uptake of extracellular DNA fragments released from lysed bacteria by a competent recipient cell, followed by integration of resistance genes into the genome. (Right) Transduction: Bacteriophage-mediated transfer of bacterial DNA fragments carrying resistance genes during viral replication and subsequent infection of a new host cell. These processes collectively facilitate the rapid dissemination of antimicrobial resistance across bacterial populations and environmental compartments. *HGT*, Horizontal gene transfer.

8.4.2 Coselection and environmental pressures

Coselection mechanisms allow resistance genes to persist in bacterial populations even in the absence of direct antibiotic pressure (Wales and Davies, 2015). Heavy metals, biocides, pesticides, and microplastics all serve as coselective agents that maintain resistance genes through genetic linkage or cross-resistance mechanisms (Gillieatt and Coleman, 2024). Sassi et al. (2025) showed that coselective agents such as heavy metals and microplastics aggravate resistance gene proliferation in different ecosystems.

The presence of subinhibitory concentrations of antibiotics in ecosystems creates selective pressures that favour resistant bacteria without necessarily killing susceptible strains (Chukwu et al., 2023). Zhang and Li (2020) observed that high concentrations of antibiotics promote selection and facilitate HGT. This phenomenon is especially observed in anthropogenically impacted environments where there are human activities. Such contaminated sites experience elevated selective pressures due to the accumulation of antibiotics, heavy metals, and other pollutants. Consequently, these locations function as ecological hotspots that promote the emergence, enrichment, and persistence of antibiotic resistance factors (Saima et al., 2019).

Biofilm formation protects antibiotic-resistant bacteria and promotes HGT (Michaelis and Grohmann, 2023). Within biofilms, microbial cells are embedded in a self-produced extracellular polymeric substance (EPS) matrix that allows tolerance to antimicrobial agents and environmental stressors (Shree et al., 2023). Biofilms established in water distribution systems, on household surfaces, and within soil aggregates create structured microenvironments that shield bacterial communities from disinfectants and antibiotics (Uruén et al., 2021). These spatially organised niches facilitate close cell-to-cell contact, thereby increasing the frequency of conjugation, transformation, and other mechanisms of genetic exchange (Nji et al., 2021).

8.4.3 Transmission pathways

The transmission of antibiotic resistance from environmental niches to human populations occurs through multiple exposure pathways. Direct contact with contaminated water, soil matter, or inanimate objects increases the acquisition and subsequent colonisation of antibiotic-resistant bacteria. Such interactions allow the transfer of resistant strains to the human microbiota, increasing the risk of asymptomatic carriage or opportunistic infection (Nji et al., 2021). Consumption of contaminated food or water represents another major way, particularly under inadequate sanitation infrastructure (Mohamad et al., 2025). Lepper and colleagues developed a compartmental mathematical model to examine resistance exchange between environmental and human reservoirs. The model stated that bidirectional transfer of resistance can significantly influence the epidemiology of resistant bacterial infections in human populations. Their findings showed that human resistance occurrence is highly sensitive to parameters regulating both compartments. In particular, the rate of resistance decay in the environment strongly affected outcomes. The transmission coefficient from the environment to humans also exerted a major influence on resistance dynamics (Lepper et al., 2022). Nadimpalli and colleagues demonstrated that bidirectional bacterial flow between humans and animals can accelerate the spreading of antibiotic-resistant pathogens in regions where there is weak environmental regulation and limited waste management. Their findings demonstrated that inadequate environmental control measures can intensify cross-sectoral transmission dynamics, thereby requiring the need for integrated One Health interventions (Nadimpalli et al., 2023). The investigation in Cambodia revealed the cooccurrence of identical mobile genetic elements and conserved transposon regions across human and animal isolates. This pattern of shared genetic structures is rarely reported in high-income countries (Nadimpalli et al., 2023).

8.5 Prevalence and distribution patterns

The occurrence of antibiotic resistance varies widely across different ecosystems. It differs between ecological niches, geographic regions, and socioeconomic settings (Zhuang et al., 2021). Despite these differences, evidence shows that environmental compartments remain contaminated with resistant bacteria and resistance genes (Huemer et al., 2020). In low- and middle-income countries, the burden is particularly high. This is largely due to inadequate water and sanitation infrastructure, limited wastewater treatment capacity, and weak monitoring and enforcement of regulations related to antibiotic distribution (Rathinavelu et al., 2024). Uncontrolled consumption in human medicine, veterinary practice, and agriculture also increases selective pressure, accelerating the environmental amplification and dissemination of resistance factors (Ur Rahman and Mohsin, 2019). A study conducted in South–South Nigeria reported that 86% of analysed environmental samples have bacterial isolates resistant to at least one antibiotic, while 58% have multidrug-resistant phenotypes (Diala et al., 2025). Molecular characterisation revealed the presence of resistance genes like blaTEM (47%), tetM (38%), and ermB (31%). These genes were detected across diverse habitats such as household water sources, sewage drainage systems, and clinical effluents, indicating the extensive environmental distribution of resistance elements. The overlap of resistance genes identified in both environmental and clinical isolates indicates ongoing genetic exchange between these compartments. This convergence supports the hypothesis of dynamic environmental–clinical connectivity mediated by HGT and communal selective pressures (Diala et al., 2025). Nji and colleagues reported 72% for ampicillin resistance and 67% for tetracycline resistance among commensal *E. coli* strains recovered from healthy populations in low- and middle-income countries (Nji et al., 2021). The high resistance burden in commensal organisms is epidemiologically significant. Commensal *E. coli* function as reservoirs of antibiotic resistance genes within the gut microbiota. Through HGT, these genes can move into pathogenic bacteria. This process contributes to the emergence and spread of clinically significant resistant infections (Moura de Sousa et al., 2023).

The distribution of specific resistance genes varies by environmental compartment and geographic region. In agricultural surface waters, sulphonamide resistance genes dominate, reflecting the use of sulphonamides in livestock production (Zhao et al., 2023). In household environments, β-lactamase genes, particularly blaCMY-2, blaACT/MIR, and blaOXA-48, are most reported (Schages et al., 2020). The presence of carbapenemase genes and mobile colistin resistance genes in environmental samples is concerning because these antibiotics are considered last-resort treatments (Schages et al., 2020; Snyman et al., 2021). Temporal trends suggest that the abundance of ARGs is increasing in environments impacted by human activities. Zhang et al. observed that ARG abundance increases significantly in areas affected by livestock farming and other anthropogenic activities (Li et al., 2022). This spatial pattern displayed the role of human activities in propelling the environmental burden of antibiotic resistance.

8.6 Challenges in tackling environmental antibiotic resistance

8.6.1 Surveillance and monitoring gaps

One of the most significant challenges related to environmental antibiotic resistance is the absence of comprehensive surveillance systems (Table 8.2). Alwis et al. stated that monitoring of AMR present in the environment remains

TABLE 8.2 Major challenges in tackling environmental antibiotic resistance.

Category	Regulatory concerns	Impact on AMR control	Spatial and sectoral variability	References
Surveillance gaps	Underdeveloped and fragmented monitoring systems; lack of integration with clinical/veterinary surveillance; AMR not recognised as a drinking water quality parameter	Incomplete understanding of resistance patterns; inability to track trends; delayed detection of emerging threats	More severe in LMICs; retail and some environmental compartments are understudied	Alwisi et al. (2025); Maldonado et al. (2025); Mohamad et al. (2025); Schages et al. (2020)
Methodological issues	Sampling heterogeneity, lack of standardised protocols, incomparable data across studies, and varying detection methods	Inability to compare results; difficulty establishing universal guidelines; challenges in meta analysis	Global issue affecting research comparability	Alwisi et al. (2025); Ndukwe and Ibrahim (2025); Schages et al. (2020)
Infrastructure limitations	Conventional WWTPs are not designed for ARG/ARB removal; inadequate sanitation; limited access to safe drinking water; insufficient laboratory capacity	Continued release of resistance determinants into the environment; environmental transmission pathways remain open	Particularly acute in LMICs, rural areas are more affected	Alwisi et al. (2025); Nadimpalli et al. (2023); Swarthout et al. (2022)
Policy/governance	Integration gaps in One Health frameworks; inadequate regulatory oversight of agricultural antibiotic use; siloed sectoral approaches; limited environmental stakeholder involvement	Uncoordinated interventions, continued environmental contamination, and insufficient resource allocation to environmental monitoring	Varies by jurisdiction; stronger in high-income countries but still inadequate	Alwisi et al. (2025); Ndukwe and Ibrahim (2025)
Resource constraints	Technical and financial limitations; high cost of advanced treatment technologies; limited funding for environmental monitoring; insufficiently trained personnel	Inability to implement comprehensive surveillance; delayed adoption of effective interventions; persistent infrastructure gaps	Most severe in LMICs; affects implementation of advanced technologies	Ndukwe and Ibrahim (2025)
Coselection pressures	Heavy metals, pesticides, microplastics, biocides, maintaining resistance genes; subinhibitory antibiotic concentrations; complex anthropogenic pressures	Resistance persists even without direct antibiotic pressure; multiple selective pressures complicate interventions	Agricultural and industrial areas are most affected; urban environments are also impacted	Malagón-Rojas et al. (2020); Puvača et al. (2022); Sassi et al. (2025)

underdeveloped and fragmented, undermining comprehensive control efforts (Alwisi et al., 2025). Unlike clinical surveillance systems, which have been established in many countries, environmental monitoring programs are sparse, inconsistent, and often lack coordination with human and veterinary health sectors (Alwisi et al., 2025).

The scoping review by Maldonado et al. revealed an uneven distribution of AMR research across environmental compartments in Canada, with most studies focused on water and soil, and very limited attention to air, retail, and community settings (Maldonado et al., 2025). This imbalance indicates substantial surveillance gaps and creates critical blind spots in understanding environmental resistomes and their transmission dynamics. Compared with clinical and veterinary sectors, environmental reservoirs remain insufficiently investigated as contributors to AMR dissemination (Maldonado et al., 2025). Mohamad et al. reported that AMR surveillance in drinking water systems remains largely neglected, despite documented detection of antibiotic-resistant bacteria (Mohamad et al., 2025). During the study period, AMR was not incorporated as a routine drinking water quality parameter in most jurisdictions, revealing a regulatory and public health gap. This oversight underscores broader challenges in embedding environmental dimensions within AMR governance frameworks (Mohamad et al., 2025).

8.6.2 Methodological challenges

The standardisation of methods for detecting and quantifying antibiotic resistance in environmental samples remains a significant challenge. Alwisi et al. identified sampling heterogeneity and the absence of uniform protocols as major obstacles to effective environmental AMR surveillance (Alwisi et al., 2025). Mutuku et al. stated the absence of standardised methodologies for assessing antibiotic occurrence in diverse environmental ecosystems. This lack of methodological consistency leads to poor comparability between independent studies. Consequently, it restricts the development of harmonised environmental protection guidelines and regulatory frameworks (Mutuku et al., 2022). A similar methodological heterogeneity is observed in the detection of AMR genes. Different investigations target distinct gene markers and apply variable quantification approaches, including culture-based, molecular, and metagenomic techniques. Such inconsistencies limit data reproducibility and constrain robust cross-regional risk assessment and surveillance efforts (Mutuku et al., 2022).

Environmental samples have highly diverse microbial communities with dynamic population structures. They contain variable concentrations of antibiotics, antibiotic metabolites, and cocontaminants such as heavy metals and biocides (Murray et al., 2024). In addition, the physicochemical properties such as pH, organic matter content, salinity, and particulate load can interfere with extraction efficiency and detection sensitivity. These factors may introduce matrix effects, inhibit amplification reactions, or reduce analytical accuracy (Likpalimor et al., 2025). Zhang et al. stated the need for standardised detection methodologies and harmonised analytical protocols. They also stress the importance of developing comprehensive and curated databases to strengthen environmental AMR surveillance and comparative research (Li et al., 2022).

8.6.3 Policy and governance issues

The integration of environmental considerations into AMR control policies remains inadequate despite growing recognition of the environmental dimensions of resistance (Bhatia, 2024). Ponzo et al. stated that although the One Health approach recognises the interrelationships between animals, humans, and their environment, implementation challenges persist, particularly regarding environmental data integration, laboratory capacity, and multisectoral coordination (Ponzo et al., 2024).

Regulatory policies that control antibiotic use in agriculture, pharmaceutical manufacturing, and aquaculture differ greatly between countries (Luthman et al., 2024). In many regions, these regulations are not strictly implemented or monitored (Malik et al., 2023), thus weakening efforts to control the spread of AMR. Puvača et al. (2022) stated the importance of identifying high-risk environments where antibiotic misuse or discharge is more likely to occur. They recommend applying appropriate regulatory frameworks in these zones to reduce the development and spread of AMR. However, they also point out that estimating the global economic burden of antibiotic resistance is challenging. Also, establishing effective and coordinated governance systems across countries is difficult (Puvača et al., 2022). Ethical issues related to data sharing and privacy present another challenge in establishing comprehensive surveillance systems (Parsonage et al., 2017). The lack of environmental stakeholders and policymakers in AMR governance structures represents a significant gap. Mustapha et al. stated that the environmental contribution to antibiotic resistance has been largely unexplored, with a significant focus historically placed on clinical isolates rather than environmental bacteria. Including environmental perspectives in policy development is essential for comprehensive AMR control (Taneja and Sharma, 2019).

8.7 Intervention strategies and their effectiveness

Several intervention strategies have been proposed to mitigate the environmental dissemination of antibiotic-resistant bacteria and antibiotic resistance genes. Enhanced wastewater treatment is an important control point. Municipal WWTPs with tertiary treatment systems can control AMR spread. Advanced technologies such as membrane bioreactors, advanced oxidation processes, and UV disinfection have shown efficacy in reducing ARB and ARG loads (Mutuku et al., 2022). Complementary approaches, including phage therapy, vaccines, nanoparticles, and natural bioactive compounds, can also be explored. However, nanoparticle-based strategies remain largely preclinical due to toxicity concerns (Mutuku et al., 2022). At the household level, optimised hygiene practices, including high-temperature automated dishwashing and laundering, significantly reduce resistant bacterial populations, although complete elimination is not always achieved (Schages et al., 2020). In agricultural systems, landscape-level interventions such as nonagricultural buffer zones and larger water bodies can attenuate ARG dissemination (Zhao et al., 2023). Antibiotic stewardship policies, including restrictions on veterinary antimicrobial use under frameworks such as the EU One-Health Action Plan, are essential but face implementation challenges (Nadimpalli et al., 2023). Strengthening water, sanitation, and hygiene (WASH) infrastructure, particularly in low- and middle-income countries, is fundamental to limiting environmental transmission pathways (Nji et al., 2021).

8.8 The One Health perspective

The One Health approach provides a conceptual framework for overcoming antibiotic resistance as an interconnected problem among human, animal, and environmental health domains (Fig. 8.3) (Danasekaran, 2024). This integrated perspective recognises that interventions targeting only one sector are insufficient and that comprehensive strategies must

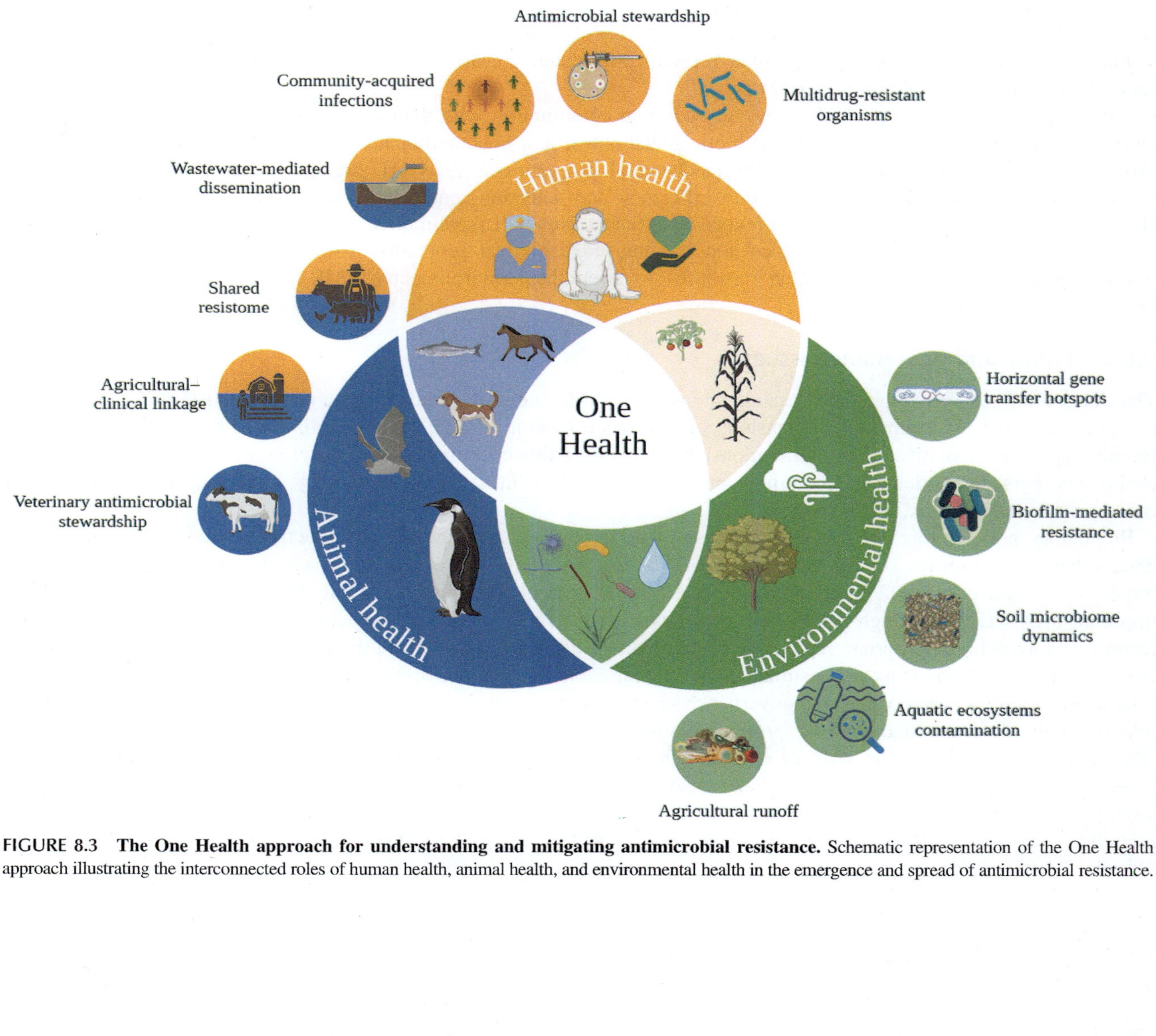

FIGURE 8.3 **The One Health approach for understanding and mitigating antimicrobial resistance.** Schematic representation of the One Health approach illustrating the interconnected roles of human health, animal health, and environmental health in the emergence and spread of antimicrobial resistance.

be adopted for understanding the complex interactions between these domains (Ponzo et al., 2024; Puvača et al., 2022; Sarkar et al., 2023). Alwisi et al. (2025) identify integration gaps within One Health frameworks, including persistent implementation barriers regarding environmental data integration, laboratory capacity, and multisectoral coordination. The siloed nature of human, veterinary, and environmental health sectors creates institutional barriers to integrated surveillance and coordinated interventions. Singh et al. (2022) propose expanding the ambit of the One Health Approach to AMR to include the 'reduction of antimicrobial contamination of the environment' as a seventh domain of activity. This expansion would clearly recognise the central role of environmental pathways in AMR dissemination. It would ensure that environmental determinants and intervention points are systematically integrated and prioritised within global AMR surveillance, mitigation, and policy frameworks (Singh et al., 2022). The success of the One Health strategy depends on strong governance systems that support coordinated collaboration across sectors, standardised surveillance methods, and fair allocation of resources (Oltean et al., 2025). Effective implementation requires the integration of data from human, animal, and environmental health systems to ensure consistent and informed decision-making. Niegowska et al. emphasise the importance of quantitative risk assessment models that combine data from food and feed chains with information from environmental reservoirs. Such integrated approaches enhance evidence-based strategies for prevention and control (Niegowska and Wögerbauer, 2022).

8.9 Future directions and recommendations

Reducing antibiotic resistance in different ecosystems requires a coordinated strategy that combines surveillance, focused research, regulatory reform, and infrastructure development (Oliveira et al., 2024). Environmental surveillance systems must be integrated with human and veterinary monitoring programs to allow cross-sectoral comparison and analysis of long-term trends (Alwisi et al., 2025; Mutuku et al., 2022). Metagenomic-based surveillance, supported by bioinformatics tools, allows culture-independent detection of microbial communities, resistance genes, and emerging pathogens. This approach improves understanding of how resistance spreads across environmental pathways (Pillay et al., 2022).

Several important knowledge gaps exist, such as the direction of transmission of resistance genes between animals and humans not clearly understood (Swarthout et al., 2022). The factors regulating the cooccurrence of antibiotic resistance genes require further investigation (Zhao et al., 2023). In addition, the role of waterways in transferring ARGs across different land-use systems needs further study (Zhao et al., 2023). The connection between antibiotic contamination present in the ecosystem and the development of resistant bacteria should be examined through long-term ecological studies. Such research would strengthen predictive models of resistance evolution and spread (Lepper et al., 2022; Mutuku et al., 2022). Advanced wastewater treatment technologies and other mitigation measures must be examined under real-world conditions to determine their scalability and practical effectiveness (Sassi et al., 2025). Regulatory frameworks should incorporate AMR indicators into environmental quality standards, including those for drinking water, soil, and food production systems (Kaiser et al., 2022). Stronger regulation of antibiotic use in agriculture and aquaculture is also required, particularly regarding medically important antibiotics (Puvača et al., 2022). Improving laboratory capacity and strengthening water and sanitation infrastructure are prerequisites where resources are limited (Alwisi et al., 2025; Nadimpalli et al., 2023).

8.10 Conclusion

The presence of antibiotic resistance in different ecosystems contributes to the global AMR crisis, requiring urgent, coordinated action. Households, water distribution systems, agro-ecosystems, soil, and air act as interconnected niches for the emergence, persistence, and dissemination of antibiotic-resistant bacteria and resistance genes. Uncoordinated surveillance systems, lack of methodological standardisation, limited infrastructure, and weak policy integration restrict effective control, particularly in resource-limited regions. Evidence supports advanced wastewater treatment, improved sanitation, and responsible antibiotic stewardship as important mitigation strategies. Sustainable control demands a One Health structure that integrates environmental monitoring, strengthened regulatory integration, and multisectoral collaboration to interrupt ecological transmission routes and preserve antibiotic efficacy.

Acknowledgements

Dipayan Das would like to thank The Assam Royal Global University for continuous support and motivation, and for providing the necessary infrastructure to conduct the present review. The author would also like to acknowledge BioRender for providing a platform to design images.

Declaration of competing interest

The authors report no declarations of interest.

AI disclosure

The authors declare that AI-assisted tools, including Grammarly and ChatGPT, were used for language editing, grammar correction, and improving the clarity and readability of the manuscript. These tools were not used to generate original data, conduct analyses, interpret results, or develop scientific conclusions. After using this tool/service, the author(s) reviewed and edited the content as needed and take(s) full responsibility for the content of the publication.

References

Abebe, T.A., Gebreyes, D.S., Abebe, B.A., Yitayew, B., 2024. Antibiotic-resistant bacteria and resistance-genes in drinking water source in north Shoa zone, Amhara region, Ethiopia. Frontiers in Public Health 12. Available at https://doi.org/10.3389/fpubh.2024.1422137, http://journal.frontiersin.org/journal/public-health/section/public-health-education-and-promotion#archive.

Alwisi, N., Alsharif, N., Charfi, K., Waqas, M., 2025. Integrating environmental surveillance into one health responses to AMR: challenges, solutions and global perspectives. Journal of Medical and Health Studies 6 (4), 01–15. Available at https://doi.org/10.32996/jmhs.2025.6.4.1.

Aminov, R.I., 2011. Horizontal gene exchange in environmental microbiota. Frontiers in Microbiology 2 (JULY). Available at https://doi.org/10.3389/fmicb.2011.00158, http://www.frontiersin.org/Journal/DownloadFile.ashx?pdf=1&FileId=116&articleId=11234&Version=1&ContentTypeId=21&FileName=fmicb-02-00158.pdf.

Aslam, B., Khurshid, M., Arshad, M.I., Muzammil, S., Rasool, M., Yasmeen, N., et al., 2021. Antibiotic resistance: one health one world outlook. Frontiers in Cellular and Infection Microbiology 11. Available at https://doi.org/10.3389/fcimb.2021.771510, http://www.frontiersin.org/Cellular_and_Infection_Microbiology/archive.

Bhatia, R., 2024. Environmental aspects of antimicrobial resistance in India: current progress & way forward. Indian Journal of Medical Research 159 (1), 10–15. Available at https://doi.org/10.4103/ijmr.ijmr_2104_23, http://www.ijmr.org.in/.

Bruni, E., Simonetti, G., Bovone, B., Casagrande, C., Castellani, F., Riccardi, C., et al., 2020. Evaluation of bioaerosol bacterial components of a wastewater treatment plant through an integrate approach and in vivo assessment. International Journal of Environmental Research and Public Health 17 (1), 273. Available at https://doi.org/10.3390/ijerph17010273.

Caneschi, A., Bardhi, A., Barbarossa, A., Zaghini, A., 2023. The use of antibiotics and antimicrobial resistance in veterinary medicine, a complex phenomenon: a narrative review. Antibiotics 12 (3), 487. Available at https://doi.org/10.3390/antibiotics12030487.

Cave, R., Cole, J., Mkrtchyan, H.V., 2021. Surveillance and prevalence of antimicrobial resistant bacteria from public settings within urban built environments: challenges and opportunities for hygiene and infection control. Environment International 157, 106836. Available at https://doi.org/10.1016/j.envint.2021.106836.

Cella, E., Giovanetti, M., Benedetti, F., Scarpa, F., Johnston, C., Borsetti, A., et al., 2023. Joining forces against antibiotic resistance: the one health solution. Pathogens 12 (9), 1074. Available at https://doi.org/10.3390/pathogens12091074.

Chukwu, K.B., Abafe, O.A., Amoako, D.G., Ismail, A., Essack, S.Y., Abia, A.L.K., 2023. Impact of environmental sub-inhibitory concentrations of antibiotics, heavy metals, and biocides on the emergence of tolerance and effects on the mutant selection window in *E. coli*. Microorganisms 11 (9). Available at https://doi.org/10.3390/microorganisms11092265, www.mdpi.com/journal/microorganisms.

Czatzkowska, M., Wolak, I., Harnisz, M., Korzeniewska, E., 2022. Impact of anthropogenic activities on the dissemination of ARGs in the environment—a review. International Journal of Environmental Research and Public Health 19 (19), 12853. Available at https://doi.org/10.3390/ijerph191912853.

Danasekaran, R., 2024. One health: a holistic approach to tackling global health issues. Indian Journal of Community Medicine 49 (2), 260–263. Available at https://doi.org/10.4103/ijcm.ijcm_521_23, http://www.ijcm.org.in.

Diala, C.G., Edobor, O., Edim, P.D., Oraekwe, C.N., Okenwa, C.R., Onyema, A.M., et al., 2025. Antimicrobial resistance (AMR) in low-resource settings: a molecular and public health investigation into the role of environmental reservoirs and over-the-counter antibiotic use in South-South Nigeria. Path of Science 11 (8), 5008. Available at https://doi.org/10.22178/pos.121-15.

Duarte, A.C., Rodrigues, S., Afonso, A., Nogueira, A., Coutinho, P., 2022. Antibiotic resistance in the drinking water: old and new strategies to remove antibiotics, resistant bacteria, and resistance genes. Pharmaceuticals 15 (4). Available at https://doi.org/10.3390/ph15040393, https://www.mdpi.com/1424-8247/15/4/393/pdf.

Galgano, M., Pellegrini, F., Catalano, E., Capozzi, L., Del Sambro, L., Sposato, A., et al., 2025. Acquired bacterial resistance to antibiotics and resistance genes: from past to future. Antibiotics 14 (3). Available at https://doi.org/10.3390/antibiotics14030222, http://www.mdpi.com/journal/antibiotics.

Ghaly, T.M., Chow, L., Asher, A.J., Waldron, L.S., Gillings, M.R., 2017. Evolution of class 1 integrons: mobilization and dispersal via food-borne bacteria. PLoS One 12 (6). Available at https://doi.org/10.1371/journal.pone.0179169, http://journals.plos.org/plosone/article/file?id=10.1371/journal.pone.0179169&type=printable.

Gillieatt, B.F., Coleman, N.V., 2024. Unravelling the mechanisms of antibiotic and heavy metal resistance co-selection in environmental bacteria. FEMS Microbiology Reviews 48 (4). Available at https://doi.org/10.1093/femsre/fuae017, https://academic.oup.com/femsre/pages/About.

Huemer, M., Mairpady Shambat, S., Brugger, S.D., Zinkernagel, A.S., 2020. Antibiotic resistance and persistence—Implications for human health and treatment perspectives. EMBO Reports 21 (12). Available at https://doi.org/10.15252/embr.202051034, http://embor.embopress.org/.

Kaiser, R.A., Taing, L., Bhatia, H., 2022. Antimicrobial resistance and environmental health: a water stewardship framework for global and national action. Antibiotics 11 (1). Available at https://doi.org/10.3390/antibiotics11010063.

Karwowska, E., 2024. Antibiotic resistance in the farming environment. Applied Sciences 14 (13), 5776. Available at https://doi.org/10.3390/app14135776.

Larsson, D.G.J., Flach, C.F., 2022. Antibiotic resistance in the environment. Nature Reviews. Microbiology 20 (5), 257–269. Available at https://doi.org/10.1038/s41579-021-00649-x, http://www.nature.com/nrmicro/index.html.

Lepper, H.C., Woolhouse, M.E.J., van Bunnik, B.A.D., 2022. The role of the environment in dynamics of antibiotic resistance in humans and animals: a modelling study. Antibiotics 11 (10). Available at https://doi.org/10.3390/antibiotics11101361, http://www.mdpi.com/journal/antibiotics.

Li, S., Zhang, C., Li, F., Hua, T., Zhou, Q., Ho, S.-H., 2021. Technologies towards antibiotic resistance genes (ARGs) removal from aquatic environment: a critical review. Journal of Hazardous Materials 411, 125148. Available at https://doi.org/10.1016/j.jhazmat.2021.125148.

Likpalimor, O.N., Zhang, S., Adagba, F.I., Fatoye, S., Adega, K.J., Njue, J., 2025. Mechanisms of microplastics sorption of antibiotics and impacts on aquatic ecosystems for sustainable development goals. Discover Environment 3 (1). Available at https://doi.org/10.1007/s44274-025-00431-3, https://link.springer.com/journal/44274.

Li, S., Ondon, B. S., Ho, S.-H., Jiang, J., & Li, F. (2022). Antibiotic resistant bacteria and genes in wastewater treatment plants: From occurrence to treatment strategies. Science of the Total Environment, 838(Part 4), 156544. https://doi.org/10.1016/j.scitotenv.2022.156544.

Liu, G., Thomsen, L.E., Olsen, J.E., 2022. Antimicrobial-induced horizontal transfer of antimicrobial resistance genes in bacteria: a mini-review. Journal of Antimicrobial Chemotherapy 77 (3), 556–567. Available at https://doi.org/10.1093/jac/dkab450, http://jac.oxfordjournals.org/.

Luthman, O., Robb, D.H.F., Henriksson, P.J.G., Jørgensen, P.S., Troell, M., 2024. Global overview of national regulations for antibiotic use in aquaculture production. Aquaculture International 32 (7), 9253–9270. Available at https://doi.org/10.1007/s10499-024-01614-0, https://www.springer.com/journal/10499.

Malagón-Rojas, J.N., Parra Barrera, E.L., Lagos, L., 2020. From environment to clinic: the role of pesticides in antimicrobial resistance. Revista Panamericana de Salud Publica/Pan American Journal of Public Health 44 (1). Available at https://doi.org/10.26633/RPSP.2020.44, https://iris.paho.org/handle/10665.2/52311.

Maldonado, M.P., Ofori-Darko, D., Nichols, V., French, J., Spence, K., Reid-Smith, R.J., et al., 2025. Investigating the occurrence of antimicrobial resistance in the environment in Canada: a scoping review. Canadian Journal of Microbiology 71, 1–13. Available at https://doi.org/10.1139/cjm-2024-0189.

Malik, H., Singh, R., Kaur, S., Dhaka, P., Singh Bedi, J., Gill, J.P.S., et al., 2023. Review of antibiotic use and resistance in food animal production in WHO South-East Asia Region. Journal of Infection and Public Health 16, 172–182. Available at https://doi.org/10.1016/j.jiph.2023.11.002.

Mann, A., Nehra, K., Rana, J.S., Dahiya, T., 2021. Antibiotic resistance in agriculture: perspectives on upcoming strategies to overcome upsurge in resistance. Current Research in Microbial Sciences 2, 100030. Available at https://doi.org/10.1016/j.crmicr.2021.100030.

Manyi-Loh, C., Mamphweli, S., Meyer, E., Okoh, A., 2018. Antibiotic use in agriculture and its consequential resistance in environmental sources: potential public health implications. Molecules (Basel, Switzerland) 23 (4), 795. Available at https://doi.org/10.3390/molecules23040795.

Michaelis, C., Grohmann, E., 2023. Horizontal gene transfer of antibiotic resistance genes in biofilms. Antibiotics 12 (2), 328. Available at https://doi.org/10.3390/antibiotics12020328.

Mohamad, Z.A., Jamil, N.M., Bakon, S.K., 2025. From Rivers to Tap: the Spread of Antimicrobial Resistance in Drinking Water Systems. IntechOpen. Available at 10.5772/intechopen.1009585.

Moura de Sousa, J., Lourenço, M., Gordo, I., 2023. Horizontal gene transfer among host-associated microbes. Cell Host & Microbe 31 (4), 513–527. Available at https://doi.org/10.1016/j.chom.2023.03.017.

Murray, L.M., Hayes, A., Snape, J., Kasprzyk-Hordern, B., Gaze, W.H., Murray, A.K., 2024. Co-selection for antibiotic resistance by environmental contaminants. npj Antimicrobials and Resistance 2 (1). Available at https://doi.org/10.1038/s44259-024-00026-7.

Muteeb, G., Rehman, M.T., Shahwan, M., Aatif, M., 2023. Origin of antibiotics and antibiotic resistance, and their impacts on drug development: a narrative review. Pharmaceuticals 16 (11), 1615. Available at https://doi.org/10.3390/ph16111615.

Mutuku, C., Gazdag, Z., Melegh, S., 2022. Occurrence of antibiotics and bacterial resistance genes in wastewater: resistance mechanisms and antimicrobial resistance control approaches. World Journal of Microbiology and Biotechnology 38 (9). Available at https://doi.org/10.1007/s11274-022-03334-0.

Nadimpalli, M.L., Stegger, M., Viau, R., Yith, V., de Lauzanne, A., Sem, N., et al., 2023. Plugging the leaks: antibiotic resistance at human–animal interfaces in low-resource settings. Frontiers in Ecology and the Environment 21 (9), 428–434. Available at https://doi.org/10.1002/fee.2639, http://onlinelibrary.wiley.com/journal/10.1002/(ISSN)1540-9309.

Nava, A.R., Daneshian, L., Sarma, H., 2022. Antibiotic resistant genes in the environment-exploring surveillance methods and sustainable remediation strategies of antibiotics and ARGs. Environmental Research 215 (Pt 1), 114212. Available at https://doi.org/10.1016/j.envres.2022.114212.

Ndukwe, N.N., Ibrahim, H.I., 2025. Antibiotic resistance genes in freshwater environment: sources, fate, ecological impact and clinical relevance. Fudma Journal Of Sciences 8 (6), 217–233. Available at https://doi.org/10.33003/fjs-2024-0806-3094.

Niegowska, M., Wögerbauer, M., 2022. Improving the risk assessment of antimicrobial resistance (AMR) along the food/feed chain and from environmental reservoirs using qMRA and probabilistic modelling. EFSA Journal 20 (Suppl 1). Available at https://doi.org/10.2903/j.efsa.2022.e200407.

Nji, E., Kazibwe, J., Hambridge, T., Joko, C.A., Larbi, A.A., Damptey, L.A.O., et al., 2021. High prevalence of antibiotic resistance in commensal *Escherichia coli* from healthy human sources in community settings. Scientific Reports 11 (1). Available at https://doi.org/10.1038/s41598-021-82693-4, www.nature.com/srep/index.html.

Oliveira, M., Antunes, W., Mota, S., Madureira-Carvalho, Á., Dinis-Oliveira, R.J., Dias da Silva, D., 2024. An overview of the recent advances in antimicrobial resistance. Microorganisms 12 (9), 1920. Available at https://doi.org/10.3390/microorganisms12091920.

Oltean, H.N., Lipton, B., Black, A., Snekvik, K., Haman, K., Buswell, M., et al., 2025. Developing a one health data integration framework focused on real-time pathogen surveillance and applied genomic epidemiology. One Health Outlook 7 (1). Available at https://doi.org/10.1186/s42522-024-00133-5.

Parsonage, B., Hagglund, P.K., Keogh, L., Wheelhouse, N., Brown, R.E., Dancer, S.J., 2017. Control of antimicrobial resistance requires an ethical approach. Frontiers in Microbiology 8. Available at https://doi.org/10.3389/fmicb.2017.02124, https://www.frontiersin.org/articles/10.3389/fmicb.2017.02124/full.

Pillay, S., Calderón-Franco, D., Urhan, A., Abeel, T., 2022. Metagenomic-based surveillance systems for antibiotic resistance in non-clinical settings. Frontiers in Microbiology 13. Available at https://doi.org/10.3389/fmicb.2022.1066995.

Ponzo, E., De Gaetano, S., Midiri, A., Mancuso, G., Giovanna, P., Giuliana, D., et al., 2024. The antimicrobial resistance pandemic is here: implementation challenges and the need for the one health approach. Hygiene 4 (3), 297–316. Available at https://doi.org/10.3390/hygiene4030024, www.mdpi.com/journal/hygiene.

Puvača, N., Vapa Tankosić, J., Ignjatijević, S., Carić, M., Prodanović, R., 2022. Antimicrobial resistance in the environment: review of the selected resistance drivers and public health concerns. Journal of Agronomy, Technology and Engineering Management (JATEM 5 (5), 793–802. Available at https://doi.org/10.55817/cscq3326.

Rathinavelu, S., Uluseker, C., Sonkar, V., Thatikonda, S., Nambi, I.M., Kreft, J.U., 2024. Mapping the scarcity of data on antibiotics in natural and engineered water environments across India. Frontiers in Antibiotics 3. Available at https://doi.org/10.3389/frabi.2024.1337261, https://www.frontiersin.org/journals/antibiotics.

Saima, S., Fiaz, M., Zafar, R., Ahmed, I., Arshad, M., 2019. Dissemination of antibiotic resistance in the environment. Antibiotics and Antimicrobial Resistance Genes in the Environment: Volume 1 in the Advances in Environmental Pollution Research Series. Elsevier, Pakistan, pp. 99–116., http://www.sciencedirect.com/science/book/9780128188828. Available at 10.1016/B978-0-12-818882-8.00006-1.

Sarkar, S., Kamle, M., Bharti, A., Kumar, P., 2023. Antibiotic-resistant bacteria risks and challenges for human health and environment: an overview. World Journal of Environmental Biosciences 12 (2), 26–34. Available at https://doi.org/10.51847/qdfn29z7ps.

Sassi, A., Basher, N.S., Kirat, H., Meradji, S., Ibrahim, N.A., Idres, T., et al., 2025. The role of the environment (Water, Air, Soil) in the emergence and dissemination of antimicrobial resistance: a one health perspective. Antibiotics 14 (8). Available at https://doi.org/10.3390/antibiotics14080764, http://www.mdpi.com/journal/antibiotics.

Schages, L., Lucassen, R., Wichern, F., Kalscheuer, R., Bockmühl, D., McBain, A.J., 2020. The household resistome: frequency of β-lactamases, class 1 integrons, and antibiotic-resistant bacteria in the domestic environment and their reduction during automated dishwashing and laundering. Applied and Environmental Microbiology 86 (23). Available at https://doi.org/10.1128/aem.02062-20.

Shree, P., Singh, C.K., Sodhi, K.K., Surya, J.N., Singh, D.K., 2023. Biofilms: understanding the structure and contribution towards bacterial resistance in antibiotics. Medicine in Microecology 16. Available at https://doi.org/10.1016/j.medmic.2023.100084, www.journals.elsevier.com/medicine-in-microecology.

Singh, A.K., Kaur, R., Verma, S., Singh, S., 2022. Antimicrobials and antibiotic resistance genes in water bodies: pollution, risk, and control. Frontiers in Environmental Science 10. Available at https://doi.org/10.3389/fenvs.2022.830861, journal.frontiersin.org/journal/environmental-science.

Snyman, Y., Whitelaw, A.C., Barnes, J.M., Maloba, M.R.B., Newton-Foot, M., 2021. Characterisation of mobile colistin resistance genes (mcr-3 and mcr-5) in river and storm water in regions of the Western Cape of South Africa. Antimicrobial Resistance and Infection Control 10 (1). Available at https://doi.org/10.1186/s13756-021-00963-2, http://www.aricjournal.com/.

Swarthout, J.M., Chan, E.M.G., Garcia, D., Nadimpalli, M.L., Pickering, A.J., 2022. Human colonization with antibiotic-resistant bacteria from nonoccupational exposure to domesticated animals in low- and middle-income countries: a critical review. Environmental Science and Technology 56 (21), 14875–14890. Available at https://doi.org/10.1021/acs.est.2c01494, http://pubs.acs.org/journal/esthag.

Taneja, N., Sharma, M., 2019. Antimicrobial resistance in the environment: the Indian scenario. Indian Journal of Medical Research 149 (2), 119–128. Available at https://doi.org/10.4103/ijmr.IJMR_331_18, http://www.ijmr.org.in/.

Tayyaba, U., Ahmed, S., 2022. Epidemiology and prevalence of beta-lactamases and recent resistance pattern in gram-negative bacteria from environmental reservoirs. Beta-Lactam Resistance in Gram-Negative Bacteria: Threats and Challenges. Springer Nature, India, pp. 219–236., https://link.springer.com/book/10.1007/978-981-16-9097-6. Available at 10.1007/978-981-16-9097-6_14.

Tokuda, M., Shintani, M., 2024. Microbial evolution through horizontal gene transfer by mobile genetic elements. Microbial Biotechnology 17 (1). Available at https://doi.org/10.1111/1751-7915.14408.

Ur Rahman, S., Mohsin, M., 2019. The under reported issue of antibiotic-resistance in food-producing animals in Pakistan. Pakistan Veterinary Journal 39 (3), 323–328. Available at https://doi.org/10.29261/pakvetj/2019.037, http://www.pvj.com.pk/.

Uruén, C., Chopo-Escuin, G., Tommassen, J., Mainar-Jaime, R.C., Arenas, J., 2021. Biofilms as promoters of bacterial antibiotic resistance and tolerance. Antibiotics 10 (1), 1–36. Available at https://doi.org/10.3390/antibiotics10010003, https://www.mdpi.com/2079-6382/10/1/3/pdf.

Velazquez-Meza, M.E., Galarde-López, M., Carrillo-Quiróz, B., Alpuche-Aranda, C.M., 2022. Antimicrobial resistance: one health approach. Veterinary World 15 (3), 743–749. Available at https://doi.org/10.14202/vetworld.2022.743-749, http://www.veterinaryworld.org/Vol.15/March-2022/27.pdf.

Vinayamohan, P.G., Pellissery, A.J., Venkitanarayanan, K., 2022. Role of horizontal gene transfer in the dissemination of antimicrobial resistance in food animal production. Current Opinion in Food Science 47. Available at https://doi.org/10.1016/j.cofs.2022.100882, http://www.journals.elsevier.com/current-opinion-in-food-science/.

Wales, A.D., Davies, R.H., 2015. Co-selection of resistance to antibiotics, biocides and heavy metals, and its relevance to foodborne pathogens. Antibiotics 4 (4), 567–604. Available at https://doi.org/10.3390/antibiotics4040567, http://www.mdpi.com/2079-6382/4/4/567/pdf.

Xi, C., Zhang, Y., Marrs, C.F., Ye, W., Simon, C., Foxman, B., et al., 2009. Prevalence of antibiotic resistance in drinking water treatment and distribution systems. Applied and Environmental Microbiology 75 (17), 5714–5718. Available at https://doi.org/10.1128/AEM.00382-09, http://aem.asm.org/cgi/reprint/75/17/5714.

Yan, Y., Zhao, X., Liang, X., Xue, Y., Niu, Q., Li, D., et al., 2025. Global wastewater microbiome reveals core bacterial community and viral diversity with regional antibiotic resistance patterns. mSystems 10 (10). Available at https://doi.org/10.1128/msystems.01428-24, https://journals.asm.org/journal/msystems.

Yu, D., Ryu, K., Zhi, S., Otto, S.J.G., Neumann, N.F., 2022. Naturalized *Escherichia coli* in wastewater and the co-evolution of bacterial resistance to water treatment and antibiotics. Frontiers in Microbiology 13. Available at https://doi.org/10.3389/fmicb.2022.810312, https://www.frontiersin.org/journals/microbiology#.

Zeng, Y., Feng, R., Huang, C., Liu, J., Yang, F., 2025. Antibiotic resistance genes in agricultural soils: a comprehensive review of the hidden crisis and exploring control strategies. Toxics 13 (4), 239. Available at https://doi.org/10.3390/toxics13040239.

Zhang, T., Li, B., 2020. Antibiotic resistance in water environment: frontiers of fundamental research, risk assessment and control strategies. Chinese Science Bulletin 65 (24), 2543–2554. Available at https://doi.org/10.1360/tb-2020-0110.

Zhao, B., van Bodegom, P.M., Trimbos, K.B., 2023. Antibiotic resistance genes in interconnected surface waters as affected by agricultural activities. Biomolecules 13 (2). Available at https://doi.org/10.3390/biom13020231, http://www.mdpi.com/journal/biomolecules.

Zhuang, M., Achmon, Y., Cao, Y., Liang, X., Chen, L., Wang, H., et al., 2021. Distribution of antibiotic resistance genes in the environment. Environmental Pollution 285. Available at https://doi.org/10.1016/j.envpol.2021.117402, https://www.journals.elsevier.com/environmental-pollution.

Chapter 9

Navigating the complexities of antibiotic resistance diagnosis

Rajkumari Elizabeth[1], Jayalaxmi Wangkheimayum[2], Manabendra Dutta Choudhury[3], and Amitabha Bhattacharjee[2]

[1]*Advanced Level Institutional Biotech Hub, Assam University, Silchar, Assam, India,* [2]*Department of Microbiology, Assam University, Silchar, Assam, India,* [3]*Rabindranath Tagore University, Hojai, Assam, India*

9.1 Introduction

Antibiotic resistance (AMR) is a global health issue when bacteria develop resistance to antibiotics, rendering these essential drugs ineffective (Maqbool and Willie, 2024). The ability of pathogens and other microorganisms to resist the effects of an antibiotic to which they were previously susceptible, enabling the organisms to survive and flourish, is known as antimicrobial resistance. AMR is difficult to prevent as microorganisms can be able to change their genetic makeup to lessen the lethal effects of antibiotic therapy (Yamin et al., 2023). The emergence of antibiotic-resistant bacteria has posed a critical challenge to the treatment of infectious diseases, leading to a gradual increase in the frequency of nosocomial infections. Multidrug-resistant (MDR) bacteria have been linked to the disseminate of resistance genes between and among bacterial species, and the antibiotics abuse, including excessive and irrelevant antibiotics' usage across clinical and agricultural domains, has been repeatedly described as a significant factor leading to the bacterial resistance's emergence, which poses a serious threat to clinical treatment and demands more attention (Chinemerem Nwobodo et al., 2022; Maqbool and Willie, 2024). Notably, antimicrobial drugs are not confined to human use. They are also being administered to treat or prevent infectious diseases or being used as a poultry growth promoter, animal husbandry, and aquaculture. In addition, antimicrobial agents are used in agriculture. However, the emergence of resistant microbes has rendered some once-effective antibiotics ineffective, a natural phenomenon amplified by antibiotic exposure (Singh et al., 2024).

India is among the countries with the highest rates of antimicrobial resistance worldwide. Antimicrobial consumption (AMC) is the main cause of antimicrobial resistance (AMR), according to several studies that showed an increase in antibiotic use among Indians (Van Boeckel et al., 2014; Klein et al., 2018). The rising prevalence of AMR in frequent fatal pathogens, including *Escherichia coli*, *Staphylococcus aureus*, and others, is also confirmed by surveillance data from the National Centre for Disease Control (NCDC) and the Indian Council of Medical Research (ICMR) (AMRSN_Annual_Report, 2022; NARS-Net, 2023).

Over 70% of India's population is employed in the agriculture and animal husbandry industry, which contributes significantly to the country's economy (FAO, 2022). India is the world's greatest milk producer, with an annual production of 221.06 million tonnes of milk (DAHDF.20th Livestock Census, 2019), 129.60 billion eggs, 9.29 million tonnes of meat, and 33.13 million kg of wool (DAHDF, Basic Animal Husbandry Statistics, 2022). However, farmers are using antibiotics as growth boosters due to rising demand. The increasing prevalence of AMR has been highlighted by numerous studies showing antibiotic-resistant bacteria and antibiotic residues in animal-derived foods (Gersema and Helling, 1986; Alagawany et al., 2018). The high occurrence of infectious diseases, incompatible infection prevention control (IPC) practices, a burdened public health system, easy access to antibiotics without prescriptions, a lack of standardised, effective surveillance platforms to track drug-resistant patterns and consumption, a lack of knowledge and awareness, and a lack of laboratory resources for disease-based diagnosis are challenges that contribute to the AMR burden in India. Notably, India was the largest user of antimicrobials among the BRICS (Brazil, Russia, India, and China) countries in 2010. This further contributes to the challenges of AMR in India (Singh et al., 2024). Bacteria develop different mechanisms of antibiotic resistance through metabolic, genetic, and structural changes that counter the effects of

Antimicrobial Resistance in the 21st Century. DOI: https://doi.org/10.1016/B978-0-443-44748-8.00006-3

antibiotics, rendering them ineffective for treating infections. To implement appropriate treatment strategies and prevent the spread of antibiotic-resistant bacteria, it is essential to promptly and accurately diagnose them (Yamin et al., 2023). An effective point-of-care diagnostic test can reduce inappropriate antibiotic use by enabling early pathogen detection and rapid antimicrobial susceptibility testing (Sharma et al., 2021).

The current topic outlines possible ways to navigate the challenges in AMR diagnosis, future perspectives for diagnosing infections more effectively and precisely, and highlights the necessity of coordinated, global collaboration and a multisectoral approach to address AMR for a more resilient and healthier future.

9.2 Mechanisms of antibiotic resistance and factors contributing to antibiotic resistance

Mechanisms of AMR are complex processes by which bacteria develop mechanisms to shield themselves against the effects of antibiotics, rendering them ineffective in treating. Different genetic, metabolic, and structural changes within bacterial cells are involved in these processes. Developing a key to mitigate AMR requires an understanding of these complex mechanisms (Reygaert, 2018).

Microbial organisms have inherent mechanisms of resistance that existed before antimicrobial agents were discovered. The substantial use of antimicrobial drugs in animals and humans has exerted selective pressure, driving the development of resistant isolates (Aworh et al., 2021). There are several mechanisms by which bacteria develop resistance to antibiotics. Some of the most frequent and main mechanisms of resistance against antibiotics shown by bacteria include mutations in the target sites of the antibiotics thereby reducing the antibiotics binding affinity, by the production of efflux pumps that pumps out the antibiotics from the bacterial cell, formation of biofilm, horizontal transfer of the resistance genes, production of enzymes that are capable of modifying or degrading antibiotics, cell permeability alterations, etc. (Reygaert, 2018). The resistance may be acquired or intrinsic; acquired resistance mechanisms include drug target modification, drug inactivation, and drug efflux, whereas limiting uptake, drug inactivation, and efflux are examples of intrinsic mechanisms. Gram-negative bacteria use different processes than Gram-positive bacteria due to structural differences, etc. Gram-positive bacteria lack the ability to use some forms of drug efflux mechanisms and are less likely to restrict the uptake of a medication because they lack an LPS outer membrane. In contrast, Gram-negative bacteria use all four main mechanisms (Chancey et al., 2012; Mahon et al., 2014). Nevertheless, new resistance mechanisms frequently emerge in bacteria and spread globally, putting the treatment of common infectious diseases at risk. Due to its propensity to spread globally and the ensuing limited treatment options, AMR is a serious issue in healthcare. Rapid cross-border spread of resistant diseases jeopardises international health interventions. The misuse, overuse, or underuse of antibiotics hastens the development of AMR (Sodhi and Singh, 2022; Shree et al., 2023; Sodhi et al., 2023; Anwar et al., 2020). Variety of factors contribute to the indiscriminate use of antibiotics, which encourages AMR. These factors include patients' noncompliance with recommended treatment and demand, prescribers' irrational use of antibiotics in human medicine, drug advertising, dispensing physicians, and antibiotic use in agriculture, poor antibiotic quality, insufficient surveillance, and susceptibility testing. Doctors and prescribers may be strongly influenced by patient demand, even when they are aware of the patient's diagnosis, which may lead to antimicrobial and AMR (Fig. 9.1) (Acharya and Wilson, 2019).

9.3 Global impact of antibiotic resistance

The worldwide implications of AMR include a complex and widespread crisis that significantly impacts economies, healthcare systems, and people's general well-being. Because AMR reduces the effectiveness of responses to infectious disease outbreaks, it threatens global health security. International health initiatives and preparedness measures can be undermined by the rapid cross-border spread of resistant diseases. This difficulty poses serious problems for the healthcare industry, greatly affecting patient care and results as well as the general well-being of many global populations (Aslam et al., 2021).

AMR is a worldwide issue that calls for a 'One Health' approach. Antimicrobials used to treat animal infectious diseases may be the same or comparable to those used in people, necessitating a comprehensive, multisectoral (One Health) approach to address the growing issue of AMR. Humans, animals, and the environment can all harbour resistant bacteria that can move from one nation to another. Geographical or human/animal boundaries are not recognised by AMR (FAO, 2018). Numerous important organisations, like the Centres for Disease Control and Prevention (CDC), the Infectious Diseases Society of America, the World Economic Forum, and the World Health Organisation (WHO), have declared AMR to be a 'global public health concern' (Michael et al., 2014; Spellberg et al., 2016). To reduce the emergence and

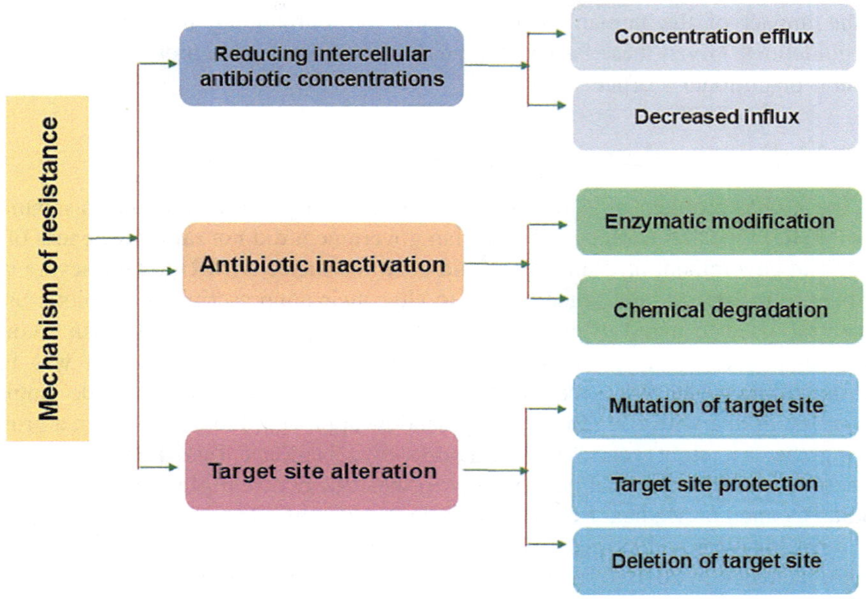

FIGURE 9.1 **Mechanism of antibiotic resistance.** Brief representation of the mechanism of antimicrobial resistance.

spread of AMR, the Food and Agriculture Organisation (FAO), the World Organisation for Animal Health (OIE), and the World Health Organisation (WHO) collaborate. A strategic action plan on AMR has been used to support a Global Action Plan (GAP) on AMR. The Tripartite cooperation has been leading the worldwide effort to combat antimicrobial resistance (AMR) and has started conducting national self-assessments to track the implementation of its National Action Plan (NAP) on AMR (WHO, 2017). The Tripartite promotes participation from multiple sectors. AMR has been recognised as a global health priority by the United Nations General Assembly (UNGA). AMR is only the fourth global health concern that the UNGA has officially addressed, making it an unprecedented move. Leading the effort to reduce AMR, particularly in the food and agricultural industries, is the FAO. It is at the forefront of helping member nations develop their national action plans and implementing cutting-edge public awareness and surveillance techniques in crop farming, aquaculture, and livestock production. Strengthening governance over antimicrobial use and AMR in food and agriculture is one of the main goals of the FAO action plan on AMR (FAO, 2018).

Increased morbidity and mortality, longer hospital stays, higher healthcare costs, disruption of medical interventions, decreased efficacy of essential medications, economic consequences, threat to food security, etc., are some of the effects of AMR (Yadav and Kapley, 2021; Romandini et al., 2021). Developing nations frequently have less access to healthcare services and a greater burden of infectious diseases. Public health in these areas is impacted by AMR, which makes managing infectious disease outbreaks and healthcare infrastructure more difficult (Maqbool and Willie, 2024).

The World Health Assembly requested that the WHO develop a global strategy to address AMR (Hoffman et al., 2015). In 2015 President Barack Obama of the United States instructed the National Security Council to create a comprehensive national action plan to combat AMR, following the advice of the US President's Council of Advisors on Science and Technology (Landers and Kavanagh, 2016; Ventola, 2015). The World Health Organisation (WHO) reports that numerous nations supported the Global Action Plan (GAP) on antimicrobial resistance (AMR) during the 2015 World Health Assembly. This plan aims to manage AMR by implementing multisectoral national action plans using a One Health approach. India has created National Action Plans (NAPs) that describe surveillance strategies to combat AMR. It should be noted that AMR prevention has greatly improved as a result of these strategies. Nevertheless, despite significant efforts to address AMR, it is still a problem that limits the antibiotics' efficacy in treating illnesses. Antibiotic-resistant bacteria must be diagnosed promptly and accurately to apply appropriate treatment and prevent the spread of resistant bacterial strains. Targeted antibiotic prescriptions and accurate, evidence-based diagnosis are now problematic for general practitioners. Some of the common difficulties encountered in diagnosing AMR include diversity in AMR mechanisms, uncertainty about the prevalence of AMR genes, shifting epidemiology of AMR due to travel and ethnic background-associated carriage, the impact of the human microbiota and its confounding influence, and ignorance of national antibiotic-prescribing guidelines. Given these issues, one possible solution to this problem is to implement point-of-care diagnostics at the general practitioner's office.

9.4 National guidelines and market perspectives

For almost a decade, the global problem of AMR remained on the periphery of political discourse despite the World Health Organisation's (WHO) repeated warnings. The Indian government did not raise awareness of this important issue till the middle of 2016. The government proclaimed antimicrobial resistance (AMR) to be a severe concern, highlighting the factors that contribute to resistance, and urged people to alter their nonprescription antibiotic use habits. The Indian government emphasised the need to avoid using antibiotics without a prescription and finish recommended treatments after realising the risks associated with self-medication. The government's announcement was followed by the first interministerial report on antimicrobial resistance (AMR) in India, which highlighted the wide-ranging difficulties in the livestock and health sectors. In line with Kingdon's policy window approach, India has shown a growing commitment to combating AMR during the last eight years. The formation of a National Task Force, the adoption of the Jaipur declaration, and the inclusion of antimicrobial containment in the 12th five-year plan marked initial steps (Hegde et al., 2024). ICMR: The Indian Council of Medical Research played a significant role by convening the 'Chennai Declaration' in 2012, which provides a roadmap for addressing AMR challenges (Ghafur et al., 2013). Public education campaigns like the 'Red Line' campaign, launched in 2016 to raise awareness of antibiotic overuse, are examples of the Prime Minister's political commitment. The intersectoral coordination committee, technical advisory group, and core working group on AMR are the three governance mechanisms established by the Ministry of Health and Family Welfare of India (MoHFW) in September 2016. The National Action Plan on Antimicrobial Resistance (NAP-AMR), which sets strategic priorities for addressing AMR and reinforcing India's leadership in the global effort, was developed under the direction of the core working group. The strategic goals of NAP-AMR align with the global action plan to mitigate AMR (GAP-AMR), adopted by the 68th World Health Assembly in May 2015. However, it also highlights India's unique role by prioritising a strategy suited to Indian conditions, emphasising the nation's leadership in the battle against AMR through multiple collaborations. Six strategic priorities are outlined by NAP-AMR (India: national action plan on antimicrobial resistance) (NAPAMR, 2017) such as improving knowledge and comprehension of AMR through effective training, education, and communication; strengthening information and evidence by monitoring AMR; decreasing infection rates through the use of efficient infection control and preventive strategies; improving the responsible application of antimicrobial agents in food, animals, and health; promoting funding for AMR-related projects, studies, and inventions; enhancing India's leading position in AMR through international, national, and subnational partnerships. Every strategic priority has distinct interventions and emphasis areas, as well as roles and deadlines for short-, medium-, and long-term periods (India: national action plan on antimicrobial resistance) (NAPAMR, 2017). In response to the challenges posed by AMR, India has put into practice important methods that focus on public education, rational use of antibiotics, the management of substandard medications, and the creation of novel medications and vaccines (NAPAMR, 2017). The ICMR made a substantial contribution that demonstrated the country's proactive approach to addressing this important global health concern by establishing Research Network and AMRSN: Antimicrobial Resistance Surveillance in 2013 (Majumder et al., 2020). The acronym AMS was first coined in 2007 by the Infectious Diseases Society of America as a comprehensive strategy to combat the threat that AMR poses to world health. To improve patient outcomes, AMS strategically optimises the choice, dosage, mode of administration, and length of antimicrobial therapy. These concerted efforts greatly enhance patient care, lower infection rates, and slow up the expansion of infections that are resilient to treatment (Doron and Davidson, 2011). The ICMR actively supported initiatives, such as an AMS curriculum and awareness and education workshops, because it recognised the urgency of AMS in healthcare institutions. To provide crucial information for successful national treatment plans, efforts were expanded to monitor antimicrobial use and resistance trends. Despite improvements, formal AMS procedures remain lacking in many Indian institutions, underscoring the need for a multidisciplinary approach. AMS programmes improve the quality and safety of patient care, which leads to higher rates of infection cure, fewer treatment failures, and more accurate therapy and prophylactic prescriptions (Mathew et al., 2020). In 2018 the ICMR launched AMSP (Antimicrobial Stewardship Programme) guidelines to help Indian hospitals establish their programmes. The AMS team should include a multidisciplinary team of doctors with expertise in infectious illnesses, a clinical microbiologist, a pharmacist, and infection control personnel, according to Section 15 of the guidelines' recommendations for quality improvement. This team focuses on capacity development, including infrastructure for monitoring, reporting, and audit capabilities, as well as reliable information systems, and it obtains separate funding within the hospital (Hegde et al., 2024). The Indian Government introduced Schedule H1 under the

Drugs and Cosmetics Rules [Rules for Selling of Drugs under Schedule H1 (pib. gov.in)] to address the problem of the over-the-counter antibiotics' sale in India. This regulation aims to reduce the use and prevent the overuse and misuse of antibiotics by making it mandatory that antibacterial drugs can be supplied to the public only on a valid prescription. Moreover, the Central Drugs Standard Control Organisation (CDSCO), a National Regulatory Body, has advised various medical departments and professional bodies to raise awareness of the importance of following standard treatment guidelines when prescribing antibiotics [AMRMiscircular.pdf (cdsco.gov.in)]. Within healthcare facilities, guidelines for hospital infection control and judicious antibiotic use were also issued by ICMR [icmr.nic. in].in, 2019]. Nevertheless, nonjudicious antibiotic use and the sale of over-the-counter antibiotics remain widespread nationwide despite these efforts (Singh et al., 2024).

9.5 Therapeutic challenges of antibiotic resistance

Resistance to antibiotics poses serious therapeutic challenges, including treatment failures, an increase in severe illness (morbidity) and death (mortality), longer hospital stays, and higher healthcare costs, as there are fewer effective medications available, particularly against multidrug-resistant (MDR) superbugs like VRE (vancomycin-resistant enterococci). The lack of developed antibiotic research, the need for improved infection treatment in biofilms, and the requirement for improved stewardship, diagnostics, and novel treatments such as phage therapy or immune boosters due to the rapid expansion of bacterial resistance are key issues (Ahmed et al., 2024). The emergence of antibiotic-resistant bacteria has made treating clinical infectious diseases extremely difficult, leading to an increase in the incidence of nosocomial infections (Zhu et al., 2022). AMR is rising globally since bacterial infections are the main causes of disease and mortality. The ability to treat common infectious diseases is threatened by the frequent appearance of novel resistance mechanisms and their global transmission. Because AMR can spread globally and reduce treatment options, it is a serious healthcare issue (Ugwu et al., 2020). Food-borne illnesses, gonorrhoea, TB, pneumonia, and other infections become untreatable when the medications used to treat them become ineffective. Swiftly, the world is entering into a postantibiotic future where common illnesses and minor injuries could become lethal once more if swift and effective action is not taken. The WHO recently ranked AMR among the top ten global public health threats to humanity (Chavada et al., 2023).

Despite knowledge of the consequences of antibiotic overuse leading to resistance, the implementation of antimicrobial stewardship programmes in healthcare and revised animal husbandry policies lags significantly. The pipeline for developing antibiotic drugs is unable to keep up with the ongoing evolution of MDR pathogens, which exacerbates these problems. With few financial incentives, pharmaceutical companies are increasingly abandoning expensive antibiotic research. Furthermore, whereas policy extensions to fund antibiotic development represent advancements, short-term fixes appear unlikely given the length of phase trials (Ahmed et al., 2024). Despite organisations such as the WHO, CDC, and UN acknowledging the borderless threats posed by AMR, international coordination on surveillance and stewardship protocols remains fragmented, further impeding containment efforts. Novel resistance variables can evolve locally and spread globally due to uneven access to high-quality diagnostics and antibiotic supervision across nations. Localised success may be continuously undermined and negated by patches of poor care. In the end, the distinct 'tragedy of the commons' character of AMR necessitates fair, collaborative international action and shared accountability. However, agreement on legally binding international laws and the funding sources required to promote antimicrobial stewardship and innovation globally is still hampered by geopolitical difficulties. AMR affects populations worldwide and transcends geographical boundaries. Infections that were previously treatable have now transformed into deadly health issues. The lack of effective antimicrobial drugs increases the risk of common medical treatments, such as organ transplants, chemotherapy, and surgery. AMR poses significant financial difficulties for governments, healthcare systems, and society at large, in addition to its detrimental effects on human health (Dadgostar, 2019; Ferri et al., 2017). Extended hospitalisations, increased medical consultations, and the necessity for high-cost drugs as a last resort greatly increase the financial burden of treating resistant infections (Ahmed et al., 2024).

9.6 Challenges in diagnosis

The lack of quick, inexpensive, and precise point-of-care diagnostics is the main obstacle to diagnosing AR, leading to delays in detecting resistant infections and improper empirical antibiotic administration. The time-consuming nature of conventional techniques, the high cost of molecular tools (such as WGS), the variety and emergence of resistance mechanisms, and the inadequate laboratory infrastructure in low-resource settings are among the main problems (Panda, 2025). For general treatment, there is a severe shortage of rapid, practical diagnostic techniques that can quickly identify specific resistance mechanisms or differentiate between bacterial and viral diseases. Misclassification of infectious illnesses is one of the main causes of antibiotic abuse in underdeveloped nations. Viral diseases like dengue and upper respiratory tract infections are frequently misdiagnosed as bacterial infections, leading to the prescription of improper and ineffective medications. Contemporary diagnostic technologies, such as molecular diagnostics, including multiplex polymerase chain reaction (PCR) panels, and point-of-care tests, such as procalcitonin and C-reactive protein (CRP), can help reduce these errors. CRP-guided antibiotic prescribing can reduce unnecessary antibiotic use by up to 50% without compromising patient outcomes, according to studies from Kenya and Vietnam (Do et al., 2023). In a similar way, molecular panels provide early de-escalation or focused therapy by quickly identifying causal infections and resistance markers. However, due to cost, logistical challenges, and the need for laboratory equipment, the adoption of such technologies remains limited in many underdeveloped nations. Furthermore, the effectiveness of diagnostic treatments depends not only on their availability but also on their incorporation into frameworks for infection prevention and antimicrobial stewardship. Broad-spectrum antibiotic use and resistance rates have been decreased in programmes that integrate diagnostic stewardship with prescriber education and feedback loops (Rahman et al. 2020). Various mechanisms of AMR in primary healthcare settings are influenced by several factors, including the accessibility and use of antibiotics by patients and general practitioners within a specific area and community. Consequently, detection of a variety of AMR, including both intrinsic and acquired resistances, may be diverse, and additionally, this resistance can easily spread between other bacterial species. Moreover, a single bacterial isolate may harbour multiple resistance mechanisms, leading to multiresistant microorganisms or even pan-resistant strains. Such a diversity of antimicrobial resistance mechanisms means that it is unlikely to come about a point-of-care diagnostic that can detect all the different AMRs described (Panda, 2025). Alternatively, a 'targeted' strategy would be needed, in which diagnostics focus on the antimicrobial resistance genuinely present in a specific geographical area (Mitsakakis et al., 2018).

9.7 Strategy for tackling drug resistance in pathogenic organisms

Several strategies have been brought up to address AMR in Gram-negative bacteria. The simultaneous administration of a Gram-positive targeting antibiotic along with enhancing agents may potentially broaden their efficacy to encompass Gram-negative strains. These potentiators might either interfere with outer membrane biogenesis or undermine outer membrane integrity through physical disruption, thus facilitating increased antibiotic influx. Moreover, compounds that inhibit efflux pumps could reduce the efflux of antibiotics from the cell, leading to higher intracellular drug concentrations. Additionally, β-lactamase inhibitors can prevent the hydrolysis of β-lactam antibiotics, thereby reinstating their effectiveness against resistant bacteria. This comprehensive strategy offers a flexible approach to overcoming significant resistance mechanisms in Gram-negative pathogens, potentially expanding therapeutic options for addressing multidrug-resistant infections. The outer membrane of Gram-negative bacteria serves as a selective barrier that restricts the entry of antibiotics, thereby complicating the treatment of infections caused by these bacteria compared to those caused by Gram-positive bacteria. Disruptors of the outer membrane can either hinder its biogenesis or physically compromise its integrity, leading to an increased influx of antibiotics (Klobucar et al., 2022). Numerous antibiotics fail to combat Gram-negative bacteria due to the action of efflux pumps. One potential approach to address this issue is to employ a synergistic combination of antibiotics alongside efflux pump inhibitors. The development of drug-like efflux pump inhibitors (EPIs) offers a promising strategy for resensitizing bacteria to antibiotics, potentially extending the effectiveness and applicability of current antibacterial therapies (Phan et al., 2023).

In a recent investigation, some notable compounds derived from asymmetric imidazole-4,5-dicarboxamide derivatives were identified as inhibitors of the AcrAB-TolC efflux pump following extensive screening. These compounds exhibited a synergistic effect when combined with levofloxacin and oxacillin, leading to a reduction in the minimum inhibitory concentrations (MICs) of these antibiotics by up to 4-fold in a wild-type strain of *E. coli* (Phan et al., 2023).

Many β-lactamase inhibitors have been recognised, including clavulanic acid, tazobactam, and sulbactam. These inhibitors demonstrate efficacy against Class A serine β-lactamases (SBLs), yet they do not work against Class B metallo-β-lactamases (MBLs), Class C SBLs, or Class A carbapenemases. Additionally, diazabicyclooctane-based β-lactamase inhibitors such as avibactam and relebactam have been identified. While avibactam is effective against Class A, Class C, and certain Class D SBLs, there is a developing resistance to this inhibitor, indicating that its future efficacy may be jeopardised by the potential rise of new β-lactamases (Ehmann et al., 2012; Blizzard et al., 2014; Doi and Chambers, 2015; Benson and Nahata, 1988).

There is an immediate requirement for solutions to the ongoing threat posed by Antimicrobial Resistance, along with the related risks, health issues, fatalities, and financial losses. Although numerous new antibiotics are currently being developed, data released by the WHO indicate they are still expected to be effective against the most critical forms of AMR. The primary biological activities attributed to antibiotics have been their direct repressing and killing effects (Berti et al., 2020). However, it is increasingly evident that antimicrobials act in concert with the host's innate immunity to exert significant indirect effects that facilitate bacterial clearance, potentially leading to quicker and more pronounced outcomes and thereby decreasing the likelihood of resistance development in bacteria (Berti et al., 2020). Research indicates that native endogenous host defence peptides have evolved to protect the host by inhibiting pathogenic organisms from inducing infection, exhibiting antimicrobial mechanisms akin to those of peptide antimicrobials used in clinical settings, such as colistin and daptomycin (Rima et al., 2021). While extensive scientific investigations have been conducted on the synergistic, additive, neutralising, and antagonistic interactions of antibiotics, there remains a lack of understanding regarding the interplay between antibiotics and the innate immune system (Yeh et al., 2009; Tartor et al., 2022). Although the direct in vitro action of an antibiotic is thought to render it effective against bacteria, additional effects of antibiotics on bacteria have been linked to enhanced immune optimisation and host immunological responses. This includes how antibiotics interact with various immune systems and virulence factors to affect the host's response (Watson et al., 2020).

In recent years, researchers have begun to focus their investigations on methods to eradicate antibiotic-resistant bacteria without necessarily creating new antibiotics. This approach is particularly compelling given the costs and challenges of discovering and developing new antibiotics. Each type of antibacterial agent operates through a unique mechanism that targets specific areas of bacterial cells to eliminate them. Furthermore, it is widely accepted that the physiological effects of antibiotics, such as changes in membrane permeability, alterations in cell shape, or molecular processes like the obstruction of essential cellular pathways, influence the efficacy of these antibiotics and their methods of bacterial eradication (Lade and Kim, 2021). After reaching their intended targets, all antibiotic treatments exhibit a common secondary effect, as shown in studies by Calhoun et al. and Hong et al. The targeted bacteria are reportedly compelled to produce 'reactive oxygen species (ROS),' commonly known as free radicals, which, if not swiftly neutralised, can inflict significant damage on bacterial DNA and proteins. This implies that all bactericidal antibiotics, irrespective of their mechanisms of action, ultimately kill bacteria through a shared secondary effect on the bacterial cell (Patel et al., 2026; Hong et al., 2019).

9.8 Current technologies for automation and rapid antimicrobial susceptibility testing

9.8.1 Traditional methods for assessing antimicrobial susceptibility

Bacterial organisms can be cultured and isolated within 24 hours to 1 week following specimen collection. Traditional phenotypic antimicrobial susceptibility testing (AST) methods are performed on pure-culture isolates and may take up to 24 hours to yield results. The gold standard methods for phenotypic AST include broth and agar dilution techniques, while other phenotypic AST methods include Kirby–Bauer (KB) disc diffusion and concentration gradient test strips, such as the E-tests (Balouiri et al., 2016; Brennan-Krohn et al., 2017).

The Kirby–Bauer disc diffusion assay evaluates bacterial susceptibility by determining the diameter of the inhibition zone surrounding the antibiotic disk. In general, a standard inoculum of 0.5 McFarland (1.5×10^8 CFU/mL) is utilised to inoculate a Muller–Hinton Agar (MHA) plate. Antibiotic discs, infused with a specific antibiotic concentration, are placed on the MHA plate and incubated for 18–24 hours. After incubation, the diameter of the inhibition zone, also referred to as the clearing around the antibiotic disc, is measured. The interpretations of the disk zone diameter (e.g. S=susceptible, I=intermediate, or R=resistant) are made based on clinical breakpoints set by the Clinical and Laboratory Standards Institute (CLSI, 2024), the United States Food and Drug Administration (FDA), or the European Committee on Antimicrobial Susceptibility Testing (EUCAST).

In agar and broth dilution assays, a standardised bacterial inoculum is introduced into agar or broth that contains antimicrobial agents, which are serially diluted in two-fold increments. The plates or broth are incubated overnight, and bacterial growth is evaluated by observing the presence of colonies or turbidity. The minimal inhibitory concentration (MIC) is the lowest concentration of an antimicrobial agent that effectively inhibits bacterial growth in broth or agar. The Epsilometer test, commonly referred to as the E-test, involves an antibiotic strip with a predetermined gradient of an antibacterial agent extending from one end to the other. Positioning of this strip is performed on an agar plate inoculated and incubated overnight. Identification of MIC is at the point where an ellipse-shaped inhibitory zone intersects with the E-test strip (Jorgensen and Ferraro, 2009). Interpretations of the results (e.g. S, I, or R) are made according to the guidelines established by CLSI, EUCAST, or the FDA.

9.8.2 Rapid phenotypic antimicrobial susceptibility testing methods

In recent years, there has been a notable rise in the availability of commercial technologies that facilitate rapid AST. These technologies employ advanced. While all methods have a runtime of under eight hours, there are general differences among the platforms, including the technology utilised, its applicability to Gram-positive or Gram-negative organisms. Most platforms are designed for positive blood cultures, except one: NGP Test (Selux Next-Generation Phenotyping, USA), which is designed for both microbial and blood cultures. The Selux NGP Assay (SeluxDX, Boston, MA, USA) necessitates the use of various instruments for testing, contingent upon whether the specimen is a blood culture or a bacterial isolate. For AST directly from positive blood cultures, the blood culture is introduced into the Selux Separator, which effectively isolates the bacteria from the blood and prepares an AST-ready McFarland equivalent to be transferred into the Inoculator instrument, where the sample will be allocated into the corresponding antibiotic panels. In cases where testing is conducted from direct colonies, users are required to prepare a McFarland inoculum prior to bringing the sample preparation carrier to the Inoculator instrument. Following the inoculation of bacteria, the plates are subsequently placed into the Analyser instrument, where AST is executed. Once adequate growth is observed in the wells, viability and surface area assays are conducted. Viability assays employ fluorescent markers to assess the metabolic activity of the bacterial population (Hattab et al., 2024).

Most rapid phenotypic AST technologies, except for the PhenoTest USA, are unable to detect organisms' detection. Organism identification will require an alternative method, such as direct identification/rapid molecular identification using matrix-assisted laser desorption/ionisation time-of-flight mass spectrometry (MALDI-TOF). The PhenoTest offers numerous advantages. First, it is capable of both the rapid identification of organisms in blood cultures and rapid antimicrobial susceptibility testing. Identification is conducted within 1.5 hours of fluorescence in situ hybridisation. AST is completed in 7 hours using the morphokinetic method. Comparing the kinetic and morphological alterations of bacteria treated with antibiotics with the untreated determines the minimum inhibitory concentrations. The PhenoTest can identify bacteria with an accuracy rate of 87%–100% (Hattab et al., 2024).

The LifeScale test, developed by Affinity Biosensors in Santa Barbara, CA, runs for less than 5 hours. This instrument utilises microfluidic sensors to generate a 'Population Profile' that corresponds to the bacterial mass present in the sample. Subsequently, the collected data is processed by an 'AI Predictor' that can identify the organism's antimicrobial MIC (Hattab et al., 2024).

The ASTar (Q-linea, Uppsala, Sweden) operates with a runtime of under 7 hours and employs high-speed time-lapse microscopy at various intervals to capture images of bacterial growth in prefilled cultured chambers containing different antibiotics. These images are utilised to calculate bacterial biomass, which is essential for determining the Minimum Inhibitory Concentration (MIC). Common errors identified through several studies were associated with amoxicillin/clavulanic acid and piperacillin/tazobactam (Hattab et al., 2024).

Several commercial systems have optimised and partially automated the monitoring of AST cultures. Systems such as Vitek and Microscan conduct automated turbidity assessments for multiwell liquid cultures. The BD Phoenix system utilises a redox indicator to enhance the identification of organism growth. These systems can achieve turnaround times as brief as 4 hours for identification and susceptibility testing, and 6 to 8 hours for testing. The CE-marked Alfed 60 ASTTM system (Alifax, Italy) employs advanced laser light-scattering technology to detect bacterial growth in liquid culture broth and delivers antimicrobial susceptibility results directly from positive blood culture bottles within 4–6 hours. These broth dilution-based systems utilise preprepared AST cards that include positive controls and wells with progressively increasing antibiotic concentrations. This method facilitates ongoing growth tracking and can analyse MIC patterns across a wide range of organisms by leveraging comprehensive databases. Pathogen detection (ID) is typically the initial phase of AST (Hattab et al., 2024).

The VITEK REVEAL (bioMerieux, Mountain View, CA, USA), which operates with a runtime of less than 7 hours, employs sensors to identify alterations in the volatile organic compounds released by bacteria during their growth. This platform offers the benefit of real-time monitoring of Minimum Inhibitory Concentrations (MICs) and allows for testing of up to 177 drug/pathogen combinations. The VITEK REVEAL has obtained CE-IVD approval and recently achieved US FDA 510(k) clearance as of June 2024 (Table 9.1) (Hattab et al., 2024).

9.9 Future perspectives in diagnosis and combating antibiotic resistance

Conventional methods in infection control and antibiotic management are essential elements of society's strategy to fight resistance. However, the World Economic Forum report highlights that AMR and the decline of the antibiotic research and development pipeline are worsening persistently, despite our continual efforts in these areas (Howell, 2013). To create

TABLE 9.1 Rapid phenotypic antimicrobial susceptibility testing methods.

Test (manufacturer)	Organisms	Functionality	Runtime	Performance	Technology	Reference
PhenoTestBC [Accelerate Diagnostics Inc., (Tucson, AZ, USA)]	Gram-positive and Gram-negative bacteria	Identification and antimicrobial susceptibility testing	ID- 2 hr AST- 7 hr	Higher accuracy for Enterobacterales	Morphokinetic cellular analysis and fluorescence in situ hybridisation	Hattab S et al., 2024
LifeScale [Affinity Biosensors (Santa, Barbara, CA, USA)]	Gram-negative bacteria	Antimicrobial susceptibility testing only	5 h	Higher accuracy	Microfluidic sensor and resonant frequency to determine bacterial concentration and mass distribution	
ASTar [Q-linea (Uppsala, Sweden)]	Gram-negative bacteria	Antimicrobial susceptibility testing only	6 h	Lower performance in amoxicillin/ clavulanic acid and piperacillin/ tazobactam testing	Time-lapse imaging of bacterial growth	
VITEK REVEAL [bioMerieux (Mountain view, CA, USA)]	Gram-negative bacteria	Antimicrobial susceptibility testing only	5 h	Higher accuracy	Colorimetric sensors reacting to volatile organic compounds due to bacterial metabolism during growth	
Selux Next-Generation Phenotyping (NGP)Test [SeluxDX (Boston, MA, USA)]	Gram-positive and Gram-negative bacteria	Antimicrobial Susceptibility Testing only	6–7 h	Higher accuracy	Fluorescent growth indicator using a viability and surface-binding assay	
Alfred [Alifax (Padova, Italy)]	Gram-positive and Gram-negative bacteria	Antimicrobial susceptibility testing only	4–7 h	Gram-negative bacteria perform slightly better than Gram-positive bacteria, lower performance in piperacillin tazobactam testing	Light scattering to detect bacterial growth	

effective countermeasures with enduring impact, innovative concepts that enhance conventional methods will be necessary. Bacteria developed AMR primarily through adaptation over countless years of antibiotic exposure. Second, it is not solely 'improper' antibiotic usage that fosters resistance. Instead, the rate at which resistance proliferates is influenced by microbial contact with all antibiotics, regardless of whether they were prescribed correctly. Consequently, even if all misuses of antibiotics were stopped, antibiotic-resistant infections would still arise. Future strategies, all of which necessitate increased societal investment in fundamental and applied research as well as policy initiatives (Bhullar et al., 2012). These initiatives focus on preventing infections from arising, promoting innovative economic structures that drive investment in antiinfective therapies, reducing the pace of resistance to extend the efficacy of antibiotics, and finding new methods to combat microbes without fostering resistance.

Conventional infection-control strategies need to be supported by innovative methods that can effectively sanitise environmental surfaces, individuals, and food. We also require technology that enables intensive healthcare without the need to implant foreign substances such as plastic or metal (Casadevall and Pirofski, 2003; Lin et al., 2012). Enhancements in population health and healthcare delivery systems can decrease hospital and skilled nursing facility admissions, consequently lowering infections. An advanced method of stewardship is creating therapies that do not induce resistance. In real-world settings, various factors hinder the broad adoption of new antibiotics, including high costs and insufficient comparative data against older drugs, as comparative research has either not occurred or consisted solely of noninferiority studies. Furthermore, for effective antibiotic treatment in severely ill patients, tailored guidelines for observational treatment have been developed based on infection type.

9.10 Conclusion

The current unit has outlined the complex mechanisms of antimicrobial resistance, challenges in therapeutics, as well as in the diagnosis of antimicrobial resistance in healthcare settings. It brought attention to the rising prevalence of AMR, the shortcomings of conventional techniques, and the growing significance of rapid molecular diagnostics, and emphasised how crucial accurate and timely detection is for directing targeted treatment, reducing unnecessary antibiotic use, and enhancing patient outcomes. This topic highlighted the importance of implementing precise, evidence-based practices, surveillance systems, current national guidelines and perspectives, and the development of rapid detection methods for AMR. Furthermore, it outlined possible ways to navigate the challenges in AMR diagnosis, future perspectives for diagnosing infections more effectively and precisely, and underscores the necessity of coordinated, global collaboration and a multisectoral approach to address AMR for a more resilient and healthier future. Henceforth, to combat the growing threat of antimicrobial resistance and preserve the efficacy of currently available antimicrobial agents, investment in rapid, cost-effective diagnostic strategies will be crucial.

AI disclosure

Grammarly Institutional software was used for English editing. After using this tool/service, the author(s) reviewed and edited the content as needed and take(s) full responsibility for the content of the publication.

References

Alagawany, M., Abd El-Hack, M.E., Farag, M.R., Sachan, S., Karthik, K., Dhama, K., 2018. The use of probiotics as eco-friendly alternatives for antibiotics in poultry nutrition. Environmental Science and Pollution Research 25 (11), 10611–10618. Available at https://doi.org/10.1007/s11356-018-1687-x.

Ahmed, S.K., Hussein, S., Qurbani, K., Ibrahim, R.H., Fareeq, A., Mahmood, K.A., Mohamed, M.G., 2024. Antimicrobial resistance: Impacts, challenges, and future prospects. Journal of Medicine, Surgery, and Public Health 2, 100081. Available at https://doi.org/10.1016/j.glmedi.2024.100081.

Acharya, K.P., Wilson, R.T., 2019. Antimicrobial resistance in Nepal. Frontiers in Medicine 6. Available at https://doi.org/10.3389/fmed.2019.00105.

Annual Report: Antimicrobial Resistance Surveillance and Research Network, Jan 2022-Dec 2022. Available online at: AMRSN_Annual_Report_2022.pdf (icmr.nic. in). (2022), Available from: https://www.icmr.gov.in/icmrobject/custom_data/pdf/resource-guidelines/AMRSN_Annual_Report_2022.pdf.

Anwar, M., Iqbal, Q., Saleem, F., 2020. Improper disposal of unused antibiotics: an often overlooked driver of antimicrobial resistance. Expert Review of Anti-Infective Therapy 18 (8), 697–699. Available at https://doi.org/10.1080/14787210.2020.1754797, http://www.tandfonline.com/loi/ierz20.

Aslam, B., Khurshid, M., Arshad, M.I., Muzammil, S., Rasool, M., Yasmeen, N., et al., 2021. Antibiotic resistance: one health one world outlook. Frontiers in Cellular and Infection Microbiology 11. Available at https://doi.org/10.3389/fcimb.2021.771510, http://www.frontiersin.org/Cellular_and_Infection_Microbiology/archive.

Aworh, M.K., Kwaga, J.K.P., Hendriksen, R.S., Okolocha, E.C., Thakur, S., 2021. Genetic relatedness of multidrug resistant *Escherichia coli* isolated from humans, chickens and poultry environments. Antimicrobial Resistance and Infection Control 10 (1). Available at https://doi.org/10.1186/s13756-021-00930-x, http://www.aricjournal.com/.

Balouiri, M., Sadiki, M., Ibnsouda, S.K., 2016. Methods for in vitro evaluating antimicrobial activity: a review. Journal of Pharmaceutical Analysis 6 (2), 71–79. Available at https://doi.org/10.1016/j.jpha.2015.11.005, http://www.journals.elsevier.com/journal-of-pharmaceutical-analysis.

Benson, J.M., Nahata, M.C., 1988. Clinical use of systemic antifungal agents. Clinical Pharmacy 7 (6), 424–438.

Berti, A., Rose, W., Nizet, V., Sakoulas, G., 2020. Antibiotics and innate immunity: a cooperative effort towards the successful treatment of infections. Open Forum Infectious Diseases. Available at https://doi.org/10.1093/ofid/ofaa302.

Bhullar, K., Waglechner, N., Pawlowski, A., Koteva, K., Banks, E.D., Johnston, M.D., et al., 2012. Antibiotic resistance is prevalent in an isolated cave microbiome. PLoS One 7 (4), e34953. Available at https://doi.org/10.1371/journal.pone.0034953.

Blizzard, T.A., Chen, H., Kim, S., Wu, J., Bodner, R., Gude, C., Imbriglio, J., Young, K., Park, Y.-W., Ogawa, A., Raghoobar, S., Hairston, N., Painter, R.E., Wisniewski, D., Scapin, G., Fitzgerald, P., Sharma, N., Lu, J., Ha, S., Hermes, J., Hammond, M.L., 2014. Discovery of MK-7655, a β-lactamase inhibitor for combination with Primaxin® Bioorganic & Medicinal Chemistry Letters 24, 780–785. Available at https://doi.org/10.1016/j.bmcl.2013.12.101.

Brennan-Krohn, T., Smith, K.P., Kirby, J.E., 2017. The poisoned well: enhancing the predictive value of antimicrobial susceptibility testing in the era of multidrug resistance. Journal of Clinical Microbiology 55 (8), 2304–2308. Available at https://doi.org/10.1128/JCM.00511-17, http://jcm.asm.org/content/55/8/2304.full.pdf+html.

Casadevall, A., Pirofski, L.A., 2003. The damage-response framework of microbial pathogenesis. Nature Reviews. Microbiology 1 (1), 17–24. Available at https://doi.org/10.1038/nrmicro732.

Chancey, S.T., Zähner, D., Stephens, D.S., 2012. Acquired inducible antimicrobial resistance in Gram-positive bacteria. Future Microbiology 7 (8), 959–978. Available at https://doi.org/10.2217/fmb.12.63.

Chavada, J., Muneshwar, K.N., Ghulaxe, Y., Wani, M., Sarda, P.P., Huse, S., 2023. Antibiotic resistance: Challenges and strategies in combating infections. Cureus 15 (9). Available at https://doi.org/10.7759/cureus.46013.

Chinemerem Nwobodo, D., Ugwu, M.C., Oliseloke Anie, C., Al-Ouqaili, M.T.S., Chinedu Ikem, J., Victor Chigozie, U., et al., 2022. Antibiotic resistance: the challenges and some emerging strategies for tackling a global menace. Journal of Clinical Laboratory Analysis 36 (9). Available at https://doi.org/10.1002/jcla.24655, http://onlinelibrary.wiley.com/journal/10.1002/(ISSN)1098-2825.

Clinical Laboratory and Standards Institute, 2024. Performance Standards for Antimicrobial Susceptibility Testing, CLSI Guideline M1000, 34th ed.; Clinical Laboratory and Standards Institute: Wayne, PA, USA, 2024. 34, Clinical Laboratory and Standards Institute, Clinical Laboratory and Standards Institute.

Dadgostar, P., 2019. Antimicrobial resistance: Implications and costs. Infection and Drug Resistance Volume 12, 3903–3910. Available at https://doi.org/10.2147/idr.s234610.

DAHDF, 2019. 20th Livestock Census 2019. Available online at: Animal Husbandry Statistics (AHS) | Department of Animal Husbandry & Dairying (dahd.nic.in). DAHDF, DAHDF.

DAHDF, 2022. Basic Animal Husbandry Statistics, 2022. Available online at: Animal Husbandry Statistics (AHS) | Department of Animal Husbandry & Dairying (dahd. nic.in). DAHDF, DAHDF Available at http://dahd.nic.in.

Do, N.T.T., Vu, T.V.D., Greer, R.C., Dittrich, S., Vandendorpe, M., Pham, N.T., Ta, D.N., Cao, H.T., Khuong, T.V., Le, T.B.T., Duong, T.H., Nguyen, T.H., Cai, N.T.H., Nguyen, T.Q.T., Trinh, S.T., van Doorn, H.R., Lubell, Y., Lewycka, S., 2023. Implementation of point-of-care testing of C-reactive protein concentrations to improve antibiotic targeting in respiratory illness in Vietnamese primary care: a pragmatic cluster-randomised controlled trial. The Lancet Infectious Diseases 23, 1085–1094. Available at https://doi.org/10.1016/s1473-3099(23)00125-1.

Doi, Y., Chambers, H.F., 2015. Penicillins and β-Lactamase Inhibitors, Elsevier BV, p. 263, 10.1016/b978-1-4557-4801-3.00020-5. Penicillins and β-Lactamase Inhibitors1.

Doron, S., Davidson, L.E., 2011. Antimicrobial stewardship. Mayo Clinic Proceedings 86 (11), 1113–1123. Available at https://doi.org/10.4065/mcp.2011.0358, http://www.mayoclinicproceedings.com/content/86/11/1113.full.pdf+html.

Ehmann, D.E., Jahić, H., Ross, P.L., Gu, R.-F., Hu, J., Kern, G., Walkup, G.K., Fisher, S.L., 2012. Avibactam is a covalent, reversible, non–β-lactam β-lactamase inhibitor. Proceedings of the National Academy of Sciences 109, 11663–11668. Available at https://doi.org/10.1073/pnas.1205073109.

FAO, 2018. Antimicrobial resistance policy review and development framework - A regional guide for governments in Asia and the Pacific to review, update and develop policy to address antimicrobial resistance and antimicrobial use in animal production. Bangkok 64.

FAO, 2022. India at a Glance. Available online at: India at a glance | FAO in India | Food and Agriculture Organization of the United Nations. Accessed on: 22nd August 2022. India at a glance | FAO in India | Food and Agriculture Organization of the United Nations, India at a glance | FAO in India | Food and Agriculture Organization of the United Nations.

Ferri, M., Ranucci, E., Romagnoli, P., Giaccone, V., 2017. Antimicrobial resistance: a global emerging threat to public health systems. Critical Reviews in Food Science and Nutrition 57 (13), 2857–2876. Available at https://doi.org/10.1080/10408398.2015.1077192.

Gersema, L.M., Helling, D.K., 1986. The use of subtherapeutic antibiotics in animal feed and its implications on human health. Drug Intelligence and Clinical Pharmacy 20 (3), 214–218. Available at https://doi.org/10.1177/106002808602000308.

Ghafur, A., Mathai, D., Muruganathan, A., Jayalal, J.A., Kant, R., Chaudhary, D., et al., 2013. The Chennai declaration: a roadmap to tackle the challenge of antimicrobial resistance. Indian Journal of Cancer 50 (1), 71. Available at https://doi.org/10.4103/0019-509X.104065.

Hattab, S., Ma, A.H., Tariq, Z., Vega Prado, I., Drobish, I., Lee, R., et al., 2024. Rapid phenotypic and genotypic antimicrobial susceptibility testing approaches for use in the clinical laboratory. Antibiotics 13 (8), 786. Available at https://doi.org/10.3390/antibiotics13080786.

Hegde, M., Raj, S., Tikadar, D., Nyamagoud, S.B., 2024. Addressing antimicrobial Rresistance in India: A 2 stewardship roadmap. Journal of Health Science and Medical Research 20241071. Available at https://doi.org/10.31584/jhsmr.20241071.

Hoffman, S.J., Caleo, G.M., Daulaire, N., Elbe, S., Matsoso, P., Mossialos, E., et al., 2015. Strategies for achieving global collective action on antimicrobial resistance. Bulletin of the World Health Organization 93 (12), 867–876. Available at https://doi.org/10.2471/blt.15.153171.

Hong, Y., Zeng, J., Wang, X., Drlica, K., Zhao, X., 2019. Post-stress bacterial cell death mediated by reactive oxygen species. Proceedings of the National Academy of Sciences 116 (20), 10064–10071. Available at https://doi.org/10.1073/pnas.1901730116.

Howell, L., editor, 2013. World Economic Forum. 2013. Global risks 2013, eighth edition: an initiative of the Risk Response Network. 8.

India: national action plan on antimicrobial resistance (NAPAMR), 2017. WHO. https://www.who.int/publications/m/item/india-national-action-plan-on-antimicrobial-resistance-(napamr)- 2017-2021.

Jorgensen, J.H., Ferraro, M.J., 2009. Antimicrobial susceptibility testing: a review of general principles and contemporary practices. Clinical Infectious Diseases 49 (11), 1749–1755. Available at https://doi.org/10.1086/647952.

Klobucar, K., Jardine, E., Farha, M.A., MacKinnon, M.R., Fragis, M., Nkonge, B., Bhando, T., Borrillo, L., Tsai, C.N., Johnson, J.W., Coombes, B.K., Magolan, J., Brown, E.D., 2022. Genetic and chemical screening reveals targets and compounds to potentiate gram-positive antibiotics against gram-negative bacteria. ACS Infectious Diseases 8, 2187–2197. Available at https://doi.org/10.1021/acsinfecdis.2c00357.

Klein, E.Y., Van Boeckel, T.P., Martinez, E.M., Pant, S., Gandra, S., Levin, S.A., et al., 2018. Global increase and geographic convergence in antibiotic consumption between 2000 and 2015. Proceedings of the National Academy of Sciences 115 (15). Available at https://doi.org/10.1073/pnas.1717295115.

Lade, H., Kim, J.-S., 2021. Bacterial targets of antibiotics in methicillin-resistant *Staphylococcus aureus*. Antibiotics 10 (4), 398. Available at https://doi.org/10.3390/antibiotics10040398.

Landers, T., Kavanagh, K.T., 2016. Is the presidential advisory council on combating antibiotic resistance missing opportunities? American Journal of Infection Control 44 (11), 1356–1359. Available at https://doi.org/10.1016/j.ajic.2016.07.008, http://www.journals.elsevier.com/ajic-american-journal-of-infection-control/.

Lin, L., Tan, B., Pantapalangkoor, P., Ho, T., Baquir, B., Tomaras, A., et al., 2012. Inhibition of LpxC protects mice from resistant *Acinetobacter baumannii* by modulating inflammation and enhancing phagocytosis. mBio 3 (5). Available at https://doi.org/10.1128/mBio.00312-12, http://mbio.asm.org/content/3/5/e00312-12.full.pdf+html.

Mahon, C.R., Lehman, D.C., Manuselis, G., 2014. Antimicrobial agent mechanisms of action and resistance, In: Textbook of Diagnostic Microbiology, St. Louis: Saunders, pp. 254–273. 5.

Majumder, M.A.A., Rahman, S., Cohall, D., Bharatha, A., Singh, K., Haque, M., et al., 2020. Antimicrobial stewardship: fighting antimicrobial resistance and protecting global public health. Infection and Drug Resistance 13, 4713–4738. Available at https://doi.org/10.2147/IDR.S290835, https://www.dovepress.com/antimicrobial-stewardship-fighting-antimicrobial-resistance-and-protec-peer-reviewed-article-IDR.

Maqbool, M., Willie, M.M., 2024. 'Antibiotic armageddon: navigating the resistance crisis', J qual healthcare eco. Journal of Quality in Health Care & Economics 7 (1), 1–16. Available at https://doi.org/10.23880/jqhe-16000363.

Mathew, P., Ranjalkar, J., Chandy, S.J., 2020. Challenges in implementing antimicrobial stewardship programmes at secondary level hospitals in India: An exploratory study. Frontiers in Public Health 8. Available at https://doi.org/10.3389/fpubh.2020.493904, http://journal.frontiersin.org/journal/public-health/section/public-health-education-and-promotion#archive.

Michael, C.A., Dominey-Howes, D., Labbate, M., 2014. The antimicrobial resistance crisis: causes, consequences, and management. Frontiers in Public Health 2. Available at https://doi.org/10.3389/fpubh.2014.00145, http://journal.frontiersin.org/article/10.3389/fpubh.2014.00145/pdf.

Mitsakakis, K., Kaman, W.E., Elshout, G., Specht, M., Hays, J.P., 2018. Challenges in identifying antibiotic resistance targets for point-of-care diagnostics in general practice. Future Microbiology 13 (10), 1157–1164. Available at https://doi.org/10.2217/fmb-2018-0084, http://www.futuremedicine.com/loi/fmb.

National AMR Surveillance Network (NARS-Net) Annual Report 1 January 2022- 31 December 2022. Available online at: National AMR Surveillance Network (NARS-Net) Annual Report 2023: Ministry of Health and Family Welfare (ncdc. gov.in). (2023). Available at http://ncdc.gov.in.

Patel, P., Wermuth, H.R., Calhoun, C., Hall, G.A., 2026. Antibiotics. StatPearls Publishing.

Panda, P.K., 2025. Wrong diagnosis — Wrong antimicrobials: Rise in antimicrobial resistance in developing countries. IDCases 41, e02313. Available at https://doi.org/10.1016/j.idcr.2025.e02313.

Rahman, M., et al., 2020. Effect of culture-driven stewardship in Bangladesh hospital. International Journal of Antimicrobial Agents 55 (5), 105910.

Phan, T.-V., Huynh, P.N.H., Nguyen, V.-T.-V., Nguyen, T.-P., Vu, T.-T., Thi Vo, C.-V., Le, M.-T., Nguyen, B.G.D., Truong, P., Thai, K.-M., 2023. Discovery of a novel AcrAB-TolC pump inhibitor using the multistep virtual screening, synthesis and biological evaluation of asymmetric imidazole-4,5-dicarboxamide derivatives. New Journal of Chemistry 47, 20718–20722. Available at https://doi.org/10.1039/d3nj01603j.

Reygaert, W.C., 2018. An overview of the antimicrobial resistance mechanisms of bacteria. AIMS Microbiology 4 (3), 482–501. Available at https://doi.org/10.3934/microbiol.2018.3.482.

Rima, M., Rima, M., Fajloun, Z., Sabatier, J.-M., Bechinger, B., Naas, T., 2021. Antimicrobial peptides: a potent alternative to antibiotics. Antibiotics 10, 1095. Available at https://doi.org/10.3390/antibiotics10091095.

Romandini, A., Pani, A., Schenardi, P.A., Pattarino, G.A.C., De Giacomo, C., Scaglione, F., 2021. Antibiotic resistance in pediatric infections: global emerging threats, predicting the near future. Antibiotics 10 (4). Available at https://doi.org/10.3390/antibiotics10040393, https://www.mdpi.com/2079-6382/10/4/393/pdf.

Sharma, M., Gangakhedkar, R.R., Bhattacharya, S., Walia, K., 2021. Understanding complexities in the uptake of indigenously developed rapid point-of-care diagnostics for containment of antimicrobial resistance in India. BMJ Global Health 6 (9), e006628. Available at https://doi.org/10.1136/bmjgh-2021-006628.

Shree, P., Singh, C.K., Sodhi, K.K., Surya, J.N., Singh, D.K., 2023. Biofilms: Understanding the structure and contribution towards bacterial resistance in antibiotics. Medicine in Microecology 16, 100084. Available at https://doi.org/10.1016/j.medmic.2023.100084.

Singh, V.P., Jha, D., Rehman, B.U., Dhayal, V.S., Dhar, M.S., Sharma, N., 2024. A mini-review on the burden of antimicrobial resistance and its regulation across one health sectors in India. Journal of Agriculture and Food Research 15, 100973. Available at https://doi.org/10.1016/j.jafr.2024.100973.

Spellberg, B., Srinivasan, A., Chambers, H.F., 2016. New societal approaches to empowering antibiotic stewardship. JAMA - Journal of the American Medical Association 315 (12), 1229–1230. Available at https://doi.org/10.1001/jama.2016.1346, http://jama.jamanetwork.com/article.aspx?articleid=2498636.

Sodhi, K.K., Singh, C.K., 2022. Recent development in the sustainable remediation of antibiotics: A review. Total Environment Research Themes 3-4, 100008. Available at https://doi.org/10.1016/j.totert.2022.100008.

Sodhi, K.K., Singh, C.K., Kumar, M., Singh, D.K., 2023. Whole-genome sequencing of Alcaligenes sp. strain MMA: insight into the antibiotic and heavy metal resistant genes. Frontiers in Pharmacology 14. Available at https://doi.org/10.3389/fphar.2023.1144561.

Tartor, Y.H., Elmowalid, G.A., Hassan, M.N., Shaker, A., Ashour, D.F., Saber, T., 2022. Promising anti-biofilm agents and phagocytes enhancers for the treatment of *Candida albicans* biofilm–associated infections. Frontiers in Cellular and Infection Microbiology 12. Available at https://doi.org/10.3389/fcimb.2022.807218, http://www.frontiersin.org/Cellular_and_Infection_Microbiology/archive.

Ugwu, M.C., Shariff, M., Nnajide, C.M., Beri, K., Okezie, U.M., Iroha, I.R., et al., 2020. Phenotypic and molecular characterization of β -lactamases among enterobacterial uropathogens in southeastern Nigeria. Canadian Journal of Infectious Diseases and Medical Microbiology 2020. Available at https://doi.org/10.1155/2020/5843904, http://www.hindawi.com/journals/cjidmm/.

Van Boeckel, T.P., Gandra, S., Ashok, A., Caudron, Q., Grenfell, B.T., Levin, S.A., et al., 2014. Global antibiotic consumption 2000 to 2010: an analysis of national pharmaceutical sales data. The Lancet Infectious Diseases 14 (8), 742–750. Available at https://doi.org/10.1016/s1473-3099(14)70780-7.

Ventola, C.L., 2015. The antibiotic resistance crisis: part 1: causes and threats. Medi Media USA Inc, undefined P and T 40 (4), 277–283. Available at http://www.ptcommunity.com/system/files/pdf/ptj4004277.pdf.

Watson, K., Russell, C.D., Baillie, J.K., Dhaliwal, K., Fitzgerald, J.R., Mitchell, T.J., et al., 2020. Developing novel host-based therapies targeting microbicidal responses in macrophages and neutrophils to combat bacterial antimicrobial resistance. Frontiers in Immunology 11. Available at https://doi.org/10.3389/fimmu.2020.00786, https://www.frontiersin.org/journals/immunology#.

Yadav, S., Kapley, A., 2021. Antibiotic resistance: global health crisis and metagenomics. Biotechnology Reports 29, e00604. Available at https://doi.org/10.1016/j.btre.2021.e00604.

Yamin, D., Uskoković, V., Wakil, A.M., Goni, M.D., Shamsuddin, S.H., Mustafa, F.H., et al., 2023. Current and future technologies for the detection of antibiotic-resistant bacteria. Diagnostics 13 (20). Available at https://doi.org/10.3390/diagnostics13203246, http://www.mdpi.com/journal/diagnostics/.

Yeh, P.J., Hegreness, M.J., Aiden, A.P., Kishony, R., 2009. Drug interactions and the evolution of antibiotic resistance. Nature Reviews. Microbiology 7 (6), 460–466. Available at https://doi.org/10.1038/nrmicro2133.

Zhu, Y., Huang, W.E., Yang, Q., 2022. Clinical perspective of antimicrobial resistance in bacteria. Infection and Drug Resistance 15, 735–746. Available at https://doi.org/10.2147/IDR.S345574, https://www.dovepress.com/getfile.php?fileID=78731.

Chapter 10

Next-gen antibiotics: breakthroughs in discovery and development

Birson Ingti[1], Anupama Moirangthem[1,2], and Ramkrishna Roy[1,3]

[1]*Department of Microbiology, Royal School of Biosciences, The Assam Royal Global University, Guwahati, Assam, India,* [2]*School of Health Science and Technology, Medhavi Skills University, Singtam, Sikkim, India,* [3]*Department of Microbiology, Suri Vidyasagar College, Birbhum, West Bengal, India*

10.1 Introduction

Antibiotics, known to be a miracle drug, have been saving millions of lives from various bacterial infections for ages. But unfortunately, their universal effectiveness has been compromised in recent years due to the rapid emergence of resistance against the available antibiotics (Golkar et al., 2014). At present, many bacterial pathogens exhibit resistance to multiple antibiotic classes, limiting treatment options and increasing the risk of serious or long-lasting infections. Thus it has become a major global concern affecting the routine care in hospitals and community settings. Antibiotic resistance has been linked not only to the overuse and abuse of these drugs but also to the inability of the pharmaceutical industry to develop new antibiotics due to diminished financial incentives and difficult regulatory constraints (Gould and Bal, 2013; Sengupta et al., 2013; Wright, 2014). As a consequence, existing antibiotics are relied upon, and the extensive and improper utilisation of them in humans, animals, or crops leads to the development of drug-resistant bacteria that evolve under significant selection pressure (Terreni et al., 2021). Beyond the clinical and agricultural sites, the environmental release of these antibiotic agents also occurs through household consumption, hospitals, and medication manufacturing companies, thus resulting in the pollution of drinking water. As a result, resistance to an antibiotic emerges shortly after its approval for clinical use. Two contributing aspects to this phenomenon are (1) the usage of antibiotics, which kill susceptible bacteria while allowing resistant bacteria to persist, and (2) the activation of latent resistance genes in bacteria due to antibiotic pressure (Lee et al., 2013). As antibiotic resistance is spreading among the bacterial population continuously, the consequences are also rapidly seen in clinical settings. The emergence of bacterial infections caused by resistant bacteria can be severe, resulting in longer illnesses owing to delayed antibiotic treatment, heightened infection risk for surgical patients, higher fatality rates, and elevated expenditures. The transmission of antibiotic-resistant bacteria among patients within a healthcare institution frequently happens via inter- or intrahospital patient transfers and may extend to the community if not effectively restricted (Farr et al., 2001). So, the resistant bacteria pose a significant global health challenge, which are increasingly involved in several bacterial diseases, include *Enterococcus* spp., *Staphylococcus aureus*, *Klebsiella pneumoniae*, *Acinetobacter baumannii*, *Pseudomonas aeruginosa*, and *Enterobacter* spp. (Tacconelli et al., 2018). Countries potentially affected by antimicrobial resistance (AMR) include India, Nigeria, Indonesia, and Russia, since they have reported elevated incidence of malaria, HIV, or tuberculosis, hence likely to experience increased resistance to existing therapies (Brüssow, 2024). The disparity in antibiotic resistance among countries correlates with differences in antibiotic usage, wherein inappropriate and excessive application fosters the emergence of antibiotic resistance (Bell et al., 2014). The declining pipeline of new antibiotics, where development and commercialisation have slowed since the 1990s, exacerbates this situation (Lewis, 2012; Singer et al., 2020). According to WHO estimates, the rise in antibiotic resistance might result in 10 million deaths by 2050, making this problem one of the biggest public health challenge (Pulingam et al., 2020). AMR not only affects humans, but it also causes an impact on plants and animals, leading to a decrease in agricultural products and compromising food security. It causes a burden in the national economy as well (Kim et al., 2023). Therefore there is an obvious need for newer antibiotics or alternative medicines to combat the resistant bacteria. In recent years, efforts to find more effective methods have accelerated. Alternative treatment approaches such as bacteriophage therapy, antimicrobial peptides, antivirulence strategies, and

Antimicrobial Resistance in the 21st Century. DOI: https://doi.org/10.1016/B978-0-443-44748-8.00015-4

nanotechnology-based delivery systems have become more popular. These methods try to either completely get around resistance processes or work with regular antibiotics. This review describes the new developments and techniques in discovering newer drugs to curtail AMR.

10.2 Current burden of priority pathogens

The WHO's global priority pathogen list has been released recently, highlighting the urgent need for antibiotics for research and development. The priority pathogens have been grouped into different categories, such as the critical priority group, which includes the mostly gram-negative bacteria that are predominantly found in immunocompromised patients who are dependent on the blood catheters and ventilators. The medium and high priority pathogens include the gram-positive and gram-negative bacteria, which are the causal organisms of common infections (Tacconelli et al., 2018). These microorganisms have developed different resistant mechanisms to withstand the conventional antibiotic therapies, making it more complicated in the treatment process of bacterial infections (Table 10.1). This list is an urgent call for finding newer drugs to fight against these bacteria that remain reason for increased mortality and morbidity. The priority list of pathogens is listed in Table 10.2.

10.3 Evolution of antibiotic discovery: from past to present

The inception of the modern 'antibiotic era' is often attributed to Paul Ehrlich and Alexander Fleming. Ehrlich's concept of a 'magic bullet', which selectively targets only pathogenic bacteria while preserving the host, was predicated on the finding that aniline and other synthetic dyes, newly available at that time, could selectively stain certain microbes but not others (Abraham and Chain, 1940). Ehrlich argued that chemical compounds could be synthesised that would be able to exert their full action exclusively on the parasite harboured within the organism. Paul Ehrlich discovered an arsenic derivative known as arsphenamine that targeted *Treponema pallidum*, which causes syphilis, in 1909. In and around 1928, which is also generally known as the starting point of the modern antibiotic era, Fleming discovered Penicillin, followed by its purification, and later on started with the mass production after a decade. In 1930 Gerhard Domagk discovered the antibiotic effects of sulphanilamide, a molecule synthesised 22 years before by Paul Gelmo. This antibiotic was marketed under the name Protonsil in 1935 and was used by soldiers during World War II (Lewis, 2013). Another important milestone in the history of antibiotics was the discovery of the first aminoglycoside antibiotic in 1943, the Streptomycin from *Actinomyces griseus* (now *Streptomyces griseus*), by Albert Schatz and Selman Waksal (Jones et al., 1944; Schatz et al., 1944). This new antibiotic was found to be effective not only against gram-positive strains but also against gram-negative bacteria, as well as against *Mycobacterium tuberculosis*. Later, around 1945, tetracyclines were discovered, which were isolated from *Streptococcus aureofaciens*. Another original antibiotic came up on the list of antibiotics in and around 1947, which was proved to be broad-spectrum antibiotic, namely chloramphenicol, a member of the phenylpropanoids, discovered initially by John Ehrlich (from *Streptomyces venezuelae*) (Ehrlich et al., 1947). The search for newer antibiotics continued with the discovery of erythromycins in 1950 (McGuire et al., 2021) and vancomycin in 1958 (Griffith, 1981). By 1962, nalidixic acid was synthesised and was found to be active against Gram-negative bacteria such as *Escherichia coli*, *Proteus mirabilis*, and *Shigella flexneri*(Lesher et al., 1962). Around this time, cephalosporins also underwent modifications in their structure. In the 1970s and 1980s, carbapenems appeared, beta-lactamase inhibitors were designed to confer protection on beta-lactam antibiotics, and fluoroquinolones entered clinical practice. Imipenem, azithromycin, clarithromycin, and third-generation cephalosporins became established treatments (Aminov, 2010). After a lengthy gap of innovation, one new class of antibiotic, teixobactin, was introduced in 2015 (Jad et al., 2015). Teixobactin is a novel type of antibiotic that works in a way that is similar to penicillins, but it is different because it cannot attach to the lipids in peptidoglycan (Fiers et al., 2017). As of December 2020, 43 new antibiotics were being developed, per the pipeline assessment. According to the WHO, there is a significant and increasing global trend in the use of phytonutrients and herbal remedies to address a variety of health issues in different national healthcare systems (Abdallah et al., 2023). As an alternative to antimicrobial drugs, there has been an increase in the screening of medicinal plants as a source worldwide. Medicinal plants have been known to have therapeutic properties for millennia, and many of the drugs today are plant-derived natural products or their derivatives (Kinghorn et al., 2011; Newman and Cragg, 2012). Numerous secondary metabolites, such as saponines, tannins, alkaloids, alkenyl phenols, glycoalkaloids, flavonoids, sesquiterpenes, lactones, terpenoids, and phorbol esters, are present in medicinal plants that exhibit antimicrobial properties (Lewis and Ausubel, 2006). Medicinal plants have been transformed into one of the oldest sciences in countries such as China, Greece, Egypt, and India. The introduction of novel antimicrobials elicits an instantaneous response to the emergence of AMR in targeted bacteria, which utilise many recognised ways to circumvent the effects of chemotherapeutic drugs such

TABLE 10.1 Major classes of antibiotics and resistance mechanisms.

Antibiotic class	Primary target	Representative drugs	Major resistance mechanisms	References
β-Lactams	Cell wall synthesis (penicillin-binding proteins)	Penicillins, Cephalosporins, Carbapenems	Production of β-lactamase, altered PBPs, reduced permeability, and efflux pumps	Bush and Bradford (2016)
Glycopeptides	Cell wall synthesis (D-Ala-D-Ala termini)	Vancomycin, Teicoplanin	Target modification (D-Ala-D-Lac), cell wall thickening	Arthur and Courvalin (1993)
Aminoglycosides	Protein synthesis (30 S ribosomal subunit)	Gentamicin, amikacin, streptomycin	Enzymatic drug modification, reduced uptake, target mutation,	Krause et al. (2016)
Macrolides	Protein synthesis (50 S ribosomal subunit)	Erythromycin, azithromycin	Efflux pump, Ribosomal methylation, enzymatic inactivation	Fyfe et al. (2016)
Fluoroquinolones	DNA gyrase and topoisomerase IV	Ciprofloxacin, levofloxacin	efflux pumps, target site mutations, reduced permeability	Hooper and Jacoby (2015)
Tetracyclines	Protein synthesis (30 S ribosomal subunit)	Tetracycline, doxycycline	Efflux pumps, ribosomal protection proteins	Chopra and Roberts (2001)
Oxazolidinones	Protein synthesis (50 S ribosomal subunit)	Linezolid	Target site mutation, methylation of rRNA	Hashemian et al. (2018)
Polymyxins	Outer membrane disruption	Colistin, polymyxin B	Lipid A modification, reduced membrane binding	Poirel et al. (2017)

TABLE 10.2 The WHO Bacterial Priority pathogen list, 2024, and its clinical significance.

Priority	Pathogen	Resistance	Clinical Significance
Critical priority	*Acinetobacter baumannii*	Carbapenem-resistant	Hospital-acquired infections
	Enterobacterales (e.g. *Escherichia coli, Klebsiella pneumoniae*)	Third-generation cephalosporin-resistant	Common causes of urinary tract infection, bloodstream, and intra-abdominal infections
	Enterobacterales	Carbapenem-resistant	Rapid global spread of carbapenemases, High mortality
	Mycobacterium tuberculosis	Rifampicin-resistant	Drug-resistant TB threatens global TB control efforts
High priority	*Salmonellatyphi*	Fluoroquinolone-resistant	Treatment of typhoid fever compromised
	Shigella spp.	Fluoroquinolone-resistant	Causes severe diarrhoea, especially in children
	Enterococcus faecium	Vancomycin-resistant (VRE)	complicated nosocomial infections
	Pseudomonas aeruginosa	Carbapenem-resistant	Main cause of severe infections in the immunocompromised patients and ICU
	Nontyphoidal *Salmonella*	Fluoroquinolone-resistant	Related to food-borne infections with invasive potential
	Neisseria gonorrhoeae	Resistant to 3rd-generation cephalosporins and/or fluoroquinolones	compromises the treatment of gonorrhoea
	Staphylococcus aureus	Methicillin-resistant (MRSA)	Major cause of bloodstream, skin, and surgical site infections
Medium priority	Group A *Streptococcus*	Macrolide-resistant	Responsible for Pharyngitis, invasive soft-tissue infections
	Streptococcus pneumoniae	Macrolide-resistant	Causes community-acquired pneumonia, meningitis
	Haemophilus influenzae	Ampicillin-resistant	Causes respiratory and invasive infections
	Group B *Streptococcus*	Penicillin-resistant	Common cause of neonatal sepsis and meningitis

Source: From the WHO Bacterial Priority pathogen list, 2024

as antibiotics (Abdallah et al., 2023). One of the early phytocompound which have shown activity against a wide range of bacteria, especially **gram-positive bacteria, is berberine, an isoquinoline quaternary alkaloid which has been derived from *Berberis vulgaris, Coptis chinensis, Coptis rhizome, Coptis japonica, Phellodendron amurense, Phellodendron chinense schneid,* and *Hydrastis canadensis* (Imanshahidi and Hosseinzadeh, 2008). This phytocompound has been used in both Ayurveda and Chinese medicine to treat various ailments (Tillhon et al., 2012). Catechins, a flavonoid compound, have also been extensively studied and are known for their antimicrobial activity. They are especially found in green teas. Catechin is found to inhibit bacteria such as *Shigella* (Vijaya et al., 1995), *Streptococcus mutans* (Sakanaka et al., 1989), and *Vibrio cholerae* O1 (Borris, 1996). The other phytocompounds that have antibacterial properties are given in Table 10.3.

10.4 Limitations of conventional antibiotic discovery

There are challenges that often limit antibiotic discovery, the most important being the permeability barriers in Gram-negative bacterial species that often compromise the penetration of antibiotics. Second, these bacteria use efflux pumps to eliminate the antibiotics that are able to enter through the cell membrane (Duffey et al., 2024). Moreover, there is a need

TABLE 10.3 Some medicinal plants having antibacterial properties.

Scientific name	Common name	Phytocompound	Class	Activity	Reference
Glycyrrhiza glabra	Licorice	Glabrol	Phenolic compound	*Staphylococcus aureus, Mycobacterium tuberculosis*	Wahab et al. (2021)
Allium cepa	Onion	Allicin	Sulfoxide	Bacteria	Vohora et al. (1973)
Withania somniferum	Ashwagandha	Withafarin A	Lactone	Bacteria	Khanchandani et al. (2019)
Berberis vulgaris	Barberry	Berberine	Alkaloid	Bacteria	El Zahar et al. (2022)
Ocimum basilicum	Basil	Essential oils	Terpenoid	*Salmonella*	Wan et al. (1998)
Vaccinium spp.	Blueberry	Fructose	Monosaccharide	*Escherichia coli*	Rodríguez-Pérez et al. (2016)
Rhamnus purshiana	Cascaras agrada	Tannins	Polyphenols, Anthraquinones	Bacteria	Nigussie et al. (2021)
Matricaria chamomilla	Chamomile	Anthemic acid	Phenolic acid	*Salmonella typhimurium*	Hamburger and Hostettmann (1991)
Erythroxylum coca	Coca	Cocaine	Alkaloid	*Gram-positive cocci*	Cervantes-Alagón et al. (2023)
Eucalyptus globulus	Eucalyptus	Globulol	Sesquiterpine	Bacteria	Tan et al. (2008)
Allium sativum	Garlic	Allicin, ajoene	Sulphoxide, Sulfated terpenoids	Bacteria	Naganawa et al. (1996)
Centella asiatica	Gotu kola	Asiatocoside	Terpenoid	*M. leprae*	Ramli et al. (2020)
Camellia sinensis	Green tea	Catechin	Flavonoid	*Shigella Vibrio cholerae*	Vijaya et al. (1995), Toda et al. (1992)
Cassia angustifolia	Senna	Rutin	Flavonoid	*E. coli, Klebsiella pneumoniae*	Ahmed et al. (2016)
Thymus vulgaris	Thyme	Thymol	Phenol	Bacteria	Borugă et al. (2014)
Curcuma longa	Turmeric	Curcumin	Terpenoids	Bacteria	Adamczak et al. (2020)

for better chemical libraries that have a broader spectrum with minimal cytotoxicity and interference with less membrane lysis properties. There is also a need for chemicals that target gram-negative organisms (Silver, 2011).

10.5 Emerging strategies for antibiotic discovery

The growing threat of AMR has led to the creation of new, nontraditional methods for finding and developing new antibacterial agents that go beyond the usual 'kill-or-cure' approach to antibiotics. These strategies often prioritise neutralising the pathogen or augmenting the host's immune response rather than solely focusing on bacterial viability. Fig. 10.1 displays the different strategies to combat the antibiotic-resistant bacteria.

10.5.1 Targeting bacterial virulence

This technique involves antivirulence therapy that targets the neutralisation of the pathogen's virulence factors.

10.5.1.1 Quorum sensing inhibition

Quorum sensing is a mechanism used by different bacteria to communicate with one another for the regulation of gene expression. It also plays an important role in biofilm formation, antibiotic resistance, and the production of virulence factors. One of the ways to curtail the antibiotic resistance of bacteria is to inhibit the quorum-sensing, which can be achieved by certain mechanisms such as synthesising the signalling molecules, degrading the signalling molecules, and hindering the binding of receptor protein by signalling molecules in quorum-sensing pathways (Defoirdt et al., 2004; Lade et al., 2014). So, quorum sensing includes mainly two types: (1) a small molecule inhibitor which may be extracted either from natural sources or by chemical synthesis (Kalia, 2013; Defoirdt et al., 2013), (2) quorum quenching enzymes, such as acylases, lactonases targeting acyl-homoserine lactones (AHLs) as signalling molecules, AI-2 kinases targeting furanosyl borate ester (autoinducer 2, AI-2) as signalling molecules, etc. Quorum-sensing inhibition mainly includes two types: One type is small molecule quorum-sensing inhibitors, either extracted from natural resources or obtained by chemical synthesis, and the other type is quorum quenching enzymes, mainly including acylases, lactonases targeting acyl-homoserine lactones (AHLs) as signalling molecules, AI-2 kinases targeting furanosyl borate ester (autoinducer 2, AI-2) as signalling molecules, etc. (Dong and Zhang, 2005; Czajkowski and Jafra, 2009; Fetzner, 2015)

10.5.1.2 Phage-antibiotic synergy

Due to the tremendous increase in antibiotic resistance in recent years, interest has been shown in the use of the combination of bacteriophages along with antibiotics known as phage-antibiotic synergy (PAS) (Liu et al., 2022). The first study of PAS was conducted in 2007, where it was found that in the presence of sublethal concentrations of certain

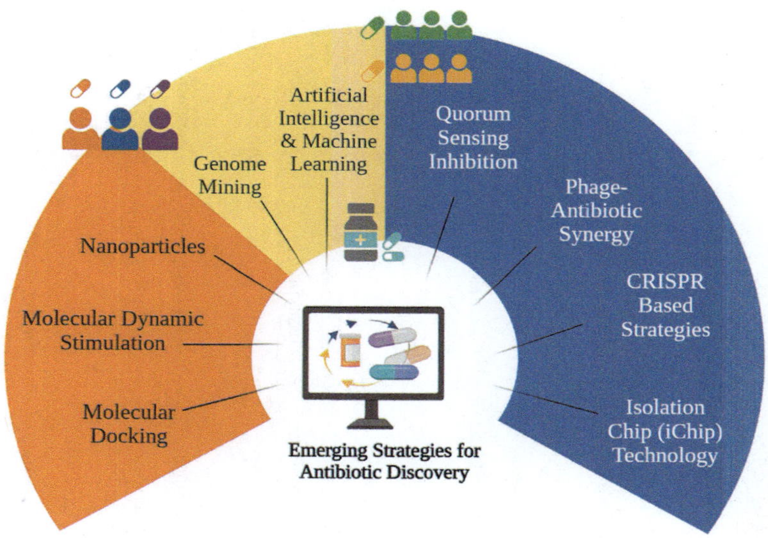

FIGURE 10.1 **Emerging strategies for antibiotic discovery**. Shows all the possible strategies to discover antibiotics.

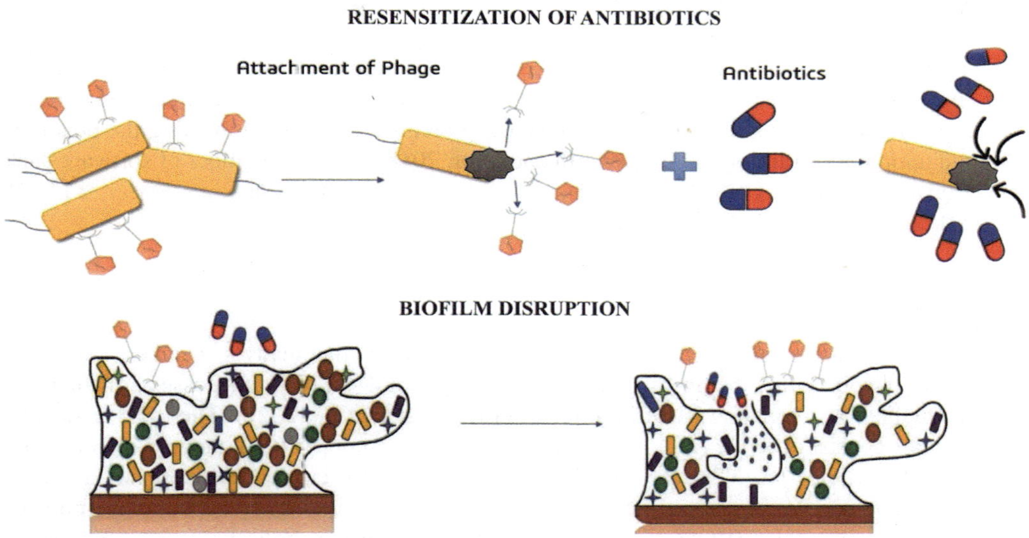

FIGURE 10.2 This diagram demonstrates the benefits of phage-antibiotic synergy (PAS) in combating resistant bacterial infections. The mechanism how phage-antibiotic synergy works.

antibiotics, phage production increased (Comeau et al., 2007). PAS plays a very important role in addressing infections caused by antibiotic-resistant strains. Bacterial growth and replication are targeted and suppressed when phages and antibiotics work in synergy, thereby helping to curtail bacterial proliferation and antibiotic resistance. For instance, when streptomycin is considered for the PAS, then it will target the 16 S rRNA and bind to inhibit bacterial protein synthesis and Podovirus, namely LUZ7, requiring LPS components for infection. Thus this approach, using the combination of phages and antibiotics, will target the different bacterial receptors, which will curtail the proliferation of bacteria and antibiotic resistance in infection (Diallo and Dublanchet, 2022). Not only this, but PAS also has potential in eliminating biofilms. Traditional antibiotic therapy alone is compromised in eliminating biofilms. However, the combination of phage and antibiotic increases the permeability of the antibiotics with the help of the phage enzyme in degrading the biofilm matrix (Łusiak-Szelachowska et al., 2020), and the mechanism of action is represented in Fig. 10.2.

10.5.2 In silico drug discovery

In silico studies have been playing a very important role in combating AMR. In silico methods for identifying novel compounds against AMR may accelerate the drug development and design process. Structural bioinformatics, high-throughput virtual screening, molecular docking, and molecular dynamics simulations can facilitate the identification of prospective therapeutic compounds. The structure–activity relationship (SAR) can further inform the creation of more effective therapeutic compounds with reduced adverse effects. Moreover, researchers have investigated novel concepts and methodologies to enhance the detection of active biomolecules that may function independently or synergistically with traditional antibiotics (Muteeb et al., 2023).

10.5.2.1 Molecular docking

It is the computational method of predicting a preferred ligand (drug molecule) conformation and orientation to its biological target, especially a protein. Docking provides ideas on the interactions between ligands and targets, which in turn helps the researchers predict the binding affinity and efficacy of potential drug compounds, and the most appropriate ones are selected. Molecular docking is particularly helpful in lead discovery, which finds intriguing leads by screening libraries of compounds against known pharmacological targets. High-throughput screening and improved precision made possible by advancements in docking algorithms have made it possible to explore enormous chemical libraries. There are several molecular docking tools available, including ArgusDock, DOCK, FRED, eHITS, AutoDock, and FTDock. Scoring methods are used in molecular modelling which are used in ranking the affinity of ligands to bind to the active site of a receptor. The scoring is to determine the most tightly bound ligand to the target macromolecule (Wadood et al., 2013).

10.5.2.2 *Molecular dynamics simulation*

Molecular dynamics simulation involves the simulation of every atom's molecular motion based on Newton's equations of motion. It also involves the increase in speed and location of every atom by increase in time. It may be used to capture biomolecular processes, ligand binding, conformational change, protein folding, etc., unveiling the positions of all of the atoms at femtosecond temporal resolution (Hollingsworth and Dror, 2018). A larger number of software packages for MD of biomolecules are available, such as GROMACS (Hess et al., 2008), AMBER (Salomon-Ferrer et al., 2013), NAMD (Phillips et al., 2020), and CHARMM (Brooks et al., 2009).

10.5.3 Nanoparticles

To address the failure of traditional antibiotics, one of the many alternatives that many researchers have shown interest in is nanotechnology. The nanoparticles, being of tiny size ranging from 1 to 100 nm, possess distinctive physical and chemical properties suitable for combating bacterial infection (Makabenta et al., 2021). The most extensively researched metal-based nanoparticles consist of silver (Ag), gold (Au), zinc (Zn), and copper (Cu), along with metal oxides like zinc oxide (ZnO). Silver nanoparticles (AgNPs) are found to combat (MRSA), carbapenem-resistant *K. pneumoniae*, and vancomycin-resistant *Enterococcus faecium* effectively. These nanoparticles disrupt bacterial membranes, induce the production of reactive oxygen species (ROS), and interfere with bacterial metabolic processes (Fontoura et al., 2023). Even the gold nanoparticles (AuNPs), in conjugation with antibiotics such as ciprofloxacin, surpass resistance mechanisms such as efflux pumps and reduced permeability, thereby enhancing the efficacy against multidrug-resistant Gram-negative bacteria. Gold nanoparticles (AuNPs) also hold promise, particularly in combination with antibiotics. Ciprofloxacin-conjugated AuNPs, for example, have shown enhanced efficacy against multidrug-resistant Gram-negative bacteria by bypassing resistance mechanisms like efflux pumps and reduced permeability (Zaman and Manghwar, 2024). Not only these nanoparticles, but selenium nanoparticles (SeNPs) also act against the *Escherichia coli and P. aeruginosa* by disrupting responses to oxidative stress and bacterial bioenergetics (Namdar et al., 2024). Lipid nanoparticles have also shown a great efficacy in overcoming barriers in treating biofilm-associated infections such as prosthetic joint infection and endocarditis (Radwan et al., 2024; Cao et al., 2024), by encapsulating it with rifampin and vancomycin that enhances the protection of the antibiotics from degradation and improves the penetration of antibiotics in the biofilm (Salama et al., 2024). Another preclinical study has shown the effectiveness of mesoporous silica nanoparticles (MSNs) against *P. aeruginosa* and *S. aureus*, mainly responsible for forming biofilm, when coloaded with antibiotics and adjuvants such as lysozyme (Su et al., 2024).

10.5.4 Genome mining

The phrase 'genome mining' refers to all bioinformatics investigations employed to identify both the biosynthetic pathways of bioactive natural compounds and their potential functional and chemical interactions (Ziemert et al., 2016). Genome mining entails the identification of previously uncharacterised natural product biosynthetic gene clusters within the genomes of sequenced organisms, the sequence analysis of the enzymes encoded by these gene clusters, and the experimental identification of the products of the gene clusters (Trivella and de Felicio, 2018). The primary use of genome mining is to identify novel biosynthetic gene clusters (BGCs). The BGCs encode two classes of enzymes: polyketide synthases (PKS) and nonribosomal peptide synthases (NRPS), which are the primary biosynthetic pathways responsible for the production of natural products (Timmermans et al., 2017). So, computational technologies are currently employed to 'mine' genomes, anticipate the structures of natural products, and identify methods to stimulate their expression in the laboratory.

10.5.5 Harnessing iChip technology

The urgent need to find new drugs to replace antibiotics that are resistant has become more and more important. Soil, which has 99% of all microbial species, is one of the main places where new antibiotic makers get their ideas. Microbes in soil make antimicrobial chemicals, but many of them cannot be grown in labs because standard methods do not work well in environments that are like their natural ones (Choi et al., 2015). One of the latest developments in strategy building for finding a way to discover newer drugs is the use of Uncultured Soil Technology (UST), or in situ incubation. The isolation chip (iChip) technique developed by D. Nichols in 2010 led to the discovery of a groundbreaking antibiotic, Teixobactin (Berdy et al., 2017; Zhao et al., 2023).

10.5.6 Clustered regularly interspaced short palindromic repeat-based strategies

Clustered regularly interspaced short palindromic repeat (CRISPR) technology has emerged as one of the most potential candidates to combat AMR in recent years (Gholizadeh et al., 2020). This revolutionary genetic engineering tool, initially derived from the bacterial immune system, offers exceptional precision in targeting and modifying pathogen genomes. Originally derived from the bacterial immune system, this groundbreaking genetic engineering technique provides an accuracy in identifying and modifying the genomes of pathogens. CRISPR has the capability of activating or inactivating sequence-specific DNA targets to combat antibiotic resistance genetically (Mayorga-Ramos et al., 2023; Tao et al., 2022). A 'scissors plasmid' was developed utilising a temperature-sensitive shuttle plasmid combined with guide RNA (gRNA) to lead Cas9 to specific DNA sequences, enabling precision DNA editing crucial for targeted antibacterial strategies. A study also showed the use of the Cas9 nuclease from type II CRISPR systems alongside a 20-nucleotide guide RNA (crRNA) to selectively eliminate particular bacterial species, thereby illustrating targeted action against antibiotic-resistant pathogens achieved by employing bacteriophages to deliver the CRISPR Type II system while preserving nontarget microorganisms (Bikard et al., 2014). Another study showed the introduction of the PRESA (Phage-Delivered Resistance Eradication with Subsequent Antibiotic) treatment, which integrates CRISPR-Cas9 with economical antibiotics (Liu et al., 2020). This approach blocked the transfer of antibiotic-resistant plasmids and prevented the emergence of resistant mutants, thereby restoring the efficacy of inexpensive antibiotics.

10.5.7 Artificial intelligence and machine learning in combating antimicrobial resistance

Machine learning and artificial intelligence methodologies have arisen to tackle the escalating issue of antibiotic resistance. These technologies facilitate the analysis of bacterial genomes, the prediction of resistance, the monitoring of epidemic patterns, and the discovery of novel antibacterial treatments or vaccines (Stokes et al., 2020; Anahtar et al., 2021). Machine learning excels in identifying factors contributing to resistance, including resistance-associated genes, resistance-associated alleles, and treatment circumstances essential for enhancing treatments. Numerous studies currently employ machine learning algorithms to forecast antibiotic resistance by analysing gene mutations, the presence or absence of genes, and antibiotic sensitivities as training data. Numerous investigations employ rule-based or machine learning methodologies to forecast the antibiotic resistance status of bacteria, utilising treatment history, patient demographics, and genomic data (Ren et al., 2022; Khaleci et al., 2020). Artificial intelligence has proven essential in expediting the identification of new antimicrobial drugs, including antimicrobial peptides (AMPs) (Wan et al., 2024; Wong et al., 2023). Antimicrobial peptides (AMPs) are essential elements of innate host defence, demonstrating potential in traversing bacterial membranes and suppressing microbial proliferation. Computational designs can forecast novel antimicrobial peptides (AMPs) from genomic sequences, offering insights into the chemical characteristics and biological functions of AMP sequences. For instance, deep learning models trained on spider transcriptome data found two peptides with significant antibacterial efficacy against many pathogens, including *Bacillus subtilis*, *Escherichia coli*, *P. aeruginosa*, *S. aureus*, and *Staphylococcus epidermidis* (Li et al., 2022; Wang et al., 2022).

10.5.8 Metagenomics

The approach of metagenomics includes the analysis of genomic data of samples from the environment without needing to isolate or culture the organism first (Roh and Villatte, 2008; Cowan et al., 2015). They used next-generation sequenching approach to understand the microbial community and functional components related to AMR (Schloss and Handelsman, 2003). The two most important metagenomic approaches are 16 S rDNA, which is a sequence- and function-driven screening of metagenomic libraries of certain traits. Another alternative to whole genome sequencing, which is superior to 16 s rDNA, is the shotgun whole genome sequencing. This method targets the diversities of microbial community, both structurally and functionally, by gene identification, complete bacterial genome reconstruction, and metabolic pathways (Chen and Pachter, 2005; De, 2019). So, this approach will not only help in detecting a wide range of species, but it will also help in identifying potential antibiotic-resistant genes.

10.6 Breakthrough molecules and drug classes

Antibiotic treatment is crucial for controlling bacterial infections, but it is greatly hindered by the global prevalence of multidrug-resistant (MRD) bacteria (Dai et al., 2025). Therefore the future of modern medicine depends upon effective antibiotics. The world produced more than 20 novel classes of antibiotics between 1930 and 1962, and since then, only two new classes of antibiotics have been marketed. It is clear that novel classes of antibiotics will be needed for long-term treatment, but how many? In this respect, it is assumed that the world will need a further 20 novel classes of antibiotics to strengthen modern medicine during the next 50 years. Therefore novel classes of antibiotics are urgently needed for the future to combat the Superbug. The two new classes of antibiotics were introduced into the market, namely oxazolidinone (Linezolid by Pfizer) and cyclic lipopeptide (Daptomycin by Cubist). This class of antibiotics inhibits protein synthesis and is active against gram-positive bacteria, such as MRSA. A study reported a novel antibiotic class named Tethered Macrocyclic Peptide (MCP), which targets the lipopolysaccharide transporter. The mechanisms of action of this MCP molecule involve blocking the transport of bacterial lipopolysaccharide from the inner membrane to its destination on the outer membrane, through inhibition of the LptB2FGC complex. This chemical class, named as Tethered MCP, represents a promising treatment paradigm for patients with invasive infections due to CRAB. The study also identifies LptB2FGC as a tractable target for antimicrobial drug development (Zampaloni et al., 2024). Another novel chemical agent, Indole-2-carboxylates, has recently been described, which has high efficacy against multidrug-resistant bacteria (Van Goethem et al., 2024).

10.7 Future perspectives

Much more research is needed to address the future of AMR, improve antibiotic development, and thus expand the spectrum of medicines used in the fight against MDR infections. The various lessons learnt from the golden age of antibiotic discovery, changes in AMR that followed, and examination of the most recent developments in genome editing, gene regulation, and systems biology can spark creative antibiotic discovery initiatives in a wide range of directions. The increasingly popular strategies include nanoparticles, immunotherapy, antisense RNA, resistance modulation, and the elimination of drug-resistant plasmids(Iskandar et al., 2022). Prevalent research is also being conducted on antibiotic resistance breakers that can be conjugated to other ineffective antibiotics or coadministered with them, irrespective of whether they have direct antibacterial action. In this context, research is being carried out on substances used to reverse AMR, such as efflux pump inhibitors, membrane permeabilizers, and modifying-enzyme inhibitors. Potential future strategies to combat AMR involve the extraction and modification of naturally existing antimicrobial peptides from eukaryotic and prokaryotic organisms, as well as antibiotic hybridisation (Khardori et al., 2020). Even though the majority of such strategies, if authorised, would probably be used in combination with antibiotics for the foreseeable future, they do offer hope for effective ways of fighting AMR (Theuretzbacher and Piddock, 2019).

10.8 Conclusion

AMR has been one major concern in treating various ailments that increases the mortality rate. The traditional way to combat the resistant microbes has been greatly challenged and is insufficient to tackle the resistant organisms. As the resistant bacteria are evolving rapidly, it necessitates finding alternatives or enhancing the existing antibiotics to combat them. The current focus of the next generation of antibiotics is shifting from the conventional discovery process to combat the increasing problem of AMR. The evolution of genomics, artificial intelligence, structure-based drug design, novel formulations, and unconventional sources of antibiotics, including synthesised molecules, antimicrobial peptides, phage therapy, and host-directed therapies, is changing the face of antibiotics. Thus an interdisciplinary approach to address the problem of AMR is imperative for enhancing infection control measures.

Acknowledgement

The corresponding author would like to thank the Assam Royal Global University for providing the necessary infrastructure to carry out the present work. The authors would also like to thank Biorender for preparing the graphical illustrations.

Funding

This research received no external funding.

Data availability statement

Not applicable.

Conflicts of interest

The authors declare no conflicts of interest.

AI disclosure

During the preparation of this work, the author(s) used Quillbot and Grammarly to better English. After using this tool/ service, the author(s) reviewed and edited the content as needed and take(s) full responsibility for the content of the publication.

References

Abdallah, E.M., Alhatlani, B.Y., de Paula Menezes, R., Martins, C.H.G., 2023. Back to nature: medicinal plants as promising sources for antibacterial drugs in the post-antibiotic era. Plants 12 (17). Available at https://doi.org/10.3390/plants12173077, www.mdpi.com/journal/plants.

Abraham, E.P., Chain, E., 1940. An enzyme from bacteria able to destroy penicillin. Nature 146 (3713), 837. Available at https://doi.org/10.1038/146837a0 837.

Adamczak, A., Ożarowski, M., Karpiński, T.M., 2020. Curcumin, a natural antimicrobial agent with strain-specific activity. Pharmaceuticals 13 (7), 153. Available at https://doi.org/10.3390/ph13070153.

Ahmed, S.I., Hayat, M.Q., Tahir, M., Mansoor, Q., Ismail, M., Keck, K., et al., 2016. Pharmacologically active flavonoids from the anticancer, antioxidant and antimicrobial extracts of *Cassia angustifolia* Vahl. BMC Complementary and Alternative Medicine 16 (1). Available at https://doi.org/10.1186/s12906-016-1443-z, http://www.biomedcentral.com/bmccomplementalternmed/.

Aminov, R.I., 2010. A brief history of the antibiotic era: lessons learned and challenges for the future. Frontiers in Microbiology 1. Available at https://doi.org/10.3389/fmicb.2010.00134, http://www.frontiersin.org/Journal/DownloadFile.ashx?pdf=1&FileId=1478&articleId=8040&Version=1&ContentTypeId=21&FileName=fmicb-01-00134.pdf.

Anahtar, M.N., Yang, J.H., Kanjilal, S., 2021. Applications of machine learning to the problem of antimicrobial resistance: an emerging model for translational research. Journal of Clinical Microbiology 59 (7). Available at https://doi.org/10.1128/JCM.01260-20, https://journals.asm.org/doi/10.1128/JCM.01260-20.

Arthur, M., Courvalin, P., 1993. Genetics and mechanisms of glycopeptide resistance in enterococci. Antimicrobial Agents and Chemotherapy 37 (8), 1563–1571. Available at https://doi.org/10.1128/aac.37.8.1563.

Bell, B.G., Schellevis, F., Stobberingh, E., Goossens, H., Pringle, M., 2014. A systematic review and meta-analysis of the effects of antibiotic consumption on antibiotic resistance. BMC Infectious Diseases 14 (1). Available at https://doi.org/10.1186/1471-2334-14-13, http://www.biomedcentral.com/1471-2334/14/13.

Berdy, B., Spoering, A.L., Ling, L.L., Epstein, S.S., 2017. In situ cultivation of previously uncultivable microorganisms using the ichip. Nature Protocols 12 (10), 2232–2242. Available at https://doi.org/10.1038/nprot.2017.074.

Bikard, D., Euler, C.W., Jiang, W., Nussenzweig, P.M., Goldberg, G.W., Duportet, X., et al., 2014. Exploiting CRISPR-Cas nucleases to produce sequence-specific antimicrobials. Nature Biotechnology 32 (11), 1146–1150. Available at https://doi.org/10.1038/nbt.3043.

Borris, R.P., 1996. Natural products research: perspectives from a major pharmaceutical company. Journal of Ethnopharmacology 51 (1-3), 29–38. Available at https://doi.org/10.1016/0378-8741(95)01347-4, www.elsevier.com/locate/jethpharm.

Borugă, O., Jianu, C., Mişcă, C., Goleţ, I., Gruia, A.T., Horhat, F.G., 2014. Thymus vulgaris essential oil: chemical composition and antimicrobial activity. Journal of Medicine and Life 7, 56–60.

Brooks, B.R., Brooks, C.L., Mackerell, A.D., Nilsson, L., Petrella, R.J., Roux, B., et al., 2009. CHARMM: the biomolecular simulation program. Journal of Computational Chemistry 30 (10), 1545–1614. Available at https://doi.org/10.1002/jcc.21287.

Brüssow, H., 2024. The antibiotic resistance crisis and the development of new antibiotics. Microbial Biotechnology 17 (7). Available at https://doi.org/10.1111/1751-7915.14510.

Bush, K., Bradford, P.A., 2016. β-Lactams and β-lactamase inhibitors: an overview. Cold Spring Harbor Perspectives in Medicine 6 (8), a025247. Available at https://doi.org/10.1101/cshperspect.a025247.

Cao, Y.H., Cai, W.J., He, X.W., Song, H.L., Gao, J., Yang, Y.L., et al., 2024. A review of advances & potential of applying nanomaterials for biofilm inhibition. npj Clean Water 7 (1). Available at https://doi.org/10.1038/s41545-024-00423-5, https://www.nature.com/npjcleanwater/.

Cervantes-Alagón, S.L., Padilla-Cáceres, T.C., Caballero-Apaza, L.M., Flores-Atencio, R., Mamani-Zambrano, R., 2023. Antibacterial effect of the ethanolic extract of *Curcuma longa* L. and *Erythroxylum coca* var. Coca at different concentrations against *Streptococcus mutans*: comparative study in vitro. Smart Innovation, Systems and Technologies 353, 530–539. Available at https://doi.org/10.1007/978-3-031-31007-2_50, https://www.springer.com/series/8767.

Chen, K., Pachter, L., 2005. Bioinformatics for whole-genome shotgun sequencing of microbial communities. PLoS Computational Biology 1 (2), e24. Available at https://doi.org/10.1371/journal.pcbi.0010024.

Choi, E.J., Nam, S.-J., Paul, L., Beatty, D., Kauffman, C.A., Jensen, P.R., et al., 2015. Previously uncultured marine bacteria linked to novel alkaloid production. Chemistry & Biology 22 (9), 1270–1279. Available at https://doi.org/10.1016/j.chembiol.2015.07.014.

Chopra, I., Roberts, M., 2001. Tetracycline antibiotics: mode of action, applications, molecular biology, and epidemiology of bacterial resistance. Microbiology and Molecular Biology Reviews 65 (2), 232–260. Available at https://doi.org/10.1128/mmbr.65.2.232-260.2001.

Comeau, A.M., Tétart, F., Trojet, S.N., Prère, M.-F., Krisch, H.M., 2007. Debbie fox, phage-antibiotic synergy (PAS): β-Lactam and quinolone antibiotics stimulate virulent phage growth. PLoS One 2 (8), e799. Available at https://doi.org/10.1371/journal.pone.0000799.

Cowan, D.A., Ramond, J.B., Makhalanyane, T.P., De Maayer, P., 2015. Metagenomics of extreme environments. Current Opinion in Microbiology 25, 97–102. Available at https://doi.org/10.1016/j.mib.2015.05.005, http://www.elsevier.com/locate/mib.

Czajkowski, R., Jafra, S., 2009. Quenching of acyl-homoserine lactone-dependent quorum sensing by enzymatic disruption of signal molecules. Acta Biochimica Polonica 56 (1), 1–16. Available at https://doi.org/10.18388/abp.2009_2512, http://www.actabp.pl/pdf/1_2009/1.pdf.

Dai, C., Liu, Y., Lv, F., Cheng, P., Qu, S., 2025. An alternative approach to combat multidrug-resistant bacteria: new insights into traditional Chinese medicine monomers combined with antibiotics. Advanced Biotechnology 3 (1). Available at https://doi.org/10.1007/s44307-025-00059-7.

De, R., 2019. Metagenomics: aid to combat antimicrobial resistance in diarrhea. Gut Pathogens 11 (1). Available at https://doi.org/10.1186/s13099-019-0331-8.

Defoirdt, T., Boon, N., Bossier, P., Verstraete, W., 2004. Disruption of bacterial quorum sensing: an unexplored strategy to fight infections in aquaculture. Aquaculture (Amsterdam, Netherlands) 240 (1-4), 69–88. Available at https://doi.org/10.1016/j.aquaculture.2004.06.031.

Defoirdt, T., Brackman, G., Coenye, T., 2013. Quorum sensing inhibitors: how strong is the evidence? Trends in Microbiology 21 (12), 619–624. Available at https://doi.org/10.1016/j.tim.2013.09.006.

Diallo, K., Dublanchet, A., 2022. Benefits of combined phage–antibiotic therapy for the control of antibiotic-resistant bacteria: a literature review. Antibiotics 11 (7), 839. Available at https://doi.org/10.3390/antibiotics11070839.

Dong, Y.H., Zhang, L.H., 2005. Quorum sensing and quorum-quenching enzymes. Journal of Microbiology 43, 101–109.

Duffey, M., Jumde, R.P., da Costa, R.M.A., Ropponen, H.K., Blasco, B., Piddock, L.J.V., 2024. Extending the potency and lifespan of antibiotics: inhibitors of gram-negative bacterial efflux pumps. ACS Infectious Diseases 10 (5), 1458–1482. Available at https://doi.org/10.1021/acsinfecdis.4c00091, http://pubs.acs.org/journal/aidcbc.

Ehrlich, J., Bartz, Q.R., Smith, R.M., Joslyn, D.A., Burkholder, P.R., 1947. Chloromycetin, a new antibiotic from a soil actinomycete. Science (New York, N.Y.) 106 (2757), 417. Available at https://doi.org/10.1126/science.106.2757.417 417.

El-Zahar, K.M., Al-Jamaan, M.E., Al-Mutairi, F.R., Al-Hudiab, A.M., Al-Einzi, M.S., Mohamed, A.A.Z., 2022. Antioxidant, antibacterial, and antifungal activities of the ethanolic extract obtained from berberis vulgaris roots and leaves. Molecules (Basel, Switzerland) 27 (18). Available at https://doi.org/10.3390/molecules27186114, http://www.mdpi.com/journal/molecules.

Farr, B.M., Salgado, C.D., Karchmer, T.B., Sherertz, R.J., 2001. Can antibiotic-resistant nosocomial infections be controlled? The Lancet Infectious Diseases 1 (1), 38–45. Available at https://doi.org/10.1016/s1473-3099(01)00020-2.

Fetzner, S., 2015. Quorum quenching enzymes. Journal of Biotechnology 201, 2–14. Available at https://doi.org/10.1016/j.jbiotec.2014.09.001.

Fiers, W.D., Craighead, M., Singh, I., 2017. Teixobactin and its analogues: a new hope in antibiotic discovery. ACS Infectious Diseases 3 (10), 688–690. Available at https://doi.org/10.1021/acsinfecdis.7b00108, http://pubs.acs.org/journal/aidcbc.

Fontoura, I., Veriato, T.S., Raniero, L.J., Castilho, M.L., 2023. Analysis of capped silver nanoparticles combined with imipenem against different susceptibility profiles of *Klebsiella pneumoniae*. Antibiotics 12 (3), 535. Available at https://doi.org/10.3390/antibiotics12030535.

Fyfe, C., Grossman, T.H., Kerstein, K., Sutcliffe, J., 2016. Resistance to macrolide antibiotics in public health pathogens. Cold Spring Harbor Perspectives in Medicine 6 (10). Available at https://doi.org/10.1101/cshperspect.a025395.

Gholizadeh, P., Köse, Ş., Dao, S., Ganbarov, K., Tanomand, A., Dal, T., et al., 2020. How CRISPR-Cas system could be used to combat antimicrobial resistance. Infection and Drug Resistance 13, 1111–1121. Available at https://doi.org/10.2147/IDR.S247271, https://www.dovepress.com/getfile.php?fileID=57487.

Golkar, Z., Bagasra, O., Pace, D.G., 2014. Bacteriophage therapy: a potential solution for the antibiotic resistance crisis. The Journal of Infection in Developing Countries 8 (02), 129–136. Available at https://doi.org/10.3855/jidc.3573.

Gould, I.M., Bal, A.M., 2013. New antibiotic agents in the pipeline and how hey can help overcome microbial resistance. Virulence 4 (2), 185–191. Available at https://doi.org/10.4161/viru.22507, http://www.landesbioscience.com/journals/virulence/2012VIRULENCE0008R.pdf?nocache=899198480.

Griffith, R.S., 1981. Introduction to vancomycin. Reviews of Infectious Diseases 3, S200. Available at https://doi.org/10.1093/clinids/3.Supplement_2.S200.

Hamburger, M., Hostettmann, K., 1991. 7. Bioactivity in plants: the link between phytochemistry and medicine. Phytochemistry 30 (12), 3864–3874. Available at https://doi.org/10.1016/0031-9422(91)83425-K.

Hashemian, S.M.R., Farhadi, T., Ganjparvar, M., 2018. Linezolid: a review of its properties, function, and use in critical care. Drug Design, Development and Therapy 12, 1759–1767. Available at https://doi.org/10.2147/DDDT.S164515, https://www.dovepress.com/getfile.php?fileID=42664.

Hess, B., Kutzner, C., Spoel, D.V.D., Lindahl, E., 2008. GRGMACS 4: algorithms for highly efficient, load-balanced, and scalable molecular simulation. Journal of Chemical Theory and Computation 4 (3), 435–447. Available at https://doi.org/10.1021/ct700301qGermany, http://pubs.acs.org/doi/pdf/10.1021/ct700301q.

Hollingsworth, S.A., Dror, R.O., 2018. Molecular dynamics simulation for all. Neuron 99 (6), 1129–1143. Available at https://doi.org/10.1016/j.neuron.2018.08.011, http://www.cell.com/neuron/home.

Hooper, D.C., Jacoby, G.A., 2015. Mechanisms of drug resistance: quinolone resistance. Annals of the New York Academy of Sciences 1354 (1), 12–31. Available at https://doi.org/10.1111/nyas.12830, http://www.blackwellpublishing.com/0077-8923.

Imanshahidi, M., Hosseinzadeh, H., 2008. Pharmacological and therapeutic effects of Berberis vulgaris and its active constituent, berberine. Phytotherapy Research 22 (8), 999–1012. Available at https://doi.org/10.1002/ptr.2399.

Iskandar, K., Murugaiyan, J., Hammoudi Halat, D., Hage, S.E., Chibabhai, V., Adukkadukkam, S., et al., 2022. Antibiotic discovery and resistance: the chase and the race. Antibiotics 11 (2), 182. Available at https://doi.org/10.3390/antibiotics11020182.

Jad, Y.E., Acosta, G.A., Naicker, T., Ramtahal, M., El-Faham, A., Govender, T., et al., 2015. Synthesis and biological evaluation of a teixobactin analogue. Organic Letters 17 (24), 6182–6185. Available at https://doi.org/10.1021/acs.orglett.5b03176, http://pubs.acs.org/journal/orlef7.

Jones, D., Metzger, H.J., Schatz, A., Waksman, S.A., 1944. Control of gram-negative bacteria in experimental animals by streptomycin. Science (New York, N.Y.) 100 (2588), 103–105. Available at https://doi.org/10.1126/science.100.2588.103.

Kalia, V.C., 2013. Quorum sensing inhibitors: an overview. Biotechnology Advances 31 (2), 224–245. Available at https://doi.org/10.1016/j.biotechadv.2012.10.004.

Khaledi, A., Weimann, A., Schniederjans, M., Asgari, E., Kuo, T.-H., Oliver, A., et al., 2020. Predicting antimicrobial resistance in *Pseudomonas aeruginosa* with machine learning-enabled molecular diagnostics. EMBO Molecular Medicine 12 (3). Available at https://doi.org/10.15252/emmm.201910264.

Khanchandani, N., Shah, P., Kalwani, T., Ardeshna, A., Dharajiya, D., 2019. Antibacterial and antifungal activity of Ashwagandha (*Withania somnifera* L.): a review. Journal of Drug Delivery and Therapeutics 9 (5-s), 154–161. Available at https://doi.org/10.22270/jddt.v9i5-s.3573.

Khardori, N., Stevaux, C., Ripley, K., 2020. Antibiotics: from the Beginning to the future: Part 2. The Indian Journal of Pediatrics 87 (1), 43–47. Available at https://doi.org/10.1007/s12098-019-03113-0.

Kim, C., Holm, M., Frost, I., Hasso-Agopsowicz, M., Abbas, K., 2023. Global and regional burden of attributable and associated bacterial antimicrobial resistance avertable by vaccination: modelling study. BMJ Global Health 8 (7), e011341. Available at https://doi.org/10.1136/bmjgh-2022-011341.

Kinghorn, A.D., Pan, L., Fletcher, J.N., Chai, H., 2011. The relevance of higher plants in lead compound discovery programs. Journal of Natural Products 74 (6), 1539–1555. Available at https://doi.org/10.1021/np200391c.

Krause, K.M., Serio, A.W., Kane, T.R., Connolly, L.E., 2016. Aminoglycosides: an overview. Cold Spring Harbor Perspectives in Medicine 6 (6). Available at https://doi.org/10.1101/cshperspect.a027029, http://perspectivesinmedicine.cshlp.org/content/6/6/a027029.full.pdf.

Lade, H., Paul, D., Kweon, J.H., 2014. Quorum quenching mediated approaches for control of membrane biofouling. International Journal of Biological Sciences 10 (5), 547–562. Available at https://doi.org/10.7150/ijbs.9028, http://www.ijbs.com/v10p0547.pdf.

Lee, C.R., Cho, I.H., Jeong, B.C., Lee, S.H., 2013. Strategies to minimize antibiotic resistance. International Journal of Environmental Research and Public Health 10 (9), 4274–4305. Available at https://doi.org/10.3390/ijerph10094274, http://www.mdpi.com/1660-4601/10/9/4274/pdf.

Lesher, G.Y., Froelich, E.J., Gruett, M.D., Bailey, J.H., Brundage, R.P., 1962. 1,8-Naphthyridine derivatives. a new class of chemotherapeutic agents. Journal of Medicinal and Pharmaceutical Chemistry 5 (5), 1063–1065. Available at https://doi.org/10.1021/jm01240a021.

Lewis, K., Ausubel, F.M., 2006. Prospects for plant-derived antibacterials. Nature Biotechnology 24 (12), 1504–1507. Available at https://doi.org/10.1038/nbt1206-1504.

Lewis, K., 2012. Recover the lost art of drug discovery. Nature 485 (7399), 439–440. Available at https://doi.org/10.1038/485439a.

Lewis, K., 2013. Platforms for antibiotic discovery. Nature Reviews Drug Discovery 12 (5), 371–387. Available at https://doi.org/10.1038/nrd3975.

Li, C., Sutherland, D., Hammond, S.A., Yang, C., Taho, F., Bergman, L., et al., 2022. AMPlify: attentive deep learning model for discovery of novel antimicrobial peptides effective against WHO priority pathogens. BMC Genomics 23 (1). Available at https://doi.org/10.1186/s12864-022-08310-4.

Łusiak-Szelachowska, M., Weber-Dąbrowska, B., Górski, A., 2020. Bacteriophages and lysins in biofilm control. Virologica Sinica 35 (2), 125–133. Available at https://doi.org/10.1007/s12250-019-00192-3, http://www.springerlink.com/content/120859/.

Liu, C., Hong, Q., Chang, R.Y.K., Kwok, P.C.L., Chan, H.-K., 2022. Phage–antibiotic therapy as a promising strategy to combat multidrug-resistant infections and to enhance antimicrobial efficiency. Antibiotics 11 (5), 570. Available at https://doi.org/10.3390/antibiotics11050570.

Liu, H., Li, H., Liang, Y., Du, X., Yang, C., Yang, L., et al., 2020. Phage-delivered sensitisation with subsequent antibiotic treatment reveals sustained effect against antimicrobial resistant bacteria. Theranostics 10 (14), 6310–6321. Available at https://doi.org/10.7150/thno.42573.

Makabenta, J.M.V., Nabawy, A., Li, C.H., Schmidt-Malan, S., Patel, R., Rotello, V.M., 2021. Nanomaterial-based therapeutics for antibiotic-resistant bacterial infections. Nature Reviews Microbiology 19 (1), 23–36. Available at https://doi.org/10.1038/s41579-020-0420-1, http://www.nature.com/nrmicro/index.html.

Mayorga-Ramos, A., Zúñiga-Miranda, J., Carrera-Pacheco, S.E., Barba-Ostria, C., Guamán, L.P., 2023. CRISPR-Cas-based antimicrobials: design, challenges, and bacterial mechanisms of resistance. ACS Infectious Diseases 9 (7), 1283–1302. Available at https://doi.org/10.1021/acsinfecdis.2c00649.

McGuire, R.J., Yu, S.C., Payne, P.R.O., Lai, A.M., Vazquez-Guillamet, M.C., Kollef, M.H., et al., 2021. A pragmatic machine learning model to predict carbapenem resistance. Antimicrobial Agents and Chemotherapy 65 (7). Available at https://doi.org/10.1128/AAC.00063-21, https://journals.asm.org/doi/10.1128/AAC.00063-21.

Muteeb, G., Rehman, M.T., Pani, B., Khan, R.H., 2023. Editorial: novel drug-designing approaches to combat antimicrobial resistance. Frontiers in Molecular Biosciences 10. Available at https://doi.org/10.3389/fmolb.2023.1342702 Available from: journal.frontiersin.org/journal/molecular-biosciences.

Naganawa, R., Iwata, N., Ishikawa, K., Fukuda, H., Fujino, T., Suzuki, A., 1996. Inhibition of microbial growth by ajoene, a sulfur-containing compound derived from garlic. Applied and Environmental Microbiology 62 (11), 4238–4242. Available at https://doi.org/10.1128/aem.62.11.4238-4242.1996.

Namdar, N., Nayeri Fasaei, B., Shariati, P., Joghataei, S.M., Arpanaei, A., 2024. *Mesoporous silica* nanoparticles co-loaded with lysozyme and vancomycin for synergistic antimicrobial action. Scientific Reports 14 (1). Available at https://doi.org/10.1038/s41598-024-78922-1.

Newman, D.J., Cragg, G.M., 2012. Natural products as sources of new drugs over the 30 years from 1981 to 2010. Journal of Natural Products 75 (3), 311–335. Available at https://doi.org/10.1021/np200906s.

Nigussie, G., Melak, H., Endale, M., 2021. Traditional medicinal uses, phytochemicals, and pharmacological activities of genus rhamnus: a review. Journal of the Turkish Chemical Society, Section A: Chemistry 8 (3), 899–932. Available at https://doi.org/10.18596/jotcsa.929188, https://dergipark.org.tr/en/download/article-file/1737984.

Phillips, J.C., Hardy, D.J., Maia, J.D.C., Stone, J.E., Ribeiro, J.V., Bernardi, R.C., et al., 2020. Scalable molecular dynamics on CPU and GPU architectures with NAMD. The Journal of Chemical Physics 153 (4). Available at https://doi.org/10.1063/5.0014475.

Poirel, L., Jayol, A., Nordmann, P., 2017. Polymyxins: antibacterial activity, susceptibility testing, and resistance mechanisms encoded by plasmids or chromosomes. Clinical Microbiology Reviews 30 (2), 557–596. Available at https://doi.org/10.1128/cmr.00064-16.

Pulingam, T., Thong, K.L., Appaturi, J.N., Nordin, N.I., Dinshaw, I.J., Lai, C.W., et al., 2020. Synergistic antibacterial actions of graphene oxide and antibiotics towards bacteria and the toxicological effects of graphene oxide on human epidermal keratinocytes. European Journal of Pharmaceutical Sciences 142, 105087. Available at https://doi.org/10.1016/j.ejps.2019.105087.

Radwan, I.T., El-Sherbiny, I.M., Selim, A.M., Metwally, N.H., 2024. Design, synthesis of some novel coumarins and their nanoformulations into lipid-chitosan nanocapsule as unique antimicrobial agents. Scientific Reports 14 (1). Available at https://doi.org/10.1038/s41598-024-79861-7, https://www.nature.com/srep/.

Ramli, S., Xian, W.J., Abd Mutalib, N.A., 2020. A review: antibacterial activities, antioxidant properties and toxicity profile of centella asiatica. EDUCATUM Journal Of Science, Mathematics And Technology 7 (1), 39–47. Available at https://doi.org/10.37134/ejsmt.vol7.1.5.2020.

Ren, Y., Chakraborty, T., Doijad, S., Falgenhauer, L., Falgenhauer, J., Goesmann, A., et al., 2022. Prediction of antimicrobial resistance based on whole-genome sequencing and machine learning. Bioinformatics (Oxford, England) 38 (2), 325–334. Available at https://doi.org/10.1093/bioinformatics/btab681.

Rodríguez-Pérez, C., Quirantes-Piné, R., Uberos, J., Jiménez-Sánchez, C., Peña, A., Segura-Carretero, A., 2016. Antibacterial activity of isolated phenolic compounds from cranberry (*Vaccinium macrocarpon*) against *Escherichia coli*. Food & Function 7 (3), 1564–1573. Available at https://doi.org/10.1039/c5fo01441g.

Roh, C., Villatte, F., 2008. Isolation of a low-temperature adapted lipolytic enzyme from uncultivated micro-organism. Journal of Applied Microbiology 105 (1), 116–123. Available at https://doi.org/10.1111/j.1365-2672.2007.03717.x.

Sakanaka, S., Kim, M., Taniguchi, M., Yamamoto, T., 1989. Antibacterial substances in Japanese green tea extract against streptococcus mutans, a cariogenic bacterium. Agricultural and Biological Chemistry 53 (9), 2307–2311. Available at https://doi.org/10.1080/00021369.1989.10869665.

Salama, S.A., Essam, D., Tagyan, A.I., Farghali, A.A., Khalil, E.M., Abdelaleim, Y.F., et al., 2024. Novel composite of nano zinc oxide and nano propolis as antibiotic for antibiotic-resistant bacteria: a promising approach. Scientific Reports 14 (1). Available at https://doi.org/10.1038/s41598-024-70490-8, https://www.nature.com/srep/.

Salomon-Ferrer, R., Case, D.A., Walker, R.C., 2013. An overview of the Amber biomolecular simulation package. Wiley Interdisciplinary Reviews: Computational Molecular Science 3 (2), 198–210. Available at https://doi.org/10.1002/wcms.1121.

Schatz, A., Bugle, E., Waksman, S.A., 1944. Streptomycin, a substance exhibiting antibiotic activity against gram-positive and gram-negative bacteria. Proceedings of the Society for Experimental Biology and Medicine 55 (1), 66–69. Available at https://doi.org/10.3181/00379727-55-14461.

Schloss, P.D., Handelsman, J., 2003. Biotechnological prospects from metagenomics. Current Opinion in Biotechnology 14 (3), 303–310. Available at https://doi.org/10.1016/S0958-1669(03)00067-3, http://www.elsevier.com/locate/copbio.

Sengupta, S., Chattopadhyay, M.K., Grossart, H.P., 2013. The multifaceted roles of antibiotics and antibiotic resistance in nature. Frontiers in Microbiology 4. Available at https://doi.org/10.3389/fmicb.2013.00047, http://www.frontiersin.org/Journal/DownloadFile.ashx?pdf=1&FileId=17019&articleId=38490&Version=1&ContentTypeId=21&FileName=fmicb-04-00047.pdf.

Silver, L.L., 2011. Challenges of antibacterial discovery. Clinical Microbiology Reviews 24 (1), 71–109. Available at https://doi.org/10.1128/CMR.00030-10, http://cmr.asm.org/cgi/reprint/24/1/71.

Singer, A.C., Kirchhelle, C., Roberts, A.P., 2020. Inter)nationalising the antibiotic research and development pipeline. The Lancet Infectious Diseases 20 (2), e54. Available at https://doi.org/10.1016/S1473-3099(19)30552-3, http://www.journals.elsevier.com/the-lancet-infectious-diseases.

Stokes, J.M., Yang, K., Swanson, K., Jin, W., Cubillos-Ruiz, A., Donghia, N.M., et al., 2020. A deep learning approach to antibiotic discovery. Cell 180 (4), 688. Available at https://doi.org/10.1016/j.cell.2020.01.021, https://www.sciencedirect.com/journal/cell.

Su, Z., Xu, D., Hu, X., Zhu, W., Kong, L., Qian, Z., et al., 2024. Biodegradable oxygen-evolving metalloantibiotics for spatiotemporal sono-metalloimmunotherapy against orthopaedic biofilm infections. Nature Communications 15 (1). Available at https://doi.org/10.1038/s41467-024-52489-x.

Tacconelli, E., Carrara, E., Savoldi, A., Harbarth, S., Mendelson, M., Monnet, D.L., et al., 2018. Discovery, research, and development of new antibiotics: the WHO priority list of antibiotic-resistant bacteria and tuberculosis. The Lancet Infectious Diseases 18 (3), 318–327. Available at https://doi.org/10.1016/S1473-3099(17)30753-3, http://www.journals.elsevier.com/the-lancet-infectious-diseases.

Tan, M., Zhou, L., Huang, Y., Wang, Y., Hao, X., Wang, J., 2008. Antimicrobial activity of globulol isolated from the fruits of *Eucalyptus globulus* Labill. Natural Product Research 22 (7), 569–575. Available at https://doi.org/10.1080/14786410701592745.

Tao, S., Chen, H., Li, N., Liang, W., 2022. The application of the CRISPR-cas system in antibiotic resistance. Infection and Drug Resistance Volume 15, 4155–4168. Available at https://doi.org/10.2147/idr.s370869.

Terreni, M., Taccani, M., Pregnolato, M., 2021. New antibiotics for multidrug-resistant bacterial strains: latest research developments and future perspectives. Molecules (Basel, Switzerland) 26 (9), 2671. Available at https://doi.org/10.3390/molecules26092671.

Theuretzbacher, U., Piddock, L.J.V., 2019. Non-traditional antibacterial therapeutic options and challenges. Cell Host & Microbe 26 (1), 61–72. Available at https://doi.org/10.1016/j.chom.2019.06.004.

Tillhon, M., Guamán Ortiz, L.M., Lombardi, P., Scovassi, A.I., 2012. Berberine: new perspectives for old remedies. Biochemical Pharmacology 84 (10), 1260–1267. Available at https://doi.org/10.1016/j.bcp.2012.07.018.

Timmermans, M.L., Paudel, Y.P., Ross, A.C., 2017. Investigating the biosynthesis of natural products from marine proteobacteria: a survey of molecules and strategies. Marine Drugs 15 (8). Available at https://doi.org/10.3390/md15080235, http://www.mdpi.com/1660-3397/15/8/235/pdf.

Toda, M., Okubo, S., Ikigai, H., Suzuki, T., Suzuki, Y., Hara, Y., et al., 1992. The protective activity of tea catechins against experimental infection by *Vibrio cholerae* O1. Microbiology and Immunology 36 (9), 999–1001. Available at https://doi.org/10.1111/j.1348-0421.1992.tb02103.x.

Trivella, D.B.B., de Felicio, R., 2018. The tripod for bacterial natural product discovery: genome mining, silent pathway induction, and mass spectrometry-based molecular networking. mSystems 3 (2). Available at https://doi.org/10.1128/msystems.00160-17.

Van Goethem, M.W., Marasco, R., Hong, P.Y., Daffonchio, D., 2024. The antibiotic crisis: on the search for novel antibiotics and resistance mechanisms. Microbial Biotechnology 17 (3). Available at https://doi.org/10.1111/1751-7915.14430, http://onlinelibrary.wiley.com/journal/10.1111/(ISSN)1751-7915.

Vijaya, K., Ananthan, S., Nalini, R., 1995. Antibacterial effect of theaflavin, polyphenon 60 (*Camellia sinensis*) and *Euphorbia hirta* on *Shigella* spp. — a cell culture study. Journal of Ethnopharmacology 49 (2), 115–118. Available at https://doi.org/10.1016/0378-8741(95)90039-x.

Vohora, S.B., Rizwan, M., Khan, J.A., 1973. Medicinal uses of common Indian vegetables. Planta Medica 23 (4), 381–393. Available at https://doi.org/10.1055/s-0028-1099460.

Wadood, A., Ahmed, N., Shah, L., Ahmad, A., Hassan, H., Shams, S., 2013. In-silico drug design: an approach which revolutionarised the drug discovery process. OA Drug Design and Delivery 1 (1). Available at https://doi.org/10.13172/2054-4057-1-1-1119.

Wahab, S., Annadurai, S., Abullais, S.S., Das, G., Ahmad, W., Ahmad, M.F., et al., 2021. Glycyrrhiza glabra (Licorice): a comprehensive review on its phytochemistry, biological activities, clinical evidence and toxicology. Plants 10 (12). Available at https://doi.org/10.3390/plants10122751, https://www.mdpi.com/2223-7747/10/12/2751/pdf.

Wan, F., Wong, F., Collins, J.J., de la Fuente-Nunez, C., 2024. Machine learning for antimicrobial peptide identification and design. Nature Reviews Bioengineering 2 (5), 392–407. Available at https://doi.org/10.1038/s44222-024-00152-x.

Wan, J., Wilcock, A., Coventry, M.J., 1998. The effect of essential oils of basil on the growth of *Aeromonas hydrophila* and *Pseudomonas fluorescens*. Journal of Applied Microbiology 84 (2), 152–158. Available at https://doi.org/10.1046/j.1365-2672.1998.00338.xAustralia, https://academic.oup.com/jambio.

Wang, G., Vaisman, I.I., van Hoek, M.L., 2022. Machine learning prediction of antimicrobial peptides. Methods in Molecular Biology 2405, 1–37. Available at https://doi.org/10.1007/978-1-0716-1855-4_1, http://www.springer.com/series/7651.

Wong, F., de la Fuente-Nunez, C., Collins, J.J., 2023. Leveraging artificial intelligence in the fight against infectious diseases. Science (New York, N.Y.) 381 (6654), 164–170. Available at https://doi.org/10.1126/science.adh1114, https://www.science.org/doi/10.1126/science.adh1114.

Wright, G.D., 2014. Something old, something new: revisiting natural products in antibiotic drug discovery. Canadian Journal of Microbiology 60 (3), 147–154. Available at https://doi.org/10.1139/cjm-2014-0063.

Zaman, W., Manghwar, H., 2024. Nanoparticles and their biological applications: recent advances in 2022–2023. Microorganisms 12 (3), 489. Available at https://doi.org/10.3390/microorganisms12030489.

Zampaloni, C., Mattei, P., Bleicher, K., Winther, L., Thäte, C., Bucher, C., et al., 2024. A novel antibiotic class targeting the lipopolysaccharide transporter. Nature 625 (7995), 566–571. Available at https://doi.org/10.1038/s41586-023-06873-0, https://www.nature.com/nature/.

Zhao, J., Shakir, Y., Deng, Y., Zhang, Y., 2023. Use of modified ichip for the cultivation of thermo-tolerant microorganisms from the hot spring. BMC Microbiology 23 (1). Available at https://doi.org/10.1186/s12866-023-02803-2.

Ziemert, N., Alanjary, M., Weber, T., 2016. The evolution of genome mining in microbes-a review. Natural Product Reports 33 (8), 988–1005. Available at https://doi.org/10.1039/c6np00025h, http://www.rsc.org/Publishing/Journals/NP/index.asp.

Chapter 11

Beyond antibiotics: exploring new therapies for resistance challenges

Tuhina Banerjee and Pue Rakshit
Department of Microbiology, Institute of Medical Sciences, Banaras Hindu University, Varanasi, Uttar Pradesh, India

11.1 Introduction

As the era of the miracle compound named "antibiotics" fades amidst the global surge of antimicrobial resistance (AMR), management of infectious diseases becomes a real challenge (Laxminarayan et al., 2013). In response, a new frontier in antimicrobial therapy is emerging, one that moves beyond the traditional paradigm of microbial eradication. Contemporary strategies aim not merely to kill pathogens, but to disarm, reprogramme, and outsmart them by targeting virulence, enhancing host defences, restoring microbial balance, and preventing infection altogether (Allen et al., 2014). These next-generation, nonantibiotic approaches offer renewed hope by minimising selective pressure for resistance while retaining clinical efficacy against already resistant organisms (Czaplewski et al., 2016). This chapter examines innovative therapeutic modalities that are poised to redefine infection management beyond antibiotics. These approaches reflect a strategic transition from broad-spectrum antimicrobial pressure to targeted, evolution-aware interventions for durable infectious disease control.

11.2 Downfall of the magic bullets: extent of the problem

Antibiotics, widely regarded as the 'magic bullets' of modern medicine, represent one of the most transformative therapeutic breakthroughs of the 20th century. Their discovery and widespread clinical application revolutionised the treatment of infectious diseases, dramatically reducing morbidity and mortality and enabling advances in surgery, oncology, transplantation, and intensive care (Aminov, 2010). For decades, antibiotics provided reliable and effective control over bacterial infections that were once considered fatal, fundamentally reshaping public health and life expectancy worldwide.

However, the extraordinary success of antibiotics has been undermined by their sustained overuse and misuse across human medicine, veterinary practice, agriculture, and environmental settings. Inappropriate prescribing, suboptimal dosing, self-medication, and extensive nontherapeutic use in animal husbandry have collectively exerted intense selective pressure on bacterial populations (Martin, 2018). As a result, bacteria have rapidly evolved and disseminated a wide array of resistance mechanisms, resulting in enzymatic drug inactivation, target modification, reduced permeability, and active efflux. The ability of bacteria to exchange resistance determinants via horizontal gene transfer has further accelerated the global spread of AMR (Mishra et al., 2025). The extent of the problem can be estimated by the fact shown in Fig. 11.1, where it is evident that AMR genes were discovered almost in parallel with the discovery of antibiotics with respect to carbapenemases (Nordmann et al., 2011). Additionally, the wide diversity of resistance genes against a particular group of antibiotics has also been highlighted. The emergence and proliferation of multidrug-resistant and pan-resistant pathogens, often referred to as 'superbugs', have transformed AMR into a silent but pervasive pandemic. Infections caused by resistant organisms are associated with prolonged illness, increased risk of treatment failure, and significantly higher mortality rates compared with susceptible infections (Bonomo et al., 2018). A list of common pathogens recently prioritised by the World Health Organisation (WHO) has been provided in Fig. 11.2.

From a healthcare perspective, AMR imposes a substantial economic burden, leading to longer hospital stays, increased need for expensive or toxic second-line therapies, and greater utilisation of healthcare resources (Murray et al., 2022).

Antimicrobial Resistance in the 21st Century. DOI: https://doi.org/10.1016/B978-0-443-44748-8.00023-3

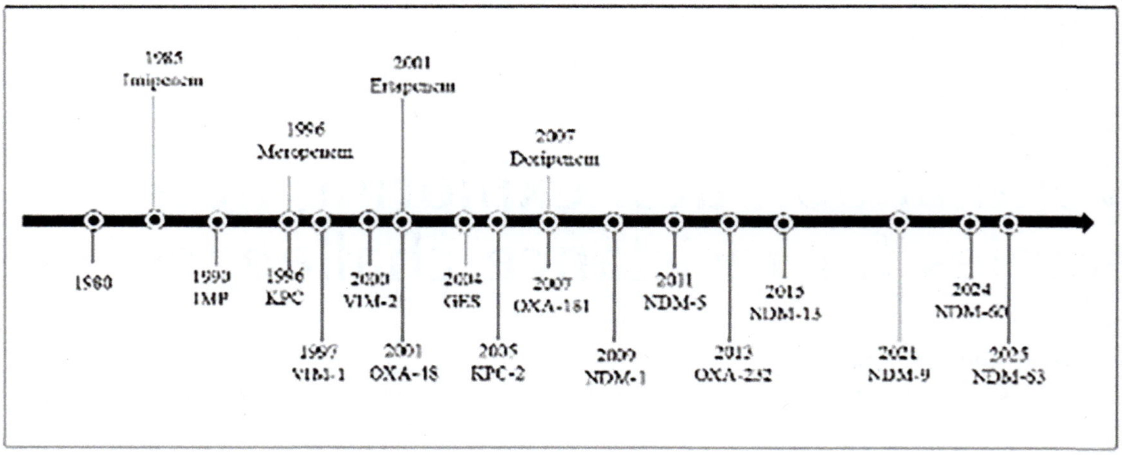

FIGURE 11.1 Carbapenems and the global emergence of carbapenemases.
Timeline depicting the introduction of key carbapenem antibiotics and the subsequent global emergence of major carbapenemases, highlighting the progressive evolution of carbapenem resistance from 1980.

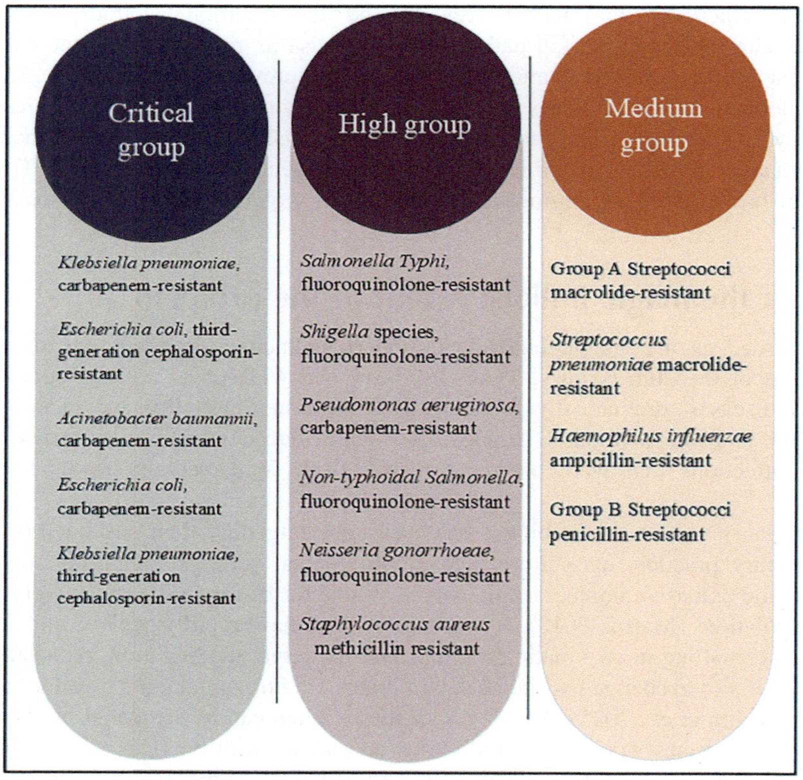

FIGURE 11.2 WHO priority bacterial pathogens driving the global antimicrobial resistance crisis.
WHO priority bacterial pathogens categorised into critical, high, and medium groups based on the urgency of new antibiotic development.

Beyond clinical settings, AMR threatens food security, animal health, and environmental stability, highlighting its complex and interconnected nature as shown in Fig. 11.3. Critically, the rise of AMR jeopardises many routine and life-saving medical procedures that rely on effective antimicrobial prophylaxis and treatment (Robinson et al., 2016). Without

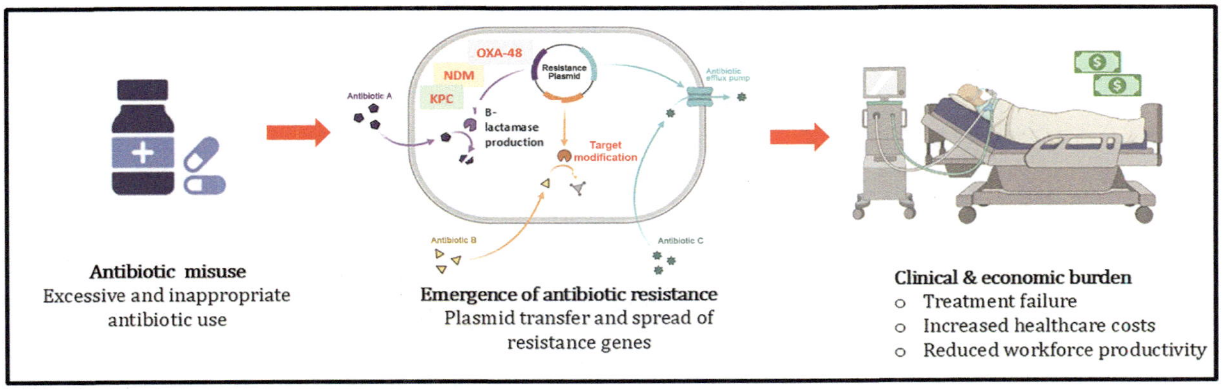

FIGURE 11.3 Complex and interconnected nature of antimicrobial resistance in various environments.
Antibiotic misuse accelerates the emergence and spread of resistant bacteria through gene transfer, which leads to treatment failure, prolonged illness, and increased healthcare burden.

urgent and coordinated action, the world risks entering a postantibiotic era in which minor infections and routine injuries could once again become life-threatening. Addressing this global crisis requires sustained commitment to antimicrobial stewardship, infection prevention, surveillance, and innovation. Preserving the efficacy of existing antibiotics while accelerating the development of alternative and next-generation therapies is essential to mitigate the growing threat posed by AMR (Tacconelli et al., 2018).

Among the most promising frontiers in antimicrobial innovation are precision-targeted therapies that exploit pathogen-specific vulnerabilities. Bacteriophage therapy and phage-derived lytic enzymes, such as endolysins and depolymerases, offer highly specific antibacterial activity by selectively targeting and eliminating pathogenic bacteria without disrupting the commensal microbiota (Abedon et al., 2011). Similarly, clustered regularly interspaced short palindromic repeats (CRISPR) based antimicrobials represent a transformative approach by enabling sequence-specific targeting of AMR genes, virulence determinants, or essential genomic regions (Bikard et al., 2014). By selectively cleaving resistance-conferring plasmids or chromosomal loci, these systems can re-sensitise bacteria to existing antibiotics or induce targeted bacterial killing, thereby minimising collateral damage and resistance spread. Antivirulence strategies constitute another innovative avenue by focusing on disarming pathogens rather than eradicating them. By inhibiting quorum sensing, toxin production, adhesion, or biofilm formation, antivirulence agents attenuate bacterial pathogenicity and enhance host immune clearance while exerting minimal selective pressure (Rasko and Sperandio, 2010; Dickey et al., 2017). Simultaneously, host-directed therapies focus on enhancing innate and adaptive immune responses or modulating host–pathogen interactions to improve infection outcomes. Such approaches offer the added advantage of reduced susceptibility to resistance, as they do not directly target bacterial viability (Zumla et al., 2016). The individual alternatives are being described briefly with respect to their mechanism, application, and challenges.

11.3 Bacteriophages

Bacteriophages are obligate intracellular viruses that exclusively parasitise prokaryotic hosts, primarily bacteria, by adsorbing to specific surface receptors and injecting their genome. Discovered over a century ago, these highly specific agents exhibit remarkable host specificity, enabling them to selectively infect and lyse pathogenic bacteria while sparing beneficial commensal microbiota, thereby preserving microbial homoeostasis (Sulakvelidze et al., 2001; Abedon et al., 2011). The antibacterial activity of lytic bacteriophages is mediated through highly specific interactions between phage surface receptors and bacterial cell surface structures, followed by intracellular replication and host cell lysis. This targeted mechanism reduces collateral damage to nontarget organisms and minimises dysbiosis, a common consequence of antibiotic therapy (Kortright et al., 2019) shown in Fig. 11.4.

A randomised controlled trial evaluating topical bacteriophage therapy for burn wound infections caused by *Escherichia coli* and *Pseudomonas aeruginosa* had demonstrated good safety and tolerability but lower than expected efficacy (Jault et al., 2019). Another study had shown phage therapy resulting in a significant reduction in bacterial load and clinical symptoms, with no serious adverse effects reported (Wright et al., 2009). A trial from Georgia had shown that phage therapy was safe but not superior to antibiotics, though it showed comparable efficacy with fewer microbiome

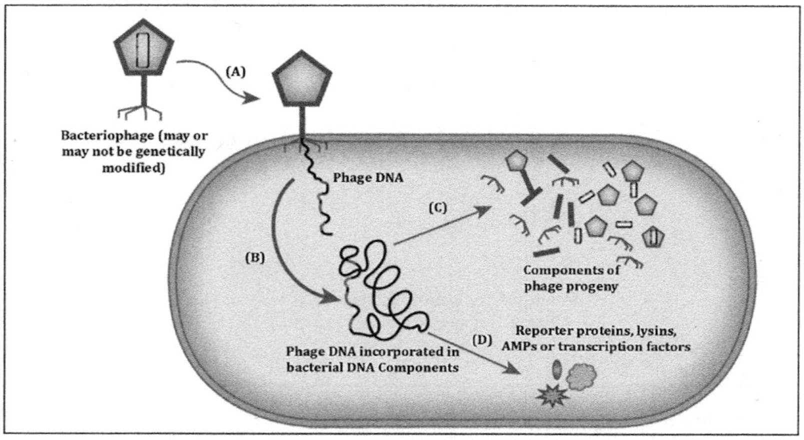

FIGURE 11.4 Principle of Bacteriophage therapy. (A) Natural or genetically modified phages selectively targets bacteria. (B) The phage genome is integrated into the bacterial genome for further replication. (C) Translation into progeny phage particles, which assemble and burst the cell. (D) Expression of reporter proteins, toxins, lysins, and antimicrobial peptides resulting in host cell damage.

disturbances when applied in urinary tract infection patients (Leitner et al., 2021). Another study from India had shown that bacteriophage therapy significantly reduced bacterial load and mortality in infected mice, showing efficacy comparable to or superior to antibiotics. It established that phages were safe, host-specific, and effective against multidrug-resistant bacteria, supporting their potential as an alternative antimicrobial strategy (Nath et al., 2019).

Importantly, phage therapy exerts lower selective pressure for resistance compared to antibiotics, as phage–bacteria coevolution drives continuous adaptation, allowing phages to evolve alongside resistant bacterial variants. Moreover, phages can be engineered or selected to overcome bacterial defence systems, further enhancing their therapeutic potential (Dedrick et al., 2019). Beyond direct bacterial killing, bacteriophages exhibit additional advantages that strengthen their role in combating multidrug-resistant (MDR) infections. Phages can penetrate and disrupt bacterial biofilms, a major contributor to chronic and device-associated infections, either directly or through the production of depolymerising enzymes (Chan et al., 2013). Phage-derived proteins, such as endolysins and polysaccharide depolymerases, have also shown potent antibacterial activity and can be developed as standalone therapeutics or used synergistically with antibiotics (Fischetti, 2018). Such combination strategies have demonstrated the ability to restore antibiotic susceptibility in resistant strains and improve treatment outcomes. Advances in genomics, synthetic biology, and bioinformatics have further accelerated the development of phage-based therapeutics. Whole-genome sequencing enables precise characterisation of phage genomes to ensure safety by excluding lysogenic or toxin-encoding genes, while bioengineering approaches allow the modification of phages for enhanced host range, stability, and delivery (Pires et al., 2016). Despite these advantages, challenges related to regulatory approval, large-scale production, immune clearance, and personalised phage selection remain significant barriers to widespread clinical implementation (Abedon, 2014).

Nevertheless, the precision, adaptability, and evolving nature of bacteriophages position them as powerful tools in next-generation antimicrobial strategies. Continued interdisciplinary research, coupled with robust clinical trials and supportive regulatory frameworks, will be essential to fully realise the potential of phage therapy in addressing the global threat posed by multidrug-resistant pathogens.

11.4 Antimicrobial peptides

Antimicrobial peptides (AMPs) are short, naturally occurring peptides produced by microorganisms as part of the innate immune system that protect against bacteria, fungi, viruses, and parasites (Hancock and Sahl, 2006). The antimicrobial activity of AMPs is primarily mediated through interactions with microbial cell membranes. Due to their positive charge, AMPs preferentially associate with negatively charged bacterial membranes, leading to membrane destabilisation, pore formation, and loss of membrane integrity. This results in rapid cell death through leakage of intracellular contents and dissipation of membrane potential. Several mechanistic models, including the barrel-stave, toroidal pore, and carpet models, have been proposed to explain AMP–membrane interactions (Matsuzaki, 2009). Importantly, membrane-targeted mechanisms reduce the likelihood of resistance development, as substantial alterations in membrane composition or charge impose significant fitness costs on bacteria. In addition to membrane disruption, many AMPs exhibit intracellular modes of action. After translocation across the membrane, AMPs can inhibit critical cellular processes such as DNA and RNA synthesis, protein translation, enzymatic activity, and cell wall biosynthesis (Nguyen, 2011). Certain peptides also interfere with quorum sensing, suppress virulence factor expression, and inhibit biofilm formation, thereby attenuating pathogenicity rather than solely exerting bactericidal effects. Moreover, AMPs possess immunomodulatory properties, including chemotactic activity, enhancement of phagocytosis, neutralisation of endotoxins, and regulation of inflammatory responses, which collectively contribute to improved host defence and tissue repair (Mookherjee, 2007). AMPs demonstrate broad-spectrum activity against Gram-positive and Gram-negative bacteria, fungi, viruses, and protozoa, including multidrug-resistant and biofilm-associated pathogens (Hancock and Sahl, 2006). Their rapid bactericidal activity and multitargeted mechanisms make them particularly effective against persistent bacterial populations. Furthermore, synergistic interactions between AMPs and conventional antibiotics have been reported, enabling dose reduction and restoration of antibiotic susceptibility (Makovitzki, 2006). Representative natural, antibiotic-derived, and synthetic AMPs, along with their sources and mechanism of action, are presented in Table 11.1.

AMPs such as pexiganan have shown broad-spectrum antibacterial activity and good safety in treating infected diabetic foot ulcers, though efficacy was comparable to standard antibiotics. They had effective fungicidal activity with low host toxicity, supporting their use against resistant fungal infection. (Hancock and Sahl, 2006) Another clinical study showed that a modified histatin-derived AMP significantly reduced pathogenic load and improved clinical status (Wang et al., 2018). A 4-week randomised clinical trial on oral spray containing P-113 demonstrated lower counts of *streptococci* and reduced dental plague, suggesting utility in oral microbial control (Wu et al., 2024).

Despite their therapeutic promise, clinical application of AMPs has been challenged by issues such as proteolytic instability, cytotoxicity, and limited bioavailability. Advances in peptide engineering, including sequence optimisation, cyclisation, incorporation of nonnatural amino acids, and development of peptidomimetics, have significantly improved stability, selectivity, and pharmacokinetic properties (Hancock and Sahl, 2006). Novel formulation and delivery strategies, such as nanoparticle-based systems and targeted delivery platforms, have further enhanced their translational potential.

11.5 Clustered regularly interspaced short palindromic repeats

The rapid global rise of AMR has created an urgent need for innovative therapeutic strategies that move beyond conventional antibiotics. In this context, CRISPR-based technologies have emerged as highly promising gene-targeted weapons capable of selectively combating resistant bacterial pathogens (Citorik et al., 2014). Originally discovered as an adaptive immune system in bacteria and archaea, CRISPR systems can be repurposed and engineered to precisely target defined DNA sequences, offering unprecedented specificity in antimicrobial intervention (Marraffini, 2015).

Unlike traditional antibiotics, which exert broad-spectrum pressure and indiscriminately affect both pathogenic and beneficial bacteria, CRISPR-based antimicrobials can be designed to specifically recognise and cleave AMR genes, virulence determinants, or essential genomic loci. By targeting resistance genes such as *mecA* or *van* clusters, CRISPR–cas nucleases can either kill resistant bacteria outright through lethal DNA damage or selectively eliminate resistance-conferring plasmids, thereby re-sensitising bacteria to existing antibiotics. This precision significantly reduces collateral damage to the commensal microbiota and limits the selective pressures that drive the emergence of new resistance mechanisms (Bikard et al., 2014; Yosef et al., 2015).

A critical aspect of CRISPR-based antimicrobial application lies in effective delivery. Several innovative delivery platforms have been explored, including bacteriophages, conjugative plasmids, and nanoparticle-based systems. Phage-mediated delivery, in particular, combines the natural host specificity of bacteriophages with the precision of CRISPR systems, enabling targeted eradication of multidrug-resistant strains (Bikard et al., 2014; Citorik et al., 2014). Conjugative plasmids can disseminate CRISPR constructs within bacterial populations, facilitating the removal of resistance genes carried on mobile genetic elements (Gomaa et al., 2014). Despite its transformative potential, the clinical translation of CRISPR-based antimicrobials faces several challenges. These include optimising delivery efficiency in complex microbial communities, mitigating host immune responses against CRISPR components or delivery vectors, preventing the evolution of CRISPR resistance, and addressing ethical and regulatory concerns related to gene-editing technologies. Additionally, large-scale manufacturing and standardisation remain important hurdles for real-world application (Zakeri and Lu, 2013; Mayorga-Ramos et al., 2023).

In a phase 1 first-in-human trial, a single-dose CRISPR-Cas9 gene-editing therapy (CTX310) safely produced a durable ~50% reduction in cholesterol and triglyceride levels in adults with resistant lipid disorders, highlighting the potential of CRISPR beyond rare diseases (Frangoul et al., 2021). In a phase 3 clinical trial of exagamglogene autotemcel (exa-cel), a

TABLE 11.1 List of representative antimicrobial peptide (AMPs): sources, mechanism of action, and therapeutic relevance.

Category AMP	Peptide/Compound	Source	Antimicrobial activity	References
Natural	Magainin	Frog skin (*Xenopus laevis*)	Membrane permeabilization of bacteria	Zasloff (2002)
Natural	Nisin	*Lactococcus lactis*	Cell wall synthesis inhibition, food, and clinical use	Cotter et al. (2005)
Natural	Defensins (α, β)	Human innate immunity	Pore formation, bacterial killing, antiviral activity	Ganz (2003)
Natural	LL-37	Human cathelicidin	Broad-spectrum activity, membrane disruption, immunomodulation	Hancock and Sahl (2006)
Antibiotic derived	Vancomycin	Glycopeptide antibiotic	Inhibit peptidoglycan synthesis	Levine (2006)
Natural	Melittin	Honeybee venom	Membrane lysis, antibiofilm activity	Raghuraman and Chattopadhyay (2007)
Antibiotic derived	Omiganan	Indolicidin analogue	Membrane disruption, topical infections	Fritsche et al. (2008)
Antibiotic derived	Daptomycin	Lipopeptide antibiotic	Membrane depolarisation of Gram-positive bacteria	Baltz (2009)
Antibiotic derived	Polymyxin B	Cyclic peptide antibiotic	Outer membrane disruption of gram-negative bacteria	Velkov et al. (2010)
Antibiotic derived	Colistin (Polymyxin E)	Cyclic peptide antibiotic	Last resort drug for multidrug-resistant gram-negative pathogens	Gupta et al. (2009)
Synthetic	Brilacidin	Defensin mimetic	Membrane-targeted, anti-MRSA, antibiofilm	Scott and Tew (2017)
Synthetic	LTX-109	Synthetic peptidomimetic	Rapid bactericidal action; nasal decolonisation	Saravolatz et al. (2012)

CRISPR-based cell therapy for sickle cell disease, 97% of the patients remained free from severe Vaso-occlusive crisis for ≥12 months after treatment, highlighting a transformative therapeutic outcome (Esrick et al., 2021).

CRISPR faces several critical challenges that limit its widespread clinical and translational application. These include off-target DNA cleavage leading to mutations, difficulties in achieving safe and efficient in vivo delivery of CRISPR components, and immune responses against commonly used Cas proteins. Nevertheless, CRISPR-based antimicrobials represent a paradigm shift in the fight against AMR. Their adaptability, precision, and ability to directly target the genetic basis of resistance position them as powerful next-generation therapeutics. When integrated with conventional antibiotics and antimicrobial stewardship strategies, CRISPR technologies hold immense promise for reshaping future approaches to managing and controlling multidrug-resistant bacterial infections (Barrangou and Doudna, 2016; Pursey et al., 2018).

11.6 Agents for modulation of gut microbiome

11.6.1 Probiotics, prebiotics, symbiotics and postbiotics

Agents that modulate the gut microbiome are increasingly recognised as a viable alternative to antibiotics, offering targeted regulation of microbial communities without promoting AMR (Ananthakrishnan et al., 2018). Probiotics are defined as live microorganisms that, when administered in adequate amounts, confer a health benefit to the host (Hill et al., 2014). Their mechanism includes competitive inclusion of pathogens, production of antimicrobial substances, enhancement of gut barrier integrity, and modulation of host immune response. Prebiotics are selectively fermentable, nondigestible substrates that stimulate the growth and activity of beneficial gut microbes. They act by enriching commercial bacteria, increasing short-chain fatty acid production, lowering gut pH, and suppressing pathogen colonisation. Symbiotics are combinations of probiotics and prebiotics designed to synergistically improve the survival, colonisation, and functional activity of beneficial microbes in the gut (Swanson et al., 2020). The prebiotic components support the administered probiotic strain, enhancing metabolic activity and ecological fitness within the intestinal environment. Postbiotics, in contrast, are nonviable microbial cells, cell components, or metabolic products that exert biological activity in the host. Their mechanisms include antiinflammatory effects, immune modulation, and inhibition of pathogenic bacteria while offering advantages of improved safety, stability, and standardisation compared to live microbes (Aguilar-Toalá et al., 2018; Salminen et al., 2021).

A randomised controlled trial (Hickson et al., 2007) had demonstrated that the probiotic drink containing *Lactobacillus casei*, *L. bulgaricus*, and *Streptococcus thermophilus* significantly reduced antibiotic-associated diarrhoea and *Clostridioides difficile* infection (CDI) in hospitalised patients. In another double-blind randomised trial, it was shown that a symbiotic combination of *Bifidobacterium* spp. with galacto-oligosaccharides modulated gut microbiota composition and improved immune markers in healthy adults (Krumbeck et al., 2018).

These microbial modulation agents have broad clinical and preventive applications. They span gastrointestinal infections, antibiotic-associated diarrhoea, inflammatory bowel disease, metabolic disorders, and immune-mediated conditions. Probiotics and symbiotics are widely used for the prevention and management of infectious diarrhoea and the restoration of microbiota after antibiotic therapy. Prebiotics support long-term gut microbial resilience. Postbiotics are emerging as safer alternatives in immunocompromised individuals (Ritchie and Romanuk, 2012). Collectively, these approaches represent microbial-centred strategies that reduce reliance on conventional antibiotics by enhancing colonisation resistance, host immunity, and microbial homoeostasis.

11.6.2 Faecal microbial transplantation

Faecal microbiota transplantation (FMT) is a therapeutic intervention in which a minimally processed stool preparation from a rigorously healthy donor is administrated to a recipient with the aim of restoring a dysbiotic gut microbiome (Cammarota et al., 2017). It is based on the principle that reconstitution of a diverse and functionally balanced microbial community can correct disease associated alternations in gut ecology that are not effectively addressed by conventional antimicrobial therapy (van Nood et al., 2013).

The therapeutic effect of FMT is mediated through the reestablishment of microbial diversity and colonisation resistance against pathogenic organisms. Transplanted commensal microbes competitively inhibit pathogen growth, restore normal metabolic functions such as short-chain fatty acid and secondary bile acid production, and reinforce intestinal barrier integrity. In addition, FMT modules host immune responses by promoting regulatory immune pathways and reducing pro-inflammatory signalling, thereby contributing to long-term microbiome stability (Khoruts and Sadowsky, 2016).

A randomised controlled trial in patients with recurrent CDI showed that FMT delivered via duodenal infusion achieves a cure rate of ~81%, significantly superior to vancomycin therapy (van Nood et al., 2013). In another randomised trial, FMT administered by colonoscopy resulted in higher remission rates of recurrent CDI compared to standard antibiotic treatment (Cammarota et al., 2015). A multicentre randomised trial demonstrated that intensive, multidonor FMT induced clinical remission and microbiome restoration in patients with ulcerative colitis (Paramsothy et al., 2017).

FMT is an established and highly effective treatment for *Clostridioides difficile* infection, where it demonstrates superior cure rates compared to antibiotics. Investigational and emerging applications include inflammatory bowel disease, irritable bowel syndrome, metabolic and liver disorders, and modulation of therapeutic responses in oncology and immunotherapy. While clinical potential is expanding, concerns related to donor screening, safety, standardisation, and regulatory frameworks continue to shape the controlled and evidence-based use of FMT. Overall, this represents a next-generation antimicrobial therapy that shifts treatment from pathogen eradication to restoration of a resilient, health-promoting gut microbiome.

11.7 Natural compounds

Natural compounds are bioactive chemical substances derived from natural sources such as plants, microorganisms, marine organisms, and animals that have emerged as a rich and largely untapped reservoir of antimicrobial agents with significant potential to counteract the growing threat of MDR pathogens (Newman and Cragg, 2020). Their structural complexity and functional diversity distinguish them fundamentally from conventional antibiotics and provide novel opportunities for antimicrobial intervention in the postantibiotic era (Atanasov et al., 2021).

A prominent mechanism involves perturbation of bacterial membrane integrity, leading to increased permeability, dissipation of membrane potential, and leakage of intracellular contents. This membrane-centric activity enables rapid bacterial killing and limits the emergence of resistance, as extensive membrane remodelling is metabolically costly for pathogens (Wang et al., 2016). In addition, numerous natural products interfere with intracellular processes, including DNA replication, protein synthesis, and energy metabolism, further enhancing their antimicrobial efficacy. (Lewis, 2020) Beyond direct bactericidal effects, natural compounds play a critical role in modulating bacterial communication and pathogenicity. Several plant- and marine-derived molecules have been shown to inhibit quorum sensing, thereby suppressing coordinated behaviours such as virulence factor expression and biofilm formation. Since quorum-sensing inhibition attenuates pathogenicity without directly killing bacteria, it imposes reduced selective pressure for resistance development. Similarly, many natural products disrupt biofilm architecture or prevent biofilm establishment, addressing one of the most formidable barriers to effective antimicrobial therapy in chronic and device-associated infections (Flemming et al., 2016). Another important mechanism involves the modulation of bacterial resistance determinants. Certain flavonoids, alkaloids, and phenolic compounds function as efflux pump inhibitors, increasing intracellular antibiotic concentrations and restoring susceptibility in resistant strains. This ability to reverse resistance has positioned natural compounds as valuable adjuvants in combination therapies. Synergistic interactions between natural products and existing antibiotics have been widely reported, resulting in enhanced antimicrobial activity, reduced effective doses, and expanded antibacterial spectra (Caesar and Cech, 2019).

In a phase 1 clinical safety trial of a natural product with antibacterial activity applied to humans, no serious adverse effects were observed following topical application, supporting the tolerability of plant-based antimicrobial mixture in early human testing. This patch test trial established safety groundwork for further clinical evaluation of natural compounds as antimicrobial alternatives to conventional antibiotics (Bruce et al., 2022).

Natural antimicrobial compounds have wide-ranging therapeutic and preventive applications, including management of skin and soft tissue infections, gastrointestinal and respiratory infections, wound healing, and use as adjuncts to conventional antibiotics to combat multidrug-resistant pathogens.

11.8 Nanomaterials

Nanomaterials are materials with at least one dimension in the nanoscale range (1–100 nm) that exhibit distinct physicochemical and biological properties compared with their bulk counterparts. In antimicrobial research, nanomaterials include metal and metal oxide nanoparticles, polymeric nanostructures, lipid-based nanocarriers, carbon-based nanomaterials, and nanocomposites. Their higher surface area, adjustable surface chemistry, and ability to be functionalized enable enhanced interaction with microbial cells and improved delivery of antimicrobial agents, positioning them as promising tools against MDR infections (Pelgrift and Friedman, 2013; Makabenta et al., 2021).

Nanomaterials exert antimicrobial effects through multiple, often simultaneous mechanisms, which limit the development of resistance. These mechanisms include disruption of bacterial cell membranes, induction of reactive oxygen species, release of antimicrobial metal ions, inhibition of enzymatic activity, and damage to DNA and proteins. Many nanomaterials can penetrate and destabilise biofilm, interfere with quorum sensing, and enhance intracellular delivery of antimicrobials. The multitargeted mode of action distinguishes nanomaterials from conventional antibiotics that typically act on a single cellular pathway (Morones et al., 2005; Slavin et al., 2017).

A nanoparticle-based liposomal formulation of amikacin had demonstrated improved sputum culture conversion in a patient with refractory nontuberculous mycobacterial lung disease (Griffith et al., 2018). Chitosan-based nanoparticles have shown safety and enhanced antimicrobial efficacy in early phase clinical studies for oral and wound infections (Fathi et al., 2020).

Nanomaterials have a wide range of applications in clinical and biomedical antimicrobial strategies, including wound dressing, implant and catheter coating, topical formulations, drug delivery systems, and oral and periodontal therapies. As nanocarriers, they improve the stability, bioavailability, and controlled release of antibiotics and AMPs while reducing systemic toxicity. Nanomaterial-based coatings and formulations have shown strong antibiofilm activity and enhanced tissue integration (Rai et al., 2009; Wang et al., 2018). Furthermore, nano-enabled combination therapies demonstrate synergistic antimicrobial effects, revitalising the efficacy of existing antibiotics against MDR pathogens.

11.9 Future directions

Recent advances in genomics, metagenomics, and transcriptomics have profoundly transformed antimicrobial research by enabling comprehensive profiling of resistance determinants, virulence factors, and microbial community dynamics. Whole-genome sequencing has become an indispensable tool for real-time surveillance of AMR, facilitating the identification of emerging high-risk clones and informing targeted therapeutic strategies. Complementing these advances, artificial intelligence and machine learning–driven drug discovery platforms have accelerated the identification of novel antimicrobial targets, optimised lead compound selection, and predicted resistance evolution, thereby improving the efficiency and precision of therapeutic design. Collectively, these emerging innovations reflect a paradigm shift toward mechanism-informed, adaptable, and sustainable antimicrobial strategies that integrate molecular precision with systems-level understanding. While challenges related to clinical validation, regulatory frameworks, scalability, and cost remain, the convergence of biotechnology, computational science, and genomics holds considerable promise. The integration of these novel approaches equips antimicrobial innovation to outpace superbug adaptation and establish a robust postantibiotic therapeutic landscape. Among the most promising alternatives is bacteriophage therapy, which exploits the natural ability of phages to specifically infect and lyse bacterial cells. Phages offer high specificity, minimising collateral damage to the commensal microbiota, and can be rapidly adapted to target evolving bacterial strains. Similarly, CRISPR-based antimicrobials enable precise targeting of resistance genes, virulence determinants, or essential genomic regions, allowing selective bacterial killing to existing antibiotics. Antivirulence therapies further expand the therapeutic arsenal by disarming pathogens through inhibition of toxins, adhesion factors, or quorum-sensing pathways, thereby reducing disease severity without imposing strong selective pressure for resistance. Host-directed approaches, including immune modulation and microbiome-based interventions, represent another critical dimension of next-generation therapies. Immunotherapeutics such as monoclonal antibodies and vaccines enhance host defence mechanisms and reduce reliance on antibiotics. Microbiome-based strategies, including probiotics, faecal microbiota transplantation, and engineered microbial consortia, restore colonisation resistance and suppress pathogen overgrowth, contributing to both infection prevention and resistance mitigation. Resistance-modifying adjuvants, such as efflux pump inhibitors and β-lactamase inhibitors, complement these approaches by restoring the efficacy of existing antibiotics.

Despite their considerable promise, the clinical translation of next-generation therapies faces substantial barriers. Regulatory uncertainty remains a major challenge, as many of these approaches do not fit within traditional drug approval frameworks. Safety concerns, including off-target effects, immunogenicity, and unintended ecological consequences, must be carefully addressed. Efficient delivery systems, scalability of production, and standardisation of complex biological products also pose significant technical hurdles. Furthermore, limited long-term efficacy and safety data hinder widespread clinical adoption and integration into standard treatment guidelines.

11.10 Conclusion

The escalating burden of AMR underscores the urgent need to move beyond conventional antibiotic-centric therapies. As the discovery of new antibiotic classes continues to lag behind the pace of resistance, next-generation therapies that extend beyond traditional bactericidal or bacteriostatic approaches are gaining increasing attention. Precision-guided, mechanism-driven strategies ranging from phage and CRISPR-based interventions to antivirulence agents, AMPs, microbiome modulation, and immunoprophylaxis collectively redefine infection control by targeting pathogenicity, resistance, and host–microbe dynamics. Outsmarting microbes rather than merely killing them will define the future of infection control in the postantibiotic era.

References

Abedon, Kuhl, Blasdel, Kutter, 2011. Phage treatment of human infections. Bacteriophage 1 (2), 66–85. Available at https://doi.org/10.4161/bact.1.2.15845.

Abedon, S.T., 2014. Phage therapy: eco-physiological pharmacology. Scientifica 2014. Available at https://doi.org/10.1155/2014/581639.

Aguilar-Toalá, J.E., Garcia-Varela, R., Garcia, H.S., Mata-Haro, V., González-Córdova, A.F., Vallejo-Cordoba, B., et al., 2018. Postbiotics: an evolving term within the functional foods field. Trends in Food Science & Technology 75, 105–114. Available at https://doi.org/10.1016/j.tifs.2018.03.009.

Allen, R.C., Popat, R., Diggle, S.P., Brown, S.P., 2014. Targeting virulence: can we make evolution-proof drugs? Nature Reviews. Microbiology 12 (4), 300–308. Available at https://doi.org/10.1038/nrmicro3232, http://www.nature.com/nrmicro/index.html.

Aminov, R.I., 2010. A brief history of the antibiotic era: lessons learned and challenges for the future. Frontiers in Microbiology 1. Available at https://doi.org/10.3389/fmicb.2010.00134, http://www.frontiersin.org/Journal/DownloadFile.ashx?pdf=1&FileId=1478&articleId=8040&Version=1&ContentTypeId=21&FileName=fmicb-01-00134.pdf.

Ananthakrishnan, A.N., Bernstein, C.N., Iliopoulos, D., Macpherson, A., Neurath, M.F., Raja Ali, R.A., et al., 2018. Environmental triggers in IBD: a review of progress and evidence. Nature Reviews Gastroenterology and Hepatology 15 (1), 39–49. Available at https://doi.org/10.1038/nrgastro.2017.136.

Atanasov, A.G., Zotchev, S.B., Dirsch, V.M., Orhan, I.E., Banach, M., Rollinger, J.M., et al., 2021. Natural products in drug discovery: advances and opportunities. Nature Reviews. Drug Discovery 20 (3), 200–216. Available at https://doi.org/10.1038/s41573-020-00114-z, http://www.nature.com/nrd/index.html.

Baltz, R.H., 2009. Daptomycin: mechanisms of action and resistance, and biosynthetic engineering. Current Opinion in Chemical Biology 13 (2), 144–151. Available at https://doi.org/10.1016/j.cbpa.2009.02.031.

Barrangou, R., Doudna, J.A., 2016. Applications of CRISPR technologies in research and beyond. Nature Biotechnology 34 (9), 933–941. Available at https://doi.org/10.1038/nbt.3659.

Bikard, D., Euler, C.W., Jiang, W., Nussenzweig, P.M., Goldberg, G.W., Duportet, X., et al., 2014. Exploiting CRISPR-cas nucleases to produce sequence-specific antimicrobials. Nature Biotechnology 32 (11), 1146–1150. Available at https://doi.org/10.1038/nbt.3043, http://www.nature.com/nbt/index.html.

Bonomo, R.A., Burd, E.M., Conly, J., Limbago, B.M., Poirel, L., Segre, J.A., et al., 2018. Carbapenemase-producing organisms: a global scourge. Clinical Infectious Diseases 66 (8), 1290–1297. Available at https://doi.org/10.1093/cid/cix893, http://cid.oxfordjournals.org/content/by/year.

Bruce, J., Oyedemi, B., Parsons, N., Harrison, F., 2022. Phase 1 safety trial of a natural product cocktail with antibacterial activity in human volunteers. Scientific Reports 12 (1). Available at https://doi.org/10.1038/s41598-022-22700-4.

Caesar, L.K., Cech, N.B., 2019. Synergy and antagonism in natural product extracts: when 1 + 1 does not equal 2. Natural Product Reports 36 (6), 869–888. Available at https://doi.org/10.1039/c9np00011a.

Cammarota, G., Ianiro, G., Tilg, H., Rajilić-Stojanović, M., Kump, P., Satokari, R., et al., 2017. European consensus conference on faecal microbiota transplantation in clinical practice. Gut 66 (4), 569–580. Available at https://doi.org/10.1136/gutjnl-2016-313017.

Cammarota, G., Masucci, L., Ianiro, G., Bibbò, S., Dinoi, G., Costamagna, G., Gasbarrini, A., 2015. Randomised clinical trial: Faecal microbiota transplantation by colonoscopy vs vancomycin for the treatment of recurrent *Clostridioides difficile* infection. Gastroenterology 149 (3), 547–555.e3. Available at https://doi.org/10.1053/j.gastro.2015.04.010.

Chan, B.K., Abedon, S.T., Loc-Carrillo, C., 2013. Phage cocktails and the future of phage therapy. Future Microbiology 8 (6), 769–783. Available at https://doi.org/10.2217/fmb.13.47.

Citorik, R.J., Mimee, M., Lu, T.K., 2014. Sequence-specific antimicrobials using efficiently delivered RNA-guided nucleases. Nature Biotechnology 32 (11), 1141–1145. Available at https://doi.org/10.1038/nbt.3011, http://www.nature.com/nbt/index.html.

Cotter, P.D., Hill, C., Ross, R.P., 2005. Bacteriocins: developing innate immunity for food. Nature Reviews Microbiology 3 (10), 777–788. Available at https://doi.org/10.1038/nrmicro1273.

Czaplewski, L., Bax, R., Clokie, M., Dawson, M., Fairhead, H., Fischetti, V.A., et al., 2016. Alternatives to antibiotics-a pipeline portfolio review. The Lancet Infectious Diseases 16 (2), 239–251.

Dedrick, R.M., Guerrero-Bustamante, C.A., Garlena, R.A., Russell, D.A., Ford, K., Harris, K., et al., 2019. Engineered bacteriophages for treatment of a patient with a disseminated drug-resistant *Mycobacterium abscessus*. Nature Medicine 25 (5), 730–733. Available at https://doi.org/10.1038/s41591-019-0437-z, http://www.nature.com/nm/index.html.

Dickey, S.W., Cheung, G.Y.C., Otto, M., 2017. Different drugs for bad bugs: antivirulence strategies in the age of antibiotic resistance. Nature Reviews. Drug Discovery 16 (7), 457–471. Available at https://doi.org/10.1038/nrd.2017.23, http://www.nature.com/nrd/index.html.

Esrick, E.B., Lehmann, L.E., Biffi, A., Achebe, M., Brendel, C., Ciuculescu, M.F., et al., 2021. Post-transcriptional genetic silencing of BCL11A to treat sickle cell disease. New England Journal of Medicine 384 (3), 205–215. Available at https://doi.org/10.1056/NEJMoa2029392, http://www.nejm.org/medical-index.

Fathi, M., Akbari, B., Taheriazam, A., 2020. Chitosan nanoparticles for antimicrobial drug delivery: clinical perspectives. International Journal of Biological Macromolecules 164, 397–414. Available at https://doi.org/10.1016/j.ijbiomac.2020.07.084.

Fischetti, V.A., 2018. Development of phage lysins as novel therapeutics: a historical perspective. Viruses 10 (6). Available at https://doi.org/10.3390/v10060310, http://www.mdpi.com/journal/viruses/.

Flemming, H.C., Wingender, J., Szewzyk, U., Steinberg, P., Rice, S.A., Kjelleberg, S., 2016. Biofilms: an emergent form of bacterial life. Nature Reviews. Microbiology 14 (9), 563–575. Available at https://doi.org/10.1038/nrmicro.2016.94, http://www.nature.com/nrmicro/index.html.

Frangoul, H., Altshuler, D., Cappellini, M.D., Chen, Y.S., Domm, J., Eustace, B.K., et al., 2021. CRISPR-Cas9 gene editing for sickle cell disease and β-thalassemia. New England Journal of Medicine 384 (3), 252–260. Available at https://doi.org/10.1056/NEJMoa2031054, http://www.nejm.org/medical-index.

Fritsche, T.R., Rhomberg, P.R., Sader, H.S., Jones, R.N., 2008. Antimicrobial activity of omiganan pentahydrochloride tested against contemporary bacterial pathogens commonly responsible for catheter-associated infections. The Journal of Antimicrobial Chemotherapy 61 (5), 1092–1098. Available at https://doi.org/10.1093/jac/dkn074.

Ganz, T., 2003. Defensins: antimicrobial peptides of innate immunity. Nature Reviews Immunology 3 (9), 710–720. Available at https://doi.org/10.1038/nri1180.

Gomaa, A.A., Klumpe, H.E., Luo, M.L., Selle, K., Barrangou, R., Beisel, C.L., 2014. Programmable removal of bacterial strains by use of genome-targeting CRISPR-cas systems. mBio 5 (1). Available at https://doi.org/10.1128/mBio.00928–13, http://mbio.asm.org/content/5/1/e00928-13.full.pdf.

Griffith, D.E., Eagle, G., Thomson, R., Aksamit, T.R., Hasegawa, N., Morimoto, K., et al., 2018. Complex (CONVERT). A prospective, open-label, randomized study. American Journal of Respiratory and Critical Care Medicine 198 (12), 1559–1569. Available at https://doi.org/10.1164/rccm.201807–1318OC.

Gupta, S., Govil, D., Kakar, P.N., Prakash, O., Arora, D., Das, S., et al., 2009. Colistin and polymyxin B: a re-emergence. Indian Journal of Critical Care Medicine Peer-Reviewed, Official Publication of Indian Society of Critical Care Medicine 13 (2), 49–53. Available at https://doi.org/10.4103/0972-5229.56048.

Hancock, R.E.W., Sahl, H.G., 2006. Antimicrobial and host-defense peptides as new anti-infective therapeutic strategies. Nature Biotechnology 24 (12), 1551–1557. Available at https://doi.org/10.1038/nbt1267.

Hickson, M., D'Souza, A.L., Muthu, N., Rogers, T.R., Want, S., Rajkumar, C., et al., 2007. Use of probiotic *Lactobacillus* preparation to prevent diarrhoea associated with antibiotics: randomised double blind placebo controlled trial. British Medical Journal 335 (7610), 80–83. Available at https://doi.org/10.1136/bmj.39231.599815.55

Hill, C., Guarner, F., Reid, G., Gibson, G.R., Merenstein, D.J., Pot, B., et al., 2014. Expert consensus document: the international scientific association for probiotics and prebiotics consensus statement on the scope and appropriate use of the term probiotic. Nature Reviews Gastroenterology and Hepatology 11 (8), 506–514. Available at https://doi.org/10.1038/nrgastro.2014.66, http://www.nature.com/nrgastro/index.html.

Jault, P., Leclerc, T., Jennes, S., Pirnay, J.-P., Que, Y.-A., Resch, G., et al., 2019. Efficacy and tolerability of a cocktail of bacteriophages to treat burn wounds infected by Pseudomonas aeruginosa (PhagoBurn): a randomised, controlled, double-blind phase 1/2 trial. The Lancet Infectious Diseases 19 (1), 35–45. Available at https://doi.org/10.1016/S1473-3099(18)30482-1.

Khoruts, A., Sadowsky, M.J., 2016. Understanding the mechanisms of faecal microbiota transplantation. Nature Reviews Gastroenterology & Hepatology 13 (9), 508–516. Available at https://doi.org/10.1038/nrgastro.2016.98.

Kortright, K.E., Chan, B.K., Koff, J.L., Turner, P.E., 2019. Phage therapy: a renewed approach to combat antibiotic-resistant bacteria. Cell Host & Microbe 25 (2), 219–232. Available at https://doi.org/10.1016/j.chom.2019.01.014.

Krumbeck, J.A., Rasmussen, H.E., Hutkins, R.W., Clarke, J., Shawron, K., Keshavarzian, A., et al., 2018. Probiotic Bifidobacterium strains and galactooligosaccharides improve intestinal barrier function in obese adults but show no synergism when used together as synbiotics. Microbiome 6 (1). Available at https://doi.org/10.1186/s40168-018-0494-4, http://www.microbiomejournal.com/.

Laxminarayan, R., Duse, A., Wattal, C., Zaidi, A.K.M., Wertheim, H.F.L., Sumpradit, N., et al., 2013. Antibiotic resistance-the need for global solutions. The Lancet Infectious Diseases 13 (12), 1057–1098. Available at https://doi.org/10.1016/S1473-3099(13)70318-9.

Leitner, L., Ujmajuridze, A., Chanishvili, N., Goderdzishvili, M., Chkonia, I., Rigvava, S., et al., 2021. Intravesical bacteriophages for treating urinary tract infections in patients undergoing transurethral resection of the prostate: a randomised, placebo-controlled, double-blind clinical trial. The Lancet Infectious Diseases 21 (3), 427–436. Available at https://doi.org/10.1016/S1473-3099(20)30330-3, http://www.journals.elsevier.com/the-lancet-infectious-diseases.

Levine, D.P., 2006. Vancomycin: a history. Clinical Infectious Diseases 42 (1), S5. Available at https://doi.org/10.1086/491709.

Lewis, K., 2020. The science of antibiotic discovery. Cell 181 (1), 29–45. Available at https://doi.org/10.1016/j.cell.2020.02.056.

Makabenta, J.M.V., Nabawy, A., Li, C.H., Schmidt-Malan, S., Patel, R., Rotello, V.M., 2021. Nanomaterial-based therapeutics for antibiotic-resistant bacterial infections. Nature Reviews Microbiology 19 (1), 23–36. Available at https://doi.org/10.1038/s41579-020-0420-1, http://www.nature.com/nrmicro/index.html.

Makovitzki, A., 2006. Ultrashort antibacterial and antifungal lipopeptides. Proceedings of the National Academy of Sciences 103 (43), 15997–16002. Available at https://doi.org/10.1073/pnas.0606129103.

Marraffini, L.A., 2015. CRISPR-Cas immunity in prokaryotes. Nature 526 (7571), 55–61. Available at https://doi.org/10.1038/nature15386, http://www.nature.com/nature/index.html.

Martin, R.M., 2018. Antimicrobial resistance history and perspectives. American Society for Clinical Laboratory Science 31 (2), 71–74. Available at https://doi.org/10.29074/ascls.2018000430.

Matsuzaki, K., 2009. Control of cell selectivity of antimicrobial peptides. Biochimica et Biophysica Acta - Biomembranes 1788 (8), 1687–1692. Available at https://doi.org/10.1016/j.bbamem.2008.09.013.

Mayorga-Ramos, A., Zúñiga-Miranda, J., Carrera-Pacheco, S.E., Barba-Ostria, C., Guamán, L.P., 2023. CRISPR-Cas-based antimicrobials: design, challenges, and bacterial mechanisms of resistance. ACS Infectious Diseases 9 (7), 1283–1302. Available at https://doi.org/10.1021/acsinfecdis.2c00649.

Mishra, S., Paul, R., Singh, K., Bagdi, R., 2025. From penicillin to superbugs: the history, mechanisms, and consequences of antibiotic resistance on pharmaceutical innovation. Journal of Population Therapeutics and Clinical Pharmacology. Available at https://doi.org/10.53555/186aae87.

Mookherjee, N., 2007. Cationic host defence peptides: innate immune regulatory peptides as a novel approach for treating infections. Cellular and Molecular Life Sciences 64 (7-8), 922–933. Available at https://doi.org/10.1007/s00018-007-6475-6.

Morones, J.R., Elechiguerra, J.L., Camacho, A., Holt, K., Kouri, J.B., Ramírez, J.T., et al., 2005. The bactericidal effect of silver nanoparticles. Nanotechnology 16 (10), 2346–2353. Available at https://doi.org/10.1088/0957-4484/16/10/059.

Murray, C.J.L., et al., 2022. Global burden of bacterial antimicrobial resistance in 2019: a systematic analysis. The Lancet. 399 (10325), 629–655. Available at https://doi.org/10.1016/s0140-6736(21)02653-2.

Nath, G., Janam, R., Kumar, R., Gangwar, M., 2019. Bacteriophage therapy: an alternative to antibiotics—an experimental study in mice. Annals of the National Academy of Medical Sciences (India) 55, 151–158. Available at https://doi.org/10.1055/s-0039-1698545.

Newman, D.J., Cragg, G.M., 2020. Natural products as sources of new drugs over the nearly four decades from 01/1981 to 09/2019. Journal of Natural Products 83 (3), 770–803. Available at https://doi.org/10.1021/acs.jnatprod.9b01285, http://pubs.acs.org/journal/jnprdf.

Nguyen, L.T., 2011. The expanding scope of antimicrobial peptide structures and their modes of action. Trends in Biotechnology 29 (9), 464–472. Available at https://doi.org/10.1016/j.tibtech.2011.05.001.

Nordmann, P., Naas, T., Poirel, L., 2011. Global spread of carbapenemase-producing Enterobacteriaceae. Emerging Infectious Diseases 17 (10), 1791–1798. Available at https://doi.org/10.3201/eid1710.110655.

Paramsothy, S., Kamm, M.A., Kaakoush, N.O., Walsh, A.J., van den Bogaerde, J., Samuel, D., et al., 2017. Multidonor intensive faecal microbiota transplantation for active ulcerative colitis: A randomised placebo-controlled trial. The Lancet 389 (10075), 1218–1228. Available at https://doi.org/10.1016/S0140-6736(17)30182-4.

Pelgrift, R.Y., Friedman, A.J., 2013. Nanotechnology as a therapeutic tool to combat microbial resistance. Advanced Drug Delivery Reviews 65 (13-14), 1803–1815. Available at https://doi.org/10.1016/j.addr.2013.07.011.

Pires, D.P., Cleto, S., Sillankorva, S., Azeredo, J., Lu, T.K., 2016. Genetically engineered phages: a review of advances over the last decade. Microbiology and Molecular Biology Reviews 80 (3), 523–543. Available at https://doi.org/10.1128/MMBR.00069-15, http://mmbr.asm.org/content/80/3/523.full.pdf+html.

Pursey, E., Sünderhauf, D., Gaze, W.H., Westra, E.R., van Houte, S., 2018. CRISPR-Cas antimicrobials: challenges and future prospects. PLoS Pathogens 14 (6), e1006990. Available at https://doi.org/10.1371/journal.ppat.1006990.

Raghuraman, H., Chattopadhyay, A., 2007. Melittin: a membrane-active peptide with diverse functions. Bioscience Reports 27 (4-5), 189–223. Available at https://doi.org/10.1007/s10540-006-9030-z.

Rai, M., Yadav, A., Gade, A., 2009. Silver nanoparticles as a new generation of antimicrobials. Biotechnology Advances 27 (1), 76–83. Available at https://doi.org/10.1016/j.biotechadv.2008.09.002.

Rasko, D.A., Sperandio, V., 2010. Anti-virulence strategies to combat bacteria-mediated disease. Nature Reviews Drug Discovery 9 (2), 117–128. Available at https://doi.org/10.1038/nrd3013.

Ritchie, M.L., Romanuk, T.N., 2012. A meta-analysis of probiotic efficacy for gastrointestinal diseases. PLoS One 7 (4). Available at https://doi.org/10.1371/journal.pone.0034938Canada, http://www.plosone.org/article/fetchObjectAttachment.action?uri=info%3Adoi%2F10.1371%2Fjournal.pone.0034938&representation=PDF.

Robinson, T.P., Bu, D.P., Carrique-Mas, J., Fèvre, E.M., Gilbert, M., Grace, D., et al., 2016. Antibiotic resistance is the quintessential One Health issue. Transactions of the Royal Society of Tropical Medicine and Hygiene 110 (7), 377–380. Available at https://doi.org/10.1093/trstmh/trw048.

Salminen, S., Collado, M.C., Endo, A., Hill, C., Lebeer, S., Quigley, E.M.M., et al., 2021. The international scientific association of probiotics and prebiotics (ISAPP) consensus statement on the definition and scope of postbiotics. Nature Reviews Gastroenterology and Hepatology 18 (9), 649–667. Available at https://doi.org/10.1038/s41575-021-00440-6, http://www.nature.com/nrgastro/index.html.

Saravolatz, L.D., Pawlak, J., Johnson, L., Bonilla, H., Saravolatz, L.D., Fakih, M.G., et al., 2012. In vitro activities of LTX-109, a synthetic antimicrobial peptide, against methicillin-resistant, vancomycin-intermediate, vancomycin-resistant, daptomycin-nonsusceptible, and linezolid-nonsusceptible *Staphylococcus aureus*. Antimicrobial Agents and Chemotherapy 56 (8), 4478–4482. Available at https://doi.org/10.1128/AAC.00194-12.

Scott, R.W., Tew, G.N., 2017. Mimics of host defense proteins: Strategies for translation to therapeutic applications. Current Topics in Medicinal Chemistry 17 (5), 576–589. Available at https://doi.org/10.2174/1568026616666160713130452.

Slavin, Y.N., Asnis, J., Häfeli, U.O., Bach, H., 2017. Metal nanoparticles: understanding the mechanisms behind antibacterial activity. Journal of Nanobiotechnology 15 (1). Available at https://doi.org/10.1186/s12951-017-0308-z, http://www.jnanobiotechnology.com/start.asp.

Sulakvelidze, A., Alavidze, Z., Morris, J., 2001. Bacteriophage therapy. Antimicrobial Agents and Chemotherapy 45 (3), 649–659. Available at https://doi.org/10.1128/AAC.45.3.649-659.2001.

Swanson, K.S., Gibson, G.R., Hutkins, R., Reimer, R.A., Reid, G., Verbeke, K., et al., 2020. The international scientific association for probiotics and prebiotics (ISAPP) consensus statement on the definition and scope of synbiotics. Nature Reviews Gastroenterology and Hepatology 17 (11), 687–701. Available at https://doi.org/10.1038/s41575-020-0344-2, http://www.nature.com/nrgastro/index.html.

Tacconelli, E., Carrara, E., Savoldi, A., Harbarth, S., Mendelson, M., Monnet, D.L., et al., 2018. Discovery, research, and development of new antibiotics: the WHO priority list of antibiotic-resistant bacteria and tuberculosis. The Lancet Infectious Diseases 18 (3), 318–327. Available at https://doi.org/10.1016/S1473-3099(17)30753-3.

van Nood, E., Vrieze, A., Nieuwdorp, M., Fuentes, S., Zoetendal, E.G., de Vos, W.M., et al., 2013. Duodenal infusion of donor feces for recurrent *Clostridioides difficile* infection. New England Journal of Medicine 368 (5), 407–415. Available at https://doi.org/10.1056/NEJMoa120503.

Velkov, T., Thompson, P.E., Nation, R.L., Li, J., 2010. Structure–activity relationships of polymyxin antibiotics. Journal of Medicinal Chemistry 53 (5), 1898–1916. Available at https://doi.org/10.1021/jm900999h.

Wang, G., Mishra, B., Lau, K., Lushnikova, T., Golla, R., Wang, X., 2016. Antimicrobial peptides in 2014. Pharmaceuticals 9 (3), 56. Available at https://doi.org/10.3390/ph9030056.

Wang, H., Ai, L., Zhang, Y., Cheng, J., Yu, H., Li, C., et al., (2018). The effects of antimicrobial peptide Nal-P-113 on inhibiting periodontal pathogens and improving periodontal status. BioMed Research International, 2018, Article ID 8146935, pp. 1–10. https://doi.org/10.1155/2018/8146935.

Wang, H., Ai, L., Zhang, Y., Cheng, J., Yu, H., Li, C., et al., 2018. The effects of antimicrobial peptide Nal-P-113 on inhibiting periodontal pathogens and improving periodontal status. BioMed Research International 2018, 8146935. Available at https://doi.org/10.1155/2018/1805793, http://www.hindawi.com/journals/biomed/.

Wright, A., Hawkins, C.H., Änggård, E.E., Harper, D.R., 2009. A controlled clinical trial of a therapeutic bacteriophage preparation in chronic otitis due to antibiotic-resistant *Pseudomonas aeruginosa*; a preliminary report of efficacy. Clinical Otolaryngology 34 (4), 349–357. Available at https://doi.org/10.1111/j.1749-4486.2009.01973.x.

Wu, Y.F., Han, B.C., Lin, W.Y., Wang, S.Y., Linn, T.Y., Hsu, H.W., et al., 2024. Efficacy of antimicrobial peptide P113 oral health care products on the reduction of oral bacteria number and dental plaque formation in a randomized clinical assessment. Journal of Dental Sciences 19 (4), 2367–2376. Available at https://doi.org/10.1016/j.jds.2024.07.011, https://www.sciencedirect.com/science/journal/19917902.

Yosef, I., Manor, M., Kiro, R., Qimron, U., 2015. Temperate and lytic bacteriophages programmed to sensitize and kill antibiotic-resistant bacteria. Proceedings of the National Academy of Sciences 112 (23), 7267–7272. Available at https://doi.org/10.1073/pnas.1500107112.

Zakeri, B., Lu, T.K., 2013. Synthetic biology of antimicrobial discovery. ACS Synthetic Biology 2 (7), 358–372. Available at https://doi.org/10.1021/sb300101g.

Zasloff, M., 2002. Antimicrobial peptides of multicellular organisms. Nature 415 (6870), 389–395. Available at https://doi.org/10.1038/415389a.

Zumla, A., Rao, M., Wallis, R.S., Kaufmann, S.H.E., Rustomjee, R., Mwaba, P., et al., 2016. Host-directed therapies for infectious diseases: current status, recent progress, and future prospects. The Lancet Infectious Diseases 16 (4), e47. Available at https://doi.org/10.1016/S1473-3099(16)00078-5, http://www.journals.elsevier.com/the-lancet-infectious-diseases.

Chapter 12

Guarding health: the role of vaccines and preventative measures

Noor Afsha[1], Ramzan Ahmed[2], Kaustuvmani Patowary[3], Tapan Kumar Mohanta[4], Nuntaree Chaichanawongsaroj[2], and Laishram Shantikumar Singh[3]

[1]*Department of Biochemistry, Oriental Institute of Science and Technology, Bardhaman, West Bengal, India,* [2]*Department of Transfusion Medicine and Clinical Microbiology, Faculty of Allied Health Sciences, Chulalongkorn University, Bangkok, Thailand,* [3]*Department of Microbiology, Faculty of Science, Assam down town University, Guwahati, Assam, India,* [4]*Department of Genetics Research, Institute for Medical Research and Consultation, Imam Abdulrahman Bin Faisal University, Dammam, Saudi Arabia*

12.1 Introduction

The most compelling and transformative approach in public health is to prevent disease rather than to treat it. Vaccination is one of the most effective and affordable public health interventions (WHO, 2024). Vaccination plays a powerful role in safeguarding global health. Over the last century, vaccines have significantly reduced the impact of infectious diseases, increased life expectancy, and enhanced the healthcare systems that treat these diseases (Shattock et al., 2024). A recent analysis published in *The Lancet (2024)* indicates that vaccines have prevented about 154 million deaths in the last 50 years, with the majority occurring among children under 5, highlighting the significant health and human benefits of vaccination. This analysis highlights that vaccines are not only medical measures but also essential components of public health (Haeuser et al., 2025; Shattock et al., 2024).

Early vaccination practices called variolation began in Asia and Africa, where people were exposed to mild forms of the disease to build immunity. This early technique was refined by the work of scientists like Edward Jenner, who introduced the smallpox vaccine in the late 1700s, and Louis Pasteur, who later developed vaccines for anthrax and rabies (Ahamed VP et al., 2024). These milestones became the foundation of modern vaccinology, a field that brings together immunology, epidemiology, and microbiology to create and apply preventive strategies against infectious diseases (Battles and Gilmour, 2022). Over the past few decades, vaccines have helped to control or eradicate major infectious threats like polio, smallpox, diphtheria, and measles. The routine childhood vaccination programme has been crucial in achieving these outcomes, ensuring that millions of children are protected from preventable diseases starting from infancy (Zhou et al., 2024).

The impact of vaccination extends beyond protecting just one person to include broader public health benefits. When a large population is immunised, herd immunity is established, which reduces disease transmission. This safeguards people who cannot get vaccinated, such as infants, immunosuppressed individuals, and older adults (Mirza et al., 2025). This indirect immunity is particularly critical for pathogens that can spread quickly when vaccination coverage decreases. Measles remains a prominent example because it can still cause outbreaks in areas where not enough people are vaccinated. This observation emphasises the importance of maintaining strong and consistent vaccination programmes, even for long-used vaccines (Orenstein and Ahmed, 2017).

In addition to their direct clinical effects, widespread immunisation programmes use vaccines. Immunisation programmes have significant socioeconomic advantages. Disease prevention helps to reduce hospitalisation rates, lower healthcare expenditures, and minimise productivity losses due to illness (Rodrigues and Plotkin, 2020). As evidence, US surveillance data from 1994 to 2023 demonstrate that routine childhood immunisation prevented millions of hospitalisations and saved billions in healthcare costs, thereby minimising productivity losses. Worldwide, vaccines enhance the economic stability of communities by preventing outbreaks that could otherwise strain health systems, disrupt education, and reduce workforce capacity. These broad and long-term benefits demonstrate the prioritisation of

Antimicrobial Resistance in the 21st Century. DOI: https://doi.org/10.1016/B978-0-443-44748-8.00009-9

immunisation by global health organisations such as the World Health Organisation (WHO), UNICEF, CDC, and GAVI, which coordinate large-scale vaccine procurement, distribution, and surveillance (Fine et al., 2011).

Vaccination is a rapidly advancing field, persistently evolving in response to new scientific discoveries, emerging pathogens, and changing societal demands. Despite major achievements, critical challenges persist. New infectious disease threats, increased global mobility, environmental shifts, and the resurgence of previously controlled diseases continue to threaten public health (Chan et al., 2020). The COVID-19 pandemic highlights both the potential and the limitations of vaccine science, demonstrating the demand for rapid vaccine development, equal access, and effective public health systems. Meanwhile, the growth of misinformation and vaccine hesitancy has become a major obstacle to vaccination efforts, showing that scientific programmes must be complemented by enduring trust and engagement (Shattock et al., 2024).

The biological basis of vaccines lies in their ability to train the immune system to recognise and respond to specific pathogens. Vaccines mimic natural infection, stimulating both innate and adaptive immunity without causing disease. They can be composed of live attenuated organisms, inactivated pathogens, subunit components, recombinant proteins, nucleic acids, or viral vectors (Pulendran and Ahmed, 2011). This diversity allows for tailored approaches depending on the pathogen, population, and health context. Understanding these principles is essential to appreciating how vaccines achieve their protective effects and why booster doses or multidose schedules may be necessary for long-term immunity (Plotkin, 2014).

In this chapter, we explore the biological foundations of vaccines, key milestones in vaccine development, and the broader spectrum of preventive healthcare strategies. We also examine policies, emerging vaccine technologies, societal challenges, and future directions in global immunisation. Through this comprehensive perspective, the chapter underscores that vaccination, combined with robust preventive health measures, remains essential to protecting global health now and in the decades ahead. By highlighting the scientific, historical, social, and economic dimensions of immunisation, this chapter provides a framework for understanding why vaccines are indispensable tools in the ongoing effort to guard human health.

12.2 The science of immunisation

12.2.1 Overview of the immune system and immune response

The immune system represents an intricate and well-organised network of cells, tissues, and signalling pathways that defend the body from infectious agents such as viruses, bacteria, parasites, and fungi. It works through two linked components, innate immunity and adaptive immunity. They play a special task for host protection, pathogen elimination, and building long-term immunity. This process is important to understanding how vaccines mimic natural infections to generate immunity without inducing disease (Pulendran et al., 2021).

12.2.1.1 Innate immunity: the body's first line of defence

Innate immunity offers rapid, nonspecific protection within minutes of encountering a pathogen (Murphy and Weaver, 2023). Its major components include the following:

- Physical and chemical barriers—skin, mucosal epithelium, antimicrobial peptides
- Cellular components—macrophages, neutrophils, dendritic cells, natural killer (NK) cells
- Soluble mediators—complement proteins, cytokines, interferons, chemokines.

Innate immune cells use pattern recognition receptors (PRRs) such as *Toll-like receptors (TLRs)* to identify conserved microbial structures known as pathogen-associated molecular patterns (PAMPs) (Plotkin, 2014; Rappuoli et al., 2024). Activation of these receptors triggers inflammatory responses, enhances antigen uptake, and boosts activation of the adaptive immune system (Seder and Hill, 2000).

12.2.1.2 Adaptive immunity: tailored and long-lasting protection

Adaptive immunity emerges more gradually but provides highly specific and long-lasting protection (Pulendran et al., 2021). It contains

- **B cells**, which differentiate into plasma cells producing antibodies
- **CD4+ T helper cells**, which regulate immunisation
- **CD8+ cytotoxic T cells**, which kill infected cells.

A main feature of adaptive immunity is its capability to form immune memory, allowing a faster and more robust response upon exposure to the same pathogen. This immunological principle forms the scientific foundation of vaccination (Orenstein et al., 2023).

12.2.2 How vaccines work: active versus passive immunity

The process of vaccination helps the immune system to recognise and fight diseases before they appear (Plotkin, 2005). This form of defence is mainly through active immunity, whereas passive immunity is through preformed antibodies. While both protect, they work differently and last for different lengths of time (Iwasaki and Medzhitov, 2015).

12.2.2.1 Active immunity

Active immunity is a resistance to infection provided by the production of antibodies by the individual's own immune system in response to direct exposure to an antigen. Vaccines stimulate natural infection by using nonpathogenic components like inactivated pathogens, subunit proteins, or mRNA without causing illness (Iwasaki and Medzhitov, 2015). This controlled exposure activates both innate and adaptive immune responses, which leads to the formation of long-lasting immunological memory. Therefore, memory B and T cells persist for years, and active immunity usually provides prolonged protection and can prevent reinfection upon later exposure to the actual pathogen (Nabel, 2013).

12.2.2.2 Passive immunity

Passive immunity occurs by the transfer of preformed antibodies directly, rather than by active production in the host. This is immediate protection but only temporary protection because no immune memory is established (Chen and Flies, 2013). Passive Immunity is very important for patients exposed to rabies or toxin-mediated diseases (such as tetanus) and for immunocompromised individuals. Examples of passive immunity include antibodies given through the placenta or in breast milk and through therapeutic preparations, such as immunoglobulin injections or monoclonal antibodies. Passive immunity typically lasts only several weeks to months (Casadevall et al., 2004).

Whereas active immunity provides long-lasting, often lifelong protection by way of immunological memory, passive immunity is shorter-lived but critical when fast-acting defence is needed. While vaccines are based on the induction of active immunity, passive immunity still provides a relevant complementary approach in urgent situations or for populations that cannot be immunised (Krammer, 2020).

12.2.3 Types of vaccines

The field of vaccinology has witnessed considerable advancements over the past hundred years, expanding from traditional whole virus vaccines to more advanced molecular and genetic vaccine platforms. Conventional vaccine platforms, including live and killed virus vaccines, have long served as a cornerstone for disease prevention. Nevertheless, safety considerations, immunological persistence, and limitations associated with vaccine production prompted innovators to focus on subunit vaccines, viral vector vaccines, DNA vaccines, and mRNA vaccines (Table 12.1) (Ahmad, 2023).

12.2.3.1 Live attenuated vaccines

Live attenuated vaccines are made by attenuation, where pathogens are rendered weaker but still possess immunogenic and nonvirulent properties. These vaccines closely resemble natural infections and thus produce excellent humoral and cell-mediated immunity (WHO, 2021). Examples include MMR and Varicella vaccines. However, a disadvantage might be seen with immunosuppressed people (Ahmad, 2023).

12.2.3.2 Inactivated/killed vaccines

Inactivated vaccines have pathogens that have been rendered inactive using heat, radiation, or chemicals. These vaccines are not capable of replication, making them relatively safer for immunosuppressed patients. Although they provide excellent antibody response, they need booster doses and adjuvants for inducing immunity. Examples include the inactivated polio vaccine and the Hepatitis A vaccine (Copier and Dalgleish, 2010).

12.2.3.3 Subunit, recombinant, and conjugate vaccines

These vaccines comprise essential antigens like proteins or polysaccharides with immune response capabilities. Contemporary methods have led to better safety (Rappuoli et al., 2024). These vaccines are not PAMPs and therefore

TABLE 12.1 Major types of modern vaccines: features and examples.

Sl. No	Vaccine type	What it contains/how it is made	Main immune response and features	Typical examples	References
1.	Live attenuated	Weakened but still replicating pathogen that has lost its ability to cause severe disease.	Very strong humoral and cellular immunity; long-lasting protection but not ideal for severe immunosuppression	MMR, Varicella vaccines	WHO (2021), Ahmad (2023)
2.	Inactivated/killed	Whole pathogen inactivated by heat, chemicals, or radiation so it cannot replicate	Good antibody responses, generally safer for immunocompromised patients but often need boosters and adjuvants	Inactivated polio vaccine, Hepatitis A vaccine	Copier and Dalgleish (2010)
3.	Subunit/ recombinant/ conjugate	Purified antigens such as proteins or polysaccharides, sometimes chemically linked to carriers	High safety; focus on specific antigens; usually require adjuvants to enhance immunogenicity	Hepatitis B and HPV vaccines	Rappuoli et al. (2024), Copier and Dalgleish (2010)
4.	Toxoid	Bacterial toxins chemically inactivated so they are nontoxic but still immunogenic	Induce strong anti-toxin antibody responses; often require multiple doses and boosters	Diphtheria and tetanus vaccines	Rumfield et al. (2020)
5.	mRNA	Synthetic mRNA encoding antigen, delivered in lipid nanoparticles into host cells	Strong antibody and T-cell responses; rapid design and update for emerging pathogens	COVID-19	Rumfield et al. (2020)
6.	Viral vector	Harmless carrier virus engineered to deliver antigen-encoding genes	Potent cellular immunity, useful for emerging infections but can be limited by preexisting vector immunity	Adenoviral COVID-19 vaccines	Hancock et al. (2018)
7.	DNA	Plasmid DNA encoding antigen taken up by host cells	Induces both humoral and cellular immunity; stable and easy to manufacture but still transitioning to widespread human use	Several veterinary and investigational human vaccines	Rappuoli et al. (2024), Hancock et al. (2018)
8.	Nanoparticle/VLP	Nanoparticles or virus-like particles that mimic viral structure without genetic material	Repetitive antigen display gives very strong B-cell and antibody responses, promising for next-generation vaccines	Licensed HPV and some HBV VLP vaccines; experimental influenza and HIV VLPs	Kumar et al. (2021)
9.	Peptide	Short antigenic epitopes synthesised chemically	Highly specific and easy to produce but low immunogenicity, need potent adjuvants or carriers	Cancer and infectious disease peptide vaccines in clinical trials	Malonis et al. (2019)

rely on adjuvants. Examples include vaccines for Hepatitis B and Human papillomavirus (HPV). A wide range of vaccines has been developed, each using different strategies to stimulate protective immunity. These approaches include classical whole-pathogen vaccines and newer platforms based on purified components or genetic material (Copier and Dalgleish, 2010).

12.2.3.4 *Toxoid vaccines*

Toxoid vaccines rely on bacterial toxins that have been made ineffective through chemicals and are used to build immunity against toxoid diseases. These vaccines produce high levels of antibodies but may need several doses. Examples include diphtheria and tetanus (Rumfield et al., 2020).

12.2.3.5 *mRNA vaccines*

mRNA vaccines offer a revolutionary breakthrough as they allow for the direct introduction of synthetic mRNA coding for antigens into host cells. Lipid nanoparticles (LNPs) protect mRNA and enhance its uptake. mRNA vaccines have shown excellent response capabilities against T cells and antibodies, and, more notably, they can be swiftly updated. The COVID-19 vaccine developed by Pfizer and Moderna would never have been possible without mRNA vaccine capabilities (Rumfield et al., 2020).

12.2.3.6 *Viral vector vaccines*

Viral vector vaccines rely on benign viruses as vectors for delivering specific antigens encoded within genetic materials. These vaccines have an efficient cellular response and are useful for emerging infections. The adenoviral vector platform, as seen with vaccines from AstraZeneca and Johnson & Johnson, emerged as a prominent tool during the pandemic. Existing immunity against virus vectors could be a limitation (Hancock et al., 2018).

12.2.3.7 *DNA vaccines*

DNA vaccines employ plasmids carrying genes coding for specific antigens. By these, cells in the host produce the antigen of interest (Rappuoli et al., 2024). DNA vaccines have several benefits, including stability and easy production. Also, they stimulate both humoral and cellular immunity. Despite heavy research on DNA vaccines, they are still at the stage of transitioning for general use among humans (Hancock et al., 2018).

12.2.3.8 *Nanoparticle and virus-like particle vaccines*

Nanoparticles and VLPs replicate the structural organisation seen in viruses but do not have any genomic material. Their repeated display of an antigen stimulates B cells more, leading to high-affinity antibodies. Examples include successful HPV and HBV vaccines. Novel self-assembling nanoparticle platforms are being readied for production against influenza, HIV, and universal coronaviruses (Kumar et al., 2021).

12.2.3.9 *Peptide vaccines*

Peptide vaccines rely on short epitopes as immunogens and have high specificity. Although peptide vaccines are safe and easy to produce, they have low immunogenicity and thus need strong adjuvants or carriers (Kumar et al., 2021).

12.2.4 Adjuvants and their role in vaccine efficacy

Vaccine adjuvants are components added to vaccine preparations with the aim of enhancing, shaping, or prolonging the specific immune response to an antigen (Lan et al., 2024). Whereas early vaccines depended mostly on the employment of whole pathogens, which were inherently strong stimulators of immunity, modern vaccines, which include the broad use of purified protein subunits, recombinant antigens, and mRNA vaccines, lack the inherent immunostimulatory signals normally needed for robust activation of the immune system. Adjuvants thus play an important role in enabling such vaccines to induce protective, long-lasting immunity (Reed et al., 2009).

Purified and highly targeted antigens have better safety profiles but are usually less immunogenic. If administered without any further stimulus, they may elicit weak antibody responses, fail to activate T-cell-mediated immunity, require repeated booster doses, and induce short-lived immune memory (Marciani, 2024).

Adjuvants represent important components in the design of vaccines, because highly purified antigens, while much safer, are usually only weak immunogens. By engaging innate immune pathways and enhancing antigen uptake, adjuvants increase the magnitude, speed, and breadth of antibody responses, promote effective T-cell immunity, and engender more durable immune memory (Ellebedy et al., 2020). They enable antigen dose sparing, facilitate the development of novel vaccines, and enhance overall vaccine efficacy and safety. The key advantages of vaccine adjuvants are illustrated in Fig. 12.1 (Essink et al., 2025).

12.2.4.1 Types of common adjuvants

Vaccine adjuvants enhance immune responses by modulating antigen presentation and innate immune activation. Aluminium salts, commonly referred to as alum, can help retain vaccine antigens for longer periods of time and stimulate a greater production of antibodies than do antigens presented alone (O'Hagan et al., 2020). Oil-in-water emulsions, such as MF59 and AS03, can be utilised to stimulate both the local uptake of antigen and to attract immune effector cells into the vicinity of where the vaccine antigen will reside (O'Hagan et al., 2020). Toll-like receptor (TLR) agonists activate pattern recognition receptors (PRR) to initiate robust cytokine-induced adaptive immune responses against specific antigens associated with the respective TLR (Pulendran et al., 2021). Saponin-containing adjuvants (e.g., QS-21 and Matrix-M) have also been shown to augment both antibody and T-cell responses through improved antigen cross-presentation (Cox and Brokstad, 2020). LNPs that act as delivery mechanisms for mRNA vaccines also have inherent immunostimulatory properties (Hou et al., 2021). Collectively, these adjuvant platforms represent a series of disparate yet complementary approaches labelled to maximise vaccine efficacy (Fig. 12.2).

12.2.4.1.1 Aluminium salts (Alum)

Aluminium salts, commonly referred to under the generic term alum, represent the first licensed vaccine adjuvant and have been continuously used since 1926. Through nearly a century, the relevance of alum has been maintained because of its excellent safety profile in conjunction with its proven ability to enhance immune responses (Gogoi et al., 2025; Reed et al., 2009). Its primary mechanism involves promoting strong humoral immunity. Hence, it is very effective in stimulating robust antibody production. Alum is also reported to bias the immune system towards a Th2-dominant response, supporting Ab-mediated protection over cell-mediated immunity. Because of these characteristics, aluminium-based adjuvants remain widely incorporated into various vaccines, including those targeting hepatitis A, hepatitis B, and diphtheria-tetanus-pertussis. Their long history of safe and effective use underlines their continued importance in modern immunisation programmes (Marciani, 2024).

12.2.4.1.2 Oil-in-water emulsions (MF59, AS03)

Oil-in-water emulsion adjuvants represent some of the most advanced adjuvant systems used in modern vaccines and include MF59 and AS03 (Ellebedy et al., 2020). These formulations contain microscopic droplets of the biodegradable oil squalene, which serves to enhance antigen delivery and stimulate stronger immune activation. MF59 induces robust antibody responses through recruitment of immune cells such as monocytes and dendritic cells to the injection site, while AS03, which includes α-tocopherol, further enhances innate signalling and antigen presentation. Emulsions are widely

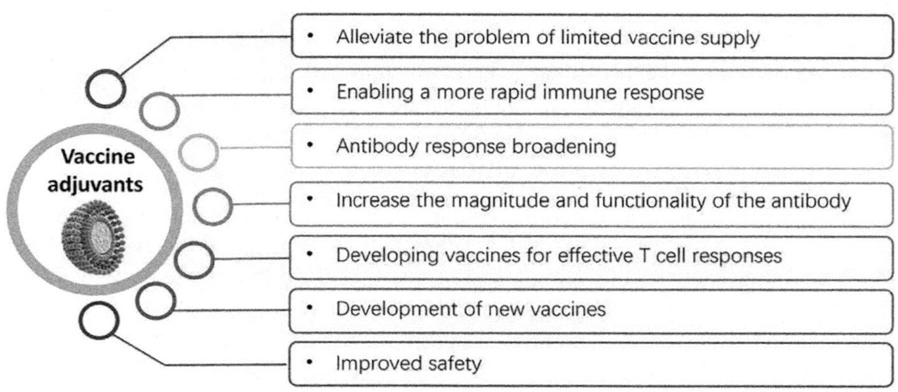

FIGURE 12.1 **Potential advantages of vaccine adjuvants.** Potential benefits of vaccine adjuvants centres include alleviating limited vaccine supply, enabling rapid immune responses, broadening antibody responses, increasing antibody magnitude and functionality, promoting effective T-cell responses, supporting development of new vaccines, and improving vaccine safety.
Cui, Y., Ho, M., Hu, Y., Shi, Y., 2024. Vaccine adjuvants: current status, research and development, licensing, and future opportunities. Journal of Materials Chemistry B, 12(17), 4118–4137. https://doi.org/10.1039/d3tb02861e.

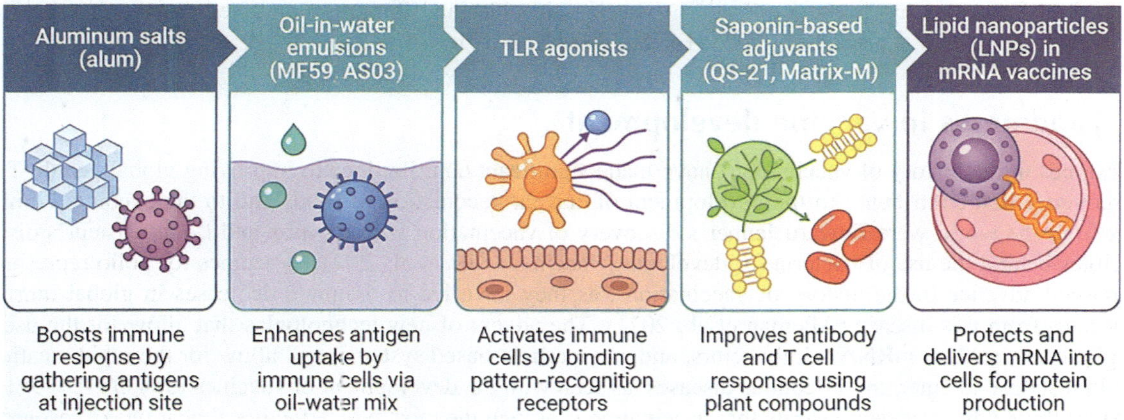

FIGURE 12.2 **Types of adjuvants vaccines.** Major classes of vaccine adjuvants and delivery systems, including aluminium salts (alum), oil-in-water emulsions (MF59, AS03), toll-like receptor (*TLR*) agonists, saponin-based adjuvants (QS-21, Matrix-M), and lipid nanoparticles (*LNPs*) used in mRNA vaccines, indicating principal immunostimulatory or delivery functions.

used in influenza vaccines and enhance immunogenicity, particularly in older adults, and support dose-sparing strategies, making them supporting tools for pandemics (Essink et al., 2025; O'Hagan et al., 2020).

12.2.4.1.3 TLR agonists

TLR agonists are adjuvants that mimic and activate Toll-like receptors on innate immune cells. By activating Toll-like receptors such as TLR4, TLR7/8, and TLR9, they stimulate dendritic cell maturation, cytokine secretion, and subsequent antigen presentation. A prominent compound within these adjuvants is Monophosphoryl Lipid A, commonly known as MPLA (Facciolà et al., 2022). It acts as a Toll-like receptor 4 agonist and has been recognised as a safe adjuvant within the licensed adjuvant system AS04, which has been incorporated into various vaccines for strains like HPV and Hepatitis B. Another prominent category would be CpG oligonucleotides, which work as Toll-like receptor 9 agonists and stimulate prominent Th1 responses (Kayesh et al., 2023; Kayraklioglu et al., 2021).

12.2.4.1.4 Saponin-based adjuvants (QS-21, Matrix-M)

Saponin adjuvants, including QS-21 and Matrix-M, are obtained from plants and have shown excellent immune responses with the production of antibodies and stimulation of T cells (Keech et al., 2020; Stertman et al., 2023). QS-21 stimulates Th1 and cytotoxic T cells, while Matrix-M adjuvants form nanoparticles that have shown less toxicity with excellent immunostimulatory effects. The adjuvants have been employed in modern vaccine technologies, including the Novavax COVID-19 vaccine (Hou et al., 2021).

12.2.4.1.5 Lipid nanoparticles in mRNA vaccines

LNPs are integral components that help deliver mRNA and safeguard it against destruction as well as deliver it inside host cells. LNPs are made up of ionisable lipids, phospholipids, cholesterol, and PEG lipids. These ingredients help stabilise the mRNA and promote efficient entry into cells (Hou et al., 2021; Hariri et al., 2025). Once inside cells, LNPs release the mRNA, and then the mRNA translates into an encoded antigen, inducing an efficient immune response. LNPs not only help stabilise mRNAs but also work as an adjuvant with low cytotoxic effects, boosting efficient B- and T-cell responses. The technological advantages have led to successful COVID-19 mRNA vaccines like Pfizer and Moderna (Krammer, 2020).

12.2.4.1.6 Role of adjuvants in contemporary vaccine development

Adjuvants are essential components of modern vaccines as they improve and direct immune responses, making vaccines more efficient. They enable dose sparing for vaccine use in outbreaks, stimulate responses associated with T cells, and are essential for controlling infections that are either complex or persistent, and allow for establishment and retention of immune memory (Essink et al., 2025; Reed et al., 2009). Moreover, adjuvants optimise advancements based on personalised adjuvants to improve the efficacy of vaccines in older and immunosuppressed patients because they have weaker immune responses (Laera et al., 2023; Ellebedy et al., 2020).

12.3 Landmarks in vaccine development

Key milestones in the history of vaccinations have made significant contributions to improving global health. Two major early discoveries that contributed to the development of current vaccination methods and to the eventual elimination of smallpox from the world were Edward Jenner's discovery of vaccination with cowpox and Louis Pasteur's discovery of the principles behind the use of bacteria for developing vaccines (Kim et al., 2021). Vaccines for polio represent another very important advance in the history of vaccinations, as they have led to dramatic decreases in global morbidity and mortality rates from this disease (Alleman et al., 2021). The advent of new technologies that allow for the use of novel vaccine platforms, such as mRNA, viral vectors, and nanoparticle-based systems, will allow for the rapid creation of new vaccines in response to emerging infectious diseases, as seen with the development of vaccines for COVID-19 (Hou et al., 2021). All of these advances demonstrate that vaccination is among the most cost-effective allocations of resources within public health and has a life-saving effect (Fig. 12.3) (Shattock et al., 2024).

12.3.1 Early breakthroughs: from smallpox to polio

The history of vaccination began with some pivotal developments that have had an enduring influence on modern immunology. The first major success story would be the invention of the smallpox vaccine by Edward Jenner in 1796, which formed the conceptual platform for immunisations that would eventually culminate in the global eradication of smallpox in 1980 (WHO, 2021). Recent historical reviews have brought forth an understanding that these formative experiences, including methods of variolation, Jenner's cowpox vaccine, and Pasteur's subsequent attenuation methods for vaccines, have provided the cornerstone on which all advances in vaccine research would be founded (Montero et al., 2023). During the 20th century, there were more advances, like the diphtheria toxoid vaccine and the IPV, which significantly brought down instances of disease prevalence on a global platform. These experiences served as a catalyst for the subsequent establishment of EPI by the World Health Organisation's efforts in 1974 and had a significant impact on expanding vaccine availability on a global scale (Cruz et al., 2025). The history of vaccination began with some pivotal developments that have had an enduring influence on modern immunology (Fig. 12.3).

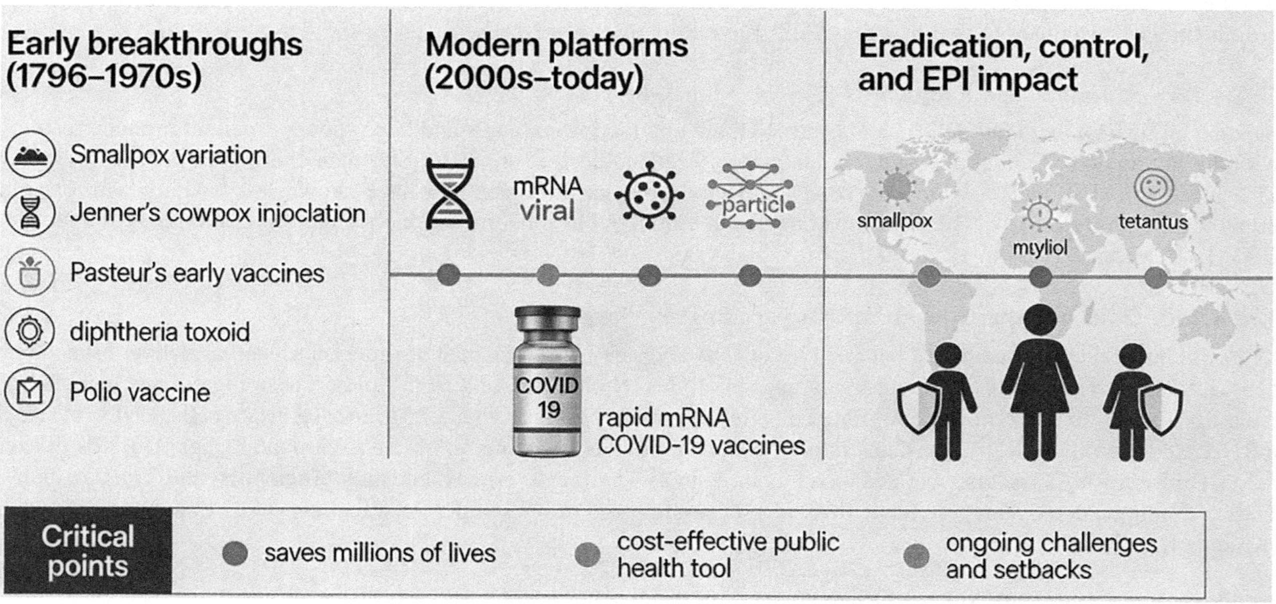

FIGURE 12.3 **Landmarks in vaccine development, highlighting early breakthroughs, modern platforms, global health impact, and persistent scientific and societal challenges.** Historical evolution of vaccines from early breakthroughs to modern vaccine platforms, highlighting milestones in vaccination development, the emergence of mRNA and viral technologies, rapid COVID-19 vaccine development, and the public health impact of immunisation through disease eradication, control, and EPI outcomes. *EPI*, Expanded Programme on Immunisation.

12.3.2 Advances in modern vaccine technology

The past two decades have witnessed an uncharted advancement in vaccine research and development. Technologies propelled by molecular biology, genomics, and nanotechnology have revolutionised vaccine research. Platforms like DNA vaccines, mRNA vaccines, viral vector vaccines, and nanoparticle-protein vaccines have completely transformed the ways and means of vaccine research (Rappuoli et al., 2024). A major stimulus for these advancements came with the onset of the COVID-19 pandemic, when mRNA LNP vaccines took less than a year to develop and enter clinical practice, thereby unearthing the promises offered by novel and modular vaccine technologies. Contemporary literature underscores that recent advances in reverse vaccinology, structural vaccinology, and computational antigen discovery have brought refinement in vaccine research and development methods and approaches (Poria et al., 2024).

12.3.3 Success stories: eradication and control of infectious diseases

Immunisation has brought about several of the greatest advances for global health. Since the eradication of smallpox, immunisation campaigns have eliminated and controlled diseases like polio, measles, and tetanus worldwide (Shattock et al., 2024; WHO, 2024). A 2025 review on behalf of the 50th anniversary of 'EPI' emphasises that childhood immunisation has eliminated millions of deaths per year and made significant progress in lowering morbidity caused by infections like whooping cough, diphtheria, and rubella (Chacon-Cruz et al., 2025). Moreover, extension efforts on EPI on new vaccines like Hib, pneumococcal conjugate, and HPV have made significant progress in lowering childhood deaths and life expectancy with cancer control as well (WHO, 2025).

12.3.4 Challenges in vaccine development and distribution

Despite making rapid progress, several challenges remain with regard to vaccine development, distribution, and adoption. From a scientific perspective, there are constantly emerging and re-emerging pathogens. These include mpox virus, L1, L2, and L3 avian influenza virus strains, as well as SARS-like viruses. However, recent research on mpox vaccines illustrates that, despite making rapid progress, there are still challenges with vaccine production and distribution on a global scale (Zhai et al., 2025; Garcia-Atutxa et al., 2024). Apart from these scientific challenges, there are several societal and logistical challenges as well. These include vaccine hesitancy, vaccine myths, and global immunisation goals. Several papers published in 2025 have pointed out that several lessons were drawn from COVID-19 pandemic research (Shafaati et al., 2025; WHO, 2025). Improving global surveillance and decentralising production and distribution would be an essential step towards rapid global vaccine distribution in times of global health crises (Fig. 12.3).

12.4 Preventative health practices aside from vaccines

Though immunisation is one of the most effective means of preventing diseases, it is more effective when it is combined with other means of preventative health practices (Mapingure et al., 2024; WHO, 2023). Nonpharmaceutical practices like good hygiene and sanitation practices, nutritional support, health promotion, early screening, as well as management of the environment, constitute an important means of minimising the burden of diseases, especially in developing nations (Fig. 12.4). This is because such practices are involved in addressing the social factors that influence health practices (Kumar et al., 2025). Table 12.2 summarises the main focus and key contributions of vaccination and each nonvaccine preventive strategy described in this section (Fig. 12.4).

12.4.1 Hygiene and sanitation

Personal and environmental hygiene, or sanitation, is one of the key principles of preventing diseases, especially those caused by the faeco-oral route of transmission. Having access to clean drinking water, toilet facilities, and observing good personal hygiene practices is referred to by the acronym WASH (Water, Sanitation, and Hygiene), which can help combat the spread of diarrhoea, respiratory, and tropical diseases (Mapingure et al., 2024).

Current research reveals that poor WASH conditions continue to significantly influence childhood mortality rates, as it is one of the major factors contributing to childhood illnesses, especially among children who are below five years old (Kumar et al., 2025). This is because cleanliness through proper WASH practices is effective in preventing the spread of diseases by ensuring that the environmental factors that affect the efficiency of other preventive strategies are optimised (Mapingure et al., 2024).

TABLE 12.2 Fundamental features of vaccine and non-vaccine preventive strategies.

Sl.No	Strategy	Main focus	Key points (2–3 phrases)	References
1.	Hygiene and sanitation (WASH)	Water, sanitation, hygiene	Reduces diarrhoea and other infections, especially protects children under five, and supports effectiveness of other interventions.	Mapingure et al. (2024)
2.	Nutritional interventions	Diet and micronutrients	Corrects undernutrition and deficiencies. It strengthens immune function, improves outcomes of infections and chronic disease.	de Hoop et al. (2025), Norris et al. (2023); Ozawa et al. (2017)
3.	Health education and behaviour change	Knowledge and practices	Promotes handwashing, safe feeding, care-seeking, uses community- and culture-specific messages for sustained change.	Shafique et al. (2025), Xie et al. (2024), Ogutu et al. (2024)
4.	Screening and early diagnosis	Early detection of NCDs	Identifies cancer, diabetes, cardiovascular risk earlier, enables timely treatment, and reduces mortality	Abbasi, 2020; Givler and Givler, 2023, Al-Abdulla et al. (2024), Mangayarkarasi, (2025)
5.	Vector control and environmental management	Mosquitoes and other vectors	Uses nets, insecticides, habitat management, lowers malaria, dengue, and other vector-borne diseases. It benefits from One Health approaches	Kumar et al. (2025), Kalitsilo, (2025)

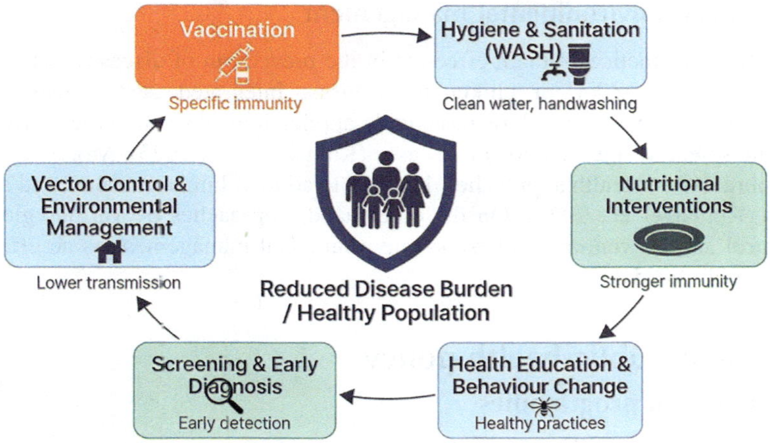

FIGURE 12.4 **Integrated preventive health strategies alongside vaccination.** The diagram shows how vaccination and nonpharmaceutical practices (WASH, nutrition, health education, screening, and vector control) act together to reduce overall disease burden and support a healthy population. Integrated public health interventions contributing to reduced disease burden and a healthy population, illustrating the complementary roles of vaccination, hygiene and sanitation (WASH), nutritional interventions, health education and behaviour change, screening and early diagnosis, and vector control with environmental management.

12.4.2 Nutritional interventions

Proper nutrition is critical for ensuring immune competency and protecting against both communicable and noncommunicable diseases. Malnutrition, including undernutrition, vitamin deficiencies, or obesity, is known to weaken immune systems. Nutrition-sensitive interventions such as supplementation, dietary diversification, or social protection instruments have proven effective in improving nutritional situations and tackling disparities in health (de Hoop et al., 2025).

Systematic reviews show that there is evidence that interventions related to food insecurity, including community-based nutritional interventions and produce prescription schemes, improve dietary quality and enhance positive health outcomes (Norris et al., 2023). Nutritional interventions complement vaccine services for the prevention of diseases by bolstering the immune systems of the host through healthy growth and development (Ozawa et al., 2016).

12.4.3 Health education and behavioural changes

Health promotion or health education, coupled with efforts directed at behaviour change, is geared towards empowering people to make healthy behaviour choices that minimise risks for diseases. Social Behaviour Change Communication approaches have been cited to make a great contribution to improving behaviour in various health matters, such as infant feeding, handwashing practices, and seeking preventive healthcare (Shafique et al., 2025).

More recent evidence suggests that the key to effective health education is to render it culture-specific, community-based, and embedded in health or social programmes (Xie et al., 2024). Such interventions, through their role of promoting behavioural change through education, improve the long-term effectiveness of health prevention strategies that extend only to biomedicine (Ogutu et al., 2024).

12.4.4 Screening and early diagnosis programmes

Screening programmes are important elements of preventive care with the capability of detecting diseases in their earlier stages, which are more easily treated. Such screens for early diagnosis are important for noncommunicable diseases such as cancer, diabetes, and cardiovascular disease. Early detection is important in reducing mortality rates for such diseases (Givler and Givler, 2023).

Research evidence suggests that screening efforts in the population have enhanced the earlier risk factor detection of diseases and made it easier for earlier clinical intervention (Al-Abdulla et al., 2024). In places such as India, the need for tackling cancer by developing more screening or diagnostic facilities for cancer is recognised as a key step for minimising the risk of the disease, as well as for ensuring survival rates (Mangayarkarasi et al., 2025).

12.4.5 Vector control and environmental management

Vector control and management practices remain effective in the prevention of diseases caused by transmission through insects, such as malaria, dengue, Zika, or chikungunya virus. Integrated vector management practices such as environmental manipulation, biological control, or insecticide applications have shown effectiveness in managing vector density and controlling the spread of the mentioned diseases (Kalitsilo et al., 2025). Moreover, recent literature suggests that it is imperative to embrace One Health approaches that seek to address human, animal, and environmental interactions in vector control efforts (Kumar et al., 2025). On the other hand, approaches by various global bodies, including the Centres for Disease Control and Prevention, underscore environmental management as an effective solution for vector-borne diseases (Kalitsilo et al., 2025).

12.5 Vaccination and public health policy

12.5.1 Global immunisation programmes

Global immunisation programmes coordinated by the World Health Organisation (WHO), UNICEF, and GAVI, the Vaccine Alliance, form the backbone of worldwide vaccination efforts (WHO, 2025). These efforts aim to ensure that everyone who needs life-saving vaccinations has access to them, especially in low- and middle-income nations. According to WHO, routine immunisation is currently estimated to save 3.5–5 million lives each year, but more than 14 million children were 'zero-dose' in 2023, despite continued challenges in vaccine access equity (Fig. 12.5) (WHO, 2025).

12.5.2 National immunisation schedules and strategies

National immunisation schedules are developed based on global advice that is adapted into country-specific immunisation platforms with regard to the epidemiological context, health system, and needs of the people (WHO, 2025). A significant number of countries are engaging in regular updates of their immunisation schedule to add newly developed vaccine products, along with boosters, to ensure high coverage of core childhood vaccinations. More recent evidence emerges that through proper governance, funding, and alignment of immunisation services with primary healthcare systems, universal

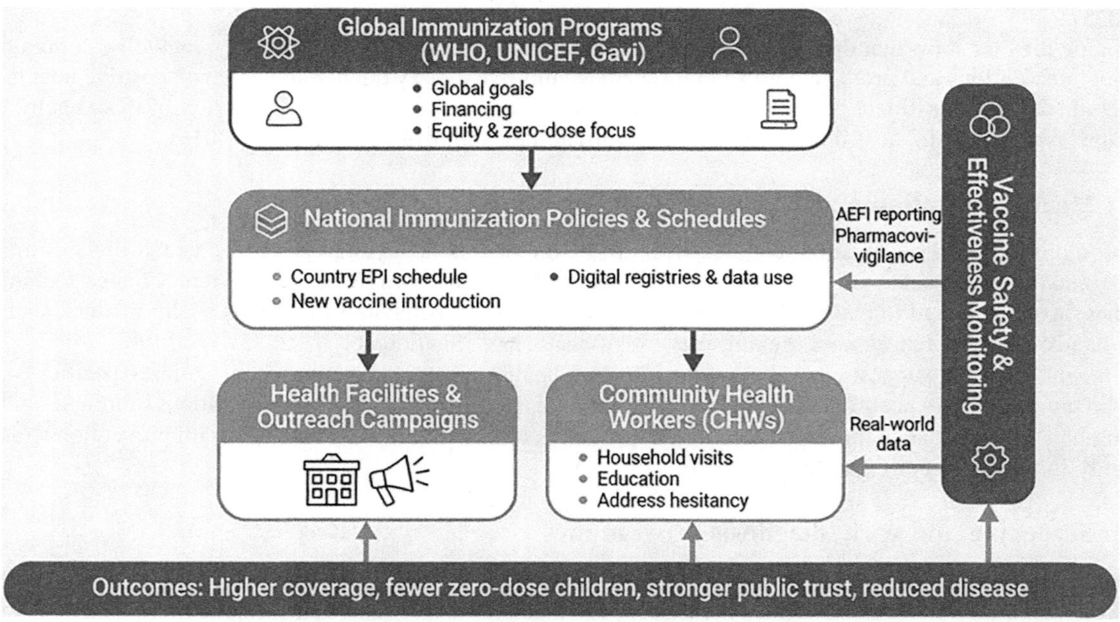

FIGURE 12.5 **Vaccination and public health policy framework showing how global programmes, national immunisation policies, community delivery (health facilities and community health workers), and vaccine safety monitoring interact to increase coverage and reduce.** Multilevel framework of immunisation programme implementation showing interactions among global immunisation programmes, national immunisation policies and schedules, health facilities and outreach campaigns, community health workers, and vaccine safety and effectiveness monitoring, collectively contributing to improved vaccine coverage and reduced disease burden.

immunisation coverage can be achieved (Chen et al., 2024). Digital immunisation registers and data-based monitoring platforms are becoming more prominent for monitoring gaps in immunisation coverage.

12.5.3 Role of Community Health Workers and outreach campaigns

Community Health Workers (CHWs) are crucial for ensuring enhanced vaccine acceptance by serving as connectors between health systems and communities. CHW engagement in vaccine delivery, educational efforts, and support has been confirmed to make a big difference in improving vaccine coverage in both rural and disadvantaged areas of society (Ogutu et al., 2024). CHW-led outreach campaigns are also effective in dealing with vaccine hesitancy by helping to build trust, dispelling misconceptions, and using appropriate communication approaches (Xie et al., 2024). Such community-based efforts are increasingly recognised as important elements of immunisation systems that ensure sustainability.

12.5.4 Monitoring vaccine safety and efficacy vaccine

Safety monitoring is a key cornerstone of ensuring public confidence in immunisations. Pharmacovigilance platforms monitor Adverse Events Following Immunisation (AEFI) cases. Such monitoring is important for overcoming vaccine hesitancy and misinformation (Nasr et al., 2025). New methodologies, such as digital platforms for monitoring vaccine safety in real-world studies, have become effective in monitoring vaccine safety after their licensure to ensure that urgent policy actions taken by international institutions, like the WHO, are timely, as happened with the COVID-19 outbreak (WHO, 2025).

12.6 Misinformation and hesitancy about vaccines

Vaccine hesitancy represents one of the most pressing public health challenges worldwide, identified by the World Health Organisation as one of the top ten threats to global health (Brumbaugh et al., 2025). While vaccines are among the most effective interventions for reducing morbidity and mortality from infectious diseases, the persistence of vaccine hesitancy and misinformation continues to undermine immunisation efforts on a global scale (Brumbaugh et al., 2025; Annandale et al., 2025).

The COVID-19 pandemic has dramatically intensified this issue, leading to outbreaks of preventable diseases fuelled by a surge in vaccine hesitancy and misinformation surrounding vaccination (Annandale et al., 2025; Ruggeri et al., 2024). Vaccine hesitancy is defined as the 'delay in acceptance or refusal of vaccination despite the availability of vaccination services'. It is a complex phenomenon influenced by a variety of interrelated factors that extend beyond mere concerns about vaccine safety (Annandale et al., 2025; Ruggeri et al., 2024).

Key contributors to vaccine hesitancy include fears about side effects and uncertainties regarding potential long-term effects. Misinformation, particularly as it spreads rapidly through social media platforms like Facebook, YouTube, and WhatsApp, plays a critical role in shaping public perception (Ingebretsen Carlson et al., 2025; Saddi et al., 2025). Additionally, low trust in institutions such as government bodies, pharmaceutical companies, and healthcare systems significantly affects vaccine acceptance (Saddi et al., 2025).

Other influencing factors include conspiracy theories, political ideologies, cultural beliefs, and structural barriers like limited access to immunisation services (e.g. transportation issues and language barriers). Social media platforms profoundly affect the dissemination of information; their algorithms tend to promote emotionally charged narratives, bypass traditional gatekeeping mechanisms, and create echo chambers where users are primarily exposed to information that aligns with their existing beliefs (Ruggeri et al., 2024). Exposure to vaccine misinformation through social media has been shown to significantly decrease vaccination intentions, and once individuals are exposed to misinformation, rectifying these misconceptions becomes increasingly challenging.

Research in the scientific literature indicates that fear-based tactics are ineffective and may even be counterproductive (Saddi et al., 2025). Instead, effective strategies involve educating trusted healthcare providers, who serve as the primary source of information for the majority of patients, on techniques such as motivational interviewing and fostering transparent dialogue (Whitehead et al., 2023). Preemptively addressing misinformation has proven to be more effective than correcting misconceptions after they have arisen (PAHO, 2025; Labbé et al., 2025).

Promoting a general scientific consensus on vaccine safety and efficacy, along with targeted educational and motivational counselling, can significantly enhance vaccination rates compared to standard care. Additionally, community support through vaccine advocates and addressing structural barriers to vaccine access are crucial for increasing willingness to receive vaccinations (Labbé et al., 2025; Tuckerman et al., 2022).

To restore public trust, it is essential to ensure institutional accountability, maintain transparency in vaccine decision-making, report adverse events accurately, and deliver clear messaging from health organisations (Brumbaugh et al., 2025; Ruggeri et al., 2024; Carosella et al., 2024). Approaches to combat vaccine hesitancy should be multifaceted, evidence-based, and take into account the complex factors that influence vaccine acceptance (Whitehead et al., 2023; PAHO, 2025). While no single strategy can entirely eliminate hesitancy, the most effective approaches involve trusted communication with healthcare providers, transparency from institutions, community engagement, structural improvements in vaccine accessibility, and framing messages that counter misinformation while reinforcing trust based on established scientific consensus (Brumbaugh et al., 2025; Labbé et al., 2025).

12.7 Emerging infectious diseases

Emerging infectious diseases (or EIDs) are typically new or rapidly spreading infections that occur as a result of zoonotic spillover incidents. Such spillover processes can be further complemented by a complex interplay of environmental changes and human behaviours (e.g., urbanisation, deforestation, and climate change) which modify the environments and interactions between wildlife and humans (Ellwanger and Chies, 2021; Mettenleiter et al., 2023). As the threats of antimicrobial resistance grow, these new infectious diseases pose global threats to health. Examples include Ebola, Severe Acute Respiratory Syndrome (SARS), and more recently COVID-19. Each of the diseases has demonstrated just how quickly a pathogen moves and disrupts societies, economies, and healthcare systems around the globe. While vaccination has been deemed a cornerstone of prevention and control for infectious diseases, many traditional vaccine development platforms struggle to scale up to new threats when they emerge quickly. These traditional methods are slow to implement, and these delays are expected to delay the advent of efficient vaccines during outbreaks (Li et al., 2022; Abbasi, 2020). Fortunately, new vaccines have revolutionised how an infection control system is controlled, allowing for immediate supply and treatment of vaccines in the most devastating disease events. The COVID-19 vaccine is one of the most famous examples of such efforts, which in this period have occurred in months, from genome sequencing to launching in the open to the entire population. The development of new modalities, notably mRNA-based technology, viral vectors, nanoparticle delivery technology, and methods made vaccines effective and adaptable, and this was achieved (Li et al., 2022; Okuyama, 2023). However, these upcoming vaccine platforms that use the technology are also known for their flexibility. They can be adaptively turned on or off to work on new pathogens that emerge later on. With these vaccines, aimed at conserved viral epitopes, they could help in time and extend a wide range of efficacy against a number of related viruses and maybe the readiness of the community for further outbreaks. Yet research and development work has been concentrated on a few points around vaccine durability, cross-resistance capability against different strains, and ease of use. The above actions are essential to improve global vaccination coverage and for the protection of human populations globally against emerging infectious diseases (Okuyama, 2023; Szabó et al., 2022). The progress in vaccine technology and development has yielded many potential advantages, but problems often prevent the timely implementation of life-saving interventions. Issues of fair access to vaccines, vaccine hesitancy among some populations, and restrictive policies can challenge the implementation of comprehensive vaccine programmes. More stringent global surveillance, sharing of knowledge between countries on an open basis, and provision of flexible production networks that can be rapidly modified in response to new threats are required to overcome these barriers (Szabó et al., 2022; Mettenleiter et al., 2023). So, whatever the newly developed vaccines are able to accomplish is just a weapon and should be part of an offensive public health strategy aimed at responding to emerging infectious diseases. Part of these efforts has to be invested in global health security, where communities are prepared not just when they develop to respond to the emergence of outbreaks but also when they are able to organise and respond successfully (Markotter et al., 2023). We can also leverage collaboration and innovation in vaccine development and distribution to help us adapt to the ubiquitous challenge of evolving infectious diseases.

12.8 Integrating vaccines within a holistic preventive health perspective

A holistic preventative health framework recognises that vaccines are important to be embedded within broader, interlinked health measures that address the human, animal, environmental, and social determinants of health (Rodrigues and Plotkin, 2020; Wang et al., 2025). Crucially, approaches such as One Health, integrated immunisation with health and nutrition services, and global research and implementation strategies are shaping how countries and global partners work together to prevent disease effectively in the 21st century (Basheer et al., 2025).

12.8.1 The One Health approach

The One Health approach emphasises that human health is linked with animal health and ecosystem integrity, forming the backbone of frameworks for infectious disease prevention and control, including diseases preventable by vaccines. One Health encourages multisectoral cooperation across human medicine, veterinary science, environmental science, and public health to tackle disease threats jointly (Kelly et al., 2022). WHO says that most infectious diseases arising at the human–animal–environment interface, most recently including many vaccine-preventable diseases, require integrated surveillance and prevention systems that include vaccines as part of a larger disease control system (Basheer et al., 2025). A recent comprehensive review notes that effective prevention and preparedness for vector-borne diseases, where vaccines may play an increasing role, requires multisectoral, multilevel governance that embodies One Health principles by integrating data, resources, and actions across sectors (Robbiati et al., 2024).

12.8.2 Linking vaccination with other preventative interventions

Integrating vaccination programmes with other preventive health services, nutrition, maternal and child health, hygiene, and mobile outreach, for example, improves both coverage and health outcomes (Rodrigues and Plotkin, 2020; WHO, 2024). This is further supported by the evidence from implementation research studies in Pakistan, demonstrating that routine immunisation integrated into broader health and nutrition services, including maternal, newborn, and child healthcare with community mobilisation, significantly improved full vaccination coverage in conflict-affected areas of Pakistan and reduced the number of 'zero-dose' children (Khan et al., 2024). The integrated outreach approach, a bringing together of immunisation with the broader health service package, intensified community engagement and health education, and mobile health teams, is one good way preventive services could synergise to reach underserved populations and strengthen health systems (Khan et al., 2024).

Similarly, regional policy research also puts a strong accent on realising innovative, sustainable solutions for reducing zero-dose vaccination. Children who have never received any routine immunisation depend on approaches that creatively integrate immunisation with primary healthcare activities, community planning, and innovations in local service delivery (Abdel Majeed et al., 2025).

12.8.3 Role of global collaboration and research networks

The integration of vaccines within holistic health frameworks is done through global collaboration and research networks. International partnerships, such as the WHO-led Immunisation Agenda 2030, are very keen on enhancing equity gaps in routine immunisation and linking vaccines with other health priorities such as nutrition and WASH (Robbiati et al., 2024). Implementation research supported by UNICEF has documented strategies that improve vaccine delivery outcomes such as microplanning, workforce development, digital tools, and governance frameworks that align immunisation with broader health systems and policies (WHO, 2024). The efforts under the umbrella of One Health also improve global capacity in the integrated detection and response of diseases through interlinking human, animal, and environmental health information to strengthen the detection of outbreaks and sharing knowledge on surveillance and coordination of the interventions between sectors. This allows for a strategic approach to vaccination, proactive in nature and supported by evidence (Oyugi et al., 2025).

12.9 Ethical, social, and economic dimensions of vaccination

Immunisation programmes have increasingly become entrenched beyond their medical efficacy and have become entrenched in ethics, social acceptance, and economic considerations (Fieselmann et al., 2022; Rodrigues and Plotkin, 2020). The aforementioned aspects have to be considered to promote equitable delivery of vaccines, social acceptance, and sustainability of gains derived from vaccination programmes. Current realities in global health have reiterated the need for ethics and sustainability in vaccination programmes (Dronova et al., 2022; Santangelo et al., 2024).

12.9.1 Access to vaccines and equity

Equal access to vaccines continues to pose a major challenge in global health, especially in low- and middle-income countries (LMICs) (Wouters et al., 2021). Inequality in vaccine distribution is affected by financial capabilities, logistics, geopolitical issues, and intellectual property rights (Altindis, 2022). The unequal distribution of vaccines during the COVID-19 outbreak led to a delayed coverage of vaccination in all affected regions, thus intensifying inequalities in all health aspects. Initiatives have been put in place globally, such as COVAX, to address inequity in vaccine access.

However, inequities in vaccine access have remained a major challenge in all health systems in all regions because of a lack of capacity in local production, health systems, and consideration of vulnerable populations such as children, elderly people, and people with weak immune systems (Wouters et al., 2021; WHO, 2023).

12.9.2 Ethical issues in vaccine mandates and clinical trials

Mandates of vaccines present a multifaceted challenge in terms of ethics, including autonomy, safety, and proportionality. Mandates can greatly raise vaccination rates and slow the spread of diseases, but they have to be placed in an ethical context (Fieselmann et al., 2022; Shachar and Reiss, 2020). Ethical considerations recommend mandates when nonmandatory tools have failed, when a serious risk is involved, and when exemptions are carefully considered. Further, conducting trials ethically, especially in terms of consent, population representation in trials, and access in regions where successful tests have proven safe and effective, is critical in vaccine trials (Canning et al., 2022). Present-day ethics directing trials include a focus on community participation and ethics globally, especially in regions with less access to medical technology (Jalilian et al., 2023).

12.9.3 Economic benefits of vaccination and disease prevention

Vaccination programmes are considered among the most cost-effective methods of medical intervention. Apart from lowering healthcare spending, vaccination programmes boost economies by preventing lost production and curtailing excess disability (Giubilini, 2021). Cost-effectiveness analysis has established a high return on investment in vaccination programmes, especially vaccination programmes targeting children. Present studies have indicated a return of multiple dollars for every dollar invested in vaccination programmes by precluding treatment spending and maintaining industrial production (Ozawa et al., 2016).

12.10 Future perspectives

Vaccination is a continuously and rapidly evolving field, fed by technological advances, computational innovation, and an ever-improved understanding of immunology and public health dynamics (Kong, 2025). Hence, the combination of personalised science with artificial intelligence, nanotechnology, and improved social systems holds great promise for the future of vaccines and preventive health (Lazarus et al., 2024).

12.10.1 Personalised and precision vaccines

Personalised and precision vaccines apply these advances in genomics and computational biology to tailor immunisations to individual immune profiles and disease risks. Currently, AI tools are applied to the identification of neoantigens and the prediction of immunogenic epitopes to optimise vaccines for specific patient profiles in therapeutic and cancer vaccines (Sahin and Türeci, 2018). This advances vaccine efficacy by aligning antigen selection with each individual's particular immune landscape, a strategy that perhaps in future years will be extended to infectious diseases. Personalised vaccines may lead to enhanced responses among subjects with variable immune competence, such as the elderly or immunocompromised, and specialise in complex conditions, including tumours with neoantigen identification (Kong, 2025).

12.10.2 Artificial intelligence and nanotechnology in vaccine design

The integration of AI and nanotechnology is revolutionising the vaccine development pipeline by enabling faster, more efficient design and delivery systems (Ong et al., 2023; He et al., 2024). AI algorithms facilitate antigen discovery, epitope prediction, and optimisation of mRNA sequences, increasing speed and accuracy in vaccine formulation. Nanotechnology supports such efforts with engineered delivery systems, including but not limited to LNPs and virus-like particles, which protect antigens and enhance cellular uptake and immune activation. Next-generation platforms exploit nanoscale carriers to improve vaccine stability, targeting, and dose efficiency, expanding the scope of immunisation strategies (Irvine and Read, 2020).

12.10.3 Strengthening public trust and health infrastructure

Despite scientific advances, public confidence and robust health infrastructure remain foundational to vaccine success. Trust in vaccines and public health institutions influences uptake and long-term programme sustainability. Large multicountry surveys show fluctuating confidence in vaccination and underscore the continued need for culturally sensitive communication, transparent data sharing, and community engagement to address hesitancy and misinformation (Lazarus et al., 2024). Strengthening health systems, integrating immunisation with routine primary care, and expanding digital health services are crucial for effective vaccine delivery, surveillance, and response to emerging threats (Brumbaugh et al., 2025).

12.10.4 Towards a disease-free future

The long-term view of vaccinology includes the development of broad-spectrum and universal vaccines that would protect against whole classes of pathogens, proactively forestall pandemics, and ensure sustained global cooperation (Erbelding et al., 2018). The attempt to formulate vaccines that grant immunity against multiple strains or future viruses, for instance, a wider coronavirus vaccine, showcases progress towards preemptive and universal approaches. Such innovations, along with equitable access to vaccines and integrated public health strategies, support the aspiration towards a reduced or disease-free world. However, this future calls for the continued investment in research, cross-disciplinary collaboration, and policies that ensure fair vaccine access globally (Pulendran et al., 2021).

12.11 Conclusion

Vaccination remains one of the most effective pillars of preventive public health, contributing substantially to disease control, reduced mortality, and improved population health. Its value extends beyond individual protection to broader community benefits through reduced transmission, outbreak prevention, and strengthened health security. As a cost-effective and scalable intervention, immunisation continues to play a central role in advancing public health goals and supporting resilient health systems.

At the same time, sustaining these gains requires recognition that vaccination alone is not sufficient. Long-term disease prevention is strengthened when immunisation is integrated with complementary interventions such as hygiene and sanitation, nutrition, maternal and child health, surveillance, health education, and vector control. Such integrated and multisectoral approaches reinforce the broader preventive framework needed to address both existing and emerging health threats.

Despite remarkable progress, persistent challenges, including inequities in access, health system limitations, vaccine hesitancy, and evolving infectious risks, continue to hinder universal protection. Addressing these barriers requires coordinated action involving scientific innovation, community engagement, strong health systems, policy support, and global partnerships to ensure equitable and sustained delivery of preventive interventions.

Looking forward, prevention must remain central to global health strategies, with vaccination serving as a cornerstone within a broader framework of primary healthcare and public health preparedness. Strengthening immunisation programmes while aligning them with other preventive measures will be essential for reducing disease burden, protecting vulnerable populations, and advancing health equity. Ultimately, the future of global health depends not only on responding to disease but also on sustained investment in prevention as the foundation of healthier, more resilient populations.

AI disclosure

During the preparation of this work, the author(s) used Quillbot for grammar check. After using this tool/service, the author(s) reviewed and edited the content as needed and take(s) full responsibility for the content of the publication.

References

Abbasi, J., 2020. COVID-19 and mRNA vaccines - first large test for a new approach. JAMA - Journal of the American Medical Association 324 (12), 1125–1127. Available at https://doi.org/10.1001/jama.2020.16866, http://jama.jamanetwork.com/journal.aspx.

Abdel Majeed, Z., Ahmad, N., Mere, O., Farrukh, S., Mohsni, E., Derow, M., et al., 2025. Innovative and sustainable solutions for reducing 'zero-dose' vaccination: how can the region respond? Frontiers in Public Health 13. Available at https://doi.org/10.3389/fpubh.2025.1652977.

Ahmad, S., 2023. A comprehensive review on types of vaccines: from classic to cutting-edge. Vaccines & Vaccination Open Access 8 (2), 1–6. Available at https://doi.org/10.23880/vvoa-16000164.

Ahamed VP, A., Choudhari, S.G., Mudey, A., Joshi, A., 2024. Jenner's Legacy: how Edward Jenner's Smallpox Vaccine Changed Public Health. Cureus 16 (9). Available at https://doi.org/10.7759/cureus.68594.

Al-Abdulla, S.A., Haj Bakri, A., Mansaray, M.A., Al-Kuwari, M.G., 2024. Assessing the impact of annual health screenings in identifying noncommunicable disease risk factors within Qatar's primary health care corporation Qatari registered population. Frontiers in Public Health 12. Available at https://doi.org/10.3389/fpubh.2024.1305636, http://journal.frontiersin.org/journal/public-health/section/public-health-education-and-promotion#archive.

Alleman, M.M., Jorba, J., Henderson, E., Diop, O.M., Shaukat, S., Traoré, M.A., et al., 2021. Update on vaccine-derived poliovirus outbreaks—worldwide, January 2020–June 2021. Morbidity and Mortality Weekly Report 70 (49), 1691–1699. Available at https://doi.org/10.15585/MMWR.MM7049A1, http://www.cdc.gov/mmwr/volumes/70/wr/mm7049a1.htm?s_cid=mm7049a1_w.

Altindis, E., 2022. Inequitable COVID-19 vaccine distribution and the intellectual property rights prolong the pandemic. Expert Review of Vaccines 21 (4), 427–430. Available at https://doi.org/10.1080/14760584.2022.2014819.

Annandale, G., Kola-Palmer, S., Duke, É., 2025. The complex landscape of vaccine hesitancy and hesitant adopters: quantitative predictors and thematic insights into COVID-19 vaccine attitudes. Human Vaccines & Immunotherapeutics 21 (1). Available at https://doi.org/10.1080/21645515.2025.2511350.

Basheer, A., Tran, M., Khan, B., Jentner, W., Wendelboe, A., Vogel, J., et al., 2025. Comprehensive review of one health systems for emerging infectious disease detection and management. One Health 21, 101253. Available at https://doi.org/10.1016/j.onehlt.2025.101253.

Battles, H.T., Gilmour, R.J., 2022. Beyond mortality: survivors of epidemic infections and the bioarchaeology of impairment and disability. Bioarchaeology International 6 (1-2), 23–40. Available at https://doi.org/10.5744/bi.2021.0003, https://journals.upress.ufl.edu/bioarchaeology/issue/view/331.

Brumbaugh, K.(Q.), Gellert, F., Mokdad, A.H., 2025. Understanding vaccine hesitancy: insights and improvement strategies drawn from a multi-study review. Vaccines 13 (10), 1003. Available at https://doi.org/10.3390/vaccines13101003.

Canning, A.G., Watson, K.E., McCreedy, K.E., Olawepo, J.O., 2022. Ethics and effectiveness of US COVID-19 vaccine mandates and vaccination passports: a review. Journal of Research in Health Sciences 22 (2). Available at https://doi.org/10.34172/jrhs.2022.81, http://jrhs.umsha.ac.ir/index.php/JRHS/article/view/7494/pdf.

Carosella, E.A., Su, M., Testa, M.A., Arzilli, G., Conni, A., Savoia, E., 2024. COVID-19 vaccine hesitancy and misinformation endorsement among a sample of native Spanish-speakers in the US: a cross-sectional study. Healthcare 12 (15), 1545. Available at https://doi.org/10.3390/healthcare12151545.

Casadevall, A., Dadachova, E., Pirofski, L.A., 2004. Passive antibody therapy for infectious diseases. Nature Reviews Microbiology 2 (9), 695–703. Available at https://doi.org/10.1038/nrmicro974.

Chan, J.F.W., Yuan, S., Kok, K.H., To, K.K.W., Chu, H., Yang, J., et al., 2020. A familial cluster of pneumonia associated with the 2019 novel coronavirus indicating person-to-person transmission: a study of a family cluster. The Lancet 395 (10223), 514–523. Available at https://doi.org/10.1016/S0140-6736(20)30154-9, http://www.journals.elsevier.com/the-lancet/.

Chen, L., Flies, D.B., 2013. Molecular mechanisms of T cell co-stimulation and co-inhibition. Nature Reviews Immunology 13 (4), 227–242. Available at https://doi.org/10.1038/nri3405.

Chen, S., Rodewald, L.E., Du, A.H., Tang, S., 2024. Advancing the national immunization program in an era of achieving universal vaccine coverage in China and beyond. Infectious Diseases of Poverty 13 (1). Available at https://doi.org/10.1186/s40249-024-01192-6, https://idpjournal.biomedcentral.com.

Copier, J., Dalgleish, A., 2010. Whole-cell vaccines: a failure or a success waiting to happen? Current Opinion in Molecular Therapeutics 12 (1), 14–20.

Cox, R.J., Brokstad, K.A., 2020. Not just antibodies: B cells and T cells mediate immunity to COVID-19. Nature Reviews Immunology 20 (10), 581–582. Available at https://doi.org/10.1038/s41577-020-00436-4, http://www.nature.com/nri/index.html.

Cruz, E., Reyes, E., Ruiz, Y., 2025. Vaccination: a legacy of success and persistent challenges, 50 years since the launch of the expanded programme on immunization. Medical Research Archives 13 (3). Available at https://doi.org/10.18103/mra.v13i3.6311.

de Hoop, T., Molotsky, A., Walcott, R., Gaitán-Rossi, P., Hernández-Cordero, S., Laar, A., et al., 2025. The role of nutrition-sensitive interventions in improving nutritional outcomes: findings from a systematic review and meta-analysis. International Journal for Equity in Health 24 (1). Available at https://doi.org/10.1186/s12939-025-02596-y.

Dronova, M., Biundo, E., Chicoye, A., Cookson, R., Devlin, N., Doherty, M., et al., 2022. P2 The value of vaccination: capturing the impact of vaccination on health equity in health economic analysis. Value in Health 25 (12), S1. Available at https://doi.org/10.1016/j.jval.2022.09.014.

Ellebedy, A.H., Nachbagauer, R., Jackson, K.J.L., Dai, Y.N., Han, J., Alsoussi, W.B., et al., 2020. Adjuvanted H5N1 influenza vaccine enhances both cross-reactive memory B cell and strain-specific naive B cell responses in humans. Proceedings of the National Academy of Sciences of the United States of America 117 (30), 17957–17964. Available at https://doi.org/10.1073/pnas.1906613117, https://www.pnas.org/content/pnas/117/30/17957.full.pdf.

Ellwanger, J.H., Chies, J.A.B., 2021. Zoonotic spillover: understanding basic aspects for better prevention. Genetics and Molecular Biology 44 (1). Available at https://doi.org/10.1590/1678-4685-GMB-2020-0355, https://www.scielo.br/j/gmb/a/TTzyffs6tcX37QwCr7fQqQp/?format=pdf&lang=en.

Erbelding, E.J., Post, D.J., Stemmy, E.J., Roberts, P.C., Augustine, A.D., Ferguson, S., et al., 2018. A universal influenza vaccine: the strategic plan for the national institute of allergy and infectious diseases. Journal of Infectious Diseases 218 (3), 347–354. Available at https://doi.org/10.1093/infdis/jiy103, http://jid.oxfordjournals.org/content/current.

Essink, B.J., Vermeulen, W., Andrade, C., de Rooij, R., Isakov, L., Casula, D., et al., 2025. A randomised phase 2 immunogenicity and safety study of a MF59-adjuvanted quadrivalent subunit inactivated cell-derived influenza vaccine (aQIVc) in adults aged 50 years and older. Vaccine 51, 126791. Available at https://doi.org/10.1016/j.vaccine.2025.126791.

Facciolà, A., Visalli, G., Laganà, A., Di Pietro, A., 2022. An overview of vaccine adjuvants: current evidence and future perspectives. Vaccines 10 (5), 819. Available at https://doi.org/10.3390/vaccines10050819.

Fieselmann, J., Annac, K., Erdsiek, F., Yilmaz-Aslan, Y., Brzoska, P., 2022. What are the reasons for refusing a COVID-19 vaccine? A qualitative analysis of social media in Germany. BMC Public Health 22 (1). Available at https://doi.org/10.1186/s12889-022-13265-y.

Fine, P., Eames, K., Heymann, D.L., 2011. Herd immunity": a rough guide. Clinical Infectious Diseases 52 (7), 911–916. Available at https://doi.org/10.1093/cid/cir007.

Garcia-Atutxa, I., Mondragon-Teran, P., Huerta-Saquero, A., Villanueva-Flores, F., 2024. Advancements in monkeypox vaccines development: a critical review of emerging technologies. Frontiers in Immunology 15. Available at https://doi.org/10.3389/fimmu.2024.1456060.

Haeuser, E., Byrne, S., et al., 2025. Global, regional, and national trends in routine childhood vaccination coverage from 1980 to 2023 with forecasts to 2030: a systematic analysis for the Global Burden of Disease Study 2023. The Lancet 406 (10500), 235–260. Available at https://doi.org/10.1016/S0140-6736(25)01037-2.

Giubilini, A., 2021. Vaccination ethics. British Medical Bulletin 137 (1), 4–12. Available at https://doi.org/10.1093/bmb/ldaa036.

Givler, D.N., Givler, A., 2023. Health screening. StatPearls [Internet. StatPearls Publishing, Treasure Island (FL). Available at https://pubmed.ncbi.nlm.nih.gov/28613785/.

Gogoi, H., Mani, R., Bhatnagar, R., 2025. Re-inventing traditional aluminum-based adjuvants: insight into a century of advancements. International Reviews of Immunology 44 (2), 58–81. Available at https://doi.org/10.1080/08830185.2024.2404095, http://www.tandfonline.com/loi/iiri20.

Hancock, G., Hellner, K., Dorrell, L., 2018. Therapeutic HPV vaccines. Best Practice & Research Clinical Obstetrics & Gynaecology 47, 59–72. Available at https://doi.org/10.1016/j.bpobgyn.2017.09.008.

Hariri, A., Mirian, M., Zarepour, A., Khosravi, A., Iravani, S., Zarrabi, A., 2025. Lipid nanoparticles driving mRNA vaccine innovations: from concept to clinic. Applied Materials Today 44, 102786. Available at https://doi.org/10.1016/j.apmt.2025.102786.

He, Y., Xiang, Z., Mobley, H.L.T., 2024. AI-driven vaccine design. Briefings in Bioinformatics. 25 (1), bbad476. Available at https://doi.org/10.1093/bib/bbad476.

Hou, X., Zaks, T., Langer, R., Dong, Y., 2021. Lipid nanoparticles for mRNA delivery. Nature Reviews Materials 6 (12), 1078–1094. Available at https://doi.org/10.1038/s41578-021-00358-0.

Ingebretsen Carlson, J., Puppo, F., Roca-Umbert, A., Folkvord, F., Lupiáñez-Villanueva, F., 2025. An experimental study on the effects of electronic medicinal product information on vaccine hesitancy. Scientific Reports 15 (1). Available at https://doi.org/10.1038/s41598-025-96092-6.

Irvine, D.J., Read, B.J., 2020. Shaping humoral immunity to vaccines through antigen-displaying nanoparticles. Current Opinion in Immunology 65, 1–6. Available at https://doi.org/10.1016/j.coi.2020.01.007, http://www.journals.elsevier.com/current-opinion-in-immunology/.

Iwasaki, A., Medzhitov, R., 2015. Control of adaptive immunity by the innate immune system. Nature Immunology 16 (4), 343–353. Available at https://doi.org/10.1038/ni.3123.

Jalilian, H., Amraei, M., Javanshir, E., Jamebozorgi, K., Faraji-Khiavi, F., 2023. Ethical considerations of the vaccine development process and vaccination: a scoping review. BMC Health Services Research 23 (1). Available at https://doi.org/10.1186/s12913-023-09237-6.

Kalitsilo, L., Oronje, R., Mbeye, N.M., 2025. Vector-borne disease control and management in irrigation areas: a neglected critical phenomenon in malawi. Tropical Medicine and Infectious Disease 10 (9), 251. Available at https://doi.org/10.3390/tropicalmed10090251.

Kayesh, M.E.H., Kohara, M., Tsukiyama-Kohara, K., 2023. TLR agonists as vaccine adjuvants in the prevention of viral infections: an overview. Frontiers in Microbiology 14. Available at https://doi.org/10.3389/fmicb.2023.1249718, https://www.frontiersin.org/journals/microbiology#.

Kayraklioglu, N., Horuluoglu, B., Klinman, D.M., 2021. CpG Oligonucleotides as vaccine adjuvants. Methods in Molecular Biology 2197, 51–85. Available at https://doi.org/10.1007/978-1-0716-0872-2_4, http://www.springer.com/series/7651.

Keech, C., Albert, G., Cho, I., Robertson, A., Reed, P., Neal, S., et al., 2020. Phase 1–2 Trial of a SARS-CoV-2 recombinant spike protein nanoparticle vaccine. New England Journal of Medicine 383 (24), 2320–2332. Available at https://doi.org/10.1056/nejmoa2026920.

Kelly, B.J., Leader, A.E., Henrikson, N.B., 2022. Editorial: promising strategies for vaccine messages: the message and the source. Frontiers in Public Health 10. Available at https://doi.org/10.3389/fpubh.2022.971283, http://journal.frontiersin.org/journal/public-health/section/public-health-education-and-promotion#archive.

Khan, A.M., Ahmed, I., Jawwad, M., Islam, M., Tahir, R., Anwar, S., et al., 2024. A low-cost, integrated immunization, health, and nutrition intervention in conflict settings in Pakistan—the impact on zero-dose children and polio coverage. Pathogens 13 (3), 185. Available at https://doi.org/10.3390/pathogens13030185.

Kim, J., Eygeris, Y., Gupta, M., Sahay, G., 2021. Self-assembled mRNA vaccines. Advanced Drug Delivery Reviews 170, 83–112. Available at https://doi.org/10.1016/j.addr.2020.12.014.

Kong, H., 2025. Advances in personalized cancer vaccine development: AI applications from neoantigen discovery to mRNA formulation. BioChem 5 (2), 5. Available at https://doi.org/10.3390/biochem5020005.

Krammer, F., 2020. SARS-CoV-2 vaccines in development. Nature 586 (7830), 516–527. Available at https://doi.org/10.1038/s41586-020-2798-3.

Kumar, A.U., Kadiresen, K., Gan, W.C., Ling, A P.K., 2021. Current updates and research on plant-based vaccines for coronavirus disease 2019. Clinical and Experimental Vaccine Research 10 (1), 13–23. Available at https://doi.org/10.7774/cevr.2021.10.1.13, http://www.ecevr.org/Synapse/Data/PDFData/9995CEVR/cevr-10-13.pdf.

Kumar, D., Ghosh, N., Matta, G., Lahariya, C., 2025. Role of safe water, sanitation, hygiene practices for child health: a review. Preventive Medicine: Research & Reviews 2 (Suppl 1), S27. Available at https://doi.org/10.4103/pmrr.pmrr_9_23.

Labbé, S., Bacon, S.L., Wu, N., Ribeiro, P.A.B., Boucher, V.G., Stojanovic, J., et al., 2025. Addressing vaccine hesitancy: a systematic review comparing the efficacy of motivational versus educational interventions on vaccination uptake. Translational Behavioral Medicine 15 (1). Available at https://doi.org/10.1093/tbm/ibae069, https://academic.oup.com/tbm/issue.

Laera, D., HogenEsch, H., et al., 2023. Aluminum adjuvants—'Back to the Future. Pharmaceutics 15 (7), 1884. Available at https://doi.org/10.3390/pharmaceutics15071884.

Lan, J., Feng, D., He, X., Zhang, Q., Zhang, R., 2024. Basic properties and development status of aluminum adjuvants used for vaccines. Vaccines 12 (10), 1187. Available at https://doi.org/10.3390/vaccines12101187.

Lazarus, J.V., White, T.M., Wyka, K., Ratzan, S.C., Rabin, K., Larson, H.J., et al., 2024. Influence of COVID-19 on trust in routine immunization, health information sources and pandemic preparedness in 23 countries in 2023. Nature Medicine 30 (6), 1559–1563. Available at https://doi.org/10.1038/s41591-024-02939-2, https://www.nature.com/nm/.

Li, M., Wang, H., Tian, L., Pang, Z., Yang, Q., Huang, T., et al., 2022. COVID-19 vaccine development: milestones, lessons and prospects. Signal Transduction and Targeted Therapy 7 (1). Available at https://doi.org/10.1038/s41392-022-00996-y.

Malonis, R.J., Lai, J.R., Vergnolle, O., 2019. Peptide-Based Vaccines: Current Progress and Future Challenges. Chemical Reviews 120 (6), 3210–3229. Available at https://doi.org/10.1021/acs.chemrev.9b00472.

Mangayarkarasi, V., Durairaj, E., Ramanathan, V., 2025. Enhancing cancer screening and early diagnosis in india: overcoming challenges and leveraging emerging technologies. Cureus 17 (2). Available at https://doi.org/10.7759/cureus.78808.

Mapingure, M., Makota, R.B., Chingombe, I., Moyo, E., Dzinamarira, T., Moyo, B., et al., 2024. Water, sanitation, and hygiene–specific risk factors of recent diarrheal episodes in children aged under 5 years: analysis of secondary data from the multiple indicator cluster survey (MICS 2019). IJID Regions 12, 100417. Available at https://doi.org/10.1016/j.ijregi.2024.100417.

Marciani, D.J., 2024. Vaccine adjuvants: from empirical to a more rational drug design. Exploratory Research and Hypothesis in Medicine. Available at https://doi.org/10.14218/erhm.2024.00002.

Markotter, W., Mettenleiter, T.C., Adisasmito, W.B., et al., 2023. Prevention of zoonotic spillover: From relying on response to reducing the risk at source. PLOS Pathogens 19 (10), e1011504. Available at https://doi.org/10.1371/journal.ppat.1011504.

Mettenleiter, T.C., Markotter, W., Charron, D.F., et al., 2023. The one health high-level expert panel (OHHLEP). One Health Outlook 5 (1), 18. Available at https://doi.org/10.1186/s42522-023-00085-2.

Mirza, I., Lemango, E.T., Lindstrand, A., 2025. Expanded programme on immunization (EPI): a legacy of 50 years and the road ahead. Vaccines 13 (6), 649. Available at https://doi.org/10.3390/vaccines13060649.

Montero, D.A., Vidal, C.R.M., Velasco, J., Carreño, L.J., Torres, J.P., Benachi O, M.A., et al., 2023. Two centuries of vaccination: historical and conceptual approach and future perspectives. Frontiers in Public Health 11. Available at https://doi.org/10.3389/fpubh.2023.1326154, http://journal.frontiersin.org/journal/public-health/section/public-health-education-and-promotion#archive.

Murphy, K.M., Weaver, C., Berg, L.J., 2023. Janeway's immunobiology 2023, 10th ed. W. W. Norton & Company.

Nabel, G.J., 2013. Designing tomorrow's vaccines. New England Journal of Medicine 368 (6), 551–560. Available at https://doi.org/10.1056/NEJMra1204186, http://www.nejm.org/doi/pdf/10.1056/NEJMra1204186.

Nasr, R., Swaidan, E., Hachem, S., Yazbek, N., Rizk, M., Abdel Rahman, A., et al., 2025. Immunization safety monitoring: addressing vaccine hesitancy and enhancing coverage in crisis-affected regions—insights from Lebanon, Ukraine, and Sudan. Therapeutic Advances in Vaccines and Immunotherapy 13. Available at https://doi.org/10.1177/25151355251380220.

Norris, K., Jilcott Pitts, S., Reis, H., Haynes-Maslow, L., 2023. A systematic literature review of nutrition interventions implemented to address food insecurity as a social determinant of health. Nutrients 15 (15), 3464. Available at https://doi.org/10.3390/nu15153464.

Ogutu, E.A., Ellis, A.S., Hester, K.A., Rodriguez, K., Sakas, Z., Jaishwal, C., et al., 2024. Success in vaccination programming through community health workers: a qualitative analysis of interviews and focus group discussions from Nepal, Senegal and Zambia. BMJ Open 14 (4). Available at https://doi.org/10.1136/bmjopen-2023-079358, http://bmjopen.bmj.com/content/early/by/section.

O'Hagan, D.T., Lodaya, R.N., Lofano, G., 2020. The continued advance of vaccine adjuvants – 'we can work it out' Seminars in Immunology 50. Available at https://doi.org/10.1016/j.smim.2020.101426, http://www.elsevier.com/inca/publications/store/6/2/2/9/4/5/index.htt.

Okuyama, R., 2023. mRNA and adenoviral vector vaccine platforms utilized in COVID-19 vaccines: technologies, ecosystem, and future directions. Vaccines 11 (12), 1737. Available at https://doi.org/10.3390/vaccines11121737.

Ong, E., et al., 2023. Artificial intelligence for vaccine design and development. Trends in Biotechnology 41 (8), 1021–1035.

Orenstein, W.A., Ahmed, R., 2017. Simply put: vaccination saves lives. Proceedings of the National Academy of Sciences 114 (16), 4031–4033. Available at https://doi.org/10.1073/pnas.1704507114.

Orenstein, W.A., Offit, P.A., Edwards, K.M., Plotkin, S.A., 2023. Plotkin's vaccines, 8th ed. Travel Medicine and Infectious Disease 55 Elsevier, 102635. Available at https://doi.org/10.1016/j.tmaid.2023.102635 accessed on October 15, 2025.

Oyugi, B., Kallander, K., Shahabuddin, A.S.M., 2025. Strengthening primary health care through implementation research: strategies for reaching zero-dose children in low- and middle-income countries' immunization programs. Vaccines 13 (10). Available at https://doi.org/10.3390/vaccines13101040, http://www.mdpi.com/journal/vaccines.

Ozawa, S., Clark, S., Portnoy, A., Grewal, S., Brenzel, L., Walker, D.G., 2016. Return on investment from childhood immunization in low- and middle-income countries, 2011–20. Health Affairs 35 (2), 199–207. Available at https://doi.org/10.1377/hlthaff.2015.1086.

Pan American Health Organization, 2025. Combating false information on vaccines: A guide for health workers 2025. Pan American Health Organization. Available at https://www.paho.org/en/documents/combating-false-information-vaccines-guide-health-workers.

Plotkin, S., 2014. History of vaccination. Proceedings of the National Academy of Sciences 111 (34), 12283–12287. Available at https://doi.org/10.1073/pnas.1400472111.

Plotkin, S.A., 2005. Vaccines: past, present and future. Nature Medicine 11 (S4), S5. Available at https://doi.org/10.1038/nm1209.

Poria, R., Kala, D., Nagraik, R., Dhir, Y., Dhir, S., Singh, B., et al., 2024. Vaccine development: current trends and technologies. Life Sciences 336, 122331. Available at https://doi.org/10.1016/j.lfs.2023.122331.

Pulendran, B., Arunachalam, P.S., O'Hagan, D.T., 2021. Emerging concepts in the science of vaccine adjuvants. Nature Reviews Drug Discovery 20 (6), 454–475. Available at https://doi.org/10.1038/s41573-021-00163-y, http://www.nature.com/nrd/index.html.

Pulendran, B., Ahmed, R., 2011. Immunological mechanisms of vaccination. Nature Immunology 12 (6), 509–517. Available at https://doi.org/10.1038/ni.2039.

Rappuoli, R., Alter, G., Pulendran, B., 2024. Transforming vaccinology. Cell 187 (19), 5171–5194. Available at https://doi.org/10.1016/j.cell.2024.07.021, https://www.sciencedirect.com/science/journal/00928674.

Reed, S.G., Bertholet, S., Coler, R.N., Friede, M., 2009. New horizons in adjuvants for vaccine development. Trends in Immunology 30 (1), 23–32. Available at https://doi.org/10.1016/j.it.2008.09.006.

Robbiati, C., Milano, A., Declich, S., Dente, M.G., 2024. One Health prevention and preparedness to vector-borne diseases: how should we deal with a multisectoral, multilevel and multigroup governance? One Health Outlook 6 (1). Available at https://doi.org/10.1186/s42522-024-00114-8.

Rodrigues, C.M.C., Plotkin, S.A., 2020. Impact of vaccines; health, economic and social perspectives. Frontiers in Microbiology 11. Available at https://doi.org/10.3389/fmicb.2020.01526, https://www.frontiersin.org/journals/microbiology#.

Ruggeri, K., Vanderslott, S., Yamada, Y., Argyris, Y.A., Većkalov, B., Boggio, P.S., et al., 2024. Behavioural interventions to reduce vaccine hesitancy driven by misinformation on social media. BMJ 384, e076542. Available at https://doi.org/10.1136/bmj-2023-076542.

Rumfield, C.S., Roller, N., Pellom, S.T., Schlom, J., Jochems, C., 2020. Therapeutic vaccines for HPV-associated malignancies. ImmunoTargets and Therapy 9, 167–200. Available at https://doi.org/10.2147/ITT.S273327, https://www.dovepress.com/getfile.php?fileID=62192.

Shachar, C., Reiss, R., 2020. When Are Vaccine Mandates Appropriate? AMA J Ethics 22 (1), 36–42. Available at doi:10.1001/amajethics.2020.36.

Saddi, F.C., Peckham, S., Vasconcelos, A.M.N., 2025. Intersectoral and pro-equity approaches in health policy. Annals of Global Health 91 (1). Available at https://doi.org/10.5334/aogh.4863, https://annalsofglobalhealth.org/articles/4863/files/68be8aa1321ef.pdf.

Shafaati, M., Forghani, S., Shahsavand, D.A., 2025. Current advances and challenges in mpox vaccine development: a global landscape. Ther Adv Vaccines Immunother. 27 (13). Available at https://doi.org/10.1177/25151355251314339.

Sahin, U., Türeci, Ö., 2018. Personalized vaccines for cancer immunotherapy. Science 359 (6382), 1355–1360. Available at https://doi.org/10.1126/science.aar7112.

Santangelo, O.E., Provenzano, S., Di Martino, G., Ferrara, P., 2024. COVID-19 vaccination and public health: addressing global, regional, and within-country inequalities. Vaccines 12 (8), 885. Available at https://doi.org/10.3390/vaccines12080885.

Seder, R.A., Hill, A.V.S., 2000. Vaccines against intracellular infections requiring cellular immunity. Nature 406 (6797), 793–798. Available at https://doi.org/10.1038/35021239.

Shafique, K., Ahmer, Z., Choudhury, S.R., Safdar, N.F., Alam, S.M., Wenndt, A.J., et al., 2025. Effect of integrating social and behavior change communication strategies in nutrition-sensitive social-protection programs on specific nutritional outcomes: a systematic review. Nutrition Reviews 84 (4). Available at https://doi.org/10.1093/nutrit/nuaf063.

Shattock, A.J., Johnson, H.C., Sim, S.Y., Carter, A., Lambach, P., Hutubessy, R.C.W., et al., 2024. Contribution of vaccination to improved survival and health: modelling 50 years of the expanded programme on immunization. The Lancet 403 (10441), 2307–2316. Available at https://doi.org/10.1016/S0140-6736(24)00850-X, https://www.sciencedirect.com/science/journal/01406736.

Stertman, L., Palm, A.K.E., Zarnegar, B., Carow, B., Lunderius Andersson, C., Magnusson, S.E., et al., 2023. The Matrix-M™ adjuvant: a critical component of vaccines for the 21st century. Human Vaccines and Immunotherapeutics 19 (1). Available at https://doi.org/10.1080/21645515.2023.2189885, http://www.tandfonline.com/loi/khvi20.

Szabó, G.T., Mahiny, A.J., Vlatkovic, I., 2022. COVID-19 mRNA vaccines: platforms and current developments. Molecular Therapy 30 (5), 1850–1868. Available at https://doi.org/10.1016/j.ymthe.2022.02.016, https://www.journals.elsevier.com/molecular-therapy.

Tuckerman, J., Harper, K., Sullivan, T.R., Fereday, J., Couper, J., Smith, N., et al., 2022. Using provider–parent strategies to improve influenza vaccination in children and adolescents with special risk medical conditions: a randomised controlled trial protocol. BMJ Open 12 (2), e053838. Available at https://doi.org/10.1136/bmjopen-2021-053838.

Wang, Q., Leung, K., Jit, M., Wu, J.T., Lin, L., 2025. Global socioeconomic inequalities in vaccination coverage, supply, and confidence. Npj Vaccines 10 (1). Available at https://doi.org/10.1038/s41541-025-01143-8.

Whitehead, H.S., French, C.E., Caldwell, D.M., Letley, L., Mounier-Jack, S., 2023. A systematic review of communication interventions for countering vaccine misinformation. Vaccine 41 (5), 1018–1034. Available at https://doi.org/10.1016/j.vaccine.2022.12.059, www.elsevier.com/locate/vaccine.

World Health Organization, 2021. A brief history of vaccination. Available at https://www.who.int/news-room/spotlight/history-of-vaccination/a-brief-history-of-vaccination.

World Health Organization, 2023. Global vaccine market report 2023. Available at https://www.who.int/publications/i/item/B09022.

World Health Organization, 2024. Global immunization efforts have saved at least 154 million lives over the past 50 years, April 24. (2024). Available at https://www.who.int/news/item/24-04-2024-global-immunization-efforts-have-saved-at-least-154-million-lives-over-the-past-50-years.

World Health Organization, 2025. Increases in vaccine-preventable disease outbreaks threaten years of progress. Available at https://www.who.int/news/item/24-04-2025-increases-in-vaccine-preventable-disease-outbreaks-threaten-years-of-progress–warn-who–unicef–Chen.

Wouters, O.J., Shadlen, K.C., Salcher-Konrad, M., Pollard, A.J., Larson, H.J., Teerawattananon, Y., et al., 2021. Challenges in ensuring global access to COVID-19 vaccines: production, affordability, allocation, and deployment. The Lancet 397 (10278), 1023–1034. Available at https://doi.org/10.1016/S0140-6736(21)00306-8, http://www.journals.elsevier.com/the-lancet/.

Xie, Y.J., Liao, X., Lin, M., Yang, L., Cheung, K., Zhang, Q., et al., 2024. Community engagement in vaccination promotion: systematic review and meta-analysis. JMIR Public Health and Surveillance 10, e49695. Available at https://doi.org/10.2196/49695.

Zhai, Y., Han, Y., Wang, W., Tan, W., 2025. Advancements in mpox vaccine development: a comprehensive review of global progress and recent data. Biomedical and Environmental Sciences 38 (2), 248–254. Available at https://doi.org/10.3967/bes2024.121, https://www.sciencedirect.com/science/journal/08953988.

Zhou, F., Jatlaoui, T.C., Leidner, A.J., Carter, R.J., Dong, X., Santoli, J.M., et al., 2024. Health and economic benefits of routine childhood immunizations in the era of the vaccines for children program United States, 1994 2023. Morbidity and Mortality Weekly Report 73 (31), 682–685. Available at https://doi.org/10.15585/mmwr.mm7331a2, https://www.cdc.gov/mmwr/volumes/73/wr/mm7331a2.htm?s_cid=mm7331a2_w.

Chapter 13

Preserving efficacy: a comprehensive approach to antimicrobial stewardship

Ajanta Sharma, Suranjana Chaliha Hazarika, and Bornali Sarmah Dutta
Department of Microbiology, Gauhati Medical College & Hospital, Guwahati, Assam, India

13.1 Introduction

Antimicrobial stewardship (AMS) is defined as 'the optimal selection, dosage, and duration of antimicrobial therapy aimed to achieve the best possible clinical outcomes while minimising adverse effects for patients and reducing the risk of developing antimicrobial resistance' (Gerding, 2001). World Health Organization (WHO) describes 'an antimicrobial stewardship programme (ASP) as a coordinated, system level approach within healthcare settings aimed at ensuring the judicious and effective use of antibiotics through evidence based strategies and interventions' (World Health Organization, 2019a,b). The education and capacity building of healthcare professionals is pivotal, as these front line workers are instrumental in safeguarding and restoring the efficacy of antimicrobial medications (World Health Organization WHO, 2021). ASP represents a highly cost-effective approach to promoting the rational use of antimicrobials, improving clinical outcomes, and mitigating the burden of antimicrobial resistance (AMR) as well as healthcare-associated infection (HAI).

The term 'Antimicrobial stewardship' was initially coined by McGowan and Gerding in 1996, focusing on the association between antimicrobial use and resistance and calling for robust epidemiologic and molecular studies to guide appropriate use (McGowan and Gerding, 1996). In 1997 the Society for Healthcare Epidemiology of America (SHEA) and Infectious Diseases Society of America (IDSA) issued guidelines to prevent AMR in hospitals (Shlaes et al., 1997). In India, the Indian Council of Medical Research (ICMR) established the Antimicrobial Resistance Surveillance and Research Network (AMRSN) in 2013, aligning with the Chennai Declaration and global AMR-control initiatives (Ghafur et al., 2013).

13.1.1 Antimicrobial resistance in bacteria

The discovery of antibiotics once led the healthcare community to believe that infectious diseases had been effectively controlled. However, the widespread emergence of multidrug-resistant bacteria has reversed this perception, presenting a growing threat to global health.

Antimicrobial agents are commonly classified according to their mechanism of action. The major groups include agents that inhibit cell wall synthesis, disrupt cell membrane integrity, inhibit protein synthesis, inhibit nucleic acid synthesis, and interfere with bacterial metabolic pathways. Despite the availability of antimicrobial agents with diverse mechanisms of action, effective control of infectious organisms has not been achieved (Reygaert, 2018). Inappropriate use and poor stewardship of antimicrobial agents have significantly contributed to the emergence and spread of AMR. Several factors have accelerated this problem, including increased consumption of antimicrobial drugs in both humans and animals, as well as inappropriate prescribing practices (Reygaert, 2018).

For many years, antibiotics have been extensively used in food-producing animals for the treatment and prevention of diseases and, in some cases, for growth promotion. Animal feed often contains antibiotics at concentrations ranging from subtherapeutic to full therapeutic levels, and many of these agents belong to the same antimicrobial classes used in human medicine.

AMR is a quintessential one health challenge, driven by the interconnected use of antimicrobials across human health, veterinary practice, agriculture, and environmental systems. In many low- and middle-income countries, including India,

Antimicrobial Resistance in the 21st Century. DOI: https://doi.org/10.1016/B978-0-443-44748-8.00020-8

antibiotics are extensively used not only for treating human infections but also in livestock for therapeutic, prophylactic, and, in some settings, growth promotion purposes. This widespread use promotes the selection of antimicrobial-resistant organisms in animals, with resistance patterns often reflecting the types and quantities of antibiotics administered. These resistant organisms can subsequently be transmitted to humans through multiple pathways, particularly via the food chain, such as consumption of contaminated meat, food, or water containing animal faecal material as well as through direct contact with animals (Landers et al., 2012; Wegener, 2012). In addition, inadequate wastewater treatment and improper disposal of pharmaceutical effluents contribute to environmental reservoirs of resistance, further amplifying transmission through contaminated water and soil. These interconnected pathways underscore the urgent need for integrated stewardship strategies that extend beyond healthcare settings to include regulation of veterinary antibiotic use, strengthened cross-sectoral surveillance, and robust environmental control measures, in alignment with the one health approach (Landers et al., 2012; Wegener, 2012).

The increasing prevalence of AMR has reduced available treatment options, leading to higher morbidity and mortality. Consequently, infections are becoming more severe, requiring more intensive therapy, prolonged illness, extended hospital stays, and significantly increased healthcare costs. Having outlined the burden of AMR and the role of AMS programmes, it is important to understand the underlying resistance mechanisms. These mechanisms explain resistance patterns and form the basis for rational antibiotic use, linking clinical practice with effective stewardship. Fig. 13.1 describes natural and acquired antibiotic resistance, and Fig. 13.2 displays the organisms intrinsically resistant to different antibiotics.

13.1.2 General mechanism of antimicrobial resistance

AMR develops through several molecular mechanisms that allow microorganisms to survive exposure to antimicrobial agents. These mechanisms include enzymatic drug inactivation, such as β-lactamase production that hydrolyses β-lactam antibiotics; alteration of antimicrobial targets, for example, modification of penicillin-binding proteins or ribosomal binding sites; reduced permeability or increased efflux of drugs, where bacteria limit antibiotic entry or actively pump drugs out via efflux pumps; and bypass of metabolic pathways, allowing organisms to circumvent inhibited biochemical processes (Reygaert, 2018). Gram-negative bacteria utilise all four major mechanisms of AMR, whereas Gram-positive bacteria less frequently employ reduced drug uptake due to the absence of an outer lipopolysaccharide (LPS) membrane and have more limited efflux capabilities for certain antimicrobial agents (Chancey et al., 2019; Mahon et al., 2014) These mechanisms may arise through genetic mutations or horizontal gene transfer via plasmids, transposons, or integrons, facilitating rapid dissemination of resistance among bacterial populations (Davies and Davies, 2010). Understanding the mechanisms of AMR provides the scientific basis for designing effective stewardship interventions aimed at preserving antibiotic efficacy. Fig. 13.3 describes antibiotic targets and the general mechanism of antibiotic resistance.

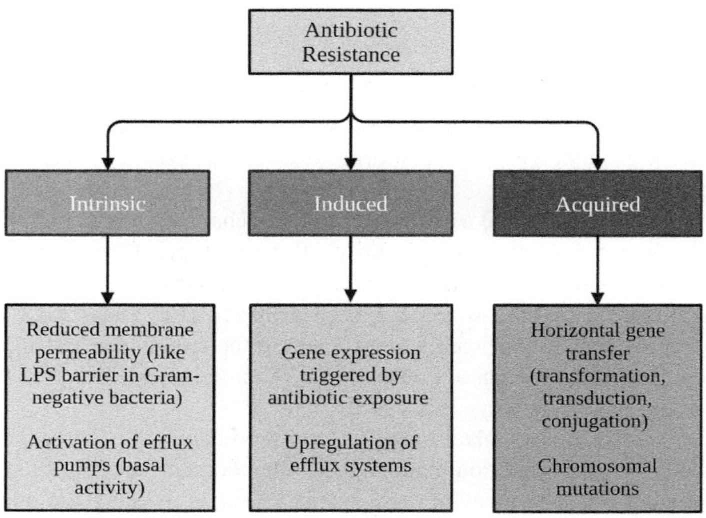

FIGURE 13.1 **Pathways of antibiotic resistance in bacteria.** Schematic representation of the three major pathways of antibiotic resistance in bacteria.

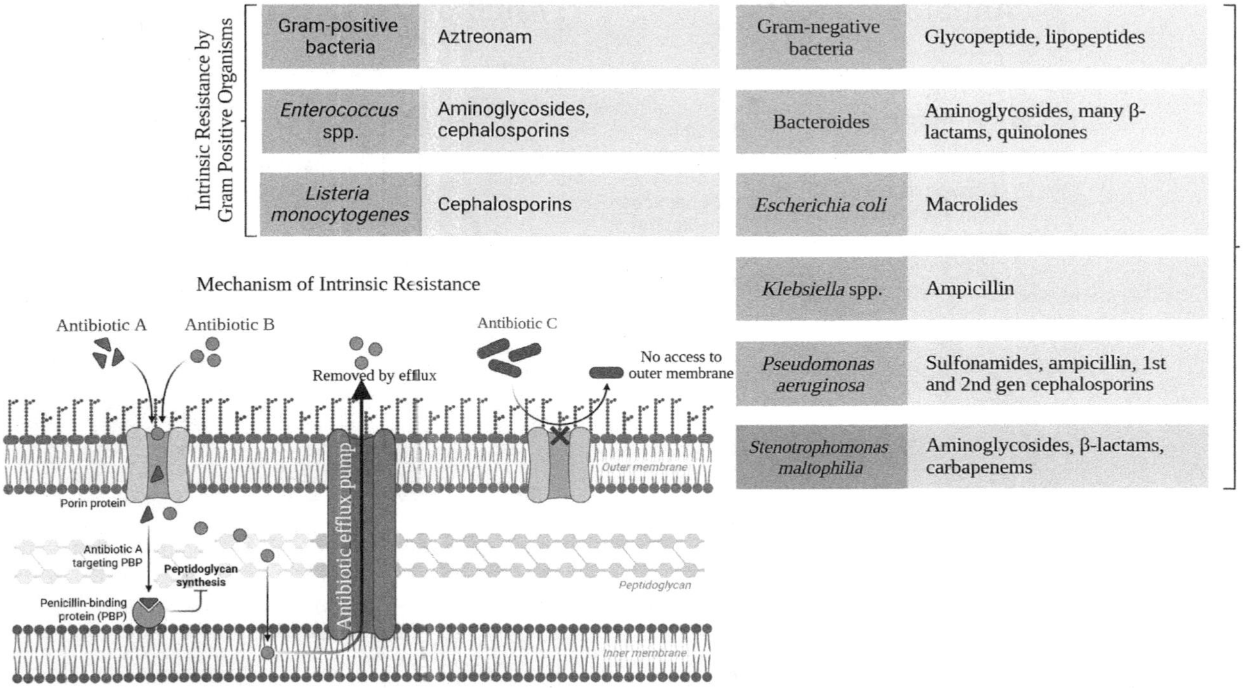

FIGURE 13.2 **Intrinsic antibiotic resistance: from cellular barriers to drug-specific profiles.** Illustration depicting intrinsic antibiotic resistance across Gram-positive and Gram-negative bacteria, highlighting representative organisms and their inherent resistance profiles. The figure also demonstrates mechanistic features of intrinsic resistance, including limited drug entry through porins, active efflux pump-mediated antibiotic removal, and restricted permeability of the outer membrane.

13.1.3 Genetic mechanisms of antimicrobial resistance

Bacterial pathogens demonstrate immense genetic plasticity, enabling rapid evolution through mutations, horizontal gene transfer (HGT), and gene expression alteration in response to antibiotic stress. The survival of the fittest mechanism ensures that resistant strains survive, passing on resistance genes to future generations (Munita and Arias, 2016).

From an evolutionary perspective, bacteria use two major genetic strategies to adapt to the antibiotic 'attack': (1) mutations in gene(s) often associated with the mechanism of action of the compound and (2) acquisition of foreign DNA coding for resistance determinants through horizontal gene transfer (HGT). While mutational resistance is helpful for fine-tuning existing cellular machinery, horizontal gene transfer is the real game changer – it is essentially the bacterial version of downloading a 'resistance patch' instantly from a neighbour (Munita and Arias, 2016).

13.1.3.1 Mutational resistance

Mutational resistance is a major evolutionary process where spontaneous mutations in bacterial genes provide a survival advantage under antibiotic selection pressure. These mutations, often costly to fitness, are maintained due to continued antibiotic presence, allowing resistant strains to outcompete susceptible ones (Munita and Arias, 2016).

13.1.3.2 Horizontal gene transfer

It is the movement of genetic material between unicellular and/or multicellular organisms other than by the vertical transmission of DNA from parent to offspring. It drives the rapid evolution and adaptation by allowing bacteria to acquire new traits such as antibiotic resistance (Burmeister, 2015).

Mechanisms of horizontal gene transfer are as follows:

1. Transformation: Bacteria take up DNA from their environment
2. Conjugation: Bacteria directly transfer genes to another cell
3. Transduction: Bacteriophages (bacterial viruses) move genes from one cell to another

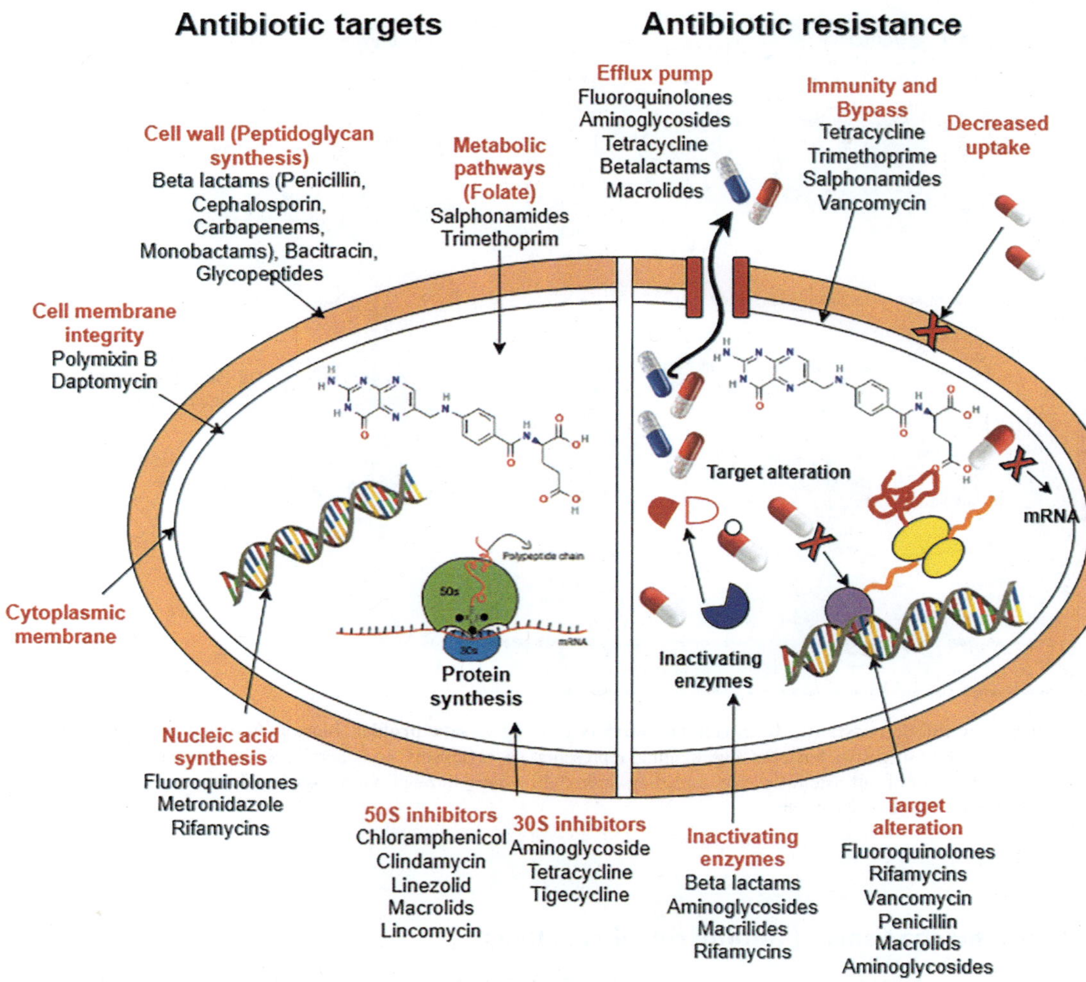

FIGURE 13.3 **Target of action of different antibiotics and the general mechanism of antibiotic resistance.** Target of action of different antibiotics and general mechanism of antibiotic resistance.

As a general rule, conjugation uses mobile genetic elements (MGEs) as vehicles to share valuable genetic information, although direct transfer from chromosome to chromosome has also been well characterised (Manson et al., 2010). The most important MGEs are plasmids and transposons, both of which play a crucial role in the development and dissemination of AMR among clinically relevant organisms. The mechanisms of horizontal transfer of antibiotic resistance genes are presented in Fig. 13.4.

Finally, one of the most efficient mechanisms for accumulating AMR genes is represented by integrons, which are site-specific recombination systems capable of recruiting open reading frames in the form of mobile gene cassettes. Integrons provide an efficient and rather simple mechanism for the addition of new genes into bacterial chromosomes, along with the necessary machinery to ensure their expression: a robust strategy of genetic interchange and one of the main drivers of bacterial evolution (Manson et al., 2010). Beyond molecular mechanisms, AMR has significant clinical and public health implications globally.

13.1.4 Global burden of antimicrobial resistance and its public health implications

AMR is among the top global public health threats, compromising the effectiveness of antimicrobial therapy and endangering human, animal, and environmental health (World Health Organization, 2019a,b). It occurs when

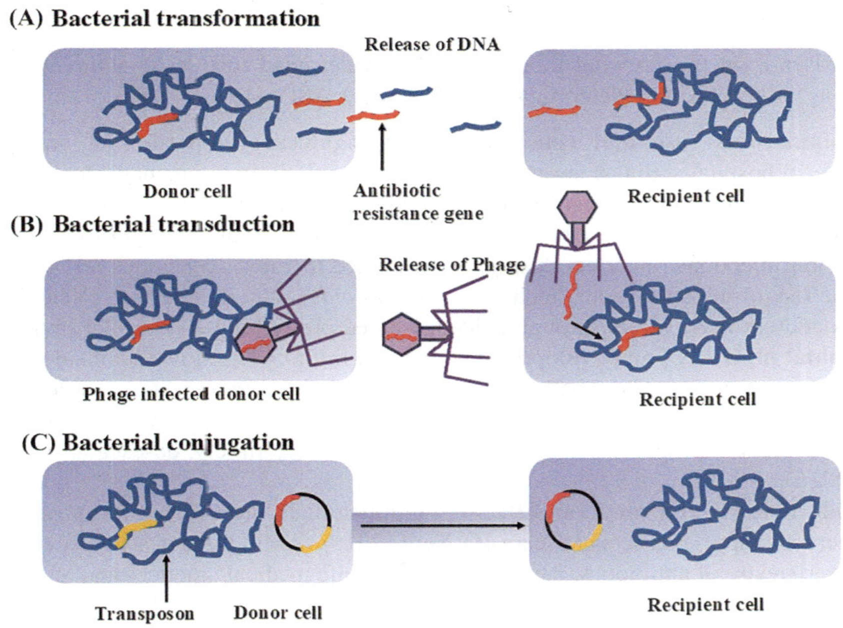

FIGURE 13.4 **Horizontal gene transfer. (A) Bacterial transformation, (B) Bacterial transduction, (C) Bacterial conjugation.** Molecular mechanism (horizontal gene transfer) of antibiotic resistance.

microorganisms develop mechanisms that render antimicrobial agents ineffective, leading to treatment failure and increased disease transmission.

In 2019 bacterial AMR was directly responsible for an estimated 1.27 million deaths and contributed to 4.95 million deaths globally (Murray et al., 2022). The economic burden is substantial, with projections of over US$1 trillion in additional healthcare costs by 2050 and US$3.4 trillion annual GDP losses by 2030. The burden is disproportionately higher in low- and middle-income countries.

Key drivers of AMR include irrational and excessive use of antimicrobials in human health, veterinary practice, and agriculture, along with limited access to diagnostics and a slow pace of new antibiotic development (World Health Organization, 2023). Multidrug-resistant (MDR) Gram-negative organisms such as *Escherichia coli, Klebsiella pneumoniae, Pseudomonas aeruginosa*, and *Acinetobacter baumannii* pose major therapeutic challenges due to limited treatment options and higher mortality (Cerceo et al., 2016; Zagaliotis et al., 2022).

Recognising the urgency, global initiatives such as the WHO Global Action Plan on AMR and the Global Antimicrobial Resistance and Use Surveillance System (GLASS) were established to strengthen surveillance and guide interventions within a one health framework (WHO GLASS Report, 2022). To support these goals, ASPs were established in institutions, led by infectious disease physicians and pharmacists, to promote judicious antimicrobial use and limit multidrug resistance while optimising patient outcomes (MacDougall and Polk, 2005).

Given the significant clinical and economic burden of AMR, the implementation of structured AMS programmes has become essential to optimise antibiotic use and mitigate its impact.

13.1.5 Goals of antimicrobial stewardship

ASP aims to ensure appropriate use of antimicrobials to enhance clinical effectiveness and limit the emergence of resistance (Doron and Davidson, 2011). The goals include

1. Promoting appropriate antimicrobial use through the '5 Ds' – right drug, dose, route, duration, and de-escalation
2. Reducing inappropriate use and overprescribing
3. Minimising adverse drug effects and toxicity
4. Slowing the emergence and spread of resistance
5. Reducing healthcare costs
6. Improving clinical outcomes, including reduced mortality and faster recovery

13.1.6 The need for antimicrobial stewardship

The need for AMS arises from the growing threat of AMR, widespread misuse of antimicrobials, and the declining effectiveness of existing drugs. The key determinants necessitating ASPs are as follows:

1. **Inappropriate antimicrobial use and consequences:** A significant proportion of antimicrobial prescriptions, estimated at 30–50% in hospital settings, are inappropriate, particularly in the management of respiratory and urinary tract infections, leading to increased resistance, adverse drug events, and healthcare costs (Centre for Disease Control and Prevention CDC, 2019; Magill et al., 2014; van de Sande-Bruinsma et al., 2008; Fridkin et al., 2014). Misuse, such as prescribing for viral infections, incorrect dosing, or prolonged therapy, accelerates resistance, increases healthcare costs, and raises the risk of adverse events, including *Clostridioides difficile* infection (Ventola, 2015; Laxminarayan et al., 2013). ASPs address these issues through evidence-based prescribing and regular review of therapy.
2. **Limited antimicrobial pipeline:** The development of new antibiotics has slowed considerably over the past three decades, while resistance among key pathogens such as *K. pneumoniae*, *P. aeruginosa,* and *A. baumannii* continues to rise (World Health Organization, 2020). Stewardship efforts are essential to preserve the effectiveness of existing antimicrobials through rational selection, dosing, and timely de-escalation (Holmes et al., 2016; Spellberg et al., 2016).
3. **Safeguarding modern medical practices:** Effective antimicrobials are critical for procedures such as surgery, transplantation, chemotherapy, and neonatal care (O'Neill, 2016). AMR threatens the safety of these interventions. AMS ensures appropriate use of antibiotics, thereby supporting safe medical practices and reducing HAIs (Davey et al., 2017; Dyar et al., 2015).

In response to these challenges, several national initiatives have been undertaken to operationalise AMS within the Indian healthcare system.

13.2 Antimicrobial stewardship programme implementation in India

The ICMR established the Antimicrobial Resistance Surveillance and Research Network (ICMR-AMRSN) in 2013 to strengthen India's response to AMR by generating high-quality evidence to guide ASPs which primarily aim to ensure the safe, rational, and optimal use of antibiotics by promoting the right drug, dose, duration, and route of administration, while also seeking to reduce adverse drug events, minimise healthcare costs, and prevent unintended ecological consequences without compromising quality of care (MacDougall and Polk, 2005; Indian Council of Medical Research, 2018). The network has enabled standardised surveillance, strengthened laboratory capacity, and promoted multidisciplinary collaboration. It has also supported regulatory actions, including restrictions on critical antimicrobials such as colistin and the expansion of surveillance for veterinary pathogens under a One Health framework.

Importantly, ICMR-AMRSN data align with the WHO Global Antimicrobial Resistance Surveillance System (GLASS), enabling global comparability. National treatment guidelines can also be harmonised with the WHO AWaRe classification (categorises antibiotics into Access, Watch, and Reserve groups to guide appropriate use and stewardship prioritisation) to guide rational prescribing and formulary decisions. This integration strengthens both national policy and global surveillance efforts.

Effective implementation of AMS programmes relies on a set of core elements that provide a structured framework for action.

13.3 Core elements of antimicrobial stewardship programme

ASPs are built on a set of core elements that provide the essential framework for optimising antimicrobial use and improving patient outcomes. These elements define the fundamental principles required for effective implementation across healthcare settings (World Health Organization, 2019a,b; Centre for Disease Control and Prevention CDC, 2019). In alignment with the objectives of ASPs, the following core components are included, which are presented in Fig. 13.5.

a. **Leadership commitment:** Strong institutional support is critical for the success of ASPs. This includes allocation of adequate financial, human, and information technology resources, as well as formal endorsement of stewardship activities as a healthcare priority.
b. **Accountability and responsibility:** Clear designation of leadership roles, typically involving a physician and a pharmacist, ensures responsibility for programme outcomes. A multidisciplinary team, including microbiologists, infection control professionals, and nursing staff, supports coordinated implementation.

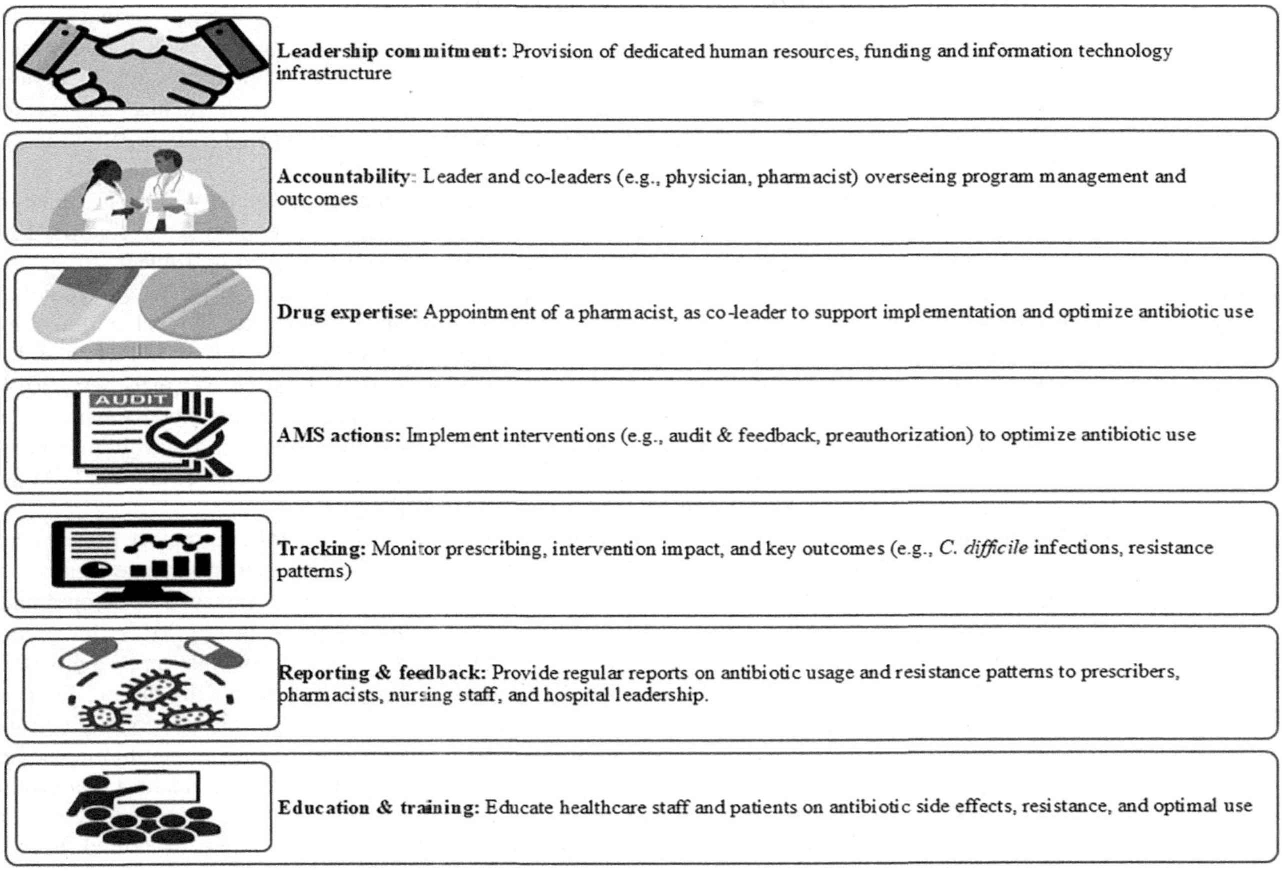

FIGURE 13.5 **ASP implementation.** Core components of ASP implementation. *ASP*, Antimicrobial stewardship programme.

c. **Drug expertise:** Access to expertise in antimicrobial therapy, particularly through trained clinical pharmacists or infectious disease specialists, is essential for guiding appropriate antibiotic selection, dosing, and monitoring.
d. **Action:** ASPs should implement targeted, evidence-based interventions tailored to local needs and resistance patterns. These may include guideline development, formulary restrictions, and decision-support tools.
e. **Monitoring and feedback:** Regular tracking of antimicrobial use, *C. difficile* rates, resistance patterns, and clinical outcomes, combined with timely feedback to prescribers, is essential to improve prescribing practices and evaluate stewardship interventions.
f. **Education and training:** Continuous education of healthcare professionals is essential to ensure appropriate antimicrobial prescribing and adherence to stewardship principles. Training should focus on resistance patterns, guideline-based therapy, and safe prescribing practices.

Core elements of ASP are classified as basic (fewer resources) or advanced (more resources), varying by facility size, needs, and context. All facilities should review the checklist to assess existing ASP practices and identify areas for improvement as per the Centre for Disease Control and Prevention (Centre for Disease Control and Prevention CDC, 2019). While core elements define the foundational principles of stewardship, their successful execution depends on well-established institutional components and infrastructure.

13.4 Major components of antimicrobial stewardship programme

Effective implementation of ASPs requires a robust institutional framework supported by appropriate infrastructure, multidisciplinary collaboration, and access to diagnostic and data systems.

a. **Administrative and organisational support:** Institutional commitment should be reflected through formal policies, integration of stewardship into quality and safety frameworks, and sustained allocation of resources such as dedicated human, financial, and IT resources, including specialist staff, funding for interventions, and systems to monitor antibiotic use and resistance. Embedding ASPs within organisational priorities ensures long-term sustainability that aligns with patient safety and infection control (Centre for Disease Control and Prevention CDC, 2019; World Health Organization WHO, 2020; Dellit et al., 2007).
b. **Multidisciplinary team approach:** A coordinated team comprising clinicians, pharmacists, microbiologists, infection control specialists, epidemiologists, and information technology personnel is essential for comprehensive stewardship activities (Pulcini et al., 2019). The roles and responsibilities of the AMS team are presented in Table 13.1.
c. **Microbiology and laboratory support:** Timely and accurate diagnostic services are central to ASPs. Microbiology laboratories enable pathogen identification, antimicrobial susceptibility testing, and development of local antibiograms, which guide empirical and targeted therapy. Advanced tools, such as automated blood cultures, MALDI-TOF, and PCR panels, enhance early detection and support prompt therapeutic decisions (Pulcini et al., 2019; Wagner et al., 2021).
d. **Pharmacological support and dose optimisation:** Therapeutic drug monitoring and pharmacokinetic/pharmacodynamic (PK/PD) principles are critical for optimising antibiotic dosing, particularly for drugs with narrow therapeutic indices. Individualised dosing improves efficacy while minimising toxicity and resistance (Dellit et al., 2007; Abdul-Aziz et al., 2020; Roberts and Lipman, 2009; World Health Organization, 2019a,b).
e. **Health information systems and data management:** Integration of electronic health records, laboratory systems, and pharmacy databases enables real-time tracking of antimicrobial use, resistance patterns, and resistance trends, and provides systematic feedback to prescribers (The Core Elements of Hospital Antibiotic Stewardship Programmes (Centre for Disease Control and Prevention CDC, 2019, Karanika et al., 2016; Howard et al., 2015). Integrating Electronic Health Records (EHRs), Laboratory Information Systems (LIS), and Pharmacy Information Systems (PIS) allow accurate tracking of indications, dose, and duration. Standardised metrics such as defined daily dose (DDD) and days of therapy (DOT) facilitate monitoring, benchmarking, and evaluation of stewardship interventions. Electronic dashboards, alerts, and periodic summary reports support prescribers and leadership in optimising antibiotic use and guiding stewardship decisions (Centre for Disease Control and Prevention CDC, 2019; World Health Organization WHO, 2020; Karanika et al., 2016; Howard et al., 2015).
f. **Capacity building:** Institutions should support ongoing professional development through structured training programmes aligned with AMS goals. In addition, patient education is essential to promote appropriate antibiotic use by improving awareness about when antibiotics are indicated, the importance of completing prescribed courses, and

TABLE 13.1 Roles and responsibilities of the AMS team members in stewardship activities.

AMS team members	Responsibilities
Infectious disease physician	They offer expert guidance on infectious disease diagnosis and management.
linical pharmacist with infectious disease training	Clinical pharmacists review prescriptions, optimise dosages, monitor therapy, provide drug information, and manage adverse effects
Clinical microbiologist	They report resistance patterns and help optimise antibiotic selection
Infection control professionals	They are responsible for adherence to infection control practices to prevent the spread of resistant pathogens
Hospital epidemiologist	They are responsible for monitoring trends of hospital infection and provide data to guide prevention strategies and antimicrobial use
Informatics professionals	They manage information systems to track antimicrobial use and resistance, supporting effective stewardship decisions
Infectious disease-trained nurse	They coordinate clinicians, pharmacists, and microbiologists to ensure optimal antimicrobial use, monitor administration, oversee timely specimen collection, and promote guideline adherence
Administrative head	The administrative head ensures that AMS is a hospital priority, providing policy support, resources, and accountability structures, while promoting coordination, staff engagement, and compliance with antimicrobial guidelines

the risks associated with self-medication and inappropriate use for viral infections (Centre for Disease Control and Prevention CDC, 2019). Building on these foundational components, ASPs employ a range of targeted strategies to optimise antibiotic use in clinical practice.

13.5 Antimicrobial stewardship strategies

The strategies include a preprescription or front-end approach and a postprescription or back-end approach, usually a combination of both.

13.5.1 Preprescription (front-end) approach

Preprescription strategies are essential components of ASPs to ensure that antimicrobials are prescribed appropriately, effectively, and safely. These strategies generally focus on guiding clinical decision making before antimicrobials are even prescribed. Front-end approach involves interventions before or at the time of prescribing. Like education, training, guidelines, formulary restrictions, and decision-support systems. The goal is to prevent misuse of antimicrobials by ensuring that prescribers are informed and guided from the start. The different components are as follows:

a. **Clinical decision-support (CDS) systems:** It helps healthcare providers make informed, evidence-based decisions, while prescribing antimicrobials. These systems can be incorporated into the electronic health record (EHR) to provide real-time guidance based on individualised patient data, such as allergies, renal function, and microbiological results. CDS can alert the prescriber about the recommended antibiotic, dosage, or duration based on national guidelines and local antibiograms (Forrest et al., 2014; Parzen-Johnson et al., 2021).
b. **Diagnostic stewardship**: Rapid diagnostic tests (RDTs) play a significant role in forming preprescription decisions. With more accurate and faster microbiological testing, healthcare providers can identify pathogens earlier and more accurately, allowing for more targeted therapy. For example, PCR-based testing can be used for detecting specific bacterial or viral pathogens in a matter of hours. The use of multiplex PCR for respiratory infections can help differentiate bacterial from viral infections, guiding clinicians to avoid unnecessary antibiotic use (Pickens et al., 2020; Lee et al., 2020).
c. **Education:** Training programmes should be held regularly for healthcare providers that focus on diagnostic stewardship, appropriate indications for antibiotic use, and alternative therapies and provide patient education to improve understanding of when antibiotics are needed and when they are not (Kakkar et al., 2020).
d. **Antibiotic guideline and policies:** Antibiograms, summarising local pathogen resistance patterns, are essential for guiding empiric antibiotic therapy (Kollef and Scott, 2016). Regularly updating and disseminating antibiograms helps prescribers choose initial therapy with the highest likelihood of success, for example targeting *Escherichia coli* resistance profiles. Hospitals may start broad-spectrum antibiotics for severe infections and de-escalate therapy once microbiology results are available (Worldwide Antimicrobial Resistance National/International Network Group WARNING Collaborators, 2023).
e. **Formulary restriction and preauthorization**: Controlled access to selected broad-spectrum or high-risk antibiotics ensures their use is justified and aligned with stewardship principles. Restricted antibiotic policies require ID specialist approval for high-resistance or broad-spectrum drugs, preventing overuse and ensuring therapy is guided by clinical and microbiological data. This approach reduces AMR risk, preserves drug efficacy, and promotes targeted therapy (Huang et al., 2022).
f. **Empiric therapy guidelines with tailored de-escalation:** Initiate broad-spectrum antibiotics for severe infections, followed by narrowing to targeted agents based on microbiological results to reduce AMR. For example, therapy may start with meropenem and be de-escalated to ceftriaxone if susceptibility is confirmed (Kaye, 2012).
g. **Multidisciplinary collaboration**: Effective AMS is contingent upon a multidisciplinary team, including physicians, pharmacists, microbiologists, and infection control specialists, to guide preprescription decisions. Daily ASP rounds help optimise antibiotic choice and therapy duration (Giamarellou et al., 2023).
h. **Preprescription review by infectious disease specialists:** For high-risk or complex patients, preprescription review by an ID specialist ensures empiric therapy is guided by local resistance patterns and patient history, optimising antibiotic selection (Umpleby et al., 2022).

13.5.1.1 Advantage of the front-end (preprescription) approach

Restricted antibiotic policies provide immediate control overprescribing, minimising inappropriate use, improving guideline compliance, reducing AMR, and lowering costs. Direct interaction between prescribers and stewardship teams also promotes rational antibiotic use and education (Indian Council of Medical Research, 2018; Reed et al., 2013; Koulenti, et al., 2025).

13.5.1.2 Limitations of the front-end approach

Front-end antibiotic restrictions may delay therapy in urgent cases, require high staffing resources, and face clinician resistance due to perceived interference with autonomy. Risks include undertreatment in emergencies and scalability challenges in large hospitals (Doron and Davidson, 2011; Reed et al., 2013; Koulenti, et al., 2025).

13.5.2 Postprescription (back-end) approach

These strategies involve review and optimisation of antimicrobial therapy after initiation, based on clinical response and diagnostic data. Interventions include audit and feedback, de-escalation, dose adjustment, and culture review to ensure appropriate antimicrobial use.

a. **De-escalation of therapy:** ASP involves narrowing empiric broad-spectrum antibiotics to targeted therapy guided by culture and sensitivity results, minimising resistance and microbiota disruption (Pulcini et al., 2019; Baur et al., 2017). This approach reduces collateral damage, secondary infections (e.g. *C. difficile*), and selective pressure for MDROs, without increasing mortality, while improving patient outcomes and shortening hospital stay. Clinically, broad-spectrum therapy is initiated empirically, followed by review at 48–72 hours to tailor treatment based on microbiology, clinical response, and biomarkers (Worldwide Antimicrobial Resistance National/International Network Group WARNING Collaborators, 2023). De-escalation strategies are associated with reduced mortality, shorter hospital stays, and no increase in treatment failure, as demonstrated in observational studies and ICU-based cohorts.
b. **Dose optimisation:** It involves tailoring of antibiotic regimens to patient-specific factors – such as renal function, body weight, infection site/severity, and MIC using PK/PD principles to maximise efficacy, minimise toxicity, and reduce resistance (Baur et al., 2017; Roberts and Lipman, 2009). Proper dosing ensures therapeutic drug levels, prevents subtherapeutic exposure that promotes resistance, and improves outcomes, including faster recovery, lower mortality, fewer adverse events, and decreased AMR in hospitals (Baur et al., 2017; Roberts and Lipman, 2009).
c. **Route optimisation:** Antibiotic regimens should be adjusted according to patient-specific factors such as renal function, body weight, and infection severity, using PK/PD principles to ensure optimal therapeutic exposure. It involves transitioning from IV to oral therapy when appropriate. The benefits include shorter hospital stays, lower healthcare costs, reduced risk of IV related complications, and improved patient satisfaction and adherence (Centre for Disease Control and Prevention CDC, 2019). Early intravenous-to-oral switch has been associated with a shorter hospital stay (9 days vs 13 days in the control group without switch), though statistically not significant, lower healthcare costs without compromising clinical outcomes (Mertz et al., 2009).
d. **Duration optimisation and discontinuation:** Antibiotic therapy should be limited to the shortest effective duration and discontinued when infection is unlikely to reduce adverse effects and resistance. It prevents unnecessary exposure to antibiotics, which can cause adverse drug reactions, increase healthcare costs, and drive AMR (MacDougall and Polk, 2005; Barlam et al., 2016; Schwartz et al., 2023).
e. **Audit and feedback:** Systematic review of antimicrobial prescriptions followed by structured, data-driven feedback to prescribers is a key stewardship intervention. It improves adherence to guidelines, promotes de-escalation, and reduces inappropriate antibiotic use (Schwartz et al., 2023; Chung et al., 2013). Prospective audit-and-feedback interventions have been shown to increase appropriate antibiotic use by 19% (95% CI 16% to 22%) and improve guideline adherence in both ICU and general ward settings (Davey et al., 2017). It improves guideline adherence for common infections, curbs broad-spectrum antibiotic use, promotes de-escalation, and reduces inappropriate antibiotic use. Continuous audit cycles are more effective than one time interventions (Chung et al., 2013; Xu et al., 2025). All these approaches maintain clinician autonomy while systematically reducing inappropriate antibiotic use.

13.5.2.1 Advantages of the back-end (postprescription) approach

Advantages of the back-end approach are that it preserves clinical autonomy, unlike front-end (preprescription) restrictive methods; back-end review is perceived as collaborative rather than punitive (Xu et al., 2025). There is improved clinical appropriateness through appropriateness of therapy, increased de-escalation rates, adherence to treatment guidelines, and better patient outcomes. This approach supports diagnostic stewardship by aligning therapy with culture results, imaging, biomarkers (e.g. procalcitonin), and clinical evolution; thus it strengthens diagnostic accuracy. Back-end interventions integrate renal function, weight, comorbidities, and drug–drug interactions, optimising PK/PD target attainment and allowing personalised therapy (Gopishetty, 2023; Nielsen et al., 2024).

13.5.2.2 Limitations of back-end (postprescription) approach

It includes delayed intervention, as inappropriate empiric therapy may already be given, and dependence on clinician cooperation, with potential resistance if autonomy is perceived as threatened (Xu et al., 2025; Gopishetty, 2023; Langford et al., 2020). It is resource-intensive, requiring trained pharmacists, ID specialists, and microbiologists, and has a limited impact on initial empiric therapy. Effectiveness depends on timely review and communication, but infrastructure, staffing, and EHR variability can hinder consistent implementation, and prescribers may only partially adopt recommendations (Xu et al., 2025; Gopishetty, 2023; Langford et al., 2020).

13.5.2.3 Case illustrations in antimicrobial stewardship programme

Presented below are selected case illustrations documenting ASP interventions at Gauhati Medical College, Guwahati, Assam, India. These cases exemplify the practical application of stewardship principles in clinical settings, emphasising rational antimicrobial use and its impact on patient care and AMR containment.

Case 1: Stewardship in ICU – de-escalation

An elderly patient in the ICU with septic shock is started empirically on meropenem and vancomycin.

Stewardship decision: Blood cultures were suggested, which grew *Escherichia coli* sensitive to ceftriaxone. The patient was stabilised clinically.

Action: De-escalated therapy from meropenem to ceftriaxone and discontinued vancomycin.

Rationale: Targeted therapy reduces selection pressure, toxicity, and cost without compromising outcomes.

Case 2: Surgical prophylaxis optimisation

A patient undergoing elective colorectal surgery is prescribed ceftriaxone for 5 days postoperatively.

Stewardship decision: Guidelines recommend a single preoperative dose (with or without ≤24-hour continuation).

Action: Antibiotic use was limited to perioperative prophylaxis only.

Rationale: Prolonged surgical prophylaxis beyond 24 hours has not shown additional benefit and is associated with increased risk of resistance and adverse events, as demonstrated in multiple randomised and guideline-based studies.

Case 3: Avoiding antibiotics in viral infection

A child presents with fever, sore throat, and cough. No bacterial features; viral infection suspected.

Stewardship decision: Avoid empirical antibiotic prescription.

Action: The patient was provided symptomatic treatment and education.

Rationale: Reduces unnecessary antibiotic exposure and prevents the emergence of resistance.

Case 4: IV to oral switch

A hospitalised patient with community-acquired pneumonia improves after 3 days of IV antibiotics and is clinically stable.

Stewardship decision: Assess eligibility for oral therapy.

Action: Switch from IV to oral antibiotics, and early discharge was planned.

Rationale: Reduces hospital stay, cost, and IV-related complications.

Case 5: Antibiotic time out (48–72 hours review)

A patient started on broad-spectrum antibiotics for suspected sepsis shows no evidence of bacterial infection after 72 hours.

Stewardship decision: Reassess need for antibiotics.

Action: Discontinued antibiotics.

Rationale: Prevents unnecessary exposure and reduces the risk of adverse effects and resistance.

Case 6: Dose optimisation with renal impairment

An elderly patient with chronic kidney disease is prescribed standard-dose vancomycin.

Stewardship decision: Evaluate renal function and need for dose adjustment.

Action: Dose was adjusted using therapeutic drug monitoring.

Rationale: Prevents toxicity while maintaining therapeutic efficacy.

13.6 Policies and interventions in antimicrobial stewardship

13.6.1 Implementing evidence based antimicrobial guidelines

Evidence-based antimicrobial guidelines optimise antibiotic use, reduce AMR, improve patient outcomes, and ensure cost-effective care. Development begins with a multidisciplinary Guideline Development Group (GDG), including ID specialists, pharmacists, microbiologists, epidemiologists, clinicians, and public health experts (Patterson et al., 2016). Using frameworks like PICO (Population/patient, intervention, comparison, and outcome), topics such as empiric therapy, diagnostic criteria, de-escalation, duration, and prophylaxis are addressed (Walia et al., 2019). High-quality evidence is gathered from Randomised Controlled Trials (RCTs), cohort studies, metaanalyses, surveillance data, and PK/PD studies and assessed via the Grading of Recommendations Assessment, Development, and Evaluation (GRADE) system (Rupali et al., 2019). Recommendations consider evidence quality, clinical benefit versus harm, local resistance, feasibility, cost-effectiveness, and patient preferences (Yasmina et al., 2023). Draughts are reviewed by external experts and stakeholders to improve clarity and uptake (Jain and Chavan, 2024). Guidelines are updated every 2–5 years or sooner with major evidence changes and implemented through education, organisational support, and IT tools (EHR integration, apps, pocket guides) (Rajendran et al., 2025; Shrestha et al., 2025). Effectiveness is monitored via audits and feedback (World Health Organization WHO, 2021), and barriers, such as clinician habits, limited resources, or patient expectations, are addressed through leadership support, training, local data, and awareness campaigns. Successful implementation reduces inappropriate antibiotic use, AMR rates, hospital stays, adverse events, and healthcare costs (World Health Organization, 2014).

13.6.2 Specific stewardship interventions

Targeted AMS interventions (ASPs) are evidence driven strategies designed to optimise antibiotic choice, dosing, and treatment duration, thereby minimising resistance development and enhancing clinical outcomes. The Broad interventions are prescribed as follows:

a. **Antibiotic time-outs (ATOs) in ASP:** ATOs are structured reassessments of ongoing antimicrobial therapy, typically performed 48–72 hours after initiation, and are endorsed by CDC, IDSA, and WHO (Patterson et al., 2016; Walia et al., 2019; Rupali et al., 2019; Yasmina et al., 2023; Jain and Chavan, 2024; Rajendran et al., 2025; Shrestha et al., 2025). ATOs ensure therapy is appropriate, necessary, and optimised, reducing broad-spectrum use, AMR, toxicity, and healthcare costs. Key steps include (1) reassessing need for antibiotics – confirm if infection is bacterial, viral, or noninfectious, (2) reviewing diagnostic/microbiology data – culture results, biomarkers (e.g. procalcitonin), imaging, and clinical updates to refine therapy, (3) optimising therapy – adjust dose, route, duration, and de-escalate when possible, and (4) documenting decisions – enhances communication and supports audit and feedback.
   ATO promotes de-escalation, aligns therapy with culture results, reduces unnecessary antibiotic exposure, and encourages prescriber behaviour change. Pharmacists and ID specialists often lead ATO discussions, with EHR alerts prompting timely review. Challenges include time constraints, unclear role ownership, variable lab turnaround, and the need for clinician education. ATO is a low-cost, high-impact ASP intervention suitable for both high- and low-resource settings (Patterson et al., 2016; Walia et al., 2019; Rupali et al., 2019; Yasmina et al., 2023; Rajendran et al., 2025).
b. **Preauthorisation:** Preauthorisation is an AMS strategy in which certain antibiotics, usually broad-spectrum or high-risk agents, require approval from an infectious disease (ID) specialist or the ASP team before they can be prescribed. This ensures that these powerful drugs are used only when clinically appropriate, preventing misuse, slowing resistance, and conserving their effectiveness. Commonly restricted antibiotics include carbapenems, fluoroquinolones, linezolid, daptomycin, and newer anti-MRSA or antiGram-negative agents (Barlam et al., 2016).
c. **Prospective audit and feedback:** Review of antibiotic therapy by external experts has been proven to be highly effective in optimising treatment for severely ill patients and those receiving broadspectrum or combination antimicrobial therapy. Prospective audit and feedback differ from an antibiotic 'time out' as the audits are conducted by personnel outside the treating team (Valiquette et al., 2007; Dancer et al., 2013).
d. **Pharmacy-driven interventions:** Pharmacists play a central role in these programmes because of their specialised knowledge in pharmacotherapy, dosing, drug–drug interactions, and infection management. Pharmacy-driven interventions have become a core component of AMS protocols globally and are associated with significant improvements in antibiotic appropriateness, resistance patterns, and clinical outcomes (World Health Organization, 2019a,b; World Health Organization, 2014). The pharmacy-driven interventions include (1) intravenous-to-oral switch when clinically appropriate, (2) dose optimisation, including renal adjustments and therapeutic drug monitoring (e.g. vancomycin and aminoglycosides), automatic alerts for therapeutic duplication, (3) time-sensitive stop orders, particularly for preoperative antibiotic prophylaxis, (4) prevention of drug-to-drug interactions, and (5) review of antibiotics in *C. difficile* and invasive infection cases to promote de-escalation.
e. **Infection and syndrome-specific interventions:** These interventions focus on (1) optimising diagnosis and treatment for common clinical conditions to improve outcomes while minimising unnecessary antibiotic use. For example, in community-acquired pneumonia, efforts aim to enhance diagnostic accuracy and ensure appropriate antibiotic selection

and duration, thereby avoiding overtreatment. (2) In urinary tract infections, unnecessary cultures should be avoided, and asymptomatic bacteriuria should generally not be treated except in specific situations, reducing inappropriate antibiotic exposure. (3) For skin and soft tissue infections, stewardship emphasises selecting the appropriate antimicrobial spectrum and limiting duration to what is clinically required. (4) Additionally, for culture proven invasive infections, management should be guided by microbiological results, ensuring targeted therapy with the correct agent, dose, and duration. Collectively, these targeted approaches help reduce AMR, improve patient safety, and promote rational antibiotic use. The antibiotic stewardship intervention strategies are presented in Fig. 13.6.

f. **Tracking antibiotic use and resistance:** Overuse and inappropriate prescribing of antibiotics drive AMR, making routine monitoring essential to detect irrational use and emerging resistance early (Dancer et al., 2013; Delpy et al., 2024). Effective ASPs require feedback loops to evaluate the impact of audits, guidelines, and other interventions, while monitoring data support National AMR action plans and One Health strategies across human, animal, and environmental sectors (Indian Council of Medical Research, 2018; McEwen and Collignon, 2018). Comprehensive surveillance should include antibiotic consumption tracking (e.g. volumes, defined daily doses, and classes of antibiotics), prescription audits linked to clinical indications, and robust AMR surveillance through culture, sensitivity testing, and regularly updated antibiograms using standardised breakpoints. Monitoring MDRO trends and integrating data across sectors strengthens stewardship efforts. National and institutional ASP committees should periodically review these data and align practices with WHO AWaRe and national guidelines to ensure an effective monitoring model (Indian Council of Medical Research, 2018; Saleem et al., 2025; Sanchez et al., 2016).

13.7 Collaboration between antimicrobial stewardship and infection control

Beyond individual interventions, collaboration with infection prevention and control programmes further strengthens the impact of AMS efforts. Coordination between the ASP and ICP is essential to enhance patient outcomes, reduce AMR, and manage HAIs. While ASP focuses on optimising antimicrobial use, ICP works to prevent infection transmission. Their synergy creates a comprehensive, integrated approach to infection prevention and resistance control.

13.7.1 Key areas of collaboration

a. **AMR and infection surveillance:** Both ASP and ICP depend on surveillance data – ICP tracks HAIs, while ASP analyses antimicrobial use and resistance trends. Collaboration allows timely detection and response to emerging pathogens and resistance patterns. For instance, rising resistance in a bacterial strain can prompt ICP to enhance

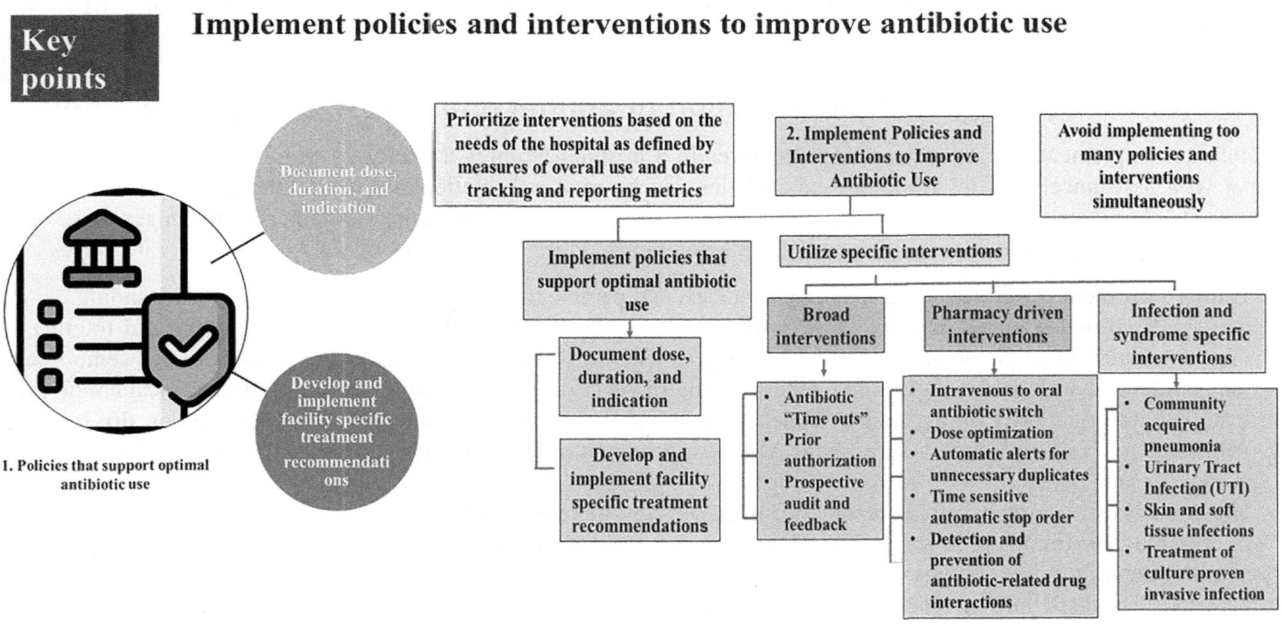

FIGURE 13.6 **Antibiotic stewardship.** Intervention strategies of antibiotic stewardship.

isolation measures and ASP to adjust treatment protocols to prevent further resistance (McElligott et al., 2017; Assi et al., 2021).

b. Optimising antibiotic use and prevention of infection: Effective infection control measures, such as hand hygiene, environmental cleaning, and isolation protocols, reduce the need for antimicrobial use. Collaboration between ICP and ASP ensures fewer infections requiring antibiotics and promotes correct, judicious antimicrobial use, minimising overuse and misuse (Hamed et al., 2024; Krishnamoorthy et al., 2025; Schuts et al., 2016).

c. Capacity building and training of healthcare staff: Both ASP and ICP require ongoing education for healthcare providers. While ASP focuses on appropriate antibiotic prescribing practices, ICP educates staff on the prevention and control of infections. Joint educational sessions that emphasise both aspects- appropriate antibiotic use and infection prevention ensure that all staff understand how their actions impact both infection rates and AMR (Indian Council of Medical Research, 2018; Knobloch et al., 2021).

d. Outbreak management: During outbreaks of MDROs such as carbapenem-resistant Enterobacteriaceae (CRE), close collaboration between ICP and ASP is crucial. ICP leads containment measures (e.g. cohorting), while ASP guides appropriate antibiotic selection to manage cases and prevent further resistance development (MacDougall and Polk, 2005; Davey et al., 2017).

e. Data integration and shared goals: Integrated data systems strengthen ASP-ICP collaboration by providing real-time information on infection rates, antibiotic use, and resistance patterns. Shared data help identify priority areas, guide targeted interventions, and refine both infection control practices and antibiotic prescribing (Cornistein et al., 2023; Büchler et al., 2024).

13.7.2 Outcomes of collaboration

a. Reduction in HAIs: Strong ASP-ICP collaboration helps lower HAIs by combining effective infection prevention with appropriate antibiotic use, reducing both infection rates and resistance risks (Zhou et al., 2021).

b. Decreased antibiotic resistance: Coordinated ASP-ICP efforts limit the spread of resistant organisms and ensure appropriate antibiotic use, helping slow the emergence of new resistance patterns (Liu et al., 2024).

c. Improved patient outcomes: ASP-ICP synergy enhances patient safety by reducing infections and ensuring effective, appropriate antibiotic therapy. Stewardship limits unnecessary broad-spectrum use, lowering risks such as *C. difficile* infection, toxicity, and prolonged hospital stay, while improving treatment success and recovery rates through measures like audit and feedback and preauthorisation (Manning et al., 2018; Dyary et al., 2023). The collaboration between ASP and ICP is vital for addressing AMR and improving healthcare safety. By aligning goals, sharing data, and coordinating interventions, these programmes reduce infections, optimise antibiotic use, and enhance clinical outcomes of patients, creating a comprehensive approach to infection management (Manning et al., 2018; Dyary et al., 2023).

13.8 Challenges in the implementation of antimicrobial stewardship programme

Meticulous implementation of stewardship measures in a hospital requires a dedicated team of experts, the healthcare along with assistance from microbiology laboratories and hospital information systems (Rolfe et al., 2021). However, despite its clear benefits, the implementation of ASP is still limited and encounters significant challenges in resource-constrained healthcare settings. Challenges in implementing ASP are depicted in Fig. 13.7.

a. Limited awareness and training: Clinicians, nurses, and paramedics often have an inadequate understanding of AMS and AMR, which hampers effective ASP implementation (Wu et al., 2022). The absence of structured training and ongoing professional development leads junior doctors to adopt suboptimal prescribing practices from seniors rather than evidence-based guidance. Limited understanding of AMS principles, antibiotic spectrums, and local antibiograms results in irrational prescribing, such as unnecessary broad-spectrum use or inappropriate combinations, driving AMR. Additionally, poor awareness of available ASP resources and guidelines further widens this knowledge gap (Lazure et al., 2022).

b. Human resource constraints: In resource-limited settings, there is a shortage of infectious disease physicians, clinical microbiologists, trained pharmacists, and data managers needed for a functional ASP team. Various studies reported it as a barrier in low- and middle-income country settings (Wu et al., 2022).

c. Limited diagnostic and microbiology support: There may be limited availability of rapid diagnostics, delayed culture reports, and suboptimal laboratory infrastructure, and poor integration of microbiology data with clinical decision-making. This has been noted across reviews and qualitative studies (Abbas, 2024).

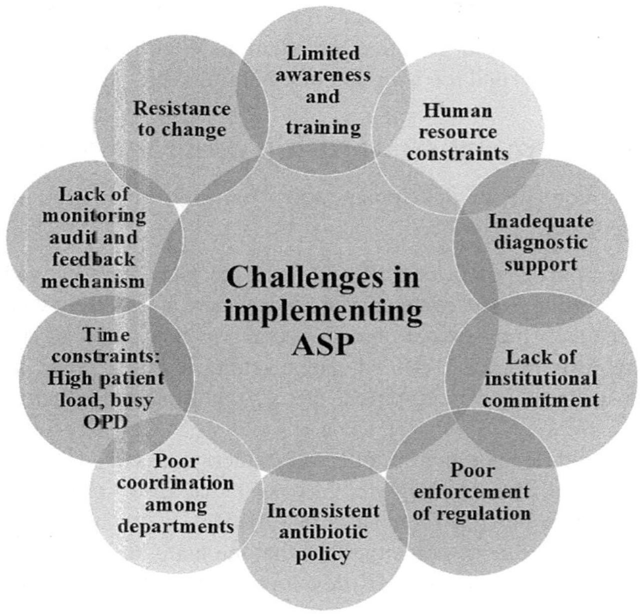

FIGURE 13.7 Challenges in the implementation of ASP. Significant challenges encountered in healthcare settings are depicted in the figure. *ASP*, Antimicrobial stewardship programme.

d. Lack of administrative commitment: A lack of governance, inconsistent policies, and inadequate resources often undermine the sustainability of ASPs (Zacchaeus et al., 2023). Limited administrative support, insufficient budget allocation, and low prioritisation of stewardship, especially in resource-constrained settings with competing demands, remain major barriers to effective implementation.

e. Weak regulatory enforcement: Many facilities lack facility-specific antibiotic guidelines or local antibiogram-based policies (Zacchaeus et al., 2023).

f. Inconsistent antibiotic policy: Absence of standard treatment guidelines (STGs) or poor adherence to existing guidelines and unregulated over-the-counter antibiotic availability (especially in low- and middle-income country settings) (Zacchaeus et al., 2023).

g. Lack of interdepartmental coordination: Lack of interdisciplinary communication between clinicians, pharmacy, microbiology, nursing, and hospital administration is a hindrance to implementation of ASP (Zacchaeus et al., 2023; Mathew et al., 2020).

h. Time constraints: High patient load and busy OPDs and wards make detailed prescription review or de-escalation rounds difficult (Zacchaeus et al., 2023).

i. Lack of audit-and-feedback mechanism: Without routine audits, prescription reviews, and feedback loops, stewardship interventions lose impact (Lazure et al., 2022). Compliance monitoring and tracking are vital to evaluate the impact of stewardship interventions, identify areas for improvement, and ensure adherence to established guidelines (Mathew et al., 2020). Compliance monitoring in ASPs includes tracking antibiotic use and resistance trends, supported by tools like the CDC's National Healthcare Safety Network (NHSN) Antibiotic Use (AU) module (process and outcome measurement). Regular audit-and -feedback cycles strengthen adherence to guidelines and enable timely correction of prescribing practices (performance feedback) (Mathew et al., 2020). Leadership commitment and defined accountability structures ensure consistent monitoring, backed by adequate financial and IT support (accountability) (Giamarellou et al., 2023). Strong compliance with stewardship guidelines improves patient safety by reducing *C. difficile* infections, mortality, and healthcare costs (patient safety) (Hwang and Kwon, 2021).

j. Resistance to change: Clinicians may perceive ASP as interference with clinical autonomy. Habitual prescribing patterns and defensive medicine (fear of patient deterioration) hinder adherence (Abbas, 2024).

In resource-poor settings, efforts to combat AMR are further weakened by vulnerable health systems marked by limited resources and poor governance (Pereira, 2020). WHO outlines six key health-system building blocks- leadership and governance, human resources, health information systems, essential medicines, funding, and service delivery (Akilimali et al., 2023). Strengthening these components, particularly infrastructure, human resources, and laboratory capacity, is essential for effective AMR control (World Health Organization WHO, 2010). Addressing these implementation challenges requires innovative approaches and the adoption of emerging technologies to strengthen stewardship practices.

13.9 Transforming antimicrobial stewardship programme through innovation

Innovation is a fundamental prerequisite to developing new strategies, tools, and treatments to stay ahead of the evolutionary challenge posed by AMR.

a. **Rapid diagnostics:** Innovative rapid diagnostic tests (RDTs) and molecular assays, such as MALDI-TOF mass spectrometry and multiplex PCR, enable faster pathogen detection and resistance pattern analysis, allowing timely intervention with targeted therapy and thereby reducing the use of unnecessary broad-spectrum antibiotics (Dixit et al., 2021).
b. **Data analytics and artificial intelligence (AI):** AI and machine learning (ML) are used to analyse vast clinical datasets to predict resistance patterns, guide empirical therapy choices, and enhance surveillance systems (Campos et al., 2022).
c. **Novel therapeutics:** Research and development in alternative therapies, including engineered antimicrobial peptides, phage therapy, and nanotechnology-based drug delivery systems, offer promising avenues to combat resistant infections when conventional antibiotics fail (Pennisi et al., 2025).
d. **One health approach:** Innovative, multisectoral strategies that integrate human, animal, and environmental health are necessary to address the complex dynamics of AMR development and dissemination across different ecosystems, as presented in Fig. 13.8 (Jacobowski et al., 2025).

Effective ASPs rely on core elements that can be systematically tracked using checklists. These tools help teams align expectations, support internal audits, and ensure consistent implementation across facilities. When widely adopted, checklists enable more reliable monitoring of ASP performance, and evidence shows they contribute to optimised antibiotic use and improved patient care (Velazquez-Meza et al., 2022; van Daalen et al., 2017a,b; de Vries et al., 2010).

13.10 Outcomes and impact of antimicrobial stewardship

ASPs effectively decrease inappropriate antibiotic use, improve clinical outcomes, and lower healthcare costs. They decrease overall antibiotic consumption, shorten hospital stays, and reduce complications from resistant infections, making them essential for public health.

a. **Reduction in inappropriate antibiotic use:** ASPs have consistently demonstrated significant improvements in antibiotic use, clinical outcomes, and healthcare costs across diverse healthcare settings. Metaanalyses and multicenter studies report reductions in overall antimicrobial consumption ranging from 19% to 36%, with decreases of up to 30%–32% in broad-spectrum antibiotic use, particularly cephalosporins and carbapenems (Braun et al., 2020). They improve prescribing appropriateness by addressing unclear indications, incorrect drug choices, and excessive duration. Controlled antibiotic use slows the spread of AMR (Braun et al., 2020).
b. **Reduction in healthcare costs:** Lower antibiotic use directly cuts drug expenditures and reduces costs associated with longer hospital stays and complications. Cost savings have been consistently reported across diverse healthcare settings implementing ASPs (Alawi and Darwesh, 2016; Lee et al., 2018). From an economic perspective, ASPs contribute to substantial cost savings through reduced drug expenditure and shorter hospital stays (Karanika et al., 2016; Rathish et al., 2019).
c. **Better clinical outcome:** An effective AMSP can reduce infection and mortality rates, decrease treatment failures, and improve the accuracy of both therapeutic and prophylactic antibiotic prescribing (Braun et al., 2020). ASPs are also associated with improved clinical outcomes, including reductions in mortality by >10% in severe infections and ICU settings, as well as decreased treatment failures and improved appropriateness of therapy (Lindsay et al., 2019; Okumura, et al., 2015). Implementation of stewardship interventions has been shown to reduce the incidence of HAIs, including *C. difficile* infection, by up to 50% in some hospital-based studies (Karanika et al., 2016; Lindsay et al., 2019; Okumura et al., 2015).
d. **Improved patient satisfaction:** ASPs contribute to enhanced patient satisfaction by promoting shorter hospital stays, optimising clinical outcomes, and reducing treatment-related complications (Abdel Hadi et al., 2024).
e. **Enhanced insight into antimicrobial usage:** Through systematic surveillance, audit, and feedback mechanisms, ASPs enable healthcare facilities to better understand antimicrobial prescribing patterns and local resistance trends, thereby informing more rational and targeted therapy (Shah and Shah, 2023).

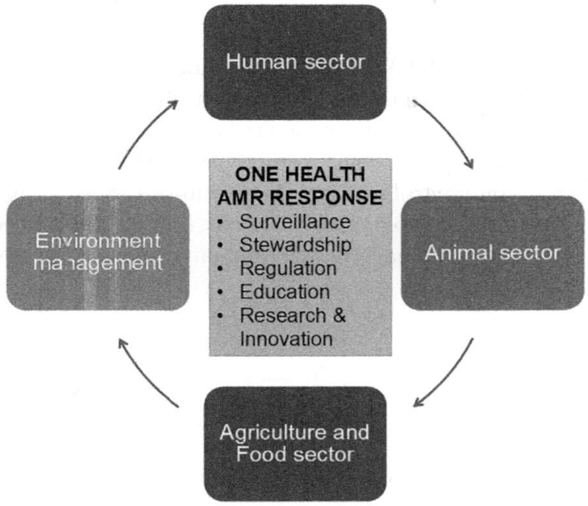

FIGURE 13.8 **One health AMR response showing multisectoral strategies that integrate human, animal, and environmental health necessary to address the complex dynamics of AMR.** One Health framework for AMR mitigation, illustrating coordinated actions – surveillance, stewardship, regulation, education, and research – across human, animal, agriculture/food, and environmental sectors. *AMR*, Antimicrobial resistance.

f. **Reduced environmental impact:** By minimising inappropriate antimicrobial use and promoting responsible disposal practices, ASPs help decrease environmental contamination with antimicrobial residues and limit the emergence and spread of AMR in environmental reservoirs Shah and Shah, 2023).

Additionally, stewardship programmes have been linked to decreased emergence of MDROs and improved adherence to evidence-based prescribing practices. These benefits are particularly evident in high-risk settings such as intensive care units, where optimised antimicrobial use has a direct impact on patient survival and resource utilisation.

13.11 Future directions

Future directions for ASPs include integrating advanced technologies such as AI, machine learning, and rapid diagnostics to enable faster, more precise, and personalised antimicrobial decisions. AI can support rapid diagnostic interpretation and tailored therapy, while point-of-care tests and emerging CRISPR-based diagnostics offer rapid, portable, and accessible testing, especially valuable in resource-limited settings (Hassan et al., 2025).

Despite advances in infectious disease rapid diagnostic tests (RDTs) and their use to guide patient-level therapy, RDT utilisation within AMS programmes (ASPs) varies widely across low-, middle-, and high-income countries. Insights from a panel of seven infectious disease experts from Colombia, Japan, Nigeria, Thailand, the United Kingdom, and the USA, along with findings from a literature review, were used to evaluate the role and value of RDTs in ASPs (Moore et al., 2023). Genomic monitoring may also be applied for real-time tracking of resistance at the genetic level, helping healthcare systems identify emerging resistant strains and adjust treatment protocols based on local resistance patterns. Additionally, global collaboration remains essential in combating AMR worldwide.

13.11.1 Technological advancements for rapid diagnostic testing integration

Technological advancements are increasingly transforming rapid diagnostic testing (RDT) by integrating cutting-edge tools to enhance speed, accuracy, and accessibility. Artificial intelligence (AI) and machine learning improve diagnostic precision, optimise guide RNA selection, and support personalised therapy, while Clustered Regularly Interspaced Short Palindromic Repeats (CRISPR)-based diagnostics enable rapid, sensitive pathogen detection and multiplexed testing for comprehensive disease profiling. Digital and point-of-care diagnostics, including Internet of Things (IoT)-enabled molecular devices and portable tools, enhance efficiency and accessibility, particularly in resource-limited settings, by providing real-time data for remote monitoring and early warning alerts. Additionally, nanotechnology-enabled biosensors facilitate instantaneous data transmission to cloud-based systems, supporting immediate analysis, reporting, and public health responses (Pennisi et al., 2025; Shah and Shah, 2023; Ramesh et al., 2022).

13.12 Conclusion

Taken together, these elements highlight the critical role of AMS in addressing the growing threat of AMR. ASPs are vital to optimise antibiotic use and improve patient outcomes, reducing healthcare costs, and combating the global challenge posed by AMR. By integrating multidisciplinary teams, evidence-based guidelines, pre and postprescription strategies like de-escalation and audits, rapid diagnostics, and robust surveillance, ASPs minimise inappropriate prescribing, curb resistant pathogens, and enhance infection control collaboration. Future success hinges on overcoming implementation barriers through leadership commitment, technological innovations like AI and CRISPR diagnostics, and a one health approach to sustain antibiotic efficacy across human, animal, and environmental sectors.

Acknowledgement

The authors would like to acknowledge BioRender for designing diagrams.

References

Abbas, S., 2024. The challenges of implementing infection prevention and antimicrobial stewardship programs in resource-constrained settings. Antimicrob Steward Healthc Epidemiol 4 (1), e45. Available at https://doi.org/10.1017/ash.2024.35.

Abdel Hadi, H., Eltayeb, F., Al Balushi, S., Daghfal, J., Ahmed, F., Mateus, C., 2024. Evaluation of hospital antimicrobial stewardship programs: implementation, process, impact, and outcomes, review of systematic reviews. Antibiotics (Basel) 13 (3), 253. Available at https://doi.org/10.3390/antibiotics13030253.

Abdul-Aziz, M.H., Alffenaar, J.C., Bassetti, M., Bracht, H., Dimopoulos, G., Marriott, D., et al., 2020. Antimicrobial therapeutic drug monitoring in critically ill adult patients: a position paper. Intensive Care Medicine 46 (6), 1127–1153. Available at https://doi.org/10.1007/s00134-020-06050-1.

Akilimali, A., Awuah, W.A., Cakwira, H., Mirindi, M.K., Masimango, G., Amani, T., et al., 2023. Antimicrobial resistance in Democratic Republic of Congo: the way forward. International Journal of Surgery: Global Health 6. Available at https://doi.org/10.1097/GH9.0000000000000150 e0150.

Alawi, M.M., Darwesh, B.M., 2016. A stepwise introduction of a successful antimicrobial stewardship program. Experience from a tertiary care university hospital in Western, Saudi Arabia. Saudi Medical Journal 37, 1350–1358. Available at https://doi.org/10.15537/smj.2016.12.15739.

Assi, M., Abbas, S., Nori, P., Doll, M., Godbout, E., Bearman, G., et al., 2021. Infection prevention and antimicrobial stewardship program collaboration during the COVID-19 Pandemic: a window of opportunity. Current Infectious Disease Reports 23 (10), 15. Available at https://doi.org/10.1007/s11908-021-00759-w.

Barlam, T.F., Cosgrove, S.E., Abbo, L.M., MacDougall, C., Schuetz, A.N., Septimus, E.J., et al., 2016. Implementing an antibiotic stewardship program: guidelines by the infectious diseases Society of America and the Society for Healthcare Epidemiology of America. Clinical Infectious Diseases 62 (10), e51–e77. Available at https://doi.org/10.1093/cid/ciw118.

Baur, D., Gladstone, B.P., Burkert, F., Carrara, E., Foschi, F., Döbele, S., et al., 2017. Effect of antibiotic stewardship on the incidence of infection and colonisation with antibiotic-resistant bacteria and *Clostridium difficile* infection: a systematic review and meta-analysis. The Lancet Infectious Diseases 17 (9), 990–1001. Available at https://doi.org/10.1016/S1473-3099(17)30325-0.

Braun, B.I., Chitavi, S.O., Suzuki, H., Soyemi, C.A., Puig-Asensio, M., 2020. Culture of safety: impact on improvement in infection prevention process and outcomes. Current Infectious Disease Reports 22 (12), 34. Available at https://doi.org/10.1007/s11908-020-00741-y.

Büchler, A.C., Haddad Galas, M., Buetti, N., 2024. Challenges and success stories of the implementation of infection control and antimicrobial stewardship strategies: proceedings of the 5th Global Ministerial Summit on Patient Safety, 2023. Antimicrobial Resistance and Infection Control 13, 16. Available at https://doi.org/10.1186/s13756-023-01344-7.

Burmeister, A.R., 2015. Horizontal gene transfer. Evolution, Medicine, and Public Health 2015 (1), 193–194. Available at https://doi.org/10.1093/emph/eov018.

Campos, A.F., Arantes, T., Cambiais, A.M.V.B., Cury, A.P., Tiroli, C.G., Rossi, F., et al., 2022. Impact of an antimicrobial stewardship program intervention associated with the rapid identification of microorganisms by MALDI-TOF and detection of resistance genes in icu patients with Gram-Negative bacteremia. Antibiotics 11 (9), 1226. Available at https://doi.org/10.3390/antibiotics11091226.

Centre for Disease Control and Prevention (CDC). The Core Elements of Hospital Antibiotic Stewardship Programs: 2019. Atlanta, GA: US Department of Health and Human Services, CDC; 2019. Available at https://www.cdc.gov/antibiotic-use/hcp/core-elements/hospital.html?CDC_AAref_Val=https://www.cdc.gov/antibiotic-use/core-elements/hospital.html. (accessed 15.09.25).

Cerceo, E., Deitelzweig, S.B., Sherman, B.M., Amin, A.N., 2016. Multidrug resistant Gram-negative bacterial infections in the hospital setting: overview, implications for clinical practice, and emerging treatment options. Microbial Drug Resistance 22 (5), 412–431. Available at https://doi.org/10.1089/mdr.2015.0220.

Chancey, S.T., Zähner, D., Stephens, D.S., 2019. Acquired inducible antimicrobial resistance in Gram-positive bacteria. Future Microbiology 7, 959–978. Available at https://doi.org/10.2217/fmb.12.63.

Chung, G.W., Wu, J.E., Yeo, C.L., Chan, D., Yang Hsu, L., 2013. Antimicrobial stewardship: a review of prospective audit and feedback systems and an objective evaluation of outcomes. Virulence 4 (2), 151–157. Available at https://doi.org/10.4161/viru.21626.

Cornistein, W., Santonato, D., Novau, P.A., Fabbro, L.G., Jorge, M.F., Malvicini, M.A., et al., 2023. Synergy between infection control and antimicrobial stewardship programs to control carbapenem-resistant Enterobacterales. Antimicrob Steward Healthcare Epidemiology 3 (1), e162. Available at https://doi.org/10.1017/ash.2023.439.

van Daalen, F.V., Lagerburg, A., de Kort, J., Sànchez Rivas, E., Geerlings, S.E., 2017a. Implementation of an antibiotic checklist increased appropriate antibiotic use in the hospital on Aruba. International Journal of Infectious Diseases 59, 14–21. Available at https://doi.org/10.1016/j.ijid.2017.03.019.

van Daalen, F.V., Prins, J.M., Opmeer, B.C., Boermeester, M.A., Visser, C.E., van Hest, R.M., et al., 2017b. Effect of an antibiotic checklist on length of hospital stay and appropriate antibiotic use in adult patients treated with intravenous antibiotics: a stepped wedge cluster randomized trial. Clinical Microbiology and Infection 23 (7), e1–485. Available at https://doi.org/10.1016/j.cmi.2017.01.019 e8.

Dancer, S.J., Kirkpatrick, P., Corcoran, D.S., Christison, F., Farmer, D., Robertson, C., 2013. Approaching zero: temporal effects of a restrictive antibiotic policy on hospital-acquired *Clostridium difficile*, extended-spectrum beta-lactamase-producing coliforms and meticillin-resistant *Staphylococcus aureus*. International Journal of Antimicrobial Agents 41 (2), 137–142. Available at https://doi.org/10.1016/j.ijantimicag.2012.10.013.

Davey, P., Marwick, C.A., Scott, C.L., Charani, E., McNeil, K., Brown, E., et al., 2017. Interventions to improve antibiotic prescribing practices for hospital inpatients. Cochrane Database of System Review(2), CD003543. Available at https://doi.org/10.1002/14651858.CD003543.pub4.

Davies, J., Davies, D., 2010. Origins and evolution of antibiotic resistance. Microbiology and Molecular Biology Reviews 74 (3), 417–433. Available at https://doi.org/10.1128/MMBR.00016-10.

Dellit, T.H., Owens, R.C., McGowan, J.E., Gerding, D.N., Weinstein, R.A., Burke, J.P., et al., 2007. Infectious Diseases Society of America and the Society for Healthcare Epidemiology of America guidelines for developing an institutional program to enhance antimicrobial stewardship. Clinical Infectious Diseases 44 (2), 159–177. Available at https://doi.org/10.1086/510393.

Delpy, L., Astbury, C.C., Aenishaenslin, C., 2024. Integrated surveillance systems for antibiotic resistance in a one health context: a scoping review. BMC Public Health 24, 1717. Available at https://doi.org/10.1186/s12889-024-19158-6.

Dixit, D., Ranka, R., Panda, P.K., 2021. Compliance with the 4Ds of antimicrobial stewardship practice in a tertiary care centre. JAC Antimicrobial Resistance 3 (3), dlab135. Available at https://doi.org/10.1093/jacamr/dlab135.

Doron, S., Davidson, L.E., 2011. Antimicrobial stewardship. Mayo Clinic Proceedings. Mayo Clinic 86 (11), 1113–1123. Available at https://doi.org/10.4065/mcp.2011.0358.

Dyary, H.O., Faraj, G.J., Saeed, N.M., 2023. History, current situation, and future perspectives on antibiotics and antibiotic resistance. In: In: Abbas, R.Z., Saeed, N.M., Younus, M., Aguilar-Marcelino, L., Khan, A. (Eds.), One Health Triad 2. Unique Scientific Publishers, Faisalabad, Pakistan, pp. 109–118. Available at 10.47278/book.oht/2023.49.

Dyar, O.J., Pagani, L., Pulcini, C., 2015. Strategies and challenges of antimicrobial stewardship in long-term care facilities. Clinical Microbiology and Infection 23 (5), 327–332. Available at https://doi.org/10.1016/j.cmi.2014.09.005.

Forrest, G.N., Van Schooneveld, T.C., Kullar, R., Schulz, L.T., Duong, P., Postelnick, M., et al., 2014. Use of electronic health records and clinical decision support systems for antimicrobial stewardship. Clinical Infectious Diseases 59 (suppl_3), S122–S133. Available at https://doi.org/10.1093/cid/ciu565.

Fridkin, S., Baggs, J., Fagan, R., Magill, S., Pollack, L.A., Malpiedi, P., et al., 2014. Vital signs: improving antibiotic use among hospitalized patients. MMWR. Morbidity and Mortality Weekly Report 63, 194–200.

Gerding, D.N., 2001. The search for good antimicrobial stewardship. The Joint Commission Journal on Quality Improvement 27 (8), 403–404. Available at https://doi.org/10.1016/s1070-3241(01)27034-5.

Ghafur, A., Mathai, D., Muruganathan, A., Jayalal, J.A., Kant, R., Chaudhary, D., et al., 2013. The Chennai declaration: a roadmap to tackle the challenge of antimicrobial resistance. Indian Journal of Cancer 50, 71–73. Available at https://doi.org/10.4103/0019-509X.104065.

Giamarellou, H., Galani, L., Karavasilis, T., Ioannidis, K., Karaiskos, I., 2023. Antimicrobial stewardship in the hospital setting: a narrative review. Antibiotics (Basel) 12 (10), 1557. Available at https://doi.org/10.3390/antibiotics12101557.

Gopishetty, T., 2023. Antibiotics transition from intravenous to oral use and their bioavailability in teritary care hospital. International Journal of Pharmaceutical Research and Applications 8 (1), 2120–2126. Available at https://doi.org/10.35629/7781-080121202126.

Hamed, N.M., Deif, O.A., El-Zoka, A.H., 2024. The impact of enhanced cleaning on bacterial contamination of the hospital environmental surfaces: a clinical trial in critical care unit in an Egyptian hospital. Antimicrobial Resistance and Infection Control 13, 138. Available at https://doi.org/10.1186/s13756-024-01489-z.

Hassan, Y.M., Mohamed, A.S., Hassan, Y.M., El-Sayed, W.M., 2025. Recent developments and future directions in point-of-care next-generation CRISPR-based rapid diagnosis. Clinical and Experimental Medicine 25, 33. Available at https://doi.org/10.1007/s10238-024-01540-8.

Holmes, A.H., Moore, L.S., Sundsfjord, A., Steinbakk, M., Regmi, S., Karkey, A., et al., 2016. Understanding the mechanisms and drivers of antimicrobial resistance. The Lancet 387 (10014), 176–187. Available at https://doi.org/10.1016/S0140-6736(15)00473-0.

Howard, P., Pulcini, C., Levy Hara, G., West, R.M., Gould, I.M., Harbarth, S., et al., 2015. An international cross-sectional survey of antimicrobial stewardship programmes in hospitals. The Journal of Antimicrobial Chemotherapy 70 (4), 1245–1255. Available at https://doi.org/10.1093/jac/dku497.

Huang, L.J., Chen, S.J., Hu, Y.W., Chun-Yu, L., Ping-Fng, W., Shu-Mei, S., et al., 2022. The impact of antimicrobial stewardship program designed to shorten antibiotics use on the incidence of resistant bacterial infections and mortality. Scientific Reports 12, 913. Available at https://doi.org/10.1038/s41598-022-04819-6.

Hwang, S., Kwon, K.T., 2021. Core elements for successful implementation of antimicrobial stewardship programs. Infection & Chemotherapy 53 (3), 421–435. Available at https://doi.org/10.3947/ic.2021.0093.

Indian Council of Medical Research. Antimicrobial Stewardship Program Guideline. Available at https://www.icmr.gov.in/icmrobject/custom_data/pdf/resource-guidelines/AMSP_0.pdf. (15.11.2018) Accessed on 30 May 2025.

Jacobowski, A.C., Boleti, A.P.A., Cruz, M.V., Santos, K.F.D.P., de Andrade, L.R.M., Frihling, B.E.F., et al., 2025. Combating antimicrobial resistance: innovative strategies using peptides, nanotechnology, phages, quorum sensing interference, and CRISPR-Cas systems. Pharmaceuticals 18 (8), 1119. Available at https://doi.org/10.3390/ph18081119.

Jain, M.Y., Chavan, S., 2024. The role and impact of antimicrobial stewardship programs in healthcare. World Journal of Pharmaceutical and Medical Research 10 (10), 113–126.

Kakkar, A.K., Shafiq, N., Singh, G., Ray, P., Gautam, V., Agarwal, R., et al., 2020. Antimicrobial stewardship programs in resource constrained environments: understanding and addressing the need of the systems. Frontiers in Public Health 8, 140. Available at https://doi.org/10.3389/fpubh.2020.00140.

Karanika, S., Paudel, S., Grigoras, C., Kalbasi, A., Mylonakis, E., 2016. Systematic review and meta-analysis of clinical and economic outcomes from computerized antimicrobial stewardship interventions. Antimicrobial Agents and Chemotherapy 60 (8), 4840–4852. Available at https://doi.org/10.1128/AAC.00825-16.

Kaye, K.S., 2012. Antimicrobial de-escalation strategies in hospitalized patients with pneumonia, intra-abdominal infections, and bacteremia. Journal of Hospital Medicine (Online) 7 (Suppl 1), S13–S21. Available at https://doi.org/10.1002/jhm.983 PMID: 23737333.

Knobloch, M.J., McKinley, L., Keating, J., Safdar, N., 2021. Integrating antibiotic stewardship and infection prevention and control programs using a team science approach. American Journal of Infection Control 49 (8), 1072–1074. Available at https://doi.org/10.1016/j.ajic.2021.01.020.

Kollef, M.H., Scott, T.M., 2016. Editorial commentary: antimicrobial de-escalation: what's in a name? Clinical Infectious Diseases 62 (8), 1018–1020. Available at https://doi.org/10.1093/cid/civ1201.

Koulenti, D., Roger, C., Lipman, J., 2025. Antibiotic dosing optimization in critically ill patients. Intensive Care Medicine 51, 603–606. Available at https://doi.org/10.1007/s00134-025-07822-3.

Krishnamoorthy, Y., Govindan, D., Karunakaran, M., Parveen, M., Manohar, A., Al-Tawfiq, J.A., 2025. A systematic review of the effectiveness and cost-effectiveness of implementing antimicrobial stewardship program at tertiary care facilities in India. IJID Reg 15, 100666. Available at https://doi.org/10.1016/j.ijregi.2025.100666.

Landers, T.F., Cohen, B., Wittum, T.E., Larson, E.L., 2012. A review of antibiotic use in food animals: perspective, policy, and potential. Public Health Reports 127 (1), 4–22. Available at https://doi.org/10.1177/003335491212700103.

Langford, B.J., Nisenbaum, R., Brown, K.A., Chan, A., Downing, M., 2020. Antibiotics: easier to start than to stop? Predictors of antimicrobial stewardship recommendation acceptance. Clinical Microbiology and Infection 26 (12), 1638–1643. Available at https://doi.org/10.1016/j.cmi.2020.07.048.

Laxminarayan, R., Duse, A., Wattal, C., Zaidi, A.K., Wertheim, H.F., Sumpradit, N., et al., 2013. Antibiotic resistance- the need for global solutions. The Lancet Infectious Diseases 13 (12), 1057–1098. Available at https://doi.org/10.1016/S1473-3099(13)70318-9.

Lazure, P., Augustyniak, M., Goff, D.A., Villegas, M.V., Apisarnthanarak, A., Péloquin, S., 2022. Gaps and barriers in the implementation and functioning of antimicrobial stewardship programmes: results from an educational and behavioural mixed methods needs assessment in France, the United States, Mexico and India. JAC Antimicrob Resist 4 (5), dlac094. Available at https://doi.org/10.1093/jacamr/dlac094. Erratum in: JAC Antimicrob Resist. 2023 Jan 21;5(1):dlad010. Available at https://doi.org/10.1093/jacamr/dlad010.

Lee, C.C., Chang, J.C., Mao, X.W., Hsu, W.T., Chen, S.Y., Chen, Y.C., et al., 2020. Combining procalcitonin and rapid multiplex respiratory virus testing for antibiotic stewardship in older adult patients with severe acute respiratory infection. Journal of the American Medical Directors Association 21 (1), 62–67. Available at https://doi.org/10.1016/j.jamda.2019.09.020.

Lee, C.F., Cowling, B.J., Feng, S., et al., 2018. Impact of antibiotic stewardship programmes in Asia: a systematic review and meta-analysis. The Journal of Antimicrobial Chemotherapy 73, 844–851. Available at https://doi.org/10.1093/jac/dkx492.

Lindsay, P.J., Rohailla, S., Taggart, L.R., Lightfoot, D., Havey, T., Daneman, N., et al., 2019. Antimicrobial stewardship and intensive care unit mortality: a systematic review. Clinical Infectious Diseases 68 (5), 748–756. Available at https://doi.org/10.1093/cid/ciy550.

Liu, L., Liu, B., Li, L., Li, Y., Zhou, X., Li, Q., 2024. Impact of antimicrobial stewardship and infection prevention and control programmes on antibiotic usage and *A. baumannii* resistance: A 2016-2023 multicentre prospective study. Infection and Drug Resistance 18, 679–692. Available at https://doi.org/10.2147/IDR.S505133.

MacDougall, C., Polk, R.E., 2005. Antimicrobial stewardship programs in health care systems. Clinical Microbiology Reviews 18, 638–656. Available at https://doi.org/10.1128/CMR.18.4.638-656.2005.

Magill, S.S., Edwards, J.R., Beldavs, Z.G., Dumyati, G., Janelle, S.J., Kainer, M.A., et al., 2014. Prevalence of antimicrobial use in US acute care hospitals, May-September 2011. The Journal of the American Medical Association 312, 1438–1446. Available at https://doi.org/10.1001/jama.2014.12923.

Mahon, C.R., Lehman, D.C., Manuselis, G., 2014. Antimicrobial agent mechanisms of action and resistance. Textbook of Diagnostic Microbiology. Saunders, St. Louis, pp. 254–273.

Manning, M.L., Septimus, E.J., Ashley, E.S.D., Cosgrove, S.E., Fakih, M.G., Schweon, S.J., et al., 2018. Antimicrobial stewardship and infection prevention-leveraging the synergy: a position paper update. American Journal of Infection Control 46 (4), 364–368. Available at https://doi.org/10.1016/j.ajic.2018.01.001.

Manson, J.M., Hancock, L.E., Gilmore, M.S., 2010. Mechanism of chromosomal transfer of *Enterococcus faecalis* pathogenicity island, capsule, antimicrobial resistance, and other traits. Proceedings of the National Academy of Sciences of the United States of America 107, 12269–12274.

Mathew, P., Ranjalkar, J., Chandy, S.J., 2020. Challenges in implementing antimicrobial stewardship programmes at secondary level hospitals in India: an Exploratory Study. Frontiers in Public Health 8, 493904. Available at https://doi.org/10.3389/fpubh.2020.493904.

McElligott, M., Welham, G., Pop-Vicas, A., Taylor, L., Crnich, C.J., 2017. Antibiotic stewardship in nursing facilities. Infectious Disease Clinics of North America 31 (4), 619–638. Available at https://doi.org/10.1016/j.idc.2017.07.008.

McEwen, S.A., Collignon, P.J., 2018. Antimicrobial resistance: a one health perspective. Microbiology Spectrum 6 (2). Available at https://doi.org/10.1128/microbiolspec.arba-0009-2017.

McGowan, J.E., Gerding, D.N., 1996. Does antibiotic restriction prevent resistance? New Horizons 4 (3), 370–376.

Mertz, D., Koller, M., Haller, P., Lampert, M.L., Plagge, H., Hug, B., et al., 2009. Outcomes of early switching from intravenous to oral antibiotics on medical wards. The Journal of Antimicrobial Chemotherapy 64 (1), 188–199. Available at https://doi.org/10.1093/jac/dkp131.

Moore, L.S.P., Villegas, M.V., Wenzler, E., Rawson, T.M., Oladele, R.O., Doi, Y., et al., 2023. Rapid diagnostic test value and implementation in antimicrobial stewardship across low-to-middle and high-income countries: a mixed-methods review. Infectious Diseases and Therapy 12 (6), 1445–1463. Available at https://doi.org/10.1007/s40121-023-00815-z.

Munita, J.M., Arias, C.A., 2016. Mechanisms of antibiotic resistance. Microbiology Spectrum 4. Available at https://doi.org/10.1128/microbiolspec.vmbf-0016-2015.

Murray, C.J., Ikuta, K.S., Sharara, F., Swetschinski, L., Robles Aguilar, G., et al., 2022. Global burden of bacterial antimicrobial resistance in 2019: a systematic analysis. Lancet 399, 629–655. Available at https://doi.org/10.1016/S0140-6736(21)02724-0.

Nielsen, N.D., Dean III, J.T., Shald, E.A., Conway Morris, A., Povoa, P., Schouten, J., et al., 2024. When to stop antibiotics in the critically ill? Antibiotics 13 (3), 272. Available at https://doi.org/10.3390/antibiotics13030272.

Okumura, L.M., Silva, M.M., Veroneze, I., 2015. Effects of a bundled antimicrobial stewardship program on mortality: a cohort study. The Brazilian Journal of Infectious Diseases 19 (3), 246–252. Available at https://doi.org/10.1016/j.bjid.2015.02.005.

O'Neill J. Tackling drug-resistant infections globally: final report and recommendations. The Review on Antimicrobial Resistance. May 2016. Available at https://amr-review.org/sites/default/files/160518_Final%20paper_with%20cover.pdf.

Parzen-Johnson, S., Kronforst, K.D., Shah, R.M., Whitmer, G.R., Scardina, T., Meg Chandarraju, R.N., et al., 2021. Use of the electronic health record to optimize antimicrobial prescribing. Clinical Therapeutics 43 (10), 1681–1688. Available at https://doi.org/10.1016/j.clinthera.2021.09.009.

Patterson, J., Tapper, M.L., Huskins, W.C., Janssen, L., 2016. Healthcare infection control practices advisory committee. Antibiotic Stewardship Statement for Antibiotic Guidelines – The Recommendations of the Healthcare Infection Control Practices Advisory Committee (HICPAC). Available at https://www.cdc.gov/hicpac/media/pdfs/antibiotic-stewardship-statement-508.pdf.

Pennisi, F., Pinto, A., Ricciardi, G.E., Signorelli, C., Gianfredi, V., 2025. The role of artificial intelligence and machine learning models in antimicrobial stewardship in public health: a narrative review. Antibiotics (Basel) 14 (2), 134. Available at https://doi.org/10.3390/antibiotics14020134.

Pereira, N.M.D., 2020. Antimicrobial stewardship programmes. Pediatric Oncall 17 (3), 73–76. Available at https://doi.org/10.7199/ped.oncall.2020.37.

Pickens, C., Wunderink, R.G., Qi, C., Mopuru, H. Donnelly, H., Powell, K., et al., 2020. A multiplex polymerase chain reaction assay for antibiotic stewardship in suspected pneumonia. Diagnostic Microbiology and Infectious Disease 98 (4), 115179. Available at https://doi.org/10.1016/j.diagmicrobio.2020.115179.

Pulcini, C., Binda, F., Lamkang, A.S., Trett, A., Charani, E., Goff, D.A., et al., 2019. Developing core elements and checklist items for global hospital antimicrobial stewardship programmes: a consensus approach. Clinical Microbiology and Infection 25 (1), 20–25. Available at https://doi.org/10.1016/j.cmi.2018.03.033.

Rajendran, R., Chitrapady, S., Tejashree, M.U., Haritha, K., Jayaraj, M.B., Nair, S., et al., 2025. Antimicrobial stewardship programs in primary and secondary care settings in India: current challenges, facilitators, perceptions, and impact – a scoping review. BMC Infectious Diseases 25, 1539. Available at https://doi.org/10.1186/s12879-025-11851-0.

Ramesh, M., Janani, R., Deepa, C., Rajeshkumar, L., 2022. Nanotechnology-enabled biosensors: a review of fundamentals, design principles, materials, and applications. Biosensors (Basel) 13 (1), 40. Available at https://doi.org/10.3390/bios13010040.

Rathish, B., Wilson, A., Warrier, A., Babu, R., Prakash, S., 2019. How to save half a million dollars: an antimicrobial stewardship program in a tertiary care center. Cureus 11 (7), e5121. Available at https://doi.org/10.7759/cureus.5121.

Reed, E.E., Stevenson, K.B., West, J.E., Bauer, K.A., Goff, D.A., 2013. Impact of formulary restriction with prior authorization by an antimicrobial stewardship program. Virulence 4 (2), 158–162 Available at https://doi.org/10.4161/viru.21657.

Reygaert, W.C., 2018. An overview of the antimicrobial resistance mechanisms of bacteria. AIMS Microbiology 4 (3), 482–501. Available at https://doi.org/10.3934/microbiol.2018.3.482.

Roberts, J.A., Lipman, J., 2009. Pharmacokinetic issues for antibiotics in the critically ill patient. Critical Care Medicine 37 (3), 840–851. Available at https://doi.org/10.1097/CCM.0b013e3181961bff.

Rolfe Jr, R., Kwobah, C., Muro, F., Ruwanpathirana, A., Lyamuya, F., et al., 2021. Barriers to implementing antimicrobial stewardship programs in three low- and middle-income country tertiary care settings: findings from a multi-site qualitative study. Antimicrobial Resistance and Infection Control 10 (1), 60. Available at https://doi.org/10.1186/s13756-021-00929-4.

Rupali, P., Palanikumar, P., Shanthamurthy, D., Peter, J.V., Kandasamy, S., Zacchaeus, N.G.P., 2019. et. al. Impact of an antimicrobial stewardship intervention in India: Evaluation of post-prescription review and feedback as a method of promoting optimal antimicrobial use in the intensive care units of a tertiary-care hospital. Infection Control and Hospital Epidemiology 40 (5), 512–519. Available at https://doi.org/10.1017/ice.2019.29.

Saleem, Z., Sheikh, S., Godman, B., Haseeb, A., Afzal, S., Qamar, M.U., et al., 2025. Increasing the use of the WHO AWaRe system in antibiotic surveillance and stewardship programmes in low- and middle-income countries. JAC Antimicrobial Resistance 7 (2), dlaf031. Available at https://doi.org/10.1093/jacamr/dlaf031.

Sanchez, G.V., Fleming-Dutra, K.E., Roberts, R.M., Hicks, L.A., 2016. Core elements of outpatient antibiotic stewardship. MMWR: Recommendations and Reports 65 (6), 1–12. Available at https://doi.org/10.15585/mmwr.rr6506a1.

van de Sande-Bruinsma, N., Grundmann, H., Verloo, D., Tiemersma, E., Monen, J., Goossens, H., et al., 2008. Antimicrobial drug use and resistance in Europe. Emerging Infectious Diseases 14, 1722–1730. Available at https://doi.org/10.3201/eid1411.070467.

Schuts, E.C., Hulscher, M.E.J.L., Mouton, J.W., Verduin, C.M., Stuart, J.W.T.C., Overdiek, H.W.P.M., 2016. et. al. Current evidence on hospital antimicrobial stewardship objectives: a systematic review and meta-analysis. The Lancet Infectious Diseases 16 (7), 847–856. Available at https://doi.org/10.1016/S1473-3099(16)00065-7. Epub 2016 Mar 3. Erratum in: Lancet Infect Dis. 2016;16(7):768. Available at https://doi.org/10.1016/S1473-3099(16)30165-7.

Schwartz, K.L., Xu, A.X.T., Alderson, S., Bjerrum, L., Brehaut, J., Brown, B.C., 2023. et. al. Best practice guidance for antibiotic audit and feedback interventions in primary care: a modified Delphi study from the Joint programming initiative on antimicrobial resistance: primary care antibiotic audit and feedback network (JPIAMR-PAAN). Antimicrobial Resistance and Infection Control 12 (1), 72. Available at https://doi.org/10.1186/s13756-023-01279-z.

Shah, A.M., Shah, S.V., 2023. Antimicrobial stewardship program: fighting antimicrobial resistance in India. National Journal of Physiology, Pharmacy and Pharmacology 13 (09), 1800–1805. Available at https://doi.org/10.5455/njppp.2023.13.07376202305082023.

Shlaes, D.M., Gerding, D.N., John, J.J.F., Craig, W.A., Bornstein, D.L., Duncan, R.A., et al., 1997. Society for Healthcare epidemiology of America and Infectious Diseases Society of America Joint Committee on the prevention of antimicrobial resistance: guidelines for the prevention of antimicrobial resistance in hospitals. Clinical Infectious Diseases 25, 584–599. Available at https://doi.org/10.1086/513766.

Shrestha, Y., Kularia, S., Doddasamiah, P., Devarasan, T., Thennarasi, V., Jadhav, V.H., 2025. Challenges and strategies for implementing antimicrobial stewardship programs in India: insights from multiple stakeholders. Journal of Young Pharmacists 17 (3), 669–678. Available at https://doi.org/10.5530/jyp.20250133.

Spellberg, B., Bartlett, J.G., Gilbert, D.N., 2016. The future of antibiotics and resistance. New England Journal of Medicine 368 (4), 299–302. Available at https://doi.org/10.1056/NEJMp1215093.

Umpleby, H., Dushianthan, A., Catton, T., Saeed, K., 2022. Antimicrobial stewardship programmes focused on de-escalation: a narrative review of efficacy and risks. Journal of Emergency and Critical Care Medicine 6, 23. Available at https://jeccm.amegroups.org/article/view/7515.

Valiquette, L., Cossette, B., Garant, M.P., Diab, H., Pepin, J., 2007. Impact of a reduction in the use of high-risk antibiotics on the course of an epidemic of *Clostridium difficile*-associated disease caused by the hypervirulent NAP1/027 strain. Clinical Infectious Diseases 45 (Suppl 2), S112–S121. Available at https://doi.org/10.1086/519258.

Velazquez-Meza, M.E., Galarde-López, M., Carrillo-Quiróz, B., Alpuche-Aranda, C.M., 2022. Antimicrobial resistance: one health approach. Veterinary World 15 (3), 743–749. Available at https://doi.org/10.14202/vetworld.2022.743-749.

Ventola, C.L., 2015. The antibiotic resistance crisis: part 1 causes and threats. Pharmacy and Therapeutics 40 (4), 277–283.

de Vries, E.N., Prins, H.A., Crolla, R.M., den Outer, A.J., van Andel, G., van Helden, S.H., et al., 2010. Effect of a comprehensive surgical safety system on patient outcomes. The New England Journal of Medicine 363 (20), 1928–1937. Available at https://doi.org/10.1056/NEJMsa0911535.

Wagner, J.L., Markovich, K.C., Barber, K.E., Stover, K.R., Biehle, L.R., 2021. Optimizing rapid diagnostics and diagnostic stewardship in Gram-negative bacteremia. Pharmacotherapy 41, 676–685. Available at https://doi.org/10.1002/phar.2606.

Walia, K., Ohri, V.C., Madhumathi, J., Ramasubramanian, V., 2019. Policy document on antimicrobial stewardship practices in India. The Indian Journal of Medical Research 149 (2), 180–184. Available at https://doi.org/10.4103/ijmr.IJMR_147_18.

Wegener, H.C., 2012. Antibiotic resistance-linking human and animal health. Institute of Medicine (US). Improving Food Safety Through a One Health Approach: Workshop Summary. National Academies Press (US), Washington (DC), pp. 331–349. A15. Available at https://www.ncbi.nlm.nih.gov/books/NBK114485/.

World Health Organisation. Global Antimicrobial Resistance and Use Surveillance System (GLASS) Report 2022. Available at https://iris.who.int/server/api/core/bitstreams/e5cc3da6-f46e-4355-a352-8d16820e0dd1/content. ;no date (accessed 30.05.23).

World Health Organization. Antimicrobial stewardship programmes in health-care facilities in low- and middle-income countries: A WHO practical toolkit. Geneva: World Health Organization; 2019a. 22 October 2019. Available at https://www.who.int/publications/i/item/9789241515481.

World Health Organization (WHO). Promoting antimicrobial stewardship to tackle antimicrobial resistance. Available at https://www.who.int/europe/activities/promoting-antimicrobial-stewardship-to-tackle-antimicrobial-resistance. Copenhagen: WHO Regional Office for Europe; 2021 (accessed 11.0326).

World Health Organization. Antimicrobial resistance: key facts. Geneva: World Health Organization; 21 November 2023. Available at https://www.who.int/news-room/fact-sheets/detail/antimicrobial-resistance.

World Health Organization (WHO). 2019 antibacterial agents in clinical development: an analysis of the antibacterial clinical development pipeline. 15 January 2020. Available at https://www.who.int/publications/i/item/9789240000193. 2020.

World Health Organization. Antimicrobial stewardship: a competency-based approach: Available from OpenWHOReviewer. JAC Antimicrob Resist. 2019b;1(1):dlz007. Available at https://doi.org/10.1093/jacamr/dlz007.

World Health Organization. WHO Handbook for Guideline Development, 2nd ed. Geneva: WHO; 18 December 2014. Available at https://www.who.int/publications/i/item/9789241548960.

World Health Organization (WHO). Monitoring the building blocks of health systems: A Handbook of Indicators and their measurement strategies. Geneva, Switzerland: World Health Organization; 2010. Available at https://www.technet-21.org/en/resources/guidance/monitoring-the-building-blocks-of-health-systems-a-handbook-of-indicators-and-their-measurement-strategies.

Worldwide Antimicrobial Resistance National/International Network Group (WARNING) Collaborators, 2023. Ten golden rules for optimal antibiotic use in hospital settings: the WARNING call to action. World Journal of Emergency Surgery: WJES 18, 50. Available at https://doi.org/10.1186/s13017-023-00518-3.

Wu, S., Tannous, E., Haldane, V., Ellen, M.E., Wei, X., 2022. Barriers and facilitators of implementing interventions to improve appropriate antibiotic use in low- and middle-income countries: a systematic review based on the consolidated framework for implementation research. Implementation Science 17, 30. Available at https://doi.org/10.1186/s13012-022-01209-4.

Xu, A.X.T., Brown, K., Schwartz, K.L., Aghlmandi, S., Alderson, S., Brehaut, J.C., 2025. et. al. Audit and feedback interventions for antibiotic prescribing in primary care: a systematic review and meta-analysis. Clinical Infectious Diseases 80 (2), 253–262. Available at https://doi.org/10.1093/cid/ciae604.

Yasmina, N.A., Krishnan, D., Lal, P.M.L., Sebin, T.K., 2023. Antibiotic stewardship: improving clinical outcomes by optimization of antibiotic use. Ijppr. Human 29 (1), 46–52.

Zacchaeus, N.G.P., Palanikumar, P., Alexander, H., et al., 2023. Establishing an effective antimicrobial stewardship program at four secondary-care hospitals in India using a hub-and-spoke model. Antimicrobial Stewardship & Healthcare Epidemiology 3 (1), e99. Available at https://doi.org/10.1017/ash.2023.17.

Zagaliotis, P., Michalik-Provasek, J., Gill, J.J., Walsh, T.J., 2022. Therapeutic bacteriophages for Gram-negative bacterial infections in animals and humans. Pathogens and Immunity 7 (2), 1. Available at https://doi.org/10.20411/pai.v7i2.516.

Zhou, S., Nagel, J.L., Kaye, K.S., LaPlante, K.L., Albin, O.R., Pogue, J.M., 2021. Antimicrobial stewardship and the infection control practitioner: a natural alliance. Infectious Disease Clinics of North America 35 (3), 771–787. Available at https://doi.org/10.1016/j.idc.2021.04.011.

Chapter 14

Global strategies for antibiotic governance: navigating policy challenges

Saurav Sarkar[1,*], Pratima Sarkar[2,*], Diksha Tarafder[3], and Swarnab Sengupta[4]

[1]*Department of Pharmaceutical Technology, University of North Bengal, Siliguri, West Bengal, India,* [2]*Department of Microbiology, University of North Bengal, Siliguri, West Bengal, India,* [3]*Department of Microbiology, North Bengal Medical College and Hospital, Siliguri, West Bengal, India,* [4]*Department of Biomedical Sciences, College of Medicine, Florida State University, Tallahassee, FL, United States*

14.1 Introduction

14.1.1 The global threat of antibiotic resistance

Antibiotic resistance is one of the threatening global health problems and a major challenge of the 21st century (Fig. 14.1. Formerly known as 'miracle drugs', antibiotics revolutionised medicine by turning once fatal infections into treatable conditions. But what made antibiotics less efficient over time was the fact that they were working against bacteria, as the latter would develop defences to make drugs invalid (Talebi Bezmin Abadi et al., 2019). The World Health Organisation's (WHO) Global Antibiotic Resistance Surveillance Report 2025 reveals an alarming reality: One in six bacterial infections worldwide is resistant to frontline antibiotic treatments, with the rate climbing to one in three in regions like South and Southeast Asia (Doron and Broom, 2026). More than 40% of the observed antibiotic-pathogen pairings are currently showing resistance and growing at an annual rate of 5%–15%. There is a significantly high level of resistance among *Escherichia coli, Klebsiella pneumoniae,* and *Acinetobacter spp.* towards the major antibiotics, which include third-generation cephalosporins and fluoroquinolones, now exceed 70% in some countries (Beckley and Wright, 2021).

Antibiotic resistance does not end at the hospital or the clinic; it is in every part of society, from animals and agriculture to factories and the environment. The scale of the threat is well understood by the WHO, which describes resistance as 'a global equity and health systems crisis', disproportionately impacting low- and middle-income countries (LMICs) with precarious healthcare systems, inadequate diagnostic ability, and poor antibiotic access (Davis et al., 2022). Without rapid intervention, AMR-attributable deaths are expected to surge from 1.14 million in 2021 to approaching 1.9 million annually by 2050, amounting to up to almost 39 million preventable deaths within a quarter of a century. Economically, resistance adds billions to healthcare costs yearly and could cause the global economy $100 trillion by 2050 in lost productivity and surplus medical spending (Naghavi et al., 2024).

These are the most critical bacterial pathogens, including *Enterococcus faecium, Staphylococcus aureus, K. pneumoniae, Acinetobacter baumannii,* and *Pseudomonas aeruginosa*, termed the 'ESKAPE pathogens', which account for more than half of hospital-related infections worldwide. The emergence of multidrug-resistant (MDR) and extensively drug-resistant (XDR) strains means that first-line treatments are no longer working, with physicians being forced to rely on drugs of last resort such as carbapenems and colistin. But even these 'safety net' antibiotics are quickly becoming useless, threatening to revert the world back to a preantibiotic era when minor cuts and common infections such as strep throat could be fatal (Al Marjani et al., 2020).

* Authors contributed equally.

Antimicrobial Resistance in the 21st Century. DOI: https://doi.org/10.1016/B978-0-443-44748-8.00018-X

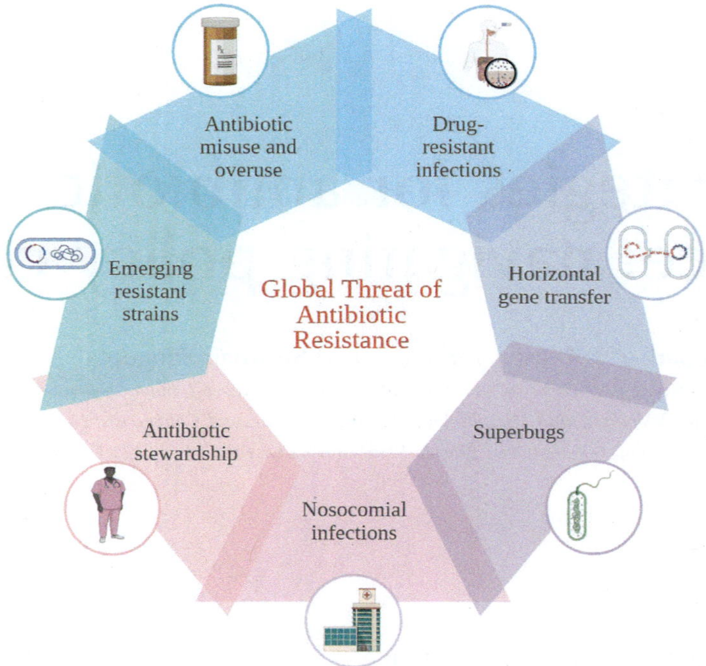

FIGURE 14.1 Representation of the global threat of antibiotic resistance, indicating drivers and consequences. Factors driving the global rise of antibiotic resistance and their interrelated impacts on public health.

14.1.1.1 Understanding the scope: resistance drivers and consequences

Resistance to antibiotics arises mainly from the misuse and overuse of these medicines in human medicine, animal husbandry as well as agriculture, along with lack of hygiene, sanitation, and infection control (Holmes et al., 2016).

14.1.1.1.1 Misuse in healthcare

An estimated 50% of all antibiotics prescribed globally are unnecessary or inappropriately targeted. Overuse for viral infections and lack of diagnostic testing as well as suboptimal courses of therapy can establish these resistant organisms. Hospital-acquired infections (HAIs) such as bloodstream infections, ventilator-associated pneumonia, and surgical site infections now increasingly involve MDR pathogens (Irfan et al., 2022).

14.1.1.1.2 Misuse in agriculture and animal husbandry

Approximately 70% of antibiotics used worldwide are consumed by production animals, frequently not to treat infections, but rather to promote growth and prevent disease among healthy animals. These methods produce reservoirs of resistant bacteria in food animals, which can be transmitted to humans via food, water, and environmental exposure (McKernan et al., 2021).

14.1.1.1.3 Environmental spread

Antibiotic residues escape from pharmaceutical waste, hospital effluents, and agricultural runoffs to soil and water systems, creating resistance hotspots that can survive for years. These environmental reservoirs allow resistance genes to disseminate among bacterial species via horizontal gene transfer (Fletcher, 2015).

The burden of antibiotic resistance goes beyond health. The Global Research on Antimicrobial Resistance (GRAM) Project calculates that since 1990, antibiotic-resistant infections have already killed more than 36 million people, equivalent to about three people dying every minute. Resistance exacerbates health inequality: in poor countries, where there is no access to new antibiotics, resistance means 'impossible-to-treat infections' rather than lengthy treatment regimens (Naghavi et al., 2024).

14.1.1.2 The case for coordinated multisectoral action

The scope of antibiotic resistance is worldwide, and no nation can combat it alone. Skilled professionals on a global scale now understand that AMR is a 'One Health' issue (meaning, it affects all humans, animals, and the environment). One Health approach brings sectors together in a united, inclusive cooperation to address an issue, with shared responsibility and accountability, so that no sector is left behind. This comprehensive approach is at the heart of the Global Action Plan on Antimicrobial Resistance, adopted by all WHO Member States in 2015 (Chen et al., 2025a, b, c).

14.1.1.2.1 The 'One Health' architecture

Antibiotic-resistant bacteria circulate freely across ecosystems, as the One Health theory acknowledges. The antibiotics that are often used on farms to stimulate growth can generate resistant bacteria in animals, which then spread through meat, water, and waste. And antibiotics discarded in human waste end up back in the environment, creating a loop of resistance that no single sector can break. Hence, successful CMR control requires cooperation of the ministries of health, agriculture, environment, and education (Häsler et al., 2013).

14.1.1.2.2 Global coordination and stakeholder partnerships

Several global organisations lead the AMR governance effort:

a) **World Health Organisation (WHO):** Oversees the Global Antimicrobial Resistance and Use Surveillance System (GLASS), which now includes data from 127 countries (World Health Organization, 2022).
b) **Food and Agriculture Organisation (FAO) and World Organisation for Animal Health (WOAH):** Develop guidelines for antibiotic use in the farming sector, promoting stewardship and phasing out nontherapeutic use (Prasarnphanich et al., 2025).
c) **United Nations Environment Programme (UNEP):** Focuses on the environmental dimensions of resistance, including pharmaceutical waste control (World Health Organization, 2021a).
d) **Global Coordination Mechanisms:** The Quadripartite Alliance, comprising WHO, FAO, WOAH, and UNEP, ensures integration of policies across sectors through shared evidence and accountability frameworks (Joshi et al., 2021).

14.1.1.2.3 Economic justification for collective action

Antibiotic resistance causes a serious economic burden. Treating resistant infections can cost 20 times as much as treating susceptible ones, with longer hospital stays and more prolonged recovery times, not to mention the higher level of care required. According to the OECD, small expenditures on infection prevention, surveillance, and stewardship equal to $2 a year per person could save approximately 24 million lives and future medical costs of billions by 2050 (Ogawa et al., 2018).

14.1.2 Need for coordinated multisectoral action

a) **Strengthening surveillance and data sharing:** Robust data systems are central to effective governance. The GLASS platform has allowed for standardised resistance monitoring, but less than half of the world's low-income countries report full data. WHO's ambition to have universal AMR surveillance coverage in 2030 would rely significantly on capacity building for diagnostics, laboratory infrastructure, and electronic reporting systems. Data-driven insights are vital for monitoring local trends, guiding policies, and assessing intervention impacts (Chen et al., 2025a, b, c).
b) **Optimising antibiotic stewardship:** Health systems should have stewardship programmes in place that balance access and control. Stewardship activities involve prescriber education, prescription review, and infection prevention policies. Government programmes such as India's National Action Plan on AMR and the EU AMR Control Directive have shown that evidence-based regulation can decrease overwhelming prescribing by 10%–20% in five years (Häsler et al., 2013).
c) **Addressing agricultural and veterinary practices:** Preventing misuse in livestock farming calls for incentives as well as enforcement. Countries such as the Netherlands and Denmark halved their use of veterinary antibiotics with better hygiene, vaccinations, and surveillance. At the policy level, reform must ban antibiotic use for growth promotion and bring all veterinary drugs under prescription only (Sanga et al., 2024a, b).
d) **Environmental governance:** Environmental regulation is still the least developed aspect of AMR governance. Industrial waste from drug manufacturing plants, particularly in South Asia, is spewing high concentrations of antibiotics into rivers and creating resistance hotspots. Improvements in waste management and the implementation of international environmental standards for antibiotic pollution discharge may be important initiatives (Häsler et al., 2013).

14.1.2.1 *Policy challenges in coordinated action*

Despite global commitment, major challenges hinder progress:

a) **Fragmented governance:** Many nations lack interministerial coordination, leading to sectoral silos that hinder policy implementation (Weldon et al., 2025).
b) **Funding gaps:** Only one-third of national action plans have secured sustainable funding for AMR interventions (Weldon et al., 2025).
c) **Limited surveillance capacity:** Diagnostic and reporting inequalities create blind spots, particularly in sub-Saharan Africa and parts of Asia (Do et al., 2023).
d) **Pharmaceutical market failures:** The global pipeline for new antibiotics remains critically underfunded. From 2017 to 2025, only a handful of new antibiotics reached the market, while resistance to existing ones grew faster than innovation (Anyanwu et al., 2023).
e) **Inequitable access:** Ironically, the same countries suffering from drug resistance also lack access to effective next-generation antibiotics, creating a dual crisis of overuse and undertreatment (Kirchhelle et al., 2020).

14.1.2.2 *Global governance in action: emerging models*

Projects like CARB-X (Combating Antibiotic-Resistant Bacteria Biopharmaceutical Accelerator) demonstrate the value in international cooperation. CARB-X has given out more than $450 million in funding to antibiotic research since 2016, with at least three treatments and diagnostics progressing through the market globally. Moreover, WHO's One Health Global Leaders Group (GLG), founded in 2020, ensures high-level political engagement and accountability across nations. On the regional level, Europe's EARS-Net (European Antimicrobial Resistance Surveillance Network) connects 30 countries' information to provide real-time comparisons and recommendations for an appropriate response. The networks are also being established in Africa and Asia, with funding provided by the Fleming Fund and the Global Antibiotic Research and Development Partnership (Luepke and Mohr, 2017). Antibiotic resistance is not a distant or isolated threat; it is a slow-moving pandemic. First tested in 2025, the evidence is clear that resistance is spreading faster than the applied efforts to stop it. The WHO's call for immediate, united, and multisectoral action also highlights the importance of integrated policy-making, sustainable financing, and equitable access to diagnostics and treatment. The way forward is to bolster global surveillance (GLASS), implement strong stewardship programmes, discourage nontherapeutic use of antibiotics in animal husbandry, and invest in environmental management. Ultimately, combating antibiotic resistance is not just an important health issue; it is a matter of global security. All sectors and countries must work together, guided by science, solidarity, and a common purpose of preserving antibiotics for future generations as one of the most effective life-saving medicines (Salam et al., 2023).

14.2 Current landscape of antibiotic resistance

14.2.1 Global overview of resistance trends

In 2025 the WHO surveillance report found more than 23 million lab-confirmed infections reported in 104 countries that covered almost 70% of the world population. Alarmingly, it found that one in six bacterial infections around the world are now resistant to standard antimicrobial treatments, and in some parts – namely South and Southeast Asia – that rises to one in three for urinary tract infections. The resistance rates in Gram-negative bacteria such as *E. coli, K. pneumoniae,* and *Acinetobacter* spp. already >40%–70% against most critical antibiotics such as third-generation cephalosporins and fluoroquinolones. The rise of methicillin-resistant Staphylococcus aureus (MRSA), affecting 27% of infections, further complicates clinical management (Mazhar et al., 2025).

The WHO categorises antibiotics under the AWaRe framework:

a) **Access drugs:** first-line, lower resistance-risk antibiotics such as ampicillin and cotrimoxazole.
b) **Watch drugs:** broad-spectrum antibiotics, including third-generation cephalosporins and fluoroquinolones, are associated with high resistance potential.
c) **Reserve drugs:** last-resort treatments such as carbapenems (Mudenda et al., 2023).

Emerging evidence suggests increasing reliance upon Watch and Reserve groups with risk-averse consequences for their future usefulness, particularly in lower income settings where surveillance and diagnostic capacity is poor. As a result, resistance is greatest in countries where healthcare systems are fragmented, and the supply of antibiotics is unregulated and infection control infrastructures are bad (Akhmetova et al., 2025).

AMR-related deaths are expected to increase by 70% by 2050, with up to 10 million AMR-linked annual total worldwide without corrective action. Over and above the loss of life, the financial implications of AMR are projected to be in 12 figures globally, closing in on US$100 trillion, making this not just a health but also a multisectoral development catastrophe (Tang et al., 2023).

14.2.1.1 *Resistance trends in healthcare*

The healthcare sector is the epicentre for antibiotic resistance, and misuse/overuse of antibiotics is common. WHO estimates that 50% of antibiotic prescriptions from primary health centres are unnecessary or inappropriate, often issued without diagnostic confirmation. In other parts of the world, antibiotics are routinely given for viral infections such as the flu and infections in the upper respiratory tract, despite their having no effect (Asghar et al., 2024).

14.2.1.2 *Drivers of resistance in healthcare*

Empirical prescribing without laboratory testing fuels broad-spectrum antibiotic use.

Incomplete treatment courses enable bacteria to adapt and develop resistance mechanisms.

Due to failure of first-line therapies the use of last-resort drugs sich as carbapenems and polymyxins are increasing .

Lack of infection prevention measures results in HAIs caused by multidrug-resistant organisms (MDROs) (Cocker et al., 2024).

Infections resulting from resistant pathogens, including *E. coli, P. aeruginosa,* and MRSA, have resulted in extended hospital stays, escalating costs and higher mortality rates. Bloodstream infections resistant to fluoroquinolones are now three times more likely to lead to mortality than those susceptible, for example. Emerging carbapenem-resistant Enterobacteriaceae (CRE) and Acinetobacter species have forced clinicians to use more toxic, less-effective agents such as colistin (Abban et al., 2023). Hospital antibiotic stewardship programmes (ASPs) developed in regions of the world such as Europe and North America have achieved reductions of up to 25% in inappropriate antibiotic use. But their penetration is limited in LMICs because of funding shortfalls and diagnostic infrastructure deficiencies (Ababneh et al., 2021).

14.2.1.3 *Resistance in agriculture and animal husbandry*

The agriculture industry, most notably the livestock and poultry industries, is the source of over half of antibiotic use around the globe for nontherapeutic uses such as growth promotion and disease prevention in somewhat healthy animals. In the developing world, where regulation is weak, antibiotics are frequently sold over the counter or added to feed for livestock without professional oversight (Kasimanickam et al., 2021).

14.2.2 Misuse in livestock and poultry production

a) Growth promotion: Continuous low-dose antibiotic use to enhance animal weight gain (Islam et al., 2024).
b) Prophylactic use: Routine administration in crowded environments to prevent disease outbreaks.
c) Metaphylactic use: Treating entire herds when only a few animals show symptoms.

These practices encourage the selection of resistant bacteria in animal microbiota. The resistant strains and resistance genes are subsequently transmitted to humans through direct contact, food-borne ingestion (via the faecal-oral route), and environmental pathways such as water and soil contamination from animal waste. In countries with industrialised agriculture, such as the United States and China, broad-spectrum antibiotics from drugs, including tetracyclines, penicillins, and macrolides, continue to be widely given to animals. 80% of antibiotic use in the United States is still linked to the system of animal production, even after some regulatory reforms. In the European Union, tighter rules led to 25–50% decreases in veterinary antibiotics used after growth-promoter antibiotics were banned in 2006 (Gwenzi et al., 2021).

i) Pig and poultry farming: Early weaning practices among piglets (at 21 days) are a source of intestinal infections, including postweaning diarrhoea, inducing high needs for antibiotics. Also, in the case of intensive poultry farming, high stocking densities and lack of ventilation lead to outbreaks of disease, thereby driving a demand for prophylactic use of antibiotics. Such misuse drives cross-species resistance genes (mecA, tetM, and blaCTX-M), responsible for MRSA and extended-spectrum β-lactamase (ESBL) production (Canibe et al., 2022).

The environmental implications of agricultural misuse are also profound. Contaminations of water bodies and soils with antibiotic residues, including resistant bacteria produced by livestock activities, create environmental reservoirs of resistance that survive beyond the point at which antibiotics are administered. These environmental 'hotspots' promote horizontal gene transfer, via which resistance can spread without antibiotic exposure (Singh et al., 2024).

ii) Resistance in aquaculture and crop production: Antibiotics are widely used in aquaculture and crop production, especially in Asia and Latin America. Fish farms frequently use antibiotics, such as oxytetracycline and florfenicol, for mass prophylaxis, which are then released into coastal marine ecosystems through the poor control of wastewater. Hence, the use of antibiotic-treated manure as crop fertiliser also brings resistant genes such as sul1 and qnrS into farmland soils, threatening food safety and ecosystem stability (Preena et al., 2020).

14.2.3 Cross-sectoral and regional variations

Resistance trends vary significantly across regions:

a) Europe and North America have demonstrated gradual declines in antibiotic consumption and certain resistance rates through strict governance and stewardship programmes (Chen et al., 2025a–c).
b) Asia and Africa, in contrast, exhibit rapid increases in multidrug-resistant infections, reflecting unregulated access to antibiotics, self-medication, and limited surveillance capacity.
c) Latin America faces growing resistance in community infections, with high levels of *E. coli* and *K. pneumoniae* resistance to third-generation cephalosporins and fluoroquinolones.

Socioeconomic disparities in surveillance and healthcare quality have led experts to term AMR an equity and health-systems crisis. Inefficient wastewater systems, poor sanitation, and environmental contamination further amplify regional inequalities.

14.2.4 Economic and environmental consequences

Antibiotic resistance has severe macroeconomic repercussions. AMR is estimated by the OECD to erode as much as 3.8% of global GDP by 2050, predominantly through productivity losses and healthcare costs. Farmers suffer economic losses due to reduced productivity in antibiotic-regulated environments, and countries have to pay the price of enhanced biosecurity implementation and closed export markets for food contaminated with these bacteria. Antibiotic residues remain in the environment, such as soil and water, for months, causing breakdown of microbial ecology, facilitating the transfer of resistance genes between different species (Aslam et al., 2024). The environmental survivors, including resistant Enterococcus and Pseudomonas species, serve as a long-term reservoir that maintains resistance even after policy change. The worldwide antibiotic-resistance profile in 2025 depicts an accelerating crisis in three sectors: human health, agriculture, and the environment. Resistance trends do not appear to be stabilising, and the findings of *E. coli, K. pneumoniae, S aureus,* and *Acinetobacter spp.* lead the surge. Excessive medication without diagnostics continues to have a status as a major challenge in the health sector. In agriculture and husbandry, over 50% of antibiotics are used nontherapeutically, a practice that induces resistant strains capable of jumping species barriers and ecosystems. Environmental fallout of antibiotic abuse in humans and animals amplifies the crisis further. Tackling this global threat demands a coordinated 'One Health' response that integrates human, animal, and ecosystem health within shared authority. Policy priorities should include increasing surveillance, making prescription-only use a requirement, improving husbandry, and regulating pharmaceutical waste. Without immediate action, antibiotic resistance will not merely challenge modern medicine; it will redefine its limits (Naheed et al., 2025). Effective governance and management of antibiotic use are central to global efforts against antimicrobial resistance (AMR). Harmonised action of international organisations, governments, and local agencies is the building block for a holistic approach towards antibiotic stewardship. The world models (individual patient levels). Similar to the national models, there are some successful global governance systems that aligned policies, monitored, and reduced overuse over decades (Anderson et al., 2019).

14.3 Governance and management of antibiotic use

The good governance of antibiotics focuses on controlling, monitoring, and promoting the optimal use of antibiotics with due regard to equitable access. A governance framework combines stewardship programmes, national laws, global coordination, and data monitoring. Good governance is positively associated with reduced resistance, highlighting accountability, successful regulation, and investment in the R&D of new antimicrobials. Antimicrobial stewardship programmes (ASPs) are pivotal components of antibiotic management at institutional and national levels. ASPs are designed to promote the appropriate prescribing of antibiotics, curb overuse (suboptimal dosing), and enhance patient care. They are based on empirical therapy review, prescription audits, and restriction of access to broad-spectrum agents.

Some countries, including India, formulated national AMS guidelines by the Indian Council of Medical Research (ICMR) addressing the need for evidence-based antibiotic treatment, formulary control, and management tools for prescribers. The European Union also promotes antibiotic stewardship through the European Centre for Disease Prevention and Control's (ECDC) strategic guidelines, which stipulate, as a requirement for member states, to establish national action plans that incorporate infection control measures, restrict veterinary use of antibiotics, and build public reporting systems (Moran, 2019).

14.3.1 Role of international organisations

International organisations have been instrumental in setting norms, frameworks, and accountability structures for antibiotic governance.

a) **World Health Organisation (WHO):** In 2015 the WHO launched the Global Action Plan on AMR and manages a global Antimicrobial Resistance and Use Surveillance System (GLASS) to support data-based decision-making across its Member States. WHO also cochairs the One Health GLG with FAO, OIE, and UNEP to advance coordinated cross-sector collaboration.
b) **Food and Agriculture Organisation (FAO) and World Organisation for Animal Health (WOAH):** FAO and WHO are working together to develop policies on how to cut antibiotic use in farming animal management. Their frameworks include monitoring antibiotics. Their frameworks include monitoring antibiotic residues in food production, phasing out antibiotic growth promoters, and developing alternatives such as vaccines and probiotics.
c) **United Nations Environment Programme (UNEP):** Since joining the Quadripartite Alliance in 2022, UNEP's and has responsibility for the environmental aspects of AMR and how soil and water are affected by pharmaceutical waste products as well as residues from agriculture.
d) **Global Coordination Mechanisms:** The UN Interagency Coordination Group on AMR (IACG), formed in 2017, has informed the current model of governance, advocating for three interconnected platforms: The Independent Panel on Evidence for Action, Global Leaders Group, and Multistakeholder Partnership Platform. This guarantees accountability, science-based leadership, and inclusivity between sectors.

Together, these organisations promote a 'One Health' approach that binds human, animal, agricultural, and environmental policies into a unified global framework (Padiyara et al., 2018).

14.3.2 Contributions of national governments and local agencies

National governments are tasked with translating international policies into tangible programmes of action through the development of laws, monitoring systems, and educational programmes. They control antibiotic production, enforce prescription-only laws, and watch over national surveillance systems.

High-income countries like the United Kingdom, Sweden, and the Netherlands have achieved a reduction of antibiotic use by means of prescription control and public campaigns. The UK's 'Antimicrobial Resistance Strategy 2019–2024' established measurable targets and decreased antibiotic prescribing in primary care by 9% by 2022.

Middle- and low-income countries encounter distinct governance challenges, such as lax regulation and over-the-counter access to antibiotics. Governments, therefore, are working together with some entities, including the WHO and the World Bank, aimed at national capacities development through National Action Plans (NAPs). Sierra Leone and Malawi, for example, had cost-effective operational AMR plans in place that linked to public awareness and infection control programmes.

Application of these measures at the local level is carried out by municipal authorities, including hospital infection committees, veterinary inspection services, and consumer education. They also gather regional data for national AMR reporting platforms and encourage community-level participation in responsible antibiotic use (World Health Organization, 2021b).

14.3.3 Case studies of successful governance models

14.3.3.1 European Union and European Centre for Disease Prevention and Control (2000–25)

The EU's cross-sectoral governance represents a model of coordinated stewardship (Delhomme and van Os, 2025). European Surveillance of Antimicrobial Consumption Network (ESAC-Net) and European Antimicrobial Resistance Surveillance Network (EARS-Net) deliver harmonised data to the ECDC in relation to qualifying regulatory systems, which resulted in a more than 25% drop in veterinary antibiotic usage since 2011. In addition, the EU banned antibiotics for growth promotion in livestock in 2006, establishing a global example.

14.3.3.2 Netherlands: integrated stewardship success

The Netherlands achieved one of the lowest antibiotic consumption rates in Europe through government-supported 'One Health' integration (Kallen et al., 2019). Its 2015 AMR Strategy integrated agriculture and medical regulation, resulting in a reduction of veterinary antibiotic use by 64% between 2009 and 2020. Public–private collaboration between farmers, scientists, and government improved transparency and accountability across sectors.

14.3.3.3 India: national action plan on antimicrobial resistance

India's National Action Plan (2017–21) unified the ministries of health, environment, and agriculture into a single AMR task force (Ranjalkar and Chandy, 2019). Antibiotic surveillance in ICMR programmes was extended to 20 tertiary care hospitals, and a stewardship programme was implemented in Urban as well as rural facilities. In 2023 India set up an AMR Surveillance Network of 32 laboratories to monitor resistant strains and sensitise people about the rational use of antibiotics.

14.3.3.4 Scandinavian models (Sweden and Denmark)

Sweden's government-led campaign since the 1990s has abolished antibiotic growth promoters in farming, relying on vaccination and hygiene instead (Björkman et al., 2021). Likewise, Denmark's 'Yellow Card Initiative' tracks veterinary sales of antibiotics and punishes overuse, slashing consumption by more than 50% in a decade. These initiatives demonstrate the impact of transparency in data collection and financial accountability in driving compliance.

14.3.3.5 Sub-Saharan Africa: Sierra Leone, Malawi, Burkina Faso

WHO's 2022 case series also highlighted that governance reform is possible in a context with limited resources. Sierra Leone had developed a costed national AMR plan to attract donor funding; the National Coordinating Committee in Malawi brought ministries, NGOs, and civil society together to focus on hospital-based infection control. Burkina Faso successfully arrived at safer antimicrobial prescription by means of its AMR Committee and surveillance strengthening.

Control and regulation of antibiotic use are built on coordinated actions across international, national, and local levels. The strategic frameworks and monitoring tools of international organisations (such as the WHO, FAO, WOAH, and UNEP) are adapted by national governments to their own particular circumstances. These policies are implemented by national, regional, and local agencies through regulation, monitoring, information, education, and stewardship. Case studies from Europe, South Asia, and Africa show that success requires multilayered governance, transparency, and sustainable financing. Moving forward, the most effective global governance model remains one that harmonises global standards with localised implementation under a unified One Health agenda (Häsler et al., 2013).

14.4 Policy challenges in antibiotic governance

14.4.1 Regulatory fragmentation/National Action Plan

One of the global concerns of the 21st century is antimicrobial resistance. The widespread use of antibiotics after the recent global outbreak of COVID-19 came with the speedy emergence of AMR. To overcome complex health challenges rising due to the increasing occurrence of antimicrobial resistance, a holistic approach known as the 'One Health' approach has been implicated. The One Health approach provides a strong framework while promoting collaboration between multisectoral disciplinaes, including human health, animal health, and environmental health. It provides a unifying platform for healthcare professionals, veterinarians, researchers, policymakers, and environmentalists to collectively work together in combating the escalating threat of antimicrobial resistance. This preserves the effectiveness of the antibiotic for future use (Danasekaran, 2024).

Antimicrobial resistance has been one of the global concerned to threat public health. To tackle the problem of AMR in India, the Ministry of Health and Family Welfare (MoHFW) in 2017 officially released the NAP-AMR which aligns with the global action plan on AMR. Aligning with the strategic objective of GAP-AMR published by WHO, the NAP-AMR was implemented in India, focusing on the one health approach. The stakeholders responsible for the implementation of

NAP-AMR include MoHFW (NCDC, ICMR, CDSCO, FSSAI, NIHFW, CHEB), states, UTs, 11 other ministries (Fisheries, Animal Husbandry and Dairying (DoF, DADH), Agriculture and Farmers Welfare (ICAR), Environment, Forest and Climate Change (CPCB), Science and Technology (DBT, CSIR), AYUSH, Chemical and Fertilisers (DoP), Jal Shakti, Education, Food Processing Industries, Information and Broadcasting (Barik et al., 2023). To overcome the issue of AMR in India, NAP-AMR 1.0 was implemented from 2017 to 2021.

14.4.1.1 The Implementation was based on six collaborative strategies involving multiple stakeholders:

a) Establishing effective communication, education, and training to improve awareness and understanding of AMR
NCDC: The National Centre for Disease Control developed material media to create awareness among healthcare professionals on the judicious use of antibiotics.
FSSAI: Food Safety and Standards Authority of India created awareness among concerned stakeholders by conducting training programmes on AMR.
Dept of Animal Husbandry uses social media platforms to create awareness for annual husbandry forums for judicious use of antibiotics in dairy, poultry, sheep, and goats.
NIPER: National Institute of Pharmaceutical Education and Research raises awareness on sanitation and hygiene on AMR at grassroot level.
The AYUSH Ministry Department conducts patient awareness sessions on AMR.

b) Strengthen and knowledge and evidence through surveillance
NCDC and the national reference laboratory established the National AMR surveillance network. The objective is to strengthen the state medical colleges for AMR surveillance. ICMR initiated the antimicrobial resistance research and surveillance network (AMRSN). The goal of ICMR-AMRSN includes
  i) To monitor the trend of antimicrobial susceptibility patterns of clinically important microorganisms in the healthcare sector.
  ii) To understand the molecular pattern of resistance.
  iii) To establish real-time data analysis tool to enable different stakeholders to execute the data in real time. Aggregated data from different stakeholders are then submitted to the Global AMR surveillance system.
  INFAAR: Indian Network for Fisheries and Animal Antimicrobial Resistance, in collaboration with ICAR and with technical support of FAO. The main objective of INFAAR includes a better understanding of AMR by establishing AMR surveillance specific to the livestock and fisheries department. ICMR and DBT have developed many biorepositories for AMR priority pathogens to guide research for the development of new antibiotics.

c) Reduce the incidence through effective infection prevention and control
Swachh Bharat Abhiyan was introduced on 2 October 2014 by our Hon'ble Prime Minister to establish a cleanliness drive all over India. Following the Swachata guidelines, MoHFW introduced KAYAKALP on 15 May 2015 to provide improved healthcare facilities, including infection control.
Good farm practices: the Department of Animal Husbandry and Dairying (DAHO) has developed biosecurity manual for good poultry farm practices. Under the National animal disease control programme, DAHD has implemented the world's largest vaccination programme for livestock diseases such as Brucellosis. Mobile veterinary unit is another strategy undertaken by the Government of India to provide services to animal owners and farmers at their doorstep.
AYUSH: Proper implementation of biomedical waste management is carried out at healthcare facilities for infection control.

d) Optimise the use of antimicrobial agents in health, animal, and food
Antimicrobial Stewardship Programme (AMSP) is a wide healthcare strategy that promotes optimal use of antibiotics. ICMR has developed a guidance document on AMSP, which can be used by healthcare facilities to strengthen their AMSP.
Schedule H and H1: Under the Drugs and Cosmetics Rules, 1945 in India, updated by the Central Drug Standard Control Organization Schedule H and H1 has been implemented to control the sale over the counters. According to this Schedule H and H1, drugs can only be bought from the pharmacy on the prescription of a registered medical practitioner.
FSSAI has published a notification in 2018 to prohibit the use of 19 antibiotics and veterinary drugs at any stage of food processing. ICAR has provided technical guidelines to DAHD to issue advisories to ensure judicious use of antibiotics in food-producing animals.

e) Promote investment for AMR activities research and innovation

ICMR-NIRBI, in collaboration with the United Nations Environment Programme, first published their report on 'Priorities for the environmental dimension of AMR in India'. To tackle the problem of AMR, ICMR has focused on the following areas of research:

1. Origin, spread, and prevention of AMR
2. Development of a novel antibiotic
3. Estimate the inherent cost of treating antimicrobial resistance in India
4. Validate available diagnostics for their utility in the field for improving the diagnosis of bacterial infection, along with drug susceptibility.

BIRAC (Biotechnology Industry Research Assistance), in collaboration with DBT, has identified new approaches to fill gaps in the area for the development of new antibiotics. In 2018 DBT launched a programme focusing on the area:

1. Research and innovation for therapeutics
2. Development of biorepository
3. Research and innovation for diagnosis
4. National priority list for AMR

By promoting research and development, the Department of Animal Husbandry and Dairying, in collaboration with the AYUSH ministry, has introduced the concept of Ayurveda into veterinary science. To promote research and development, the Department of Pharmaceuticals proposed a new scheme named PRIP (Promotion of Research and Innovation in Pharma Med Tech Sector). According to this scheme, R&D department of the pharma academic institute is strengthened. The R&D sector of ICAR has focused on the development of phytochemicals, which can be a potential alternative to antibiotics.

f) Strengthen India's leadership, commitment, and collaboration on AMR at the international, national, and subnational level

India has taken the lead in several international collaborations on AMR. Partnership of ICMR and Global Antibiotic Research and Development leads to strengthening clinical trials, epidemiological study, and facilitating the development of new drugs. One of the members of the global antimicrobial resistance R&D hub is DBT, which collectively works to leverage the existing resources that support research and development addressing the global AMR concern. US-CDC and NCDC collaboratively developed SOP for AMR surveillance. NCDC, in partnership with WHO and USAID, has supported states in the development of a state action plan. Seven states in India have launched their State Action Plan on containment of AMR, and efforts are being made to support the remaining states to launch their State Action Plan.

14.4.2 Challenges faced during the implementation of various policies

There are several barriers to the implementation of NAP in lower and middle-income countries. The spread of AMR can be correlated with two main factors, that is, economic, market-driven barriers and public engagement or awareness gaps. These contributing factors are more complicated by the number of overlapping barriers organised into the following themes.

14.4.2.1 Challenges in human healthcare sectors

a) Overuse and misuse of antibiotics: The complications related to antibiotic resistance might worsen due to the inadequate competence of healthcare professionals (Akunne et al., 2025). Symptom ambiguity is another reason for the misuse of antibiotics. Symptoms of some viral infections often mix with bacterial infection; as a result, medical practitioners often prescribe antibiotics prior to a clear diagnosis, which is problematic. Viral infections are untreatable with antibiotics. However, patients infected with viral infections are often prescribed antibiotics due to delayed diagnosis, which ultimately leads to commensal microbiota dysbiosis. This promotes the commercial microorganism to become antibiotic resistant, which may later transfer to harmful bacteria by horizontal gene transfer.

b) Self-medication or accessibility: Inappropriate self-medication is often encouraged due to easy access to antibiotics over the counter without a prescription. Another major contributor to antibiotic resistance is failure to complete the full duration and dosage of the antibiotic prescribed.

c) **Poor infection control:** HAIs often occur due to poor infection control in the facility. An immunocompromised patient gets this infection caused by antibiotic-resistance microorganism. Without appropriate infection control actions, germs can spread from one person to another through surface contact, such as bed rails or the hand of a healthcare worker. Use of medical devices such as catheters and ventilators can be a potential pathway for the spread of germs. Antibiotic-resistant bacteria can cause infection at the community level if healthcare settings fail to prevent the spread of infection.

14.4.2.2 *Challenges in the pharmaceutical industries*

The three major challenges faced by pharmaceutical companies, which ultimately lead to the rise in antibiotic-resistant microorganisms, include:

a) **Economic challenges:** The development of new drugs and the necessity of conserving to prevent resistance place many pharmaceutical companies in conflicting situations. No company would wish to spend billions of dollar lot of resources, and decades of time just to discover an antibiotic that is reserved for last-resort use (Akunne et al., 2025). This greatly affects the potential profitability of different pharma companies. The pharmaceutical antibiotic R&D department thus gets stuck in a loop, where development of new antibiotics drives resistance; on the other hand, emergence of antibiotic resistance compels the necessity to develop new antibiotics.
b) **Scientific challenges:** In 2024 the updated format of the WHO's bacterial priority pathogen list (BPPL) states the urgent need for development of Nobel antibiotics (Sati et al., 2025). Based on certain criteria, WHO has classified pathogens as critical, high, and medium. Among the critical group, three pathogens (including *K. pneumoniae, Acinetobacter baumanni, and P. aeruginosa*) have played a persistent role due to associated mortality and present a massive threat in hospital settings because of the ability to become resistant to almost or class of antibiotics. These three pathogens have tended to acquire resistance against the carbapenem class of antibiotics, which are considered a last resort, by certain mechanisms, including changes in bacterial physiology, chromosomal mutation, and modification of the drug target (Muntean et al., 2022). The changes acquired to become antibiotic resistant are one of the major scientific challenges faced by the R&D sectors in the development of novel antibiotics.
c) **Regulatory challenges:** Lack of good clinical trial proof has been a large obstacle for the pharmaceutical company. Telithromycin and trovafloxacin are a few antibiotics which was initially approved by the FDA but later were withdrawn from the market due to their life-threatening hepatotoxicity (Gargate et al., 2025). To prevent the recurrence of such an incident. Regulatory agencies have introduced much more stringent guidelines for the approval of novel drugs used for commercial purposes.
d) **Complex inclusion criteria:** Due to strict eligibility criteria provided by the regulatory agency, enroling participants for a clinical trial along with their consent has been one of the major struggles for conducting such a trial. This further results in low participant enrolment.

14.4.2.3 *Challenges in animal care sectors*

a) **Pastoralism:** The pastoral setting is practised in many countries worldwide. The main idea of this system includes the constant movement of livestock in search of grass and water. In such a system, the livestock owners often get access to antibiotics easily for their livestock from the local market. As a result, overuse, misuse, or prolonged use of this antibiotic leads to the development of antibiotic resistance, endangering not only animal health but also human health (Abdi Hussein et al., 2023).
b) **Use of antibiotics as prophylactics and growth promoters:** Small doses of antibiotics are administered in livestock to prevent the occurrence of disease. On the other hand, antibiotics are also administered at lower doses for a prolonged period as growth promoters, which is another factor responsible for antibiotic resistance in livestock (Abdi Hussein et al., 2023).
c) **Lack of access to veterinary services:** One of the critical safeguards to animal health is easy access to veterinary services. Due to a lack of access to veterinary services in pastoral settings, the pastoralist often relies on self-medication. This accelerates the development of antibiotic resistance.
d) **Lack of hygiene and sanitation measures:** One of the major contributors to accelerating the dissemination of antibiotic resistance is poor husbandry practices.
e) **Constrained access to vaccines:** Livestock diseases can be prevented by vaccinating the livestock, which simultaneously decreases the need for antibiotic use. However, overpriced vaccines become a limitation to being easily accessed by the pastoralist. It is crucial to ensure that livestock are vaccinated to prevent the rise of antibiotic-resistant bacteria.

14.4.2.4 Challenges in environmental care sectors

Contamination of water bodies: Wastewater from the healthcare facilities, irrigation runoff, and urban stormwater are some major sources that lead to the development of antibiotic-resistant bacteria in water bodies (Meradji et al., 2025). Water plays a significant role as an ARB reservoir in the environment. In agriculture and livestock systems, antibiotics are usually used to manage bacterial diseases and as a growth promoter. They collectively result in the environmental resistome.

Soil contamination: Soil is a natural hotspot for antibiotic-resistant bacteria (ARB). Anthropogenic pressure is the major cause for the introduction of large amounts of ARB into the soil through manure, sludge, compost, and wastewater.

14.5 Science policy integration

An estimation given by a comprehensive screening process concludes that the burden of antimicrobial resistance is potentially larger than any of the major diseases in the majority of countries (Mestrovic et al., 2022). The estimated mortality projects it to reach at least 10 million deaths if not less on an annual basis if it remains unaddressed (Farinola and Ayodeji, 2025). Despite many National and International efforts by different organisations worldwide since the start of the Global Action Plan in 2010, the results have been disappointing. The fundamental issue is the lack of interaction between the scientific evidence and the institutional and political policymakers, resulting in the failure of implementation. This disconnection between the scientific scholars and the policy makers is prevalently known as the science-policy interaction gap (Reichmann and Wieser, 2022). The main idea of science-policy integration with respect to AMR is to bridge the gap between the two and ensure evidence-based findings so that they can be well translated into policies for dealing with the in-ground situations of AMR (Cousins et al., 2023).

14.5.1 Advantages of science-policy integration in antimicrobial resistance governance

Among the most persuasive benefits of this interaction is evidence-based policymaking. It is also referred to as the 'problem-solving model of policymaking'. The idea of this model is to identify a problem and solve it by filling the knowledge gap using research evidence. This way of making the policy decision is observed as a logical process which comprises selection, synthesis, and critical evaluation of the research evidence from which the most appropriate answer to the problem will emerge and can be implemented (Malbon and Parkhurst, 2023). Mainly, six domains are needed for evidence-based policymaking. They are governance, standards and routinised processes, collective action and support, leadership and commitment, resources, culture (Steel et al., 2025). Sweden reported a significant reduction in penicillin-resistant pneumococcal infections through evidence-based policies. There are certain fundamental challenges to evidence-based policymaking. This includes financial constraints leading to resource limitations, especially in low- and middle-income countries; research evidence being less applicable or ineffective implementation of policy due to unregulated sales of antibiotics; and poor infection control (Skoog Ståhlgren et al., 2022).

Another critical advantage of Science-Policy interaction is Multisectoral coordination and cross-disciplinary collaboration (Sworowska-Baranowska, 2022). Coordinated action across all the sectors, like health, animal, plant, and human, is important to effectively deal with AMR. Although each of these sectors are operated and regulated by different governance structures and funding bodies like World Health Organisation (WHO), Food and Agriculture Organization of the United Nations (FAO), World Organisation for Animal Health (WOAH), United Nations Environment Programme (UNEP), but according to One Health scheme multiple sectors need to implement coordinated solutions (Sanga et al., 2024a, b). There are certain bodies, like the Consolidated Framework for Implementation Research (CFIR), which provide a well-studied approach for identifying the barriers and enabling collaboration across multiple sectors. A multiministerial project in South Korea based on AMR reported to receive 92.2% performance rate. This was achieved by using the One Health core principles, which include multisectoral collaboration and data sharing, despite the persisting challenges like surveillance coverage and long-term governance. Besides this, they applied the evaluation framework across the four domains—building of infrastructure, surveillance across multiple sectors, avoiding the spreading of AMR, translation, and dissemination of outcomes (Jeong et al., 2025; Na et al., 2025).

Yet another advantage of science-policy intervention is its potential to bridge the knowledge translation gap between the various research findings and proper policy implementation (Khomsi et al., 2024). Whenever the alignment between the research findings, policy needs, and managers fails, it results in a critical barrier, but this intervention ensures that the scientific discoveries can be implemented in real-world settings for maximum effectiveness. This is very important when dealing with AMR, especially in low- and middle-income countries where there is a high demand for effective policy approaches. The intervention of science and policy also strengthens the AMR interventions by implementation of policy designs and strategies with evidence of what works in context to real life and why (Suazo-Galdames et al., 2025).

14.5.2 Barriers of science-policy integration in antimicrobial resistance governance

The primary reason for the integration gap in science-policy is not a knowledge gap or lack of political commitment. Panels or specific mechanisms are needed for effective conversion of scientific evidence into governance policy, but this seems to be lacking in many countries. This is the very reason for the establishment of the Scientific Panel on Evidence for Action (SPEA), which plays a role in outlining policy options based on the insights of scientific evidence (Ruckert et al., 2025). This problem of misalignment is more prevalently seen in low- and middle-income countries where research on AMR is being conducted in large numbers, but the mechanism of reviewing this research and translating it into policy remains confined only to high-income countries. Another barrier in this integration process is that, in spite of AMR being one of the serious global threats, this issue has failed to attract political support and the necessary resources. This is why the governance of AMR is still being done at ground level, 'below the political radar'. Lastly, the lack of feedback and adaptive mechanisms has added to this ongoing problem. The policies governing AMR need constant evaluation and monitoring from time to time to ensure that the policies have not become ineffective and static. Without the feedback mechanism, policymakers will not know the reason behind the failure of existing policy, thus making it difficult to engage in adaptive policy methods (Rony et al., 2023;Sulis et al., 2022;Pokharel et al., 2024).

14.5.3 Adaptive strategies for dynamic antimicrobial resistance patterns

Recent studies related to antimicrobial resistance show that resistance in organisms is a dynamic process caused by the continuous evolution of mechanisms, shifting of pathogen prevalence with time, dissemination of resistance genes across space and time, etc. In other words, antibiotic resistance is not a static phenomenon. Advanced spectroscopic analysis using Raman spectroscopy (RS), Fourier-transform infra-red spectroscopy (FTIR), and Nuclear Magnetic Resonance (NMR) combined with other algorithms can help accelerate and understand the early detection of AMR (Al-Shaebi et al., 2025). In one of the recent studies, an FTIR spectral study was conducted *Staphylococcus aureus* exposed to different antibiotics like trimethoprim, oxacillin, and azithromycin in successive order. As soon as the resistance patterns were checked post-24-hour antibiotic exposure, early resistance patterns were detected. This increased the model performance accuracy by 96%, depending on factors like the type of antibiotic and the components. This demonstrates that adaptation to resistance occurs in real time, which can be detected by appropriate monitoring approaches and thus requires adaptive approaches to manage the antimicrobials (Bonner et al., 2025). Another study reported the increased percentage of carbapenem-resistant *K. pneumoniae* resistance, providing gene *bla*-KPC dramatically increasing from 16.61% to 58.46% and a simultaneous decrease of *bla*-NDM from 81.48% to 38.46% just within a year from 2021 to 2022. From this study, it can be concluded that a rapid shift of genes in pathogenic organisms is occurring in response to the exposure of antibiotics, and this is why static approaches of infection control will soon be outdated (Yao et al., 2023). A few adaptive strategies are mentioned below.

14.5.3.1 Real-time surveillance framework

Surveillance system is one of the effective adaptive governance methods which can be capable of detecting resistance patterns that are presently emerging and generate immediate responses like policy changes and clinical responses, before the resistance becomes part of persistent conditions. This surveillance system can be enhanced by (1) integration and access to microbiological data generated during pathogenesis, clinical data and also epidemiological level data to generate adaptive policy responses, (2) establishing a stronger collaboration across various sectors like public healthcare, animal healthcare as well as environment using One Health approaches, (3) interconnection and collaboration of surveillance network at an international level by methods like global reporting, and (4) access to whole genome sequencing data of the pathogen to the research and development section for conducting study on antimicrobial and vaccines, to the health officials for required policy changes and infection control (Oltean et al., 2025). Complementary systems like wastewater-based epidemiology can also serve as a surveillance system to detect antimicrobial resistance. By combining the analysed data obtained after sampling the hospital wastewater with agricultural and municipal waste, early warning systems for effective AMR monitoring with enhanced early warning systems can be built (Balcázar, 2025).

14.5.3.2 Multisectoral collaboration and adaptive governance frameworks

As already mentioned earlier, the One Health approach is the foundation for adaptive governance frameworks bridging across the various sectors for collaboration and data sharing. Many countries have adapted this approach with certain modifications depending upon their needs, aiming to generate an AMR response (Faijue et al., 2024). Uganda reported adaptive governance for multisectoral collaboration using a model comprising mainly four components. They are sharing of leadership responsibilities, collaborative financing, cross-sectoral combined actions, feedback mechanisms, and regular interval evaluation. Due to insufficient coordination, lack of a proper strategy, knowledge gap, lack of awareness, and proper funding constraints, the final implementation process was difficult. Therefore a unified policy like the National AMR framework is necessary with active participation of people from different sectors and systemic monitoring for adaptive policy implementation (Ssennyonjo et al., 2022).

14.5.3.3 Localised strategies for implementation

Policymaking should be flexible. It must have modified policies that are unique in terms of economy, healthcare facilities, and cultural beliefs in that region. Before planning, it is important to know the tax and trade systems, partnership between public and private, agreement of sharing benefits, etc. (Ugwu et al., 2025). In Argentina, surveillance was conducted in two different phases to understand the major themes based on which AMR can be dealt with: (1) cultural factors and political factors influencing AMR progress, (2) fragmented governance, (3) access to antibiotic and providing information regarding its challenges, (4) AMR awareness, (5) implementation challenges of National plan, and (6) external factors influencing AMR like international pressure and internal factors like proper funding (Allel et al., 2024). Education approaches on AMR at the community level provide a foundation for adaptive governance frameworks. Healthcare facilities combining traditional and conventional medical practices demonstrate adaptive frameworks engaging diverse health systems. Discussions on AMR approaches revealed challenges, a lack of scientific validation of traditional approaches, and misunderstandings. These educational approaches with healthcare interventions can serve as a foundation for appropriate AMR educational efforts (Mitchell et al., 2022).

14.5.3.4 Digital innovation and data-driven antimicrobial resistance governance

Sophisticated applications like artificial intelligence and machine learning are some important tools for the prediction of antibiotic resistance (Sakagianni et al., 2023). Machine learning, along with data obtained by phylogenetic analysis of the pathogen, can be used for the determination of the phylogeny for improving the model performance, besides the prediction of antibiotic resistance (Yurtseven et al., 2023). An example of such a language model is AMR-Policy GPT. It is a large database containing 146 countries' policy guidelines, thus demonstrating how the latest technologies can support evidence-based development of policy. This model can search and summarise data from global document databases like national action plans, implementation guidelines, and technical reports (Chen et al., 2025a, b, c).

14.6 Support for alternative therapies

One of the major health challenges faced globally is the emergence of antibiotic-resistant bacteria (ARB). Antibiotics have proven to be a boon to human society. The rapid growth of ARB and the lack of new antibiotics have become a major concern of the 21st century. To combat the rise of antibiotic resistance and maintain the efficacy of existing antibiotics, new alternative therapeutic approaches should be introduced Table 14.1.

14.6.1 Antimicrobial peptides

AMPs are short-chain amino acid sequences (15–150) that have a valuable therapeutic approach against antibiotic-resistant bacteria. Study shows that the synergistic action of AMPs and antibiotics disrupt bacterial cell membranes, thus increasing the effectiveness of antibiotics. In a study, lactoferrin did not display any antimicrobial effect. However, the combination of Lactoferrin and chloramphenicol shows a remarkable synergistic action against the isolates producing extended-spectrum beta-lactamase (Aziz, 2023). AMPs also target biofilm formation. A combination of AMP GVF27 and ciprofloxacin targets the biofilm formed by *Burkholderia cepacia* (Bosso et al., 2022). Enhancing the efficacy of nalidixic acid, peptide Cecropin A has been shown to treat recurrent urinary tract infection by disrupting uropathogenic *E. coli* biofilm (Sun et al., 2025). Thus AMPs disrupt the integrity of bacterial cell membrane allowing the antibiotic to penetrate the bacterial cells more efficiently and target the intracellular processes responsible for developing antibiotic resistance.

TABLE 14.1 Summary of different alternative therapeutics to combat antibiotic resistance.

Alternative therapeutics	Agent used for therapy	Mode of action	Reference
Antimicrobial peptide (AMPs)	Cec4	Eradicates the biofilm formation of carbapenem-resistant *A. baumannii* by interfering with quorum sensing	Liu et al. (2020)
RNA based	sRNA NcS25	Regulates the membrane transport across outer membrane porin BCAL3473 and facilitates antibiotic uptake in *Burkholderia cenocepacia*	Sass and Coenye (2023)
Aptamers	R8-su12	Inhibits the biofilm formation in *Streptococcus suis*	Matchawong et al. (2022)
Phage therapy	phage Sb-1	Synergistic action with oxacillin inhibits the cell wall synthesis in MRSA	
Antibodies	Raxibacumab	Neutralises anthrax toxin	Skoura et al. (2020)
Metal ions and compounds	Ruthenium	Ruthenium polypyridine complexes with triphenylamine affect the integrity of bacterial membrane	Jiang et al. (2022)
Phytochemical	Berberine	Synergistic action of berberine and ciprofloxacin downregulates the quorum-sensing genes in *Salmonella* strains	Shi et al. (2018)
Probiotics	silymarin-enriched *Lactobacillus*	Suppresses gut microbiota related to secondary bile acid synthesis	Haghshenas et al. (2023)

14.6.2 RNA-based therapy

One of the strategies which is actively being researched as a potential alternative to conventional antibiotics is RNA silencing. This technique is based on the natural regulatory mechanism of bacteria. In this mechanism, a single mRNA (antisense sequence) contains cis and trans sequences that are complementary and, upon binding, can reversibly block translation. In this alternative therapy, synthetic antisense sequences complementary to the gene responsible for antibiotic resistance are constructed. These sRNAs then bind to the complementary sequence and block the translation of those enzymes that enable bacteria to resist antibiotics. In a study, M13 bacteriophage was used to deliver sRNA, which silences the expression of chloramphenicol acetyl transferase and kanamycin phosphotransferase (Abdullah et al., 2022).

14.6.3 Aptamers

Aptamers are single-stranded oligonucleotide sequences (DNA or RNA) that can be used as a flexible tool for therapeutic goals. The aptamers usually show their antimicrobial effect by binding to specific proteins that are responsible for developing resistance mechanisms in bacteria. *Staphylococcus aureus* targets the host epithelial tissue and cell membrane by producing the virulence factor alpha-toxin. Studies show that Aptamers (AT-27, AT-33, AT-36, and AT-49) prevent host tissue damage by targeting the toxin (Alomran et al., 2022). Aptamers and Ampicillin complex show a synergistic effect on biofilm formation (Ning et al., 2022). One of the advantages of using an aptamer over antibody therapy is higher affinity and specificity against various targets.

14.6.4 Phage therapy

The global surge in the increase of antibiotic-resistant demands the development of novel therapeutic approaches to overcome antibiotic resistance. Bacteriophage therapy has gained interest as an alternative therapeutic approach. This therapy uses bacteriophages to use the lytic cycle to target the bacterial pathogen in a patient's system. In recent studies, PP1131 (phage cocktail made of 12 anti-*P. aeruginosa* phages) is used to treat burn wound infection in patients caused by *P. aeruginosa* (Zhang, 2023). It was observed that there was 1.5 log reduction in *P. aeruginosa* density in ventilator-associated pneumonia when treated with a cocktail of five phages comprising three Myoviridae Pseudomonas phages and two Podoviridae Pseudomonas phages (Khan et al., 2023). Urinary tract infection is one of the most common bacterial infections affecting almost 150 million people each year. Recurrent UTIs are prevalent among children and females. A 58-year-old renal transplant patient suffering from recurrent UTI caused by *K. pneumoniae* strain with extended-spectrum beta-lactamase (ESBL) was successfully treated using a combination of meropenem and bacteriophage (Krawczyk et al., 2022).

14.6.5 Antibodies

Even before the widespread use of antibiotics, serum antibodies efficiently treated infectious diseases. Antibodies are advantageous over antibiotics because they reduce clearance rate, lower toxicity, and increase longevity (Seixas et al., 2022). Trellis Biosciences developed a monoclonal antibody which inhibited a plane formed by *Staphylococcus aureus* and *P. aeruginosa* by targeting DNA B II Protein (Conway et al., 2024). Suvratoxumab is a monoclonal antibody designed to target the toxin produced by *Staphylococcus aureus*. The antibodies are mainly used to treat ventilator-associated pneumonia (Piscaglia et al., 2025). Bezlotoxumab, approved by the FDA in 2016, is used to prevent the recurrence of *Clostridium difficile* infection by neutralising Toxin B (Thandavaram et al., 2022). In 2016 Obitoxaximab was approved for clinical use. Its neutralisers the protective antigen component of anthrax toxin and prevents respiratory and track infections caused by *Bacillus anthracis*. Though they are potential therapeutic agents, challenges exist in their development because of high manufacturing costs.

14.6.6 Metal ions and compounds

Though not in clinical use, metalloantibiotic complexes and nanoparticles show promising activity against AR bacteria. Altering metal haemostasis to combat infection has become a strategic approach. P. aeruginosa infection can be treated by forming a copper -ciprofloxacin complex (Golden et al., 2024). Cu increases the efficacy of nalidixic acid to show broader antibacterial activity. AuNPs conjugated berberine achieves lower MIC compared to free berberine, along with higher biofilm eradication and increased ROS, leading to membrane disruption in methicillin-resistance *Staphylococcus aureus* (Yang et al., 2025). Synthetic AuNPs also showed efficient antituberculosis activity. Bacterial membrane disruption, targeting enzymatic mechanisms, interference with DNA replication, and generation of oxidative stress are some of the mechanisms by which AuNPs shows its antibacterial activity.

14.6.7 Phytochemical

Secondary metabolites obtained from plants show efficient antibacterial activity. They are becoming increasingly popular because of their lower cost, lower side effect, and higher efficacy. Extract of *Calpurnia aurea* showed an activity profile against the tested strains. *S.aureus* was found to be most vulnerable, whereas *E. coli* was the most sensitive. However, the compound from the most active fractional extract was not identified (Geletu et al., 2022). *Launaea arborescens* and *Limonium* sp. extracts shorten the bacterial activity. Ethanolic and aqueous extracts of *Limonium* sp. reduce the growth of *E. coli* by 82.85% and 78.57%, respectively. Chloric acid extract of *Limonium* sp. showed the highest antibacterial activity against *Staphylococcus aureus*. Aqueous extract of *Launaea arborescens* inhibited the growth of *E. coli* by 78.57% and *S. aureus* by 68.42% (Lakhdar et al., 2023). Phytochemicals fight MDR by damaging the cell membrane, interfering with quorum sensing, inhibiting efflux pumps that expel drugs, and increasing the sensitivity of the existing drugs. Thus interfering with the mechanism which makes the bacteria resistance against antibiotic.

14.6.8 Probiotics

In contrast to antibiotics, which kill the pathogen directly, probiotics are living microorganism which produces antimicrobials metabolite and compete with the pathogen for nutrients. As a result, it indirectly prevents infection. *Lactobacillus* and *Bifidobacterium* are most commonly used strains for human consumption. *Pediococcus acidilactici* inhibits *P.aeruginosa* by targeting its motility and biofilm-forming mechanism. By producing bacteriocin-like substances, probiotics have been found to reduce the formation by MRSA (Singh et al., 2025). *Lactobacillus brevis* and *Bifidobacterium bifidum* show significant inhibitory effect against *S. aureus* biofilm formation. The synergistic action of prebiotics and probiotics has enhanced antirotavirus responses (Díaz et al., 2022). Fengycin produced by Bacillus sp. has also shown antifungal activity. Inhibition of quorum sensing is another target to prevent *S. aureus* infection (Gao et al., 2026).

14.7 Strengthening public–private partnership

The establishment of a partnership between the public and private sectors has played an important role in successfully implementing the action plan at the global and national levels. Public–Private Partnership (PPPs) include crucial alliances between governments, industries, academia, and NGOs to combat the global threat of antimicrobial resistance. These sectors work together for management of R&D cost and risk, strengthening the surveillance system, improving the antibiotic stewardship programme, and supporting research for the development of new antibiotics and diagnostic tools.

In low- and middle-income countries (LMICs), research and development are usually undervalued by pharmaceutical companies, governments, including other stakeholders involved in the development of new antibiotics. This is because of enormous cost, high risk of failure, low return on investment, and long development timelines. To share this financial burden and risk, PPPs pool funds from diverse sources.

To overcome the market failure and to support R&D, a global nonprofit funding agency, CARB-X (Combating Antibiotic-Resistant Bacteria Biopharmaceutical Accelerator), was established in 2016 by Boston University. It is funded by the United States government (through BARDA and NIAID, within the United States Department of Health and Human Services), the UK government (through the Global AMR Innovation Fund), GAMRIF (in the Department of Health and Social Care), the Wellcome Trust, and the Bill and Melinda Gates Foundation. These funding agencies mainly focus on bacteria on the Priority Pathogens list published by the WHO and CDC. The annual report of CARB-X 2024 states that since its foundation in 2016, CARB-X has supported 104 R&D projects in 13 countries.

To support the second objective of GAP-AMR, 'Strengthen knowledge through surveillance and research'. The Global Antimicrobial Resistance and Use Surveillance System (GLASS) was launched by WHO on 22 October 2015. Surveillance is an essential tool for monitoring the spread of AMR at local, national, and global levels (Nath et al., 2024). GLASS publishes its annual reports by featuring both global reports for antimicrobial resistance and antimicrobial consumption data (Ajulo et al., 2024). GLASS-AMR is based on two objectives:

a) provides a standard approach for the collection, analysis, and sharing of AMR data by countries and enables the documentation of the status of existing or newly developed AMR surveillance systems.

b) promotes a shift from a surveillance approach based solely on laboratory data to a system that includes epidemiological, clinical, and population-level data.

PPPs have played an important role in strengthening the AMR surveillance. The strategy in PPPs for AMR surveillance includes the following.

14.7.1 Development of digital tool for data compilation and sharing

a) WHONET: WHONET is a software developed by the World Health Organisation for the management and analysis of microbiology laboratory data with a particular focus on antimicrobial resistance surveillance. Applications of WHONET are prescribed as follows:

1. To understand the antibiotic-resistance pattern of a particular geographic region.
2. Used for the development of the local antibiogram guideline.
3. Enables the clinician to analyse the local antibiogram pattern and prescribe the most appropriate antibiotic.
4. It supports research by focusing on the development of new antimicrobials against the bacteria in the global Priority Pathogen list.
5. Used to rationalise antibiotic use in local hospitals.
6. The length of hospital stays and the cost of treatment increase due to inappropriate use of antibiotics. WHONET software provides clinicians with pathogen-directed therapy, thus reducing the chance of failure and cost of treatment.
7. It also alarms us with future preparedness for any outbreak.

b) FAO-ATLASS: Food and Agriculture Organisation of the United Nations piloted the FAO-ATLASS tool in 2016 for assessing AMR surveillance in the food and agriculture sectors. The FAO-ATLASS tool is divided into two different modules, namely the surveillance module and the laboratory module.

c) Surveillance module: This module is designed for the national-level authorities working at veterinary and animal healthcare services, food safety authorities, environmental monitoring agencies, national reference laboratories, policy makers, and stakeholders involved or expected to be involved in AMR surveillance. The five core domains, which are assessed by this surveillance module, include Governance, Data Production Laboratories, Data Collection and Analysis, Communication, and Sustainability.

d) Laboratory module: This module is designed for the national-level laboratory involved in AMR surveillance. The laboratory module acts as a technical backbone of AMR surveillance practices, integrating the data to ensure the quality assurance policies. In doing so, it establishes a functional network between microbiological practice and national surveillance governance (Keck et al., 2023).

e) i-AMRSS: Collection, validation, and analysis of Antimicrobial Susceptibility Testing (AMST) data are centralised in a web-based open-source tool developed by the Indian Council of Medical Research (ICMR). AMR data of clinical isolates are routinely shared with these sites by the clinical microbiology laboratory. Consumption of antibiotics has been controlled due to the formation of a local antibiogram, which helps the medical practitioner to prescribe appropriate medicine to treat the patient effectively. The standardised data collected through i-AMRSS is valuable to monitor outbreaks and infection control practices. Annual data are published online. The National Coordinating Centre for India annually reports anonymised data to the Centre for Disease Control (NCDC), New Delhi, which is then reported to the Global Antimicrobial Resistance Surveillance System. Being an online tool, it can be easily integrated into the laboratory. Data are entered through built-in integrated form or using Excel. The i-AMRSS portal provides standards to the laboratory and hospital to follow the nationwide AMST guidelines. It provides reliability, timelessness, and standard operating procedures for the collection and management of AMR data.

14.7.2 Strengthening funding sources for antimicrobial resistance surveillance

14.7.2.1 Fleming fund

The fund is named after the famous scientist Alexander Fleming. To tackle the antibiotic resistance UK's Department of Health and Social Care supports 30 different countries in Asia and Africa to strengthen their AMR surveillance system. So far, 170 National Action Plans have been developed with the support of the Fleming Fund. It provides technical assistance and funding to bacteriology diagnostics to improve AMR reporting and data analysis (Burki, 2025).

14.7.2.2 The pandemic fund

During the outbreak, one of the major problems faced by the lower–middle-income countries is the financial crisis. To support the LMICs during an outbreak, the World Bank has come up with the multilateral financing mechanism known as 'The pandemic fund'. The main aim of this funding agency is to strengthen the infrastructure for AMR surveillance. Thus supporting preparedness and response capacities at the local, regional, and global level. To date, the pandemic funding has awarded financial grants to 75 countries and 47 projects to support pandemic prevention, preparedness, and response (Boyce et al., 2023).

14.7.2.3 Antimicrobial resistance multipartner trust fund

Aligning with the Quadripartite framework (FAO, UNEP, WHO, and WOAH), the AMR Multipartner Trust Funds contribute to advancing the National Action Plan of LMICs. Since 2019, AMR MPTF has supported 14 countries, including Bangladesh, Cambodia, Ethiopia, Ghana, and Indonesia, and from 2026, seven additional countries are expected to receive multiyear funding. The funding supports four global projects, which include

1. Supporting AMR-relevant legislation by developing a 'One Health' approach
2. Setting up an integrated surveillance programme
3. Raising awareness on AMR
4. Developing AMR monitoring and evaluation

14.7.2.4 Global fund to fight AIDS, tuberculosis, and malaria

Focusing on a few diseases, international financing and partnership organisation aim to provide research grants to countries epidemic to HIV, tuberculosis, and malaria. 70 million lives have been saved, 25.6 million people, by providing them with antiretroviral therapy for HIV. It has supported 7.4 million people to treat tuberculosis, and 160 million insecticide-treated mosquito nets were distributed.

14.7.3 International and education campaign

The emergence and spread of AMR are mainly due to a lack of awareness. For example, antibiotic overprescribing, self-medication, not using antimicrobial as prescribed, and incomplete or short-duration antibiotic courses are some of the causes that lead to the survival of the most resistant bacteria and result in recurrent infection. Does controlling AMR require behavioural changes within the population? These behavioural changes can be achieved through an awareness-raising campaign, education, and training

14.7.3.1 World Antimicrobial Resistance Awareness Week

WAAW is a global campaign initiated by the Quadripartite framework, which is celebrated every year from 18 to 24 November. This week aims to encourage the public healthcare workers, animal health professionals, the youth civil society, and the media about the best practices and handle the antibiotic with care. Over the week, activities like webinars, workshops, panel discussions, poster presentations, and many other celebrations are organised. During this week, many campaigns are organised, such as

a. **Go Blue for AMR:** The 'Go Blue for AMR' campaign was initiated by the Quadripartite framework to symbolise unity in the fight against AMR infection. The different stakeholders under the One Health mission illuminate their workplace, building, or infrastructure with blue light. Individual teams and members are encouraged to wear blue clothing. This symbolises the unifying commitment of people working in the healthcare sector, animal care sector, and environmental care sector, as well as the general public, for the responsible use of antibiotics towards combating AMR.
b. **Handle with care campaign:** In 2021 alone, AMR directly caused 1.14 million death, and if action is not taken, it is estimated that 39 million lives might be affected by 2050. 'Handle with care' stamps are a unifying visual identity for the global AMR campaign. Healthcare professionals, patient services, and animal care services often wear t-shirts with printed stamps to stand united against the development of antimicrobial resistance. The core message of this campaign highlights that antimicrobials should not be overused or misused, as they are precious resources.
c. **Spread awareness, stop resistance:** In 2021 the theme for the WAAW was 'Spread awareness, stop resistance'. This campaign aimed to create awareness among the general public, healthcare workers, farmers, animal health professionals, and policymakers to avoid the emergence of AMR and to encourage best practices to effectively use antibiotics.

a) **Preventing antimicrobial resistance together:** AMR affects all four sectors, affecting human, animal, plant, and environmental health. Addressing AMR becomes more challenging because of the exchange of genetic material between microbes, conveying disease resistance between them. Therefore a multisectoral approach between people, animals, plants, and their shared environment is required to address the problem of AMR. By implementing multisectoral programmes, policies, legislation, and research, AMR can be addressed effectively. Each sector must play its part, and everyone has a role to play.
b) **Educate Advocate, Act Now:** In 2024 the theme for WAAW was 'Educate Advocate, Act Now'. Drug-resistant pathogens pose a threat to human health, animal health, food production, and the environment. The aim of the theme is based on three principles of the multisectoral department.
c) Educate the multisectoral stakeholders on AMR.
d) Advocate for bold commitment.
e) Take concrete action in response to AMR.

14.7.3.2 ReAct campaign

As AMR has become a global threat to the world, a powerful coalition of communities and civil society groups is needed. The campaign 'From People to Leaders: Act on AMR NOW!', spearheaded by ReAct and the Pan American Health Organisation (PAHO), is going ahead to conduct the crucial High-Level Meeting on AMR at the United Nations General Assembly in New York. The campaign addresses the role of communities and civil society organisations in the global AMR response, arguing that their involvement is vital for effective strategies and should be a priority for global leaders.

14.7.3.3 The antibiotic guardian campaign

AMR can be addressed by behavioural changes and by increasing the knowledge among the community. The campaign targets the healthcare professionals and members of the public to become Antibiotic Guardians (AG). The aim is to provide knowledge to healthcare professionals through an online platform. The AG then aims to increase awareness about the threat of AMR to public health and encourage appropriate antimicrobial use. The campaign extends the EAAD (European Antibiotic Awareness Day) goal of raising awareness and seeks to change behaviour through the use of pledges and support for collective action or responsibility (identifying as an AG, visiting the website and being part of a group of AGs) by addressing the intention-behaviour gap in relation to the appropriate use of antibiotics (Mitchell et al., 2023).

14.7.3.4 The Centres for Disease Control and Prevention's campaign to prevent antimicrobial resistance

The CDC's Campaign to Prevent Antimicrobial Resistance was launched in 2002 and focuses on reducing resistance in healthcare settings through four core strategies:

a) **Prevent infection:** Protecting the individual from high-risk infection by vaccination. Vaccination drives should be carried out effectively, targeting healthcare workers, the elderly, individuals with chronic conditions, and pregnant women. This reduces the emergence of a global pandemic.
b) **Diagnose and treat infection effectively:** Potential experts should diagnose the pathogen by culturing and calculating the minimal inhibitory concentration to generate a local antibiogram, which helps the medical practitioner to prescribe appropriate medicine to effectively treat the patient and prevent the emergence of AMR.
c) **Use antimicrobials wisely:** Antibiotics should be used judiciously. Follow the Right Drug, Right Dose, Right Time rule, that is prescribing the most effective antibiotic at the correct concentration for the right duration.
d) **Prevent transmission:** Proper hygiene practices, proper handling practices in the laboratory, and effective vaccination drives can prevent the transmission of infection. The campaign targets hospital-based clinicians, later expanding to include surgeons, paediatricians, and long-term care facilities. The campaign provides educational materials, including slide sets, posters, and pocket cards for clinical staff (Ashiru-Oredope et al., 2025).

14.8 Balancing innovation, regulation, and education

Addressing antimicrobial resistance requires a deliberate balance between innovation, regulation and knowledge, each reinforcing a coherent policy framework. Innovation is the soul of progress in today's world. Development of new products and services or any other processes that can improve the quality of life is a result of innovation (Kolawole et al., 2023). The key factors of the innovation include research and development, market forces, globalisation, and government support. With respect to antibiotic resistance, many governmental and nongovernmental organisations such as the WHO, UN, G20, and EU have designed several policies (Badau, 2021). These policy responses encourage the innovation of antibiotics, which are in the form of public incentives. These incentives can be of two different types. The first type is the push incentive, which aims to help the firms in drug development by cutting down the R&D costs, whereas pull incentives, on the other hand, are generally seen as payments given to the developers as compensation against risks and costs to accomplish R&D outputs (Anderson et al., 2023a, b). Besides these, scientists are also investigating a range of other alternatives to antibiotics. Antimicrobial peptides (AMPs) are one such alternative, which are broad-spectrum antimicrobials which have multivalent actions on bacterial targets, due to which resistance against them is difficult to occur (Erdem Büyükkiraz and Kesmen, 2022). Phage Therapy is another method which has a strong potential due to the specificity of killing targets (Hibstu et al., 2022). Besides the research and development section, market forces, globalisation, and government support also play an important role in Innovation. Market demands the firms for more innovative in production, whereas globalisation has led to thorough cross-cultural collaborations and exchange of data and technology. Government plays the main regulatory role; without whose involvement, the scientific evidence would be transformed into policies that cannot be implemented (Hinterleitner et al., 2024). The need for regulation in antibiotic governance is to align the goals of the policies. In general, the regulation in technology governance is limited to the protection of consumer rights, ensuring public safety, addressing the social impacts, and also encouraging sustainable development. One of the important pillars in maintaining the balance between innovation and regulation is knowledge (Sulaeman et al., 2023). Global public awareness about antibiotic resistance is very limited. It is reported that in Kosovo, out of the 100 people interviewed, 75% had only heard of antibiotic resistance and only a handful of them understood the key concepts (Pasha et al., 2025). A study revealed that there are many misconceptions related to the use of antibiotics (Chalkidou et al., 2023). Around 67% of the population understood the need to complete the entire course of medicines, whereas 29% people stop the medication once they start to feel better. Even now, a large percentage of the population reports self-medication of antibiotics or thinks it would be useful for treating viral infections and also stores the medicine for future use (Gashaw et al., 2025). Community education is effective in bridging this gap by educating people about the use, misuse, and consequences of self-medication of antibiotics (Popoola et al., 2024).

14.9 Impact of antibiotic overuse and increasing efficacy

One massively important issue giving rise to high mortality, morbidity, and long-term health issues is Antibiotic Failure. The major contributors to antibiotic failure are Biofilms, Adaptive resistance, Sepsis, compromised immune and host defences, Microbiome, etc. 65% of antibiotic failure is mainly caused by biofilm. These are microbial aggregates enclosed within a polymeric matrix and found on different surfaces. They can cause numerous infections, such as those of the lungs, throat, bladder, skin, ear, and nose. They are highly resistant to stress conditions, due to which they can easily evade the host defence mechanisms, adding to their ability to persist and lead to chronic infections (Giordano and Giannoudis, 2024). Annually, 103 million cases of HAIs caused by biofilm have been reported in the Global Report on Infections and Prevention and Control, 2022 (Mishra et al., 2024). Abscess and chronic rhinosinusitis are conditions frequently caused by biofilm formation but are not treatable with antibiotics (Shaghayegh et al., 2022). Bacteria can be both tolerant and persistent to antibiotics. In both scenarios, bacteria can again regrow when conditions turn favourable, and this condition is known as adaptive resistance. One of the life-threatening conditions caused by uncontrolled infection leading to multiple organ dysfunction is called sepsis (Kumar et al., 2024). It has been reported that annually, sepsis causes around 49.5 million cases and 11 million deaths pre-COVID-19 (Chen et al., 2025a, b, c). Even though antibiotics are generally used for sepsis treatment but increased mortality rate due to multiple organ failure is an indication of their decreased and limited effectiveness in treating this condition. Failure of immune cells gives rise to immunocompromised conditions. Immunodeficiencies in the host can be of two varying types: Primary immunodeficiency in patients is observed with an altered immune response, mainly caused by a single gene disorder. A recent study shows that primary immunodeficiency causes 6 million cases annually (Griffith et al., 2016) and a total of around 4000 deaths (Cassini et al., 2019). These individuals show recurring infections and are best treated with antibiotics, despite which they succumb to infections within the first year of their life. On the other hand, individuals who are affected by other illnesses like asthma, cystic fibrosis, and AIDS etc or under immunosuppressive therapy due to transplants have secondary immunodeficiency. These are mainly acquired conditions, and the patients suffer from chronic infections. In patients suffering from cystic fibrosis, it is commonly observed that despite the intensive treatment with antibiotics, patients succumb to chronic infections caused by *P. aeruginosa*, *Staphylococcus aeruginosa*, *Burkholderia spp*., etc. Besides these causes of antibiotic failure, understanding the dynamics of the host and the microbiome has also been proven to be equally important. Administration of prolonged antibiotic treatment, besides the aim to eliminate pathogens, profoundly impacts the gut microbiome. Many patients report antibiotic-associated diarrhoea, caused by the colonisation of opportunistic pathogens like *Clostridium difficile*. This condition leads to colitis, fever, and abdominal pain. The disruption of the healthy gut microbiome and the growth of opportunistic pathogens promote antibiotic-resistant strains, contributing to antibiotic failure. Addressing the issue of antibiotic failure demands an extremely comprehensive approach that delves beyond the discovery of a new antibiotic. This demands a critical need for diagnostics so that antibiotics can be avoided when unnecessary. Rapid diagnostics using multiplex PCR-based assays can help distinguish between bacterial and viral infections. This will reduce the overprescription of the broad-spectrum antibiotics and also provide clinical information which can provide more therapeutic options for the treatment of the ailment. Other methods to increase the efficacy of antibiotics are by using combination antibiotic therapy, in which two or more antimicrobials are used for different purposes, which limits the chances of resistance strain formation. Antivirulence therapies use alternative therapeutic strategies for the treatment of bacteria-mediated diseases. Phage therapy is another alternative method where viruses that can infect bacteria are employed for treating the infections. Being very specific in target, they are being widely used nowadays. Finally, to ensure the long-term resilience of the antibiotic efficacy, focus needs to be given in surveillance programmes to integrate different sectors, national and international coordination, political commitment, awareness among people, and strategic plan to maintain a coordination among all as preserving antibiotic efficacy for future generations is equally important as saving nature for them as a shared responsibility.

14.10 Conclusion

Antimicrobial resistance, as mentioned earlier, is one of the major threats of the 21st century. Several factors which contribute to the spread of antibiotic resistance are overuse and misuse of antibiotics, incomplete treatment, lack of new antibiotics, education, etc. The limitations which contribute to AMR are variable national policy implementation, uneven surveillance programme, inadequate resources and finances, resulting from fragmented governance. To tackle this issue, the World Health Organisation (WHO) encouraged the states to develop their own One Health approach in 2015. Besides this, Global Action Plan (GAP) and Global Antimicrobial Resistance (GLASS) and Use Surveillance System (GLASS) were also launched, which are integral to achieving sustainable development. One Health scheme is a multisectoral approach; in simple terms, it can be defined as a collaboration across various sectors to achieve optimal health outcomes. All the sectors and stakeholders are intended to collaborate and work effectively to interconnect people, animals, plants, etc., with their shared environment. Since AMR has to be dealt with a dynamic and flexible approach, the governance framework should be adaptive. The adaptive mechanisms should consist of certain components like real-time surveillance programme. This will collect information on antibiotic susceptibility and its consumption across the various sectors, like agriculture, veterinary, hospital, and environmental. Other components include equal availability of resources, knowledge, and data sharing across all the countries, be it high-, low-, or middle-income countries.

References

Ababneh, M.A., Nasser, S.A., Rababa'h, A.M., 2021. A systematic review of antimicrobial stewardship program implementation in middle eastern countries. International Journal of Infectious Diseases 105, 746–752. Available from. Available at https://doi.org/10.1016/j.ijid.2021.03.035, https://www.journals.elsevier.com/international-journal-of-infectious-diseases.

Abban, M.K., Ayerakwa, E.A., Mosi, L., Isawumi, A., 2023. The burden of hospital acquired infections and antimicrobial resistance. Heliyon 9 (10) Available from. Available at https://doi.org/10.1016/j.heliyon.2023.e20561, http://www.journals.elsevier.com/heliyon/.

Abdullah, S., Ain, Q., Jalil, A., Khan, D., Khan, A., Qasim, M., et al., 2022. Silencing of curlin protein via M13 phagemid-mediated synthetic sRNA expression reduces virulence in the avian pathogenic *E. coli* (APEC). Current Microbiology 79 (4) Available from. Available at https://doi.org/10.1007/s00284-022-02791-y, link.springer.de/link/service/journals/00284/index.htm.

Ajulo, S., Awosile, B., Negatie, B.A., 2024. Global antimicrobial resistance and use surveillance system (GLASS 2022): investigating the relationship between antimicrobial resistance and antimicrobial consumption data across the participating countries. PLoS One 19 (2), e0297921. Available at https://doi.org/10.1371/journal.pone.0297921.

Akhmetova, K., Makalkina, L., Pivina, L., Lim, L., Aukenov, N., Boranbayev, K., et al., 2025. Trends in 'Watch' and 'Reserve' antibiotic use in primary care in Kazakhstan: the imperative for enhancing stewardship strategies. Antibiotics 14 (10). Available at https://doi.org/10.3390/antibiotics14100963.

Akunne, O.Z., Emmanuel, B.N., Saidu, U.F., Akinnawo, A.S., Adeyemi, N.A., Egoh, L.C., et al., 2025. Antibiotic resistance: the growing threats and potential solutions. American Journal of Pharmacotherapy and Pharmaceutical Sciences 4, 13. Available at https://doi.org/10.25259/ajpps_2025_013.

Al Marjani, M.F., Hasan, R.N., Khadam, Z.A., Al-Saryi, N.A., Al Rahhal, A.H., 2020. 'ESKAPE bacteria and antimicrobial resistance. PalArch s Journal of Archaeology of Egypt/Egyptology 17.

Allel, K., Fernandez-Miyakawa, M., Gaze, W., Petroni, A., Corso, A., Luna, F., et al., 2024. Opportunities and challenges in antimicrobial resistance policy including animal production systems and humans across stakeholders in Argentina: a context and qualitative analysis. BMJ Open 14 (6), e082156. Available at https://doi.org/10.1136/bmjopen-2023-082156.

Alomran, N., Chinnappan, R., Alsolaiss, J., Casewell, N.R., Zourob, M., 2022. Exploring the utility of ssDNA aptamers directed against snake venom toxins as new therapeutics for snakebite envenoming. Toxins 14 (7), 469. Available at https://doi.org/10.3390/toxins14070469.

Al-Shaebi, Z., Akdeniz, M., Ahmed, A.O., Altunbek, M., Aydin, O., 2025. Breakthrough solution for antimicrobial resistance detection: surface-enhanced raman spectroscopy-based on artificial intelligence. Advanced Materials Interfaces 12 (14) Available from. Available at https://doi.org/10.1002/admi.202300664, http://onlinelibrary.wiley.com/journal/10.1002/(ISSN)2196-7350.

Anderson, M., Schulze, K., Cassini, A., Plachouras, D., Mossialos, E., 2019. A governance framework for development and assessment of national action plans on antimicrobial resistance. The Lancet Infectious Diseases 19 (11), e371. Available from. Available at https://doi.org/10.1016/S1473-3099(19)30415-3, http://www.journals.elsevier.com/the-lancet-infectious-diseases.

Anderson, M., Panteli, D., van Kessel, R., Ljungqvist, G., Colombo, F., Mossialos, E., 2023a. Challenges and opportunities for incentivising antibiotic research and development in Europe. The Lancet Regional Health - Europe 33. Available at https://doi.org/10.1016/j.lanepe.2023.100705.

Anderson, M., Panteli, D., Mossialos, E., 2023b. How can the EU support sustainable innovation and access to effective antibiotics? Policy options for existing and new medicines'. Available from: https://eurohealthobservatory.who.int/publications/i/how-can-the-eu-support-sustainable-innovation-and-access-to-effective-antibiotics-policy-options-for-existing-and-new-medicines.

Anyanwu, M.U., Jaja, I.F., Okpala, C.O.R., Njoga, E.O., Okafor, N.A., Oguttu, J.W., 2023. Mobile colistin resistance (mcr) gene-containing organisms in poultry sector in low- and middle-income countries: epidemiology, characteristics, and one health control strategies. Antibiotics 12 (7) Available from. Available at https://doi.org/10.3390/antibiotics12071117, http://www.mdpi.com/journal/antibiotics.

Asghar, A., Khalid, A., Baqar, Z., Hussain, N., Saleem, M.Z., Sairash, et al., 2024. An insights into emerging trends to control the threats of antimicrobial resistance (AMR): an address to public health risks. Archives of Microbiology 206 (2). Available at https://doi.org/10.1007/s00203-023-03800-9.

Ashiru-Oredope, D., Onwunle, M., Gilham, E., Charlesworth, J., 2025. 'Engaging health workers, students and the public to take action to tackle AMR: 10 years of the antibiotic guardian campaign' JAC-Antimicrobial Resistance 7 (Supplement 1). Available at https://doi.org/10.1093/jacamr/dlae217.002.

Aslam, B., Asghar, R., Muzammil, S., Shafique, M., Siddique, A.B., Khurshid, M., et al., 2024. AMR and sustainable development goals: at a crossroads. Globalization and Health 20 (1) Available from. Available at https://doi.org/10.1186/s12992-024-01046-8, https://globalizationandhealth.biomedcentral.com/.

Aziz, R.A., 2023. Study of the synergistic effect of proteins produced from *Saccharomyces cerevisiae* with lactoferrin against multi resistant diarrheal bacteria. Iraqi Journal of Market Research and Consumer Protection 15 (1), 45–53. Available at https://doi.org/10.28936/jmracpc15.1.2023.(4).

Badau, E., 2021. A one health perspective on the issue of the antibiotic resistance. Parasite (Paris, France) 28, 16. Available at https://doi.org/10.1051/parasite/2021006.

Balcázar, J.L., 2025. Wastewater-based epidemiology as a complementary tool for antimicrobial resistance surveillance: overcoming barriers to integration. Bioessays: News and Reviews in Molecular, Cellular and Developmental Biology 47 (8) Available from. Available at https://doi.org/10.1002/bies.70034, http://onlinelibrary.wiley.com/journal/10.1002/(ISSN)1521-1878.

Barik, S.R., Pradhan, H.S., Prakash, J., Satpathy, S.K., 2023. A comparative analysis of national and state level action plans on antimicrobial resistance in India. Journal of Multidisciplinary Research in Healthcare 9 (2), 37–50. Available at https://doi.org/10.15415/jmrh.2023.92004.

Beckley, A.M., Wright, E.S., 2021. Identification of antibiotic pairs that evade concurrent resistance via a retrospective analysis of antimicrobial susceptibility test results. The Lancet Microbe 2 (10), e545. Available from. Available at https://doi.org/10.1016/S2666-5247(21)00118-X, https://www.thelancet.com/lanmic/about.

Björkman, I., Röing, M., Sternberg Lewerin, S., Stålsby Lundborg, C., Eriksen, J., 2021. Animal production with restrictive use of antibiotics to contain antimicrobial resistance in Sweden—a qualitative study. Frontiers in Veterinary Science 7. Available at https://doi.org/10.3389/fvets.2020.619030.

Bonner, M., Barrera Patiño, C.P., Borsatto, A.R., Soares, J.M., Blanco, K.C., Bagnato, V.S., 2025. FTIR-derived feature insights for predicting time-dependent antibiotic resistance progression. Antibiotics 14 (8) Available from. Available at https://doi.org/10.3390/antibiotics14080831, http://www.mdpi.com/journal/antibiotics.

Bosso, A., Gaglione, R., Di Girolamo, R., Veldhuizen, E.J.A., García-Vello, P., Fusco, S., et al., 2022. Human cryptic host defence peptide GVF27 exhibits anti-infective properties against biofilm forming members of the *Burkholderia cepacia* complex. Pharmaceuticals 15 (2), 260. Available at https://doi.org/10.3390/ph15020260.

Boyce, M.R., Sorrell, E.M., Standley, C.J., 2023. An early analysis of the World Bank's Pandemic Fund: a new fund for pandemic prevention, preparedness and response. BMJ Global Health 8 (1) Available from. Available at https://doi.org/10.1136/bmjgh-2022-011172, gh.bmj.com/.

Burki, T., 2025. UK government dismantles the fleming fund. The Lancet Microbe 6 (11), 101239. Available at https://doi.org/10.1016/j.lanmic.2025.101239.

Canibe, N., Højberg, O., Kongsted, H., Vodolazska, D., Lauridsen, C., Nielsen, T.S., et al., 2022. Review on preventive measures to reduce post-weaning diarrhoea in piglets. Animals 12 (19) Available from. Available at https://doi.org/10.3390/ani12192585, http://www.mdpi.com/journal/animals.

Cassini, A., Högberg, L.D., Plachouras, D., Quattrocchi, A., Hoxha, A., Simonsen, G.S., et al., 2019. Attributable deaths and disability-adjusted life-years caused by infections with antibiotic-resistant bacteria in the EU and the European Economic Area in 2015: a population-level modelling analysis. The Lancet Infectious Diseases 19 (1), 56–66. Available from. Available at https://doi.org/10.1016/S1473-3099(18)30605-4, http://www.journals.elsevier.com/the-lancet-infectious-diseases.

Chalkidou, A., Lambert, M., Cordoba, G., Taxis, K., Hansen, M.P., Bjerrum, L., 2023. Misconceptions and knowledge gaps on antibiotic use and resistance in four healthcare settings and five European countries—a modified Delphi study. Antibiotics 12 (9), 1435. Available at https://doi.org/10.3390/antibiotics12091435.

Chen, C., Li, S.L., So, A.D., Xu, Y.Y., Guo, Z.F., Wang, X., et al., 2025a. Using large language models to assist antimicrobial resistance policy development: integrating the environment into health protection planning. Environmental Science and Technology 59 (2), 1243–1252. Available from. Available at https://doi.org/10.1021/acs est.4c07842, http://pubs.acs.org/journal/esthag.

Chen, Z., Tian, Y., Liu, J., Ran, J., Sun, S., Zhao, S., et al., 2025b. COVID-19 and sepsis-related excess mortality in the United States during the first 3 years: a national-wide time-series study. American Journal of Epidemiology 194 (8), 2219–2229. Available from. Available at https://doi.org/10.1093/aje/kwae411, http://aje.oxfordjournals.org/.

Cheng, Y., Wang, X., Ma, Y., Chen, T., Xiong, L., Chen, P., et al., 2025c. A multi-sectoral approach to antimicrobial resistance containment: insights from one health. One Health Bulletin 5 (3), 105–112. Available at https://doi.org/10.4103/ohbl.ohbl_68_24.

Cocker, D., Birgand, G., Zhu, N., Rodriguez-Manzano, J., Ahmad, R., Jambo, K., et al., 2024. Healthcare as a driver, reservoir and amplifier of antimicrobial resistance: opportunities for interventions. Nature Reviews. Microbiology 22 (10), 636–649. Available from. Available at https://doi.org/10.1038/s41579-024-01076-4, https://www.nature.com/nrmicro/.

Conway, J., Delanois, R.E., Mont, M.A., Stavrakis, A., McPherson, E., Stolarski, E., et al., 2024. Phase 1 study of the pharmacokinetics and clinical proof-of-concept activity of a biofilm-disrupting human monoclonal antibody in patients with chronic prosthetic joint infection of the knee or hip. Antimicrobial Agents and Chemotherapy 68 (8) Available from. Available at https://doi.org/10.1128/aac.00655-24, https://journals.asm.org/doi/10.1128/aac.00655-24.

Cousins, M., Parmley, E.J., Greer, A.L., Neiterman, E., Lambraki, I.A., Graells, T., et al., 2023. Is scientific evidence enough? Using expert opinion to fill gaps in data in antimicrobial resistance research. PLoS One 18 (8) Available from. Available at https://doi.org/10.1371/journal.pone.0290464, https://journals.plos.org/plosone/article/file?id=10.1371/journal.pone.0290464&type=printable.

Danasekaran, R., 2024. One health: a holistic approach to tackling global health issues. Indian Journal of Community Medicine 49 (2), 260–263. Available at https://doi.org/10.4103/ijcm.ijcm_521_23.

Davis, A., Lembo, T., Laurie, E., Mutua, E., Loosli, K., Nthambi, M., et al., 2022. How public health crises expose systemic, day-to-day health inequalities in low- and-middle income countries: an example from East Africa. Antimicrobial Resistance and Infection Control 11 (1) Available from. Available at https://doi.org/10.1186/s13756-022-01071-5, http://www.aricjournal.com/.

Delhomme, V., van Os, C., 2025. Building the European Health Union (2019–2024): successes, limits and future perspectives. European Journal of Risk Regulation 16 (3), 942–960. Available at https://doi.org/10.1017/err.2025.10021.

Díaz, M.A., Alberto, M.R., Vega-Hissi, E.G., González, S.N., Arena, M.E., 2022. Interference in *Staphylococcus aureus* biofilm and virulence factors production by human probiotic bacteria with antimutagenic activity. Arabian Journal for Science and Engineering 47 (1), 241–253. Available from. Available at https://doi.org/10.1007/s13369-021-05934-8, https://link.springer.com/journal/13369.

Do, P.C., Assefa, Y.A., Batikawai, S.M., Reid, S.A., 2023. Strengthening antimicrobial resistance surveillance systems: a scoping review. BMC Infectious Diseases 23 (1) Available from. Available at https://doi.org/10.1186/s12879-023-08585-2, https://bmcinfectdis.biomedcentral.com/.

Doron, A., Broom, A., 2026. A World of Resistance: India and the Global Antibiotic Crisis. Harvard University Press. Available at 10.2307/jj.31913392.

Erdem Büyükkiraz, M., Kesmen, Z., 2022. Antimicrobial peptides (AMPs): a promising class of antimicrobial compounds. Journal of Applied Microbiology 132 (3), 1573–1596. Available from. Available at https://doi.org/10.1111/jam.15314, https://academic.oup.com/jambio.

Faijue, D.D., Segui, A.O., Shringarpure, K., Razavi, A., Hasan, N., Dar, O., et al., 2024. Constructing a one health governance architecture: a systematic review and analysis of governance mechanisms for one health. European Journal of Public Health 34 (6), 1086–1094. Available from. Available at https://doi.org/10.1093/eurpub/ckae124, http://eurpub.oxfordjournals.org/.

Farinola, L., Ayodeji, I.T., 2025. Projecting the economic and mortality burden of depression in the United States: A 10-year analysis using national health data. International Journal of Population Data Science 10 (1). Available at https://doi.org/10.23889/ijpds.v10i1.3046.

Fletcher, S., 2015. Understanding the contribution of environmental factors in the spread of antimicrobial resistance. Environmental Health and Preventive Medicine 20 (4), 243–252. Available at https://doi.org/10.1007/s12199-015-0468-0.

Gao, Y., Zhao, L., Zhang, D., Zhao, D., Li, Q., Jiang, H., et al., 2026. Surfactin–bacillaene copathway engineering strategy boosts fengycin production and antifungal activity in *Bacillus velezensis* HN-Q-8. Microorganisms 14 (1) Available from. Available at https://doi.org/10.3390/microorganisms14010246, www.mdpi.com/journal/microorganisms.

Gargate, N., Laws, M., Rahman, K.M., 2025. Current economic and regulatory challenges in developing antibiotics for Gram-negative bacteria. npj Antimicrobials and Resistance 3 (1). Available at https://doi.org/10.1038/s44259-025-00123-1.

Gashaw, T., Yadeta, T.A., Weldegebreal, F., Demissie, L., Jambo, A., Assefa, N., 2025. The global prevalence of antibiotic self-medication among the adult population: systematic review and meta-analysis. Systematic Reviews 14 (1) Available from. Available at https://doi.org/10.1186/s13643-025-02783-6, https://systematicreviewsjournal.biomedcentral.com/.

Geletu, U.S., Usmael, M.A., Ibrahim, A.M., 2022. Isolation, identification, and susceptibility profile of *E. coli, Salmonella,* and *S. aureus* in dairy farm and their public health implication in Central Ethiopia. Veterinary Medicine International 2022 Available from. Available at https://doi.org/10.1155/2022/1887977, http://www.hindawi.com/journals/vmi/.

Gupta, R., & Sharma, S. (2022). Role of alternatives to antibiotics in mitigating the antimicrobial resistance crisis. Indian Journal of Medical Research, 156(3), 464-477.

Giordano, V., Giannoudis, P.V., 2024. Biofilm formation, antibiotic resistance, and infection (BARI): the triangle of death. Journal of Clinical Medicine 13 (19), 5779. Available at https://doi.org/10.3390/jcm13195779.

Golden, M.M., Heppe, A.C., Zaremba, C.L., Wuest, W.M., 2024. Metal chelation as an antibacterial strategy for *Pseudomonas aeruginosa* and *Acinetobacter baumannii*. RSC Chemical Biology 5 (11), 1083–1096. Available from. Available at https://doi.org/10.1039/d4cb00175c, www.rsc.org/journals-books-databases/about-journals/rsc-chemical-biology.

Griffith, L.M., Cowan, M.J., Notarangelo, L.D., Kohn, D.B., Puck, J.M., Shearer, W.T., et al., 2016. Primary immune deficiency treatment consortium (PIDTC) update. Journal of Allergy and Clinical Immunology 138 (2), 375–385. Available from. Available at https://doi.org/10.1016/j.jaci.2016.01.051, http://www.elsevier.com/inca/publications/store/6/2/3/3/6/8/index.htt.

Gwenzi, W., Chaukura, N., Muisa-Zikali, N., Teta, C., Musvuugwa, T., Rzymski, P., et al., 2021. Insects, rodents, and pets as reservoirs, vectors, and sentinels of antimicrobial resistance. Antibiotics 10 (1), 1–42. Available from. Available at https://doi.org/10.3390/antibiotics10010068, https://www.mdpi.com/2079-6382/10/1/68/pdf.

Haghshenas, B., Kiani, A., Mansoori, S., Mohammadi-noori, E., Nami, Y., 2023. Probiotic properties and antimicrobial evaluation of silymarin-enriched *Lactobacillus bacteria* isolated from traditional curd. Scientific Reports 13 (1) Available from. Available at https://doi.org/10.1038/s41598-023-37350-3, https://www.nature.com/srep/.

Hibstu, Z., Belew, H., Akelew, Y., Mengist, H.M., 2022. Phage therapy: a different approach to fight bacterial infections. Biologics: Targets and Therapy 16, 173–186. Available from. Available at https://doi.org/10.2147/BTT.S381237, https://www.dovepress.com/getfile.php?fileID=84479.

Hinterleitner, M., Knill, C., Steinebach, Y., 2024. The growth of policies, rules, and regulations: a review of the literature and research agenda. Regulation & Governance 18 (2), 637–654. Available at https://doi.org/10.1111/rego.12511.

Holmes, A.H., Moore, L.S.P., Sundsfjord, A., Steinbakk, M., Regmi, S., Karkey, A., et al., 2016. Understanding the mechanisms and drivers of antimicrobial resistance. The Lancet 387 (10014), 176–187. Available from. Available at https://doi.org/10.1016/S0140-6736(15)00473-0, http://www.journals.elsevier.com/the-lancet/.

Abdi Hussein, H., Abdi, S.M., Ahad, A.A., 2023. Factors and challenges contributing to antimicrobial resistance in East African pastoral settings and importance of one health approach. CABI One Health. Available at https://doi.org/10.1079/cabionehealth.2023.0025.

Irfan, M., Almotiri, A., AlZeyadi, Z.A., 2022. Antimicrobial resistance and its drivers—a review. Antibiotics 11 (10) Available from. Available at https://doi.org/10.3390/antibiotics11101362, http://www.mdpi.com/journal/antibiotics.

Islam, M.A., Bose, P., Rahman, M.Z., Muktaruzzaman, M., Sultana, P., Ahamed, T., et al., 2024. A review of antimicrobial usage practice in livestock and poultry production and its consequences on human and animal health. Journal of Advanced Veterinary and Animal Research 11 (3), 675–685. Available from. Available at https://doi.org/10.5455/javar.2024.k817, bdvets.org/javar/.

Jeong, Y.-I., Lee, H.-Y., Lee, S., Jeong, G.Y., Kim, S.H., Kim, S., et al., 2025. Korea's national action plan on antimicrobial resistance: focusing on the appropriate use of antibiotics. Infection & Chemotherapy 57 (2), 203. Available at https://doi.org/10.3947/ic.2025.0028.

Jiang, L., Ma, Y., Xiong, Y., Tan, Y., Duan, X. Liao, X., et al., 2022. Ruthenium polypyridine complexes with triphenylamine groups as antibacterial agents against *Staphylococcus aureus* with membrane-disruptive mechanism. Frontiers in Chemistry 10 Available from. Available at https://doi.org/10.3389/fchem.2022.1035741, http://journal.frontiersin.org/journal/chemistry.

Joshi, M.P., Hafner, T., Twesigye, G., Ndiaye, A., Kiggundu, R., Mekonnen, N., et al., 2021. Strengthening multisectoral coordination on antimicrobial resistance: a landscape analysis of efforts in 11 countries. Journal of Pharmaceutical Policy and Practice 14 (1) Available from. Available at https://doi.org/10.1186/s40545-021-00309-8, http://www.joppp.org/content.

Kallen, M.C., Binda, F., ten Oever, J., Tebano, G., Pulcini, C., Murri, R., et al., 2019. Comparison of antimicrobial stewardship programmes in acute-care hospitals in four European countries: a cross-sectional survey. International Journal of Antimicrobial Agents 54 (3), 338–345. Available at https://doi.org/10.1016/j.ijantimicag.2019.06.005.

Kasimanickam, V., Kasimanickam, M., Kasimanickam, R., 2021. Antibiotics use in food animal production: escalation of antimicrobial resistance: where are we now in combating AMR? Medical Sciences 9 (1), 14. Available at https://doi.org/10.3390/medsci9010014.

Keck, N., Treilles, M., Gordoncillo, M., Ivette, O.L.I., Dauphin, G., Dorado-Garcia, A., et al., 2023. A systematic approach toward progressive improvement of national antimicrobial resistance surveillance systems in food and agriculture sectors. Frontiers in Veterinary Science 9 Available from. Available at https://doi.org/10.3389/fvets.2022.1057040, journal.frontiersin.org/journal/veterinary-science.

Khan, M.F., Khattak, A.A., Saleem, A., Rizwan, M., Asif, M., Alvi, I.A., 2023. Efficient reduction of *Pseudomonas aeruginosa* biofilms using the myoviridae lytic bacteriophage vBPaeM MLG. Lahore Garrison University Journal of Life Sciences 7 (03), 262–279. Available at https://doi.org/10.54692/lgujls.2023.0703275.

Khomsi, K., Bouzghiba, H., Mendyl, A., Al-Delaimy, A.K., Dahri, A., Saad-Hussein, A., et al., 2024. Bridging research-policy gaps: an integrated approach. Environmental Epidemiology 8 (1), e281. Available at https://doi.org/10.1097/ee9.0000000000000281.

Kirchhelle, C., Atkinson, P., Broom, A., Chuengsatiansup, K., Ferreira, J.P., Fortané, N., et al., 2020. Setting the standard: multidisciplinary hallmarks for structural, equitable and tracked antibiotic policy. BMJ Global Health 5 (9) Available from. Available at https://doi.org/10.1136/bmjgh-2020-003091, gh.bmj.com/.

Kolawole, T.O., Mustapha, A.Y., Mbata, A.O., Tomoh, B.O., Forkuo, A.Y., Kelvin-Agwu, M.T.C., 2023. Innovative strategies for reducing antimicrobial resistance: a review of global policy and practice. Journal of Frontiers in Multidisciplinary Research 4 (1), 25–38. Available at https://doi.org/10.54660/.ijfmr.2023.4.1.25-38.

Krawczyk, B., Wysocka, M., Michalik, M., Gołębiewska, J., 2022. Urinary tract infections caused by *K. pneumoniae* in kidney transplant recipients – epidemiology, virulence and antibiotic resistance. Frontiers in Cellular and Infection Microbiology 12 Available from. Available at https://doi.org/10.3389/fcimb.2022.861374, http://www.frontiersin.org/Cellular_and_Infection_Microbiology/archive.

Kumar, N.R., Balraj, T.A., Kempegowda, S.N., Prashant, A., 2024. Multidrug-resistant sepsis: a critical healthcare challenge. Antibiotics 13 (1) Available from. Available at https://doi.org/10.3390/antibiotics13010046, http://www.mdpi.com/journal/antibiotics.

Lakhdar, M., Abir, D., Fatna, M., 2023. Plant growth-promoting endophytic bacteria isolated from *Launaea arborescens* (characterization and molecular identification). Journal of Agriculture and Applied Biology 4 (1), 71–82. Available at https://doi.org/10.11594/jaab.04.01.08.

Liu, W., Wu, Z., Mao, C., Guo, G., Zeng, Z., Fei, Y., et al., 2020. Antimicrobial peptide Cec4 eradicates the bacteria of clinical carbapenem-resistant *Acinetobacter baumannii* biofilm. Frontiers in Microbiology 11 Available from. Available at https://doi.org/10.3389/fmicb.2020.01532, https://www.frontiersin.org/journals/microbiology#.

Luepke, K.H., Mohr, J.F., 2017. The antibiotic pipeline: reviving research and development and speeding drugs to market. Expert Review of Anti-Infective Therapy 15 (5), 425–433. Available at https://doi.org/10.1080/14787210.2017.1308251.

Malbon, E., Parkhurst, J., 2023. System dynamics modelling and the use of evidence to inform policymaking. Policy Studies 44 (4), 454–472. Available from. Available at https://doi.org/10.1080/01442872.2022.2080814, www.tandf.co.uk/journals/titles/01442872.asp.

Matchawong, A., Srisawat, C., Sangboonruang, S., Tharinjaroen, C.S., 2022. The ability of nuclease-resistant RNA aptamer against *Streptococcus suis* serotype 2, strain P1/7 to reduce biofilm formation in vitro. Molecules (Basel, Switzerland) 27 (12). Available at https://doi.org/10.3390/molecules27123894.

Mazhar, T., Sheikh, T.R., Khan, A.U., Rahman, A.U., Rehman, M.U., Shaheen, R., et al., 2025. Global trends in antimicrobial resistance: a systematic review of surveillance data. Insights-Journal of Life and Social Sciences 3 (1), 153–160. Available at https://doi.org/10.71000/3164k763.

McKernan, C., Benson, T., Farrell, S., Dean, M., 2021. Antimicrobial use in agriculture: critical review of the factors influencing behaviour. JAC-Antimicrobial Resistance 3 (4). Available at https://doi.org/10.1093/jacamr/dlab178.

Meradji, S., Basher, N.S., Sassi, A., Ibrahim, N.A., Idres, T., Touati, A., 2025. The role of water as a reservoir for antibiotic-resistant bacteria. Antibiotics 14 (8) Available from. Available at https://doi.org/10.3390/antibiotics14080763, http://www.mdpi.com/journal/antibiotics.

Mestrovic, T., Robles Aguilar, G., Swetschinski, L.R., Ikuta, K.S., Gray, A.P., Davis Weaver, N., et al., 2022. The burden of bacterial antimicrobial resistance in the WHO European region in 2019: a cross-country systematic analysis. The Lancet Public Health 7 (11), e897. Available from. Available at https://doi.org/10.1016/S2468-2667(22)00225-0, http://thelancet.com/journals/lanpub/issue/vol1no2/PIIS2468-2667(16)X0003-5.

Mishra, A., Aggarwal, A., Khan, F., 2024. Medical device-associated infections caused by biofilm-forming microbial pathogens and controlling strategies. Antibiotics 13 (7), 623. Available at https://doi.org/10.3390/antibiotics13070623.

Mitchell, J., Cooke, P., Ahorlu, C., Arjyal, A., Baral, S., Carter, L., et al., 2022. Community engagement: the key to tackling antimicrobial resistance (AMR) across a one health context? Global Public Health 17 (11), 2647–2664. Available at https://doi.org/10.1080/17441692.2021.2003839.

Mitchell, J., Hawkings, H., Latham, S., Fieroze, F., Arjyal, A., Barrington, D.J., et al., 2023. Addressing antimicrobial resistance through community engagement: a framework for developing contextually relevant and impactful behaviour change interventions. JAC-Antimicrobial Resistance 5 (6) Available from. Available at https://doi.org/10.1093/jacamr/dlad124, https://academic.oup.com/jacamr.

Moran, D., 2019. A framework for improved one health governance and policy making for antimicrobial use. BMJ Global Health 4 (5), e001807. Available at https://doi.org/10.1136/bmjgh-2019-001807.

Mudenda, S., Daka, V., Matafwali, S.K., 2023. World health organization AWaRe framework for antibiotic stewardship: where are we now and where do we need to go? An expert viewpoint. Antimicrobial Stewardship and Healthcare Epidemiology 3 (1) Available from. Available at https://doi.org/10.1017/ash.2023.164, www.cambridge.org/core/journals/antimicrobial-stewardship-and-healthcare-epidemiology.

Muntean, M., Muntean, A.-A., Preda, M., Manolescu, L., Dragomirescu, C., Popa, M.-I., et al., 2022. Phenotypic and genotypic detection methods for antimicrobial resistance in ESKAPE pathogens (Review). Experimental and Therapeutic Medicine 24 (2). Available at https://doi.org/10.3892/etm.2022.11435.

Na, S.H., Jun, I., Kang, S., Yoo, J.S., Yoon, E.J., 2025. Laying the groundwork for an inter-ministerial one health response to antimicrobial resistance in South Korea. One Health 21 Available from. Available at https://doi.org/10.1016/j.onehlt.2025.101235, https://www.sciencedirect.com/science/journal/23527714.

Naghavi, M., Vollset, S.E., Ikuta, K.S., Swetschinski, L.R., Gray, A.P., Wool, E.E., et al., 2024. Global burden of bacterial antimicrobial resistance 1990–2021: a systematic analysis with forecasts to 2050. The Lancet 404 (10459), 1199–1226. Available from. Available at https://doi.org/10.1016/S0140-6736(24)01867-1, https://www.sciencedirect.com/science/journal/01406736.

Naheed, G., Sultan, T., Athar Hussain Barvi, L.A., Mahnoor, 2025. Emerging antimicrobial resistance in companion, farm animals and poultry: a veterinary concern. Journal of Medical & Health Sciences Review 2 (2). Available at https://doi.org/10.62019/fxjemz43.

Nath, N.D., Debnath, A., Siddique, M.A.B., 2024. 'Combatting antimicrobial resistance: public health policies, community awareness and global surveillance. Clinical Epidemiology & Public Health 2 (1), 1–8.

Ning, Y., Wang, X., Chen, P., Liu, S., Hu, J., Xiao, R., et al., 2022. Targeted inhibition of methicillin-resistant *Staphylococcus aureus* biofilm formation by a graphene oxide-loaded aptamer/berberine bifunctional complex. Drug Delivery 29 (1), 1675–1683. Available from. Available at https://doi.org/10.1080/10717544.2022.2079768, https://www.tandfonline.com/loi/idrd20.

Ogawa, V.A., Tran, T.A., Shah, C.M., 2018. Understanding the Economics of Microbial Threats. National Academies Press, Washington, DC Available from. Available at https://www.ncbi.nlm.nih.gov/books/NBK534893/.

Oltean, H.N., Lipton, B., Black, A., Snekvik, K., Haman, K., Buswell, M., et al., 2025. Developing a one health data integration framework focused on real-time pathogen surveillance and applied genomic epidemiology. One Health Outlook 7 (1). Available at https://doi.org/10.1186/s42522-024-00133-5.

Padiyara, P., Inoue, H., Sprenger, M., 2018. Global governance mechanisms to address antimicrobial resistance. Infectious Diseases: Research and Treatment 11. Available at https://doi.org/10.1177/1178633718767887.

Pasha, F., Krasniqi, V., Ismaili, A., Krasniqi, S., Bahtiri, E., Qorraj Bytyqi, H., et al., 2025. Antibiotic resistance awareness in Kosovo: insights from the WHO antibiotic resistance: multi-country public awareness survey. Antibiotics 14 (6), 599. Available at https://doi.org/10.3390/antibiotics14060599.

Piscaglia, M., Scaglione, G., Genovese, C., Borgonovo, F., Brivio, F., Rampichini, F., et al., 2025. Exploring human use of monoclonal antibodies against critical bacteria: a scoping review of clinical trials. Infectious Diseases and Therapy 14 (8), 1619–1647. Available from. Available at https://doi.org/10.1007/s40121-025-01195-2, https://www.springer.com/journal/40121.

Pokharel, S., Adhikari, B., Johnson, T., Cheah, P.Y., 2024. Interventions to address antimicrobial resistance: an ethical analysis of key tensions and how they apply in low-income and middle-income countries. BMJ Global Health 9 (4) Available from. Available at https://doi.org/10.1136/bmjgh-2023-012874, gh.bmj.com/.

Popoola, O.O., Adepitan, D.S., Adeyemi, A.S., Oladeru, O.F., Yusuff, S.I., 2024. A national survey of the antibiotic use, self-medication practices, and knowledge of antibiotic resistance among graduates of tertiary institutions in Nigeria. Scientific African 23, e01978. Available at https://doi.org/10.1016/j.sciaf.2023.e01978.

Prasarnphanich, O.O., Berger, C., Abdikadir, M.I., Alfonso-Dilley, C.S., Belot, G., Boniol, M., et al., 2025. Developing a tripartite (food and agriculture organization of the United Nations; world health organization; and world organisation for animal health) tool to strengthen the workforce for effective management of zoonotic diseases. BMC Global and Public Health 3 (1) Available from. Available at https://doi.org/10.1186/s44263-025-00194-2, https://bmcglobalpublichealth.biomedcentral.com/.

Preena, P.G., Swaminathan, T.R., Kumar, V.J.R., Singh, I.S.B., 2020. Antimicrobial resistance in aquaculture: a crisis for concern. Biologia (Lahore, Pakistan) 75 (9), 1497–1517. Available from. Available at https://doi.org/10.2478/s11756-020-00456-4, http://www.springer.com/journal/11756.

Ranjalkar, J., Chandy, S.J., 2019. India's national action plan for antimicrobial resistance – an overview of the context, status, and way ahead. Journal of Family Medicine and Primary Care 8 (6), 1828. Available at https://doi.org/10.4103/jfmpc.jfmpc_275_19.

Reichmann, S., Wieser, B., 2022. Open science at the science–policy interface: bringing in the evidence? Health Research Policy and Systems 20 (1) Available from. Available at https://doi.org/10.1186/s12961-022-00867-6, http://www.health-policy-systems.com/home/.

Rony, M.K.K., Sharmi, P.D., Alamgir, H.M., 2023. Addressing antimicrobial resistance in low and middle-income countries: overcoming challenges and implementing effective strategies. Environmental Science and Pollution Research 30 (45), 101896–101902. Available from. Available at https://doi.org/10.1007/s11356-023-29434-4, https://www.springer.com/journal/11356.

Ruckert, A., Fafard, P., Hindmarch, S., Morris, A., Packer, C., Patrick, D., ... & Labonté, R. (2020). Governing antimicrobial resistance: a narrative review of global governance mechanisms. Journal of public health policy, 41(4), 515.

Ruckert, A., Hoffman, S.J., Bishop, J., Van Katwyk, S.R., Fafard, P., Poirier, M.J.P., 2025. Intergovernmental or fully independent? Designing a scientific panel on evidence for action against antimicrobial resistance. PLOS Global Public Health 5 (1) Available from. Available at https://doi.org/10.1371/journal.pgph.0004039, https://journals.plos.org/globalpublichealth/article?id=10.1371/journal.pgph.0004039.

Sakagianni, A., Koufopoulou, C., Feretzakis, G., Kalles, D., Verykios, V.S., Myrianthefs, P., et al., 2023. Using machine learning to predict antimicrobial resistance—a literature review. Antibiotics 12 (3), 452. Available at https://doi.org/10.3390/antibiotics12030452.

Salam, M.A., Al-Amin, M.Y., Salam, M.T., Pawar, J.S., Akhter, N., Rabaan, A.A., et al., 2023. Antimicrobial resistance: a growing serious threat for global public health. Healthcare (Switzerland) 11 (13) Available from. Available at https://doi.org/10.3390/healthcare11131946, www.mdpi.com/journal/healthcare/.

Sanga, V.T., Hoza, A.S., Karimuribo, E.D., 2024a. Knowledge gap among one health key actors on multi-sectoral coordination at national and sub-national levels in Tanzania. Research Directions: One Health 2. Available at https://doi.org/10.1017/one.2024.12.

Sanga, V.T., Karimuribo, E.D., Hoza, A.S., 2024b. One health in practice: benefits and challenges of multisectoral coordination and collaboration in managing public health risks: a meta-analysis. International Journal of One Health 10 (1), 26–36. Available from. Available at https://doi.org/10.14202/IJOH.2024.26-36, www.onehealthjournal.org.

Sass, A.M., Coenye, T., 2023. The small RNA NcS25 regulates biological amine-transporting outer membrane porin BCAL3473 in *Burkholderia cenocepacia*. mSphere 8 (2) Available from. Available at https://doi.org/10.1128/msphere.00083-23, https://aac.asm.org/content/by/year.

Sati, H., Carrara, E., Savoldi, A., Hansen, P., Garlasco, J., Campagnaro, E., et al., 2025. The WHO bacterial priority pathogens List 2024: a prioritisation study to guide research, development, and public health strategies against antimicrobial resistance. The Lancet Infectious Diseases 25 (9), 1033–1043. Available from. Available at https://doi.org/10.1016/S1473-3099(25)00118-5, https://www.sciencedirect.com/science/journal/14733099.

Seixas, A.M.M., Sousa, S.A., Leitão, J.H., 2022. Antibody-based immunotherapies as a tool for tackling multidrug-resistant bacterial infections. Vaccines 10 (11) Available from. Available at https://doi.org/10.3390/vaccines10111789, http://www.mdpi.com/journal/vaccines.

Shaghayegh, G., Cooksley, C., Ramezanpour, M., Wormald, P.-J., Psaltis, A.J., Vreugde, S., 2022. Chronic rhinosinusitis, *S. aureus* biofilm and secreted products, inflammatory responses, and disease severity. Biomedicines 10 (6), 1362. Available at https://doi.org/10.3390/biomedicines10061362.

Shi, C., Li, M., Muhammad, I., Ma, X., Chang, Y., Li, R., et al., 2018. Combination of berberine and ciprofloxacin reduces multi-resistant *Salmonella* strain biofilm formation by depressing mRNA expressions of luxS, rpoE, and ompR. Journal of Veterinary Science 19 (6), 808–816. Available from. Available at https://doi.org/10.4142/jvs.2018.19.6.808, http://www.vetsci.org/journal/download_pdf.php?doi=10.4142/jvs.2018.19.6.808.

Singh, B., Bhat, A., Ravi, K., 2024. Antibiotics misuse and antimicrobial resistance development in agriculture: a global challenge. Environment & Health 2 (9), 618–622. Available at https://doi.org/10.1021/envhealth.4c00094.

Singh, N., Gulhane, R.D., Singh, A., Goel, M., Udelal, P.P., Sangwan, V., et al., 2025. Exploring the antimicrobial potential of lactobacilli against early-stage and mature biofilms of *Staphylococcus aureus* and *Pseudomonas aeruginosa*. Frontiers in Chemistry 13 Available from. Available at https://doi.org/10.3389/fchem.2025.1425666, http://journal.frontiersin.org/journal/chemistry.

Skoog Ståhlgren, G., Grape, M., Edlund, C., 2022. The swedish model for prioritising research on the use of antibiotics: aligning public funding with research gaps. Health Policy (Amsterdam, Netherlands) 126 (8), 725–730. Available from. Available at https://doi.org/10.1016/j.healthpol.2022.05.012, www.elsevier.com/locate/healthpol.

Skoura, N., Wang-Jairaj, J., Della Pasqua, O., Chandrasekaran, V., Billiard, J., Yeakey, A., et al., 2020. Effect of raxibacumab on immunogenicity of Anthrax vaccine adsorbed: a phase 4, open-label, parallel-group, randomised non-inferiority study. The Lancet Infectious Diseases 20 (8), 983–991. Available from. Available at https://doi.org/10.1016/S1473-3099(20)30069-4, http://www.journals.elsevier.com/the-lancet-infectious-diseases.

Ssennyonjo, A., Criel, B., Van Belle, S., Ssengooba, F., Titeca, K., 2022. What are the tools available for the job? Coordination instruments at Uganda's national government level and their implications for multisectoral action for health. Health Policy and Planning 37 (8), 1025–1041. Available from. Available at https://doi.org/10.1093/heapol/czac047, http://heapol.oxfordjournals.org/.

Steel, A., Gallego-Perez, D.F., Ijaz, N., Gall, A., Bangpan, M., dos Santos Boeira, L., et al., 2025. Integration of traditional, complementary, and integrative medicine in the institutionalization of evidence-informed decision-making: the world health organization meeting report. Journal of Integrative and Complementary Medicine 31 (4), 388–394. Available from. Available at https://doi.org/10.1089/jicm.2024.0837, https://home.liebertpub.com/publications/journal-of-integrative-and-complementary-medicine/26.

Suazo-Galdames, I.C., Saracostti, M., Chaple-Gil, A.M., 2025. Scientific evidence and public policy: a systematic review of barriers and enablers for evidence-informed decision-making. Frontiers in Communication 10 Available from. Available at https://doi.org/10.3389/fcomm.2025.1632305, www.frontiersin.org/journals/communication#.

Sulaeman, M.M., Suhartono, E., Wibowo, S.N., 2023. Knowledge management as a strategic pillar in organisational adaptation to the dynamics of business environment change. Journal of Contemporary Administration and Management (ADMAN) 1 (3), 208–214. Available at https://doi.org/10.61100/adman.v1i3.82.

Sulis, G., Sayood, S., Gandra, S., 2022. Antimicrobial resistance in low- and middle-income countries: current status and future directions. Expert Review of Anti-Infective Therapy 20 (2), 147–160. Available from. Available at https://doi.org/10.1080/14787210.2021.1951705, http://www.tandfonline.com/loi/ierz20.

Sun, L., Jia, M., Zhu, K., Hao, Z., Shen, J., Wang, S., 2025. The efficacy of cecropin against multidrug-resistant bacteria is linked to the destabilization of outer membrane structure LPS of gram-negative bacteria. Probiotics and Antimicrobial Proteins 17 (6), 4867–4877. Available from. Available at https://doi.org/10.1007/s12602-024-10424-y, https://www.springer.com/journal/12602.

Sworowska-Baranowska, A., 2022. Knowledge pluralisation in for-common-good science: cross-disciplinary, cross-institutional and cross-sectoral research in environmental conservation in Poland. Forum Scientiae Oeconomia 10 (2), 45–72. Available from. Available at https://doi.org/10.23762/FSO_VOL10_NO2_3, http://ojs.wsb.edu.pl/index.php/fso/article/download/526/310.

Talebi Bezmin Abadi, A., Rizvanov, A.A., Haertlé, T., Blatt, N.L., 2019. World health organization report: current crisis of antibiotic resistance. BioNanoScience 9 (4), 778–788. Available from. Available at https://doi.org/10.1007/s12668-019-00658-4, http://www.springerlink.com/content/2191-1630.

Tang, K.W.K., Millar, B.C., Moore, J.E., 2023. Antimicrobial resistance (AMR). British Journal of Biomedical Science 80 Available from. Available at https://doi.org/10.3389/bjbs.2023.11387, https://www.frontierspartnerships.org/journals/british-journal-of-biomedical-science.

Thandavaram, A., Channar, A., Purohit, A., Shrestha, B., Patel, D., Shah, H., et al., 2022. The efficacy of bezlotoxumab in the prevention of recurrent clostridium difficile: a systematic review. Cureus 14 (8). Available at https://doi.org/10.7759/cureus.27979.

Ugwu, C.N., Ugwu, O.P.C., Alum, E.U., Eze, V.H.U., Basajja, M., Ugwu, J.N., et al., 2025. Sustainable development goals (SDGs) and resilient healthcare systems: addressing medicine and public health challenges in conflict zones. Medicine (United States) 104 (7), e41535. Available from. Available at https://doi.org/10.1097/MD.0000000000041535, https://journals.lww.com/md-journal/pages/default.aspx.

Weldon, I., Liddell, K., Outterson, K., 2025. Policy options for antimicrobial resistance: exploring lessons from environmental governance. The Milbank Quarterly 103 (4), 1003–1046. Available at https://doi.org/10.1111/1468-0009.70050.

World Health Organization, 2021. Antimicrobial Resistance and the United Nations Sustainable Development Cooperation Framework: Guidance for United Nations country teams. World Health Organization, Geneva (2021), Available from. Available at https://www.who.int/publications/i/item/9789240036024.

World Health Organization. (2021b). Antimicrobial resistance multi-partner trust fund: forging tripartite collaboration for urgent global and country action against antimicrobial resistance (AMR): annual progress report 2020. Rome: Food and Agriculture Organization. (2021).

World Health Organization, 2022. Global antimicrobial resistance and use surveillance system (GLASS) report 2022. Geneva: World Health Organization.

Yang, X., Wang, Y., Li, L., Tang, D., Yan, Z., Li, M.Y., et al., 2025. Berberine and its nanoformulations and extracts: potential strategies and future perspectives against multi-drug resistant bacterial infections. Frontiers in Microbiology 16. Available at https://doi.org/10.3389/fmicb.2025.1643409.

Yao, L., Wei, B., Wang, Y., Xu, B., Yang, M., Chen, X., et al., 2023. A critical role of outer membrane vesicles in antibiotic resistance in carbapenem-resistant *Klebsiella pneumoniae*. Annals of Clinical Microbiology and Antimicrobials 22 (1) Available from. Available at https://doi.org/10.1186/s12941-023-00645-4, https://ann-clinmicrob.biomedcentral.com/.

Yurtseven, A., Buyanova, S., Agrawal, A.A., Bochkareva, O.O., Kalinina, O.V., 2023. Machine learning and phylogenetic analysis allow for predicting antibiotic resistance in *M. tuberculosis*. BMC Microbiology 23 (1). Available at https://doi.org/10.1186/s12866-023-03147-7.

Zhang, Kexin, 2023. Phage therapy in the battle towards antibiotic resistance. Proceedings of SPIE - The International Society for Optical Engineering 12789. Available at https://doi.org/10.1117/12.2691862.

Chapter 15

Navigating antibiotic resistance challenges in low- and middle-income countries

Divya Varshney, Samreen, and Iqbal Ahmad
Department of Agricultural Microbiology, Aligarh Muslim University, Aligarh, Uttar Pradesh, India

15.1 Introduction

15.1.1 Antimicrobial resistance as an evolutionary and ecological crisis

Antimicrobial resistance (AMR) is one of the biggest risks to public health in the 21st century, which seriously impairs our capacity to treat bacterial diseases that were once thought to be common medical issues. AMR is fundamentally an evolutionary process that is fuelled by both anthropogenic forces and natural selection (Naghavi et al., 2024). People with genetic changes that make them resistant to antibiotics survive and reproduce when bacteria are exposed to these drugs, leading to the growth of resistant bacteria that spread among people, communities, and the environment. Although this process is a predictable result of microbial evolution, human activities such as the widespread use and abuse of antibiotics, insufficient infection control procedures, inadequate sanitation infrastructure, and the globalised movement of people and goods have significantly accelerated and magnified it (Iskandar et al., 2020).

The divide between low- and middle-income nations (LMICs) and high-income countries (HICs) is contextual rather than essentially biological. Geographically, resistance mechanisms appear the same, but the conditions that facilitate rapid transmission are very different. Multidrug-resistant (MDR) and extensively drug-resistant (XDR) pathogens thrive in LMICs due to some structural vulnerabilities, including fragmented healthcare systems, limited diagnostic capacity, lack of pharmaceutical regulation, antibiotic contamination, and the prevalence of informal healthcare sectors (Nashwan et al., 2024). As depicted in Fig. 15.1, the "One Health" framework highlights this interconnectedness, as bacteria are not limited by artificial boundaries, including those between animals, healthcare settings, and environmental sources. In low- and middle-income countries (LMICs), the forces that promote resistance are present in all three areas at the same time, due to their frequent physical and functional overlap (Laxminarayan et al., 2016).

15.1.2 Global Burden: magnitude and inequitable distribution

The impact of AMR has been assessed at alarming levels. The 2024 Global Research on AMR (GRAM) report, which compiled extensive epidemiological data, indicates that bacterial AMR caused 1.27 million direct deaths in 2019 (Naghavi et al., 2024). The substantial underestimation stemming from conservative mortality assessments is underscored by the estimated 4.95 million deaths associated with AMR, where resistance was a contributing factor, beyond those directly attributed to it. Projections for 2050 are concerning: bacterial AMR is forecasted to result in approximately 39 million deaths over the next twenty-five years if serious intervention is not implemented (Naghavi et al., 2024). According to this trajectory, AMR would rank among the top causes of early death, surpassing the prevalence of HIV/AIDS in many areas and being comparable to the current burden of tuberculosis.

A world sharply divided by health inequality is revealed by the regional distribution of this burden. With an estimated 249,000 deaths per year, Western sub-Saharan Africa has the highest age-standardised death rate owing to AMR, at 27.3 per 100,000 people. South Asia comes in second at 22.4 per 100,000, or about 280,000 deaths per year. An estimated 100,000 deaths per year in South Asia are attributed to bacterial AMR in India alone. In sharp contrast, North America reports 8.0 fatalities per 100,000, and Australasia has 6.8 deaths per 100,000, which are around three to four times lower than the highest burden regions (Naghavi et al., 2024).

Antimicrobial Resistance in the 21st Century. DOI: https://doi.org/10.1016/B978-0-443-44748-8.00007-5

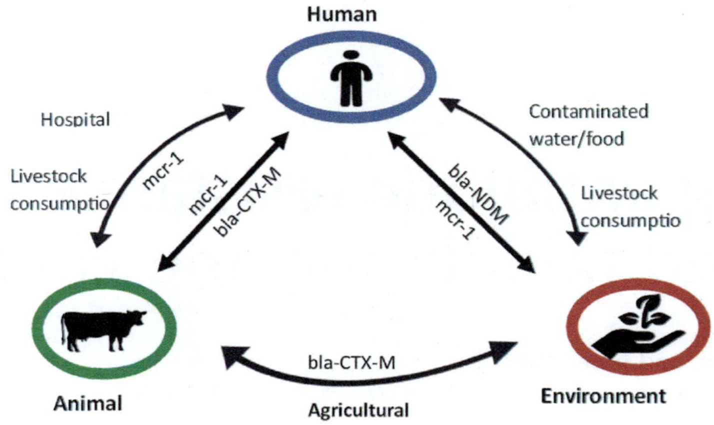

FIGURE 15.1 One health framework for antimicrobial resistance transmission. The One Health framework illustrates interconnected pathways of antimicrobial resistance transmission across human, animal, and environmental reservoirs, with dissemination of plasmid-mediated resistance genes (bla-$_{CTX-M}$, bla-NDM, mcr-1) through multiple routes.

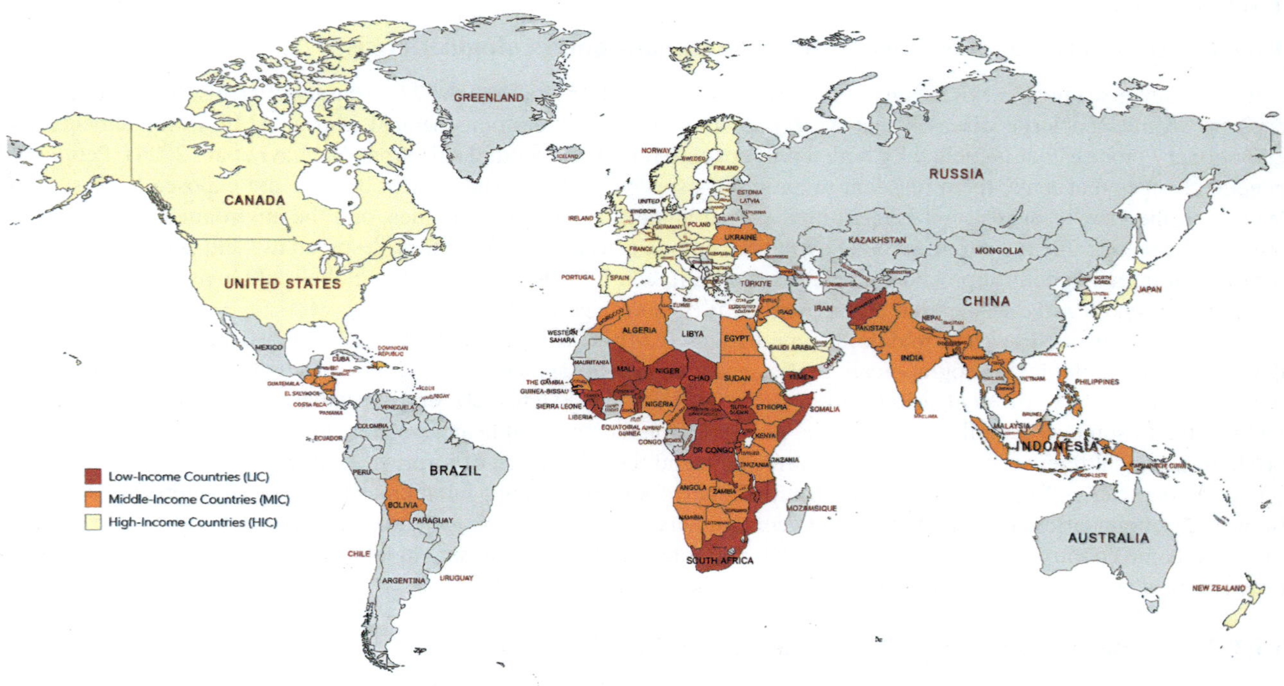

FIGURE 15.2 Income-level classification of countries worldwide. Global distribution of countries by income level. Map lines delineate study areas and do not necessarily depict accepted national boundaries.

Inadequate capacity for culture-based diagnostics and antimicrobial susceptibility testing (AST); lack of surveillance systems; prevalence of subpar and counterfeit medications; unregulated and over-the-counter antibiotic availability; contaminated water and sanitation infrastructure; and inadequate infection prevention and control (IPC) practices in healthcare facilities are just a few of the systemic factors unique to LMICs that contribute to this unequal distribution. The selection pressure for resistance becomes overpowering when both factors overlap, as they do in many LMICs.

Economic stratification is closely reflected in the global distribution of AMR vulnerability. Due to weak health systems, limited laboratory capacity, and significant gaps in water, sanitation, and hygiene (WASH) facilities, low-income countries (LICs), which are mostly found in sub-Saharan Africa and portions of South Asia, bear the brunt of the burden (Fig. 15.2). With large populations exposed to unregulated antibiotic markets, extensive agricultural antibiotic use, and environmental contamination, middle-income countries (MICs) – including populous nations like India, Pakistan,

Bangladesh, China, Brazil, Indonesia, and Nigeria – represent the epicentre of AMR dissemination. High-income nations are vulnerable to imported resistant microorganisms and diseases linked to healthcare, despite having more robust regulatory frameworks and surveillance systems (Jinks et al., 2016; Sriram et al., 2021a; Hafeez et al.,2023).

15.1.2.1 Objectives and scope

This chapter employs a comprehensive, systems-based methodology to gather the latest findings concerning antibiotic resistance within LMICs. Rather than considering AMR as an isolated clinical concern, and examine it from several interconnected perspectives:

Epidemiological characterisation: This section addresses the current prevalence and evolving patterns of resistance, encompassing both phenotypic and genotypic characteristics, in significant pathogenic species prevalent in LMIC areas.
Mechanisms and drivers: The molecular basis of resistance, encompassing β-lactamases, carbapenemases, and resistance to last-resort drugs, is examined, with particular emphasis on plasmid-mediated dissemination and horizontal gene transfer within resource-limited environments.
Environmental determinants: Environmental factors, including wastewater, agricultural practices, and aquaculture systems, represent significant, yet often overlooked, environmental reservoirs that contribute to the selection and amplification of resistance.
Socio-behavioural dimensions: Socio-behavioural aspects, such as prescription practices, patient behaviours, retail pharmacy dispensing trends, and the impact of traditional medicine, significantly influence antibiotic utilisation and the emergence of resistance.
Implementation sciences: The gap in reality between the creation of policies (National Action Plans) and their execution; methods that have proven workable in LMIC settings.
Emerging innovations: Phage therapy, bacteriocins, quick molecular diagnostics, and vaccine-based preventative techniques are examples of low-cost, context-appropriate remedies that show promise for settings with limited resources.

15.2 Epidemiology of antimicrobial resistance in low- and middle-income countries: current status and trends

15.2.1 Surveillance infrastructure and data limitations

Functional monitoring systems that can identify resistance mechanisms, monitor temporal and geographic trends, and produce useful data to guide clinical and policy decisions are necessary for a robust epidemiological characterisation of AMR. AMR surveillance in LMICs is still inadequate and fragmented. The most coordinated worldwide attempt to standardise AMR surveillance is the World Health Organization's Global Antimicrobial Resistance and Use Surveillance System (GLASS), which was created in 2015 (World Health Organization, 2024). As of 2024, however, GLASS data are still regionally biased, with significant representation from Europe and a few middle-income nations, little coverage from the poorest LMICs, limited coverage from sub-Saharan Africa, and fragmented coverage in South Asia.

There are significant infrastructure obstacles to efficient surveillance. Many LMICs lack functional diagnostic microbiology labs, and those that do frequently rely on phenotypic identification and basic culture without the ability to perform antibiotic susceptibility testing (AST) (Islam et al., 2021). Due to equipment costs (USD 300,000–500,000 for next-generation sequencing platforms), reagent costs (USD 50–150 per sample), and the need for specialised bioinformatic knowledge and reference databases, whole-genome sequencing (WGS), the gold standard for thoroughly characterising resistance mechanisms and tracking clonal dissemination, is still mainly unavailable in LMIC contexts (Wasswa et al., 2022). Because of this, surveillance data from LMICs primarily rely on phenotypic AST using broth microdilution or disc diffusion, which yields categorical classifications (susceptible, intermediate, and resistant) but hides the underlying molecular determinants and may overlook newly emerging resistance mechanisms (Islam et al., 2021).

Additionally, only the portion of bacterial illnesses that reach official healthcare institutions is captured by monitoring in LMICs, which is primarily focused on hospitals. 60–80% of illnesses in many LMICs are treated in settings related to traditional medicine, the community, or the unofficial private sector (Cai et al., 2015). Preventing the measurement of these illnesses and the documentation of their resistance patterns are critical. This deliberate bias toward hospital surveillance inflates the apparent incidence of invasive infections, whereas community-acquired resistance that develops in mostly untreated populations is underestimated.

15.2.2 Prevalence and resistance rates in major pathogens

15.2.2.1 Gram-negative organisms with the extended-spectrum β-lactamase phenotype

One of the most alarming and pervasive resistance phenomena in LMICs is the predominance of Enterobacteriaceae that produce ESBLs. Third-generation cephalosporins, which are frequently used empirically for serious infections in hospital settings with limited resources and little diagnostic capability, are resistant to ESBL enzymes, with 33% having a 95% Confidence Interval (CI) of 28%–38% of clinical *Escherichia coli* isolates producing ESBL, while 17% (95% CI: 13%–21%) create metallo-β-lactamases (MBLs), according to a thorough 2021 metaanalysis of South Asian data integrating research from Bangladesh, Pakistan, and India (Islam et al., 2021). A total of 58% of the enzymes found in ESBL-producing isolates belong to the *bla* $_{CTX\text{-}M}$ gene family, with bla-$_{CTX\text{-}M}$ being the predominant variation in 51% of instances. This enzyme spreads easily among Enterobacteriaceae and increasingly among other Gram-negative species. It is usually found on highly mobilisable IncF plasmids (Rozwandowicz et al., 2018; Islam et al., 2021).

Moreover, AmpC β-lactamases represent an important but often underrecognised resistance mechanism in Gram-negative bacteria. These enzymes may be chromosomally encoded or plasmid-mediated. These enzymes make bacteria resistant to penicillins and most cephalosporins, and they are not affected by standard β-lactamase inhibitors. Plasmid-mediated AmpC genes are common in Enterobacterales and can be spread through integrons. The expression of AmpC can be increased or turned on when exposed to β-lactams. This can lead to treatment failures and the development of additional resistance mechanisms, such as ESBLs and carbapenemases (Pospišil et al., 2023). A metaanalysis encompassing 25 African studies, which collectively analysed more than 16,000 isolates, revealed a pooled prevalence of 18.8% for AmpC-producing *E. coli* and *K. pneumoniae*. Moreover, studies published after 2020 and those conducted in Egypt indicated a higher prevalence, suggesting an increasing incidence of AmpC-mediated resistance (Tekele et al., 2025).

Clinical implications are significant: patients infected with ESBL-producing organisms have higher mortality rates, longer hospital stays, and escalation to carbapenem therapy, which is frequently unavailable in LMIC settings and costs 10–50 times more than third-generation cephalosporins (Tilahun et al., 2021). Third-generation cephalosporin-resistant Enterobacterales (3GCRE) colonised 30.2% (95% CI: 25.1–35.6%) of neonates in participating sites, according to a 2024 metaanalysis on neonatal colonisation in LMICs (Beaumont et al., 2024). This suggests that antibiotic-resistant bacteria establish themselves in the microbiota very early in life, with implications for future infection risk and treatment options.

15.2.2.2 Gram-negative organisms; carbapenem resistance and metallo-β-lactamases

A turning point in the AMR epidemic is the appearance and spread of carbapenem-resistant Gram-negative bacteria, which are mostly caused by metallo-β-lactamase (MBL) enzymes. In LMICs, carbapenems are one of the few remaining effective treatments for serious infections brought on by ESBL-producing organisms; their resistance loss eliminates effective therapeutic alternatives. In many LMIC regions, especially in South and Southeast Asia, the bla-$_{NDM\text{-}1}$ gene, which codes for New Delhi metallo-β-lactamase-1, has emerged as the most common MBL (Naghavi et al., 2024).

The epidemiology of $bla_{NDM\text{-}1}$ dissemination demonstrates the ability of horizontal gene transfer to spread resistance worldwide. The gene has spread to more than 65 nations on six continents since its discovery in New Delhi, India, in 2008 (Naghavi et al., 2024). This gene is typically found on self-transmissible plasmids, such as IncL/M and IncC/A replicons, which are easily transferred between species and even between bacterial taxa that are not closely related (Karampatakis et al., 2023). NDM-1 incidence among clinical isolates of Enterobacteriaceae in South Asia is estimated to be between 3 and 8%. Higher rates in wastewater (up to 15–20%) suggest substantial environmental selection and circulation (Larsson et al., 2018).

National or regional prevalence of carbapenem-resistant *E. coli* is 7% (95% CI 5–9%). NDM is the most common carbapenemase gene, detected in approximately 40%–50% of molecular events. Carbapenem-resistant *E. coli* occurs at frequencies between 12% and 15% in some regions of South Asia and between 2% and 5% in most of sub-Saharan Africa, although the latter region has a rapidly rising prevalence (Naghavi et al., 2024).

Carbapenem resistance is even more common in *Klebsiella pneumoniae*. Vectors of $bla_{NDM\text{-}1}$, $bla_{OXA\text{-}48}$, and other carbapenemase genes have spread throughout the world thanks to high-risk clones, including ST11, ST147, and ST258 (Karampatakis et al., 2023). The frequency of carbapenem-resistant *K. pneumoniae* in clinical isolates can reach 20–30% in some parts of South Asia and the Middle East (Naghavi et al., 2024).

15.2.2.3 Gram-positive organisms: methicillin-resistant Staphylococcus aureus

The *mecA* gene, which encodes a penicillin-binding protein with increased affinity for β-lactams, mediates methicillin resistance in *Staphylococcus aureus*, which is common in both clinical and community settings in LMICs (Idrees et al.,

2023). MRSA prevalence exhibits considerable variation across LMIC regions, a disparity due to variations in healthcare system organisation, infection control protocols, and antibiotic utilisation patterns. In the Middle East and North Africa (MENA) region, invasive MRSA rates among clinical isolates fluctuate between 9% and 67%, contingent upon the specific nation and healthcare setting (Naghavi et al., 2024). The Asia-Pacific region demonstrates detection proportions ranging from 26% to 73%, with South Asia generally reporting a higher MRSA prevalence (40–60%) compared to East Asia (25–35%). This observed divergence is likely a consequence of gaps in healthcare system maturity, infection control capabilities, and the implementation of antibiotic stewardship programmes (Lim et al., 2021). In LMICs, the prevalence of community-associated MRSA (CA-MRSA) is on the rise, particularly in regions characterised by high population density and poor sanitation (DeLeo et al., 2010). A 2024 metaanalysis examining neonatal MRSA colonisation within LMICs revealed a pooled prevalence of 2.7% (95% CI: 1.8%–3.8%) across the included studies (Beaumont et al., 2024). This finding implies that MRSA colonisation occurs early in life, which may increase the likelihood of subsequent skin and soft tissue infections.

15.3 Geographic variation and regional epidemiology

The impact of AMR is not evenly spread around the world. Instead, it is concentrated in specific areas that have unique social, economic, and infrastructure characteristics. A conceptual map of the global AMR burden shows that sub-Saharan Africa and South/Southeast Asia are the most concerning regions. There are also secondary areas of concern in the Middle East, North Africa, and Central Asia (Naghavi et al., 2024).

Western sub-Saharan Africa: At 27.3 deaths per 100,000 people per year, this area has the highest burden (Naghavi et al., 2024). Controlling enteric fever has become difficult due to high rates of fluoroquinolone resistance in *Salmonella typhi*; invasive nontyphoidal Salmonella (iNTS) with MDR phenotypes is a major source of bacteremia, especially in HIV coinfected persons (Al-Emran et al., 2016). Significant morbidity and death are caused by *S. aureus*, including both methicillin-sensitive and methicillin-resistant strains. High rates of ESBL synthesis and rising carbapenem resistance are seen in gram-negative bacteria, such as *K. pneumoniae* and *E. coli* (Naghavi et al., 2024).
South Asia (India, Bangladesh, Pakistan): Almost 280,000 deaths annually and a burden of 22.4 deaths per 100,000, this region is acknowledged as the epicentre of worldwide AMR. An estimated 100,000+ deaths worldwide per year are attributed to AMR in India alone (Naghavi et al., 2024). High levels of ESBL-producing Enterobacteriaceae with CTX-M-15 dominance, endemic circulation of NDM-1 and other MBLs, widespread colistin-resistant organisms, high rates of community-acquired resistance due to uncontrolled antibiotic availability, and severe WASH deficiencies that facilitate faecal-oral transmission of resistant organisms are the characteristics of this region (Okeke et al., 2007). High levels of antibiotics and antibiotic-resistant bacteria from sewage, untreated hospital effluent, and pharmaceutical production waste are found in the Ganges and other important rivers (Larsson et al., 2018).
Southeast Asia: Vietnam, Cambodia, and Thailand have become secondary AMR hotspots. Due to the extensive use of antibiotics in fish farming, aquaculture is a significant risk factor in this area (Reverter et al., 2020). The prevalence of CTX-M-15 ESBL is increasing, and new carbapenem resistance presents more difficulties. MDR organisms are frequently found in healthcare-associated infections (HAIs), which are caused by overcrowded hospitals with inadequate infection control capabilities (Reverter et al., 2020).
Middle East and North Africa (MENA): The load is variable but typically moderate to high, with MRSA frequency being especially noticeable. Macrolide-resistant respiratory infections, such as influenza, are prevalent. Disrupted healthcare services and displacement may increase resistance in conflict-affected communities (Naghavi et al., 2024).

15.4 Molecular mechanism of resistance in low- and middle-income country contexts

15.4.1 Genetic basics: β-lactamases, carbapenemases, and beyond

β-Lactamases constitute one of the most prevalent genetic mechanisms of AMR in Gram-negative bacteria, enabling the hydrolysis and inactivation of β-lactam antibiotics. These enzymes are encoded by a diverse group of genes located on either plasmids or chromosomes. The first plasmid-mediated β-Lactamase resistance was discovered in *E. coli* from a Greek patient, Temoneira, in 1963 and named TEM-1 (Datta and Kontomichalou, 1965). Over 2770 naturally occurring β-lactamases have been discovered. Ambler molecular classification and Bush–Jacoby–Medeiros functional classification are used to categorise β-lactamases, a diverse enzyme group. Ambler categorises β-lactamases into four groups (A, B, C, and D) based on primary sequence motifs. Classes A, C, and D use serine as the enzyme active centre, whereas Class B uses zinc. The Bush–Jacobi–Medeiros classification categorises β-lactamases into three groups based on substrate and inhibitor profiles, molecular masses, and isoelectric points (Bush and Jacoby, 2010). Bush and Bradford (2019) found that combining β-lactams with β-lactamase inhibitors effectively inhibits the hydrolytic activity of these enzymes. Some β-lactam-β-lactamase inhibitor combinations are ineffective against Gram-negative bacteria producing extended-spectrum β-lactamases (ESBLs).

ESBLs are a leading cause of antibiotic resistance in Enterobacteriaceae and antibiotic treatment failure globally. Numerous studies have examined the incidence of ESBL-producing enteric bacteria in clinical settings worldwide (Ibrahim, 2023; Bezabih et al., 2022). In recent decades, CTX-M (cefotaxime-hydrolysing β-lactamase)-type ESBLs have grown prominent in *E. coli, K. pneumoniae, Salmonella enterica* serovar Typhimurium, and *Shigella spp.* (Yu et al., 2025). These CTX-M enzymes are assumed to be derived from Kluyvera spp. chromosome-encoded genes that were transferred to plasmids by insertion sequences such as ISEcp1 and ISCR1 (Yu et al., 2025). The CTX-M family is divided into major groups based on amino acid sequence divergence (e.g. CTX-M-1, 2, 8, 9, 25, and KLUC), and more than 80 CTX-M variants have been discovered from various sources (Huang et al., 2020; Olaitan et al., 2025). Because earlier resistance mechanisms still predominate in resource-constrained environments where older antibiotics are still used, the genetic architecture of resistance in LMIC isolates differs somewhat from that in high-income nations (Rawat and Nair, 2010). Designing stewardship and diagnostic procedures that are appropriate for the situation requires an understanding of this molecular epidemiology.

15.4.2 Extended-spectrum β-lactamases

ESBLs are serine proteases that hydrolyse the β-lactam ring to permanently block β-lactam antibiotics. Approximately 50–70% of ESBL phenotypes in LMICs are dominated by the CTX-M family, particularly CTX-M-15 (Islam et al., 2021). Usually expressed on IncF plasmids carrying other resistance genes (to tetracyclines, aminoglycosides, and fluoroquinolones), CTX-M-15 produces a multidrug-resistant phenotype that affects available treatments (Gelalcha et al., 2024). Historically dominant in HICs, the SHV and TEM families are still common in older institutions in LMICs, but CTX-M variations are gradually replacing them (Rawat and Nair, 2010).

15.4.3 Metallo-β-lactamases and carbapenem resistance

MBLs are a unique type of β-lactamase that exhibits resistance to β-lactamase inhibitors and zinc-dependent catalytic activity. In LMICs, the main MBL kinds are as follows:

NDM (New Delhi Metallo-β-lactamases): Accounts for 35–50% of MBL detections in South Asia and increasingly in other LMICs (Naghavi et al., 2024), confers resistance to every β-lactam, with the exception of aztreonam, and is usually found on highly mobilizable plasmids (IncL/M, IncC), which enable quick horizontal spread (Tilahun et al., 2021).
VIM (Verona integron-encoded metallo-β-lactamases): Previously common in Europe, it is now more frequently found in LMICs (Rawat and Nair, 2010) and is frequently linked to *Pseudomonas aeruginosa* in medical environments.
IMP (imipenemase): Predominantly found in *Acinetobacter baumannii* in Asia-Pacific LMICs (Naghavi et al., 2024).

15.4.4 Serine carbapenemases (OXA-48, KPC)

Serine carbapenemases provide carbapenem resistance via a different mechanism in addition to zinc-dependent MBLs. The class D β-lactamase OXA-48 is becoming increasingly common in the Middle East, parts of Europe with Mediterranean drainage, and South Asia (Naghavi et al., 2024; Tilahun et al., 2021). *K. pneumoniae* carbapenemase, or KPC, is primarily found in the Western Hemisphere, with limited occurrences in Asia (Naghavi et al., 2024).

15.4.5 Colistin resistance mechanisms

Changes in lipopolysaccharide (LPS) structure, usually due to loss-of-function mutations in genes encoding two-component regulatory systems (pmrAB, phoPQ) or enzymes involved in LPS modification (lipid A phosphatase, pmrC), induce colistin resistance (Alba et al., 2018). These chromosomal mutations are more prevalent and have been found worldwide (Alba et al., 2018; Karampatakis, Tsergouli and Behzadi, 2023). In contrast, the plasmid-mediated *mcr-1* gene, which is mostly linked to agricultural use, produces a phosphoethanolamine transferase that alters lipid A (Laxminarayan et al., 2016). Although *mcr-1* is still the most common, several mcr variations (*mcr-2* through *mcr-10*) have now been discovered (Alba et al., 2018).

15.5 Horizontal gene transfer, plasmid mobility, and genetic elements

The natural environment plays a crucial role in the development and dissemination of antibiotic resistance in Gram-negative bacteria. It acts as a source of resistance genes and encourages the transfer of these genes across bacteria, especially when environmental stresses are present (Wellington et al., 2013). The incredibly effective mobilisation of resistance genes through horizontal gene transfer (HGT) is what distinguishes the dissemination of resistance in LMICs. In contrast to HICs, where single clones (such as ST131 *E. coli* and ST258 *K. pneumoniae*) may predominate, polyclonal spread occurs in LMIC environments when resistance genes are simultaneously acquired by numerous divergent genetic backgrounds through plasmids and mobile genetic elements (Partridge et al., 2009).

15.5.1 Plasmid-mediated transfer

Bacteria can conjugatively transfer plasmids, which are self-replicating extrachromosomal DNA elements. In LMICs, the density of bacteria in hospital effluents, wastewater treatment plant influents, and untreated sewage is extremely high, often surpassing 10^6 organisms per millilitre, creating ideal circumstances for conjugative transfer (Rizzo et al., 2013). Identical plasmids harbouring $bla\text{-}_{CTX\text{-}M}$, $bla\text{-}_{NDM}$, or *mcr-1* have been found in isolates derived simultaneously from humans, livestock, food, and wastewater within single geographic areas, according to empirical studies using molecular marker tracking (Le et al., 2016). This offers undisputed support for the occurrence of frequent and reciprocal gene transfer. Significant plasmid groups implicated in the spread of resistance determinants encompass:

Inc replicons: These elements harbour ESBL genes, fluoroquinolone resistance determinants (qnr), and aminoglycoside resistance genes (aac(3)-II). Their prevalence is notably high in South Asia, as reported by Gelalcha et al. (2024).
IncL/M replicons: Associate with NDM-1 and other MBLs. Possess high conjugative transfer rates and a broad host range (Karampatakis et al., 2023).
IncC/A replicons: Increasingly associated with NDM-1 and other carbapenemases in the Middle East and Asia (Karampatakis et al., 2023).
IncX plasmids: Characterised by a narrow host range but extremely efficient transfer among Enterobacteriaceae; frequently carry mcr-1 (Laxminarayan et al., 2016).

15.5.2 Integrons and insertion sequences

Integrons are genetic elements that can use site-specific recombination to capture and express resistance genes. In LMIC populations of Enterobacteriaceae and Acinetobacter, class 1 integrons (including the intI1 gene) are common and frequently carry cassettes expressing ESBL genes, aminoglycoside resistance, and trimethoprim resistance simultaneously. Resistance genes can be mobilised into plasmids and between chromosomal sites with the use of insertion sequences (IS elements). IS*Aba1* frequently carries genes for carbapenem resistance and is specifically linked to the Acinetobacter gene (Partridge et al., 2009).

15.6 Environmental reservoirs and amplification of resistance

The importance of environmental reservoirs in maintaining and enhancing resistance is a key difference between the epidemiology of AMR in LMICs and HICs. Antibiotic-resistant bacteria flourish in the environment, which also serves as an active selective environment where resistance genes are constantly mobilised and redistributed (Cytryn, 2013; Nesme and Simonet, 2015; Delgado-Baquerizo et al., 2022). It has been underlined that environmental compartments such as wastewater, soil, and surface waters serve not just as passive sinks but also as dynamic reservoirs, allowing AMR to persist, evolve, and spread across human–animal interfaces (Ahmad et al., 2019).

15.6.1 Wastewater and sewage

Antibiotic-resistant bacteria are found in remarkably high percentages in untreated or insufficiently treated wastewater in LMICs. Carbapenemase-producing Enterobacteriaceae have been found in sewage from major South Asian cities (Delhi, Dhaka, Karachi) at densities of 10^{-6}-10^{-8} CFU per millilitre (Larsson et al., 2018), while effluent from European cities had less than 1 CFU per millilitre (Gros et al., 2010; Aydin et al., 2018). In LMICs, hospital effluents often flow untreated directly into city sewers, serving as concentrated point sources of resistant microbes (Law et al., 2021; Wolters et al., 2022). Strong selection conditions are produced by these conditions: The high bacterial density promotes horizontal gene transfer, and the subinhibitory antibiotic doses select for resistant mutants (Larsson et al., 2018). In the Indian context,

Taneja and Sharma (2019) demonstrate that inadequate wastewater treatment and poor environmental regulations significantly contribute to the persistence of antibiotic resistance in the environment, a problem that affects many LMICs.

15.6.2 Pharmaceutical manufacturing waste

Massive amounts of antibiotics are released into the environment during production operations by bulk antibiotic producers in China, India, and other LMICs (Bengtsson-Palme et al., 2018). Ciprofloxacin concentrations of 30–40 μg/mL, above therapeutic levels, were found in surface water samples taken downstream of an active pharmaceutical ingredient (API) manufacturer in South Asia in one seminal investigation (Bengtsson-Palme et al., 2018; Lin et al., 2021). Over a vast geographic area, these ambient antibiotic concentrations favour resistant organisms, and resistant bacteria that start in these habitats may eventually spread to clinical populations (Fletcher, 2015; Bengtsson-Palme et al., 2018; Lin et al., 2021).

15.6.3 Agricultural runoff and livestock waste

In LMICs, antibiotics are frequently used in livestock to promote growth and prevent disease, especially in environments with low biosecurity (Van Boeckel et al., 2015). Antibiotic-resistant bacteria and antibiotics are both present in the faeces of treated animals; when this waste is used as manure in agricultural fields, resistance genes are introduced into the soil habitats (Van Boeckel et al., 2015; Mulchandani et al., 2023). These organisms are mobilised into surface water and groundwater by surface runoff during periods of heavy precipitation (Van Boeckel et al., 2015). Research conducted in South Asia has shown that bacteria isolated from livestock, agricultural soil, irrigation water, and vegetables irrigated with polluted water share the same resistance determinants ($bla_{\text{CTX-M}}$, *mcr-1*) (Sharan et al., 2023; Sarkar, 2023).

15.6.4 Biosolids and sludge

Bacteria and resistance genes are concentrated in wastewater treatment sludge (also known as "biosolids"). Massive quantities of resistant organisms are discharged into the soil when biosolids are applied to land, as is typical in LMICs, as a fertility amendment and disposal method (Nichols, Lake and Heaviside, 2018; Cissé, 2019; Lin et al., 2021).

15.7 Socio-behavioural and structural drivers

15.7.1 Unregulated antibiotic markers and over-the-counter availability

Antibiotics can be obtained without a prescription from retail pharmacies, private physicians, and unofficial medicine vendors in the majority of LMICs. Antibiotics were available without a prescription in most sampled retail shops in 19 out of 22 countries with sufficient data, according to a systematic analysis that synthesised data from 29 LMICs. OTC availability is more than 90% in various South Asian and sub-Saharan African nations (Sriram et al., 2021b; Hutchings et al., 2019). Numerous factors contribute to this unregulated access, including (1) lax enforcement of regulations; (2) retailers' financial incentives; (3) patient expectations and demand; (4) reliance on the unofficial private sector for primary healthcare; and (v) the perception that antibiotics are safe (Sriram et al., 2021b).

There are significant repercussions for resistance. Antibiotics are used by populations with underlying conditions that lead to poor outcomes (e.g. untreated TB patients receiving fluoroquinolones), in suboptimal doses, for suboptimal durations, and for situations that do not require them (e.g. viral respiratory infections) in the absence of prescriber oversight. Resistance is selected for by each of these patterns (Ibrahim, 2023).

15.7.2 Prescribing practices in healthcare settings

Due to diagnostic ambiguity, empirical broad-spectrum antibiotic usage is widespread in LMIC healthcare institutions. Clinicians cannot wait for culture findings before starting treatment in environments that do not have the capability for quick bacterial identification and susceptibility testing (Lalani et al., 2018). As a result, based solely on clinical symptoms, patients are started on broad-spectrum medications (carbapenems, fluoroquinolones, or combination regimens) (Lalani et al., 2018; Njoki et al., 2024). This strategy heavily selects for resistance, even if it might guarantee coverage of the causal organism (Debebe et al., 2022; Lalani et al., 2018). In most LMICs, prescribers receive little training on antimicrobial stewardship. Only 35% of the 89 LMICs evaluated in a systematic evaluation reported receiving any formal stewardship training in medical curricula. Many prescribers have had little exposure to local resistance epidemiology or evidence-based prescribing approaches due to high patient loads and inadequate diagnostic assistance (Kakkar et al., 2020; Abbas, 2024).

15.7.3 Patient behaviour and healthcare-seeking patterns

Antibiotic use trends in LMICs are significantly influenced by patient variables. Due to financial limitations, patients are forced to buy partial courses, take medicine sporadically when their symptoms get better, or put off getting treatment until their illnesses have progressed significantly. Without a prescriber evaluation, patients frequently buy antibiotics through networks of friends and family (Ocan et al., 2014; Niwandinda et al., 2020). The desire for needless parenteral antibiotics and the related injection safety hazards are driven in certain cultures by strong cultural ideas about the superiority of injections over oral drugs (Mishra, Kumar and Sharma, 1994; Memon et al., 2013). An estimated 60–80% of infections in rural LMIC areas are treated by traditional medicine practitioners and informal providers, who frequently lack training in evidence-based diagnosis and therapy (Cai et al., 2015). Leftover antibiotics from prior infections are frequently used for self-medication (Niwandinda et al., 2020).

15.8 Environmental and water challenges

15.8.1 Water, sanitation, and hygiene infrastructure

The significance of WASH in the emergence of AMR within LMICs cannot be overstated. Worldwide, 2.3 billion individuals lack access to adequate sanitation, and 2.1 billion are without safely managed drinking water; the vast majority of these people reside in LMICs (World Health Organization and UNICEF). When sanitation is inadequate, the faecal-oral transmission of microorganisms, especially those resistant to antibiotics, becomes widespread within populations. The presence of faecal markers, such as antibiotic-resistant *E. coli*, which are frequently found in water sources across sub-Saharan Africa and South Asia, serves as an indicator of significant faecal contamination (Swarthout et al., 2020; Grembi et al., 2023). Vegetables irrigated with wastewater or groundwater contaminated by sewage can transmit resistant pathogens (Grembi et al., 2023).

Unlike meat, vegetables are often consumed raw, thereby maintaining bacterial viability. This poses a significant risk in areas where diarrhoeal diseases, often associated with enteric bacteria, are endemic (Swarthout et al., 2020; Grembi et al., 2023). Furthermore, exposure to contaminated water is directly linked to the incidence of diarrhoeal illness among children in LMICs with inadequate water, sanitation, and hygiene (WASH) infrastructure; research indicates that faecal-oral transmission of antibiotic-resistant bacteria is a primary mechanism of spread in these environments (Porter et al., 2021; Llor et al., 2024). WASH interventions (including access to clean water, latrine construction, and handwashing promotion) decreased antibiotic use among children by 24%, primarily by preventing diarrhoeal disease, according to a landmark 2025 cluster-randomised controlled trial conducted in rural Bangladesh (Llor et al., 2024). This result highlights the critical role that WASH plays in preventing AMR (AMR); by lowering the incidence of infection, WASH immediately lowers the demand for drug selection (Porter et al., 2021; Llor et al., 2024).

15.8.2 Climate change and environmental factors

In LMIC settings, climate change exacerbates preexisting vulnerabilities. Bacterial growth and metabolic rates are accelerated by higher ambient temperatures, which may lead to an increase in mutation rates. Flooding introduces resistant organisms into food crops by flooding agricultural areas with sewage and wastewater (van Daalen et al., 2022, 2024). Droughts increase the concentrations of antibiotic-resistant bacteria and other contaminants in surface water (van Daalen et al., 2024). According to a 2024 analysis, climate-driven extreme weather events in LMICs were linked to a notable rise in waterborne illness and, secondarily, to an increase in the use of antibiotics and the evolution of resistance (Nashwan et al., 2024).

15.9 Policy, stewardship, and emerging solutions

15.9.1 National action plans: formulations and implementation gap

Acknowledging the worldwide concern, the WHO called on all member nations to create National Action Plans (NAPs) in its 2015 Global Action Plan on AMR. 172 nations reported having an AMR NAP in place as of 2024 (World Health Organization, 2024). But the implementation of policy has been woefully inadequate. Critical obstacles to NAP implementation in LMICs were found in a 2023 analysis. These included a lack of "ring-fenced" budgets specifically allocated to AMR activities, fragmented governance across several ministries with poor coordination, insufficient laboratory capacity for surveillance, limited stewardship expertise, and a lack of political will (Smith and Coast, 2002; Jinks et al., 2016).

A dedicated multisectoral AMR coordination body, specific budget lines for surveillance, stewardship, and infection control, regional cooperation and peer learning networks, and technical support from international organisations have all been necessary for the successful implementation of NAP in a few LMIC contexts (such as Bangladesh, Vietnam, and Morocco). Nevertheless, these all-encompassing methods continue to be the exception rather than the rule (Smith and Coast, 2002; Prestinaci, Pezzotti and Pantosti, 2015; Jinks et al., 2016).

15.9.2 Antimicrobial stewardship adapted to low- and middle-income contexts

Although it has been widely developed in HIC settings, antimicrobial stewardship – the prudent use of antibiotics to prevent needless use while assuring effective therapy – needs significant adaptation for resource-limited situations (Cox et al., 2017). Several important stewardship efforts are practical in LMICs, including

Diagnostic stewardship: Using rapid diagnostic tests, such as CRISPR-based technologies, in primary healthcare settings can help reduce the unnecessary use of antibiotics when they are initiated based solely on clinical symptoms rather than confirmed microbial diagnosis (Olatunji et al., 2024).
Antifungal rotation strategies: Antifungal rotation strategies can be effective. Rotating the use of different antifungal treatments among healthcare providers can help reduce the development of resistance in areas where multiple resistance patterns are present (Dellit, 2007).
Task-shifted stewardship: Reducing reliance on infectious disease specialists by teaching nurses and nonphysician health workers how to apply streamlined stewardship algorithms and get stewardship metrics (Karanika et al., 2016).
Paper-based decision support: Healthcare staff with limited clinical experience can use straightforward, printed clinical algorithms tailored to local resistance patterns to support evidence-based decision-making and rational antibiotic use (Karanika et al., 2016; Hwang and Kwon, 2021).

15.10 Emerging innovations with low- and middle-income country applicability

Given the declining effectiveness of current antibiotics, it is imperative to look at different strategies for the development of antibacterial agents (Ahmad et al., 2019). Several alternative approaches have been explored to combat AMR, particularly in settings where conventional antibiotics are no longer effective or easily accessible. These emerging strategies, ranging from biological interventions to low-cost preventive measures, offer realistic options for LMICs when adapted to local needs and constraints.

15.10.1 Phage therapy: from compassionate use to evidence-based

Before the development of broad-spectrum antibiotics, bacteriophage therapy, which uses lytic viruses to destroy harmful bacteria, was frequently used. However, it was mostly abandoned in affluent nations. However, phage therapy has seen a notable comeback in the setting of newly emerging pan-drug-resistant (PDR) germs against which no traditional antibiotics are effective (Wang and Yu, 2025). Interest has been reignited by recent developments in phage production, discovery, and clinical use. Over 200 peer-reviewed papers on phage therapy from the previous five years were found in a 2025 review; they included several case reports and small case series detailing the effective treatment of MDR and PDR infections (Hatem and Abdelaziz, 2024; Mafe and Büsselberg, 2025).

When conventional antibiotics were no longer effective, phage cocktails were used in India to treat patients with chronic wounds or respiratory tract infections brought on by bacteria like *Mycobacterium abscessus*, *Pseudomonas aeruginosa*, and *K. pneumoniae*. In 60%–70% of cases, clinical improvement was noted (Sulakvelidze et al., 2001; Hitchcock et al., 2023). Phage therapy has several benefits in LMIC settings, such as (1) incredibly low production costs (~USD 0.50–2.00 per treatment course compared to USD 50–500 for antibiotics); (2) pathogen specificity, which minimises disruption of normal microbiota; (3) minimal side effects (phages are obligatory lytic parasites for bacteria, not eukaryotic pathogens); (4) local production using locally circulating phages; and (5) lack of antibiotic cross-resistance (Hatem and Abdelaziz, 2024; Wang and Yu, 2025).

The main challenges include (1) a lack of data from clinical trials, (2) unclear regulations in most countries, and (3) the complex process of producing phages, which requires specialised knowledge (Wang and Yu, 2025).

15.10.2 Bacteriocins and probiotic approaches

Bacteria, especially lactic acid bacteria (LAB), create antimicrobial peptides called bacteriocins that inhibit closely related bacterial species. *Listeria monocytogenes*, *Salmonella*, *Campylobacter*, and *Staphylococcus aureus* were among the important pathogens against which a 2024 metaanalysis of bacteriocin applications revealed proven activity. In food preservation, bacteriocins have been effectively used to stop the growth of pathogens without compromising food safety or quality. Furthermore, several LAB-derived bacteriocins have been studied for potential therapeutic applications; there is growing evidence that these bacteriocins may help prevent foodborne illnesses and possibly treat some GI infections (Putri et al., 2024).

Bacteriocin-based strategies are quite popular in LMICs because they are safe, naturally occurring, and can be made locally from food-grade LAB strains using straightforward fermentation techniques. Several LAB strains isolated from traditional fermented foods in South Asia and Africa were found to have strong bacteriocins with action against regional MDR pathogens in a 2024 systematic study (Putri et al., 2024).

15.10.3 Rapid molecular diagnostics: CRISPR and isothermal amplification

Empirical broad-spectrum antibiotic use in LMICs is fundamentally hampered by the lack of quick diagnostic capability. Without the need for laboratory equipment, novel diagnostics based on CRISPR nuclease activity or isothermal amplification (loop-mediated isothermal amplification, or LAMP) can identify bacterial infections and resistance determinants in less than one to two hours. Targeted therapy based on quick pathogen and resistance detection was made possible by a 40% reduction in empirical antibiotic courses, according to a 2024 review of a CRISPR-Cas12-based diagnostic used in rural clinics in Uganda. The cost of these diagnostics has significantly decreased: LAMP-based diagnostics currently cost USD 1–3 per test, whereas CRISPR-based assays cost USD 5–10, and conventional molecular approaches cost USD 50–100 (Shanmugakani et al., 2020). Crucially, these tests can be used in distant LMIC clinics because they do not require electricity, refrigeration, or specific staff training. Simultaneously, metagenomic sequencing is progressively acknowledged as an effective diagnostic technique capable of simultaneously identifying uropathogens and AMR determinants without the need for prior cultivation. A recent systematic review and metaanalysis on metagenomics in urinary tract infections highlighted its potential to enhance pathogen identification and inform targeted treatment, especially in instances where traditional diagnostics are unsuccessful or produce delayed outcomes (Jeff et al., 2025).

15.10.4 Vaccine-based prevention

In the fight against AMR, vaccines are a potent but underutilized tool. Vaccines reduce the need for antibiotics and the pressure of antibiotic selection by avoiding infections (Date et al., 2015). An estimated 30,000+ cases of extensively drug-resistant (XDR) *Salmonella typhi* have already been avoided since the Typhoid Conjugate Vaccine (TCV) was introduced in Nepal (2018) and Pakistan (2019), negating the need for fluoroquinolone or carbapenem therapy (Guzman et al., 2006; Milligan et al., 2018). When compared to alternative therapies, mass vaccination against *S. Typhi* produces cost-effective AMR prevention in nations with high XDR prevalence, according to a 2023 health-economic analysis (Bloom et al., 2018). Further AMR prevention may be possible with vaccines being developed against common resistance-associated bacteria, such as *A. baumannii* and *K. pneumoniae* (Bloom et al., 2018; Milligan et al., 2018).

15.11 Intersectional inequalities and social dimensions of antimicrobial resistance

15.11.1 Gender and antimicrobial resistance vulnerability

The clinical AMR literature frequently ignores the manner in which the burden of AMR is unevenly distributed across gender lines. In places with insufficient piped supply, women in LMICs are disproportionately responsible for collecting water, which exposes them to contaminated water sources and resistant bacteria that colonise the gastrointestinal system (Pradipta et al., 2018). Due to cultural limitations on independent travel, household duties, and mobility issues, women in rural areas have less access to healthcare. When MDR organisms cause maternal and newborn infections (such as mastitis or bacteremia in postpartum sepsis), treatment choices are restricted due to the urgency of preventing severe sepsis and death as well as the requirement to avoid teratogenic drugs (during pregnancy) (Pradipta et al., 2018).

15.11.2 Vulnerable populations: refugees, migrants, and displaced persons

Due to congestion, poor sanitation, starvation, and restricted access to healthcare, refugee camps and informal settlements provide the perfect environment for the quick spread of resistant infections. Up to 40% of samples from some camps had *bla*-$_{CTX-M}$ ESBL-producing *E. coli*, according to a 2024 analysis of antibiotic resistance among refugees in several locales, which revealed significantly higher rates of MDR pathogen colonisation compared to nondisplaced populations. When AMR restricts treatment options, poor nutritional conditions weaken immune systems, increasing susceptibility to infection and risk of severe disease (Comelli et al., 2024).

15.12 Conclusions and future perspectives

AMR in LMICs is a result of a confluence of structural, biological, environmental, and socioeconomic issues that cannot be resolved by antibiotics and diagnostics alone. With approximately 280,000 deaths in South Asia alone each year, the unprecedented burden necessitates immediate, all-encompassing action based on solid evidence and tailored to local circumstances. The current epidemiological data showing the high frequency of ESBL-producing organisms, the emergence of carbapenem resistance, and the alarming spread of last-resort resistance mechanisms in LMIC settings have been compiled in this chapter. Coordinated policy initiatives and strategic investments are essential for mitigating the primary drivers of AMR, which encompass unregulated antibiotic markets, insufficient WASH infrastructure, environmental antibiotic contamination, and inadequate diagnostic capabilities.

In resource-constrained settings, novel technologies such as phage therapy, bacteriocins, rapid molecular diagnostics, and vaccine-based prophylaxis present significant potential. Nevertheless, these advancements will not achieve their optimum impact without concurrent investments in WASH, infection control protocols, stewardship training, and surveillance infrastructure. The period from 2025 to 2035 is critical; lacking substantial intervention, projections indicate that AMR-related mortality could exceed 39 million deaths by 2050. Conversely, sustained commitment to comprehensive, locally adapted interventions can alter this trajectory. The scientific rationale is clear; the requisite resources and political resolve are now imperative.

Acknowledgement

The author (DV) is thankful to the Council of Scientific & Industrial Research (CSIR) for awarding a Junior Research Fellowship (JRF). Samreen and IA are grateful to DBT, New Delhi, for financial support in the form of Research Associate-I and Principal/Coprincipal Investigator, respectively, under Indo-UK Environmental AMR DBT projects (SELECTAR and ResPharm 2020–2025).

AI disclosure

During the preparation of this work, the author(s) used ChatGPT (OpenAI) to use for language editing, clarity improvement, and restructuring of some sections. After using this tool/service, the author(s) reviewed and edited the content as needed and take(s) full responsibility for the content of the publication.

References

Abbas, S., 2024. The challenges of implementing infection prevention and antimicrobial stewardship programs in resource-constrained settings. Antimicrobial Stewardship & Healthcare Epidemiology 4 (1). Available at https://doi.org/10.1017/ash.2024.35.

Ahmad, I., Ahmad, S., Rumbaugh, K.P., 2019. Antibacterial drug discovery to combat MDR: natural compounds, nanotechnology and novel synthetic sources pp. 1–680. https://doi.org/10.1007/978-981-13-9871-1.

Alba, P., Leekitcharoenphon, P., Franco, A., Feltrin, F., Ianzano, A., Caprioli, A., et al., 2018. Molecular epidemiology of mcr-encoded colistin resistance in enterobacteriaceae from food-producing animals in Italy revealed through the EU harmonized antimicrobial resistance monitoring. Frontiers in Microbiology 9. Available at https://doi.org/10.3389/fmicb.2018.01217.

Al-Emran, H.M., Eibach, D., Krumkamp, R., Ali, M., Baker, S., Biggs, H.M., et al., 2016. A multicountry molecular analysis of *Salmonella enterica* serovar typhi with reduced susceptibility to ciprofloxacin in Sub-Saharan Africa. Clinical Infectious Diseases 62, s42. Available at https://doi.org/10.1093/cid/civ788, http://cid.oxfordjournals.org/content/by/year.

Aydin, S., Aydin, M.E., Ulvi, A., Kilic, H., 2018. Antibiotics in hospital effluents: occurrence, contribution to urban wastewater, removal in a wastewater treatment plant, and environmental risk assessment. Environmental Science and Pollution Research 26, 544–558. Available at https://doi.org/10.1007/s11356-018-3563-0.

Beaumont, A.L., Kermorvant-Duchemin, E., Breurec, S., Huynh, B.T., 2024. Neonatal Colonization with antibiotic-resistant pathogens in low- and middle-income countries: a systematic review and meta-analysis. JAMA Network Open 7 (11). Available at https://doi.org/10.1001/jamanetworkopen.2024.41596, https://jamanetwork.com/journals/jamanetworkopen.

Bengtsson-Palme, J., Kristiansson, E., Larsson, D.G.J., 2018. Environmental factors influencing the development and spread of antibiotic resistance. FEMS Microbiology Reviews 42 (1), 68–80. Available at https://doi.org/10.1093/femsre/fux053, http://femsre.oxfordjournals.org/.

Bezabih, Y.M., Bezabih, A., Dion, M., Batard, E., Teka, S., Obole, A., et al., 2022. Comparison of the global prevalence and trend of human intestinal carriage of ESBL-producing *Escherichia coli* between healthcare and community settings: a systematic review and meta-Analysis. JAC-Antimicrobial Resistance 4 (3). Available at https://doi.org/10.1093/jacamr/dlac048, https://academic.oup.com/jacamr.

Bloom, D.E., Black, S., Salisbury, D., Rappuoli, R., 2018. Antimicrobial resistance and the role of vaccines. Proceedings of the National Academy of Sciences of the United States of America 115 (51), 12868–12871. Available at https://doi.org/10.1073/pnas.1717157115, https://www.pnas.org/content/pnas/115/51/12868.full.pdf.

Bush, K., Bradford, P.A., 2019. Interplay between β-lactamases and new β-lactamase inhibitors. Nature Reviews. Microbiology 17 (5), 295–306. Available at https://doi.org/10.1038/s41579-019-0159-8, http://www.nature.com/nrmicro/index.html.

Bush, K., Jacoby, G.A., 2010. Updated functional classification of β-lactamases. Antimicrobial Agents and Chemotherapy 54 (3), 969–976. Available at https://doi.org/10.1128/AAC.01009-09, http://aac.asm.org/cgi/reprint/54/3/969.

Cai, J., Wang, X., Ma, A., Wang, Q., Han, X., Li, Y., et al., 2015. Factors associated with patient and provider delays for tuberculosis diagnosis and treatment in Asia: a systematic review and meta-analysis. PLoS One 10 (3), e0120088. Available at https://doi.org/10.1371/journal.pone.0120088.

Cissé, G., 2019. Food-borne and water-borne diseases under climate change in low- and middle-income countries: further efforts needed for reducing environmental health exposure risks. Acta Tropica 194, 181–188. Available at https://doi.org/10.1016/j.actatropica.2019.03.012.

Comelli, A., Gaviraghi, A., Cattaneo, P., Motta, L., Bisoffi, Z., Stroffolini, G., 2024. Antimicrobial resistance in migratory paths, refugees, asylum seekers and internally displaced persons: a narrative review. Current Tropical Medicine Reports 11 (3), 153–166. Available at https://doi.org/10.1007/s40475-024-00322-2.

Cox, J.A., Vlieghe, E., Mendelson, M., Wertheim, H., Ndegwa, L., Villegas, M.V., et al., 2017. Antibiotic stewardship in low- and middle-income countries: the same but different? Clinical Microbiology and Infection 23 (11), 812–818. Available at https://doi.org/10.1016/j.cmi.2017.07.010.

Cytryn, E., 2013. The soil resistome: the anthropogenic, the native, and the unknown. Soil Biology and Biochemistry 63, 18–23. Available at https://doi.org/10.1016/j.soilbio.2013.03.017.

Datta, N., Kontomichalou, P., 1965. Penicillinase Synthesis Controlled By Infectious R Factors In Enterobacteriaceae. Nature 208, 239–241. Available at https://doi.org/10.1038/208239a0.

Date, K.A., Bentsi-Enchill, A., Marks, F., Fox, K., 2015. Typhoid fever vaccination strategies. Vaccine 33 (3), C55. Available at https://doi.org/10.1016/j.vaccine.2015.04.028, www.elsevier.com/locate/vaccine.

Debebe, F., Goffi, A., Haile, T., Alferid, F., Estifanos, H., Adhikari, N.K.J., 2022. Predictors of ICU mortality among mechanically ventilated patients: an inception cohort study from a tertiary care center in addis Ababa, Ethiopia. Critical Care Research and Practice 2022. Available at https://doi.org/10.1155/2022/7797328.

Delgado-Baquerizo, M., Hu, H.-W., Maestre, F.T., Guerra, C.A., Eisenhauer, N., Eldridge, D.J., et al., 2022. The global distribution and environmental drivers of the soil antibiotic resistome. Microbiome 10 (1). Available at https://doi.org/10.1186/s40168-022-01405-w.

DeLeo, F., Otto, M., Kreiswirth, B., et al., 2010. Community-associated meticillin-resistant Staphylococcus aureus. The Lancet 375, 1557–1568.

Dellit, T.H., 2007. Summary of the Infectious diseases society of america and the society for healthcare Epidemiology of America guidelines for developing an institutional program to enhance antimicrobial stewardship. Infectious Diseases in Clinical Practice 15 (4), 263–264. Available at https://doi.org/10.1097/IPC.0b013e318068b1c0.

Fletcher, S., 2015. Understanding the contribution of environmental factors in the spread of antimicrobial resistance. Environmental Health and Preventive Medicine 20 (4), 243–252. Available at https://doi.org/10.1007/s12199-015-0468-0.

Gelalcha, B.D., Gelgie, A.E., Kerro Dego, O., 2024. Antimicrobial resistance and prevalence of extended-spectrum beta-lactamase-producing Klebsiella species in East Tennessee dairy farms. Microbiology Spectrum 12. Available at https://doi.org/10.1128/spectrum.03537-23.

Grembi, J.A., Lin, A., Karim, M.A., Islam, M.O., Miah, R., Arnold, B.F., et al., 2023. Effect of water, sanitation, handwashing, and nutrition interventions on enteropathogens in children 14 months old: a cluster-randomized controlled trial in Rural Bangladesh. Journal of Infectious Diseases 227 (3), 434–447. Available at https://doi.org/10.1093/infdis/jiaa549, http://jid.oxfordjournals.org/content/current.

Gros, M., Petrović, M., Ginebreda, A., Barceló, D., 2010. Removal of pharmaceuticals during wastewater treatment and environmental risk assessment using hazard indexes. Environment International 36 (1), 15–26. Available at https://doi.org/10.1016/j.envint.2009.09.002.

Guzman, C.A., Borsutzky, S., Griot-Wenk, M., Metcalfe, I.C., Pearman, J., Collioud, A., et al., 2006. Vaccines against typhoid fever. Vaccine 24 (18), 3804–3811. Available at https://doi.org/10.1016/j.vaccine.2005.07.111.

Hatem, H., Abdelaziz, R., 2024. Revolutionizing the fight against multidrug-resistant bacteria: phage and phage products as the leading armament in future: review. Journal of Advanced Veterinary Research 14 (5), 913–916. Available at https://www.advetresearch.com/index.php/AVR/article/download/1714/1256.

Hafeez, A., Dangel, W., Ostroff, S., et al., 2023. The state of health in Pakistan and its provinces and territories, 1990–2019: a systematic analysis for the Global Burden of Disease Study 2019. The Lancet Global Health 11, e229–e243.

Hitchcock, N.M., Devequi Gomes Nunes, D., Shiach, J., Valeria Saraiva Hodel, K., Dantas Viana Barbosa, J., Alencar Pereira Rodrigues, L., et al., 2023. Current clinical landscape and global potential of bacteriophage therapy. Viruses 15 (4), 1020. Available at https://doi.org/10.3390/v15041020.

Huang, Y.H., Kuan, N.L., Yeh, K.S., 2020. Characteristics of extended-spectrum β-lactamase–producing *Escherichia coli* from dogs and cats admitted to a veterinary teaching hospital in Taipei, Taiwan From 2014 to 2017. Frontiers in Veterinary Science 7. Available at https://doi.org/10.3389/fvets.2020.00395, journal.frontiersin.org/journal/veterinary-science.

Hutchings, M., Truman, A., Wilkinson, B., 2019. Antibiotics: past, present and future. Current Opinion in Microbiology 51, 72–80. Available at https://doi.org/10.1016/j.mib.2019.10.008, http://www.elsevier.com/locate/mib.

Hwang, S., Kwon, K.T., 2021. Core elements for successful implementation of antimicrobial stewardship programs. Infection and Chemotherapy 53 (3), 421–435. Available at https://doi.org/10.3947/ic.2021.0093, https://icjournal.org/DOIx.php?id=10.3947/ic.2021.0093.

Ibrahim, M.E., 2023. Risk factors in acquiring multidrug-resistant *Klebsiella pneumoniae* infections in a hospital setting in Saudi Arabia. Scientific Reports 13 (1). Available at https://doi.org/10.1038/s41598-023-38871-7, https://www.nature.com/srep/.

Idrees, M.M., Saeed, K., Shahid, M.A., Akhtar, M., Qammar, K., Hassan, J., Khaliq, T., Saeed, A., 2023. Prevalence of mecA- and mecC-Associated Methicillin-Resistant Staphylococcus aureus in Clinical Specimens, Punjab, Pakistan. Biomedicines 11, 878. Available at https://doi.org/10.3390/biomedicines11030878.

Iskandar, K., Molinier, L., Hallit, S., Sartelli, M., Catena, F., Coccolini, F., et al., 2020. Drivers of antibiotic resistance transmission in low-and middle-income countries from a "one health" perspective—a review. Antibiotics 9 (7), 1–23. Available at https://doi.org/10.3390/antibiotics9070372, https://www.mdpi.com/2079-6382/9/7/372/pdf.

Islam, K., Heffernan, A.J., Naicker, S., Henderson, A., Chowdhury, M.A.H., Roberts, J.A., Sime, F.B., 2021. Epidemiology of extended-spectrum β-lactamase and metallo-β-lactamase-producing *Escherichia Coli* in South Asia. Future Microbiology 16, 521–535. Available at https://doi.org/10.2217/fmb-2020-0193.

Jeff, C., Quarton, S., Hatton, C., Parekh, D., Thickett, D., McNally, A., et al., 2025. Metagenomics in the diagnosis and treatment of urinary tract infections: a systematic review and meta-analysis. Diagnostic Microbiology and Infectious Disease 113 (3), 116995. Available at https://doi.org/10.1016/j.diagmicrobio.2025.116995.

Jinks, T., Lee, N., Sharland, M., Rex, J., Gertler, N., Diver, M., et al., 2016. A time for action: antimicrobial resistance needs global response. Bulletin of the World Health Organization 94 (8), 558. Available at https://doi.org/10.2471/BLT.16.181743.

Kakkar, A.K., Shafiq, N., Singh, G., Ray, P., Gautam, V., Agarwal, R., et al., 2020. Antimicrobial stewardship programs in resource constrained environments: understanding and addressing the need of the systems. Frontiers in Public Health 8. Available at https://doi.org/10.3389/fpubh.2020.00140, http://journal.frontiersin.org/journal/public-health/section/public-health-education-and-promotion#archive.

Karampatakis, T., Tsergouli, K., Behzadi, P., 2023. Carbapenem-resistant *Klebsiella pneumoniae*: virulence factors, molecular epidemiology and latest updates in treatment options. Antibiotics 12 (2). Available at https://doi.org/10.3390/antibiotics12020234, http://www.mdpi.com/journal/antibiotics.

Karanika, S., Paudel, S., Grigoras, C., Kalbasi, A., Mylonakis, E., 2016. Systematic review and meta-analysis of clinical and economic outcomes from the implementation of hospital-based antimicrobial stewardship programs. Antimicrobial Agents and Chemotherapy 60 (8), 4840–4852. Available at https://doi.org/10.1128/aac.00825-16.

Lalani, H.S., Waweru-Siika, W., Mwogi, T., Kituyi, P., Egger, J.R., Park, L.P., et al., 2018. Intensive care outcomes and mortality prediction at a National Referral Hospital in Western Kenya. Annals of the American Thoracic Society 15 (11), 1336–1343. Available at https://doi.org/10.1513/AnnalsATS.201801-051OC, https://www.atsjournals.org/doi/pdf/10.1513/AnnalsATS.201801-051OC.

Larsson, D.G.J., Andremont, A., Bengtsson-Palme, J., Brandt, K.K., de Roda Husman, A.M., Fagerstedt, P., et al., 2018. Critical knowledge gaps and research needs related to the environmental dimensions of antibiotic resistance. Environment International 117, 132–138. Available at https://doi.org/10.1016/j.envint.2018.04.041, www.elsevier.com/locate/envint.

Law, A., Solano, O., Brown, C.J., Hunter, S.S., Fagnan, M., Top, E.M., et al., 2021. Biosolids as a source of antibiotic resistance plasmids for commensal and pathogenic bacteria. Frontiers in Microbiology 12. Available at https://doi.org/10.3389/fmicb.2021.606409.

Laxminarayan, R., Sridhar, D., Blaser, M., Wang, M., Woolhouse, M., 2016. Achieving global targets for antimicrobial resistance. Science 353, 874–875. Available at https://doi.org/10.1126/science.aaf9286.

Le, T.H., Ng, C., Chen, H., Yi, X.Z., Koh, T.H., Barkham, T.M.S., et al., 2016. Occurrences and characterization of antibiotic-resistant bacteria and genetic determinants of hospital wastewater in a tropical country. Antimicrobial Agents and Chemotherapy 60 (12), 7449–7456. Available at https://doi.org/10.1128/AAC.01556-16, http://aac.asm.org/content/60/12/7449.full.pdf+html.

Lim, C., Ashley, E.A., Hamers, R.L., Turner, P., Kesteman, T., Akech, S., Corso, A., Mayxay, M., Okeke, I.N., Limmathurotsakul, D., van Doorn, H.R., 2021. Surveillance strategies using routine microbiology for antimicrobial resistance in low- and middle-income countries. Clinical Microbiology and Infection. 27 (10), 1391–1399.

Lin, Z., Yuan, T., Zhou, L., Cheng, S., Qu, X., Lu, P., et al., 2021. Impact factors of the accumulation, migration and spread of antibiotic resistance in the environment. Environmental Geochemistry and Health 43 (5), 1741–1758. Available at https://doi.org/10.1007/s10653-020-00759-0.

Llor, C., Benkő, R., Bjerrum, L., 2024. Global restriction of the over-the-counter sale of antimicrobials: does it make sense? Frontiers in Public Health 12. Available at https://doi.org/10.3389/fpubh.2024.1412644, http://journal.frontiersin.org/journal/public-health/section/public-health-education-and-promotion#archive.

Mafe, A.N., Büsselberg, D., 2025. Phage therapy in managing multidrug-resistant (MDR) infections in cancer therapy: innovations, complications, and future directions. Pharmaceutics 17 (7). Available at https://doi.org/10.3390/pharmaceutics17070820, http://www.mdpi.com/journal/pharmaceutics.

Memon, K.N., Shaikh, K., Pandhiani, B.S., Usman, G., 2013. How do mothers recognize & treat pneumonia in their children at home? A study in Union Council Jhudo, district Mirpurkhas. Journal of the Liaquat University of Medical and Health Sciences 12 (3), 208–213. Available at http://lumhs.edu.pk/jlumhs/Vol12No03/pdfs/16.pdf.

Milligan, R., Paul, M., Richardson, M., Neuberger, A., 2018. Vaccines for preventing typhoid fever. Cochrane Database of Systematic Reviews 2018 (5). Available at https://doi.org/10.1002/14651858.cd001261.pub4.

Mishra, S., Kumar, H., Sharma, D., 1994. How do mothers recognize and treat pneumonia at home? Indian Pediatrics 31 (1), 15–18.

Mulchandani, R., Wang, Y., Gilbert, M., Boeckel, T.P.V., Odetokun, I.A., 2023. Global trends in antimicrobial use in food-producing animals: 2020 to 2030. PLOS Global Public Health 3 (2), e0001305. Available at https://doi.org/10.1371/journal.pgph.0001305.

Naghavi, M., Vollset, S.E., Ikuta, K.S., Swetschinski, L.R., Gray, A.P., Wool, E.E., et al., 2024. Global burden of bacterial antimicrobial resistance 1990–2021: a systematic analysis with forecasts to 2050. The Lancet 404 (10459), 1199–1226. Available at https://doi.org/10.1016/S0140-6736(24)01867-1, https://www.sciencedirect.com/science/journal/01406736.

Nashwan, A.J., Shah, H.H., Hussain, T., Rauf, S.A., Ahmed, S.K., 2024. Environmental drivers of antimicrobial resistance in low and middle-income countries: the impacts of a changing world. Environmental Health Insights 18. Available at https://doi.org/10.1177/11786302241246420, https://journals.sagepub.com/home/EHI.

Nesme, J., Simonet, P., 2015. The soil resistome: a critical review on antibiotic resistance origins, ecology and dissemination potential in telluric bacteria. Environmental Microbiology 17 (4), 913–930. Available at https://doi.org/10.1111/1462-2920.12631France, http://onlinelibrary.wiley.com/journal/10.1111/(ISSN)1462-2920.

Nichols, G., Lake, I., Heaviside, C., 2018. Climate change and water-related infectious diseases. Atmosphere 9 (10), 385. Available at https://doi.org/10.3390/atmos9100385.

Niwandinda, F., Lukyamuzi, E.J., Ainebyona, C., Ssebunya, V.N., Murungi, G., Atukunda, E.C., 2020. Patterns and practices of self-medication among students enrolled at mbarara university of science and technology in Uganda. Integrated Pharmacy Research and Practice 9, 41–48. Available at https://doi.org/10.2147/iprp.s237940.

Njoki, C., Simiyu, N., Kaddu, R., Mwangi, W., Sulemanji, D., Oduor, P., et al., 2024. Epidemiology, clinical characteristics, and outcomes of 4546 adult admissions to high-dependency and ICUs in Kenya: a multicenter registry-based observational study. Critical Care Explorations 6 (2), e1036. Available at https://doi.org/10.1097/cce.0000000000001036.

Ocan, M., Bwanga, F., Bbosa, G.S., Bagenda, D., Waako, P., Ogwal-Okeng, J., et al., 2014. Patterns and predictors of self-medication in Northern Uganda. PLoS One 9 (3), e92323. Available at https://doi.org/10.1371/journal.pone.0092323.

Okeke, I.N., Aboderin, O.A., Byarugaba, D.K., Ojo, K.K., Opintan, J.A., 2007. Growing problem of multidrug-resistant enteric pathogens in Africa. Emerging Infectious Diseases 13 (11), 1640–1646. Available at https://doi.org/10.3201/eid1311.070674, http://www.cdc.gov/eid/content/13/11/pdfs/1640.pdf.

Olaitan, M.O., Orababa, O.Q., Shittu, R.B., Obunukwu, G.M., Kade, A.E., Arowolo, M.T., et al., 2025. Prevalence of ESBL-producing *Escherichia coli* in sub-Saharan Africa: a meta-analysis using a One Health approach. One Health 20. Available at https://doi.org/10.1016/j.onehlt.2025.101090, https://www.sciencedirect.com/science/journal/23527714.

Olatunji, A.O., Olaboye, J.A., Maha, C.C., Kolawole, T.O., Abdul, S., 2024. Next-Generation strategies to combat antimicrobial resistance: integrating genomics, CRISPR, and novel therapeutics for effective treatment. Engineering Science & Technology Journal 5 (7), 2284–2303. Available at https://doi.org/10.51594/estj.v5i7.1344.

Partridge, S.R., Tsafnat, G., Coiera, E., Iredell, J.R., 2009. Gene cassettes and cassette arrays in mobile resistance integrons: review article. FEMS Microbiology Reviews 33 (4), 757–784. Available at https://doi.org/10.1111/j.1574-6976.2009.00175.x.

Porter, G., Kotwani, A., Bhullar, L., Joshi, J., 2021. Over-the-counter sales of antibiotics for human use in India: the challenges and opportunities for regulation. Medical Law International 21 (2), 147–173. Available at https://doi.org/10.1177/09685332211020786.

Pospišil, M., Car, H., Elveđi-Gašparović, V., Beader, N., Herljević, Z., Bedenić, B., 2023. Bloodstream Infections by AmpC-producing enterobacterales: risk factors and therapeutic outcome. Pathogens 12 (9), 1125. Available at https://doi.org/10.3390/pathogens12091125.

Pradipta, I.S., Forsman, L.D., Bruchfeld, J., Hak, E., Alffenaar, J.W., 2018. Risk factors of multidrug-resistant tuberculosis: a global systematic review and meta-analysis. Journal of Infection 77 (6), 469–478. Available at https://doi.org/10.1016/j.jinf.2018.10.004, http://www.elsevier.com/inca/publications/store/6/2/3/0/5/4/index.htt.

Prestinaci, F., Pezzotti, P., Pantosti, A., 2015. Antimicrobial resistance: a global multifaceted phenomenon. Pathogens and Global Health 109 (7), 309–318. Available at https://doi.org/10.1179/2047773215y.0000000030.

Putri, D.A., Lei, J., Rossiana, N., Syaputri, Y., 2024. Biopreservation of food using bacteriocins from lactic acid bacteria: classification, mechanisms, and commercial applications. International Journal of Microbiology 2024 (1). Available at https://doi.org/10.1155/ijm/8723968, https://onlinelibrary.wiley.com/journal/8472.

Rawat, D., Nair, D., 2010. Extended-spectrum ß-lactamases in gram negative bacteria. Journal of Global Infectious Diseases 2 (3), 263. Available at https://doi.org/10.4103/0974-777x.68531.

Reverter, M., Sarter, S., Caruso, D., Avarre, J.-C., Combe, M., Pepey, E., et al., 2020. Aquaculture at the crossroads of global warming and antimicrobial resistance. Nature Communications 11 (1). Available at https://doi.org/10.1038/s41467-020-15735-6.

Rizzo, L., Manaia, C., Merlin, C., Schwartz, T., Dagot, C., Ploy, M.C., Michael, I., Fatta-Kassinos, D., 2013. Urban wastewater treatment plants as hotspots for antibiotic resistant bacteria and genes spread into the environment: a review. Science of the Total Environment. 447, 345–360.

Rozwandowicz, M., Brouwer, M.S.M., Fischer, J., Wagenaar, J.A., Gonzalez-Zorn, B., Guerra, B., et al., 2018. Plasmids carrying antimicrobial resistance genes in Enterobacteriaceae. Journal of Antimicrobial Chemotherapy 73 (5), 1121–1137. Available at https://doi.org/10.1093/jac/dkx488.

Sarkar, P.R., 2023. Epidemiological trends in zoonotic diseases comparative insights from south asia and the U.S. American Journal of Interdisciplinary Studies 04 (03), 166–207. Available at https://doi.org/10.63125/wrrfmt97.

Shanmugakani, R.K., Srinivasan, B., Glesby, M.J., Westblade, L.F., Cárdenas, W.B., Raj, T., et al., 2020. Current state of the art in rapid diagnostics for antimicrobial resistance. Lab on a Chip 20 (15), 2607–2625. Available at https://doi.org/10.1039/d0lc00034e, http://pubs.rsc.org/en/journals/journal/lc.

Sharan, M., Vijay, D., Yadav, J.P., Bedi, J.S., Dhaka, P., 2023. Surveillance and response strategies for zoonotic diseases: a comprehensive review. Science in One Health 2, 100050. Available at https://doi.org/10.1016/j.soh.2023.100050.

Smith, R.D., Coast, J., 2002. Antimicrobial resistance: a global response. Bulletin of the World Health Organization 80 (2), 126–133.

Sriram, A., Kalanxhi, E., Kapoor, G., Craig, J., Balasubramanian, R., Brar, S., Criscuolo, N., Hamilton, A., Klein, E., Tseng, K., Van Boeckel, T., Laxminarayan, R., 2021. State of the World's Antibiotics 2021: A Global Analysis of Antimicrobial Resistance and Its Drivers. Center for Disease Dynamics, Economics & Policy, Washington, DC, USA, pp. 1–15.

Sriram, A., Kalanxhi, E., Kapoor, G., Craig, J., Balasubramanian, R., Brar, S., Criscuolo, N., Hamilton, A., Klein, E., Tseng, K., Van Boeckel, T., Laxminarayan, R., 2021. The State of the World's Antibiotics 2021. 8 (2), 30–34.

Sulakvelidze, A., Alavidze, Z., Morris, J., 2001. Bacteriophage therapy. Antimicrobial Agents and Chemotherapy 45 (3), 649–659. Available at https://doi.org/10.1128/AAC.45.3.649-659.2001.

Swarthout, J., Ram, P.K., Arnold, C.D., Dentz, H.N., Arnold, B.F., Kalungu, S., et al., 2020. Effects of individual and combined water, sanitation, handwashing, and nutritional interventions on child respiratory infections in Rural Kenya: a cluster-randomized controlled trial. The American Journal of Tropical Medicine and Hygiene 102 (6), 1286–1295. Available at https://doi.org/10.4269/ajtmh.19-0779.

Taneja, N., Sharma, M., 2019. Antimicrobial resistance in the environment: the Indian scenario. Indian Journal of Medical Research 149 (2), 119–128. Available at https://doi.org/10.4103/ijmr.IJMR_331_18, http://www.ijmr.org.in/.

Tekele, S.G., Mulatie, Z., Gedefie, A., Ebrahim, H., Eshetu, B., Tilahun, M., et al., 2025. Prevalence of AmpC beta-lactamase producing *Escherichia coli* and *Klebsiella pneumoniae* in Africa: a systematic review and meta-analysis. Antimicrobial Resistance and Infection Control 14 (1). Available at https://doi.org/10.1186/s13756-025-01578-7, https://aricjournal.biomedcentral.com/.

Tilahun, M., Kassa, Y., Gedefie, A., 2021. M. Ashagire, Emerging carbapenem-resistant enterobacteriaceae infection, its epidemiology and novel treatment options: a review. Infection and Drug Resistance 14, 4363–4374. Available at https://doi.org/10.2147/IDR.S337611, https://www.dovepress.com/getfile.php?fileID=75134.

Van Boeckel, T.P., Brower, C., Gilbert, M., Grenfell, B.T., Levin, S.A., Robinson, T.P., et al., 2015. Global trends in antimicrobial use in food animals. Proceedings of the National Academy of Sciences of the United States of America 112 (18), 5649–5654. Available at https://doi.org/10.1073/pnas.1503141112, http://www.pnas.org/content/112/18/5649.full.pdf.

van Daalen, K.R., Romanello, M., Rocklöv, J., Semenza, J.C., Tonne, C., Markandya, A., et al., 2022. The 2022 Europe report of the lancet countdown on health and climate change: towards a climate resilient future. The Lancet Public Health 7 (11), e942. Available at https://doi.org/10.1016/S2468-2667(22)00197-9, http://thelancet.com/journals/lanpub/issue/vol1no2/PIIS2468-2667(16)X0003-5.

van Daalen, K.R., Tonne, C., Semenza, J.C., Rocklöv, J., Markandya, A., Dasandi, N., et al., 2024. The 2024 Europe report of the Lancet Countdown on health and climate change: unprecedented warming demands unprecedented action. The Lancet Public Health 9 (7), e495. Available at https://doi.org/10.1016/S2468-2667(24)00055-0, https://www.sciencedirect.com/science/journal/24682667.

Wang, Y., Yu, Y., 2025. Phage therapy as a revitalized weapon for treating clinical diseases. Microbiome Research Reports 4 (3). Available at https://doi.org/10.20517/mrr.2025.31, https://f.oaes.cc/xmlpdf/published/article/bc8939f6662dc6fe6b0e494983364d25/mrr4031.pdf.

Wasswa, F.B., Kassaza, K., Nielsen, K., Bazira, J., 2022. MinION whole-genome sequencing in resource-limited settings: challenges and opportunities. Current Clinical Microbiology Reports 9 (4), 52–59. Available at https://doi.org/10.1007/s40588-022-00183-1, https://www.springer.com/journal/40588.

Wellington, E.M.H., Boxall, A.B.A., Cross, P., Feil, E.J., Gaze, W.H., Hawkey, P.M., et al., 2013. The role of the natural environment in the emergence of antibiotic resistance in Gram-negative bacteria. The Lancet Infectious Diseases 13 (2), 155–165. Available at https://doi.org/10.1016/S1473-3099(12)70317-1.

World Health Organization. 2024. Food and Agriculture Organization of the United Nations, United Nations Environment Programme, World Organisation for Animal Health, 2024. Implementing the Global Action Plan on Antimicrobial Resistance: First Quadripartite Biennial Report. World Health Organization, Geneva.

Wolters, B., Hauschild, K., Blau, K., Mulder, I., Heyde, B.J., Sørensen, S.J., et al., 2022. Biosolids for safe land application: does wastewater treatment plant size matters when considering antibiotics, pollutants, microbiome, mobile genetic elements and associated resistance genes? Environmental Microbiology 24 (3), 1573–1589. Available at https://doi.org/10.1111/1462-2920.15938.

Yu, W., Zhang, H., Xu, Y., Zhu, Y., Jia, P., Kang, Y., Yang, Q., 2025. In vitro activity of ceftolozane/tazobactam against ESBL-producing enterobacterales in China: SMART 2016–2019. Journal of Global Antimicrobial Resistance 42, 161–166. Available at https://doi.org/10.1016/j.jgar.2025.02.001.

Chapter 16

The antibiotic dilemma: how the pharmaceutical industry shapes our future

Varsha Shridhar[1] and Maneesh Paul Satyaseela[2]

[1]*Deparment of Microbiology and Molecular Diagnostics, Molecular Solutions Care Health LLP, Bengaluru, Karnataka, India,* [2]*Lead Co-Inventor Enmetazobactam, College of Pharmaceutical Sciences, Dayananda Sagar University, Bengaluru South, Karnataka, India*

16.1 Introduction

Antibiotics are everywhere. Far from being just the smooth little pills seen in healthcare facilities or our homes, these chemical compounds are produced by bacteria and fungi in nearly every part of the Earth's environment that you can think of – in water, in soil, in the ice of the Antarctic, in the boiling slurries around geysers, on the deep-sea floor of oceans. Why restrict them to Earth? Bacteria on the International Space Station are also known to produce antibiotics! Where there are antibiotics, there is certainly some type of antibiotic resistance being developed in some neighbouring species of microorganisms. It is a fact universally acknowledged that antibiotic production and development of resistance to antibiotics go hand-in-hand: neither is a human creation, although human activities certainly can (and do) influence both.

Why then is the focus specifically on the pharmaceutical industries in this chapter? There are four main reasons for this:

1. The economics of antibiotic production by these same industries drives the innovation (or lack of) of novel antibiotics: Antibiotic courses are short and limited, making the process of innovation and development of novel antibiotics fiscally unattractive as compared to medicines for life-long diseases or at least those of longer duration (such as hypertension, diabetes, and cardiovascular diseases). Thus the arsenal of safe and effective antibiotic compounds to combat various types of pathogens remains limited.
2. The pharmaceutical industry's influence on the use and abuse of antibiotics in the market: Their policies, guidelines, and practices have the potential to allow for rational use and stewardship of antibiotics, or for their rampant misuse or abuse, often driven by relentless marketing of unnecessary antibiotics.
3. Pharmaceutical policies direct the methods of storage, transportation, time of retention, and methods of disposal of antibiotics, all of which have a role in actively shaping clinical, therapeutic, or empirical practices and the AMR ecosystem among humans and animals (including all forms of marine life) in a region.
4. By the dint of the sheer volume of antibiotic chemicals they produce, the pharmaceutical industries have the capacity to shape the antibiotic and AMR profile of large swathes of the environment around them. This includes wastewater/ sewage. Wastewater and wastewater treatment plants are being recognised as evolutionary hubs which support the maintenance, transmission, and amplification of antibiotic resistance genes within and across bacterial species.

Pharmaceutical industries hold immense sway over market forces and policy directives that permit rational use or rampant misuse of antibiotics; they have a great capacity to harm individuals, but they also have enormous power to heal and rejuvenate the ailing wasteland of healthcare options in the background of crippling AMR.

In this chapter, four points are delved into deeper detail, namely,

1. Economics of antibiotic innovations,
2. Institutional frameworks that support and enhance antibiotic use/sales,
3. Storage, transport, retention, and disposal policies and practices for antibiotics,
4. Environmental impact of pharmaceutical industries' manufacturing practices.

Antimicrobial Resistance in the 21st Century. DOI: https://doi.org/10.1016/B978-0-443-44748-8.00008-7

16.2 The antibiotics golden age and its decline

The period from the 1940s to the 1970s marked a golden age of antibiotic discovery and development. However, the rate of new antibiotic approvals has slowed significantly. Despite the introduction of numerous novel antibiotic classes during the 'golden era' and subsequent advancements such as quinolone antimicrobials and carbapenems, several attributes have contributed to a substantial decline in new antibiotic development since the 1970s (Murugaiyan et al., 2022; Abdallah et al., 2023). The reasons encompass both scientific hurdles, which are of significant threat, and the diminishing interest of major pharmaceutical companies in antibiotic research and development due to unattractive return on investment (Alas Portillo et al., 2024). Antibiotic discovery is tougher still, as bacteria evolve rapidly under selective pressure, developing resistance even during trials. A single surviving bacterium can generate over 16 million offspring in a day, making resistance inevitable and development uniquely difficult (Blaskovich and Cooper, 2025). But a critical challenge continues to be the rising prevalence of antimicrobial resistance, which renders existing antibiotics ineffective and necessitates continuous innovation (Gargate et al., 2025). In addition to this, significant investment in preclinical research has not translated into new classes of approved antibiotics, leading many commercial developers to exit the field due to economic pressures and the inherent difficulty in identifying promising chemical starting points (Klug et al., 2021). This trend began to manifest in the 1970s and intensified through the 1990s, culminating in many large pharmaceutical companies abandoning antiinfective research by the early 2000s (Alas Portillo et al., 2024). This withdrawal has created an urgent imperative for the discovery of novel antibiotics, particularly given the escalating problem of antimicrobial resistance (Coxe and Azad, 2023). This trend, coupled with the rapid emergence of resistance to existing antimicrobial agents, has created a critical public health crisis (Bo et al., 2024). Persistent stagnation in antibiotic development, despite significant scientific advancements, highlights a critical paradox in the pharmaceutical innovation landscape (Maharramov et al., 2025; Alas Portillo et al., 2024). The success of early antibiotics in transforming medicine from a diagnostic to a therapeutic science further underscores the impact of this decline, as the pipeline for new antimicrobial agents has drastically diminished since the initial wave of discovery in the mid-20th century (Alghamdi et al., 2021). Indeed, the last truly novel class of antibiotics approved for clinical use against Gram-negative pathogens was discovered over five decades ago, leaving the global market dominated by classes identified much earlier (Farha, Tu and Brown, 2025; Fischbach and Walsh, 2009). Nearly all current clinically used antibiotics were discovered during the 'golden era', with newer drugs often being chemically tailored analogues of existing scaffolds rather than genuinely novel chemical entities (Maglangit et al., 2020). Developing a new drug is extremely challenging, that is fewer than 1 in 200 candidates reach the market, even after years of testing millions of compounds (Blaskovich and Cooper, 2025). While natural products have historically provided effective starting points for antibiotic development, the identification of novel, safe, and effective scaffolds remains a significant hurdle (Bellucci et al., 2024), underscoring the urgent need for innovative approaches to discover and develop new antibacterial agents with novel mechanisms of action to combat the growing threat of multidrug-resistant pathogens (Miethke et al., 2021; Prasad et al., 2022). There is a need for a renewed focus on fundamental research into bacterial physiology and pathogenesis to uncover vulnerabilities that can be exploited for therapeutic intervention (Walesch et al., 2023). Moreover, the historical reliance on natural products for antibiotic discovery, while successful in the past, has yielded diminishing returns, especially in terms of the rate of resistance emergence, prompting a shift towards synthetic compounds and novel pharmacophores (Parkhill and Johnson, 2024).

This overreliance on modifying existing scaffolds contributes to a faster onset of resistance, creating an urgent need for innovative approaches to discover entirely new classes of antimicrobial agents (Iskandar et al., 2022). The identification of new antibiotics with distinct mechanisms of action remains a significant challenge, with the last truly novel chemical class discovered more than three decades ago (Farha, Tu and Brown, 2025). Consequently, the development of new scaffolds should be prioritised, exploring underexplored microbial niches and repurposing existing synthetic molecule libraries to identify novel antimicrobial agents (Fischbach and Walsh, 2009).

16.3 The economics of antibiotics

Antibiotics are less profitable compared to drugs for chronic diseases due to shorter treatment durations and stewardship efforts. This section discusses the market failure in antibiotic development and the policy and economic factors, both historical and current, influencing pharmaceutical companies.

The economic landscape disincentivises investment in antibiotic research and development, creating a significant barrier to addressing the global health crisis of antimicrobial resistance (Årdal et al., 2020). This stems from various factors, including the comparatively low return on investment for antibiotics due to short treatment courses and the imperative for judicious use, alongside the substantial costs associated with developing entirely new antimicrobial classes

(Miethke et al., 2021). Furthermore, the substantial financial outlay and protracted timelines associated with clinical trials for novel antibiotics, coupled with the limited market exclusivity and lower pricing potential compared to other drug classes, further dissuade pharmaceutical companies from engaging in this vital area of drug discovery (Miethke et al., 2021). This financial disincentive is exacerbated by the high cost of drug development, with estimates suggesting that each successfully developed antibiotic would require an additional $1 billion to $2 billion in incentives. This economic reality results in a market failure where the societal need for new antibiotics is not met by the current for-profit drug development model. The lack of continuous market expansion and profitability, coupled with the increasing global burden of antimicrobial resistance, has compelled many companies to withdraw from antimicrobial development, shifting interest towards nontraditional strategies (Murugaiyan et al., 2022). In fact, 7 out of 12 antimicrobial manufacturers that have gone public since 2010 are no longer active, underscoring the severity of the market crisis. This situation is further complicated by the fact that pharmaceutical companies often hold back new antibiotics from the market to preserve them as last-resort treatments, further diminishing their immediate monetary value. This strategic withholding, while critical for resistance stewardship, paradoxically reduces the commercial viability and attractiveness of antibiotic development. This phenomenon, known as the 'antibiotic paradox', illustrates how efforts to conserve the efficacy of new antimicrobial agents inadvertently undermine their commercial potential, creating a significant impediment to sustained innovation in this field (Årdal et al., 2020). To overcome this paradox, novel economic models and policy interventions are needed to decouple the return on investment from sales volume and instead reward the societal value of novel antibiotics, thereby incentivising pharmaceutical companies to re-engage in antibiotic research and development (Gotham et al 2021). Such interventions might include market entry rewards, transferable exclusivity extensions, or subscription-based payment models to ensure predictable revenue streams for developers (Howard-Anderson et al., 2022; Årdal et al., 2020). Despite these challenges, various incentives, such as grants, prizes, and market entry rewards, are being explored to promote antibiotic innovation. For instance, financial incentives for both new and existing antibiotics are crucial to prolong their lifespan and ensure accessibility, requiring a reconfigured approach at both global and country levels (Årdal et al., 2020). These mechanisms are vital to sustain research and development activities, particularly given the inherent difficulties in commercialising new antibiotics within the existing market structure. Furthermore, global coordination and policy gaps in establishing guidelines for antibiotic efficacy tracking and quality assurance along the supply chain also hinder sustained innovation. These challenges are further compounded by disparities in regulatory frameworks and economic incentives between high-income and low- and middle-income countries, where a significant burden of antimicrobial resistance exists (Wells, Nguyen and Harbarth, 2024). Bankruptcy, despite developing a novel antibiotic, highlights the inadequacy of current incentive structures, particularly for small to medium-sized enterprises struggling with commercialisation cash flow. This case exemplifies how current 'pull' incentives often fall short in supporting commercialisation beyond drug approval, particularly for small and medium-sized enterprises that dominate the antibiotic pipeline. Specifically, plazomicin, a novel aminoglycoside developed with substantial US government biosecurity investments, failed commercialisation postapproval due to a misalignment between global health needs for extensively drug-resistant Gram-negative bacilli and the relatively small US market for these infections (Wells, Nguyen and Harbarth, 2024; Singer et al., 2019). This commercial failure underscores the urgent need for innovative economic models that decouple antibiotic development from volume-based sales and instead incentivise their societal value and stewardship (Årdal et al., 2020; McEnany and Outterson, 2024). One promising approach involves 'de-linkage' payment models, such as subscription models, which compensate developers based on the value of their antibiotics rather than the volume sold. This shift aims to provide a predictable revenue stream for developers, mitigating the commercial risk associated with low sales volumes inherent in responsible antibiotic stewardship (Wells et al., 2024). Such models, including fixed annual payments or guaranteed minimum revenues, offer enhanced cash flow visibility and stimulate the development of new antibacterial products by mitigating market risks (Wells et al., 2024; Miethke et al., 2021). Another innovative approach includes 'antibiotic bonds', where investors are repaid through guaranteed purchase contracts backed by public health funders, providing long-term cash flow support for commercialisation (Wells et al., 2024). These mechanisms are crucial for fostering a sustainable ecosystem for antibiotic development, moving beyond the traditional pharmaceutical business models that have proven inadequate for this unique market (Wells et al., 2024). The invention of Enmetazobactam offers a glimpse into potential breakthroughs that could revitalise the antibacterial pipeline, demonstrating the ongoing scientific capability to identify novel antimicrobial agents. The interesting part of this NCE is that it is the first time the USFDA has approved any invention that emerged from India (Paul-Satyaseela, 2024). Thus Enmetazobactam became the first-ever 100% India-invented antimicrobial that is now globally approved in major markets. However, the successful commercialisation and deployment of such agents still hinge on overcoming the profound financial vulnerabilities faced by small and medium-sized biotechnology enterprises, which are responsible for a significant portion of the late-stage antibiotic pipeline (Wells et al., 2024). The commercial failure of plazomicin, despite its regulatory approval and efficacy against multidrug-resistant pathogens, starkly illustrates the systemic financial challenges confronting novel antibiotic commercialisation, especially for small and medium-sized enterprises (Wells et al., 2024). This failure highlights the critical need for robust 'pull' incentives that extend beyond initial research and development to support the complex commercialisation phase, ensuring that innovative therapies reach patients (Wells et al., 2024). To maintain or increase investment in antibiotic innovation, the public sector must demonstrate a willingness to pay for new antibiotics that address unmet public health needs, potentially through national or regional access pilots. For instance, the United Kingdom and Sweden have experimented with subscription-based models, offering guaranteed payments for novel antibiotics, irrespective of volume sold, which helps stabilise revenue streams and incentivise development (Roope et al., 2024). These innovative models are designed to overcome the market failure inherent in antibiotic development, where the societal value of a new drug is not reflected in its market price, leading to insufficient investment in crucial antimicrobial research (Wells et al., 2024). Moreover, the global community must address policy gaps in tracking antibiotic efficacy and ensuring quality assurance across the supply chain, as these factors also impede sustained innovation. The disproportionate burden of antimicrobial resistance in low- and middle-income countries further necessitates equitable access policies and pricing strategies that do not exacerbate existing health disparities. The financial fragility of small and medium-sized biotechnology enterprises, which are responsible for a substantial portion of early stage and late-stage clinical research in antibiotics, further complicates this landscape, as evidenced by the high rate of bankruptcy or unprofitable exits among SME sponsors of FDA-approved antibacterials (Wells et al., 2024). This trend underscores the acute vulnerability of smaller entities, which, despite often being at the forefront of innovation, lack the financial resilience to navigate the intricate and costly path to commercialisation (Wells et al., 2024). Therefore comprehensive policy interventions are urgently required to mitigate these financial risks and foster a more sustainable antibiotic development ecosystem (Wells et al., 2024).

16.4 Institutional framework that supports and enhances antibiotic use/sales

While the economics of antibiotic innovation remain intensively challenging, the economics of antibiotic consumption and sales are no less complex. These tend to be fairly country- or region-specific. This chapter focuses on India.

As detailed in the above section, the current economic model of antibiotic sales is entirely volume-based. This naturally leads to several practices that, combined with deep structural and institutional enablers, drive the extreme misuse and irrational prescriptions of antibiotics. The structural and institutional enablers go into the heart of the economic and healthcare system of India, both for humans and animals. Some of these factors are delved into more detail below.

The expenditure by India on healthcare is around 1.8% of its total gross domestic product (GDP) (Government of India, Ministry of Finance, 2025). While this percentage has been gradually increasing over the years, it is significantly lower compared to several other countries, including the ones in strategically important blocs that India is part of: For instance, countries under the Organization for Economic Cooperation and Development (OECD) spend about 7.6% on average for and those in BRICS countries spend about 3.6% of their total GDP on average for healthcare (Majhi et al., 2024). This extremely low proportion of public spending on health drives a strong private healthcare in the country. Indeed, the private sector contribution to total healthcare is about 52% of the total healthcare expenditure in the country (Government of India, Ministry of Finance, 2025). This percentage used to be significantly higher in previous decades. Insufficient focus has been given to primary care, which is first-contact, comprehensive, contextual, continuity-based, and coordinated care, shown to be economically wiser (Shi, 2012). In recent years, with the launch of the PMJAY and Ayushman Bharat schemes, a gradual increase in investment in primary care with the hope of universal health coverage (UHC) is now in progress (Kadarpeta, Anand and Achungura, 2024; Kumar, 2023). Unfortunately, the lack of a primary care system, with low availability or access to healthcare facilities that can provide high quality and timely care at affordable rates, that is well-resourced, of high quality and can provide timely care, leads to the mushrooming of several alternate forms of healthcare which not just include private medical facilities but also informal health providers, faith healers, and pharmacies which act as pill providers. This leads to an ever-growing and nearly ungovernable network of antibiotic sellers who sell for the consumption of humans and animals both. Some other key factors are the lack of health insurance that covers outpatient (current versions of Ayushman Bharat only deal with inpatient expenditures) and the fact that over two-thirds of India's workforce (including agriculture) are daily wage workers without the ability to take days off. A strong private sector investment in healthcare has led to focussed growth of secondary and tertiary care (speciality and superspeciality hospitals) for several decades (Kumar, 2023), which are typically more profitable than local primary care clinics. This economic model promotes hospital admissions, sees sicker patients, may be linked to insurance schemes linked to admissions, and initiates more aggressive antimicrobial therapy, which also promotes unnecessary antibiotic usage. Lack of laboratory facilities in many small hospitals may induce doctors and patients to save money on microbiological culture and antibiotic sensitivity testing and initiate empirical antibiotic regimens. There is a consistent and significant trend that private facilities do prescribe more antibiotics than public facilities, regardless of age group, country, or broader health ecosystem (Ab Rahman, Teng and Sivasampu, 2016; Granlund and Zykova, 2021).

While increasing human consumption of antibiotics is certainly of concern, the antibiotic consumption in the animal sector is several times higher (Fig. 16.1). The animal husbandry sector in India consists of dairy, poultry, sheep/goat, piggeries, and fisheries. Antibiotics are used in animals primarily for three reasons: to treat infections, as metaphylaxis (a veterinary practice to mass treat animals, typically cattle, to prevent widespread infections) and as growth promoters. Global consumption of antimicrobials in food animal production was estimated at 63,151 (±1560) tonnes in 2010 and is projected to rise by 67%, to 105,596 (±3605) tonnes by 2030 (Global Antimicrobial Use in the Livestock Sector OECD Committee for Agriculture, 2015). A recent study (Fig. 16.1) estimated a massive increase of about 67% between 2010 and 2030 in animal antimicrobial use in Asia (global average estimates were 8%, with the lowest in Africa at <1%) (Mulchandani et al., 2023).

Regulations of and restrictions on antibiotic use in the animal husbandry industry exist but are poorly implemented, especially in countries where animal husbandry industries are less formalised, with a majority of small owners and less structured production systems. A ban on antibiotics would have an outsized effect on such countries and have a drastic impact on food production and on the economy. For instance, India's livestock industry is one of the largest in the world, being the largest producer of milk, second-largest producer of eggs (after China and the United States), and the third-largest producer of meat. It contributes to 5.5% of the Gross Value Add (GVA), or approximately 9.5 lakh crores in 2024, with over 87% of the Indian livestock industry being driven by small farmers (Government of India, Ministry of Fisheries, Animal Husbandry and Dairying, 2024).

Where does the pharmaceutical industry come in this messy picture of the healthcare and animal husbandry systems? On the face of it, given that the economic system is geared towards volume-based sales of pharmaceutical products, the motivations of the pharmaceutical industry could be expected to be in opposition to the motivations of antimicrobial stewardship. Indeed, interviews with pharmaceutical industry wholesale businesses in India indicate a strong distributor-doctor nexus, incentivised sales by pharmacies and informal health providers. Fig. 16.2 highlights the supply side (pharmaceutical industry) and demand side (users of antibiotics) dynamics that facilitate AMR spread and transmission (Kotwani et al., 2022). The pharmaceutical industry (which comprises the pharmaceutical drug manufacturing companies as well as wholesale merchants, retail merchants, other distributors, and medical representatives) uses a variety of methods to exploit existing loopholes in regulatory frameworks to increase drug sales. Some manufacturing companies tie up directly with hospital chains or large animal husbandry units, to directly supply them with antibiotics at a discounted price, thereby cutting the middle distributors and ensuring direct and larger amounts of profits; some manufacturing companies act as third party white-label manufacturers, allowing middle level marketers to brand and market them in whatever way

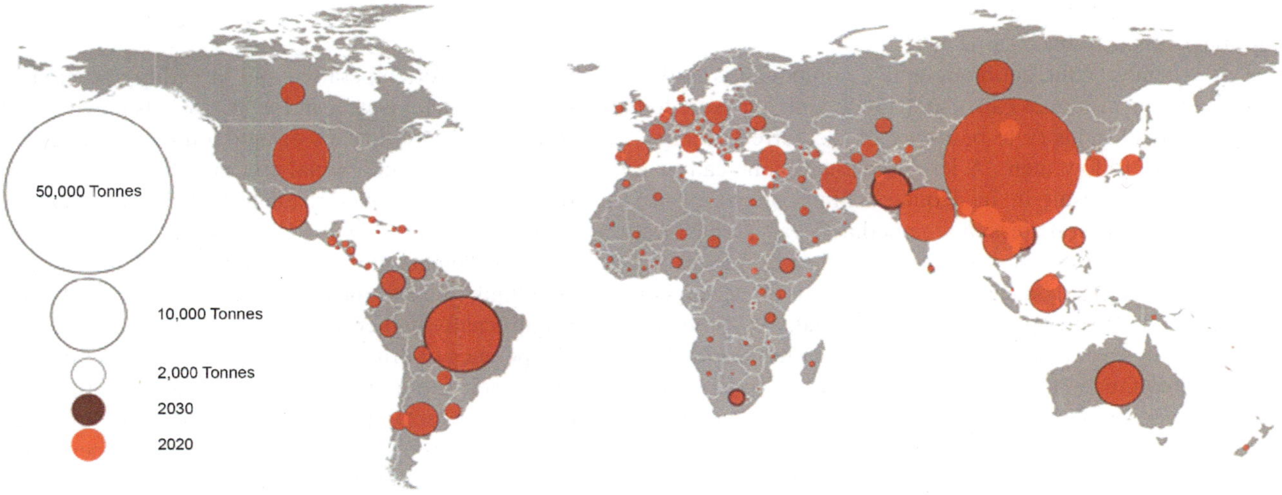

FIGURE 16.1 Global antimicrobial usage in livestock. Global trends in antimicrobial use in food-producing animals: 2020–2030. Map lines delineate study areas and do not necessarily depict accepted national boundaries.
Mulchandani, R., Wang, Y., Gilbert, M., Van Boeckel, T. P., Odetokun, I. A., 2023. Global trends in antimicrobial use in food-producing animals: 2020 to 2030. PLOS Global Public Health 3(2). https://doi.org/10.1371/journal.pgph.0001305.

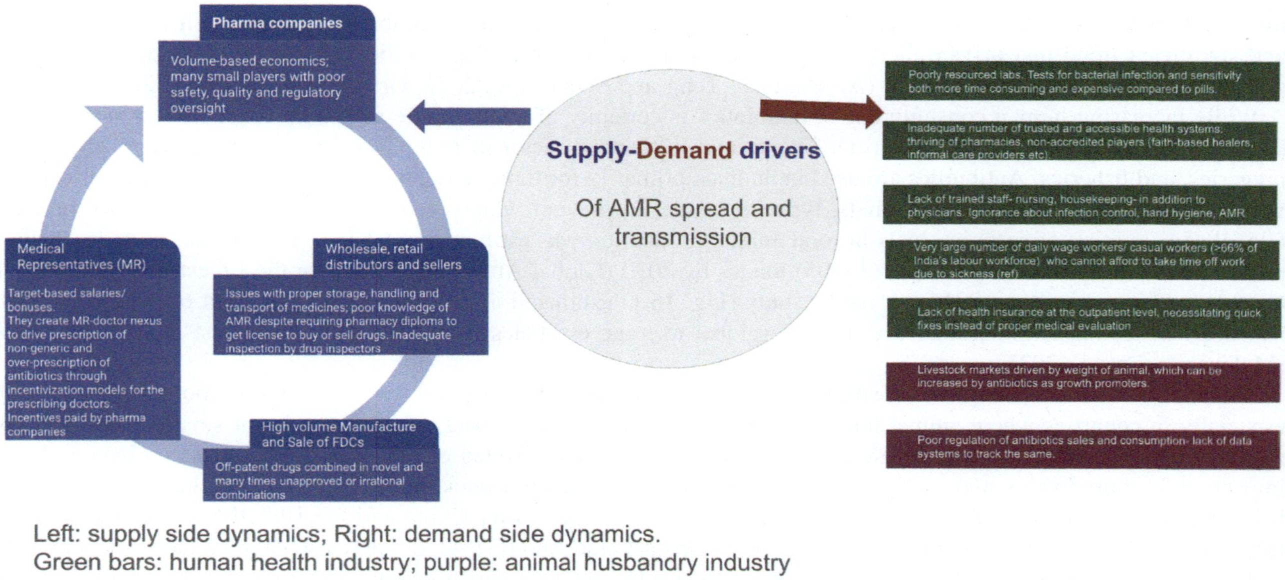

FIGURE 16.2 Drivers of AMR from supply (pharmaceutical) and demand (consumer) sides. Pictorial representation of the drivers of AMR from the pharmaceutical industry (supply) side and the health/animal husbandry industry (demand) side.
[*Picture created using insights from* (Kotwani et al., 2022; Yuktha et al., 2024; Barker et al., 2017).

they desire; others use off-patent drugs to combine them in novel, and frequently unapproved (Mehta et al., 2022) or irrational combinations (Anand et al., 2022), to create 'new' drugs (so-called fixed drug combinations or FDCs) which do not fall under the government price caps. India is the largest manufacturer and consumer of FDCs, with over 37% of the total antibiotic sales being that of FDCs (Bortone, 2021). Of these, over 41% were not recommended by the WHO (Mehta et al., 2022). FDCs not only are marketed as being more effective than single-drug regimens but also expose the individual to high levels of unnecessary antibiotics, increasing the risk of development of AMR.

16.5 Storage, transport, retention, and disposal policies and practices for antibiotics

India, unfortunately, has been in the news frequently in the past few years for the manufacture and distribution of counterfeit or substandard drugs (Khangan and Tiple, 2025; The Economic Times, 2025). But even more than manufacturing substandard drugs is the problem of incorrect storage and retention that degrades even good-quality drugs and makes them substandard.

This potency of the antibiotic active compound is critical and is sensitive to temperature and humidity variations, as well as time. Antibiotics can also degrade with exposure to harsh sunlight. Thus it is critical for the various players in the pharmaceutical industry to be aware of the correct storage, transport, and retention times of antibiotics (and indeed, all drugs) and to follow them. A less potent antibiotic can give rise to AMR due to suboptimal concentrations of the active compound that result in inadequate killing of pathogens. In countries with weaker regulatory and implementation landscapes, or in rural areas, it is remarkably easy to have poorly stored drugs or even expired drugs in the market. Storage conditions depend on the type of antibiotics: Some are stable at ambient temperature, while others need refrigeration. However, what is 'ambient or room temperature'? Most people tend to think that this means around 18°C–25°C. Ambient temperature in several parts of India, especially in summer, can be significantly higher than this. Rising temperatures and heat waves in several parts of the country and the world also affect the potency of drugs, including antibiotics. While India has clear requirements for good manufacturing practices of antibiotics, there are no mandatory standards for appropriate storage practices. There are no guidelines for maintaining medicines at the correct temperatures in retail stores, other than that they must have a refrigerator. Thus electricity issues can play havoc with refrigerator temperatures, and there are no quality control checks for temperature maintenance or the mandatory provision of a backup generator in case of electric outages. Similarly, while cold chain logistics are expected and mandatory for drugs that are to be kept at 4°C or frozen, those that can be at ambient temperature can be exposed to exceptionally high temperatures during transit and distribution. Thus given global warming and the vulnerability of several parts of the world towards unseasonably high temperatures, it is critical that our regulators, the pharma industry, and pharma retailers follow correct guidelines to ensure drug potency and reduce drug degradation. Similarly, studies indicate that end users are ill-informed about drug storage practices (Khan et al., 2022; Karuniawati et al., 2025) and efforts to educate the general public on appropriate drug storage are also critical.

Expired drugs are meant to be incinerated, as per the CDSCO guidelines (Central Drugs Standard Control Organisation, 2025). Households are expected to return unused drugs to pharmacies (World Health Organization, 2006). However, as can be expected, very few actually do (Karol et al., 2021). Knowledge of the correct disposal of antibiotics is not commonly known. Several studies (David, Jayachandran and Palatty, 2024; Zalpuri et al., 2022) indicate the flushing of drugs, including antibiotics, down drains, or discarding them into garbage bins as common disposal practices (Kumari et al., 2022). These practices are irresponsible and lead to contamination of soil and water and their ingestion by animals such as stray dogs, cattle, and vermin, which in turn can get sick or die, leading to further public health concerns.

Analysis of returned drugs in countries where such practices are more prevalent indicates that antiinfectives constitute the majority of drugs returned, and the vast majority of them are expired. It also indicates overprescribing and underutilisation, both of which are known drivers of AMR (Romanelli and Lucente, 2022). Thus the responsibilities of the pharma industry towards AMR control go far beyond merely manufacturing and pushing sales, but also in appropriate storage, retention and disposal.

16.6 Challenges faced by the pharmaceutical industry on the future of antimicrobials esp in view of waste management and novel drug discovery

The pharmaceutical industry encounters significant economic hurdles in antibiotic development due to high research costs, low returns on investment, and the necessity for stewardship programs that restrict usage to preserve efficacy, creating a market failure that discourages innovation (Salam et al., 2023; Dunachie and Pizza, 2025). To address this systemic issue, successful innovation relies on resource mobilisation, requiring highly visible funding commitments from both public and private investors to demonstrate that antimicrobial R&D is a priority area and to inspire investor confidence (Rodrigues et al., 2023). The lack of return on investment has particularly caused many major pharmaceutical companies to shift their focus to more profitable areas, such as chronic diseases and cancer treatments (Loo, Song Lai and Mansor, 2020). This redirection of research efforts has led to a significant decline in the development of new antimicrobial agents, exacerbating the global challenge of antimicrobial resistance (Dhingra et al., 2020). Consequently, academic researchers and smaller biotechnology firms are increasingly shouldering the responsibility for advancing antimicrobial discovery, often with limited resources and facing substantial developmental timelines (Yam et al., 2019). The quality, cost of goods, manufacturing, distribution, price control, etc., are all the drivers that contribute to adherence to quality products in the market. The absence of attractive market incentives and consistent funding streams further compounds these challenges, making it difficult for even promising antimicrobial candidates to advance through the rigorous and expensive stages of clinical development (Mudenda et al., 2023). Moreover, the stringent regulatory requirements and the extended timelines associated with clinical trials for antimicrobial agents further deter investment, as these factors prolong the period before any potential market entry and revenue generation (Dunachie and Pizza, 2025) This situation is further complicated by the short duration of antimicrobial treatments compared to therapies for chronic diseases, which inherently limits the potential for sustained revenue generation and reduces the economic viability of developing new antibiotics (Dhingra et al., 2020; Chaudhry and Tomar, 2017). This economic disincentive, coupled with the societal imperative to preserve the effectiveness of new antimicrobials by limiting their use, creates a challenging landscape for pharmaceutical companies (Årdal et al., 2020; Plackett, 2020). This dilemma is intensified by the preference of health regulators for generic brands and reimbursement schemes that favour the cheapest available drugs, further diminishing the profitability of novel antimicrobial products (Plackett, 2020). This creates a critical need for innovative economic models, such as 'pull' incentives and delinked payment systems, to bridge the gap between high research and development costs and the low expected returns on investment, thereby stimulating pharmaceutical companies to re-engage in antimicrobial discovery (Gargate et al., 2025; Schurer et al., 2023). To conclude, the current trajectory of antimicrobial development is unsustainable, necessitating urgent and comprehensive interventions across regulatory, economic, and scientific domains to safeguard public health against the growing threat of AMR (Gargate et al., 2025).

16.7 Environmental impact of pharmaceutical industries manufacturing practices

It is an unexpected and underappreciated fact that our wastewater systems are the biggest contributors to AMR transmission. The modern wastewater system relies on the collection of faecal material and accompanying water into pipes that are underground (and can be accessed at specific points through manholes). Underground pipes from various parts of the city, like blood vessels, capture this water and merge together to form larger pipes and then head into a wastewater treatment plant, which could be of several types but typically includes a holding tank, anaerobic chamber, chlorination, other types of treatment, and finally discharge into lakes or other water bodies. Thus this system depends on the availability and proper functioning of such underground pipes. In several lower or middle-income countries, not all faecal waste finds its way into these pipes. If the speed of urban expansion exceeds the speed of laying of pipes, or if slums grow within a city, or there are illegal constructions, then the underground sewerage network is absent in these areas. Thus the faecal matter and accompanying water end up in areas which are not meant for them – usually storm water drains meant to collect rainwater, or directly to neighbouring lakes or water bodies.

AMR bacteria and ARGs in the water: Regardless of where the dirty water (called 'black' water, denoting the presence of faecal material) flows, faecal material is rich in pathogens and rich in antimicrobial resistance genes (ARGs). This is due to the presence of the same in the guts of human beings. As humans (and other animals) get exposed to antibiotics, either through medication or through other means, gut microbiota face evolutionary pressures to develop, retain, and spread ARGs. This gets shed in faeces, which then finds its way into water systems, either formal underground ones or informal drains, lakes, and other water bodies.

In addition to human faecal presence, wastewater also contains faecal matter from animals. Livestock and animal husbandry farms may or may not have special canals to remove the faecal matter of their cattle into a separate treatment facility. In places like India, where most farms are small, it is very likely that faecal matter from animals enters open canals or drains and flows directly into lakes in an untreated manner. This same lake water is used for agricultural irrigation.

The role of the pharmaceutical industries is critical here. Policies in India to control and treat effluents from pharmaceutical manufacturing units in India to remove antibiotics (specifically the API or active pharmaceutical ingredients) before allowing their flow into neighbouring streams or lakes have been sporadic and not widely accepted. While WHO policies recommend strict management of pharmaceutical effluent, suggesting that wastewater from antibiotic manufacturing units should achieve at least 99% removal of antibiotic residues, it was not until 2020 that a recommendation by the Ministry of Environment, Forest and Climate Change (MoEF&CC) introduced a proposal to limit the concentration of antibiotic residues in treated effluent from the manufacture of more than 121 drugs (Ministry of Environment, Forests and Climate Change, 2020). However, by August 2021, this recommendation was dropped from the final rules governing pharmaceutical manufacturing. The decision to drop these recommendations is widely considered to have been due to intense lobbying by the pharmaceutical industry, citing high costs of implementation, which would naturally trickle downstream and increase the costs of the drugs themselves. Several of India's pharmaceutical manufacturing companies are considered 'small' (<Rs. 250 crore turnover), and the requirements to monitor and control the API residues in the effluents are estimated to cost between Rs. 2 and 10 crores per unit, making it impossible for such companies to take the hit (*India* Business and Trade, 2025). This is unfortunate. Several studies have categorically and without any doubt shown high levels of antibiotics in pharmaceutical effluents: a study on antibiotic residues in water samples taken from a common effluent treatment plant near Hyderabad, India, (which receives water from approximately 90 bulk drug manufacturers); nearby lakes not known to be contaminated by the treatment plant as well as wells from six nearby villages showed that antibiotic residues were present in all the wells with Ciprofloxacin, enoxacin, cetirizine, terbinafine, and citalopram detected at more than 1 μg/L in several wells. Additionally, 'Very high concentrations of ciprofloxacin (14 mg/L) and cetirizine (2.1 mg/L) were found in the effluent of the treatment plant, together with high concentrations of seven additional pharmaceuticals. Very high concentrations of ciprofloxacin (up to 6.5 mg/L), cetirizine (up to 1.2 mg/L), norfloxacin (up to 0.52 mg/L), and enoxacin (up to 0.16 mg/L) were also detected in the two lakes, which clearly shows that the investigated area has additional environmental sources of insufficiently treated industrial waste' (Fick et al., 2009).

The primordial soup that is wastewater, which contains AMR bacteria, ARGs, antibiotics, heavy metals, and a host of other organic and inorganic materials, serves as a dynamic hub for AMR evolution. Thus the effluent treatment practices of pharmaceutical companies and their policies for antibiotic disposal are important direct contributors to antibiotic levels in the environment.

Larger pharmaceutical manufacturing companies in India, whose markets are primarily global and therefore have to adhere to an international standard of manufacturing practices, have signed up for voluntary effluent treatment, as framed by the AMR Industry Alliance in 2018. This agreement has several aspects: a Common Antibiotic Manufacturing Framework to encourage manufacturers to review their processes and try to reduce their effluent concentrations of antibiotics (while a broad framework was suggested in 2018, this was followed by an Antibiotic Manufacturing Standard in 2022 which requires accredited companies to adhere to certain quality and manufacturing standards which also include effluent treatment); a Predicted No-Effect Concentration (PNEC) to scientifically define and justify the concentrations at which an antibiotic is not expected to contribute to the development of AMR (AMR Industry Alliance, 2026); and third party certification of standards for control of environmental contamination (by British Standards Institute, BSI) (AMR Industry Alliance, 2023). In recent times, several pharmaceutical companies in India are becoming compliant to these standards (e.g. India's Orchid Pharma's satisfactory PNEC values, among other qualifying criteria, enabled GARDP to choose them to partner for licensing Cefiderocol manufacturing) (GARDP and Orchid Pharma Sublicence and Technology Transfer Agreement, 2023), though it is not known as to what proportion of India's over 10,500 pharmaceutical manufacturing units follow these guidelines.

16.8 The road ahead

Promising scientific frontiers, such as phage therapy, AI-driven drug discovery, and microbiome modulation, offer hope for the future. Policy recommendations, including de-linkage models and subscription-based incentives, are essential for sustainable antibiotic development. This section emphasises the need for a global treaty on AMR, reworking financial models for pharmaceutical companies in the antibiotic business, and the reframing of the AMR-antibiotic issue from pharmaceutical industries perceived to be agents promoting AMR to agents promoting AMS towards mutual benefit to all stakeholders. In addition, surveillance in the community is essential in addition to the tertiary care hospital data alone, especially of antibiotic dispensing, the shifting of antibiotics from first line to second and third line (Paul, 2026) to establish the true rates of AMR and its influencers.

References

Ab Rahman, N., Teng, C.L., Sivasampu, S., 2016. Antibiotic prescribing in public and private practice: a cross-sectional study in primary care clinics in Malaysia. BMC Infectious Diseases 16 (1). Available at https://doi.org/10.1186/s12879-016-1530-2.

Abdallah, E.M., Alhatlani, B.Y., de Paula Menezes, R., Martins, C.H.G., 2023. Back to nature: medicinal plants as promising sources for antibacterial drugs in the post-antibiotic era. Plants 12 (17), 3077. Available at https://doi.org/10.3390/plants12173077.

Alas Portillo, M., Gómez Rodríguez, I.M., Gradmann, C., Kirchhelle, C., Leisner, J.J., Martinenghi, L.D., et al., 2024. Paradoxes of the antibiotic pipeline. Humanities and Social Sciences Communications 11 (1). Available at https://doi.org/10.1057/s41599-024-03211-1, https://www.nature.com/palcomms/.

Alghamdi, S., Berrou, I., Aslanpour, Z., Mutlaq, A., Haseeb, A., Albanghali, M., et al., 2021. Antimicrobial stewardship programmes in saudi hospitals: evidence from a national survey. Antibiotics 10 (2), 1–11. Available at https://doi.org/10.3390/antibiotics10020193, https://www.mdpi.com/2079-6382/10/2/193/pdf.

AMR Industry Alliance, 2023. AMR Alliance Follows Up Global Industry Standard With Responsible Manufacturing Certification. Available from: https://www.amrindustryalliance.org/mediaroom/amr-alliance-follows-up-global-industry-standard-with-responsible-manufacturing-certification/.

AMR Industry Alliance, 2026. Available from: https://www.amrindustryalliance.org/2026-changes-to-the-pnec/.

Anand, P., Kaur, N., Verma, V., Shafiq, N., Malhotra, S., 2022. Assessment of rationality of available fixed dose combinations of antibiotics in India. Expert Review of Anti-Infective Therapy 20 (5), 797–808. Available at https://doi.org/10.1080/14787210.2022.2015324.

Årdal, C., Balasegaram, M., Laxminarayan, R., McAdams, D., Outterson, K., Rex, J.H., et al., 2020. Antibiotic developments—economic, regulatory and societal challenges. Nature Research, Norway Nature Reviews Microbiology 18 (5), 267–274. Available at https://doi.org/10.1038/s41579-019-0293-3, http://www.nature.com/nrmicro/index.html.

Barker, A.K., Brown, K., Ahsan, M., Sengupta, S., Safdar, N., 2017. What drives inappropriate antibiotic dispensing? A mixed-methods study of pharmacy employee perspectives in Haryana, India. BMJ Open 7 (3), e013190. Available at https://doi.org/10.1136/bmjopen-2016–013190.

Bellucci, M.C., Romani, C., Sani, M., Volonterio, A., 2024. Dual antibiotic approach: synthesis and antibacterial activity of antibiotic–antimicrobial peptide conjugates. Antibiotics 13 (8), 783. Available at https://doi.org/10.3390/antibiotics13080783.

Blaskovich, M.A.T., Cooper, M.A., 2025. Antibiotics re-booted-time to kick back against drug resistance. npj Antimicrobials and Resistance 3 (1). Available at https://doi.org/10.1038/s44259-025-00096-1.

Bo, L., Sun, H., Li, Y.D., Zhu, J., Wurpel, J.N.D., Lin, H., et al., 2024. Combating antimicrobial resistance: the silent war. Frontiers in Pharmacology 15. Available at https://doi.org/10.3389/fphar.2024.1347750, http://www.frontiersin.org/Pharmacology.

Bortone, B., Jackson, C., Hsia, Y., Bielicki, J., Magrini, N., Sharland, M., et al., 2021. High global consumption of potentially inappropriate fixed dose combination antibiotics: analysis of data from 75 countries. PLoS One 16 (1), e0241899. Available at https://doi.org/10.1371/journal.pone.0241899.

Business and Trade, 2025. Compliance or closure? India's small pharma dilemma India. Available from: https://www.linkedin.com/pulse/compliance-closure-indias-small-pharma-dilemma-india-business-trade-js0lc/.

Central Drugs Standard Control Organisation, 2025. 5 26 2026 3 7 Guidelines on disposal of expired/unused drugs. Available from: https://cdsco.gov.in/opencms/resources/UploadCDSCOWeb/2018/UploadPublic_NoticesFiles/Guidance%20document%20on%20disposal.pdf.

Chaudhry, D., Tomar, P., 2017. Antimicrobial resistance: the next BIG pandemic. International Journal Of Community Medicine And Public Health 4 (8), 2632. Available at https://doi.org/10.18203/2394–6040.ijcmph20173306.

Coxe, T., Azad, R.K., 2023. Silicon versus superbug: assessing machine learning's role in the fight against antimicrobial resistance. Antibiotics 12 (11). Available at https://doi.org/10.3390/antibiotics12111604, http://www.mdpi.com/journal/antibiotics.

David, R., Jayachandran, M., Palatty, P.L., 2024. A study on household medicine wastage and disposal practices with screening of water samples for antibiotic presence in a South Indian district. International Journal of Community Medicine And Public Health 11 (7), 2787–2793. Available at https://doi.org/10.18203/2394–6040.ijcmph20241839.

Dhingra, S., Rahman, N.A.A., Peile, E., Rahman, M., Sartelli, M., Hassali, M.A., et al., 2020. Microbial resistance movements: an overview of global public health threats posed by antimicrobial resistance, and how best to counter. Frontiers in Public Health 8. Available at https://doi.org/10.3389/fpubh.2020.535668, http://journal.frontiersin.org/journal/public-health/section/public-health-education-and-promotion#archive.

Dunachie, S.J., Pizza, M., 2025. Global antimicrobial resistance—the ostrich's head is in the sand. PLoS Biology 23 (9), e3003382. Available at https://doi.org/10.1371/journal.pbio.3003382.

Farha, M.A., Tu, M.M., Brown, E.D., 2025. Important challenges to finding new leads for new antibiotics. Current Opinion in Microbiology 83. Available at https://doi.org/10.1016/j.mib.2024.102562, https://www.sciencedirect.com/science/journal/13695274.

Fick, J., Söderström, H., Lindberg, R.H., Phan, C., Tysklind, M., Larsson, D.G.J., 2009. Contamination of surface, ground, and drinking water from pharmaceutical production. Environmental Toxicology and Chemistry 28 (12), 2522–2527. Available at https://doi.org/10.1897/09-073.1.

Fischbach, M.A., Walsh, C.T., 2009. Antibiotics for emerging pathogens. Science (New York, N.Y.) 325 (5944), 1089–1093. Available at https://doi.org/10.1126/science.1176667.

GARDP and Orchid Pharma Sublicence and Technology Transfer Agreement, 2023. Available from: https://gardp.org/wp-content/uploads/2023/09/GARDP-Orchid-Manufacturing-Sublicense-Agreement_Final_Redacted-Version.pdf#page=52.

Gargate, N., Laws, M., Rahman, K.M., 2025. Current economic and regulatory challenges in developing antibiotics for Gram-negative bacteria. npj Antimicrobials and Resistance 3 (1). Available at https://doi.org/10.1038/s44259-025-00123-1.

Global Antimicrobial Use in the Livestock Sector OECD Committee for Agriculture, 2015. Working Party on Agricultural Policies and Markets. Available from: https://one.oecd.org/document/TAD/CA/APM/WP(2014)34/FINAL/en/pdf.

Gotham, D., Moja, L., van der Heijden, M., Paulin, S., Smith, I., Beyer, P., 2021. Reimbursement models to tackle market failures for antimicrobials: Approaches taken in France, Germany, Sweden, the United Kingdom, and the United States. Health Policy 125, 296–306. Available at https://doi.org/10.1016/j.healthpol.2020.11.015.

Government of India, Ministry of Finance, 2025. Share of government health expenditure in the total health expenditure has increased from 29.0 % to 48.0 % between FY15 AND FY22: economic survey 2024-25. Available from: https://www.pib.gov.in/PressReleasePage.aspx?PRID=2097868&reg=3&lang=2#:~:text=HEALTH%20EXPENDITURE:-,As%20per%20Economic%20Survey%2C%20the%20Total%20Health%20Expenditure%20(THE),of%20India's%20most%20vulnerable%20populations.

Government of India, Ministry of Fisheries, Animal Husbandry and Dairying 2024. YEAR END REVIEW 2024: DEPARTMENT OF ANIMAL HUSBANDRY AND DAIRYING. Available from: https://www.pib.gov.in/PressReleasePage.aspx?PRID=2086052&reg=3&lang=2.

Granlund, D., Zykova, Y.V., 2021. Can private provision of primary care contribute to the spread of antibiotic resistance? A study of antibiotic prescription in Sweden. PharmacoEconomics - Open 5 (2), 187–195. Available at https://doi.org/10.1007/s41669-020-00234-7, http://www.springer.com/adis/journal/41669.

Howard-Anderson, J., Dai, W., Yahav, D., Hamasaki, T., Turjeman, A., Koppel, F., et al., 2022. A desirability of outcome ranking analysis of a randomized clinical trial comparing seven versus fourteen days of antibiotics for uncomplicated gram-negative bloodstream infection. Open Forum Infectious Diseases 9 (6). Available at https://doi.org/10.1093/ofid/ofac140, https://academic.oup.com/ofid.

Iskandar, K., Murugaiyan, J., Hammoudi Halat, D., El Hage, S., Chibabhai, V., Adukkadukkam, S., et al., 2022. Antibiotic discovery and resistance: the chase and the race. Antibiotics 11 (2), 182. Available at https://doi.org/10.3390/antibiotics11020182.

Kadarpeta, R.S.R., Anand, J.S., Achungura, G., 2024. Strengthening public financing of primary healthcare in India: a perspective. Health Services Insights 17. Available at https://doi.org/10.1177/11786329241249289.

Karol, J., Marwa, G., Mcharo, S., Mwita, D., Katabalo, D., Ruganuza, A., et al., 2021. Isangula, disposal practices of expired and unused medications among households in Mwanza, Tanzania. PLoS One 16 (2), e0246418. Available at https://doi.org/10.1371/journal.pone.0246418.

Karuniawati, H., Rahman, R.A., Hossain, M.S., Wahyuni, A.S., Hakim, D.Z.'aini A., Kusuma, I.Y., 2025. Assessment of storage behavior of antibiotics and influence factors among household members in Boyolali, Indonesia: a cross-sectional study. Open Public Health Journal 18. Available at https://doi.org/10.2174/0118749445388122250619115503.

Khan, F.U., Mallhi, T.H., Khan, Q., Khan, F.U., Hayat, K., Khan, Y.H., et al., 2022. Assessment of antibiotic storage practices, knowledge, and awareness related to antibiotic uses and antibiotic resistance among household members in post-conflict areas of Pakistan: Bi-central study. Frontiers in Medicine 9. Available at https://doi.org/10.3389/fmed.2022.962657.

Khangan, V.R., Tiple, R.H., 2025. The challenge of counterfeit drugs in Indian market: a comprehensive review. Indian Journal of Pharmacy Practice 18 (4), 352–361. Available at https://doi.org/10.5530/ijopp.20250285.

Klug, D.M., Idiris, F.I.M., Blaskovich, M.A.T., von Delft, F., Dowson, C.G., Kirchhelle, C., et al., 2021. There is no market for new antibiotics: this allows an open approach to research and development. Wellcome Open Research 6, 146. Available at https://doi.org/10.12688/wellcomeopenres.16847.1.

Kotwani, A., Bhanot, A., Singal, G.L., Gandra, S., 2022. Marketing and distribution system foster misuse of antibiotics in the community: insights from drugs wholesalers in India. Antibiotics 11 (1). Available at https://doi.org/10.3390/antibiotics11010095, https://www.mdpi.com/2079-6382/11/1/95/pdf.

Kumar, A., 2023. The transformation of the Indian healthcare system. Cureus 15 (5). Available at https://doi.org/10.7759/cureus.39079.

Kumari, A., Alam, M.S., Kujur, M., Kumar, S., 2022. The pattern of disposal practices of unused and expired medications among healthcare professionals: a cross-sectional survey in Rajendra Institute of Medical Sciences, Ranchi, Jharkhand. Cureus 14 (8). Available at https://doi.org/10.7759/cureus.27555.

Loo, E., Song Lai, K., Mansor, R., 2020. Antimicrobial Usage and Resistance in Dairy Cattle Production. IntechOpen. Available at https://doi.org/10.5772/intechopen.81365.

Maglangit, F., Zhang, Y., Kyeremeh, K., Deng, H., 2020. Discovery of new antibacterial accramycins from a genetic variant of the soil bacterium, *Streptomyces* sp. MA37. Biomolecules 10 (10), 1464. Available at https://doi.org/10.3390/biom10101464.

Maharramov, E., Czikkely, M.S., Szili, P., Farkas, Z., Grézal, G., Daruka, L., et al., 2025. Exploring the principles behind antibiotics with limited resistance. Nature Research, Hungary Nature Communications 16 (1). Available at https://doi.org/10.1038/s41467-025-56934-3, https://www.nature.com/ncomms/.

Majhi, M.M., Mishra, A., Garg, S., Kumar, R., 2024. How healthy is the health budget of Amrit Kaal: 2023-24? Journal of Family Medicine and Primary Care 13 (1), 5–9. Available at https://doi.org/10.4103/jfmpc.jfmpc_398_23.

McEnany, M., Outterson, K., 2024. Changes in revenues associated with antimicrobial reimbursement reforms in Germany. Humanities and Social Sciences Communications 11 (1). Available at https://doi.org/10.1057/s41599-024-03374-x.

Mehta, A., Brhlikova, P., McGettigan, P., Pollock, A., Roderick, P., Hasan Farooqui, H., 2022. Systemic antibiotic sales and WHO recommendations, India. Bulletin of the World Health Organization 100 (10), 610–619. Available at https://doi.org/10.2471/blt.22.287908.

Miethke, M., Pieroni, M., Weber, T., Brönstrup, M., Hammann, P., Halby, L., et al., 2021. Towards the sustainable discovery and development of new antibiotics. Nature Reviews Chemistry 5 (10), 726–749. Available at https://doi.org/10.1038/s41570-021-00313-1, https://www.nature.com/natrevchem/.

Ministry of Environment, Forests and Climate Change, 2020. Draft Notification. Available from: https://embed.documentcloud.org/documents/6770398-The-following-draft-of-the-notification-which/.

Mudenda, S., Chabalenge, B., Daka, V., Mfune, R.L., Salachi, K.I., Mohamed, S., et al., 2023. Global strategies bto combat antimicrobial resistance: a one health perspective. Pharmacology & Pharmacy 14 (8), 271–328. Available at https://doi.org/10.4236/pp.2023.148020.

Mulchandani, R., Wang, Y., Gilbert, M., Van Boeckel, T.P., Odetokun, I.A., 2023. Global trends in antimicrobial use in food-producing animals: 2020 to 2030. PLOS Global Public Health 3 (2), e0001305. Available at https://doi.org/10.1371/journal.pgph.0001305.

Murugaiyan, J., Anand Kumar, P., Rao, G.S., Iskandar, K., Hawser, S., Hays, J.P., et al., 2022. Progress in alternative strategies to combat antimicrobial resistance: focus on antibiotics. Antibiotics 11 (2). Available at https://doi.org/10.3390/antibiotics11020200, https://www.mdpi.com/2079-6382/11/2/200/pdf.

Parkhill, S.L., Johnson, E.O., 2024. Integrating bacterial molecular genetics with chemical biology for renewed antibacterial drug discovery. Biochemical Journal 481 (13), 839–864. Available at https://doi.org/10.1042/BCJ20220062, https://portlandpress.com/biochemj/article/481/13/839/234662/Integrating-bacterial-molecular-genetics-with.

Paul, M.S., 2026. A brief historical perspective of antimicrobial resistance in India. ASM mBio in press.

Paul-Satyaseela, M., 2024. Invention of Enmetazobactam: An Indian Triumph in Antimicrobial Drug Discovery. ACS Infectious Diseases 11, 1–3. Available at https://doi.org/10.1021/acsinfecdis.4c00982.

Plackett, B., 2020. Why big pharma has abandoned antibiotics. Nature 586 (7830), S50. Available at https://doi.org/10.1038/d41586-020-02884-3.

Prasad, N.K., Seiple, I.B., Cirz, R.T., Rosenberg, O.S., 2022. Leaks in the pipeline: a failure analysis of gram-negative antibiotic development from 2010 to 2020. Antimicrobial Agents and Chemotherapy 66 (5). Available at https://doi.org/10.1128/aac.00054-22, https://journals.asm.org/doi/10.1128/aac.00054-22.

Rodrigues, T.D.S.G., Maxwell, A., Mercer, D.K., Lappin, O., 2023. Antimicrobial resistance: a biochemical society position statement. Biochemist 45 (1), 33–38. Available at https://doi.org/10.1042/bio_2022_148.

Romanelli, L., Lucente, F., 2022. Analysis of medicines returned to pharmacies for disposal and estimation of the cost due to medicine wasting. Exploratory Research in Clinical and Social Pharmacy 5, 100133. Available at https://doi.org/10.1016/j.rcsop.2022.100133.

Salam, M.A., Al-Amin, M.Y., Salam, M.T., Pawar, J.S., Akhter, N., Rabaan, A.A., et al., 2023. Antimicrobial resistance: a growing serious threat for global public health. Healthcare (Switzerland) 11 (13). Available at https://doi.org/10.3390/healthcare11131946, www.mdpi.com/journal/healthcare/.

Schurer, M., Patel, R., van Keep, M., Horgan, J., Matthijsse, S., Madin-Warburton, M., 2023. Recent advances in addressing the market failure of new antimicrobials: learnings from NICE's subscription-style payment model. Frontiers in Medical Technology 5. Available at https://doi.org/10.3389/fmedt.2023.1010247, www.frontiersin.org/journals/medical-technology.

Shi, L., 2012. The impact of primary care: a focused review. Scientifica 2012. Available at https://doi.org/10.6064/2012/432892.

Singer, A.C., Xu, Q., Keller, V.D.J., Leekha, S., 2019. Translating antibiotic prescribing into antibiotic resistance in the environment: a hazard characterisation case study. PLoS One 14 (9), e0221568. Available at https://doi.org/10.1371/journal.pone.0221568.

The Economic Times, 2025. Surge in fake medicines in retail markets leaving druggists distraught: Traders' body. Available from: https://economictimes.indiatimes.com/industry/healthcare/biotech/pharmaceuticals/surge-in-fake-medicines-in-retail-markets-leaving-druggists-distraught-traders-body/articleshow/118765250.cms?from=mdr.

Walesch, S., Birkelbach, J., Jézéquel, G., Haeckl, F.P.J., Hegemann, J.D., Hesterkamp, T., et al., 2023. Fighting antibiotic resistance—strategies and (pre)clinical developments to find new antibacterials. EMBO Reports 24 (1). Available at https://doi.org/10.15252/embr.202256033, http://embor.embopress.org/.

Wells, N., Nguyen, V.K., Harbarth, S., 2024. Novel insights from financial analysis of the failure to commercialise plazomicin: implications for the antibiotic investment ecosystem. Humanities and Social Sciences Communications 11 (1). Available at https://doi.org/10.1057/s41599-024-03452-0, https://www.nature.com/palcomms/.

World Health Organization, 2006. Guidelines for the safe disposal of expired drugs. Available from: https://www.emro.who.int/images/stories/pakistan/documents/pak_documents/Guidelines_for_Expired_Medicines.pdf.

Yam, E.L.Y., Yang Hsu, L., Peng-Huat Yap, E., Yeo, T.W., Lee, V., Schlundt, J., et al., 2019. Antimicrobial resistance in the Asia Pacific region: a meeting report. Antimicrobial Resistance & Infection Control 8 (1). Available at https://doi.org/10.1186/s13756-019-0654-8.

Yuktha, K.B., Munavar, K.S.A., Sarmah, P., 2024. Hemavathi, Knowledge on antibiotic use and antibiotic dispensing practices in pharmacies in Bengaluru city. IP Innovative Publication Pvt. Ltd., India Indian Journal of Microbiology Research 11 (3), 163–169. Available at https://doi.org/10.18231/j.ijmr.2024.030, https://www.ijmronline.org/journal-article-file/22880.

Zalpuri, R., Sharma, J.K., Singh, R., Rawat, L., 2022. Disposal practices and awareness of medicine waste management among general population of Delhi-National Capital Region, India. Journal of Applied Pharmaceutical Science 12 (1), 078–083. Available at https://doi.org/10.7324/japs.2021.120107, www.japsonline.com/index.php.

Chapter 17

Farming under fire: the role of agricultural practices in antibiotic resistance development

Pimily Langthasa[1], Pakiza Begum[2], and Rhitu Kotoky[3]
[1]*Department of Zoology, Gargaon College, Gargaon, Simaluguri, Sivasagar, Assam, India,* [2]*Department of Chemistry, Gargoan College, Gargaon, Simaluguri, Sivasagar, Assam, India,* [3]*Department of Biology, University of Southern Denmark, Odense, Denmark*

17.1 Introduction

The discovery of antibiotics changed everything, transforming medicine and global food production. However, decades of excessive and often improper use of antibiotics have put significant pressure on microbial communities. This has led to the emergence and spread of antibiotic resistance (ABR) (Laxminarayan et al., 2013). What once thought was a distant threat is now a real challenge, as ABR is making it increasingly difficult to treat routine infections. Thereby, it adds risks to major medical procedures such as surgeries, cancer chemotherapy, and organ transplants (Murray et al., 2022). The World Health Organisation (WHO) ranks antimicrobial resistance (AMR) among the most urgent public health threats worldwide, and reports predict that resistant infections could cause over 10 million deaths annually by 2050 (Kraker et al., 2016). This is partly due to antibiotics becoming less effective, while resistance traits spread across bacterial groups (Brown & Wright, 2016). ABR spreads more quickly than new antibiotics can be developed, leading to situations where current drugs fail against pathogens. Alarmingly, the rate of new antibiotic discovery has dropped sharply over the past two decades, and this gap between rising resistance and available treatments is widening (Brüssow, 2024). Additionally, the improper use of antibiotics in healthcare, agriculture, and the environment further increases selective pressure, allowing microbial resistance to emerge and spread (Belay et al., 2024).

The widespread and continuous discharge of antibiotics into the environment, arising from agricultural farms, hospitals, municipal waste, and industries, poses a threat to organisms in these environments. Antibiotics are not fully metabolised in humans and animals, so a large portion of these drugs ends up in the environment. Municipal wastewater, animal manure, sewage sludge, and biosolids often carry discharged antibiotics onto agricultural land (Cycoń et al., 2019). Environmental stressors like heavy metals, pesticides, and rising temperatures can also boost the selective pressure of AMR genes, even without antibiotics present (Helliwell et al., 2021; Nawaz et al., 2023). These stressors can induce broad stress responses, leading to cross-resistance to antibiotics, including the induction of efflux pumps (Katz et al., 2013). Furthermore, environmental stressors may encourage mutations in global regulatory networks or accelerate the movement of resistance genes through mobile genetic elements (MGEs). Thus conventionally managed soils often have more AMR genes than organically managed soils (Bassitta et al., 2022). In agricultural ecosystems, ABR factors move through wastewater discharge, biosolids, manure-treated soils, and food, connecting agricultural resistance to ecological and human health networks (Heuer et al., 2011; Cheng et al., 2016; Cycoń et al., 2019).

The amendment of organic material like animal manure as fertiliser introduces significant amounts of resistant bacteria and ARGs into soil systems (Cheng et al., 2016). Once in the soil, these elements can accumulate and spread, turning agricultural land into a reservoir of ABR (Heuer et al., 2011). Furthermore, even conventional agricultural practices, such as using inorganic fertilisers and pesticides, can alter the soil microbiome and promote AMR development (Ahmad, 2025; Kelbrick et al., 2023; Huang et al., 2025). Some research indicates that good drainage in agricultural systems can lessen the impact in both field and controlled lab studies (Huang et al., 2025; Szadziul et al., 2025). In contrast, other findings

Antimicrobial Resistance in the 21st Century. DOI: https://doi.org/10.1016/B978-0-443-44748-8.00016-6

suggest that while ARGs may remain in soil for a year or more, pathogenic bacteria from manure usually fail to survive long-term in soil environments (Peng et al., 2017, 2022).

Animal agriculture adds complexity to the ABR issue, working as a point of selection, amplification, and dissemination of resistant microbes and genes through multiple routes. Globally, about 73% of all medically important antibiotics are used in livestock production, typically at low doses for growth promotion or disease prevention (Tiseo et al., 2020; Magnusson, 2022). This creates ideal conditions for the growth of resistant bacteria and the spread of resistance genes. The connection is evident: Higher antibiotic use in livestock leads to higher resistance levels (Wang et al., 2023; Enshaie et al., 2025). Changes in policy, such as the European Union's ban on antibiotic growth promoters and the emergence of monitoring programmes, reflect increasing awareness. However, environmental pathways, especially those linked to manure application and leakage, are still critical yet underappreciated routes for the spread of resistance genes beyond farms.

Using antibiotics more responsibly is essential, but emerging evidence indicates that merely reducing antibiotic use may not be sufficient to reverse resistance trends (Mudenda et al., 2023). The old belief that resistance always incurs a fitness cost to bacteria is being reevaluated. In reality, bacteria often balance this with compensatory mutations and genetic coselection with other genes (Andersson & Hughes, 2010). These patterns highlight that the AMR crisis is an interconnected system rather than a series of isolated issues. This chapter discusses current knowledge on three different dimensions of agricultural ABR: (1) the contribution of farming practices, including high antibiotic use in livestock, to resistance emergence, (2) different environmental transport systems or mechanisms mediating the transfer of farm-derived resistance to human population, and (3) the epidemiological and molecular evidence linking antimicrobial consumption to adverse clinical outcomes in human medicines.

17.2 Historical context and scale of agricultural antibiotic use

The use of antibiotics in agriculture started in the late 1940s and early 1950s. The growth-promoting effect of antibiotics was discovered in the 1940s when researchers found that animals fed dried mycelia of *Streptomyces aureofaciens* containing chlortetracycline residues showed improved growth (Castanon, 2007). This led the US Food and Drug Administration to approve the use of antibiotics as feed additives without prescriptions in 1951 (Jones & Ricke, 2003). Since then, antibiotic use in agriculture has grown significantly. Recent estimates suggest that more than 67% of global consumption of antibiotics is for agricultural purposes (Van Boeckel et al., 2019). Limited historical data from 1970 is available; however, evidence from 2013 indicates that use may be stabilising or declining in some areas due to policy changes (Van Boeckel et al., 2019; Tiseo et al., 2020).

Remarkably, projections for 2030 show that terrestrial food-producing animals will account for 73.7%, aquaculture for 5.7%, and human use for 20.5% of global antibiotic consumption (Tiseo et al., 2020). Global antibiotic use could reach about 143,481 tons by 2040, representing a 29.5% increase from the 2019 baseline of approximately 110,777 tons. Alternative scenarios suggest that these projections could vary from an increase of 14.2% to a decrease of 56.8%, based on changes in livestock biomass and antibiotic use intensity (Acosta et al., 2025). Agricultural antibiotic consumption varies significantly by region, with Asia currently accounting for 67% of global agricultural antibiotic consumption hotspots. In 2010 China alone was responsible for 23% of global antimicrobial use in food animals, the highest share of any country (Van Boeckel et al., 2019). The United States (13%), Brazil (9%), India (3%), and Germany (3%) followed as major consumers (Van Boeckel et al., 2019). By 2030, China's share is expected to rise to 30%, while the United States, Brazil, India, and Mexico are projected to account for 10%, 8%, 4%, and 2%, respectively. Notably, the most rapid relative increases in antimicrobial consumption for 2030 are expected in Myanmar, Indonesia, Nigeria, Peru, and Vietnam, indicating a shift towards more intensive livestock production systems in low- and middle-income countries (Van Boeckel et al., 2019).

Governments around the world have supported the 79th United Nations General Assembly (UNGA) declaration, committing to a significant reduction in global antimicrobial use in the agrifood system by 2030 (UNGA, 2024). Additionally, 47 countries have promised to decrease antimicrobial use in agrifood systems by 30–50% by 2030 (FAO & VMD, 2022). These policies have begun to reduce agricultural antibiotic consumption in several high-income countries. China demonstrated measurable reductions in veterinary antimicrobial sales between 2013 and 2017 following targeted policy measures, and global agricultural antibiotic use fell between 2016 and 2018. These developments indicate that regulatory actions and stewardship programmes are effective in reducing consumption. They highlight the need for ongoing monitoring and coordinated policy efforts on an international scale.

17.3 Mechanisms: how agricultural practices drive antibiotic resistance

The resistant bacteria are not just passive byproducts of antibiotic use; rather, they are active players in microbial ecosystems shaped by both antimicrobial and nonantimicrobial pressures (Fig. 17.1). Such stress not only alters the

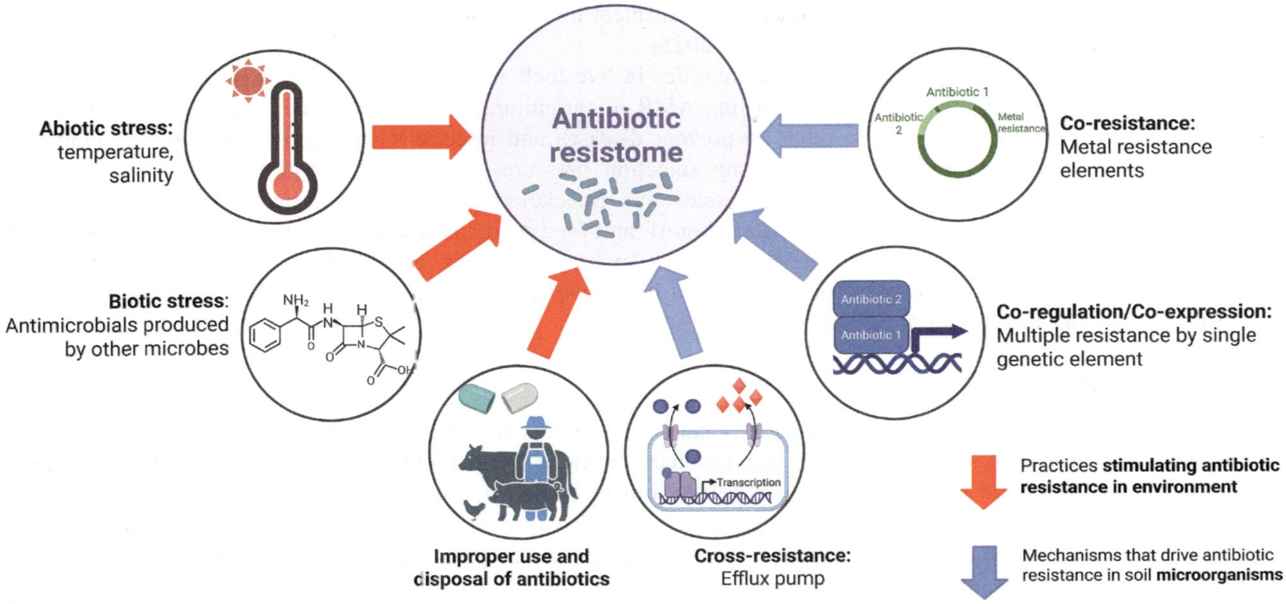

FIGURE 17.1 **Drivers of antibiotic resistance in soil microbiome.** Representation of the key stressors and mechanisms that drive antibiotic resistance in soil microorganisms. Direct stressors, abiotic (such as changes in temperature, drought or moisture fluctuations, salinity, metals, and UV exposure) and biotic (such as microbial competition) influence microbial physiology and resistance to antimicrobials. Additionally, improper disposal of antibiotics in environments further intensifies selective pressure. Together, these conditions promote key resistance pathways: coresistance, coregulation, and cross-resistance.

resistance pattern but also reshapes the entire microbiome and ecosystem. Agricultural soils, water bodies, and even animal and human microbiomes have become long-term reservoirs for resistance genes and mobile genetic elements. Even at low concentrations, antibiotic residues in soil and runoff create continuous selective pressure that keeps resistant populations stable, while altering the nutrient cycling and overall microbial diversity (Kümmerer, 2009; Cycoń et al., 2019; Li et al., 2023). At the evolutionary level, such strong selection pressure enhances adaptive processes through mutations and rapid exchange of genes (Davies & Davies, 2010). Mobile genetic elements, including plasmids, transposons, and integrons, act as shuttles transferring resistance genes across diverse species lines and communities (Allen et al., 2010; Partridge et al., 2018). Over time, these adaptations become stable and established, making it difficult to reverse the resistance (Andersson & Hughes, 2010).

Modern agriculture is one of the major contributors to the global AMR crisis, not only because of how many antibiotics it uses but also how those practices interact with complex biological and environmental systems. Although the large volume of antibiotics used is a major contributor to the problem, the deeper issue is the interplay between antibiotics, complex microbial communities, environmental stressors, and ecological networks that helps to magnify and spread the resistance far beyond the farm.

In agricultural systems, low-dose antibiotics are often used for growth promotion or disease prevention, creating conditions where only resistant microbes thrive. Subsequently, the resistant microbes dominate the microbiomes of animals, soils, and waters (Manyi-Loh et al., 2018). Additionally, abiotic stressors like heavy metals, pesticides, and disinfectants further intensify this selection, often through shared genetic mechanisms that bind resistance traits together (Bobate et al., 2023). Horizontal gene transfer (HGT) and MGEs then accelerate this resistance spread through environmental microbiomes and eventually reach towards human and other ecosystems (Zhang et al., 2022; Li et al., 2024).

17.3.1 Direct selection pressure

The strongest single mechanism driving AMR in agriculture is direct selection by antibiotics. When antibiotics are present in the gut, tissues, soil, or water, they kill susceptible bacteria and allow resistant variants to proliferate. Low-dose and prolonged exposures, typical of growth-promotion treatments, are especially influential because they do not kill bacteria outright but instead give resistant mutants evolutionary advantages (Andersson & Hughes, 2010; Nazir et al., 2025). Using antibiotics in low doses for longer time periods and at higher intensity provides microorganisms with a prolonged time to adjust and develop resistance. Even therapeutic doses can select for resistance when treatment durations are too short or when resistant subpopulations preexist (Nazir et al., 2025).

This selection pressure reshapes microbial communities in livestock animals, manure, soil, and aquatic environments and remains the most noteworthy mechanism causing AMR in agricultural practices. Subtherapeutic doses are regularly given as feed additives to boost growth rates, to prevent diseases and infections, aiming to improve feed conversion efficiency. However, this practice creates a strong selection pressure, removing susceptible bacterial strains while permitting resistant variants to survive and proliferate (Van Boeckel et al., 2019). Eventually, these resistant strains become dominant and readily exit farms via manure, runoff, and food products, contributing to environmental and human exposure. For example, commonly used antibiotic classes such as β-lactams, tetracyclines, and macrolides have been directly linked to resistance evolution in zoonotic species like *Campylobacter*, *Salmonella*, and *Escherichia coli* (Manyi-Loh et al., 2018; Van Boeckel et al., 2019).

17.3.2 Coselection and cross-resistance

Resistance is not driven by antibiotic use alone. Nonantibiotic stress in agricultural environments, particularly heavy metals, biocides, herbicides, and nanoparticles, can coselect for ARGs located on the same genetic element, meaning that selection for one resistance trait simultaneously selects others (Baker-Austin et al., 2006). Accordingly, coselection implies the simultaneous selection for multiple resistance traits, such as antibiotics, biocides, and heavy metals, when their resistance genes are located on the same MGE (Wales & Davies, 2015; Li et al., 2025). Thus the exposure to a single stress factor can maintain multiple resistance genes at once.

The occurrence of coselection can be explained through the following processes: (1) coresistance, where several resistance genes are present together on a single mobile genetic element, and selection for one resistance gene (metal resistance) can maintain other genes (ABR). Heavy metals like copper and zinc are used as feed additives and disinfectants, and the metal resistance genes often cooccur with ABR genes on the same mobile genetic elements, driving coresistance (Cheng et al., 2019; Gillieatt & Coleman, 2024). Industrial effluents, wastewater irrigation, and aquaculture runoff with heavy metals or biocides can coselect for bacteria harbouring both metal and antibiotic-resistant genes, increasing their spread and persistence in the environment (Zhu et al., 2017; Liu et al., 2022). (2) Coregulation: A coordinated expression due to exposure to any of the agents of multiple resistance genes with common promoters or regulatory elements (Lee et al., 2005; Murray et al., 2024). For example, the copper-sensing system CusRS has been shown to sense Cu^{+} and activate the efflux pump gene cluster cusCFBA and the β-lactamase gene blaCTX-M (Argudín et al., 2019; Li et al., 2025). Mercury synchronously regulates merA (mercury reductase) and the tetracycline efflux pump gene tetA through the MerR transcription factor (Li et al., 2025). Importantly, even a subinhibitory concentration of Cu^{2+} has been shown to activate cusRS, thereby increasing *Escherichia coli*'s minimum inhibitory concentration (MIC) for cefotaxime eight-fold (Wang et al., 2021).

Furthermore, another phenomenon, cross-resistance, refers to a single mechanism, such as multidrug efflux pumps, reduced permeability, or ribosomal modification, conferring resistance to multiple structurally or functionally distinct antibiotic/nonantibiotic factors. A single mechanism, such as an efflux pump or stress–response pathway, can give rise to cross-resistance, tolerance to multiple, unrelated antibiotics within the same or different classes (Gaurav et al., 2023). A common example is multidrug efflux pumps, which can extrude both heavy metals and antibiotics from the bacterial cell, providing added tolerance to several substances used in animal husbandry. Cross-resistance can develop either through a single physiological mechanism or through a genetic basis, where one gene provides tolerance to multiple stressors.

These processes operate even in the absence of antibiotics, maintaining and amplifying resistance factors across soil and water microbiomes. A recent study has shown that herbicides such as glyphosate, glufosinate, and dicamba also enrich ARGs and MGEs in soil by increasing bacterial cell permeability and enhancing plasmid transfer, connecting routine crop protection chemicals to AMR emergence (Liao et al., 2021). Studies using metagenomic investigations have also demonstrated that metal resistance genes (MRGs) and ARGs often coexist in agricultural soils (Zhu et al., 2017). The pattern of their coabundance can aid as valuable markers for surveillance and assessing risk profiling of resistomes in animal, soil, and aquatic environments. Another study found that manure from farms contains diverse antibiotic and heavy metal resistance determinants, with MRGs, specifically those providing resistance to Zn and Cu, frequently cooccur alongside multiple ARGs (Liu et al., 2022).

17.3.3 Horizontal gene transfer

HGT is a central mechanism for the spread of several antimicrobial-resistance genes throughout diversified bacterial species. It can be defined as an exchange of genetic material between two organisms that are not related by an ancestor-descendant relationship. This mechanism contributes significantly to creating novel biological traits that otherwise might take millions of years (Boto, 2009; Andam et al., 2011). The quick and broad propagation of resistance genes by HGT, followed by selective pressure from antimicrobial usage, threatens the effectiveness of these drugs (Barlow, 2009). Conjugation (plasmid transfer), transformation (uptake of environmental DNA), and transduction (bacteriophage-mediated transfer) are the main HGT modes. Some virus-like gene transfer agents further diversify the routes bacteria use to exchange genetic material (Okoń et al., 2024). These pathways allow resistance genes to spread efficiently through dense microbiomes found in animal guts, soils, and water bodies (Lerminiaux & Cameron, 2019; Borodovich et al., 2022). The connection of HGT with environmental microbiomes means resistance developed in one niche can quickly be mobilised into others, significantly accelerating evolution and spread across ecological borders (Raziq et al., 2026).

17.3.4 Persistence of resistance

Persistence of resistance refers to the long-term maintenance and stability of ARGs within bacterial populations, even when the direct selective pressure of antibiotics is reduced or absent. HGT plays a central role in this persistence by enabling resistance determinants to spread, integrate, and become established across diverse microbial communities (Barlow, 2009). In agricultural settings, this occurs through rapid exchange of genetic materials in bacteria like integrons, plasmids, transposons, and other mobile elements. The persistence results from the survival and proliferation of resistance determinants in livestock production, aquaculture, and crop protection through repeated use of antibiotics (Barlow, 2009). These resistance determinants persist within microbial groups even after decreasing the use of antibiotics. Resistance genes frequently impose low fitness costs on bacteria, allowing resistant strains to compete effectively with susceptible ones (Gillieatt & Coleman, 2024). Moreover, coselection mechanisms maintain them even without antibiotic pressure.

17.3.5 The microbiome as an amplifier

The animal and environmental microbiomes act as dense and dynamic hosts serving as active sites of exchange and reservoirs of resistance. Antibiotic exposure disrupts microbiome balance, reducing diversity and allowing resistant microbes to flourish. In animals, they can form a genetic reservoir of pathogenic species (Woolhouse et al., 2015). Similarly, soil and marine microbiomes exposed to antibiotic-laden manure and wastewater act as lasting ARG reservoirs (Sassi et al., 2025). Dense microbial populations facilitate all major HGT pathways, and interconnected gene exchange networks within soil and gut microbiomes enhance mobility of resistance traits. Thus environmental microbiomes act as genetic amplifiers that escalate both the abundance and diversity of ARGs.

17.4 Transmission pathways of agricultural to humans

The evolution of ABR in agricultural settings has consequences that extend far beyond farm boundaries via multiple transmission pathways. The resistant bacteria and resistance genes move through an interconnected ecological network and human activities, ultimately reaching human populations and the broader environment. This transmission creates a complex network linking agricultural practices, environmental changes, and human activities.

17.4.1 Through food

For most people, food is the main route of exposure to AMR microorganisms present in agricultural farms (Morel, 2018). When food-producing animals carry resistant bacteria, the microbes often colonise the animals themselves, and the food derived from that host becomes a primary source for ARB (Almansour et al., 2023). During slaughter and processing, contamination can be transferred to food products. From there, inadequate food processing or cross-contamination facilitates such resistant bacteria into humans, posing direct risks of colonisation or resistant infection (Morel, 2018). Fresh agricultural produce adds another layer of risks (Rahman et al., 2022). Agricultural crops watered with contaminated water or enriched with untreated manure take up the resistant bacteria on the surface or internally. Therefore the leafy greens and other vegetables from such farms represent a critical, and often overlooked, exposure route (Hölzel et al., 2018; Rahman et al., 2022). In an interesting study by Castello and his associates, more than $3/4^{th}$ of the tested vegetables hosted AMR strains, where $1/5^{th}$ of the isolates were multidrug resistant(Castello et al., 2023).

17.4.2 Through the environment

When livestock manure is applied to agricultural fields, it introduces high concentrations of ARB, ARGs, and residual antimicrobials into soil (Liu et al., 2022; Zhang et al., 2022). Then the resistant factors can persist, integrate into indigenous microbial communities, and be taken up by crop plants. Further water movement, like rainfall-induced runoff and leaching, can mobilise resistance determinants from fields into surface waters and groundwater (Corno et al., 2019). Such aquatic compartments then become reservoirs for ABR, often exposed to the human population through irrigation, recreation, or potable water sources (Jampani et al., 2024). Aquaculture contributes similarly, with antibiotic-laden effluents flowing directly into surrounding waters, enriching environmental resistomes.

Several studies have reported the importance of airborne transmission of ARBs (Xin et al., 2023; Wang et al., 2020). Activities such as manure spraying, livestock housing ventilation, and feed handling generate bioaerosols loaded with ARBs and ARGs. These particulates can travel considerable distances through the wind and deposit onto soil, water bodies, and residential areas. Farm workers and nearby communities are therefore often exposed via inhalation.

17.4.3 Through direct contact

People who work with animals, such as farmers, veterinarians, slaughterhouse workers, and transporters, face higher risks of exposure to ARB (Koutsoumanis et al., 2022; Panicker et al., 2025). Exposure happens in multiple ways, through skin contact, accidental ingestion, and inhalation of contaminated dust. Once colonised, such individuals with livestock-associated ARB become susceptible to further transmission within households, communities, and healthcare settings, effectively linking farm reservoirs to human populations.

17.4.4 Through wildlife and insect vectors

Wildlife and insects play a key role in the geographic spread of ARB and ARGs. Birds, rodents, flies, and other creatures that move between farms, manure heaps, or contaminated water sources can pick up resistant bacteria and carry them long distances (Radhouani et al., 2014; Furness et al., 2017). That way, they introduce ARB into previously unexposed ecosystems or agricultural sites, facilitating cross-ecosystem spread. In some cases, wildlife species carry resistance levels comparable or even higher than those of domesticated livestock, and these animals are not only vectors but also reservoirs of new resistance traits that can move back into livestock populations.

17.5 Priority antimicrobial-resistant pathogens from agricultural sources

Several resistant patterns have been observed, each originating or amplified under antibiotic selection in livestock production, as a wide range of antibiotics has been used (Fig. 17.2). These compromise the efficacy of critically important human medicines as they persist within transmission networks that transcend the artificial boundaries between veterinary and human medicine. The key pathogens described below share a common route, may emerge or disseminate under agricultural antibiotic selection and circulate across the human–animal–environment interface, and each undermines the efficacy of medicines essential for human health.

Extended-spectrum beta-lactamase (ESBL) and AmpC-producing bacteria represent one of the most urgent threats at the human–animal interface. These bacteria hydrolyse third and fourth-generation cephalosporins, critically important antibiotics in human medicine, and frequently exhibit multidrug resistance (WHO, 2018). Their global dissemination from food-producing animals has been reported to be so extensive that livestock function as reservoir populations from which resistance continuously spills over into human communities, either through whole bacterium transmission or via mobile genetic elements (Lazarus et al., 2015; Muloi et al., 2018). In a recent study by Irimaso et al. (2026), third-generation cephalosporin-resistant (3GC-R) *Enterobacterales* were isolated from conventionally kept domestic pigs and their environment in Rwanda. The isolates showed predominant harbouring of$bla_{\text{CTX-M}}$ group 1 β-lactamase genes, alongside *Klebsiella pneumoniae* isolates, all displaying extended-spectrum β-lactamase (ESBL) phenotypes, highlighting a potential public health risk in the context of extensive human–animal–environment interaction. These ESBL genes reside on mobile plasmids that move effortlessly between microorganisms; once established in agricultural reservoirs, resistance becomes a permanent resident of the shared microbial environment, and apparently colonises healthy animals (Carattoli, 2013; Rozwandowicz et al., 2018). Fluoroquinolone-resistant *Campylobacter* demonstrates how an agricultural practice systematically dismantles a life-saving drug class (Endtz et al., 1991; Engberg et al., 2001).

In 1980s the increase in quinolone resistance was observed to coincide with the increasing use of fluoroquinolones in human and veterinary medicine (Endtz et al., 1991). *Campylobacter jejuni* and *C. coli* are among the most common causes

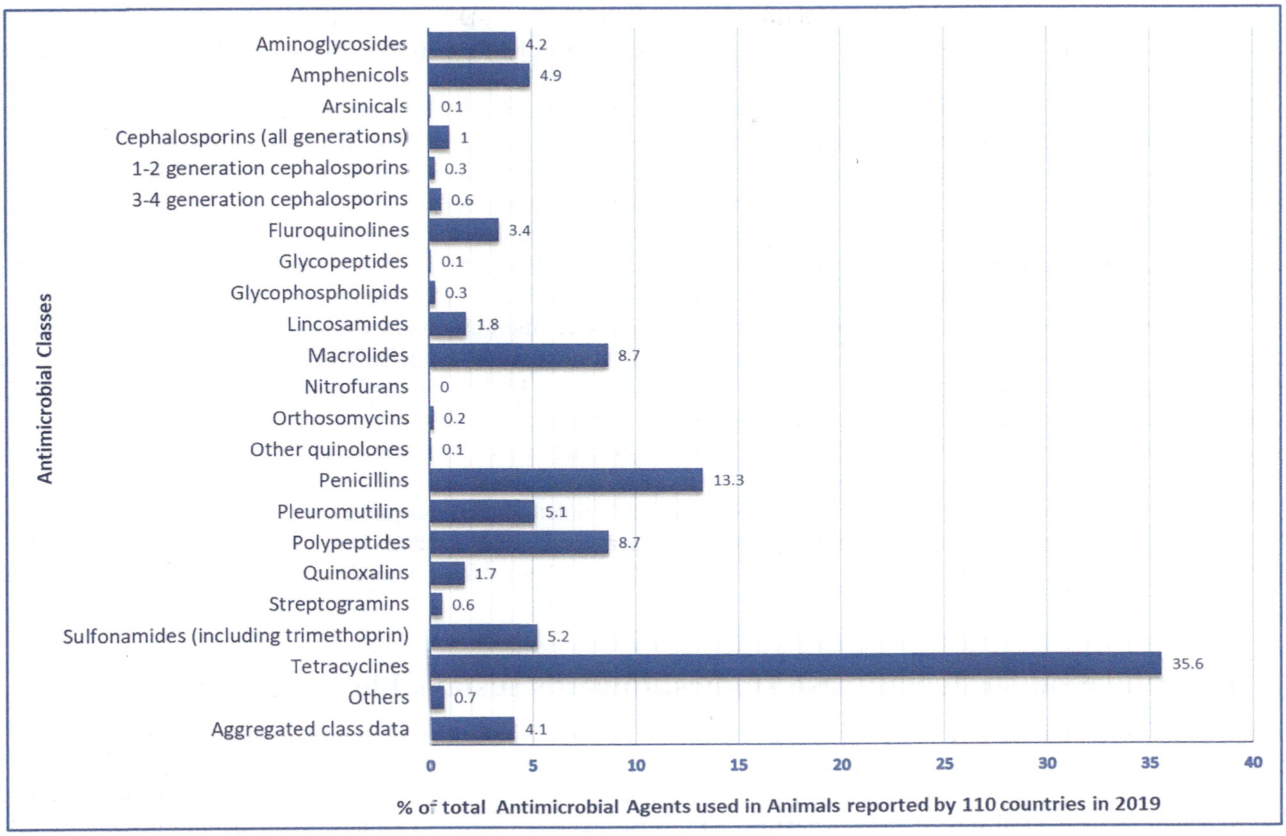

FIGURE 17.2 **Global pattern of antimicrobial use in animals.** Proportion of antimicrobial classes reported for use in animals by 110 countries in 2019.
Data from 7th report on Antimicrobial agents intended for the use in animals, WHOA.

of bacterial gastroenteritis worldwide, with poultry as the dominant reservoir and source of transmission of campylobacteriosis to humans (WHO, 2020; Kaakoush et al., 2015). Resistance arises primarily through a single point mutation in the gyrA gene that often confers a fitness advantage, where resistant strains may outcompete susceptible ones even without antibiotic selection (Luo et al., 2005; Luangtongkum et al., 2012). Once established in the host, resistant Campylobacter persists for years after drug withdrawal (Luo et al., 2005; Price et al., 2005). Importantly, such Fluoroquinolone-resistant strains have been detected in poultry faeces, carcasses, and retail meat across North America, Europe, Asia, and Africa, with prevalence exceeding 80% in some regions(Ge et al., 2003; Smith & Fratamico, 2010; Zhao et al., 2010).

Livestock-associated methicillin-resistant *Staphylococcus aureus* (LA-MRSA), sequence type (ST398), emerged in the early 2000s and has since become a global phenomenon (Voss et al., 2005; Fluit, 2012). These LA-MRSAs are carried by healthy food-producing animals and the humans who work with them (Cuny et al., 2015; Crespo-Piazuelo & Lawlor, 2021). A report from Voss showed more than 760 times greater prevalence rate of *S. aureus* among pig farmers than other patients admitted to Dutch hospitals (Voss et al., 2005). A recent study documented MRSA contamination not only in animals, or retail meats, but also in environments like shared supermarket shopping trolley handles (Yang et al., 2026). Whole-genome sequencing revealed close relatedness to previously reported animal and human isolates, with identical SCC*mec* IVa cassettes and enterotoxin genes (Yang et al., 2026). Moreover, dust from pig farms can also carry MRSA onto clothing and skin, leading to direct exposure to farmers (Angen et al., 2017).

Vancomycin-resistant enterococci (VRE) in food-producing animals are associated with unintended consequences (Miller et al., 2016; Wegener, 2003). Avoparcin, a glycopeptide growth promoter used in European livestock from the 1970s to its 1997 ban, led to the occurrence of vancomycin-resistant enterococci in food animals. However, it was observed that the levels of resistance have been markedly reduced after the termination of use (Wegener, 2003). Historical antimicrobial use in New Zealand agriculture has driven the evolution of ST108, a VRE lineage carrying a range of clinically relevant AMRs (Rushton-Green et al., 2019). The vancomycin resistance in Enterococci is due to the changes in the formation of peptidoglycan; vancomycin binds to the D-Ala-D-Ala terminus of the protein precursors to peptidoglycan. Resistance develops as this terminus is changed to D-Ala-D-lactate so that vancomycin binds with less affinity. This is coded by genotypes identified alphabetically as VanA to VanG (Levitus et al., 2025). The plasmid-based VanA and VanB genotypes are most common.

Colistin is a polymyxin antibiotic discovered more than 50 years ago, abandoned in human medicine due to nephrotoxicity, but used continuously and largely unregulated in livestock for decades (Li et al., 2006; Talat et al., 2023). Colistin was resurrected as last-line therapy; however, colistin-resistant determinants have also been discovered from human as well as animal origin, causing great concern (European medicines agency, 2016). The plasmid-mediated MCR-1 gene, reported in 2015 from animal and human *E. coli* in China, ruined the assumption that colistin resistance was exclusively chromosomal and nontransferable (Liu et al., 2016). Wang et al. reported a high prevalence of colistin resistance in *E. coli* (72.7%) and *Klebsiella pneumoniae* (66.7%) isolated from food animals in China (Wang et al., 2018). MCR genes frequently colocalise with ESBLs, carbapenemases, and fluoroquinolone resistance determinants on the same plasmids, generating drug-resistant phenotypes extensively with no reliable therapeutic options (Sun et al., 2016; Zheng et al., 2017). A study showed colistin-resistant *E. coli* isolated from healthy poultry in Algeria; 78.6% of the colistin-resistant isolates carried the mcr-1 gene, and 57.1% harboured both mcr-1 and mcr-5 (Akkari et al., 2026). Another study showed high prevalence (25%) of mcr-1-encoded colistin resistance in commensal *E. coli* from broiler chickens in Bangladesh (Ahmed et al., 2020).

These five pathogen groups differ in their microbiology and clinical manifestations but have a common shared origin: Each can emerge or amplify under agricultural antibiotic selection pressure, each circulates across the humananimal–environment interface via interconnected transmission networks, and each produce challenges to confront the consequences of how the food are produced.

17.6 Mitigation strategies: case studies, towards sustainable agriculture

The overuse of antibiotics in modern agricultural practices, manure application, and environmental contamination has created a complex network of resistance. It has made the health of animals, humans, and the environment vulnerable through the spread and emergence of resistance via contaminated food products, soil, water runoff, and dung. Regulating this situation requires unified regulatory, scientific, and fiscal involvement, aligning with sustainability goals.

17.6.1 Antibiotic stewardship in agriculture

Reducing the use of antibiotics in agriculture is important, but the effective stewardship requires actions more than just simple reduction of use. There is also a need to understand the underlying reasons for the use of antibiotics in agriculture. A PhD thesis comparing poultry production in France and Indonesia demonstrated that stewardship fails when it ignores local realities (Hibbard, 2025). In France, the focus was on veterinary prescription patterns alongside clinical outcomes. In Indonesia, researchers engaged farmers to identify barriers, highlighting that stewardship involvements fail when they disregard the economic and infrastructural realities of the local farmers (Hibbard, 2025). This suggests that any stewardship must take the local responsible factors into account to be effective.

The United Kingdom offers a more encouraging example for collaborative stewardship. Between 2014 and 2025, the UK livestock industry halved antibiotic sales through a voluntary, multisectoral approach (FAO & VMD, 2022). Success factors included building strong relationships between farmers and veterinarians, setting sector-specific targets, and transparent monitoring. Thereby, the UK pig sector achieved a 74% reduction in antibiotic use (FAO & VMD, 2022).

A cross-sectional study of 200 poultry farms in Enugu State, Nigeria (2024) revealed the scale of stewardship challenges. Where 90.5% of farmers used antibiotics, 61% for growth promotion or disease prevention, and 65% had never heard of (Ejikeugwu et al., 2025). These findings highlight that stewardship in resource-limited settings requires foundational education and regulatory enforcement alongside technical interventions.

17.6.2 Improving animal health and welfare

A direct correlation exists between animal health conditions and antibiotic requirements. Therefore strategies that enhance animal health directly reduce overall antibiotic consumption in livestock production. A review of genetic strategies for pig robustness states that high antibiotic use in intensive systems reflects disease pressure in high-density, high-stress environments. This can be removed through better biosecurity, vaccination, and management, alongside genetic approaches (Abbas et al., 2026). Functional nutritional interventions like probiotics, prebiotics, synbiotics, organic acids, dietary enzymes, and phytogenic feed additives have emerged as good alternatives to antibiotic growth promoters.

Among these, phytogenic additives rich in terpenoids and phenolics demonstrated particular promise due to their antimicrobial, antioxidant, and gut health-promoting properties (Abbas et al., 2026).

A report with 19 studies across multiple species confirmed the potential of these approaches (Dell'Anno et al., 2025). In swine, microencapsulated essential oils improved health in weaned piglets. *Alpinia oxyphylla* oil boosted immunity and gut health in fattening pigs. Probiotics increased beneficial short-chain fatty acids and daily gain. Maternal nutrition also mattered; fish oil supplementation in sows protected piglets from inflammatory damage, and one-carbon metabolites influenced antimicrobial peptide expression in offspring (Dell'Anno et al., 2025).

In poultry, a synbiotic enzyme complex improved growth and gut health in broilers. A *Bacillus-derived* postbiotic reduced mortality. An antimicrobial peptide from the Isalo scorpion enhanced performance and immunity in yellow-feathered broilers (Dell'Anno et al., 2025).

In ruminants, *Saccharomyces cerevisiae* supplementation in buffalo showed trends toward improved weight gain. Medium-chain fatty acids as silage additives reduced harmful fermentations (Dell'Anno et al., 2025).

17.6.3 Selection of better breeds

Genetics offers a preventive route to reducing antibiotic dependence. Immune competence, the ability to mount both antibody- and cell-mediated responses, is moderately heritable, with estimates of 0.35 and 0.36 in cattle (Madsen et al., 2025). However, genetic correlations between immune competence and weaning weight were unfavourable: −0.35 for antibody-mediated immune response (AMIR) and −0.26 for cell-mediated immune response (CMIR), supporting the hypothesis that mounting effective immune responses can divert resources from growth when nutrition is limited (Madsen et al., 2025). This trade-off must be managed in breeding programmes that balance productivity with disease resilience.

In pigs, genetic variation in resilience to Porcine Reproductive and Respiratory Syndrome (PRRS) has been quantified. During outbreaks, additive genetic variance for reproductive traits was highest, suggesting real potential for selective breeding (de Oliveira et al., 2025).

African indigenous cattle, meanwhile, carry genetic adaptations to tropical climates and endemic diseases like trypanosomiasis. Genomic analysis of over 1,000 cattle from 17 breeds identified 229 selective regions linked to immune function, promising targets for breeding programmes aimed at resilience (Bitew et al., 2026).

17.6.4 Manure and wastewater management

Land application of animal waste is often considered the least damaging and environmentally preferable manure benefitting agriculture. However, with the use of antibiotics to cure diseases, effective disposal of manure and wastewater has become an ardent protective tactic for minimising the use of antibiotics and restricting the dissemination of ARB in livestock production. Livestock wastes serving as reservoirs of antibiotic residues, pathogenic microorganisms, and antibiotic-resistant bacteria can thus be curbed from spreading further (He et al., 2020). Treatment practices start by storing husbandry waste in lagoons, followed by treatments, including anaerobic digestion (AD), composting, and biological wastewater treatment, which effectively reduce pathogen loads. This further facilitates residual antimicrobial breakdown, preventing their presence in the environment (Tian et al., 2021). Enhanced environmental hygiene and effective wastewater and manure disposal can reduce the risk of pathogen exposure, promote livestock health, and decrease dependence on routine antibiotics.

It has been observed that the implementation of mitigation strategies has yielded a measurable reduction in antibiotic use in agriculture. For instance, in China, the promotion of composting and anaerobic digestion alongside restrictions on nontherapeutic use resulted in significant reductions in ARGs in manure and lower antibiotic use in animals (Tian et al., 2021). Numerous dairy farms in Europe practice manure storage and handling, as well as introduce biological wastewater treatment. These practices significantly reduce ARB contamination in farm effluents, lowering the incidence of infectious diseases like mastitis and lowering the need for antibiotic treatments. This demonstrates that integrated strategies through effective wastewater management and proper manure disposal can help regulate the use of antibiotics and promote sustainable livestock production (Heuer et al., 2011).

17.6.5 Policy, regulation, and economic incentives

Regulation works best when it is backed by incentives. Many countries now restrict nontherapeutic antibiotic use and insist on stewardship programmes and surveillance (WHO, 2023). But rules alone are not enough. Farmers need support, subsidies for vaccination, biosecurity upgrades, and alternative health technologies to make the shift without sacrificing productivity.

For example, in Denmark, an inclusive countrywide accomplishment plan is in place, combining stringent regulation and compulsory reporting of antimicrobial usage and rewarding farmers for adopting low-antibiotic farming practices. The outcome of these practices includes significant reductions in the use of antibiotics in poultry and pig farming. Further, no adverse effects have been observed on animal well-being or productivity output (Hammerum et al., 2007). To tackle AMR, the National Action Plan has been launched in India. This includes promoting balanced antimicrobial use across agriculture, human and animal health, emphasising infection prevention, surveillance, and stewardship. The plan encourages reduced nontherapeutic antibiotic use in livestock and supports the adoption of preventive measures such as vaccination, biosecurity, and amended farm management (Nair et al., 2021).

17.7 Conclusion

The evidence shows that ABR arising from agricultural practices is a threat, and it's reaching our homes and hospital wards. The pathogens discussed in this chapter, ESBL-producing microorganisms, fluoroquinolone-resistant Campylobacter, livestock-associated MRSA, vancomycin-resistant enterococci, and colistin-resistant bacteria, each show resistance derived from agricultural practices. The routine use of antibiotics in animal production serves as a major driver for these resistance threats, which are transmitted through multiple pathways like the food chain, water system, soil, or atmosphere. Such transmission ultimately establishes these farm-evolved ABRs in distant environments and nontarget hosts. However, recent data suggest that global antibiotic use in livestock has passed its peak and has begun to decline. Countries, including China and the United States, have achieved measurable reductions. Countries that banned colistin use in animals saw corresponding declines in mcr-positive bacteria in livestock. In Zimbabwe, participatory training programmes helped smallholder farmers reduce prophylactic antibiotic use from 73% to below 10%. This is proof that change is possible and that the trajectory of resistance can be bent. Moving forward, a sustained and integrated response is essential. This requires phasing out the routine use of antibiotics in healthy animals, investing in biosecurity and vaccination at the farm level, and strengthening surveillance systems to track resistance across sectors. This will ensure that countries can now scale up livestock production without unnecessary use of antibiotics. The path toward sustainable agriculture is neither simple nor quick, but the tools and knowledge to navigate it already exist. What remains is the collective effort to apply them.

AI disclosure

During the preparation of this work, the author(s) used Grammerly, ChatGPT to correct grammar and minor editing. After using this tool/service, the author(s) reviewed and edited the content as needed and take(s) full responsibility for the content of the publication.

References

Abbas, M., Abbas, G., Fatima, Hashmi, A.H., Jaffery, S., Li, Y., et al., 2026. Sustainable AGP alternatives: a systems approach to non-antibiotic growth regulators standardization, synergistic formulation and environmental safety. Frontiers in Veterinary Science 12. Available at https://doi.org/10.3389/fvets.2025.1695160.

Acosta, A., Tirkaso, W., Nicolli, F., Van Boeckel, T.P., Cinardi, G., Song, J., 2025. The future of antibiotic use in livestock. Nature Communications 16 (1). Available at https://doi.org/10.1038/s41467-025-56825-7.

Ahmad, N., 2025. Agricultural Chemicals in Crops and How They Shape Soil Microbiome. Available at https://doi.org/10.20944/preprints202510.1396.v1.

Ahmed, S., Das, T., Islam, M.Z., Herrero-Fresno, A., Biswas, P.K., Olsen, J.E., 2020. High prevalence of mcr-1-encoded colistin resistance in commensal *Escherichia coli* from broiler chicken in Bangladesh. Scientific Reports 10 (1). Available at https://doi.org/10.1038/s41598-020-75608-2.

Akkari, H., Heleili, N., Ozgumus, O.B., Akarsu, N., Kiliç, A.O., 2026. Intestinal carriage of mcr-1 and mcr-5 mediated colistin-resistant *Escherichia coli* in avian species: first report from broiler breeders and laying hens in Algeria. Microbial Pathogenesis 210, 108180. Available at https://doi.org/10.1016/j.micpath.2025.108180.

Allen, H.K., Donato, J., Wang, H.H., Cloud-Hansen, K.A., Davies, J., Handelsman, J., 2010. Call of the wild: antibiotic resistance genes in natural environments. Nature Reviews. Microbiology 8 (4), 251–259. Available at https://doi.org/10.1038/nrmicro2312.

Almansour, A.M., Alhadlaq, M.A., Alzahrani, K.O., Mukhtar, L.E., Alharbi, A.L., Alajel, S.M., 2023. The silent threat: antimicrobial-resistant pathogens in food-producing animals and their impact on public health. Microorganisms 11 (9), 2127. Available at https://doi.org/10.3390/microorganisms11092127.

Andam, C.P., Fournier, G.P., Gogarten, J.P., 2011. Multilevel populations and the evolution of antibiotic resistance through horizontal gene transfer. FEMS Microbiology Reviews 35 (5), 756–767. Available at https://doi.org/10.1111/j.1574-6976.2011.00274.x.

Andersson, D.I., Hughes, D., 2010. Antibiotic resistance and its cost: is it possible to reverse resistance? Nature Reviews. Microbiology 8 (4), 260–271. Available at https://doi.org/10.1038/nrmicro2319.

Angen, Ø., Feld, L., Larsen, J., Rostgaard, K., Skov, R., Madsen, A.M., et al., 2017. Transmission of methicillin-resistant *Staphylococcus aureus* to human volunteers visiting a swine farm. Applied and Environmental Microbiology 83 (23). Available at https://doi.org/10.1128/aem.01489-17.

Argudín, M.A., Hoefer, A., Butaye, P., 2019. Heavy metal resistance in bacteria from animals. Research in Veterinary Science 122, 132–147. Available at https://doi.org/10.1016/j.rvsc.2018.11.007.

Baker-Austin, C., Wright, M.S., Stepanauskas, R., McArthur, J.V., 2006. Co-selection of antibiotic and metal resistance. Trends in Microbiology 14 (4), 176–182. Available at https://doi.org/10.1016/j.tim.2006.02.006.

Barlow, M., 2009. What antimicrobial resistance has taught us about horizontal gene transfer. Methods in Molecular Biology (Clifton, N.J.). 532, 397–411. Available at https://doi.org/10.1007/978-1-60327-853-9_23.

Bassitta, R., Nottensteiner, A., Bauer, J., Straubinger, R.K., Hölzel, C.S., 2022. Spread of antimicrobial resistance genes via pig manure from organic and conventional farms in the presence or absence of antibiotic use. Journal of Applied Microbiology 133 (4), 2457–2465. Available at https://doi.org/10.1111/jam.15717.

Belay, W.Y., Getachew, M., Tegegne, B.A., Teffera, Z.H., Dagne, A., Zeleke, T.K., et al., 2024. Mechanism of antibacterial resistance, strategies and next-generation antimicrobials to contain antimicrobial resistance: a review. Frontiers in Pharmacology 15. Available at https://doi.org/10.3389/fphar.2024.1444781, http://www.frontiersin.org/Pharmacology.

Bitew, M.K., Paguem, A., Abanda, B., Achukwi, D.M., Manchang, K.T., Preuß, S., et al., 2026. Genomic evidence of positive selection highlights functional divergence associated with environmental resilience in central and West African indigenous cattle. Scientific Reports 16 (1). Available at https://doi.org/10.1038/s41598-025-30767-y, https://www.nature.com/srep/.

Bobate, S., Mahalle, S., Dafale, N.A., Bajaj, A., 2023. Emergence of environmental antibiotic resistance: mechanism, monitoring and management. Environmental Advances 13, 100409. Available at https://doi.org/10.1016/j.envadv.2023.100409.

Van Boeckel, T.P., Pires, J., Silvester, R., Zhao, C., Song, J., Criscuolo, N.G., et al., 2019. Global trends in antimicrobial resistance in animals in low- and middle-income countries. Science (New York, N.Y.) 365 (6459). Available at https://doi.org/10.1126/science.aaw1944, https://science.sciencemag.org/content/365/6459/eaaw1944/tab-pdf.

Borodovich, T., Shkoporov, A.N., Ross, R.P., Hill, C., 2022. Phage-mediated horizontal gene transfer and its implications for the human gut microbiome. Gastroenterology Report 10. Available at https://doi.org/10.1093/gastro/goac012.

Boto, L., 2009. Horizontal gene transfer in evolution: facts and challenges. Proceedings of the Royal Society B: Biological Sciences 277 (1683), 819–827. Available at https://doi.org/10.1098/rspb.2009.1679.

Brown, E.D., Wright, G.D., 2016. Antibacterial drug discovery in the resistance era. Nature 529 (7586), 336–343. Available at https://doi.org/10.1038/nature17042, http://www.nature.com/nature/index.html.

Brüssow, H., 2024. The antibiotic resistance crisis and the development of new antibiotics. Microbial Biotechnology 17 (7). Available at https://doi.org/10.1111/1751-7915.14510.

Carattoli, A., 2013. Plasmids and the spread of resistance. International Journal of Medical Microbiology 303 (6-7), 298–304. Available at https://doi.org/10.1016/j.ijmm.2013.02.001.

Castanon, J.I.R., 2007. History of the use of antibiotic as growth promoters in european poultry feeds. Poultry Science 86 (11), 2466–2471. Available at https://doi.org/10.3382/ps.2007-00249.

Castello, A., Lo Cascio, G., Ferraro, C., Pantano, L., Costa, A., Butera, G., et al., 2023. Food risk associated with vegetable consumption, exposure to antimicrobial-resistant strains and pesticide residues. Italian Journal of Food Safety 12 (2). Available at https://doi.org/10.4081/ijfs.2023.11134.

Cheng, W., Li, J., Wu, Y., Xu, L., Su, C., Qian, Y., et al., 2016. Behavior of antibiotics and antibiotic resistance genes in eco-agricultural system: a case study. Journal of Hazardous Materials 304, 18–25. Available at https://doi.org/10.1016/j.jhazmat.2015.10.037.

Cheng, G., Ning, J., Ahmed, S., Huang, J., Ullah, R., An, B., et al., 2019. Selection and dissemination of antimicrobial resistance in agri-food production. Antimicrobial Resistance & Infection Control 8 (1). Available at https://doi.org/10.1186/s13756-019-0623-2.

Corno, G., Yang, Y., Eckert, E.M., Fontaneto, D., Fiorentino, A., Galafassi, S., et al., 2019. Effluents of wastewater treatment plants promote the rapid stabilization of the antibiotic resistome in receiving freshwater bodies. Water Research 158, 72–81. Available at https://doi.org/10.1016/j.watres.2019.04.031.

Crespo-Piazuelo, D., Lawlor, P.G., 2021. Livestock-associated methicillin-resistant *Staphylococcus aureus* (LA-MRSA) prevalence in humans in close contact with animals and measures to reduce on-farm colonisation. Irish Veterinary Journal 74 (1). Available at https://doi.org/10.1186/s13620-021-00200-7, http://www.veterinary-ireland.org/.

Cuny, C., Wieler, L.H., Witte, W., 2015. Livestock-associated MRSA: the impact on humans. Antibiotics 4 (4), 521–543. Available at https://doi.org/10.3390/antibiotics4040521, http://www.mdpi.com/2079-6382/4/4/521/pdf.

Cycoń, M., Mrozik, A., Piotrowska-Seget, Z., 2019. Antibiotics in the soil environment—degradation and their impact on microbial activity and diversity. Frontiers in Microbiology 10. Available at https://doi.org/10.3389/fmicb.2019.00338, https://www.frontiersin.org/journals/microbiology#.

Davies, J., Davies, D., 2010. Origins and evolution of antibiotic resistance. Microbiology and Molecular Biology Reviews 74 (3), 417–433. Available at https://doi.org/10.1128/MMBR.00016-10Canada, http://mmbr.asm.org/cgi/reprint/74/3/417.

Dell'Anno, M., Hejna, M., Riva, F., Rossi, L., 2025. Editorial: functional nutritional strategies as alternatives to antimicrobials. Frontiers in Veterinary Science 12. Available at https://doi.org/10.3389/fvets.2025.1711519, journal.frontiersin.org/journal/veterinary-science.

Ejikeugwu, C.P., Nwakaeze, E.A., Aniekwe, C.W., Onu, E.N., Adikwu, M.U., Eze, P.M., 2025. Investigating the factors influencing antibiotic use practices and their association with antimicrobial resistance awareness among poultry farmers in Enugu State, Nigeria. Antimicrobial Stewardship and Healthcare Epidemiology 5 (1). Available at https://doi.org/10.1017/ash.2025.10141, www.cambridge.org/core/journals/antimicrobial-stewardship-and-healthcare-epidemiology.

Endtz, H.P., Ruijs, G.J., Van Klingeren, B., Jansen, W.H., Van Der Reyden, T., Mouton, R.P., 1991. Quinolone resistance in campylobacter isolated from man and poultry following the introduction of fluoroquinolones in veterinary medicine. Journal of Antimicrobial Chemotherapy 27 (2), 199–208. Available at https://doi.org/10.1093/jac/27.2.199.

Engberg, J., Aarestrup, F.M., Taylor, D.E., Gerner-Smidt, P., Nachamkin, I., 2001. Quinolone and macrolide resistance in *Campylobacter jejuni* and *C. coli*: resistance mechanisms and trends in human isolates. Emerging Infectious Diseases 7 (1), 24–34. Available at https://doi.org/10.3201/eid0701.010104.

Enshaie, E., Nigam, S., Patel, S., Rai, V., 2025. Livestock antibiotics use and antimicrobial resistance. Antibiotics 14 (6), 621. Available at https://doi.org/10.3390/antibiotics14060621.

European medicines agency, 2016. Updated advice on the use of colistin products in animals within the European Union: development of resistance and possible impact on human and animal health. Available at https://www.ema.europa.eu/en/documents/scientific-guideline/updated-advice-use-colistin-products-animals-within-european-union-development-resistance-and-possible-impact-human-and-animal-health_en.pdf-0.

FAO, VMD, 2022. Tackling antimicrobial use and resistance in food-producing animals. Lessons learned in the United Kingdom of Great Britain and Northern Ireland. FAO; Veterinary Medicines Directorate, Rome, Italy. Available at https://doi.org/10.4060/cc0927en.

Fluit, A.C., 2012. Livestock-associated Staphylococcus aureus. Clinical Microbiology and Infection: The Official Publication of the European Society of Clinical Microbiology and Infectious Diseases 18 (8), 735–744. Available at 10.1111/j.1469-0691.2012.03846.x.

Furness, L.E., Campbell, A., Zhang, L., Gaze, W.H., McDonald, R.A., 2017. Wild small mammals as sentinels for the environmental transmission of antimicrobial resistance. Environmental Research 154, 28–34. Available at https://doi.org/10.1016/j.envres.2016.12.014, http://www.elsevier.com/inca/publications/store/6/2/2/8/2/1/index.htt.

Gaurav, A., Bakht, P., Saini, M., Pandey, S., Pathania, R., 2023. Role of bacterial efflux pumps in antibiotic resistance, virulence, and strategies to discover novel efflux pump inhibitors. Microbiology (Reading, England) 169 (5). Available at https://doi.org/10.1099/mic.0.001333.

Ge, B., White, D.G., McDermott, P.F., Girard, W., Zhao, S., Hubert, S., et al., 2003. Antimicrobial-resistant campylobacter species from retail raw meats. Applied and Environmental Microbiology 69 (5), 3005–3007. Available at https://doi.org/10.1128/aem.69.5.3005-3007.2003.

Gillieatt, B.F., Coleman, N.V., 2024. Unravelling the mechanisms of antibiotic and heavy metal resistance co-selection in environmental bacteria. FEMS Microbiology Reviews 48 (4). Available at https://doi.org/10.1093/femsre/fuae017, https://academic.oup.com/femsre/pages/About.

Hammerum, A.M., Heuer, O.E., Emborg, H.D., Bagger-Skjøt, L., Jensen, V.F., Rogues, A.M., et al., 2007. Danish integrated antimicrobial resistance monitoring and research program. Emerging Infectious Diseases 13 (11), 1632–1639. Available at https://doi.org/10.3201/eid1311.070421, http://www.cdc.gov/eid/content/13/11/pdfs/1632.pdf.

He, Y., Yuan, Q., Mathieu, J., Stadler, L., Senehi, N., Sun, R., et al., 2020. Antibiotic resistance genes from livestock waste: occurrence, dissemination, and treatment. npj Clean Water 3 (1). Available at https://doi.org/10.1038/s41545-020-0051-0.

Helliwell, R., Raman, S., Morris, C., 2021. Environmental imaginaries and the environmental sciences of antimicrobial resistance. Environment and Planning E: Nature and Space 4 (4), 1346–1368. Available at https://doi.org/10.1177/2514848620950752, journals.sagepub.com/home/ene.

Heuer, H., Schmitt, H., Smalla, K., 2011. Antibiotic resistance gene spread due to manure application on agricultural fields. Current Opinion in Microbiology 14 (3), 236–243. Available at https://doi.org/10.1016/j.mib.2011.04.009.

Hibbard, R., 2025. Redefining and understanding antimicrobial stewardship: context-based approaches applied to poultry production in Indonesia and France. Theses. Université de Toulouse. Available at https://theses.hal.science/tel-05425305 (accessed 17.02.26). HAL theses, HAL theses, (2025), Available at https://theses.hal.science/tel-05425305.

Huang, C., Zeng, Y., Yang, F., Wu, Q., Ding, Y., 2025. Exploring the relationship between farmland management and manure-derived antibiotic resistance genes and their prevention and control strategies. Antibiotics 14 (11), 1117. Available at https://doi.org/10.3390/antibiotics14111117.

Hölzel, C.S., Tetens, J.L., Schwaiger, K., 2018. Unraveling the role of vegetables in spreading antimicrobial-resistant bacteria: a need for quantitative risk assessment. Foodborne Pathogens and Disease 15 (11), 671–688. Available at https://doi.org/10.1089/fpd.2018.2501, www.liebertonline.com/fpd.

Irimaso, E., Hagenimana, V., Nzabamwita, E., Blümlinger, M., Fischer, O.W., Schwarz, L., et al., 2026. Third-generation cephalosporin-resistant enterobacterales and methicillin-resistant *Staphylococcus aureus* (MRSA) in Pigs in Rwanda. Animals 16 (1), 122. Available at https://doi.org/10.3390/ani16010122.

Jampani, M., Mateo-Sagasta, J., Chandrasekar, A., Fatta-Kassinos, D., Graham, D.W., Gothwal, R., et al., 2024. Fate and transport modelling for evaluating antibiotic resistance in aquatic environments: current knowledge and research priorities. Journal of Hazardous Materials 461, 132527. Available at https://doi.org/10.1016/j.jhazmat.2023.132527.

Jones, F.T., Ricke, S.C., 2003. Observations on the history of the development of antimicrobials and their use in poultry feeds. Poultry Science 82 (4), 613–617. Available at https://doi.org/10.1093/ps/82.4.613, https://www.journals.elsevier.com/poultry-science/.

Kaakoush, N.O., Castaño-Rodríguez, N., Mitchell, H.M., Man, S.M., 2015. Global epidemiology of campylobacter infection. Clinical Microbiology Reviews 28 (3), 687–720. Available at https://doi.org/10.1128/CMR.00006-15, http://cmr.asm.org/content/28/3/687.full.pdf.

Katz, S., Hershberg, R., Malik, H.S., 2013. Elevated mutagenesis does not explain the increased frequency of antibiotic resistant mutants in starved aging colonies. PLoS Genetics 9 (11), e1003968. Available at https://doi.org/10.1371/journal.pgen.1003968.

Kelbrick, M., Hesse, E., O' Brien, S., 2023. Cultivating antimicrobial resistance: how intensive agriculture ploughs the way for antibiotic resistance. Microbiology (Reading, England) 169 (8). Available at https://doi.org/10.1099/mic.0.001384.

Koutsoumanis, K., Allende, A., Álvarez-Ordóñez, A., Bolton, D., Bover-Cid, S., Chemaly, M., et al., 2022. Transmission of antimicrobial resistance (AMR) during animal transport. EFSA Journal 20 (10). Available at https://doi.org/10.2903/j.efsa.2022.7586.

Kraker, M.E.A., Stewardson, A.J., Harbarth, S., 2016. Will 10 million people die a year due to antimicrobial resistance by 2. PLoS Medicine 13 (11), e1002184. Available at https://doi.org/10.1371/journal.pmed.1002184.

Kümmerer, K., 2009. Antibiotics in the aquatic environment – a review – part I. Chemosphere 75 (4), 417–434. Available at https://doi.org/10.1016/j.chemosphere.2008.11.086.

Laxminarayan, R., Duse, A., Wattal, C., Zaidi, A.K.M., Wertheim, H.F.L., Sumpradit, N., et al., 2013. Antibiotic resistance—the need for global solutions. The Lancet Infectious Diseases 13 (12), 1057–1098. Available at https://doi.org/10.1016/s1473-3099(13)70318-9.

Lazarus, B., Paterson, D.L., Mollinger, J.L., Rogers, B.A., 2015. Do human extraintestinal *Escherichia coli* infections resistant to expanded-spectrum cephalosporins originate from food-producing animals? A systematic review. Clinical Infectious Diseases 60 (3), 439–452. Available at https://doi.org/10.1093/cid/ciu785.

Lee, L.J., Barrett, J.A., Poole, R.K., 2005. Genome-wide transcriptional response of chemostat-cultured *Escherichia coli* to zinc. Journal of Bacteriology 187 (3), 1124–1134. Available at https://doi.org/10.1128/JB.187.3.1124-1134.2005.

Lerminiaux, N.A., Cameron, A.D.S., 2019. Horizontal transfer of antibiotic resistance genes in clinical environments. Canadian Journal of Microbiology 65 (1), 34–44. Available at https://doi.org/10.1139/cjm-2018-0275, www.nrc.ca/cgi-bin/cisti/journals/rp/rp_desy_e?cjm.

Levitus, M., Rewane, A., Perera, T., 2025. Vancomycin-resistant enterococci. StatPearls. StatPearls Publishing, Treasure Island (FL). Available at http://www.ncbi.nlm.nih.gov/books/NBK513233/, https://www.ncbi.nlm.nih.gov/books/NBK513233/.

Li, J., Nation, R.L., Turnidge, J.D., Milne, R.W., Coulthard, K., Rayner, C.R., et al., 2006. Colistin: the re-emerging antibiotic for multidrug-resistant Gram-negative bacterial infections. The Lancet Infectious Diseases 6 (9), 589–601. Available at https://doi.org/10.1016/s1473-3099(06)70580-1.

Li, S., Ondon, B.S., Ho, S.-H., Li, F., 2023. Emerging soil contamination of antibiotics resistance bacteria (ARB) carrying genes (ARGs): new challenges for soil remediation and conservation. Environmental Research 219, 115132. Available at https://doi.org/10.1016/j.envres.2022.115132.

Li, S., Zhu, Y., Zhong, G., Huang, Y., Jones, K.C., 2024. Comprehensive assessment of environmental emissions, fate, and risks of veterinary antibiotics in China: an environmental fate modeling approach. Environmental Science & Technology 58 (12), 5534–5547. Available at https://doi.org/10.1021/acs.est.4c00993.

Li, Z., Tang, J., Wang, X., Ma, X., Yuan, H., Gao, C., et al., 2025. The environmental lifecycle of antibiotics and resistance genes: transmission mechanisms, challenges, and control strategies. Microorganisms 13 (9), 2113. Available at https://doi.org/10.3390/microorganisms13092113.

Liao, H., Li, X., Yang, Q., Bai, Y., Cui, P., Wen, C., et al., 2021. Herbicide selection promotes antibiotic resistance in soil microbiomes. Molecular Biology and Evolution 38 (6), 2337–2350. Available at https://doi.org/10.1093/molbev/msab029.

Liu, Y.Y., Wang, Y., Walsh, T.R., Yi, L.X., Zhang, R., Spencer, J., et al., 2016. Emergence of plasmid-mediated colistin resistance mechanism MCR-1 in animals and human beings in China: a microbiological and molecular biological study. The Lancet Infectious Diseases 16 (2), 161–168. Available at https://doi.org/10.1016/S1473-3099(15)00424-7, http://www.journals.elsevier.com/the-lancet-infectious-diseases.

Liu, C., Li, G., Qin, X., Xu, Y., Wang, J., Wu, G., et al., 2022. Profiles of antibiotic- and heavy metal-related resistance genes in animal manure revealed using a metagenomic analysis. Ecotoxicology and Environmental Safety 239, 113655. Available at https://doi.org/10.1016/j.ecoenv.2022.113655.

Luangtongkum, T., Shen, Z., Seng, V.W., Sahin, O., Jeon, B., Liu, P., et al., 2012. Impaired fitness and transmission of macrolide-resistant *Campylobacter jejuni* in its natural host. Antimicrobial Agents and Chemotherapy 56 (3), 1300–1308. Available at https://doi.org/10.1128/aac.05516-11.

Luo, N., Pereira, S., Sahin, O., Lin, J., Huang, S., Michel, L., et al., 2005. Enhanced in vivo fitness of fluoroquinolone-resistant *Campylobacter jejuni* in the absence of antibiotic selection pressure. Proceedings of the National Academy of Sciences 102 (3), 541–546. Available at https://doi.org/10.1073/pnas.0408966102.

Madsen, M.D., van der Werf, J.H.J., Ingham, A., Hine, B., Reverter, A., Clark, S.A., 2025. The genetic relationship between immune competence traits and micro-genetic environmental sensitivity of weight, fat, and muscle traits in Australian Angus cattle. Genetics Selection Evolution 57 (1). Available at https://doi.org/10.1186/s12711-025-00998-8, https://gsejournal.biomedcentral.com/.

Magnusson, U., 2022. Antimicrobial use and resistance in food-producing animals—how can we protect the efficacy of antibiotics for reproductive diseases? Reproduction in Domestic Animals 57 (S4), 13–20. Available at https://doi.org/10.1111/rda.14170.

Manyi-Loh, C., Mamphweli, S., Meyer, E., Okoh, A., 2018. Antibiotic use in agriculture and its consequential resistance in environmental sources: potential public health implications. Molecules (Basel, Switzerland) 23 (4), 795. Available at https://doi.org/10.3390/molecules23040795.

Miller, W.R., Murray, B.E., Rice, L.B., Arias, C., 2016. Vancomycin-resistant enterococci: therapeutic challenges in the 21st century. Infectious Disease Clinics 30 (2), 415–439. Available at https://doi.org/10.1016/j.idc.2016.02.006 Infectious Disease Clinics of North America. **30** (2) (2016), 415–439.

Morel, C., 2018. Transmission of antimicrobial resistance from livestock agriculture to humans and from humans to animals. OECD Food, Agriculture and Fisheries Papers 133. Available at https://doi.org/10.1787/fcf77850-en. OECD Food, Agriculture and Fisheries Papers. **133**.

Mudenda, S., Daka, V., Matafwali, S.K., 2023. World Health Organization AWaRe framework for antibiotic stewardship: where are we now and where do we need to go? An expert viewpoint. Antimicrobial Stewardship and Healthcare Epidemiology 3 (1). Available at https://doi.org/10.1017/ash.2023.164, www.cambridge.org/core/journals/antimicrobial-stewardship-and-healthcare-epidemiology.

Muloi, D., Ward, M.J., Pedersen, A.B., Fèvre, E.M., Woolhouse, M.E.J., van Bunnik, B.A.D., 2018. Are food animals responsible for transfer of antimicrobial-resistant *Escherichia coli* or their resistance determinants to human populations? A systematic review. Foodborne Pathogens and Disease 15 (8), 467–474. Available at https://doi.org/10.1089/fpd.2017.2411.

Murray, C.J., Ikuta, K.S., Sharara, F., Swetschinski, L., Robles Aguilar, G., Gray, A., et al., 2022. Global burden of bacterial antimicrobial resistance in 2019: a systematic analysis. The Lancet 399 (10325), 629–655. Available at https://doi.org/10.1016/S0140-6736(21)02724-0, http://www.journals.elsevier.com/the-lancet/.

Murray, L.M., Hayes, A., Snape, J., Kasprzyk-Hordern, B., Gaze, W.H., Murray, A.K., 2024. Co-selection for antibiotic resistance by environmental contaminants. npj Antimicrobials and Resistance 2 (1). Available at https://doi.org/10.1038/s44259-024-00026-7.

Nair, M., Zeegers, M.P., Varghese, G.M., Burza, S., 2021. India's national action plan on antimicrobial resistance: a critical perspective. Journal of Global Antimicrobial Resistance 27, 236–238. Available at https://doi.org/10.1016/j.jgar.2021.10.007.

Nawaz, S., Rafiq, M., Pepper, I.L., Betancourt, W.Q., Shah, A.A., Hasan, F., 2023. Prevalence and abundance of antibiotic-resistant genes in culturable bacteria inhabiting a non-polar passu glacier, karakorum mountains range, Pakistan. World Journal of Microbiology and Biotechnology 39 (4). Available at https://doi.org/10.1007/s11274-023-03532-4.

Nazir, A., Nazir, A., Zuhair, V., Aman, S., Rehman Sadiq, S.U., Hasan, A.H., et al., 2025. The global challenge of antimicrobial resistance: mechanisms, case studies, and mitigation approaches. Health Science Reports 8 (7). Available at https://doi.org/10.1002/hsr2.71077.

de Oliveira, L.F., Dunkelberger, J., Sevillano, C.A., Arirangan, S., Wedow, R., Tegtmeyer, M., et al., 2025. Genetics of reproductive performance across porcine reproductive and respiratory syndrome (PRRS) outbreak phases in purebred and crossbred sows. Genetics Selection Evolution 57 (1). Available at https://doi.org/10.1186/s12711-025-01011-y, https://gsejournal.biomedcentral.com/.

Panicker, A., Changaroth, A., Gangadharan, S.V., Yamamoto, T., Noothalapati, H., Nambudiri, M.K., et al., 2025. From farms to homes: navigating antimicrobial resistance landscapes from livestock to humans. One Health Advances 3 (1). Available at https://doi.org/10.1186/s44280-025-00084-0.

Partridge, S.R., Kwong, S.M., Firth, N., Jensen, S.O., 2018. Mobile genetic elements associated with antimicrobial resistance. Clinical Microbiology Reviews 31 (4). Available at https://doi.org/10.1128/CMR.00088-17, http://cmr.asm.org/.

Peng, S., Feng, Y., Wang, Y., Guo, X., Chu, H., Lin, X., 2017. Prevalence of antibiotic resistance genes in soils after continually applied with different manure for 30 years. Journal of Hazardous Materials 340, 16–25. Available at https://doi.org/10.1016/j.jhazmat.2017.06.059.

Peng, S., Song, D., Zhou, B., Hua, Q., Lin, X., Wang, Y., 2022. Persistence of *Salmonella* Typhimurium and antibiotic resistance. Ecotoxicology and Environmental Safety 248, 114330. Available at https://doi.org/10.1016/j.ecoenv.2022.114330.

Price, L.B., Johnson, E., Vailes, R., Silbergeld, E., 2005. Fluoroquinolone-resistant *Campylobacter* isolates from conventional and antibiotic-free chicken products. Environmental Health Perspectives 113 (5), 557–560. Available at https://doi.org/10.1289/ehp.7647.

Radhouani, H., Silva, N., Poeta, P., Torres, C., Correia, S., Igrejas, G., 2014. Potential impact of antimicrobial resistance in wildlife, environment and human health. Frontiers in Microbiology 5. Available at https://doi.org/10.3389/fmicb.2014.00023.

Rahman, M., Alam, M.-U., Khan Luies, S., Kamal, A., Ferdous, S., Lin, A., et al., 2022. Contamination of fresh produce with antibiotic-resistant bacteria and associated risks to human health: a scoping review. International Journal of Environmental Research and Public Health 19 (1), 360. Available at https://doi.org/10.3390/ijerph19010360.

Raziq, K., Saleem, R., Zafar, S., Sanaullah, T., Nazir, M.M., Ummara, U.E., et al., 2026. Environmental resistomes and antimicrobial resistance: integrating the one health framework. Naunyn-Schmiedeberg's Archives of Pharmacology 399 (2), 2081–2095. Available at https://doi.org/10.1007/s00210-025-04639-1.

Rozwandowicz, M., Brouwer, M.S.M., Fischer, J., Wagenaar, J.A., Gonzalez-Zorn, B., Guerra, B., et al., 2018. Plasmids carrying antimicrobial resistance genes in Enterobacteriaceae. Journal of Antimicrobial Chemotherapy 73 (5), 1121–1137. Available at https://doi.org/10.1093/jac/dkx488.

Rushton-Green, R., Darnell, R.L., Taiaroa, G., Carter, G.P., Cook, G.M., Morgan, X.C., et al., 2019. Agricultural origins of a highly persistent lineage of vancomycin-resistant *Enterococcus faecalis* in New Zealand. Applied and Environmental Microbiology 85 (13). Available at https://doi.org/10.1128/aem.00137-19.

Okoń, S., Zimowska, B., Rai, M., Okoń, S., Zimowska, B., Rai, M. (Eds.), 2024. Microbial Genetics. CRC Press, Boca Raton. Available at https://doi.org/10.1201/9781003328933.

Sassi, A., Basher, N.S., Kirat, H., Meradji, S., Ibrahim, N.A., Idres, T., et al., 2025. The role of the environment (Water, Air, Soil) in the emergence and dissemination of antimicrobial resistance: a one health perspective. Antibiotics 14 (8), 764. Available at https://doi.org/10.3390/antibiotics14080764.

Smith, J.L., Fratamico, P.M., 2010. Fluoroquinolone resistance in *Campylobacter*. Journal of Food Protection 73 (6), 1141–1152. Available at https://doi.org/10.4315/0362-028X-73.6.1141, http://www.ingentaconnect.com/content/iafp/jfp.

Sun, J., Yang, R.-S., Zhang, Q., Feng, Y., Fang, L.-X., Xia, J., et al., 2016. Co-transfer of blaNDM-5 and mcr-1 by an IncX3–X4 hybrid plasmid in *Escherichia coli*. Nature Microbiology 1 (12). Available at https://doi.org/10.1038/nmicrobiol.2016.176.

Szadziul, M., Goryluk-Salmonowicz, A., Popowska, M., 2025. The link between antibiotic resistance level and soil physico-chemical properties. Frontiers in Microbiology 16. Available at https://doi.org/10.3389/fmicb.2025.1584660, https://www.frontiersin.org/journals/microbiology#.

Talat, A., Miranda, C., Poeta, P., Khan, A.U., 2023. Farm to table: colistin resistance hitchhiking through food. Archives of Microbiology 205 (5). Available at https://doi.org/10.1007/s00203-023-03476-1.

Tian, M., He, X., Feng, Y., Wang, W., Chen, H., Gong, M., et al., 2021. Pollution by antibiotics and antimicrobial resistance in livestock and poultry manure in China, and countermeasures. Antibiotics 10 (5), 539. Available at https://doi.org/10.3390/antibiotics10050539.

Tiseo, K., Huber, L., Gilbert, M., Robinson, T.P., Van Boeckel, T.P., 2020. Global trends in antimicrobial use in food animals from 2017 to 2030. Antibiotics 9 (12), 918. Available at https://doi.org/10.3390/antibiotics9120918.

UNGA, UN General Assembly High-Level Meeting on antimicrobial resistance 2024. Available at: https://www.who.int/news-room/events/detail/2024/09/26/default-calendar/un-general-assembly-high-level-meeting-on-antimicrobial-resistance-2024 (Accessed: February 12, 2026). (2024).

Voss, A., Loeffen, F., Bakker, J., Klaassen, C., Wulf, M., 2005. Methicillin-resistant *Staphylococcus aureus* in pig farming. Emerging Infectious Diseases 11 (12), 1965–1966. Available at https://doi.org/10.3201/eid1112.050428.

Wales, A.D., Davies, R.H., 2015. Co-selection of resistance to antibiotics, biocides and heavy metals, and its relevance to foodborne pathogens. Antibiotics 4 (4), 567–604. Available at https://doi.org/10.3390/antibiotics4040567, http://www.mdpi.com/2079-6382/4/4/567/pdf.

Wang, R., Liu, Y., Zhang, Q., Jin, L., Wang, Q., Zhang, Y., et al., 2018. The prevalence of colistin resistance in *Escherichia coli* and *Klebsiella pneumoniae* isolated from food animals in China: coexistence of mcr-1 and bla NDM with low fitness cost. International Journal of Antimicrobial Agents 51 (5), 739–744. Available at https://doi.org/10.1016/j.ijantimicag.2018.01.023.

Wang, X., Yang, T., Jian, Su, Q., Qin, Y., Wang, S., et al., 2020. Air pollution could drive global dissemination of antibiotic resistanc. The ISME Journal 15. Available at https://doi.org/10.1038/s41396-020-00780-2.

Wang, M., Liu, X., Yang, B., Fei, Y., Yu, J., An, R., et al., 2021. Heavy metal contamination in surface sediments from lakes and their surrounding topsoils of China. Environmental Science and Pollution Research 28 (23), 29118–29130. Available at https://doi.org/10.1007/s11356-020-12091-2.

Wang, H., Qi, J.-F., Qin, R., Ding, K., Graham, D.W., Zhu, Y.-G., 2023. Intensified livestock farming increases antibiotic resistance genotypes and phenotypes in animal feces. Communications Earth & Environment 4 (1). Available at https://doi.org/10.1038/s43247-023-00790-w.

Wegener, H.C., 2003. Antibiotics in animal feed and their role in resistance development. Current Opinion in Microbiology 6 (5), 439–445. Available at https://doi.org/10.1016/j.mib.2003.09.009, http://www.elsevier.com/locate/mib.

WHO, 2018. Critically important antimicrobials for human medicine: 6th revision. Available at https://www.who.int/publications/i/item/9789241515528 (accessed 12.02.26.).

WHO, 2020. Global priority list of antibiotic-resistant bacteria to guide research, discovery, and development of new antibiotics.

Woolhouse, M., Ward, M., van Bunnik, B., Farrar, J., 2015. Antimicrobial resistance in humans, livestock and the wider environment. Philosophical Transactions of the Royal Society B: Biological Sciences 370 (1670), 20140083. Available at https://doi.org/10.1098/rstb.2014.0083.

Xin, H., Qiu, T., Guo, Y., Gao, H., Zhang, L., Gao, M., 2023. Aerosolization behavior of antimicrobial resistance in animal farms: a field study from feces to fine particulate matter. Frontiers in Microbiology 14. Available at https://doi.org/10.3389/fmicb.2023.1175265.

Yang, T.-Y., Sun, C.-X., Wang, J., You, Z., Wang, H., Yi, K., et al., 2026. Prevalence and characterization of methicillin-resistant *Staphylococcus aureus* from animals, retail meats and market shopping vehicles in Shandong, China. Foods 15 (2), 248. Available at https://doi.org/10.3390/foods15020248.

Zhang, Y., Cheng, D., Xie, J., Zhang, Y., Wan, Y., Zhang, Y., et al., 2022. Impacts of farmland application of antibiotic-contaminated manures on the occurrence of antibiotic residues and antibiotic resistance genes in soil: a meta-analysis study. Chemosphere 300, 134529. Available at https://doi.org/10.1016/j.chemosphere.2022.134529.

Zhao, S., Young, S.R., Tong, E., Abbott, J.W., Womack, N., Friedman, S.L., et al., 2010. Antimicrobial Resistance of *Campylobacter* isolates from retail meat in the United States between 2002 and 2007. Applied and Environmental Microbiology 76 (24), 7949–7956. Available at https://doi.org/10.1128/aem.01297-10.

Zheng, B., Yu, X., Xu, H., Guo, L., Zhang, J., Huang, C., et al., 2017. Complete genome sequencing and genomic characterization of two *Escherichia coli* strains co-producing MCR-1 and NDM-1 from bloodstream infection. Scientific Reports 7 (1). Available at https://doi.org/10.1038/s41598-017-18273-2.

Zhu, Y.G., Zhao, Y., Li, B., Huang, C.L., Zhang, S.Y., Yu, S., et al., 2017. Continental-scale pollution of estuaries with antibiotic resistance genes. Nature Microbiology 2. Available at https://doi.org/10.1038/nmicrobiol.2016.270, www.nature.com/nmicrobiol/.

Chapter 18

Cultural contexts and behavioural patterns: impacts on antibiotic utilisation

Payel Paul[1], Gargee Dhar Purkayastha[2], and Sourish Dasgupta[1]
[1]*Department of Botany, North Bengal St. Xavier's College, Jalpaiguri, West Bengal, India,* [2]*Sri Sri Academy, Siliguri, West Bengal, India*

18.1 Introduction

There is increasing recognition that the problem of antibiotic use lies on a continuum between individual medical need, societal norms and behaviours, and social pressure. While antibiotic use tends to be framed as a medical/technical or drug issue, there is growing evidence that its uses are embedded in everyday social practices. These practices influence how people think about, request, prescribe, and use medications. According to the World Health Organisation (WHO), the misuse of antibiotics is one of the primary drivers for antimicrobial resistance (AMR), which poses a global threat responsible for an estimated 1.27 million deaths in 2019, as well as contributing to nearly 4.95 million deaths worldwide (Murray et al., 2022; WHO, 2015). Such data position AMR as one of the leading causes of death worldwide and underscore the need to address the social and cultural systems that fuel antibiotic overuse.

Recent global studies indicate antibiotic use continues to increase, particularly in low- and middle-income countries, where access has expanded more rapidly than the ability to do so safely (Klein et al., 2019). Although this has enabled access to life-saving antibiotics, the absence of culturally specific management strategies has led to wrongful usage on a wide scale. In many cultures, antibiotics are associated with strong, fast-acting drugs that can cure a wide variety of illnesses, even viral ailments. This understanding is strengthened by lived experiences, anecdotes from the community hall, and expectations of a fast recovery that create pressure-form demands on patients to influence how physicians write prescriptions (Roope et al., 2018). In many instances, doctors are in a position where not prescribing antibiotics when requested may lead to complaints and dissatisfaction from patients, damage to the reputation and trustworthiness of the doctor, or loss of patients (which can be particularly damaging in private practice/high volume settings). This shows that antibiotic prescribing is not just a clinical decision; it is a social negotiation.

Another important factor in antibiotic utilisation is self-medication, heavily influenced by cultural acceptance and practical situations. Research from Asia, Africa, and Latin America indicates that informal access to antibiotics via pharmacies or leftover medicines in homes is still common (Ocan et al., 2015; Torres et al., 2019). Many communities see these actions as reasonable forms of care for other people, as money-saving and efficient symptom management, not misuse. In places where healthcare is difficult or expensive to access, antibiotics enter into home-based survival strategies. These lead to incomplete treatments, incorrect dosages, and wrong medicine choices, adding to continual pressure for resistant bacteria at population levels.

Antibiotic behaviour is significantly influenced by health literacy and trust in healthcare systems. Cautious use of antibiotics is attributable to limited public information regarding microbial resistance and a fear of potentially worsening conditions (McCullough et al., 2016). Behavioural research has found that many people link antibiotics with safety and responsibility, particularly for those looking after others, be it children or elderly relatives. At the same time, scepticism about formal healthcare establishments can result in people looking for informal sources, which dispense antibiotics without adequate medical scrutiny. These trust issues emphasise the need for communication and community interaction in management (Hulscher and Prins, 2017).

Insights from various disciplines increasingly point to the need for action beyond medical regulation as part of a solution to AMR. Anthropological and behavioural science insights suggest that antibiotic use is linked to deep moral obligations of care, productivity demands, and social responsibilities (Chandler, 2019). For many, antibiotics are safe amid

Antimicrobial Resistance in the 21st Century. DOI: https://doi.org/10.1016/B978-0-443-44748-8.00019-1

uncertain healthcare settings. Time pressures from economic stress, work demands, and caregiving responsibilities create desperation to find quick fixes. Thus management programs that fail to acknowledge these social realities will likely struggle for acceptability and experience backlash within communities.

International policy initiatives are increasingly acknowledging the value of management strategies that are culturally relevant. WHO and other international action plans addressing AMR argue that education, changing behaviours, and engaging the public are important complements to regulation and monitoring (WHO, 2015; Holmes et al., 2016). Research suggests that limiting access is not as effective as approaches rooted in behavioural science, such as communication training for doctors, community-centred awareness campaigns and culturally relevant messaging (Gong et al., 2019). The strengthening of the ties between public health objectives and lived experiences through sociocultural analysis in management design can facilitate sustained behavioural changes as opposed to short-term compliance.

This is key to addressing AMR as a systemic issue, instead of simply viewing it through an economic lens, if not completely disregarding the importance of antibiotic use patterns by taking the overall supply chain into account. The wrong use of antibiotics is not only about knowledge or regulation; it is also the reflection of how societies understand disease, responsibility, and care. Incorporating these insights into antimicrobial management paves the way for scientifically sound and socially acceptable actions. This method is an essential component of successfully reducing persistent inappropriate use of antibiotics and preserving the efficacy of treatments currently available.

18.2 Beliefs about illnesses and medicines in cultures

Beliefs about illness and medications play an important role in how antibiotics are used. Antibiotics are not just understood at the level of medicine; they gain social meaning that informs how communities interpret what is considered to be appropriate treatment. Social medicine and anthropology have long noted that pharmaceuticals are embedded in moral economies of care, trust, and responsibility (Kleinman, 1995; Helman, 2007), and this has been true since the early 1990s. Antibiotics in particular have come to embody the promise of modern healing, creating demands for recovery and suitable care. These meanings shape demand, prescribing practices and adherence, making it clear that antibiotic consumption cannot be divorced from its social context (Whyte et al., 2002; Chandler, 2019).

18.2.1 Antibiotics as symbols of a strong or 'effective' medicine

Antibiotics are often viewed as powerful, definitive medicines in cultures around the world. Even early sociomedical research on pharmaceuticals revealed their role as a signifier of technological authority and modernity (Van der Geest et al., 1996). Antibiotics became considered 'curative' medicines because of their association with rapid symptom relief, reinforcing the perception that they work for everyone. This perspective has persisted for generations, contributing to patients' treatment expectations today (Payer, 1996; Whyte et al., 2002). Modern studies confirm that patients have come to equate antibiotics with good care. Some patients might interpret a lack of antibiotic prescription as evidence that they are not being adequately treated (Butler et al., 1998; Roope et al., 2018). The heroic image of antibiotics is believed to contribute to pressure for these to be prescribed, particularly in outpatient care settings where a social negotiation happens between doctor and patient (Hulscher and Prins, 2017). Antibiotic prescribing is thus shaped by symbolic beliefs even within tightly monitored healthcare systems (Holmes et al., 2016).

18.2.2 Misunderstanding of infection causation

Chronic confusion about the difference between bacterial and viral infections has been noted for over three decades. Studies of the public's knowledge from the 90 s showed considerable misunderstanding regarding how antibiotics work (Belongia et al., 2002). Despite significant global awareness campaigns, many existing misconceptions persist according to surveys conducted recently (McCullough et al., 2016; WHO, 2015). The gap is not only of education but also cultural: Categories of illness are often filtered through personal experience as opposed to scientific comprehension (Helman, 2007). Antibiotics are often perceived by parents and carers as safeguards against more serious illness, particularly in the case of children (Mangione-Smith et al., 1999). Fear of complications fuels demand even when healthcare providers recommend a wait-and-see approach. Psychological insights from behavioural economics research reveal that individuals generally prefer instant perceived safety over future AMR threats (Roope et al., 2018). This illustrates how risk perception is influenced more by culture than by raw information (Chandler, 2019; Holmes et al., 2016).

18.2.3 Reinforcement of the lived experience and social narratives

Anthropological studies since the early 2000s have linked medicines to social networks and knowledge sharing through experience (Whyte et al., 2002). Factual storytelling, peer recommendations, and familial care practices all help to normalise the use of antibiotics. Anecdotes of personal recovery are powerful forms of informal evidence that tend to trump medical communications. Recent social science studies reveal that peer recommendations constitute the strongest influencer on self-medication (Ocan et al., 2015; Torres et al., 2019). Medicines serve as social goods shared within networks of reciprocity in many communities. Sharing antibiotics becomes responsible caregiving instead of misuse. These are moral economies of medicine in which drug access denotes social connection (Chandler, 2019).

18.2.4 Heavy cultural pressure around fast recovery and high visibility

Medical anthropology has long recognised that visible actions of treatment carry moral significance among many cultures (Kleinman, 1995). Sickness is perceived not merely as a biological state but as a social rupture demanding bold intervention. Antibiotics fit into cultural demands that responsible people take an active role in restoring productivity and family stability. Studies of clinical communication in the late 1990s showed that patients prefer some form of medical intervention to 'watchful waiting' (Mangione-Smith et al., 1999; Butler et al., 1998). This expectation is strongly upheld in current evidence (Roope et al., 2018). Economic strains, precarious employment, and caregiving responsibilities create increased pressures to relieve symptoms (Holmes et al., 2016; WHO, 2015). Antibiotics are tools for social stability as much as medical treatment.

18.2.5 Tension with evidence-based prescribing norms

The result of some symbolic expectations and social obligations is chronic tension with the proper application of antibiotics. Healthcare providers have learned to expect certain patient expectations over lost decades, even as they are caught between professional guidelines and such. From the 1990s, studies raised concerns about patient dissatisfaction generating defensive prescribing (Butler et al., 1998). This trend persists, even in light of newer management theories (Gong et al., 2019; Hulscher and Prins, 2017). Recent cross-disciplinary studies highlight that management of cultural meanings is more essential than constraint (Holmes et al., 2016). Direct patient-centred strategies (e.g. responding to patients' worry, understanding patients' caregiving motives, and reframing antibiotics as finite resources) are more effective than prescriptive approaches (WHO, 2015). Stricter regulations alone are insufficient to address misuse; transforming social narratives surrounding medicine is equally important (Table 18.1).

18.3 Patient expectations and prescribing pressure

Patient expectations are a powerful driver of antibiotic prescribing behaviour and an important cause of inappropriate antibiotic use. Patients in many primary care practices relate antibiotics to successful treatment. Studies have shown that demand from patients – explicitly or implicitly – often coerces clinicians to prescribe antibiotics, despite guidelines suggesting against it (Butler et al., 1998; Mangione-Smith et al., 1999). Time pressures, uncertainty regarding diagnosis, and risk of unhappiness are more responsible for underpinning this behaviour. You may also keep antibiotics because they preserve trust, prevent confrontation, or minimise the chances of complications (Little et al., 1997). Research shows that doctors frequently overestimate patients' expectations, which can contribute to unnecessary prescribing of medication. In paediatric care, the perception of parental pressure has been particularly associated with increased rates of antibiotic prescribing (Mangione-Smith et al., 1999).

New findings indicate that this cycle of expectations and prescriptions is self-reinforcing. Patients interpret antibiotics prescribed as a response to perceived expectations as confirmation that they are required, which in turn increases the likelihood of future requests (Roope et al., 2018; McCullough et al., 2016). This circumstance complicates initiatives to promote responsible antibiotic prescribing, as prescribing decisions are determined by clinical judgement and also social and relational contextual factors (Hulscher and Prins, 2017; WHO, 2015).

Noting the pressure to prescribe antibiotics, as well as the lack of efficacy in antibiotics for some conditions, the authors recommend improved clinician-patient communication and shared decision-making strategies, along with public education aimed at resetting expectations for antibiotic necessity. In the absence of addressing these behavioural determinants, responsible prescribing practices that meet acceptable standards would be difficult to achieve (Fig. 18.1).

TABLE 18.1 Sociocultural determinants shaping antibiotic use and their behavioural consequences.

Cultural factor	Underlying belief or norm	Behavioural outcome	Impact on antibiotic use	Public health implication	Key references
Antibiotics as 'strong medicine'	Pharmaceuticals symbolise powerful, modern, curative care	Preference for receiving antibiotics during consultation	Demand for antibiotics in nonbacterial illness	Reinforces overprescribing and resistance selection	Van der Geest et al. (1996); Whyte et al. (2002); Roope et al. (2018)
Misunderstanding of infection causation	All infections perceived as requiring antibiotics	Expectation of antibiotics for colds, flu, and fever	Inappropriate initiation of therapy	Weakens stewardship and increases misuse	Belongia et al. (2002); McCullough et al. (2016); WHO (2015)
Social sharing of medicines	Caregiving equals providing medicine to family members	Household antibiotic reuse and incomplete courses	Subtherapeutic exposure	Community-level resistance pressure	Whyte et al. (2002); Ocan et al. (2015); Torres et al. (2019)
Rapid recovery culture	Illness seen as social and economic disruption	Immediate treatment seeking	Premature antibiotic consumption	Normalises pharmaceutical dependence	Butler et al. (1998); Mangione-Smith et al. (1999); Holmes et al. (2016)
Peer and family narratives	Experiential knowledge trusted over formal medical advice	Self-medication based on anecdotal success	Reinforcement of irrational use	Undermines regulatory control	Chandler (2019); Torres et al. (2019)
Distrust in healthcare systems	Informal providers perceived as more accessible	Pharmacy-led or informal antibiotic access	Bypass of diagnostic oversight	Fragmented stewardship implementation	Hulscher & Prins (2017); WHO (2015)
Caregiving responsibility norms	Moral obligation to protect children/elderly	Precautionary antibiotic use	Overexposure in vulnerable populations	Increased resistance risk	Mangione-Smith et al. (1999); Roope et al. (2018)

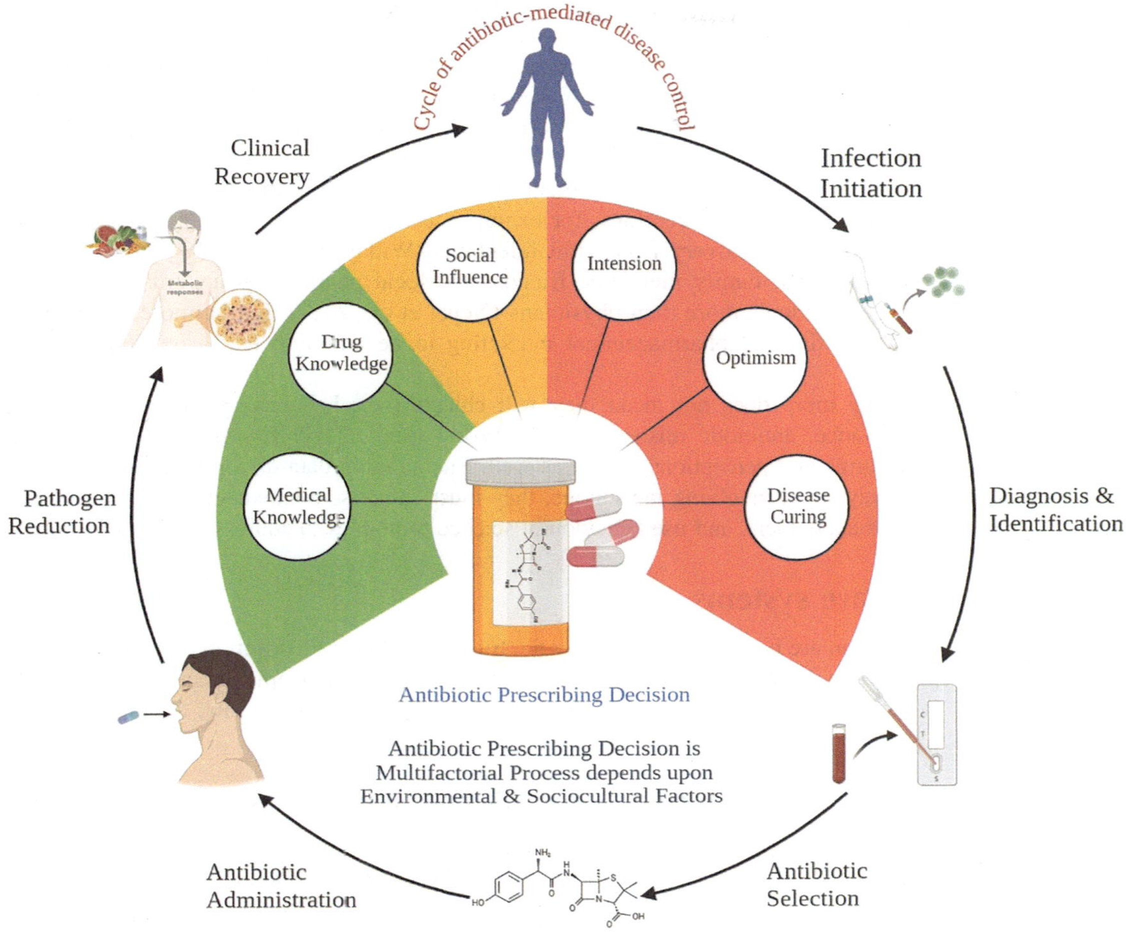

FIGURE 18.1 Cycle of antibiotic-mediated disease control showing infection progression, treatment stages, and factors influencing antibiotic prescribing decisions. The figure depicts the cyclic process of antibiotic-mediated disease control, integrating clinical progression (infection to recovery) with behavioural and knowledge-driven factors influencing antibiotic prescribing decisions. It highlights how sociocultural and environmental determinants intersect with clinical judgement to shape antibiotic use and therapeutic outcomes.

18.4 Self-medication and informal access

Self-medication with antibiotics is a major cause of inappropriate use, in particular in low- and middle-income countries, but increasingly also a problem in high-income areas. Community pharmacies, informal vendors, online markets, and leftover supplies tend to make antibiotics accessible without a prescription or any supervision (Auta et al., 2019; Morgan et al., 2011). Where regulations do exist, however, enforcement gaps and commercial incentives often facilitate the sale of nonprescription elements (Do et al., 2021).

Individuals frequently make decisions regarding self-administration based on previous therapy or based on historical advice from family members or recommendations made by pharmacists, frequently without adequate microbiological studies. In some contexts, pharmacists serve as primary care providers, especially in areas with limited healthcare access (Sakeena et al., 2018). Although this can enhance accessibility, it can also result in inappropriate choice of antibiotics, incorrect dosages, and shortened duration of treatment. Another cause contributing to this phenomenon is sharing leftover antibiotics within households, which may appear to be a sign of caring (Grigoryan et al., 2006).

Indeed, these practices accelerate the development of antibiotic resistance by exposing pathogens to low doses and incomplete courses of treatment conditions under which resistant strains flourish (Ventola, 2015; Llor and Bjerrum, 2014). At the (sub)systemic level, unregulated circulation of antibiotics goes against national regulations and hampers accurate consumption monitoring (Rather et al., 2017). So, addressing self-medication, therefore, needs to include tighter regulations and initiatives to get communities engaged on why they are accessing informal means.

18.5 Health literacy and community knowledge

Health literacy is important to antibiotic-related behaviours, including the initiation of treatment, adherence, and understanding effectiveness. A limited understanding of microbial resistance, the effectiveness of antibiotics, and how long therapy should be given is prevalent across much (André et al., 2010; Essack et al., 2016). People often switch antibiotics prematurely or insist on broader-spectrum treatments when there is no clinical need, and misunderstandings such as the belief that persistent symptoms represent treatment failures contribute to this practice.

People with low health literacy have misconceptions, for example, using the belief that antibiotics reduce fever no matter what causes it or leads to faster recovery from viral infections (Widayati et al., 2012). For example, in many communities, informal social networks, family members, friends, and local pharmacists are trusted sources of health information and may influence individuals more than physicians (Nepal et al., 2019). Furthermore, antibiotics are often incorrectly promoted in digital media and pharmaceutical marketing to be common and a quick fix, facilitating the normalisation of their use.

But just because people know more does not mean they will change their behaviour. Indeed, even where there is awareness of the risks of resistance, antibiotic misuse persists (World Bank, 2017; Essack et al., 2016), proving that behaviour can reflect cultural norms and perceptions of risk as well. Encouraging data-driven family planning and similar interventions cannot be a one-way street involving facts alone; they must comprise context-sensitive education campaigns, community dialogues with key stakeholders, and use behavioural science principles to address deeply held misconceptions.

18.6 Trust in healthcare systems

The patterns of antibiotic utilisation are heavily influenced by the trust in healthcare institutions. Trust in health providers, regulatory agencies, and public health guidance matters when it comes to whether people adhere to prescriptions, accept nonantibiotic treatment, or seek alternative care options. When trust in these systems is low, antibiotic misuse tends to increase.

18.6.1 Distrust of formal healthcare providers and reliance on informal ones

Where the public healthcare system appears as if unreachable, inefficient, or expensive, patients will typically select informal providers, private pharmacies, or simply do not go at all (Gebrekirstos et al., 2017). In these organisational situations, the antibiotics may not receive appropriate diagnostic assessment or follow-up, which promotes further use without a prescription. This issue is exacerbated in low-resource settings by weak implementation of pharmaceutical regulations (Alhomoud et al., 2018). It means antibiotics can circulate beyond the bounds of trusted gatekeeping, outside formal oversight frameworks.

18.6.2 Awareness of quality of care and expectations for antibiotics

Trust is closely linked to perceptions of care quality. Patients who distrust thorough diagnostics or clinician competence are prone to demanding antibiotics or seeking second opinions (Calnan and Rowe, 2008). Conversely, higher trust in institutions was associated with greater willingness to accept delayed prescriptions and nonantibiotic treatment strategies. Patients are more likely to comply with recommendations when doctors explain the rationale for withholding antibiotics

18.6.3 Communication and relational trust

Communication issues, linguistic, cultural or educational, can obstruct understanding of when antibiotics are appropriate and how long they must be taken. Shared decision-making and culturally competent communication that instils confidence in patients has been shown to reduce inappropriate antibiotic expectations (Schneider et al., 2021). Interactions that build trust help clinicians in explaining more explicitly the difference between viral and bacterial infections, thereby supporting efforts for responsible use. As such, trust is both a relational factor central to clinical exchanges and a structural one linked to the effectiveness of health systems.

18.7 Socioeconomic and occupational influences

Health-seeking behaviour and antibiotic use are strongly influenced by socioeconomic factors. Financial problems, job insecurity, and lack of access to healthcare often cause people to seek quick symptom relief rather than evidence-based treatment.

18.7.1 Productivity pressures and rapid recovery norms

In many places, missing work means losing income immediately, which can be especially dire for sectors that operate through informal labour without access to paid sick leave. In these settings, antibiotics represent instant cures that enable workers to get back on the job more quickly. The pressure to be productive leads to early use of antibiotics, prioritising immediate recovery over appropriate medical evaluation.

18.7.2 Cost barriers and self-treatment

People regard self-medication as an affordable alternative due to high healthcare costs and limited insurance availability (Ayukekbong et al., 2017). Antibiotics purchased directly from pharmacies often appear to be cheaper than going to a doctor, getting diagnostic tests, and then the appropriate antibiotic. Geographic challenges also exacerbate accessibility to formal healthcare services in rural or underprivileged areas, increasing dependence upon informal therapeutic practices (Sulis et al., 2022).

18.7.3 Structural inequality and antimicrobial exposure

Broader socioeconomic inequalities create different patterns of antibiotic exposure. Antibiotic use is driven by the installation of higher infection rates due to overcrowded living stations and inadequate sanitation (Collignon et al., 2018). Inadequate public health infrastructure may lead to repeated empirical antibiotic treatment without the necessary lab tests. These systemic factors reveal that antibiotic misuse is frequently driven by economic and environmental restrictions, rather than exclusively personal decisions.

18.8 Family structures and caregiving norms

Expectations around family structure and caregiving heavily influence antibiotic use, particularly for children and elderly relatives. Decisions about health are commonly made collectively by households, influenced by cultural standards around responsibility, care, and avoidance of risk.

18.8.1 Antibiotic use in high-risk populations as a precaution

For instance, caregivers often take precautionary treatment measures to stop illness from getting worse, especially among young children and older adults. Analysis indicates that parental anxiety and concern about complications increase the likelihood of antibiotic request, prescription, or initiation in response to minor/ self-limiting conditions (Cantarero-Arévalo et al., 2017; Cabral et al., 2015). In multigenerational households, older family members may be given antibiotics early because of concerns about frailty and other medical conditions. This is a conservative way to proceed because it prioritises immediate reassurance over diagnostic precision.

18.8.2 Accountability and the avoidance of delegation

In countless cultures, administering medicine is a benchmark of good care. Failure to provide treatment can also constitute neglect (Colvin et al., 2013). Antibiotics are seen as powerful, so they tend to be a tangible demonstration that suitable measures have been taken. This moral framing can induce pressure on caregivers to pursue antibiotics even when professionals recommend a watchful approach. Overuse is driven by the need to perform caregiving duties.

18.8.3 Gender roles in treatment decision-making

Gender norms are powerful determinants of health behaviours within the household. In many cultures, women are primarily responsible for children and the elderly, which makes them key decision-makers in antibiotic usage. But their decisions can be influenced by the counsel of older relatives or male heads of household, reflecting family hierarchies. Research indicates that health education for women can lead to better adherence with treatment plans as well as a decrease in the misuse of antibiotics. Stewardship efforts, therefore, need to take family power dynamics and caregiving responsibilities into account.

18.9 Interaction with traditional medicine

In much of the world, antibiotics are used within plural healthcare contexts where biomedical medicine coexists with traditional healing systems. Patients move fluidly across both paradigms, resulting in hybrid trajectories mediated by access to care, trust in the systems of treatment, cost considerations, and cultural beliefs.

18.9.1 Concurrent use and hybrid treatment practices

Patients often take antibiotics together with herbal products, spiritual therapies or folk treatments without communicating with their physicians (James et al., 2018). This combination may derive from the belief that mixing treatments both increases efficacy and decreases time to recovery. But without guidance, doing so can throw off adherence, alter the way the body metabolises medications, or delay appropriate medical evaluations. At times, advances from traditional remedies might result in patients discontinuing antibiotics prematurely, increasing the risk.

18.9.2 Sequential care-seeking behaviour

In several contexts, patients may first visit traditional healers and then go on to biomedical providers in the case that symptoms do not subside or vice versa (Green and Colucci, 2020). These stepwise pathways can delay initiation of antibiotics for true bacterial infections or cause inappropriate use of antibiotics after poor response to traditional therapies. This behaviour makes diagnosis challenging and may result in re-exposure to medications.

18.9.3 Legitimacy in culture and authority of healers

Traditional healers are highly respected sources of knowledge within their own cultures. In a lot of cases, the expertise that these individuals can give may be more powerful than what formal healthcare providers can offer, especially in rural or neglected communities (WHO, 2019). Healers sometimes serve therapeutic, spiritual, and social advisory purposes. When biomedical systems reject or disdain traditional practices, patients may abandon formal healthcare services, thereby curtailing opportunities for effective antibiotic stewardship.

18.9.4 Knowledge transmission and community beliefs

Traditional health systems disseminate knowledge and create community perspectives on illness/treatment. Concepts of 'heat', 'imbalance', or spiritual causes can apply to conditionalities and timings of antibiotic use (Langwick, 2011). Seen as an adjunct to orthodox medicine where bacteria are concerned, antibiotics can be misused if there is disagreement between biomedical and cultural pathways of disease.

18.9.5 Risks of unregulated herbal–antibiotic interactions

Research on interactions between herbal medicines and antibiotics is limited. There may, however, be concerns for altered drug processing and efficacy (Fasinu et al., 2012). Self-directed combination treatments may also result in more side effects or less adherence due to mishaps with pill schedules. This begs the need for open conversations on the traditional medicine usage between both clinicians and patients.

18.9.6 Opportunities for collaborative stewardship

Rather than considering traditional medicine a barrier, several public health initiatives have demonstrated that involving traditional healers in teaching about antimicrobials can be useful (King and Homsy, 1997). Incorporating trusted community members as awareness advocates improves trust and acceptance of appropriate antibiotic use practices. Community-based stewardship grounded in traditional healing practices generally has a better reception than purely regulatory measures (WHO, 2019).

18.10 Conclusion

To use antibiotics is not merely a clinical decision; it is also a social practice, which is influenced by beliefs, roles of care and trust in health systems and economic conditions, as well as diverse medical traditions. Home use of antibiotics, gender roles in caregiving in the family, informal access to care, integration with traditional healing, and pressure from patients all influence antibiotic prescription. Misuse frequently arises not from purposeful carelessness but rather from culturally influenced efforts to safeguard relatives, remain productive despite illness, and demonstrate visible responsiveness to sickness.

Consequently, efficient antimicrobial stewardship must extend beyond regulations and clinical practice guidelines. The approaches must embrace diverse communication strategies, community outreach, and culturally competent health literacy while accommodating better healthcare systems. Emphasising deference to traditional practices, addressing the economic difficulties, and building institutional trust are more likely to bring about lasting behaviour change. Treating antibiotic use as medical and cultural work positions us to create successful interventions for limiting AMR and preserving the ability of antibiotic treatments far into the future.

AI disclosure

The authors declare that AI- assisted tools, including Grammarly and ChatGPT, were used for language editing, grammar correction, and improving the clarity and readability of the manuscript. These tools were not used to generating original data, conduct analyses, interpret results, or develop scientific conclusions. After using this tool/service, the author(s) reviewed and edited the content as needed and take(s) full responsibility for the content of the publication.

References

Alhomoud, F., Almahasnah, R., Alhomoud, F.K., 2018. "You could lose when you misuse" - Factors affecting over-the-counter sale of antibiotics in community pharmacies in Saudi Arabia: a qualitative study. BMC Health Services Research 18 (1). Available at https://doi.org/10.1186/s12913-018-3753-y, https://bmchealthservres.biomedcentral.com/.

André, M., Vernby, A., Berg, J., Lundborg, C.S., 2010. A survey of public knowledge and awareness related to antibiotic use and resistance in Sweden. Journal of Antimicrobial Chemotherapy 65 (6), 1292–1296. Available at https://doi.org/10.1093/jac/dkq104.

Auta, A., Hadi, M.A., Oga, E., Adewuyi, E.O., Abdu-Aguye, S.N., Adeloye, D., et al., 2019. Global access to antibiotics without prescription in community pharmacies: a systematic review and meta-analysis. Journal of Infection 78 (1), 8–18. Available at https://doi.org/10.1016/j.jinf.2018.07.001, http://www.elsevier.com/inca/publications/store/6/2/3/0/5/4/index.htt.

Ayukekbong, J.A., Ntemgwa, M., Atabe, A.N., 2017. The threat of antimicrobial resistance in developing countries: causes and control strategies. Antimicrobial Resistance and Infection Control 6 (1). Available at https://doi.org/10.1186/s13756-017-0208-x, http://www.aricjournal.com/.

Belongia, E.A., Naimi, T.S., Gale, C.M., Besser, R.E., 2002. Antibiotic use and upper respiratory infections: a survey of knowledge, attitudes, and experience in Wisconsin and Minnesota. Preventive Medicine 34 (3), 346–352. Available at https://doi.org/10.1006/pmed.2001.0992.

Butler, C.C., Rollnick, S., Pill, R., Maggs-Rapport, F., Stott, N., 1998. Understanding the culture of prescribing: qualitative study of general practitioners' and patients' perceptions of antibiotics for sore throats. British Medical Journal 317 (7159), 637–642. Available at https://doi.org/10.1136/bmj.317.7159.637, http://www.bmj.com/.

Cabral, C., Lucas, P.J., Ingram, J., Hay, A.D., Horwood, J., 2015. "It's safer to ..." parent consulting and clinician antibiotic prescribing decisions for children with respiratory tract infections: an analysis across four qualitative studies. Social Science & Medicine 136-137, 156–164. Available at https://doi.org/10.1016/j.socscimed.2015.05.027.

Calnan, M., Rowe, R., 2008. Trust relations in a changing health service. Journal of Health Services Research & Policy 13 (3_suppl), 97–103. Available at https://doi.org/10.1258/jhsrp.2008.008010.

Cantarero-Arévalo, L., Hallas, M.P., Kaae, S., 2017. Parental knowledge of antibiotic use in children with respiratory infections: a systematic review. International Journal of Pharmacy Practice 25 (1), 31–49. Available at https://doi.org/10.1111/ijpp.12337, http://onlinelibrary.wiley.com/journal/10.1111/(ISSN)2042-7174.

Chandler, C.I.R., 2019. Current accounts of antimicrobial resistance: stabilisation, individualisation and antibiotics as infrastructure. Palgrave Communications 5 (1). Available at https://doi.org/10.1057/s41599-019-0263-4, www.palgrave-journals.com/palcomms/.

Collignon, P., Beggs, J.J., Walsh, T.R., Gandra, S., Laxminarayan, R., 2018. Anthropological and socioeconomic factors contributing to global antimicrobial resistance: a univariate and multivariable analysis. The Lancet Planetary Health 2 (9), e398. Available at https://doi.org/10.1016/S2542-5196(18)30186-4, https://www.journals.elsevier.com/the-lancet-planetary-health.

Colvin, C.J., Smith, H.J., Swartz, A., Ahs, J.W., de Heer, J., Opiyo, N., et al., 2013. Understanding careseeking for child illness in sub-Saharan Africa: a systematic review and conceptual framework based on qualitative research of household recognition and response to child diarrhoea, pneumonia and malaria. Social Science & Medicine 86, 66–78. Available at https://doi.org/10.1016/j.socscimed.2013.02.031.

Do, N.T.T., Vu, H.T.L., Nguyen, C.T.K., Punpuing, S., Khan, W.A., Gyapong, M., et al., 2021. Community-based antibiotic access and use in six low-income and middle-income countries: a mixed-method approach. The Lancet Global Health 9 (5), e610. Available at https://doi.org/10.1016/S2214-109X(21)00024-3, http://www.elsevier.com/journals/the-lancet-global-health/2214-109x.

Essack, S.Y., Desta, A.T., Abotsi, R.E., Agoba, E.E., 2016. Antimicrobial resistance in the WHO African region: current status and roadmap for action. Journal of Public Health 39 (1), fdw015. Available at https://doi.org/10.1093/pubmed/fdw015.

Fasinu, P.S., Bouic, P.J., Rosenkranz, B., 2012. An overview of the evidence and mechanisms of herb-drug interactions. Frontiers in Pharmacology 3. Available at https://doi.org/10.3389/fphar.2012.00069, http://www.frontiersin.org/Journal/DownloadFile.ashx?pdf=1&FileId=%201343&articleId=%2021203&Version=%201&ContentTypeId=21&FileName=%20fphar-03-00069.pdf.

Gebrekirstos, N.H., Workneh, B.D., Gebregiorgis, Y.S., Misgina, K.H., Weldehaweria, N.B., Weldu, M.G., et al., 2017. Non-prescribed antimicrobial use and associated factors among customers in drug retail outlet in Central Zone of Tigray, northern Ethiopia: a cross-sectional study. Antimicrobial Resistance and Infection Control 6 (1). Available at https://doi.org/10.1186/s13756-017-0227-7, http://www.aricjournal.com/.

Gong, C.L., Zangwill, K.M., Hay, J.W., Meeker, D., Doctor, J.N., 2019. Behavioral economics interventions to improve outpatient antibiotic prescribing for acute respiratory infections: a cost-effectiveness analysis. Journal of General Internal Medicine 34 (6), 846–854. Available at https://doi.org/10.1007/s11606-018-4467-x, http://www.springerlink.com/content/120414/.

Green, B., Colucci, E., 2020. Traditional healers' and biomedical practitioners' perceptions of collaborative mental healthcare in low- and middle-income countries: a systematic review. Transcultural Psychiatry 57 (1), 94–107. Available at https://doi.org/10.1177/1363461519894396, http://tps.sagepub.com/content/by/year.

Grigoryan, L., Haaijer-Ruskamp, F.M., Burgerhof, J.G.M., Mechtler, R., Deschepper, R., Tambic-Andrasevic, A., et al., 2006. Self-medication with antimicrobial drugs in Europe. Emerging Infectious Diseases 12 (3), 452–459. Available at https://doi.org/10.3201/eid1203.050992.

Helman, C.G., 2007. Culture, Health and Illness, 5th ed. Hodder Arnold.

Holmes, A.H., Moore, L.S.P., Sundsfjord, A., Steinbakk, M., Regmi, S., Karkey, A., et al., 2016. Understanding the mechanisms and drivers of antimicrobial resistance. The Lancet 387 (10014), 176–187. Available at https://doi.org/10.1016/S0140-6736(15)00473-0, http://www.journals.elsevier.com/the-lancet/.

Hulscher, M.E.J.L., Prins, J.M., 2017. Antibiotic stewardship: does it work in hospital practice? A review of the evidence base. Clinical Microbiology and Infection 23 (11), 799–805. Available at https://doi.org/10.1016/j.cmi.2017.07.017.

James, P.B., Wardle, J., Steel, A., Adams, J., 2018. Traditional, complementary and alternative medicine use in Sub-Saharan Africa: a systematic review. BMJ Global Health 3 (5). Available at https://doi.org/10.1136/bmjgh-2018-000895, gh.bmj.com/.

King, R., Homsy, J., 1997. Involving traditional healers in AIDS education and counselling in sub-Saharan Africa: a review. AIDS (London, England) 11.

Klein, E.Y., Tseng, K.K., Pant, S., Laxminarayan, R., 2019. Tracking global trends in the effectiveness of antibiotic therapy using the drug resistance index. BMJ Global Health 4 (2). Available at https://doi.org/10.1136/bmjgh-2018-001315, gh.bmj.com/.

Kleinman, A., 1995. Writing at the Margin: Discourse Between Anthropology and Medicine. University of California Press.

Langwick, S., 2011. Bodies, Politics, and African Healing: The Matter of Maladies in Tanzania. Indiana University Press. Available at 10.2979/5787.0.

Little, P., Gould, C., Williamson, I., Warner, G., Gantley, M., Kinmonth, A.L., 1997. Reattendance and complications in a randomised trial of prescribing strategies for sore throat: the medicalising effect of prescribing antibiotics. BMJ (Clinical Research ed.) 315 (7104), 350–352. Available at https://doi.org/10.1136/bmj.315.7104.350.

Llor, C., Bjerrum, L., 2014. Antimicrobial resistance: risk associated with antibiotic overuse and initiatives to reduce the problem. Therapeutic Advances in Drug Safety 5 (6), 229–241. Available at https://doi.org/10.1177/2042098614554919.

Mangione-Smith, R., McGlynn, E.A., Elliott, M.N., Krogstad, P., Brook, R.H., 1999. The relationship between perceived parental expectations and pediatrician antimicrobial prescribing behavior. Pediatrics 103 (4 I), 711–718. Available at https://doi.org/10.1542/peds.103.4.711.

McCullough, A.R., Parekh, S., Rathbone, J., Del Mar, C.B., Hoffmann, T.C., 2016. A systematic review of the public's knowledge and beliefs about antibiotic resistance. Journal of Antimicrobial Chemotherapy 71 (1), 27–33. Available at https://doi.org/10.1093/jac/dkv310, http://jac.oxfordjournals.org/.

Morgan, D.J., Okeke, I.N., Laxminarayan, R., Perencevich, E.N., Weisenberg, S., 2011. Non-prescription antimicrobial use worldwide: a systematic review. The Lancet Infectious Diseases 11 (9), 692–701. Available at https://doi.org/10.1016/S1473-3099(11)70054-8, https://www.sciencedirect.com/science/journal/14733099.

Murray, C.J., Ikuta, K.S., Sharara, F., Swetschinski, L., Robles Aguilar, G., Gray, A., et al., 2022. Global burden of bacterial antimicrobial resistance in 2019: a systematic analysis. The Lancet 399 (10325), 629–655. Available at https://doi.org/10.1016/S0140-6736(21)02724-0, http://www.journals.elsevier.com/the-lancet/.

Nepal, A., Hendrie, D., Robinson, S., Selvey, L.A., 2019. Knowledge, attitudes and practices relating to antibiotic use among community members of the Rupandehi District in Nepal. BMC Public Health 19 (1). Available at https://doi.org/10.1186/s12889-019-7924-5.

Ocan, M., Obuku, E.A., Bwanga, F., Akena, D., Richard, S., Ogwal-Okeng, J., et al., 2015. Household antimicrobial self-medication: a systematic review and meta-analysis of the burden, risk factors and outcomes in developing countries. BMC Public Health 15 (1). Available at https://doi.org/10.1186/s12889-015-2109-3, http://www.biomedcentral.com/bmcpublichealth.

Payer, L., 1996. Medicine and Culture: Varieties of Treatment in the United States. Owl Books, England, West Germany, and France.

Rather, I.A., Kim, B.-C., Bajpai, V.K., Park, Y.-H., 2017. Self-medication and antibiotic resistance: crisis, current challenges, and prevention. Saudi Journal of Biological Sciences 24 (4), 808–812. Available at https://doi.org/10.1016/j.sjbs.2017.01.004.

Roope, L.S.J., Tonkin-Crine, S., Butler, C.C., Crook, D., Peto, T., Peters, M., et al., 2018. Reducing demand for antibiotic prescriptions: evidence from an online survey of the general public on the interaction between preferences, beliefs and information, United Kingdom, 2015. Eurosurveillance 23 (25). Available at https://doi.org/10.2807/1560-7917.ES.2018.23.25.1700424, https://www.eurosurveillance.org/docserver/fulltext/eurosurveillance/23/25/eurosurv-23-25-2.pdf?expires=1531563262&id=id&accname=guest&checksum=9398E4C73B718AE09875E7CFEF29C5E9.

Sakeena, Bennett, A.A., McLachlan, A.J., 2018. Non-prescription sales of antimicrobial agents at community pharmacies in developing countries: a systematic review. International Journal of Antimicrobial Agents 52 (6), 771–782. Available at https://doi.org/10.1016/j.ijantimicag.2018.09.022.

Schneider, C.R., Dryhurst, S., Kerr, J., Freeman, A.L.J., Recchia, G., Spiegelhalter, D., et al., 2021. COVID-19 risk perception: a longitudinal analysis of its predictors and associations with health protective behaviours in the United Kingdom. Journal of Risk Research 24 (3-4), 294–313. Available at https://doi.org/10.1080/13669877.2021.1890637, www.tandf.co.uk/journals/titles/13669877.asp.

Sulis, G., Sayood, S., Gandra, S., 2022. Antimicrobial resistance in low- and middle-income countries: current status and future directions. Expert Review of Anti-Infective Therapy 20 (2), 147–160. Available at https://doi.org/10.1080/14787210.2021.1951705, http://www.tandfonline.com/loi/ierz20.

Torres, N.F., Solomon, V.P., Middleton, L.E., 2019. Patterns of self-medication with antibiotics in Maputo City: a qualitative study. Antimicrobial Resistance and Infection Control 8 (1). Available at https://doi.org/10.1186/s13756-019-0618-z, http://www.aricjournal.com/.

Van der Geest, S., Whyte, S.R., Hardon, A., 1996. The anthropology of pharmaceuticals: a biographical approach. Annual Review of Anthropology 25 (1), 153–178. Available at https://doi.org/10.1146/annurev.anthro.25.1.153, http://www.annualreviews.org/journal/anthro.

Ventola, C.L., 2015. The antibiotic resistance crisis: part 1: causes and threats. P & T: A Peer-reviewed Journal for Formulary Management. 40 (4), 277–283.

Whyte, S.R., Van der Geest, S., Hardon, A., 2002. Social Lives of Medicines. Cambridge University Press.

Widayati, A., Suryawati, S., de Crespigny, C., Hiller, J.E., 2012. Knowledge and beliefs about antibiotics among people in Yogyakarta City Indonesia: a cross sectional population-based survey. Antimicrobial Resistance and Infection Control 1. Available at https://doi.org/10.1186/2047-2994-1-38, http://www.aricjournal.com/content/1/1/38.

World Bank, 2017. Drug-Resistant Infections: A Threat to Our Economic Future. World Bank Group.

World Health Organization, 2015. Global Antimicrobial Resistance and Use Surveillance System (GLASS) report. WHO Press.

World Health Organization, 2019. WHO global report on traditional and complementary medicine. WHO Press.

Chapter 19

Revolutionising safety: advances in infection control and hygiene practices

Soumitra Nath
Department of Biotechnology, Gurucharan University, Silchar, Assam, India

19.1 Introduction

Hospital-acquired infection (HAI) is a major global challenge for healthcare systems, which has increased in recent years due to the COVID-19 pandemic. It has disrupted existing infection prevention and control (IPC) practices and led to the misuse/overuse of antimicrobial agents. It has been observed that HAI rates remain higher in low- and middle-income countries (LMICs) than in high-income countries. A recent metaanalysis reported that the overall prevalence of HAIs in LMICs is approximately 22%, with rates reaching up to 37% in South-East Asia (Odoom et al., 2025). Patients in LMIC hospitals face a two- to twenty-fold higher risk of acquiring HAIs compared to those in high-income settings (Allegranzi et al., 2011; Asghar et al., 2025). Thus there is an urgent need for improved IPC education, reliable access to hygiene resources, and strengthened policy and surveillance frameworks.

In recent times, IPC strategies have undergone rapid transformation due to emerging technologies and increased attention to antimicrobial resistance (AMR). Technologies such as ultraviolet-C (UV-C) disinfection robots, real-time sensor-based monitoring, artificial intelligence (AI), and machine learning (ML) are now being integrated into clinical settings to strengthen infection prevention strategies. The application of UV-C disinfection robots has been shown to significantly reduce environmental contamination on hospital surfaces, contributing to a lower incidence of HAIs in intensive care and surgical units (Casini et al., 2023). Real-time sensor-based systems enhance hand hygiene compliance by providing immediate feedback to healthcare workers, thereby significantly decreasing HAI rates (Abubeker and Baskar, 2022). AI-driven models enable early identification of potential HAI outbreaks by analysing large datasets from electronic health records and microbiology labs (Villanueva-Miranda et al., 2025). The synthesis and characterisation of antimicrobial coatings incorporating nanomaterials such as silver, copper, and titanium dioxide have demonstrated strong antimicrobial effects (Ashhari et al., 2023).

The effective implementation of these IPC practices requires integrating technological tools with adequate human resources, proper training, and effective policy development. Therefore it is essential to optimise implementation strategies, assess cost-effectiveness, and monitor long-term impact on infection transmission and resistance patterns. The World Health Organisation's 2022 Global Report on Infection Prevention and Control emphasised the need for IPC programmes with dedicated budgets, trained staff, and regular monitoring (Tomczyk *et al.*, 2022). The objective of this study is to critically examine current IPC practices in healthcare settings, identify gaps, and explore emerging technologies that could revolutionise patient safety.

19.2 Understanding hospital-acquired infections

HAIs are defined as infections that occur in patients after admission to a healthcare facility and were neither present nor incubating at the time of admission. These infections can also arise shortly after discharge, particularly in patients who underwent invasive procedures or prolonged hospital stays. Most studies define HAI as an infection occurring 48 hours after infection or within 30 days of discharge. Such infections represent a significant public health concern, resulting in prolonged hospitalisations, increased healthcare costs, and higher mortality rates.

Antimicrobial Resistance in the 21st Century. DOI: https://doi.org/10.1016/B978-0-443-44748-8.00013-0

19.2.1 Epidemiology and distribution

The global prevalence of HAI varies by region, hospital type, and patient population. It is often observed in intensive care units, tuberculosis centres, and HIV/AIDS wards. A study conducted in a large Chinese infectious disease hospital between 2017 and 2023 found HAI prevalence ranging from 0.89% to 2.52%, with the majority of cases located in respiratory, tuberculosis, and HIV/AIDS departments (Deng et al., 2024). In another study, Canadian multihospital surveillance confirms reductions in some HAI categories, including a 7% decrease in Clostridioides difficile infections (CDIs) and a 2.9% decline in Methicillin-resistant *Staphylococcus aureus* (MRSA) bloodstream infections from 2018 to 2022. However, bloodstream infections caused by vancomycin-resistant *Enterococcus* (VRE) increased by 5.9% during the same period (2024). A multicenter surveillance study from Brazil reported an overall HAI prevalence of 9.5% across 100 hospitals, with bloodstream infections and pneumonia being most common in ICUs (Alvim et al., 2020). Overall, global evidence consistently indicates a disproportionate burden of HAIs in LMICs, largely due to infrastructural limitations, inadequate surveillance systems, and poor adherence to IPC practices.

19.2.2 Risk factors and causative agents

The development of HAIs is driven by a combination of patient-related, procedural, and systemic risk factors. Older age, comorbidities, prolonged hospital stays, and the use of invasive procedures such as urinary catheters, central lines, or mechanical ventilation significantly increase infection risk. A case–control investigation in China identified antibiotic use as the primary factor (OR=7.46), while cardiovascular comorbidities were associated with an even higher risk (OR=26.69) (Deng et al., 2024). Similarly, UK hospitals reported advanced age, frequent transfers, cross-infections, prior antibiotic use, and longer institutional stays as significant risk amplifiers (Isigi et al., 2023). The common causative organisms of HAIs include mainly Gram-negative bacteria (*Klebsiella pneumoniae*), multidrug-resistant strains (MRSA and VRE), and nosocomial fungi (Deng et al., 2024).

19.2.3 Consequences for patients and health systems

HAIs continue to drive longer hospital stays, increased costs, and higher morbidity and mortality. Recent studies show that HAIs significantly extend hospital stays and increase healthcare costs (Rosenthal et al., 2023). In high-acuity settings such as ICUs, device-associated infections can double or triple mortality risk, especially in sepsis cases, with global data indicating over 24% mortality among patients with HAI-related sepsis, and rates surpassing 50% for infections acquired in the ICU (Khavandegar et al., 2025). These infections also disrupt hospital workflows, increase bed occupancy rates, and consume essential healthcare resources such as antibiotics, diagnostic testing, and staff time, particularly in LMICs, where systemic constraints are greatest (Rosenthal et al., 2023). In Australia, reducing HAI prevalence by just 1% could free up 150,000 hospital bed-days per year, which is enough to treat an additional 38,500 patients (Graves et al., 2009). The ripple effects of HAIs, therefore, go far beyond the individual, impacting entire health systems.

19.2.4 Classification and surveillance

A structured classification, proper surveillance, and adoption of innovative control strategies are essential to reduce the burden of HAIs. These infections are typically grouped into three major categories based on their source and clinical presentation: (1) device-associated, (2) procedure-associated, and (3) nondevice-associated infections (Fig. 19.1). This classification provides targeted prevention measures and enables accurate surveillance across healthcare settings. Some modern infection-tracking programmes use real-time data, laboratory-confirmed cases, and regional or international databases to monitor infection trends and AMR patterns. The World Health Organisation has recently updated its global strategy (2024–30), which emphasises effective surveillance as a core component of IPC. It also integrates advanced digital tools and molecular techniques to effectively monitor HAIs. Hospitals in the United States and Canada adopted electronic tracking and real-time reporting, as well as the management of simplified datasets, to improve national surveillance ("The Canadian Nosocomial Infection Surveillance Program: Keeping an eye on antimicrobial resistance in Canadian hospitals since 1995," 2022). The use of rapid diagnostic tools, ML-based outbreak prediction, environmental sensor monitoring, and genomic surveillance is also gaining attention and can provide early detection and a more precise response to emerging HAI threats (Kolbe-Busch et al., 2025).

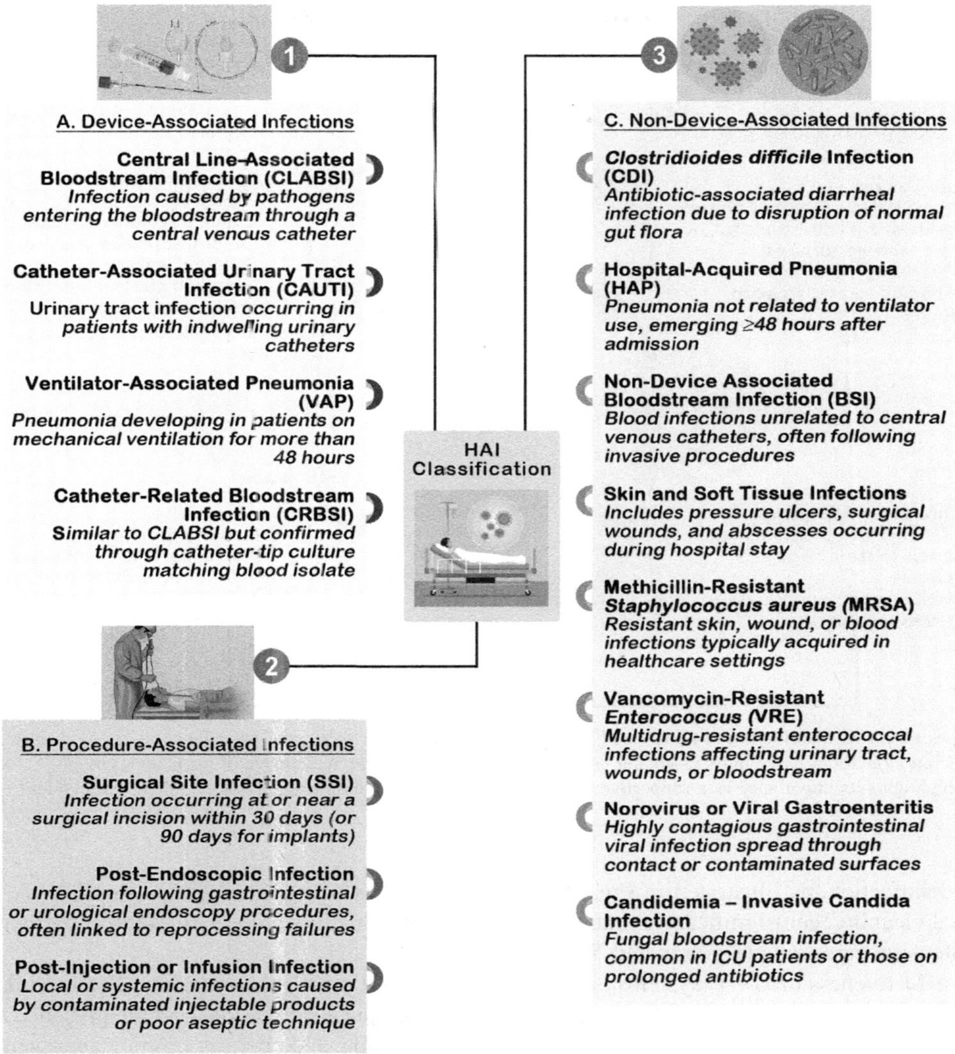

FIGURE 19.1 **Classification of HAIs based on the source and clinical manifestation.** This figure presents the classification of HAIs based on their source and clinical presentation. It groups infections into device-associated, procedure-related, and nondevice-associated categories. The classification provides a clear framework for understanding different infection types in healthcare settings. *HAIs*, Hospital-acquired infections.

19.3 Technological innovations in infection prevention and control

The growing burden of HAIs, together with the global rise of AMR, has led to increasing reliance on technological innovations in IPC. Traditional practices such as hand hygiene and environmental cleaning remain fundamental, and modern healthcare systems are increasingly incorporating advanced tools to enhance efficiency, reduce variability, and support data-driven decision-making. These technologies, however, must be evaluated not only for their antimicrobial effectiveness but also for their feasibility, cost, and compatibility with existing healthcare infrastructure. The following subsections discuss key technological interventions and their practical relevance.

19.3.1 Ultraviolet-C disinfection systems

UV-C radiation, operating within a wavelength range of 200–280 nm, has gained considerable attention as a nonchemical disinfection method in healthcare environments. UV-C induces DNA and RNA damage and effectively inactivates a wide range of pathogens, including *C. difficile*, *Acinetobacter baumannii*, and *S. aureus* (Yang et al., 2019). Different types of UV-C systems, such as mobile units, fixed ceiling-mounted devices, and robotic platforms, are increasingly used for

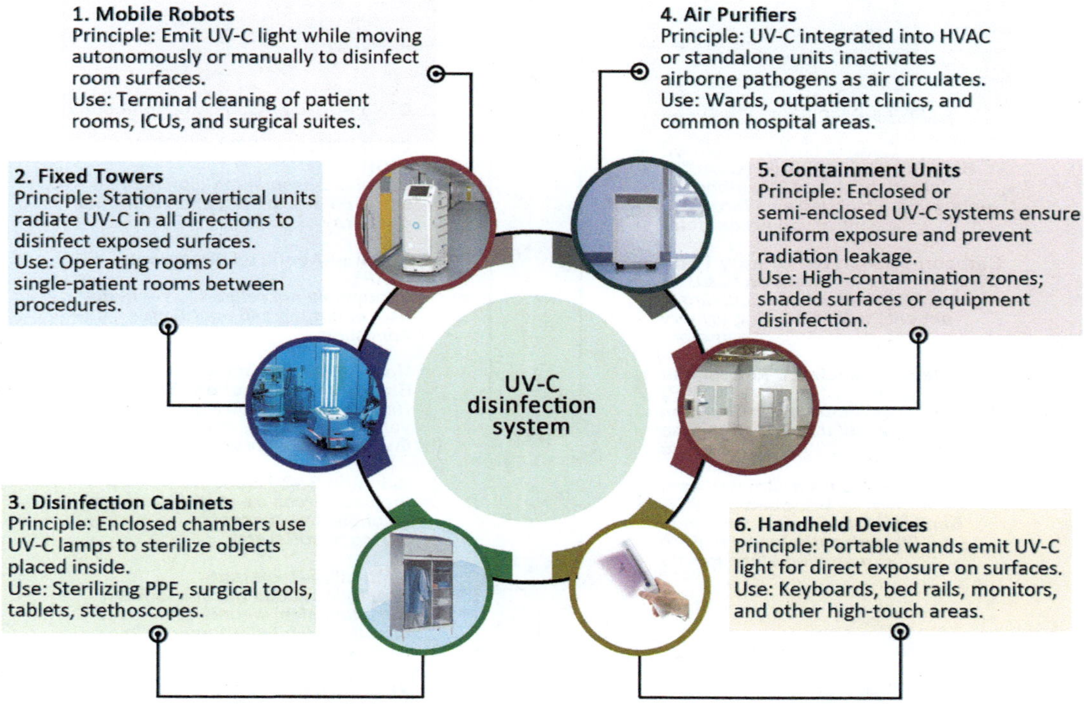

FIGURE 19.2 **Types of UV-C disinfection systems used in healthcare.** This figure shows the main types of UV-C disinfection systems used in healthcare, ranging from large room-based units to smaller portable devices. It reflects the wide range of applications of UV-C technology in infection control. The figure highlights its importance as a supportive method for improving environmental hygiene. *UV-C*, Ultraviolet-C.

surface and air disinfection, as illustrated in Fig. 19.2. Clinical studies have demonstrated that UV-C, when used as an adjunct to manual cleaning, can significantly reduce microbial contamination and infection rates, particularly in high-risk areas such as intensive care units (Koutras and Wade, 2023).

However, the effectiveness of UV-C systems is influenced by several practical constraints. Their reliance on direct line-of-sight exposure means that shadowed surfaces may not receive adequate irradiation, and the presence of organic matter can further reduce efficacy. In addition, proper positioning, exposure time, and operator training are essential for achieving optimal results. As a result, UV-C is most appropriately used as a complementary tool alongside conventional cleaning practices rather than as a standalone solution. This approach aligns with current IPC recommendations that emphasise layered strategies for infection prevention.

19.3.2 Vapour-based sterilisation systems

Vapour-based sterilisation systems using vaporised hydrogen peroxide (VHP) and ozone (O_3) provide high-level disinfection by inducing oxidative damage to microbial cells. These systems are particularly effective against resistant organisms and spores, making them valuable for terminal disinfection of isolation rooms and outbreak control (Molloy-Simard et al., 2019). Their ability to penetrate complex surfaces and enclosed spaces offers a clear advantage over conventional surface-based methods.

The operational workflow and comparative features of these systems are summarised in Table 19.1, which highlights their applications, benefits, and limitations in healthcare settings. Studies have shown that VHP systems can achieve substantial reductions in environmental contamination when used under controlled conditions (Yuan et al., 2026).

Despite their effectiveness, these technologies are associated with practical challenges. They require airtight sealing, controlled environmental parameters, and strict safety measures to prevent occupational exposure. Longer processing times and the need for specialised equipment further limit their routine use in busy clinical settings. In resource-limited healthcare systems, the high cost and technical complexity of these systems may restrict their adoption, making them more suitable for targeted, high-risk applications rather than widespread routine use.

TABLE 19.1 Applications of artificial intelligence/machine learning (*AI/ML*) in infection prevention and control (*IPC*).

Application area	Technology used	Key outcome	References
HAI surveillance	ML (Random forest)	Detected 63% more HAIs than manual method	Lukasewicz Ferreira et al. (2024)
Early outbreak detection	NLP + ML (BlueDot)	Predicted outbreaks days before CDC alerts	Adeoye et al. (2025)
Predicting stroke HAI risk	Gradient boosting	AUROC of 0.85+ for 30-day infection risk prediction	Chang et al. (2024)
Infection control compliance	AI vision systems	Real-time hygiene compliance alerts	Shrimali and Teuscher (2023)
Antimicrobial stewardship	Decision trees	Predicted resistant organisms and optimised antibiotic use	El Arab et al. (2025)
ICU sterilisation scheduling	Predictive AI scheduler	Improved throughput, reduced human error	Godbole et al. (2025)
Explainable AI in risk models	SHAP – neural networks	Improved clinician trust and intervention uptake	Hoogestraat and Wulff (2024)
Real-time environmental data	Sensors + ML	Enabled adaptive disinfection response	Arzilli et al. (2024)
ICU infection risk monitoring	XGBoost + EHR	Accurate identification of high-risk patients	Baddal et al. (2024)
AI for protocol optimisation	Multimodal ML pipelines	Improved resource use and infection rate reduction	Alhusain (2025)

19.3.3 *Nanomaterial-based*antimicrobial strategies

Nanotechnology has introduced innovative approaches to infection prevention through the development of antimicrobial coatings and materials. Metal and metal oxide nanoparticles, such as silver, copper oxide, and zinc oxide, exhibit antimicrobial activity through multiple mechanisms, including membrane disruption, oxidative stress induction, and interference with cellular processes. These properties enable their application on high-touch surfaces, medical devices, and hospital infrastructure.

The potential applications and emerging trends in nanotechnology-based IPC strategies are summarised in Table 19.2. Recent studies have demonstrated the effectiveness of these materials against a broad spectrum of microorganisms, including *S. aureus*, *Streptococcus pneumoniae*, *Escherichia coli,* and *Pseudomonas aeruginosa* (Igiebor and Omoregie, 2025).

Thus nanomaterial-based coatings are a promising solution for infection prevention, especially when combined with material-specific binding techniques and validated for biosafety. However, studies have suggested the need for 'safe-by-design' strategies to reduce ecotoxicological hazards and potential AMR (Ahonen et al., 2017). Furthermore, variability in long-term stability, potential cytotoxic effects, and lack of standardised regulatory guidelines remain key challenges that must be addressed before widespread clinical adoption.

19.3.4 Artificial intelligence/machine learning-supported sterilisation and infection prevention and control workflows

AI and ML are increasingly being applied in IPC to enhance surveillance, predict infection trends, and improve compliance monitoring. These systems can process large volumes of healthcare data in real time, enabling early detection of outbreaks and supporting evidence-based decision-making. AI and ML systems have also been used for real-time compliance monitoring, which integrates sensor- and vision-based systems for tracking hygiene practices in clinical settings (Fig. 19.3). Furthermore, AI is capable of generating personalised infection control plans based on patient history, comorbidities, and environmental exposure (El Arab et al., 2025). Some key applications of AI and ML in IPC are summarised in Table 19.1, which provides an overview of their potential roles in healthcare settings.

TABLE 19.2 Emerging trends in infection prevention and control (*IPC*).

Trend	Description	Expected outcome	References
AI and ML	Predictive surveillance, outbreak detection	Early response, risk profiling	Kubde et al. (2023)
Genomic surveillance	Whole-genome sequencing for pathogen tracking	Faster outbreak source identification	Mayhall (2012)
Environmental design	Antimicrobial materials, improved airflow	Reduction in surface contamination	Gray (2015)
IPC training and leadership	Role expansion and education reforms	Stronger frontline implementation	Qureshi et al. (2022)
Global networks	Cross-border surveillance and standardisation	Enhanced outbreak preparedness	Harbarth and Pittet (2010)
Behavioural strategies	Professional accountability, culture change	Higher adherence to IPC protocols	Gilbert and Kerridge (2020)
Patient engagement	Infection risk education, hygiene advocacy	Improved outcomes postdischarge	Fofanah and Conteh (2020)
Universal decolonisation	Routine skin/mucosa decolonisation	Reduced multidrug-resistant colonisation	Ramage et al. (2025)
Economic modelling	Data-driven cost-benefit frameworks	Smarter resource allocation	Gray (2015)
Decentralised monitoring	Mobile platforms, wearable sensors	Real-time, location-specific alerts	Kubde et al. (2023)

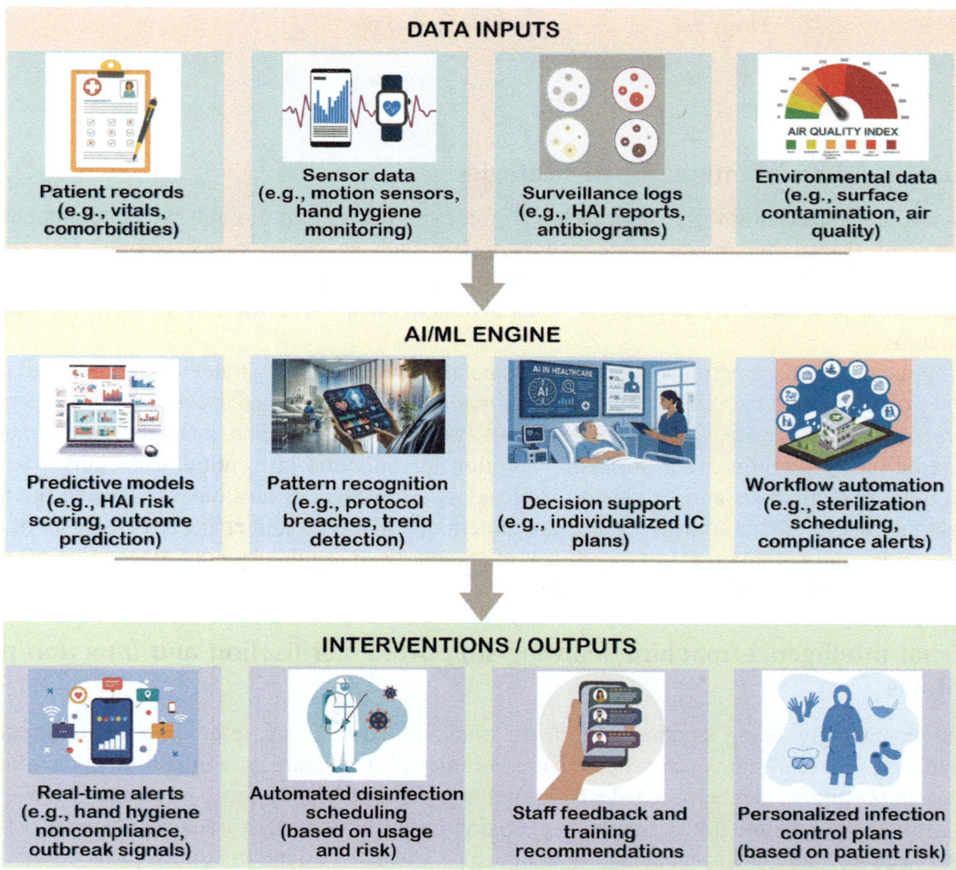

FIGURE 19.3 **Integration of AI/ML into infection control workflows showing data inputs, machine learning models, and resulting interventions in hospital settings.** This figure illustrates the integration of AI and machine learning into infection control workflows in hospitals. It presents the flow from data inputs to analysis and resulting interventions. The figure emphasises the role of these technologies in strengthening monitoring and decision-making in infection prevention. *AI*, Artificial intelligence; *ML*, machine learning.

Despite these advantages, several limitations affect their implementation. AI systems rely on high-quality, standardised datasets, which may not be available in all healthcare environments. Issues related to data privacy, algorithmic transparency, and potential bias also require careful consideration. In LMICs, limited digital infrastructure and technical expertise pose additional barriers. Consequently, the integration of AI into IPC programmes should be approached gradually, ensuring alignment with existing systems, regulatory standards, and workforce capabilities.

19.4 Real-world implementations and challenges

Despite advances in infection prevention and sterilisation technologies, their implementation in healthcare settings remains challenging. The most important barriers include inadequate infrastructure, limited funding, and technical constraints, which collectively hinder the effective translation of IPC strategies into routine clinical practice.

19.4.1 Global disparities in implementation

Data from a large national survey of Indonesian hospitals revealed that although 90% of institutions reported performing HAI surveillance, more than 57% lacked adequate microbiology laboratory capacity. Behavioural resistance and limited access to antibiograms were identified as major barriers (Supriadi et al., 2023). Similarly, in Tanzania, IPC compliance across 26 regional referral hospitals remained suboptimal. While intensive care units showed relatively higher adherence (71.4%), departments such as sterilisation and blood transfusion reported compliance levels below 50% (Bahegwa et al., 2024). These findings highlight significant disparities in IPC implementation across healthcare systems, particularly in LMICs, where infrastructure and resource limitations persist as challenges.

19.4.2 Structural and organisational barriers

Hospitals in Islamabad demonstrated poor performance in the WHO IPC assessment framework, with median scores below 200. The study revealed substantial deficiencies in essential IPC components, including the absence of functional programmes, trained personnel, and adequate supplies (Savul et al., 2020). In Georgia, although 78% of hospitals reported having IPC guidelines, only 44% utilised monitoring data for action planning. In addition, most institutions lacked structured compliance-monitoring systems, defined work plans, and sufficient budget allocation (Deryabina et al., 2021). Such systemic limitations significantly restrict the effective implementation of the WHO IPC core components and reduce the overall impact of infection control strategies.

19.4.3 Implementation gaps and sustainability challenges

Evidence from implementation studies suggests that targeted interventions can significantly improve IPC outcomes even in resource-limited settings. In Sierra Leone, a focused operational research initiative enabled three tertiary hospitals to improve IPC compliance from 'Basic' to 'Intermediate' levels within two years. This improvement was primarily driven by staff training, updated guidelines, and external evaluation mechanisms (Kamara et al., 2023).

However, the sustainability of these improvements was constrained by the absence of dedicated IPC budgets and weak HAI surveillance systems. These findings indicate that while short-term gains are achievable, long-term success depends on sustained institutional commitment, financial support, and continuous monitoring mechanisms. These persistent challenges highlight the need for comprehensive policy frameworks and system-level interventions.

19.5 Policy framework and future directions

Effective infection prevention and control requires a balanced integration of established practices with emerging technological and policy-driven approaches. Particular emphasis should be placed on adapting these technologies for LMIC settings through cost-reduction strategies, modular deployment, and integration with existing healthcare infrastructure. These approaches are considered advanced interventions and should be implemented based on resource availability and institutional capacity. Some of the key policy components for strengthening IPC systems across diverse healthcare settings are discussed below.

19.5.1 Foundations of effective infection prevention and control policy

The global burden of healthcare-associated infections necessitates policies that are grounded in strong scientific evidence and adapted to local healthcare conditions. At the core of IPC are standardised protocols, including hand hygiene, prevention of ventilator-associated pneumonia (VAP), and control of catheter-associated urinary tract infections (CAUTI). The guidelines from the Indian Society of Critical Care Medicine strongly support these measures (Mehta et al., 2014). Similarly, evidence from European hospitals shows that well-structured IPC programmes are associated with lower infection rates, although implementation challenges persist due to limited human resources and adherence (Struelens et al., 2006). These foundational measures represent essential IPC interventions that should be prioritised across all healthcare settings.

19.5.2 Antibiotic stewardship and surveillance systems

AMR is a growing global concern closely linked to hospital practices. Antibiotic stewardship programmes are designed to ensure that antibiotics are used judiciously and only when necessary. These programmes typically include prescription audits, clinician feedback, and de-escalation strategies.

Indian guidelines recommend integrating stewardship programmes with digital tools to monitor antibiotic usage (Mehta et al., 2014). Similarly, CDC guidelines emphasise structured antibiotic classification and linkage of prescription practices with clinical decision-support systems (Flynn and Guarner, 2023). The NHS EPIC Project in the United Kingdom demonstrated that continuous data collection and feedback to healthcare workers can significantly improve IPC outcomes (Pellowe et al., 2005).

19.5.3 Environmental and structural controls

The healthcare environment plays a critical role in the transmission of infections. Improvements in infrastructure, including HEPA-filtered rooms, isolation units, and controlled airflow systems, have been shown to reduce the spread of airborne pathogens (Brady et al., 2025). Recent studies further demonstrate that HEPA-filtered, positive-pressure rooms can substantially reduce airborne infections (Ouirungroj et al., 2024). Technological interventions, such as UV-C disinfection systems, have demonstrated effectiveness in reducing microbial contamination in high-risk areas (Casini et al., 2023). In addition, antimicrobial surface coatings using metals such as silver and copper have been reported to reduce pathogen survival on frequently touched surfaces (Long et al., 2023).

Water systems also represent an important source of infection. Improved water management strategies, including filtration and biofilm control, have been associated with reductions in waterborne infections (Hayward et al., 2020). However, implementing these interventions requires careful consideration of costs, maintenance, and technical feasibility, particularly in resource-limited settings.

19.5.4 Compliance, training, and behavioural change

Noncompliance among healthcare workers remains one of the most significant barriers to effective infection prevention. Studies from Zambia indicate that factors such as staff shortages, high workload, and inadequate training contribute to reduced adherence to IPC practices (Zimba et al., 2022). In Austria, compliance levels were higher among trained IPC personnel than among general clinicians, highlighting the importance of specialised training and awareness (Berktold et al., 2023).

Improving compliance requires a comprehensive approach that includes regular training programmes, performance monitoring, and institutional support. Behavioural change strategies, combined with continuous feedback and accountability mechanisms, are essential to ensure sustained adherence to IPC protocols.

19.6 Case studies of effective infection prevention and control interventions

Global case studies from healthcare environments demonstrate the impact of targeted infection control interventions on hospital safety and quality. Interventions such as hand hygiene campaigns, active surveillance, antibiotic audits, and organisational changes have led to measurable improvements in many settings. Rural hospitals in Brazil and the United Kingdom reveal that IPC success depends on strong leadership, dedicated IPC teams, adaptive policy frameworks, and continuous performance evaluation. These case-based experiences offer practical insights for replicating and scaling effective infection prevention strategies across varied settings.

A study in rural hospitals in the United States reported a 28.4% increase in hand hygiene compliance after implementing focused IPC strategies (Stevenson et al., 2014). At Kobe University Hospital in Japan, closer monitoring of antibiotic use and the implementation of rapid alert systems improved clinical outcomes over time (Shigemura et al., 2013). In Australia, a national survey covering 2005–11 revealed frequent HAI outbreaks, primarily from Norovirus and VRE. IPC strategies were inconsistently applied across hospitals, with no standardised implementation protocol.

However, hospitals with clearer internal reporting pathways showed reduced HAI recurrence (Hospital acquired infections: outbreaks and infection control interventions, a national descriptive survey. Queensland University of Technology., 2016).

Recent studies showed that economic and organisational factors strongly influence IPC outcomes. Studies have suggested that infection prevention programmes should be cost-effective and reduce both patient burden and hospital expenses (Asegu et al., 2024; Rice et al., 2023). Another study highlighted that improved infection surveillance, such as postdischarge monitoring, helps better detect and control infections, especially in resource-limited settings (Srivastav et al., 2024). In addition, the use of digital tools and AI-based surveillance systems can reduce workload, improve detection accuracy, and support cost-effective IPC strategies (Arzilli et al., 2024). Some notable case studies and their outcomes are presented in Table 19.2, which highlights differences in intervention strategies, resource utilisation, and impact across settings. These case studies highlight that successful IPC implementation depends on technological interventions, strong leadership, continuous monitoring, and adaptation to local resource availability.

19.7 Future directions and emerging trends in infection prevention and control

Healthcare systems are facing increasingly complex challenges, and IPC is undergoing a transformative change. The rise of AMR, emerging pathogens, and global patient movement has made IPC more demanding. Future IPC depends on innovative technologies as well as systemic evolution in the design, implementation, and monitoring of IPC strategies.

19.7.1 Technological integration and digital transformation

Digital health tools are becoming an important part of IPC. AI and ML are now used to track infections in real time and identify high-risk patients early. These tools can help hospitals respond faster to outbreaks and plan preventive measures more effectively. AI-powered tools have been applied to model patient movement and environmental risks, which may help facilitate timely interventions (Kubde et al., 2023). Moreover, next-generation diagnostics and molecular surveillance tools such as whole-genome sequencing (WGS) are revolutionising pathogen tracking. They help identify the source of infections and detect AMR more quickly (Oniciuc et al., 2018).

19.7.2 Infrastructure and environmental design

Hospital design is increasingly recognised as an important factor in IC. According to the WHO report, gaps in water, sanitation, and hygiene facilities remain a key factor in poor IPC in many hospitals. Studies also show that better ventilation, controlled airflow, and isolation design reduce airborne transmission and improve patient safety (Gray, 2015). In addition, the use of antimicrobial surfaces and improved cleaning protocols has been associated with lower contamination levels in hospital settings. However, infrastructure upgrades are often expensive and require long-term maintenance. The WHO has emphasised that although 71% of countries report having IPC programmes, only a small proportion meet full standards, mainly due to limited resources and infrastructure gaps (Balakrishnan, 2025). This shows that infrastructure improvement must align with sustained funding and planning.

19.7.3 Workforce transformation and interdisciplinary collaboration

A strong and well-trained workforce is essential for effective IPC. Healthcare workers, especially IPC teams, are responsible for implementing guidelines in daily practice. However, many studies in ICU settings found that even when staff are aware of IPC guidelines, factors such as heavy workload, staff shortages, and unclear responsibilities reduce compliance (Madran et al., 2025).

19.7.4 Global surveillance and policy harmonisation

The WHO Global Action Plan for IPC (2024–30) stresses the importance of unified surveillance systems and shared indicators to improve global health security (Satta et al., 2025). These systems allow countries to compare data, identify risks, and respond more effectively. Europe has made significant measure in harmonising surveillance and reporting frameworks through multicenter trials and networks like ECDC's HAI-Net (Harbarth and Pittet, 2010). However, differences in healthcare infrastructure, data quality, and technical capacity remain a challenge for applying uniform standards everywhere. Therefore future policies should focus on flexible frameworks that can be adapted to local conditions while maintaining global coordination.

19.7.5 Emphasis on behaviour change and cultural transformation

Behavioural factors remain among the most difficult barriers in IC. Even when guidelines are available, they are not always followed consistently in clinical practice. Recent research shows that compliance is affected by multiple factors, including workload, workplace culture, lack of supervision, and the perceived importance of guidelines (Madran et al., 2025). In many cases, healthcare workers may skip steps due to time pressure or routine habits. To address this, future IPC strategies must go beyond rules and focus on behaviour change. This includes regular training, peer monitoring, feedback systems, and creating a culture where IPC is seen as a shared responsibility.

19.7.6 Inclusion of patients and the public

Patient involvement is also being recognised as an important part of IC (infection control). Educated patients can support hygiene practices, follow discharge instructions, and help reduce infection risks. Recent studies highlight that patient engagement improves adherence to IPC measures and extends their impact beyond hospital settings (Arzilli et al., 2024). For example, patient awareness programmes and postdischarge monitoring can help identify infections early and prevent complications. Further studies on infection prevention are expected to move beyond hospitals and involve the wider community. This broader approach can make IPC more sustainable and effective, especially in controlling outbreaks and AMR.

19.8 Conclusion

In conclusion, IPC is undergoing a significant transformation driven by technological advancements, notably UV-C disinfection systems, nanomaterial-based coatings, and AI-driven surveillance tools. These innovations offer substantial potential to reduce HAIs by improving environmental hygiene, enhancing monitoring, and minimising human error. However, their successful implementation depends on practical considerations, notably cost, infrastructure, workforce training, and regulatory support. In resource-limited settings, particularly in LMICs, scalable and cost-effective solutions remain a priority. Strengthening IPC programmes by integrating essential practices and advanced technologies in a balanced way is critical for sustainable impact. Future efforts should focus on improving accessibility, ensuring data-driven decision-making, and developing context-specific strategies to effectively reduce the global burden of HAIs.

References

Abubeker, K.M., Baskar, S., 2022. Wireless sensor and wireless body area network assisted biosensor network for effective monitoring and prevention of non-ventilator hospital-acquired pneumonia. Frontiers in Sustainable Cities 4. Available at https://doi.org/10.3389/frsc.2022.1063067.

Adeoye, A., Onah, C.O., Orobator, E.T., Akintayo, E.A., Inuaeyen, J.U., Umoru, D.O., et al., 2025. AI and machine learning for early detection of infectious diseases in the US: opportunities and challenges. Journal of Medical Science, Biology, and Chemistry 2 (1), 54–63. Available at https://doi.org/10.69739/jmsbc.v2i1.465.

Ahonen, M., Kahru, A., Ivask, A., Kasemets, K., Kõljalg, S., Mantecca, P., et al., 2017. Proactive approach for safe use of antimicrobial coatings in healthcare settings: opinion of the cost action network AMiCI. International Journal of Environmental Research and Public Health 14 (4), 366. Available at https://doi.org/10.3390/ijerph14040366.

Alhusain, 2025. Harnessing artificial intelligence for infection control and prevention in hospitals: a comprehensive review of current applications, challenges, and future directions. Saudi Medical Journal 46.

Allegranzi, B., Nejad, S.B., Combescure, C., Graafmans, W., Attar, H., Donaldson, L., et al., 2011. Burden of endemic health-care-associated infection in developing countries: systematic review and meta-analysis. The Lancet 377 (9761), 228–241. Available at https://doi.org/10.1016/S0140-6736(10)61458-4, http://www.journals.elsevier.com/the-lancet/.

Alvim, A.L.S., Couto, B.R.G.M., Gazzinelli, A., 2020. Quality of the hospital infection control programs: an integrative review. Revista Gaucha de Enfermagem 41.

El Arab, R.A., Almoosa, Z., Alkhunaizi, M., Abuadas, F.H., Somerville, J., 2025. Artificial intelligence in hospital infection prevention: an integrative review. Frontiers in Public Health 13. Available at https://doi.org/10.3389/fpubh.2025.1547450, http://journal.frontiersin.org/journal/public-health/section/public-health-education-and-promotion#archive.

Arzilli, G., De Vita, E., Pasquale, M., Carloni, L.M., Pellegrini, M., Di Giacomo, M., et al., 2024. Innovative techniques for infection control and surveillance in hospital settings and long-term care facilities: a scoping review. Antibiotics 13 (1). Available at https://doi.org/10.3390/antibiotics13010077, http://www.mdpi.com/journal/antibiotics.

Asegu, L.M., Kitschen, A., Neuwirth, M.M., Sauerland, D., 2024. The economic burden of nosocomial infections for hospitals: evidence from Germany. BMC Infectious Diseases 24 (1). Available at https://doi.org/10.1186/s12879-024-10176-8.

Asghar, S., Atif, M., Masood, I., Khan, M., 2025. A comprehensive review of current status of infection prevention and control program in low- and middle-income countries. Infection, Disease & Health 30 (3), 260–283. Available at https://doi.org/10.1016/j.idh.2025.03.005.

Ashhari, S., Sehhat, E., Ranjbar, Z., 2023. Nanomaterial-based antibacterial and antiviral thin film coatings. ACS Symposium Series 1458, 203–250. Available at https://doi.org/10.1021/bk-2023-1458.ch007.

Baddal, B., Taner, F., Uzun Ozsahin, D., 2024. Harnessing of artificial intelligence for the diagnosis and prevention of hospital-acquired infections: a systematic review. Diagnostics 14 (5). Available at https://doi.org/10.3390/diagnostics14050484.

Bahegwa, R.P., Hokororo, J.C., Msigwa, Y.S., Ngowi, R.R., German, C.J., Marandu, L., et al., 2024. Infection prevention and control compliance during COVID-19 pandemic era: assessment of 26 regional referral hospitals in Tanzania. Discover Health Systems 3 (1). Available at https://doi.org/10.1007/s44250-024-00078-6.

Balakrishnan, V.S., 2025. Policy and investment gaps in IPC. The Lancet. Microbe.

Berktold, M., Fuchs, S., Kuppelwieser, B., Ulmer, H., Kettner, M., Thummer, L., et al., 2023. Beyond the guidelines" – Deviations in adherence to infection control measures in Tyrolean hospitals, Austria. American Journal of Infection Control 51 (4), 406–412. Available at https://doi.org/10.1016/j.ajic.2022.07.005.

Brady, G., Bennin, F., De Koning, R., Vindrola-Padros, C., Clark, S.E., Tiwari, M.K., et al., 2025. Interventions used to reduce infectious aerosol concentrations in hospitals—a review. eClinicalMedicine 79. Available at https://doi.org/10.1016/j.eclinm.2024.102990.

Casini, B., Tuvo, B., Scarpaci, M., Totaro, M., Badalucco, F., Briani, S., et al., 2023. Implementation of an environmental cleaning protocol in hospital critical areas using a UV-C disinfection robot. International Journal of Environmental Research and Public Health 20 (5), 4284. Available at https://doi.org/10.3390/ijerph20054284.

Chang, C.W., Chang, C.H., Chien, C.Y., Jiang, J.L., Liu, T.W., Wu, H.C., et al., 2024. Predictive modelling of hospital-acquired infection in acute ischemic stroke using machine learning. Scientific Reports 14 (1). Available at https://doi.org/10.1038/s41598-024-82280-3, https://www.nature.com/srep/.

Deng, B., Li, P., Liu, Y., Xie, J., Huang, Y., Sun, Q., et al., 2024. Investigation of hospital-acquired infections prevalence and analysis of influencing factors: a case study of a specialized infectious disease hospital in Chongqing, 2017–2023. Frontiers in Public Health 12. Available at https://doi.org/10.3389/fpubh.2024.1417645.

Deryabina, A., Lyman, M., Yee, D., Gelieshvilli, M., Sanodze, L., Madzgarashvili, L., et al., 2021. Core components of infection prevention and control programs at the facility level in Georgia: key challenges and opportunities. Antimicrobial Resistance & Infection Control 10 (1). Available at https://doi.org/10.1186/s13756-020-00879-3.

Flynn, C.E., Guarner, J., 2023. Emerging antimicrobial resistance. Modern Pathology 36 (9), 100249. Available at https://doi.org/10.1016/j.modpat.2023.100249.

Fofanah, B.D., Conteh, C., 2020. Patient involvement in infection prevention and control (IPC) practice: knowledge and perception study. Infection Control & Hospital Epidemiology 41 (S1), s493. Available at https://doi.org/10.1017/ice.2020.1169.

Gilbert, G.L., Kerridge, I., 2020. Hospital infection control: old problem – evolving challenges. Internal Medicine Journal 50 (1), 105–107. Available at https://doi.org/10.1111/imj.14686, http://onlinelibrary.wiley.com/journal/10.1111/(ISSN)1445-5994.

Godbole, A.A., Paras, Mehra, M., Banerjee, S., Roy, P., Deb, N., et al., 2025. Enhancing infection control in ICUS through AI: a literature review. Health Science Reports 8 (1). Available at https://doi.org/10.1002/hsr2.70288, onlinelibrary.wiley.com/journal/23988835.

Graves, N., Halton, K., Paterson, D., Whitby, M., 2009. Economic rationale for infection control in Australian hospitals. Healthcare Infection 14 (3), 81–88. Available at https://doi.org/10.1071/hi09010.

Gray, J., 2015. Infection control: beyond the horizon. Journal of Hospital Infection 89 (4), 237–240. Available at https://doi.org/10.1016/j.jhin.2014.11.004.

Harbarth, S., Pittet, D., 2010. Infection prevention research in europe: recent advances and future priorities. Infection Control & Hospital Epidemiology 31 (S1), S11. Available at https://doi.org/10.1086/655986.

Hayward, C., Ross, K.E., Brown, M.H., Whiley, H., 2020. Water as a source of antimicrobial resistance and healthcare-associated infections. Pathogens 9 (8), 1–21. Available at https://doi.org/10.3390/pathogens9080667, https://www.mdpi.com/2076-0817/9/8/667/pdf.

Hoogestraat, A.T., Wulff, A., 2024. A vision on user-centered implementation and evaluation of explainable AI for predicting hospital-onset bacteremia. Studies in Health Technology and Informatics 316, 766–770. Available at https://doi.org/10.3233/SHTI240525.

Hospital acquired infections: outbreaks and infection control interventions, a national descriptive survey, 2016. Queensland University of Technology. Queensland University of Technology, Australia, Queensland University of Technology, Australia.

Igiebor, F.A., Omoregie, W.A., 2025. Assessment of plant-mediated copper and zinc nanoparticles against selected microorganisms. BioNanoScience 15 (3). Available at https://doi.org/10.1007/s12668-025-02096-x, https://www.springer.com/journal/12668.

Isigi, S.S., Parsa, A.D., Alasqah, I., Mahmud, I., Kabir, R., 2023. Predisposing factors of nosocomial infections in hospitalized patients in the united kingdom: systematic review. JMIR Public Health and Surveillance 9 (1). Available at https://doi.org/10.2196/43743, https://publichealth.jmir.org/2023/1/e43743/PDF.

Kamara, R.Z., Kamara, I.F., Moses, F., Kanu, J.S., Kallon, C., Kabba, M., et al., 2023. Improvement in infection prevention and control compliance at the three tertiary hospitals of sierra leone following an operational research study. Leone Tropical Medicine and Infectious Disease 8 (7). Available at www.mdpi.com/journal/tropicalmed, 10.3390/tropicalmed8070378.

Khavandegar, A., Siami, Z., Rasouli, A., Nazemi, P., Gull, A., 2025. Impact of healthcare-associated infections on in-hospital outcomes during the COVID-19 era: a multicenter comparative study of 20,942 isolated microorganisms from ICU patients. Frontiers in Public Health 13. Available at https://doi.org/10.3389/fpubh.2025.1475221.

Kolbe-Busch, S., Djouela Djoulako, P.D., Stingu, C.-S., 2025. Trends in healthcare-acquired infections due to multidrug-resistant organisms at a german university medical center before and during the COVID-19 pandemic. Microorganisms 13 (2). Available at https://doi.org/10.3390/microorganisms13020274.

Koutras, C., Wade, R.L., 2023. Effectiveness of an Ultraviolet-C Disinfection System for Inactivating Clostridioides Difficile Spores in Critical Areas of a Simulated Patient Room. bioRxiv, United States. Available at https://www.biorxiv.org, 10.1101/2023.02.16.528885.

Kubde, D., Badge, A.K., Ugemuge, S., Shahu, S., 2023. Importance of hospital infection control. Cureus 15 (12). Available at https://doi.org/10.7759/cureus.50931.

Long, J., Chen, J., Chen, C., Xu, T., Gou, X., 2023. MOFs encapsulated nanorods derived CoNi@CN composites with open structure as highly efficient bifunctional catalysts for rechargeable Zn-air batteries. Journal of Colloid and Interface Science 629 (Pt A), 73–82. Available at https://doi.org/10.1016/j.jcis.2022.08.088.

Lukasewicz Ferreira, S.A., Franco Meneses, A.C., Vaz, T.A., da Fontoura Carvalho, O.L., Hubner Dalmora, C., Pressotto Vanni, D., et al., 2024. Hospital-acquired infections surveillance: the machine-learning algorithm mirrors National Healthcare Safety Network definitions. Infection Control and Hospital Epidemiology 45 (5), 604–608. Available at https://doi.org/10.1017/ice.2023.224.

Madran, B., Demir, Z.I., Yalcin, B., Ayaz, O.T., Iyikosker, K., Keskin, A., et al., 2025. Reasons for insufficient compliance with infection prevention and control measures in the intensive care unit: a qualitative study conducted in Türkiye in 2024. Antimicrobial Resistance and Infection Control 14 (1). Available at https://doi.org/10.1186/s13756-025-01544-3, https://aricjournal.biomedcentral.com/.

Mayhall, C.G., 2012. Hospital Epidemiology and Infection Control. Wolters Kluwer Health, Lippincott Williams & Wilkins.

Mehta, Y., Gupta, A., Todi, S., Myatra, S.N., Samaddar, D.P., Patil, V., et al., 2014. S. Ramasubban, guidelines for prevention of hospital acquired infections. Indian Journal of Critical Care Medicine 18 (3), 149–163. Available at https://doi.org/10.4103/0972-5229.128705, https://www.ijccm.org/lov/IJCCM.

Molloy-Simard, V., Lemyre, J.L., Martel, K., Catalone, B.J., 2019. Elevating the standard of endoscope processing: terminal sterilization of duodenoscopes using a hydrogen peroxide–ozone sterilizer. American Journal of Infection Control 47 (3), 243–250. Available at https://doi.org/10.1016/j.ajic.2018.09.009, http://www.journals.elsevier.com/ajic-american-journal-of-infection-control/.

Odoom, A., Tetteh-Quarcoo, P.B., Donkor, E.S., 2025. Prevalence of hospital-acquired infections in low- and middle-income countries: systematic review and meta-analysis. Asia Pacific Journal of Public Health 37 (5), 448–466. Available at https://doi.org/10.1177/10105395251338002.

Oniciuc, E.A., Likotrafiti, E., Alvarez-Molina, A., Prieto, M., Santos, J.A., Alvarez-Ordóñez, A., 2018. The present and future of whole genome sequencing (WGS) and whole metagenome sequencing (WMS) for surveillance of antimicrobial resistant microorganisms and antimicrobial resistance genes across the food chain. Genes 9 (5). Available at https://doi.org/10.3390/genes9050268, http://www.mdpi.com/2073-4425/9/5/268/pdf.

Ouirungroj, T., Apichai, S., Pattananandecha, T., Grudpan, K., Saenjum, C., 2024. Smart-detection approach for protein residues to evaluate the cleaning efficacy of reusable medical devices. Journal of Hospital Infection 145, 44–51. Available at https://doi.org/10.1016/j.jhin.2023.12.005.

Pellowe, C., Pratt, R., Loveday, H., Harper, P., Robinson, N., Jones, S.R.L.J., 2005. The epic project: updating the evidence base for national evidence-based guidelines for preventing healthcare-associated infections in NHS hospitals in England. A report with recommendations. Journal of Hospital Infection 59 (4), 373–374. Available at https://doi.org/10.1016/j.jhin.2004.11.007.

Qureshi, M., Chughtai, A., Seale, H., 2022. Supporting the delivery of infection prevention and control training to healthcare workers: insights from the sector. Healthcare (Switzerland) 10 (5). Available at https://doi.org/10.3390/healthcare10050936.

Ramage, G., Kean, R., MacKay, W.G., Vaccaro, A., Lopez-Ribot, J.L., Howe, M.H., 2025. Candidozyma auris: Implications for Infection Prevention and Control in the Management of This Resilient Biofilm Former. Springer Science and Business Media LLC, pp. 165–192. Available at 10.1007/978-3-031-93315-8_7.

Rice, S., Carr, K., Sobiesuo, P., Shabaninejad, H., Orozco-Leal, G., Kontogiannis, V., et al., 2023. Economic evaluations of interventions to prevent and control health-care-associated infections: a systematic review. The Lancet Infectious Diseases 23 (7), e228. Available at https://doi.org/10.1016/s1473-3099(22)00877-5.

Rosenthal, V.D., Yin, R., Lu, Y., Rodrigues, C., Myatra, S.N., Kharbanda, M., et al., 2023. The impact of healthcare-associated infections on mortality in ICU: a prospective study in Asia, Africa, Eastern Europe, Latin America, and the Middle East. American Journal of Infection Control 51 (6), 675–682. Available at https://doi.org/10.1016/j.ajic.2022.08.024, http://www.journals.elsevier.com/ajic-american-journal-of-infection-control/.

Satta, Rogers, Costa, B.e, Tartari, Santos, Bischoff, et al., 2025. Identifying indicators and targets for the new World Health Organization global framework on infection prevention and control: an e-Delphi study. Journal of Hospital Infection 165, 181–188. Available at https://doi.org/10.1016/j.jhin.2025.04.039.

Savul, S., Lalani, F.K., Ikram, A., Khan, M.A., Khan, M.A., Ansari, J., 2020. Infection prevention and control situation in public hospitals of Islamabad. Journal of Infection in Developing Countries 14 (9), 1040–1046. Available at https://doi.org/10.3855/JIDC.12779, https://www.jidc.org/index.php/journal/article/view/12779.

Shigemura, K., Osawa, K., Mukai, A., Ohji, G., Lee, J.J., Yoshida, H., et al., 2013. Infection control team activity and recent antibiograms in the Kobe University Hospital. Journal of Antibiotics 66 (9), 511–516. Available at https://doi.org/10.1038/ja.2013.36.

Shrimali, S., Teuscher, C., 2023. A novel deep learning-, camera-, and sensor-based system for enforcing hand hygiene compliance in healthcare facilities. IEEE Sensors Journal 23 (12), 13659–13670. Available at https://doi.org/10.1109/jsen.2023.3271297.

Srivastav, S., Khurana, S., Mukhopadhyay, C., Myatra, S.N., Katyal, S., Katoch, O., et al., 2024. Surveillance for surgical site infections developed during hospital stay & after discharge: a multicentric study. Indian Journal of Medical Research 160 (5), 428–437. Available at https://doi.org/10.25259/IJMR_369_2024, https://ijmr.org.in/?view-pdf=1&embedded=true&article=6661b694742757c2f60a3819dbdaedc7Jh9ieeR11bYIBg%3D%3D.

Stevenson, K.B., Searle, K., Curry, G., Boyce, J.M., Harbarth, S., Stoddard, G.J., et al., 2014. Infection control interventions in small rural hospitals with limited resources: results of a cluster-randomized feasibility trial. Antimicrobial Resistance and Infection Control 3 (1). Available at https://doi.org/10.1186/2047-2994-3-10, http://www.aricjournal.com/content/3/1/10.

Struelens, M.J., Wagner, D., Bruce, J., MacKenzie, F.M., Cookson, B.D., Voss, A., et al., 2006. Status of infection control policies and organisation in European hospitals, 2001: the ARPAC study. Clinical Microbiology and Infection 12 (8), 729–737. Available at https://doi.org/10.1111/j.1469-0691.2006.01462.x.

Supriadi, I.R., Saptawati, L., Widodo, N.H., Satari, H.I., Sitohang, G., Usman, Y., et al., 2023. SG-APSIC1108: implementation of infection prevention and control in Indonesian hospitals: identification of strengths, gaps, and challenges in current practices. Antimicrobial Stewardship & Healthcare Epidemiology 3 (S1), s17. Available at https://doi.org/10.1017/ash.2023.52.

The Canadian Nosocomial Infection Surveillance Program: Keeping an eye on antimicrobial resistance in Canadian hospitals since 1995, 2022. Canada Communicable Disease Report 48 (11/12), 506–511. Available at https://doi.org/10.14745/ccdr.v48i1112a03.

Tomczyk, S., Twyman, A., de Kraker, M.E.A., Coutinho Rehse, A.P., Tartari, E., Toledo, J.P., et al., 2022. The first WHO global survey on infection prevention and control in health-care facilities. The Lancet Infectious Diseases 22 (6), 845–856. Available at https://doi.org/10.1016/S1473-3099(21)00809-4.

Villanueva-Miranda, I., Xiao, G., Xie, Y., 2025. Artificial intelligence in early warning systems for infectious disease surveillance: a systematic review. Frontiers in Public Health 13. Available at https://doi.org/10.3389/fpubh.2025.1609615.

Yang, J.H., Wu, U.I., Tai, H.M., Sheng, W.H., 2019. Effectiveness of an ultraviolet-C disinfection system for reduction of healthcare-associated pathogens. Journal of Microbiology, Immunology, and Infection 52 (3), 487–493. Available at https://doi.org/10.1016/j.jmii.2017.08.017, http://www.elsevier.com/wps/find/journaldescription.cws_home/722895/description#description.

Yuan, Y., Sui, J., Kong, X., 2026. Effect of initial temperature and relative humidity on VHP penetration during HEPA in-situ fumigation disinfection. Energy and Built Environment 7 (1), 1–13. Available at https://doi.org/10.1016/j.enbenv.2024.08.003.

Zimba, M.M., Dorothy, C.O., Mulubwa, C., Ngoma-Hazemba, A., 2022. Infection Prevention and Control practices; an Exploratory Qualitative study of Experiences of Health Care Providers and Support Staff at the University Teaching Hospital. Zambia. medRxiv, Lusaka. Available at https://www.medrxiv.org/, 10.1101/2022.03.23.22269760.

Chapter 20

The ripple effect: understanding the economic and social consequences of antibiotic resistance

Alok Sharma[1], Pratik Kolhe[1], Biswajit Paul[2], Pardeep Kumar[3], Vishek Choudhary[4,5], Sapna A. Kondalkar[1], and Brijesh Singh Sisodia[1]

[1]*Regional Ayurveda Research Institute, Gwalior, Madhya Pradesh, India,* [2]*Central Ayurveda Research Institute, Guwahati, Assam, India,* [3]*Department of Botany, Mata Gujri College, Fatehgarh Sahib, Punjab, India,* [4]*CSIR-Institute of Himalayan Bioresource Technology, Palampur, Himachal Pradesh, India,* [5]*Academy of Scientific and Innovative Research (AcSIR), Ghaziabad, Uttar Pradesh, India*

20.1 Introduction

Before going into a detailed discussion of antimicrobial resistance (AMR), one should go through the background of antibiotics and the generation of AMR. Finally, after the past nine decades of the emergence of antibiotics, it emerges as a new challenge for healthcare, medicine, and pharmaceutical fields (Ahmad and Khan, 2019). Even in the 21st century, the majority of health risks and threats are avoided with the application of various suitable antibiotics (Teillant et al., 2015). However, unrestricted use of antibiotics can also develop AMR. The identification of AMR as a public health threat has drawn attention to the controlled and wise practice of antibiotics (WHO. Global action plan on AMR, Official Journal of the European Union. Geneva., 2016). The survival and proliferation tendency of microbes, including bacteria, protozoa, fungi, viruses, and parasites, regardless of antibiotic use, is known as AMR. Infection caused by such resistant species is hard to cure and also risks causing severe health conditions (https://www.cdc.gov/drugresistance/about/how-resistance-happens.html).

Resistance is a condition where micro-organisms are not responsive to antibiotics, generated with prolonged use of an antibiotic. However, resistance developed is not a single-day phenomenon, and it is a result of the uncontrolled and persistent use of antibiotics. The recent rise and development of AMR are addressed by both clinical and nonclinical approaches of the healthcare systems. By considering the clinical standpoint, all pathogens are initially sensitive towards antibiotics, but their uncontrolled exposure leads bacterial systems to adapt and modify their longevity either by genetic or metabolic mutations (Koch et al., 2021). In routine life, antimicrobials can also be categorised into antifungal, antibiotic, antiprotozoal, antiviral, antiseptic, disinfectant, and food preservatives that are used for different purposes. AMR is considered an evolutionary phenomenon as bactericidal action leads to species adaptation and forces natural selection, which leads to the ineffectiveness of drug (https://www.cdc.gov/drugresistance/about/how-resistance-happens.html). In the 21st century, resistant strain infection has become a severe and potential threat to global health and a silent pandemic cause (Read and Woods, 2014). Numerous interrelated elements associated with agriculture and healthcare have a simultaneous impact on the emergence and spread of AMR. In addition to that, improper waste management of clinical and pharmaceutical drugs and social factors like unregulated trading of pharmaceutical products also contribute to making AMR one of world's severe concerns (Samreen et al., 2021). According to the Cambridge dictionary, the ripple effect is a situation where one event is responsible for spreading and producing a further series of events or effects. In the present context, the ripple effect can be explained as how failure at one point of healthcare resulted in long-term side effects in different social aspects. Ripple effects explain the chain of various events that are related or can be said as a consequence of one event; for example, when the antibiotic resistance increases, patients will invest more in healthcare, and the medical expenses and their treatment will increase the cost of alternative therapy. Therefore in this chapter, a thorough discussion about the socioeconomic effects of AMR has been done.

Antimicrobial Resistance in the 21st Century. DOI: https://doi.org/10.1016/B978-0-443-44748-8.00001-4

20.2 Drivers for antimicrobial resistance

Adaptive behaviour towards external stimuli gives rise to AMR. Drivers that play a role in the evolution of AMR are categorised as intrinsic drivers and extrinsic drivers. Intrinsic determinants are related to the genetic components that decide the latency of bacterial genes, while extrinsic determinants are economic and social, as shown in Fig. 20.1 (Irfan et al., 2022; Yu et al., 2022).

20.2.1 Improper prescription

Improper prescription can be identified as a wrong selection of antibiotics, inappropriate dose and duration of antibiotic treatment, and unnecessary prescription of antibiotics. A report provided by Center for Disease Control (CDC) in 2017 mentioned that $1/3^{rd}$ of the patient population is getting prescribed antibiotics for a prolonged period and are inadequately examined (Woolhouse et al., 2016).

20.2.2 Increment in gross production

In recent times, it has been observed that the Antibiotics usage as a treatment and as an animal supplement has increased mainly in low- and middle-income countries. In the decade 2000–2010, 76% of the total antibiotic consumption was observed in countries like India, Russia, Brazil, China, and South Africa, which comprise about 73% of the total global population. Similarly, in the 2010–2021 decade, countries like Algeria, China, and some countries from west region of Africa had the maximum usage (Wang et al., 2025). Hence, the significantly higher usage of antibiotics in certain regions of the world poses the long-term threat of AMR to the people of those countries.

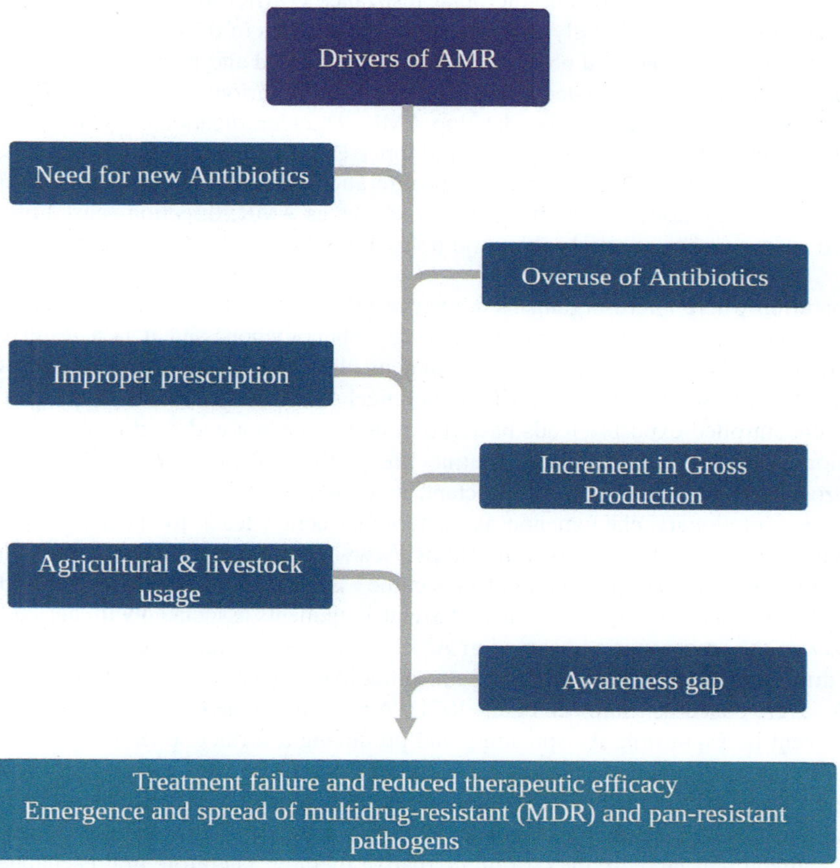

FIGURE 20.1 **Drivers for antimicrobial resistance.** Figure showing the various drivers for the antimicrobial resistance.

20.2.3 Need for new antibiotics

According to the present situation, the scarcity of therapeutics makes it difficult to manage bacterial infections like pneumonia, UTIs, drug-resistant TB, and other Gram-negative diseases effectively. Moreover, aged patients are even more prone to fatal illnesses due to the need for new medications (Chang et al., 2015). Due to an increase in the number of resistant strains, there is a need to develop new formulations or medications to combat continuously evolving bacterial infections, but ongoing research is mainly focusing on modifying the present drugs rather new innovations; hence, as a result, these new drugs failed even earlier or became resistant in a shorter time.

20.2.4 Overuse of antibiotics

In spite of continuous cautioning advice by health organisations, people are still using antibiotics on a large scale at an unprecedented rate. As per the global observation, among the people, there are also many wrong beliefs and misconceptions concerning antibiotics. Moreover, one example of a commonly known abuse pattern is the management of antibiotics without a clear warning, especially in less developed countries where healthcare facilities are in inadequate condition (Chokshi et al., 2019).

20.2.5 Agricultural and livestock usage

Antibiotics are used in livestock as a disease preventive measure, supplement, or feed additives to increase production and to regain health. However, this unregulated use further becomes one of the prominent reasons to saturate the human body, which can eventually lead to resistance (Food and Agriculture Organization of the United Nations. Antimicrobial Resistance., 2026). Especially, the poor governance of the less developed countries is the major cause of this factor to gain more, which ultimately promotes human AMR (Castro-Sánchez et al., 2016).

20.2.6 Awareness gap

To estimate the extent of the AMR load, there is a need for constant vigilance and to develop remedial options like antibiotics supervision or antimicrobial stewardship (Carter et al., 2016). Deprived enforcement and protocols and a lack of awareness can cause improper application of antibiotics and delivery of healthcare system (Lekagul et al., 2019). To get the required and optimal results from the remedial intervention, one needs to get proper vigilance data. However, the lack of such measures in the current situation across many regions of the world is also one of the major causes of AMR.

20.3 Economic impression of antimicrobial resistance

Till now, this chapter has discussed the biological and social impact of AMR worldwide, but economic parameter and their variables also have their own impact on AMR. When we are discussing economic parameters, there are terms 'cost' and 'economic influence' which cover the whole category of economic impact. Research and Development Organization (RAND) in one of their studies noted that the world's population would be between 11 million and 444 million fewer by 2050 if AMR is not addressed in the near future (Taylor et al., 2014). As per the economic definition of resistance, it is a negative externality. Externality means results (positive or negative) mentioned in terms of economic action that has an impact on a third party, which has no direct transaction. The external cost is multifaceted as it not only affects the direct consumer (A fact sheet from ReAct-Action on Antibiotic Resistance., 2012). The risk of long-term poverty, untreated sickness, and mortality is a consequence of the higher expenses of healthcare. In the end, this process results in the loss of human resources. World Bank Report released in 2014 regarding investment in the healthcare sector worldwide, which mentioned that around 9.9% of global GDP is used for the health system management (https://data.worldbank.org/indicator/SH.XPD.TOTL.ZS).

Due to the inadequate antibiotic policies and lesser knowledge in low- and middle-income countries, it is difficult to treat microbial infections that create havoc. Moreover, AMR also has an impact on socioeconomic growth. The nation's economy is dependent on labour; however, the lack of healthcare or increasing sickness directly disturbs the economy. There are countries where economical agriculture-based animal husbandry has a big share in it. So, maintenance of animal health has a relation with AMR as it is one of the main carriers of AMR, and this can have an impact on the price of animal products as well as the farming sector. The elevated price of food has a role in human nutrition and shows a negative ripple on human nutrition. According to the World Bank, AMR may significantly affect the economic performance (Pulingam et al., 2022).

In 2017 the World Bank released its report regarding the economic scenario on antibiotic resistance, which showed the economic facet of antibiotic resistance (World Bank Group. World Bank. "DRUG-RESISTANT INFECTIONS A Threat to Our Economic Future." Washington, DC., 2017).

20.3.1 GDP influence

If AMR has minimal impact, it will decrease global GDP by 1.1% but in the high-impact scenario, it would decrease by 3.8% by the year 2050. Nation-wise loss in GDP reflects 5% loss in the low-income nations by the prior 2050. Increased burden on economy because of poor productivity and elevated healthcare costs, even though every country has a variable GDP according to its economic structure, international trade, and agricultural development. Low-middle-income countries suffer the most influence on GDP due to AMR conditions (https://documents1.worldbank.org/curated/en/323311493396993758/pdf/final-report.pdf, accessed 15 January 2023).

20.3.2 Global poverty

There is a significant rise in poverty as a consequence of AMR. As a result of high-impact scenario, 26.2 million individuals in low-income nations will experience extreme poverty in 2050. Although the world in the current scenario is trying to obliterate extreme poverty, with the goal that by 2030, <3% people should become below poverty. However, the global threat of AMR could make this goal unachievable. Increased burden on economy due to AMR can cause increased household poverty on global level. When healthcare expense increases, lower income families and societies spend more than their budget compared to higher incomes and burden lends on government (Ahmed et al., 2018).

20.3.3 Influence on global trade

Global exports and animal-based outcomes would decline by 3.8% and 7.5%, respectively, while healthcare expenditures might rise from $300 billion to over $1 trillion annually by 2050. Declining health effect caused by AMR is responsible for reducing productivity of the intense laborious sector, also ongoing recession has negative impacts on global trade (Lekagul, et al., 2019).

Analysing the economic perspective of AMR on society is crucial to understand the seriousness and global economic influence caused by this. While understanding AMR and the economic approach, we should use statistical models on both the economic model and factors related to AMR. One variable of AMR is the microbial population, which can become resistant in the future. The second variable is the economic consequences derived from the damaged function. Both variables are explained as merely discovered and inexplicable. The damage function calculation in AMR is helpful to go through both the economic and health consequences (damage) due to resistant diseases (Lekagul et al., 2019).

The next variable to understand is cost–benefit analysis. It is an evolutionary model that estimates cost comparison and estimated benefits. Since AMR causes long-term damage, the selected discount rate significantly influences how significant those future harms are in today's scenario. A low discount rate reflects high intergenerational equity. Minimal changes in the discount rate can cause huge deflection in net present value (NPV). NPV is estimated from fair use of resources, analysis of current expenses, and its impact on future generations (all these are economic considerations). It is important to note that a slight variation in the current economic planning has severe results on future outcomes of NPV. Many economists, including Stern (Stern, 2016), suggested not to rely on cost–benefit analysis as it omits some important intangible factors and employs techniques to prevent the impact of AMR, that having disastrous results (Roope et al., 2019).

20.4 Clinical Implications of antimicrobial resistance on the economy

Healthcare and public health are two important pillars for the growing economy. Any change or direct stress on these factors will directly or indirectly create a hurdle in raising the economy. Clinical consequences of AMR include directly affecting public health, cost, and expenditure on healthcare facilities in many ways; some of which are mentioned in Fig. 20.2.

1. Resistant strain pathogen infection is hard to cure, and it will act as a source for secondary infection in healthy individuals. AMR makes it more difficult to treat infections caused by bacteria, fungi, and viruses.
2. Treatment for many common infections, including common cold (flu), typhoid, upper respiratory tract infections (URTIs), and urinary tract infections (UTIs), is difficult to treat because of the spread of novel resistant drivers, which can lead to fatal results.

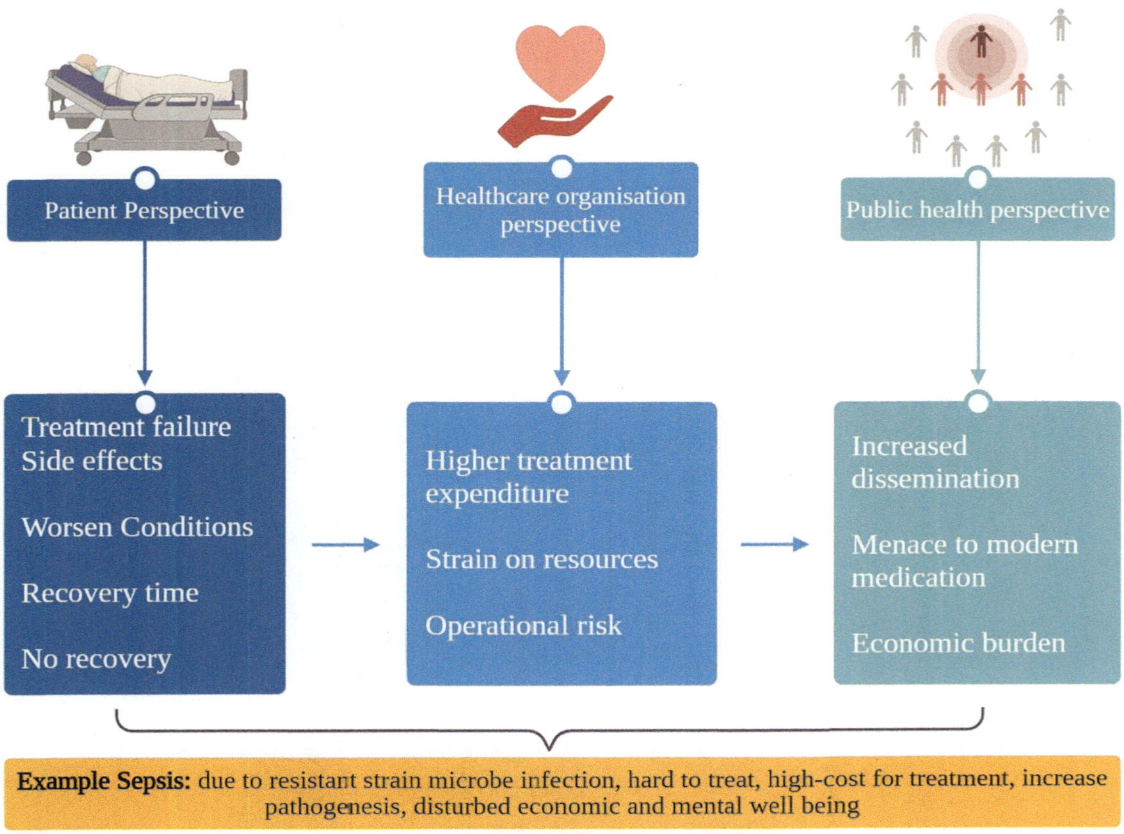

FIGURE 20.2 **Schematic presentation of clinical implications of antibiotic resistance.** Figure showing the various clinical impacts of AMR on patient, healthcare organisation, and public health level. *AMR*, Antimicrobial resistance.

3. Because of the unavailability of new medications, AMR creates a major threat to the success of cancer chemotherapy, transplant surgery, and even basic dentistry procedures.
4. Long-term necessary therapy, which costs healthcare expenses and may require expensive alternative medications (Lin et al., 2022).

These patients require further laboratory testing and have prolonged hospitalisation, and administration of second-line therapy may be expensive, toxic, and still less resultant. Therefore this will have a negative outcome on the economic and administrative on healthcare organisation (Bhatia, 2018). The WHO studied the global scenario of antibiotic resistance, which showed that different antibiotics are resistant to different Gram-positive and Gram-negative bacteria (Global Antimicrobial Resistance and Use Surveillance System GLASS Report, 2021) (Table 20.1).

20.5 Socioeconomic impact of antibiotic resistance on pandemic

Economic factors have pushed us significantly to develop the combat strategies against the microbial resistance. The AMR has both social and economic impacts; however, there is a difference between them. To understand this difference in economic and social behaviour, it is crucial to know the common element that has roles in economic behaviour and societal health (OECD. Antimicrobial Resistance Tackling the Burden in the European Union: Tackling the Burden in the European Union. OECD Publishing, 2019). The degree of steps taken to counter against AMR or the betterment of healthcare organisations is dependent on the economic model of the country. It is also important to understand whether our control measures like antimicrobial stewardship are sufficient, and how they are positively regulating the AMR, along with the acceptance of the improved approaches to tackle the human health challenges (Giacobbe et al., 2020).

During pandemic conditions, the human race was facing two twisted crises: COVID-19 and AMR. As supportive treatment was not sufficiently effective on severe cases of COVID-19, to manage these helpless conditions, clinical

TABLE 20.1 List of antibiotics showing resistance to bacteria in different regions.

Sr. no	Region	Antibiotics	Bacteria	Countries under region
1.	Eastern Mediterranean region	Cefepime, Ceftazidime, Ceftriaxone, Ciprofloxacin, Cotrimoxazole, Imipenem, Meropenem, Ampicillin, Cefotaxime, Colistin, Ertapenem, Levofloxacin	*Escherichia coli*	Bahrain
		Amikacin, Colistin, Gentamicin, Imipenem, Meropenem, Tigecycline	*Acinetobacter spp.*	
		Cefepime, Ceftazidime, Ceftriaxone, Ciprofloxacin, Cotrimoxazole, Colistin, Ertapenem, Imipenem, Meropenem, Cefotaxime, Ceftazidime, Levofloxacin	*Klebsiella. pneumoniae*	
		Ceftazidime, Ceftriaxone, Ciprofloxacin, Imipenem, Meropenem	*Salmonella spp.*	
		Azithromycin, Cefixime, Ceftriaxone, Ciprofloxacin	*Neisseria gonorrhoeae*	
		Ceftriaxone, Penicillin G	*Streptococcus pneumoniae*	
		Oxacillin	*Staphylococcus aureus*	
		Ceftriaxone, Ciprofloxacin	*Shigella spp.*	
		Cefepime, Ceftriaxone, Ceftazidime, Cotrimoxazole, Ampicillin, Ciprofloxacin, Imipenem, Ertapenem, Levofloxacin, Meropenem, Cefotaxime, Ertapenem, Colistin	*E. coli*	Lebanon
		Amikacin, Gentamicin, Imipenem, Minocycline	*Acinetobacter spp.*	
		Cefoxitin	*S. aureus*	
		Cefepime, Cefotaxime, Ceftriaxone, Ceftazidime, Cotrimoxazole, Ciprofloxacin, Imipenem, Ertapenem, Colistin, Ertapenem, Levofloxacin, Meropenem	*K. pneumoniae*	
		Cefotaxime, Ceftriaxone, Ceftazidime, Ciprofloxacin, Ertapenem, Imipenem, Levofloxacin	*Salmonella spp.*	
		Not recorded	*S. pneumoniae, N. gonorrhoeae, Shigella spp.*	
		Cefepime, Ceftazidime, Cefotaxime, Ceftriaxone, Levofloxacin, Meropenem, Imipenem, Ampicillin, Ciprofloxacin, Ertapenem	*E. coli*	Egypt
		Amikacin, Imipenem, Colistin, Gentamicin, Meropenem,	*Acinetobacter spp.*	
		Cefepime, Ceftazidime, Cefotaxime, Ceftriaxone, Imipenem, Levofloxacin, Meropenem, Ciprofloxacin, Cotrimoxazole, Doripenem, Ertapenem	*K. pneumoniae*	

		Oxacillin	*S. aureus*	
		Not recorded	*S. pneumoniae, N. gonorrhoeae, Shigella spp., S. pneumoniae*	
		Cefepime, Ceftazidime, Cefotaxime, Cotrimoxazole, Ampicillin, Ciprofloxacin, Imipenem, Ertapenem, Meropenem, Colistin, Levofloxacin	*E. coli*	Tunisia
		Ceftazidime, Cefotaxime, Ciprofloxacin, Imipenem, Ertapenem, Levofloxacin	*Salmonella spp.*	
		Cefepime, Ceftazidime, Cefotaxime, Cotrimoxazole, Colistin, Ciprofloxacin, Imipenem, Ertapenem, Meropenem, Levofloxacin	*K. pneumoniae*	
		Cefoxitin	*S. aureus*	
		Not recorded	*S. pneumoniae, N. gonorrhoeae, Shigella spp.*	
2.	European region	Imipenem, Meropenem, Ceftriaxone, Ceftazidime, Cefotaxime, Ertapenem, Levofloxacin, Ciprofloxacin, Ampicillin	*E. coli*	Bosnia and Herzegovina
		Imipenem, Ertapenem, Ceftriaxone, Ceftazidime, Cefotaxime, Levofloxacin, Ciprofloxacin, Meropenem	*K. pneumoniae*	
		Meropenem, Colistin, Imipenem, Gentamicin, Amikacin	*Acinetobacter spp.*	
		Cefotaxime, Ceftriaxone, Oxacillin, Penicillin G	*S. pneumoniae*	
		Oxacillin, Cefoxitin	*S. aureus*	
		Not recorded	*Salmonella spp., N. gonorrhoeae, Shigella spp.*	
		Ceftriaxone, Ceftazidime, Ampicillin, Levofloxacin, Ciprofloxacin, Meropenem, Imipenem, Cefotaxime, Ertapenem, Cotrimoxazole	*E. coli*	Finland
		Ceftriaxone, Ceftazidime, Cotrimoxazole, Levofloxacin, Ciprofloxacin, Meropenem, Imipenem, Cefotaxime, Ertapenem	*K. pneumoniae*	
		Ceftriaxone, Ciprofloxacin, Meropenem, Ceftazidime, Levofloxacin	*Salmonella spp.*	
		Meropenem	*Acinetobacter spp.*	
		Penicillin G	*S.pneumoniae*	
		Oxacillin	*S. aureus*	

(Continued)

TABLE 20.1 (Continued)

Sr. no	Region	Antibiotics	Bacteria	Countries under region
		Ceftriaxone	*N. gonorrhoeae*	
		Not recorded	*Shigella spp.*	
		Ampicillin, Cefotaxime, Ceftazidime, Ceftriaxone, Ciprofloxacin, Ertapenem, Imipenem, Levofloxacin, Meropenem	*E. coli*	Georgia
		Cefepime, Ceftriaxone, Ceftazidime, Colistin, Ciprofloxacin, Meropenem, Imipenem, Ertapenem, Cefotaxime, Cotrimoxazole, Levofloxacin	*K. pneumoniae*	
		Colistin, Gentamicin, Imipenem, Meropenem	*Acinetobacter spp.*	
		Cefoxitin, Oxacillin	*S. aureus*	
		Not recorded	*S. pneumoniae, Shigella spp., N. gonorrhoeae, Salmonella spp.*	
		Ceftriaxone, Ceftazidime, Cefotaxime, Ampicillin, Levofloxacin, Ciprofloxacin, Meropenem, Imipenem, Ertapenem, Cotrimoxazole	*E. coli*	Germany
		Cefotaxime, Ceftazidime, Ceftriaxone, Ciprofloxacin, Ertapenem, Imipenem, Meropenem, Levofloxacin	*Salmonella spp.*	
		Ceftriaxone, Ceftazidime, Cefotaxime, Levofloxacin, Ciprofloxacin, Meropenem, Imipenem, Ertapenem, Cefepime	*K. pneumoniae*	
		Meropenem, Imipenem, Gentamicin, Amikacin	*Acinetobacter spp.*	
		Cefotaxime, Ceftriaxone, Penicillin G	*S. pneumoniae*	
		Cefoxitin, Oxacillin	*S. aureus*	
		Colistin, Imipenem, Gentamicin, Amikacin, Meropenem	*Acinetobacter spp.*	
		Cefotaxime, Ceftazidime, Ceftriaxone, Ciprofloxacin, Levofloxacin	*Shigella spp.*	
		Not recorded	*N. gonorrhoeae*	
		Ceftazidime, Cefotaxime, Ampicillin, Ciprofloxacin, Ceftriaxone, Meropenem, Imipenem, Ertapenem, Levofloxacin	*E. coli*	Lativia
		Ceftazidime, Cefotaxime, Ciprofloxacin, Meropenem, Imipenem, Ertapenem, Ceftriaxone, Levofloxacin	*K. pneumoniae*	
		Cefotaxime, Ceftriaxone, Oxacillin, Penicillin G	*S. pneumoniae*	
		Cefoxitin, Oxacillin	*S. aureus*	

		Not recorded	*Shigella spp., Salmonella spp.,N. gonorrhoeae*	
		Ceftriaxone, Ceftazidime, Cefotaxime, Cotrimoxazole, Ampicillin, Ciprofloxacin	*E. coli*	Norway
		Ceftazidime, Cefotaxime, Cotrimoxazole, Ciprofloxacin, Meropenem, Cefepime, Ceftriaxone, Ertapenem, Imipenem	*K. pneumoniae*	
		Cefotaxime, Ceftriaxone, Oxacillin, Penicillin G	*S. pneumoniae*	
		Ceftazidime, Cefotaxime, Ciprofloxacin, Meropenem	*Salmonella spp.*	
		Meropenem, Gentamicin	*Acinetobacter spp.*	
		Cefoxitin, Oxacillin	*S. aureus*	
		Cefotaxime, Ceftazidime, Ciprofloxacin	*Shigella spp.*	
		Azithromycin, Cefixime, Ceftriaxone	*N. gonorrhoeae*	
		Cefotaxime, Ceftazidime, Cotrimoxazole, Ampicillin, Ciprofloxacin, Meropenem, Imipenem	*E. coli*	Sweden
		Ceftazidime, Cefotaxime, Ciprofloxacin, Meropenem, Imipenem	*Salmonella spp.*	
		Ceftazidime, Cefotaxime, Ciprofloxacin, Meropenem, Imipenem	*K. pneumoniae*	
		Amikacin, Meropenem, Gentamicin, Imipenem	*Acinetobacter spp.*	
		Cefotaxime, Oxacillin, Penicillin G	*S. pneumoniae*	
		Cefoxitin, Oxacillin	*S. aureus*	
		Azithromycin, Cefotaxime, Ceftazidime, Ciprofloxacin	*Shigella spp.*	
		Not recorded	*N. gonorrhoeae*	
		Cefepime, Ceftazidime, Cefotaxime, Cotrimoxazole, Ampicillin, Ciprofloxacin, Imipenem, Meropenem, Ertapenem, Ceftriaxone, Colistin, Levofloxacin	*E. coli*	Switzerland
		Cefepime, Ceftazidime, Cefotaxime, Cotrimoxazole, Ceftriaxone, Colistin, Levofloxacin, Ciprofloxacin, Meropenem, Imipenem, Ertapenem	*K. pneumoniae*	
		Ceftriaxone, Ceftazidime, Cefotaxime, Levofloxacin, Ciprofloxacin, Meropenem, Ertapenem	*Salmonella spp.*	
		Azithromycin, Cefixime, Ceftriaxone, Ciprofloxacin, Spectinomycin	*N. gonorrhoeae*	
		Azithromycin, Levofloxacin, Ceftazidime, Ciprofloxacin, Ceftriaxone	*Shigella spp.*	

(*Continued*)

TABLE 20.1 (Continued)

Sr. no	Region	Antibiotics	Bacteria	Countries under region
		Meropenem, Imipenem, Gentamicin, Amikacin	*Acinetobacter*	
		Cefoxitin, Oxacillin	*S. aureus*	
3.	South-east Asia	Cefepime, Ceftazidime, Cefotaxime, Cotrimoxazole, Ampicillin, Ciprofloxacin, Meropenem, Imipenem, Ertapenem, Colistin, Ceftriaxone, Levofloxacin, Doripenem	*E. coli*	Thailand
		Cefepime, Ceftazidime, Cefotaxime, Cotrimoxazole, Ceftriaxone, Imipenem, Colistin, Ertapenem, Levofloxacin, Ciprofloxacin, Doripenem, Meropenem	*K. pneumoniae*	
		Ceftazidime, Cefotaxime, Ceftriaxone, Meropenem, Imipenem, Ertapenem, Ciprofloxacin, Doripenem	*Salmonella spp.*	
		Meropenem, Imipenem, Gentamicin, Amikacin, Colistin, Doripenem, Tigecycline	*Acinetobacter spp.*	
		Cefotaxime, Ceftriaxone, Cotrimoxazole, Oxacillin, Penicillin G	*S. pneumoniae*	
		Ceftriaxone, Ciprofloxacin	*N. gonorrhoeae*	
		Cefoxitin, Oxacillin	*S. aureus*	
		Not recorded	*Shigella spp.*	
		Cefepime, Ceftazidime, Cefotaxime, Cotrimoxazole, Ampicillin, Ciprofloxacin, Meropenem, Imipenem, Ertapenem, Colistin, Ceftriaxone, Levofloxacin, Doripenem	*E. coli*	Indonesia
		Cefepime, Ceftazidime, Cefotaxime, Cotrimoxazole, Ciprofloxacin, Meropenem, Imipenem, Ertapenem, Colistin, Ceftriaxone, Levofloxacin, Doripenem	*K. pneumoniae*	
		Amikacin, Doripenem, Gentamicin, Imipenem, Meropenem, Tigecycline	*Acinetobacter spp.*	
		Cefotaxime, Ceftazidime, Ceftriaxone, Ciprofloxacin, Ertapenem, Meropenem	*Salmonella spp.*	
		Cefoxitin, Oxacillin	*S. aureus*	
		Ceftriaxone, Cotrimoxazole, Penicillin G	*S. pneumonia*	
		Not recorded	*Shigella spp., N. gonorrhoeae*	
		Cefepime, Ceftazidime, Ampicillin, Ciprofloxacin, Meropenem, Imipenem, Ceftriaxone, Levofloxacin	*E. coli*	Bangladesh

		Ceftazidime, Ceftriaxone, Ciprofloxacin, Imipenem	*K. pneumoniae*	
		Cefotaxime, Ceftazidime, Ceftriaxone, Ciprofloxacin, Ertapenem, Meropenem	*Salmonella spp.*	
		Not recorded	*S. pneumoniae, Acinetobacter spp., S. aureus, N. gonorrhoeae, Shigella spp.*	
		Ampicillin, Cefepime, Cefotaxime, Ceftazidime, Ceftriaxone, Ciprofloxacin, Imipenem, Levofloxacin, Meropenem	*E. coli*	Japan
		Cefepime, Ceftazidime, Cefotaxime, Imipenem, Ceftriaxone, Levofloxacin, Ciprofloxacin, Meropenem	*K. pneumoniae*	
		Cefotaxime, Ceftazidime, Ciprofloxacin, Imipenem, Meropenem	*Salmonella spp.*	
		Amikacin, Imipenem, Meropenem, Minocycline	*Acinetobacter spp.*	
		Cefotaxime, Ceftriaxone, Penicillin G	*S. pneumoniae*	
		Azithromycin, Cefixime, Ceftriaxone, Ciprofloxacin, Spectinomycin	*N.gonorrhoeae*	
		Oxacillin	*S. aureus*	
		Cefotaxime, Ciprofloxacin	*Shigella spp.*	
		Cefepime, Ceftazidime, Cefotaxime, Ampicillin, Cotrimoxazole, Colistin, Ciprofloxacin, Meropenem, Imipenem, Ertapenem,	*E. coli*	Republic of Korea
		Cefepime, Ceftazidime, Cefotaxime, Cotrimoxazole, Ciprofloxacin, Meropenem, Imipenem, Ertapenem, Colistin	*K. pneumonia*	
		Cefotaxime, Ceftriaxone, Oxacillin, Penicillin G	*S. pneumoniae*	
		Ceftazidime, Cefotaxime, Ciprofloxacin, Imipenem	*Salmonella spp.*	
		Cefoxitin	*S. aureus*	
		Not recorded	*N. gonorrhoeae, Shigella spp.*	
		Ampicillin, Cefepime, Cefotaxime, Ceftazidime, Ceftriaxone, Ciprofloxacin, Cotrimoxazole, Ertapenem, Imipenem, Meropenem	*E. coli*	Cambodia
		Amikacin, Colistin, Gentamicin, Imipenem, Meropenem	*Acinetobacter spp.*	
		Not recorded	*K. pneumoniae, S. pneumoniae, S. aureus, Salmonella spp., shigella spp.,N. gonorrhoeae*	
		Cefepime, Ceftriaxone, Ceftazidime, Cefotaxime, Cotrimoxazole, Colistin, Levofloxacin, Ciprofloxacin, Imipenem Ertapenem, Doripenem, Meropenem	*K. pneumoniae*	Philippines

(*Continued*)

TABLE 20.1 (Continued)

Sr. no	Region	Antibiotics	Bacteria	Countries under region
		Ceftriaxone, Cefotaxime, Levofloxacin, Imipenem, Ciprofloxacin, Ceftazidime, Ertapenem, Meropenem	*Salmonella spp.*	
		Azithromycin, Cefixime, Ceftriaxone, Ciprofloxacin, Spectinomycin	*N. gonorrhoeae*	
		Azithromycin, Cefotaxime, Ceftriaxone, Ciprofloxacin	*Shigella spp.*	
		Ceftriaxone, Cotrimoxazole, Oxacillin, Penicillin G,	*S. pneumoniae*	
		Cefoxitin, Oxacillin	*S. aureus*	
		Ampicillin, Cotrimoxazole, Cefepime, Ceftriaxone, Ceftazidime, Cefotaxime, Ciprofloxacin, Levofloxacin	*E. coli*	
		Not recorded	*Acinetobacter spp.*	
4.	Africa	Cefepime, Ceftazidime, Ceftriaxone, Levofloxacin, Meropenem, Ampicillin, Imipenem, Ciprofloxacin, Ertapenem, Colistin, Cotrimoxazole, Doripenem	*E. coli*	Madagascar
		Cefepime, Ceftazidime, Ceftriaxone, Ciprofloxacin, Cotrimoxazole, Imipenem, Colistin	*K. pneumoniae*	
		Ceftazidime, Ceftriaxone, Ciprofloxacin	*Shigella spp.*	
		Azithromycin, Ceftriaxone, Ciprofloxacin	*N. gonorrhoeae*	
		Amikacin, Gentamicin, Meropenem	*Acinetobacter spp.*	
		Cefoxitin	*S. aureus*	
		Not recorded	*Salmonella spp., S. pneumoniae*	
		Ampicillin, Cotrimoxazole, Meropenem, Cefepime, Cefotaxime, Ceftazidime, Ceftriaxone, Imipenem	*E. coli*	Ethiopia
		Cefepime, Cefotaxime, Ceftazidime, Ceftriaxone, Cotrimoxazole, Meropenem	*K. pneumoniae*	
		Amikacin, Gentamicin, Meropenem	*Acinetobacter spp.*	
		Not recorded	*S. aureus, S. pneumoniae, Salmonella spp., Shigella spp., N. gonorrhoeae*	
		Cefepime, Ceftazidime, Ceftriaxone, Meropenem, Imipenem, Ampicillin, Ciprofloxacin, Ertapenem, Cotrimoxazole, Doripenem, Cefotaxime	*E. coli*	South Africa
		Cefepime, Ceftazidime, Ceftriaxone, Cotrimoxazole, Ciprofloxacin, Imipenem, Meropenem, Ertapenem	*K. pneumoniae*	

		Cefotaxime, Ceftazidime, Ceftriaxone, Ciprofloxacin, Ertapenem, Imipenem,	*Salmonella spp.*	
		Amikacin, Gentamicin, Imipenem, Meropenem, Minocycline, Tigecycline	*Acinetobacter spp.*	
		Azithromycin, Cefixime, Ceftriaxone, Spectinomycin	*N. gonorrhoeae*	
		Ceftriaxone, Cotrimoxazole, Penicillin G	*S. pneumoniae*	
		Cefotaxime, Ceftazidime, Ceftriaxone, Ciprofloxacin	*Shigella spp.*	
		Oxacillin	*S. aureus*	
5.	United States of America	Ampicillin, Cefotaxime, Ceftazidime, Ceftriaxone, Ciprofloxacin, Cotrimoxazole, Imipenem, Meropenem, Ertapenem, Levofloxacin	*E. coli*	Brazil
		Ceftazidime, Ceftriaxone, Ciprofloxacin, Cotrimoxazole, Ertapenem, Imipenem, Meropenem, Levofloxacin, Cefotaxime	*K. pneumoniae*	
		Amikacin, Gentamicin, Imipenem, Meropenem, Tigecycline	*Acinetobacter spp.*	
		Cefixime, Ceftriaxone, Ciprofloxacin, Gentamicin, Spectinomycin	*N. gonorrhoeae*	
		Oxacillin	*S. aureus*	
		Not recorded	*S. pneumoniae, Salmonella spp. Shigella spp.*	

experts have to choose the available antibiotics as a curative step. Interestingly, the symptoms of COVID-19, common cold, cough, and fever were somewhat similar, which leads to the unregulated and extensive use of antibiotics in healthcare organisations and commodities, which worsens the diagnosis and therapeutic scenario (Nieuwlaat et al., 2021). Several antimicrobials are subjected to experimental randomised trials to improve infected person's health that is responsible for the AMR condition. To handle this condition, the WHO suggested the implementation of the principle of parsimony.

The observations made by Sultana and group brought attention to azithromycin's inefficacy, as it revealed that patients had previous exposure to the same antibiotic. Therefore there is still a need for an in-depth study focusing on the use and poor antimicrobial action of azithromycin (Sultana et al., 2020). During the COVID-19 pandemic, the extensive usage of antibiotics may be resulted in various coinfections and secondary infections and an enhancer for AMR (Collignon and Beggs, 2020). Therefore these data highlight the rising concerns of AMR, especially in the challenging pandemic time, which can be tackled better in the future if necessary directions are taken in future studies.

20.6 Local perspective of antimicrobial resistance

Likewise, in the rest of the world, antibiotic resistance is widespread in India also because of the different drivers explained above in this chapter. India, despite being the largest antibiotic user in absolute volume, lacks a structured system of antibiotic usage surveillance similar to those in the United States and Europe, which is aimed to create awareness regarding antibiotics. In India, the State Drug Regulatory Authorities (SDRAs) and the Central Drugs Standard Control Organization (CDSCO) are the regulatory bodies for pharmaceuticals and medical devices, which coordinate between state and central government, respectively. The majority of states' SDRAs have minimal technical capacity, but this capacity varies throughout SDRAs (Koya et al., 2022). Despite the fact that several studies have covered the resistance profiles of bacterial isolates, these still lack information regarding the usage of antibiotics and their justifications. Indian Council of Medical Research (ICMR) released an annual report for AMR that explained the current scenario of prevalent infective agents found in patients. In the report, it was very clear that *Escherichia coli, Klebsiella pneumoniae, Pseudomonas aeruginosa, Acinetobacter baumannii, and* Staphylococcus *aureus* are the prevalent pathogens that showed resistance against different antibiotics. Moreover, in recent times, there has been an increase in the antibiotic resistance ratio exerted by bacteria (ICMR Indian Council of Medical Research, 2022. AMR Research and Surveillance Network. New Delhi., 2022), and the same results have been reported by the National Centre for Disease Control (NCDC) also. This report overviewed the antibiotic resistance surveillance throughout the nation, reviewed the different preventive measures, and provided fruitful feedback. One of the major benefits of such a report is to keep track of surveillance and introspect on the actions. Interestingly, the significant drop in resistance against *E.coli* infection is also seen, which indicates the positive growth in healthcare management of the country.

In conclusion, the AMR surveillance data from NARS-NET highlight the need for coordinating efforts to improve the execution of antimicrobial stewardship policies, infection prevention, and control programmes at the state and federal levels. This target can be accomplished by creating and putting IPC programmes and facility-level antibiotic policies into practice and pushing the system to adhere to national treatment standards for infectious illnesses (National Centre for Disease Control NCDC, Directorate General of Health services, Ministry of Health & Family Welfare, Government of India, 2023. National Antimicrobial Surveillance Network NARS-Net, 2023).

20.7 Preventive, measures to control antimicrobial resistance

The correlation between AMR and COVID-19 is complicated; factors like physical distancing and personal and hand hygiene help to minimise the transmission of AMR within the surrounding (Langford et al., 2023). The on-ground, real practical conduct for AMR regulation and suggested measures showed a significant gap. The various statistical models have been applied to understand the future outcomes and what type of new approaches can be applied for AMR prevention. However, a wide gap has been seen between the statistical model expectations and onsite measures, because of the variability in the occurrence of incidence, discharge, and admission per incident, ratio of incident, ratio of infection, and point of prevalence. There can be other reformatory steps for confining antibiotic use, such as taxes and quotas, as 80% of US antibiotic market is covered by agricultural and aquacultural professionals for fodder or sanitisation purposes. So, tax imposition can reduce the unregulated use of antibiotics by various nonclinical domain personnel and help in the AMR regulation. Similarly, recently, the WHO also gave the recommendation to impose restrictions on antibiotics as an additive supplement for animals. Imposition of user fees on unregulated antibiotic use is another preventive measure; user fees can be easily imposed at any stage of antibiotic conduct from company to consumption, and such revenue can be invested for future research and development purposes. Furthermore, the rapidly evolving AMR challenges insist that we make the required corrections and amendments in the antimicrobial stewardship programme to further strengthen and empower these stewardship programmes with the aim of better patient care, a decrease in the usage of antibiotics, and more affordable medical care (Kollef and Micek, 2023).

Moreover, the present era is of technologies and artificial intelligence; therefore the future approaches should be more technologically sufficient and with aids like different biosensors-based food testing. There are lab-based POC sensors that can track antibiotics as contaminants; such hand-held, onsite sensors are helpful to governing authorities for surveillance as a point-of-care sensor (Raysyan et al., 2020; Prakashan et al., 2024).

20.8 Conclusions

AMR is a condition where the efficacy of an antibiotic to combat against pathogen is significantly reduced. It arises due to the persistent use of antibiotics. Till now, several biological mechanisms have shown how AMR affects biologically and clinically. Moreover, social and economic perspectives are also impacted, which shows how society and the global economy are being affected by this condition. Furthermore, these impacts create a ripple effect; how one factor, that is clinical emergence, is related to socioeconomic growth. The economic growth has an effect on social and public health and affects a series of events. Therefore these overall combined impacts result into negative impact on the healthcare system as it creates stress on facilities, disrupts everyday actions, and negatively hampers personal productivity. As the work-life productivity degrades, it will also halt the economic growth of a nation.

Similarly, like any other clinical condition, AMR also has long-term adverse effects on the social and economic health of the country and ultimately worldwide. Therefore the public awareness and acknowledgement regarding antibiotic applicability and misuse are critical to control these conditions. Addressing the drivers and their effects on each aspect, like social and economic, is also crucial. Furthermore, the implementation of health programmes to slow down the propagation of this pandemic condition, which can eventually abate its socioeconomic consequences, is also a way forward.

Acknowledgements

The authors are thankful to the Director General, Central Council for Research in Ayurvedic Science, Ministry of Ayush, Government of India, for providing encouragement and support. The authors also acknowledge BioRender for designing the illustrations.

Conflicts of Interest

The authors have no conflict of interest to disclose.

References

A fact sheet from ReAct-Action on Antibiotic Resistance, 2012. Available at http://www.reactgroup.org.

Ahmad, M., Khan, A.U., 2019. Global economic impact of antibiotic resistance: a review. Journal of Global Antimicrobial Resistance 19, 313–316. Available at https://doi.org/10.1016/j.jgar.2019.05.024, http://www.journals.elsevier.com/journal-of-global-antimicrobial-resistance/.

Ahmed, S.A., Barış, E., Go, D.S., Lofgren, H., Osorio-Rodarte, I., Thierfelder, K., 2018. Assessing the global poverty effects of antimicrobial resistance. World Development 111, 148–160. Available at https://doi.org/10.1016/j.worlddev.2018.06.022.

Bhatia, R., 2018. Antimicrobial resistance: threat, consequences and options. National Medical Journal of India 31 (3), 133–135. Available at https://doi.org/10.4103/0970-258X.255752, https://www.journalonweb.com/nmji/.

Carter, R.R., Sun, J., Jump, R.L.P., 2016. A survey and analysis of the American public's perceptions and knowledge about antibiotic resistance. Open Forum Infectious Diseases 3 (3). Available at https://doi.org/10.1093/ofid/ofw112, https://academic.oup.com/ofid.

Castro-Sánchez, E., Moore, L.S.P., Husson, F., Holmes, A.H., 2016. What are the factors driving antimicrobial resistance? Perspectives from a public event in London, England. BMC Infectious Diseases 16 (1). Available at https://doi.org/10.1186/s12879-016-1810-x, http://www.biomedcentral.com/bmcinfectdis/.

Chang, Q., Wang, W., Regev-Yochay, G., Lipsitch, M., Hanage, W.P., 2015. Antibiotics in agriculture and the risk to human health: how worried should we be? Evolutionary Applications 8 (3), 240–247. Available at https://doi.org/10.1111/eva.12185, http://www.blackwellpublishing.com/eva_enhanced/.

Chokshi, A., Sifri, Z., Cennimo, D., Horng, H., 2019. Global contributors to antibiotic resistance. Journal of Global Infectious Diseases 11 (1), 36–42. Available at https://doi.org/10.4103/jgid.jgid_110_18.

Collignon, P., Beggs, J.J., 2020. CON: COVID-19 will not result in increased antimicrobial resistance prevalence. JAC-Antimicrobial Resistance 2 (3). Available at https://doi.org/10.1093/jacamr/dlaa051, https://academic.oup.com/jacamr.

Food and Agriculture Organization of the United Nations, 2026. Antimicrobial Resistance. Available at https://www.fao.org/antimicrobial-resistance/key-sectors/animal-production/en/.

Giacobbe, D.R., Giani, T., Bassetti, M., Marchese, A., Viscoli, C., Rossolini, G.M., 2020. Rapid microbiological tests for bloodstream infections due to multidrug resistant Gram-negative bacteria: therapeutic implications. Clinical Microbiology and Infection 26 (6), 713–722. Available at https://doi.org/10.1016/j.cmi.2019.09.023.

Global Antimicrobial Resistance and Use Surveillance System (GLASS) Report, 2021. Available at https://www.who.int/publications/i/item/9789240027336.

ICMR Indian Council of Medical Research, 2022. Antimicrobial Resistance Research and Surveillance Network. New Delhi. Available at https://www.icmr.gov.in/icmrobject/custom_data/pdf/resource-guidelines/AMR_Annual_Report_2021.pdf.

Irfan, M., Almotiri, A., AlZeyadi, Z.A., 2022. Antimicrobial resistance and its drivers—a review. Antibiotics 11 (10). Available at https://doi.org/10.3390/antibiotics11101362.

Koch, N., Islam, N.F., Sonowal, S., Prasad, R., Sarma, H., 2021. Environmental antibiotics and resistance genes as emerging contaminants: methods of detection and bioremediation. Current Research in Microbial Sciences. 2, 100027. Available at https://doi.org/10.1016/j.crmicr.2021.100027.

Kollef, M.H., Micek, S.T., 2023. Antimicrobial Stewardship Textbook of Critical Care. Available at https://doi.org/10.1016/B978-0-323-75929-8.00114-3.

Koya, S.F., Ganesh, S., Selvaraj, S., Wirtz, V.J., Galea, S., Rockers, P.C., 2022. Consumption of systemic antibiotics in India in 2019. The Lancet Regional Health - Southeast Asia. 4. Available at https://doi.org/10.1016/j.lansea.2022.100025, https://www.thelancet.com/lansea/about.

Langford, B.J., Soucy, J.P.R., Leung, V., So, M., Kwan, A.T.H., Portnoff, J.S., et al., 2023. Antibiotic resistance associated with the COVID-19 pandemic: a systematic review and meta-analysis. Clinical Microbiology and Infection 29 (3), 302–309. Available at https://doi.org/10.1016/j.cmi.2022.12.006, https://www.journals.elsevier.com/clinical-microbiology-and-infection.

Lekagul, A., Tangcharoensathien, V., Yeung, S., 2019. Patterns of antibiotic use in global pig production: a systematic review. Veterinary and Animal Science 7. Available at https://doi.org/10.1016/j.vas.2019.100058.

Lin, T.Z., Jayasvasti, I., Tiraphat, S., Pengpid, S., Jayasvasti, M., Borriharn, P., 2022. The predictors influencing the rational use of antibiotics among public sector: a community-based Survey in Thailand. Drug, Healthcare and Patient Safety 14, 27–36. Available at https://doi.org/10.2147/DHPS.S339808, http://www.dovepress.com/drug-healthcare-and-patient-safety-journal.

National Centre for Disease Control (NCDC), 2023. Directorate General of Health services, Ministry of Health & Family Welfare, Government of India, 2023. National Antimicrobial Surveillance Network (NARS-Net). Available at https://ncdc.mohfw.gov.in/wp-content/uploads/2024/09/Final-Annual-Report-2023-06_08_2024.pdf.

Nieuwlaat, R., Mbuagbaw, L., Mertz, D., Burrows, L.L., Bowdish, D.M.E., Moja, L., et al., 2021. Coronavirus disease 2019 and antimicrobial resistance: parallel and interacting health emergencies. Clinical Infectious Diseases 72 (9), 1657–1659. Available at http://cid.oxfordjournals.org/content/by/year, 10.1093/cid/ciaa773.

OECD, 2019. Antimicrobial Resistance Tackling the Burden in the European Union: Tackling the Burden in the European Union. OECD Publishing. Available at https://doi.org/10.1787/33CBFC1C-EN https://www.oecd.org/content/dam/oecd/en/publications/reports/2019/03/antimicrobial-resistance-tackling-the-burden-in-the-european-union_e3631398/33cbfc1c-en.pdf.

Prakashan, D., Kolhe, P., Gandhi, S., 2024. Design and fabrication of a competitive lateral flow assay using gold nanoparticle as capture probe for the rapid and on-site detection of penicillin antibiotic in food samples. Food chemistry 439, 138120. Available at https://doi.org/10.1016/j.foodchem.2023.138120.

Pulingam, T., Parumasivam, T., Gazzali, A.M., Sulaiman, A.M., Chee, J.Y., Lakshmanan, M., et al., 2022. Antimicrobial resistance: prevalence, economic burden, mechanisms of resistance and strategies to overcome. European Journal of Pharmaceutical Sciences 170. Available at https://doi.org/10.1016/j.ejps.2021.106103, www.elsevier.com/locate/ejps.

Raysyan, A., Galvidis, I.A., Schneider, R.J., Eremin, S.A., Burkin, M.A., 2020. Development of a latex particles-based lateral flow immunoassay for group determination of macrolide antibiotics in breast milk. Journal of Pharmaceutical and Biomedical Analysis 189. Available at https://doi.org/10.1016/j.jpba.2020.113450, www.elsevier.com/locate/jpba.

Read, A.F., Woods, R.J., 2014. Antibiotic resistance management. Evolution, Medicine, and Public Health 2014 (1), 147. Available at https://doi.org/10.1093/emph/eou024 –147.

Roope, L.S.J., Smith, R.D., Pouwels, K.B., Buchanan, J., Abel, L., Eibich, P., et al., 2019. The challenge of antimicrobial resistance: what economics can contribute. Science 364 (6435). Available at https://doi.org/10.1126/science.aau4679, https://science.sciencemag.org/content/364/6435/eaau4679/tab-pdf.

Samreen, I., Ahmad, H.A., Malak, H.H., Abulreesh, 2021. Environmental antimicrobial resistance and its drivers: a potential threat to public health. Journal of Global Antimicrobial Resistance 27, 101–111. Available at https://doi.org/10.1016/j.jgar.2021.08.001, http://www.journals.elsevier.com/journal-of-global-antimicrobial-resistance/.

Stern, N., 2016. Economics: current climate models are grossly misleading. Nature 530 (7591), 407–409. Available at https://doi.org/10.1038/530407a.

Sultana, J., Cutroneo, P.M., Crisafulli, S., Puglisi, G., Caramori, G., Trifirò, G., 2020. Azithromycin in COVID-19 patients: pharmacological mechanism, clinical evidence and prescribing guidelines. Drug Safety 43 (8), 691–698. Available at https://doi.org/10.1007/s40264-020-00976-7, http://rd.springer.com/journal/40264.

Taylor, J., Hafner, M., Yerushalmi, E., Smith, R., Bellasio, J., Vardavas, R., et al., 2014. Estimating the economic costs of antimicrobial resistance: model and results. Estimating the economic costs of antimicrobial resistance: Model and Results 1–113. Available at https://doi.org/10.7249/RR911, https://www.rand.org/pubs/research_reports/RR911.html.

Teillant, A., Gandra, S., Barter, D., Morgan, D.J., Laxminarayan, R., 2015. Potential burden of antibiotic resistance on surgery and cancer chemotherapy antibiotic prophylaxis in the USA: a literature review and modelling study. The Lancet Infectious Diseases 15 (12), 1429–1437. Available at https://doi.org/10.1016/S1473-3099(15)00270-4, http://www.journals.elsevier.com/the-lancet-infectious-diseases.

Wang, L., Chen, H., Zhang, Y., Tian, Y., Hu, X., Wu, J., et al., 2025. Global antibiotic consumption and regional antimicrobial resistance, 2010-21: an analysis of pharmaceutical sales and antimicrobial resistance surveillance data. The Lancet. Globalization and Health 13 (11), e1880. Available at https://doi.org/10.1016/S2214-109X(25)00308-0.

WHO, 2016. Global action plan on antimicrobial resistance, Official Journal of the European Union. Geneva. Available at https://iris.who.int/server/api/core/bitstreams/1a487887-e162-46a0-8aef-802907c66070/content.

Woolhouse, M., Waugh, C., Perry, M.R., Nair, H., 2016. Global disease burden due to antibiotic resistance - State of the evidence. Journal of Global Health 6 (1). Available at https://doi.org/10.7189/jogh.06.010306, http://jogh.org/documents/issue201601/jogh-06-010306.pdf.

World Bank Group. World Bank, 2017. DRUG-RESISTANT INFECTIONS A Threat to Our Economic Future. Washington, DC. Available at https://documents1.worldbank.org/curated/en/323311493396993758/pdf/final-report.pdf.

Yu, P., Dong, P., Wang, H., 2022. Deciphering changes in the abundance of intracellular and extracellular antibiotic resistance genes and mobile genetic elements under anaerobic fermentation: driven by bacterial community. Bioresource Technology 355, 127264. Available at https://doi.org/10.1016/j.biortech.2022.127264.

Chapter 21

Target-centric drug repurposing: a next-generation approach to confront antimicrobial resistance

Indrani Gogoi[1,2], Shyamalima Saikia[1], and Pankaj Chetia[1]
[1]*Molecular Plant Taxonomy and Bioinformatics Research Laboratory, Department of Life Sciences, Dibrugarh University, Dibrugarh, Assam, India,*
[2]*Department of Botany, Dergaon Kamal Dowerah College (Autonomous), Golaghat, Assam, India*

21.1 Introduction

Target-centric drug is based on the identification of validated molecular targets essential for pathogen survival and screening existing approved drugs against these targets. Unlike traditional drug discovery, which is time-intensive and costly, this approach begins with the identification of validated molecular targets that are essential for microbial survival, virulence, or resistance. Once such targets are defined, existing approved or investigational drugs are systematically screened for potential activity against them.

21.1.1 Growing risk of antimicrobial resistance among bacteria

Antimicrobial resistance (AMR) is a long-standing evolutionary response that supports the adaptation and survival of organisms in the presence of natural antimicrobial agents found in their ecosystems. The majority of antimicrobial agents are naturally occurring substances, and over time, the microorganisms, such as viruses, bacteria, fungi, and parasites, living alongside them have developed mechanisms to counteract their effects for survival (Ahmed et al., 2024). The risk of antibiotic resistance is becoming a persistent and significant health problem worldwide. The repeated use of antibiotics, self-medication, and prolonged exposure to hospital-based infections have created significant selective pressure, leading to antibiotic resistance in microorganisms (Niederman, 2005). This has caused the evolution of intrinsic resistance, the generation of hydrolysing enzymes, the activation of efflux pumps, and a decrease in membrane permeability to acquired mechanisms of drug modification in bacteria that regulate the epidemiology of antibiotic resistance (Iredell et al., 2016). In this way, the burden of AMR among bacteria has increased steadily and created unavoidable havoc in the healthcare system. AMR has significant consequences for public health and economic well-being. In 2010, the World Health Organisation (WHO) listed antibiotic resistance as one of the three main threats to human health (Vaara, 2019). In a report by the Centres for Disease Control and Prevention (CDC) in 2019 on antibiotic resistance threats, it was stated that in the United States, more than 2.8 million antibiotic-resistant cases occur, and around 35,000 people have lost their lives (Patel and Sridhar, 2022). Similarly, Europe reports over 25,000 cases of antibiotic-resistant bacterial infections that result in death each year, costing the healthcare system more than 1.5 billion dollars (MacGowan and Macnaughton, 2017).

Multidrug-resistant (MDR) bacteria are causing serious health problems, which is considered one of the significant challenges to treating hospital-acquired infections (HAIs) and community-acquired bacterial infections in the present day (Davies, 1996; Ferreira et al., 2019). The acquisition of antibiotic resistance by ESKAPE pathogens (*Enterococcus faecium, Staphylococcus aureus, Klebsiella pneumoniae, Acinetobacter baumannii, Pseudomonas aeruginosa*, and *Enterobacter* species), alongside other critical MDR pathogens, has reduced the effectiveness of antibiotics (Boucher et al., 2009). In low- and middle-income countries (LMICs), where antibiotic stewardship and infection control measures are often limited, resistance rates of MDR strains to multiple first- and second-line agents are substantially higher (WHO, 2020). The worldwide increase in the occurrence of MDR pathogens highlights the multifactorial processes, which include

Antimicrobial Resistance in the 21st Century. DOI: https://doi.org/10.1016/B978-0-443-44748-8.00024-5

the continuous evolution and diversification of chromosomal or plasmid-encoded antibiotic resistance genes (ARGs) (Navon-Venezia et al., 2017). Bacteria have evolved to inactivate antibiotics by producing hydrolysing enzymes, such as β-lactamases and carbapenemases (Lutgring and Limbago, 2016). The acquisition of the most prevalent *K. pneumoniae* carbapenemase (KPC) type enzymes has led to the global emergence of carbapenem-resistant *K. pneumoniae* (CR-KP). This resistance phenotype is recognised as one of the major causes of life-threatening HAI (Liu et al., 2018; Indrajith et al., 2021). The possible molecular mechanisms that could account for the upregulation of efflux pumps include mutated insertion elements in the upstream region of the efflux pump gene, mutated transcriptional activator or repressor genes, and the promoter region, which directly promotes bacterial pathogenicity (Piddock, 2006). In addition to carbapenem resistance, bacteria exhibit a remarkable ability to hydrolyse other β-lactam antibiotics, which have historically formed the backbone of antibacterial therapy. This resistance is largely mediated by β-lactamases productions that hydrolyses the β-lactam ring and renders the antibiotic ineffective. Extended-spectrum β-lactamases (ESBLs), such as TEM, SHV, and CTX-M types, AmpCs, and metallo-β-lactamases (MBLs) have been widely reported in various MDR strains and confer resistance to penicillins, cephalosporins, monobactams, and even to carbapenems and β-lactamase inhibitor combinations (Drawz and Bonomo, 2010; Paterson and Bonomo, 2005).

Recognising the rapid emergence of antibiotic resistance among bacteria, the WHO has prepared the Bacterial Priority Pathogen List (BPPL), which identifies clinically important bacterial pathogens that need urgent attention in research and development for their control. The increasing rate of antibiotic resistance has raised questions about the effectiveness of market-available antibiotics, making it a global challenge (Rossolini et al., 2014). As per the estimation of the CDC, if new and effective antibiotics are unavailable until 2050, the number of deaths from MDR bacterial infections alone will equal the number of total deaths from heart and cancer diseases combined.

21.1.2 Integrated strategies for antimicrobial development

The continuous shortage of antimicrobials in antibiotic discovery pipelines and the spread of resistant pathogens necessitate alternative strategies to address these significant health issues. Novel antimicrobial interventions need to be developed as the conventional antibiotic therapy is becoming incapable of fighting against MDR bacterial pathogens. According to a report published by the WHO, while several new antibiotics are currently under development, none of them are anticipated to be effective against the most critical forms of antibiotic-resistant bacteria. This implies that addressing antibiotic resistance requires more than the discovery and development of new antibiotics or the implementation of new interventions. Instead, it necessitates the development of novel strategies that can limit the emergence and rise of resistance to existing antibiotics (Chinemerem Nwobodo et al., 2022). Recent advancements in drug discovery have led to the development of several antimicrobial strategies, including host-directed therapies, phage therapies, antimicrobial peptides, and nanotechnology (Rathored and Budhbaware, 2025). In addition to these, the employment of the CRISPR-Cas system can eradicate the antibiotic resistance mechanisms in bacterial pathogens by inducing death or by curing antibiotic resistance-expressing plasmids (Mayorga-Ramos et al., 2023).

The development of new drugs through traditional methods requires substantial investment of capital, skilled personnel, and expertise. It also requires compliance with testing regulations and manufacturing standards before introduction to the general population (Dickson and Gagnon, 2004). The discovery of traditional medicine can take up to approximately 10–15 years, and it is solely based on nature-based products (Berdigaliyev and Aljofan, 2020). Therefore to overcome such limitations, the focus has shifted towards integrated strategies for antimicrobial development, which offer a multidisciplinary response to the rising risk of AMR. Integrated target-based drug discovery, high-throughput approaches to accelerate the identification and design of drug molecules, structure-guided drug design, and artificial intelligence—driven molecule optimisation can be a good choice for novel antimicrobial development to minimise the resistance potential of bacteria (Ferreira and Carneiro, 2025). The evaluation of resistance mechanisms and their primary drivers in bacteria could serve as a crucial point to initiate the screening of potential drug targets. Such insights also enhance the ability of multiomics to facilitate the systematic identification, refinement, and prioritisation of drug candidates, as well as the prediction of their potential for drug resistance (Majumder and Panigrahi, 2025).

Drug repurposing is another emerging strategy in the field of contemporary drug discovery, and it has gained significant importance in the identification of new therapeutic applications for existing drugs. Drug repurposing offers a systematic and efficient approach to exploring approved drugs for new uses by combining various techniques, including computational modelling, network pharmacology, and multiomics analysis (Parvathaneni et al., 2019). Drug repurposing has been placed at the forefront for its reduced cost and longer approval timelines for antimicrobial discovery. This approach explores the cryptic antimicrobial activities of old drugs, thereby expanding their therapeutic potential beyond their original indications. Numerous in vitro and computational-based studies and validation on model organisms have confirmed the potential of this strategy in antimicrobial research (Farha and Brown, 2019). Thus the integrated approaches not only accelerate the discovery of effective antimicrobial targets but also contribute to searching for therapeutics that are less prone to rapid resistance developed in pathogenic bacteria.

21.2 Drug-target identification using different approaches

21.2.1 Genomics and proteomics in drug-target discovery

Genomics and proteomics-based approaches are target-based methods that are employed in the early phase of drug repurposing for the identification of a druggable target. To consider a molecule as a desirable drug target, it should exhibit standard features such as essentiality, druggability, specificity, and relevance to the metabolic pathways of pathogens (Serral et al., 2021). Genetic-driven identification and validation of druggable targets is a promising tool that was initiated in 2003 and is currently widely used to predict novel drug targets using multiple algorithms (Murthy et al., 2013). However, validation of the identified drug target is crucial as it ensures the involvement of the molecular target in the disease mechanisms (Wang et al., 2010). It can be achieved through structure–activity relationship analysis, gene manipulation, mutant generation, knockout study, overexpression analysis, and altering signalling pathways downstream of the identified target (Baxevanis and Francis Ouellette, 2002). The quality assessment of selected drug targets helps to ensure the success and efficiency of drug development. The presence of suitable binding pockets is crucial for the druggability of the target, which needs to be screened at the initial stage of drug discovery (Zhou et al., 2024). The structure of the binding pocket of the target was crucial for drug design and lead optimisation, which could expand the druggable genome and enable the novel development of target-based therapeutics (Jakubec et al., 2022). Recently, the integration of various modern approaches with traditional drug-target discovery has escalated this process.

The identification of disease-modifying targets is the initial step in drug discovery. This requires manual investigation of scientific literature and various biomedical databases to understand the association between molecular targets and disease or disease-causing pathogens and to evaluate the efficacy and commercial potential of the target (Paananen and Fortino, 2020). Being an interdisciplinary subject, integration of omics with different fields of life sciences helps revolutionise drug-target discovery. There are various approaches to use genomics and proteomics to screen novel drug targets. However, the detection of genetic variants and mutations in genes via genomics affects susceptibility and therapeutic response, enabling personalised treatment. The structural elucidation of target proteins, their functions, and interactions can be accessed through proteomics studies. Recent advancements in genomics and proteomics, including next-generation sequencing (NGS), single-cell genomics and proteomics, mass spectrometry, gene editing, and whole-genome sequencing, have aided in accelerating drug development (Dogiparthi et al., 2025).

21.2.1.1 Genome-wide association studies (GWAS)

GWAS is a powerful tool to detect disease-associated mutations. This compares the genomic materials of patients against those of people without disease in large numbers to identify any disease-linked genetic variations. GWAS analyses identify specific single-nucleotide polymorphisms (SNPs) and large numbers of loci that increase disease risk (Dogiparthi et al., 2025). Studies also reveal that GWAS has proven successful in identifying target-disease combinations and systematically evaluating historical drug development, highlighting the potential of genetic evidence in drug development (Cao and Moult, 2014; Hukerikar et al., 2024). The generation of gene-gene interactions through computational gene networks identifies influential genes within the network that participate in various metabolic pathways. The genes that exhibit overexpression or mutations are considered ideal for drug targets (You et al., 2019). Studies also reported a close association between GWAS-reported genes and drug-target genes in the gene network, which justifies their synergistic modulations in target identification (Cao and Moult, 2014). Recently, the metaanalysis of published GWAS data has uncovered disease-associated novel genetic loci and opened new avenues for drug repurposing (Pun et al., 2023).

21.2.1.2 Subtractive genomics approach

The identification of novel drug targets and elucidation of their structures and functions in pathogens is crucial for developing therapeutics. In such cases, subtractive genomics can be helpful to interpret the role of therapeutic targets for differentiation of essential genes of a pathogen from noncritical or host-homologous genes (Mudgal et al., 2025). Subtractive genomics methodology enables the systematic identification of genes vital to pathogenicity or drug resistance mechanisms. This approach mines the whole-genome DNA of pathogens to filter out potential drug targets using several online databases and servers (Shahid et al., 2019). Several pathogens, such as *Mycobacterium tuberculosis*, *Bartonella baciliformis*, *Mycoplasma genitalium*, *Brucella suis*, and many more, have already implemented subtractive genomics to find out effective drug targets against the pathogens (Hosen et al., 2014; Khan et al., 2020, 2022; Fatoba et al., 2021).

Subtractive genomics requires multiple online tools and bioinformatics software for in silico identification of genetic elements within a specific group of organisms. Microbial Genome Comparison (MGC) is one such software that enables independent and comprehensive comparison of microbial genomes, giving detailed insights into genomic similarities and differences. Some other software, mGenomeSubtractor and FindTarget, are also available to exploit the DNA blueprint of bacterial species. The mGenomeSubstractor is a BLAST-based web tool that can compare a reference bacterial genome with multiple genomes to identify species-variable accessory regions. This can efficiently perform large-scale comparative analysis in a very short period of time, which requires only about one minute to compare all 5312 coding sequences of *Escherichia coli* O157:H7 EDL933 chromosome (~5.5 Mb) against 23 completely sequenced and phylogenetically diverse *E. coli* strains (Shao et al., 2010). Thus it suggests that high-throughput subtractive genomics is able to identify strain-specific, conserved, or pathogenic genes relevant to virulence and AMR. In recent years, a large number of targets have been identified for drug-resistant bacterial pathogens using subtractive genomics (Barh et al., 2011).

21.2.2 Systems biology and network-based target identification

Systems biology is one of the most powerful approaches for large-scale screening of protein–ligand interactions for the identification of drug targets and mechanisms of action to lead compound characterisation (Yadav and Tripathi, 2018). A combination of high-throughput experimental data and computational biology has revolutionised the understanding of disease-associated cellular pathways and gene-regulatory networks, which provide critical insights into disease phenotypes, AMR mechanisms, etc. (Wang et al., 2012). Multitarget interventions of drugs have shown greater efficacy against diseases with complexities compared with the conventional mono-target drugs currently in use. Multitarget drugs may consist of a single chemical entity or a combination of multiple chemicals, and most are multicomponent formulations that act synergistically to modulate the complex pathological networks underlying disease progression (Leung et al., 2013). The mode of action of the network identification (MNI) algorithm has been developed to identify the gene targets of a drug from gene-expression data. This algorithm constructs a gene-regulatory network of cells and describes their influential effects on each other (Xing and Gardner, 2007). Systems biology-driven network analysis also facilitates the integration of diverse omics datasets and generates a holistic view of host–pathogen interactions and cellular response to drug exposure. For example, TargetTB is a drug target for *Mycobacterium tuberculosis* that was identified through interactome, reactome, and genome-scale structural analysis of the pathogen and found to be effective in addressing mycobacterial persistence and drug resistance (Raman et al., 2008). Proteome network analysis is a new approach that reveals interactions between proteins and their association with other proteins present in the cell, which could serve as an alternate drug target (Salvato et al., 2021). The Biomolecular Interaction Database (BIND) is such a freely available database of protein–protein, protein–DNA, and genetic interactions of *Saccharomyces cerevisiae*, *Drosophila melanogaster*, and *Caenorhabditis elegans* (Ouellette and Baxevanis, 2001). STRING is another notable database for protein–protein interaction prediction (Szklarczyk et al., 2019). Thus by mapping gene-regulatory networks and system biology, it becomes possible to infer mechanisms of action of drugs, predict off-target effects, and identify synergistic drug combinations.

21.2.3 Artificial intelligence in predicting novel targets

Prediction of novel drug targets involves the identification of specific proteins or pathways that modulate target regulatory mechanisms. This requires costly high-throughput screens, which are gradually replaced by AI due to the ability of complex datasets analysis, identification of efficient drug targets and biomarkers, as well as lead compound optimisation (Vora et al., 2025). AI-driven synthetic data mimics the original pattern and characteristics. Therefore these data can be used to predict biological scenarios for exploring and analysing a broad range of possibilities (Pun et al., 2023). The AI algorithms have been recently integrated into several drug discovery processes, such as structure and ligand-based virtual screening, toxicity prediction, pharmacophore modelling, peptide synthesis, drug repositioning, and reverse docking. Additionally, the development of AI has made big data analysis more effective and easier, as there are ample machine learning (ML) and deep learning (DL) approaches available, which can be useful in extracting significant features and patterns of medical data (Gupta et al., 2021). ML and DL are two important subsets of AI. Using DL algorithms, recently, a group of researchers has successfully identified and annotated AMR genes ARGs) utilising data obtained from genomic and plasmid sequences (Olatunji et al., 2024). Similarly, AI-based mobile applications have also been developed for rapid phenotypic detection analysis (Pascucci et al., 2021), which illustrates the utility of AI to make timely and evidence-based antibiotic choices for treatment. By integrating AI-guided chemical modelling, high-throughput screening, and multiomics approaches, studies have highlighted the utility of AI in investigating the role of TNFRSF10A/TRAILR1 as a major mediator of drug-resistant pancreatic ductal adenocarcinoma (PDAC) (Sharma et al., 2025). Computer-aided drug designing (CADD) is widely used in AMR-related studies, which typically investigate already available datasets to design new lead molecules. ML boosts the target prediction, off-target selection, and rapid identification of molecular functions associated with drug-target identification (Talat and Khan, 2023). AI-generated data can also help to solve questions that usually arise from data imbalance or bias. For instance, in several therapeutic areas, the underrepresentation of specific patient populations in existing datasets complicates target identification. In such cases, AI-generated synthetic data can help to explore more inclusive analysis. Furthermore, broader possibilities have been offered by AI-generated datasets to avoid data scarcity (Pun et al., 2023). The following schematic illustration in Fig. 21.1 demonstrates different approaches of multiomics, network-based analysis, phenotypic screening, and artificial intelligence that can collectively contribute to systematic identification and validation of drug targets.

21.3 Drug repurposing: a next-generation approach

Drug repurposing or drug repositioning is a promising approach for the identification of effective therapeutic agents from existing drugs (Begley et al., 2021). This is a highly efficient, low-cost, and time-saving drug discovery strategy with minimum risk of failure (Latif et al., 2022). In drug profiling or therapeutic switching, new drugs are identified from existing FDA-approved drugs and used as therapeutics for the treatment of diseases different from their original use

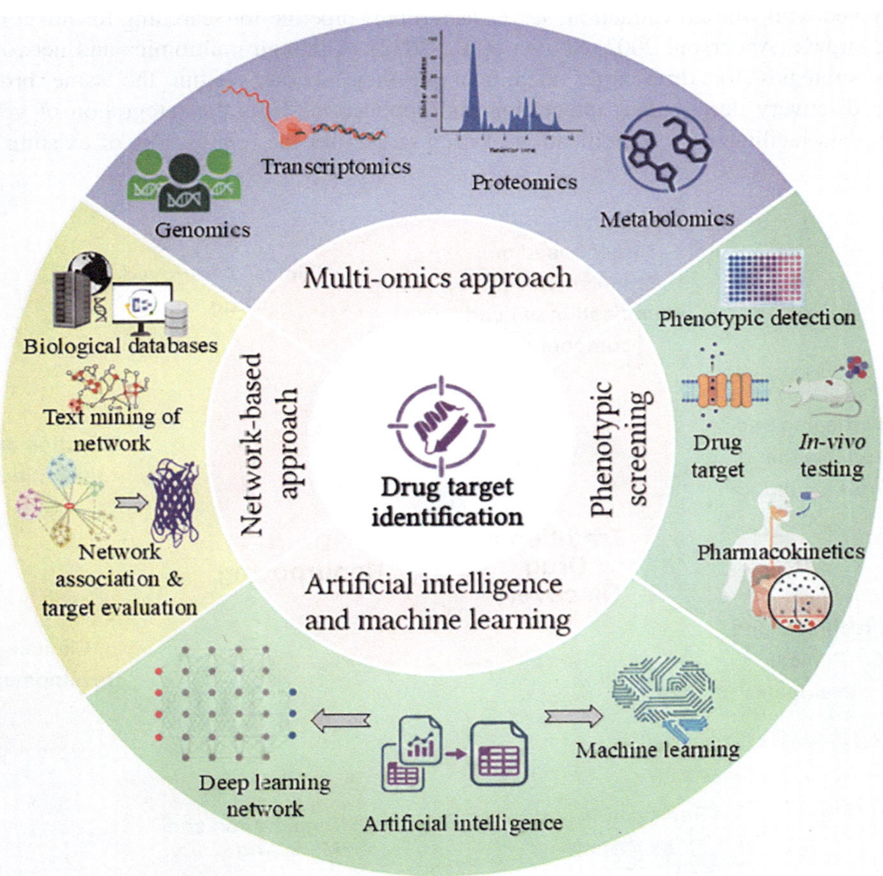

FIGURE 21.1 **Various approaches to drug-target identification** . Overview of the various approaches used in drug-target identification, including genomics-based methods, proteomics analyses, bioinformatics and network pharmacology, phenotypic screening, and systems biology approaches. These complementary strategies integrate experimental and in silico techniques to identify, validate, and prioritise biologically relevant molecular targets for therapeutic intervention.

(Projan and Shlaes, 2004; Begley et al., 2021). In 2015 the market for repositioning drugs was placed at $24.4 billion by the pharmaceutical industry, which was reported to increase to $31.1 billion in 2020 (Krishnamurthy et al., 2022). Drug repurposing offers several advantages over traditional methods, such as reducing the time required for development of the drug and lowering the risk of failure, as these repurposed drugs exhibit a favourable pharmacokinetic profile, and their preclinical trials have already been conducted (Begley et al., 2021). The reprofiled drugs are cost-effective as the investment required for different phases of clinical trials, especially phase III and regulatory phases, is comparatively higher for newly developed drugs (Krishnamurthy et al., 2022; Boyd et al., 2021). Thus repurposing of existing drugs is less risky and offers a more rapid return compared to the traditional drug discovery approach (Fig. 21.2. During repurposing, a drug has to cross three stages, including identification of a potential drug candidate against a specific target, evaluation of the effectiveness of the identified drug candidate in a preclinical model, and finally assessment of phase III clinical trials to confirm the efficiency of the drug molecule (Krishnamurthy et al., 2022).

21.3.1 Target-centric drug repurposing strategies

The traditional methods of drug discovery are time-consuming, and these require 20–25 years from identification to clinical approval of a drug, with a cost of approximately $350 million (Shulga et al., 2022). With the ever-increasing antibiotic resistance and growing demand for novel therapeutics, computational drug design is drawing greater attention due to its high efficiency in identifying drug-like molecules with high selectivity, low toxicity, and desirable pharmacokinetic properties in a very short time (Maia et al., 2020; Batool et al., 2019). Computational drug-target identification, combined with wet lab validation, serves as a robust pipeline for searching for novel medicines for known and unknown drug targets (Anderson, 2003; Shulga et al., 2022). Although multiomics and network biology are often treated as distinct strategies for drug-target identification, they operate within the same broader framework of computational drug discovery due to their integration and application. Thus the integration of genomics, proteomics, and network-driven data facilitates the identification of drug targets for the repurposing of existing drugs.

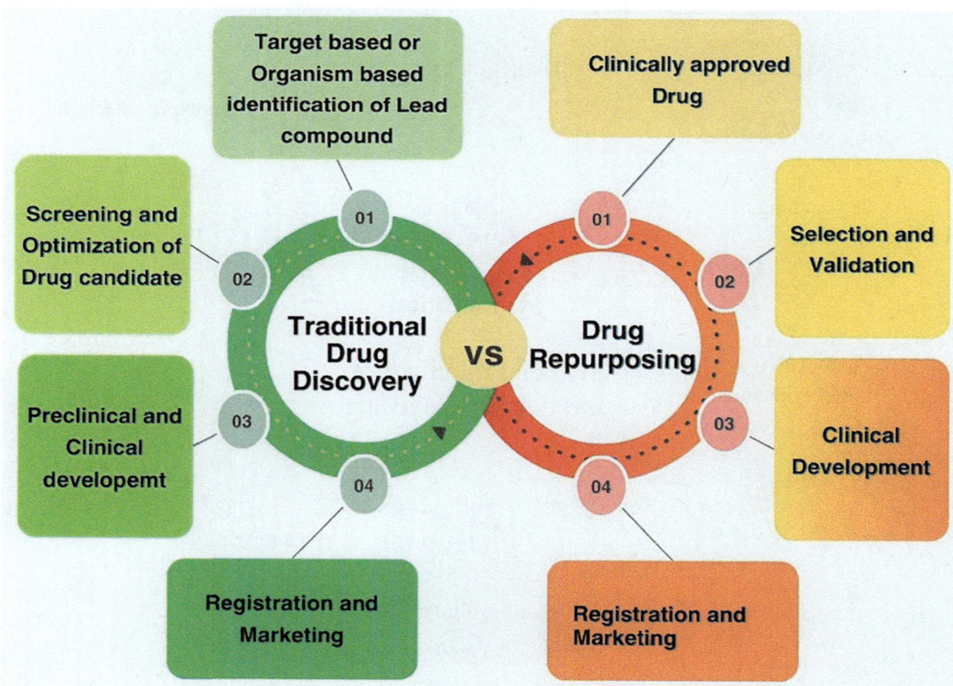

FIGURE 21.2 **Key steps of traditional drug discovery and drug repurposing pipeline** . Overview of the key stages in the pharmaceutical development process, highlighting the sequential steps of traditional drug discovery, from target identification and lead optimisation to preclinical studies and clinical trials, alongside the drug repurposing pipeline.

21.3.1.1 Computer-aided drug designing (CADD)

Natural products are the primary source of most chemical entities used in medicine, including antibiotics. The traditional methods of identifying lead molecules for the discovery and development of drugs are challenging due to the complexities involved in isolation, characterisation, and standardisation (Fortman and Mukhopadhyay, 2016). With the recent advancements in omics sciences and computational biology, CADD tools serve as a promising alternative that robustly screens large databases, identifies drug-like molecules, and predicts their pharmacokinetic properties in a short time (Table 21.1. Additionally, it reduces the costs required for the development of a drug-like compound and reduces the chances of failure in clinical trials (Oselusi et al., 2024). With the ever-increasing antibiotic resistance and growing

TABLE 21.1 Databases that are widely used in the computer-aided drug designing (*CADD*) approach for drug-target identification.

Database type	Name of the database	Description	References
Metabolic Pathway Databases	KEGG Pathway Database	An integrated database that contains information on metabolic, molecular, cellular, and regulatory pathways. This database provides detailed information at the molecular level and is widely employed for mapping metabolism, cellular processes, human diseases, genetic information processing, and drug development.	https://www.genome.jp/kegg/
	Pathway Commons	A publicly available integrated database of biological pathways. This database stores and integrates molecular interactions data from 22 public databases.	https://www.pathwaycommons.org/
	MetaCyc	A database of metabolic pathways that holds information on the primary and secondary metabolic pathways from all domains of life. Presently, it has information on 3,105 pathways, 18,973 metabolites, and 18,566 reactions.	https://metacyc.org/
	HumanCyc	It is a curated database that stores information on human metabolic pathways, enzymatic reactions, the human genome, and metabolites	https://humancyc.org/
	BioCyc	A collection of 20,043 pathways that hold information on metabolites and genomes of eukaryotes and microbes	https://biocyc.org/
	Reactome	A freely available database that stores information on biological pathways and provides tools for analysing the mechanisms and small molecules involved in the process	https://reactome.org/
Interaction Analysis Databases	BIND	The Bimolecular Interaction Network Database (BIND) is a tool for analysing protein interactions and other molecular pathways	https://bio.tools/bind
	STRING	It is a database and computational tool for the prediction of gene-gene or protein–protein interaction networks. The tool generates interaction between the genes or proteins based on information obtained from lab experiments, coexpression, genomic context prediction, automated text mining, and previous knowledge in databases.	https://string-db.org/
	BioGRID	The Biological General Repository for Interaction Datasets (BioGRID) is a freely available database for analysis of genomic and protein interactions, especially in human and model organisms. It holds more than 1,740,000 interactions at present.	https://thebiogrid.org/

demand for novel therapeutics, CADD is drawing greater attention due to its higher efficiency in identifying drug-like molecules with high selectivity, low toxicity, and desirable pharmacokinetic properties in a very short time (Maia et al., 2020). CADD, combined with wet lab validation, serves as a robust pipeline for identifying novel medicines for both known and unknown drug targets (Anderson, 2003). For the identification of the lead compounds through the CADD approach, numerous compound databases are screened, and the top hits with the highest binding affinity against the drug targets are identified through molecular docking, which are further selected for in vitro testing, preclinical, and clinical trials to receive approval from the regulatory bodies (Bruch et al., 2020). The advantage of this process lies in its ability to identify the top lead molecules supported by scoring functions, which reduces the number of compounds to be tested experimentally, thereby increasing the success rate (Shulga et al., 2022).

Additionally, a structure-based CADD approach requires minimal knowledge of compound structure; however, it identifies several lead compounds with known chemical structure against the drug targets (van Montfort et al., 2017). However, identifying promising lead compounds with a unique mode of action to combat antibiotic resistance in bacteria is challenging, as bacteria are highly adaptive organisms that continue to evolve and find ways to tolerate antimicrobial agents. Therefore integrating the CADD tool and in vitro validation can serve as a transformative solution that may accelerate the development of effective therapeutics to combat drug-resistant pathogens. Computational models can be used for repurposing nonantibiotic drugs for antimicrobial therapy using various approaches, including molecular docking, quantitative structure–activity relationship, ML, molecular dynamics simulation, as well as genomics and proteomics-based approaches.

21.3.1.2 Molecular docking

With the emerging antibiotic resistance in bacteria and the constant evolution of resistance genes, the discovery of antimicrobial agents capable of combating resistant pathogens has become increasingly complex. The Structure-based Drug Discovery (SBDD) approach is a promising alternative for the discovery of antimicrobial therapeutics, which identifies high-affinity small molecules against a druggable target with a known 3D structure (Oselusi et al., 2024). In this process, the preferred orientation of the ligands and their binding energies with active site residues of proteins are predicted using molecular docking. Molecular docking employs several scoring functions to identify the hit compounds, which are based on several parameters (Aparoy et al., 2012). It predicts the binding affinity of the protein–ligand complex, which in turn is influenced by the type of amino acid residues present in the active site, the chemical nature of the drug-like compound, molecular weight, and the number and nature of interactions (Raval and Ganatra, 2022). The pharmacological potential of the small molecules depends on various physicochemical properties such as water solubility, partition coefficient (LogP), bioconcentration factor (BCF), melting point (MP), and boiling point (BP) (Alam and Khan, 2018).

In AMR research, the SBDD tool has been widely employed for high-throughput screening of compound libraries for the identification of antimicrobial compounds with high affinity against the resistance genes (Wang et al., 2016). In drug repurposing, molecular docking is widely employed for high-throughput screening of large libraries of approved and failed drug candidates against one or multiple targets (Meng et al., 2011). Several such databases store information on various drugs to support drug repurposing. For instance, DrugBank, DrugRepoBank, and repoDB are repository of drugs and drug-like molecules that stores information on approved and failed drugs and their indications (Huang et al., 2024). These databases are robustly screened against the active site of the target using molecular docking tools to identify probable drug candidates from existing drugs. Some of the databases used as a compound library in drug repurposing are listed in Table 21.2.

21.3.1.3 MD simulation

The computational prediction of interactions between the protein–ligand complexes in molecular docking may not be stable in biological systems due to various physiological parameters such as pressure, temperature, and the presence of water molecules. These factors interfere with the bond stability and, therefore, may work differently in biological systems. Thus the interaction between the protein–ligand complexes is simulated in an artificial system resembling the biological system with specific standard parameters. In MD simulation, the protein–ligand complexes are subjected to energy minimisation, placed in a solvent system, and temperature and pressure are applied to evaluate the stability of the bonds under different conditions. In MD simulation studies, any macromolecule can be simulated alone or with small molecules. Analysis of their interactions can offer insights into the changes in the protein chains with respect to changes in physiological factors and the stability of interactions between the two molecules. In drug repurposing studies, MD simulation can be helpful to validate the interaction of the repurposed drug with a specific drug target. Molecular docking

TABLE 21.2 List of databases that store information on approved drugs and drug-like molecules.

Database	Description	Estimated size
CEMTDD	Chinese Ethnic Minority Traditional Drug for the discovery and optimisation of lead compounds	4060
DrugBank	A comprehensive database that stores information on drugs and drug targets	13,600 compounds, 25,886 annotations
ChEBI	Chemical Entities of Biological Interest is a chemical database of small chemical compounds	15,736
ChEMBL	Chemicals from the European Molecular Biology Laboratory, a curated database of bioactive molecules with drug-like properties, bringing together chemical, bioactivity, and genomic data for the development of effective new drugs	2,399,743
ChemDB	A chemical database of small molecules, founded in 2005	5 M commercially available small molecules
KEGG Compound Database	Database for chemical structures integrated with genomics	19,270
DRDOCK	A database for the repurposing of FDA-approved drugs	Approved drugs of 2016
PubChem	World's largest collection of freely accessible chemical information databases, maintained by the National Library of Medicine, National Institute of Health	-
ZINC	Free database of commercially available compounds for virtual screening, provided by Irwin and Shoichet Laboratories, University of California	Contains over 750 million purchasable compounds and over 85,190 Natural compounds
RepoDB	A database that contains information on approved and failed drugs	8931 approved drugs

and simulations enable the assessment of binding affinity, interaction stability, and molecular compatibility prior to experimental validation. In a study, a total of 8815 approved drugs were computationally screened against three selected proteins of MDR *Candida albicans*, where arbutamine-TRR1 was found to be the best interactor of YHB1 protein, highlighting its potential as an antifungal drug (Verma and Aggarwal, 2021). The primary steps involved in any molecular dynamics simulation of a protein–ligand complex are illustrated in Fig. 21.3

At the starting point, it is crucial to ensure the absence of unwanted heteroatoms in the system, and the ligands should be docked appropriately with the active site residues of the proteins. Several force fields, such as AMBER, OPLS-AA, CHARMM, and GROMOS87/96, are applied for the preparation of proteins and ligands. For ligand topology prediction, several tools such as acpype, CGenFF, Antechamber, PRODRG, Topolbuild, LigParGen, and ATB can also be used. After the generation of topology files for the system, a box with a proper size and shape is defined and visualised using tools such as VMD to monitor the changes at each step of the run. Once the solvent-filled box covers the complex, ions are added to the solvated system to neutralise any net charge. Energy minimisation in an equilibrated system is a crucial step in this process. The prepared equilibrated system is then subjected to MD simulation for a definite time, which depends on the size of the system as well as the capacity of the computational tool. Various analyses are performed to understand the changes in the simulated system, which provide detailed information on hydrogen bond formation, binding energy, solvent accessible surface area, root mean square deviation, radius of gyration, root mean square fluctuation of residues, etc.

21.3.2 Drug repurposing in combating antimicrobial resistance

Drug repurposing has recently gained so much attention as a strategic approach in identifying approved or investigational drugs for new therapeutic use beyond their original medical indication (Talat et al., 2022). AMR has been recognised as a global public health threat and is going to be responsible for 10 million deaths annually by 2050 (de Kraker et al., 2016). A

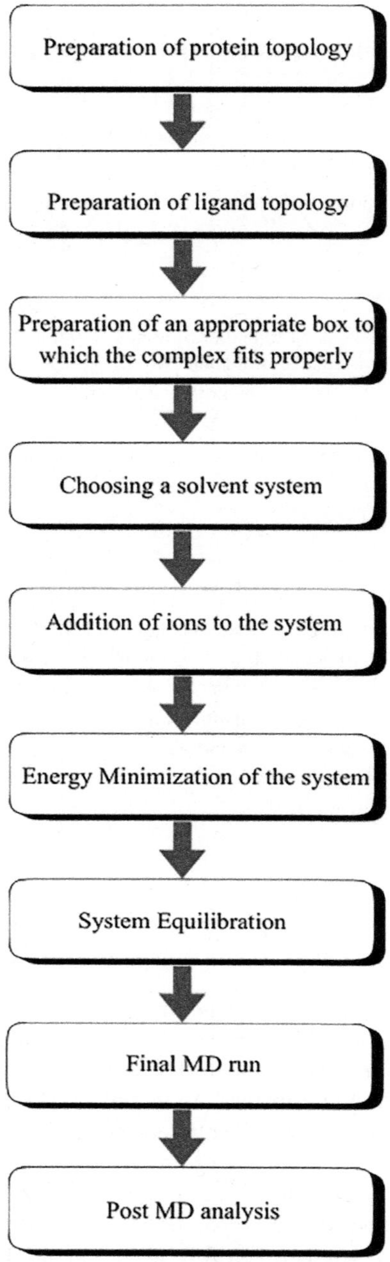

FIGURE 21.3 **Steps for molecular dynamics (*MD*) simulation of protein–ligand complex** . The workflow illustrating the standard steps involved in MD simulation of a protein–ligand complex. The process begins with protein and ligand structure preparation, followed by force field parameterisation and topology generation. The MD run is subsequently carried out to generate trajectory data, which is analysed using structural and energetic parameters such as RMSD, RMSF, hydrogen bonding, and binding free energy calculations to evaluate stability and interaction dynamics. *MD*, Molecular dynamics.

recent study has analysed the global burden of AMR from 1990 to 2021, and this also forecasted a large number of deaths by 2050 due to bacterial AMR (Naghavi et al., 2024). Therefore drug repurposing could be a significant solution for controlling the AMR threat, particularly in the current era of limited availability of effective antibiotics. Moreover, drug repurposing offers several advantages, as it bypasses extensive Phase I safety studies of drug discovery by providing preexisting safety and clinical data and allowing the drug to enter later phases of clinical trials directly. This reduces both the time and costs involved in the process of developing new therapeutics (Kaul et al., 2019). Drug repurposing also increases the potential of drugs by reducing their toxicity level and thus supports the utilisation of the drugs as a new therapeutic option (Dubey et al., 2020).

With the uncertainties of effective antibiotic development, especially against pan-drug-resistant (PDR) organisms, drug repurposing holds immense potential for the discovery of next-generation antibiotics (Pushpakom et al., 2018; Rossolini et al., 2014). Drug repurposing can serve as a rapidly emerging alternative to combat the increasing resistance in pathogens (Latif et al., 2022). The National Centre for Advancing Translational Sciences (NCATS) launched 'Discovering New Therapeutic Uses for Existing Molecules' in 2012, a drug repurposing programme to find alternative therapeutics for human diseases with urgent clinical need (Wang et al., 2012).

Although drug reprofiling is widely used in AMR research to address the growing resistance, few drugs have been repurposed and approved for the treatment of fungal and protozoal infections (Sahoo et al., 2021). For example, antimalarial drug pyrimethamine has been repurposed for the treatment of toxoplasmosis, and chloroquine has received approval for the treatment of amoebiasis (Sahoo et al., 2021). Similarly, amphotericin B has received approval for treating visceral leishmaniasis. However, no drug has received approval to date as an antibiotic for the treatment of bacterial infections. Nevertheless, few antibiotics have been employed for the treatment of other diseases, such as spiramycin for toxoplasmosis, paromomycin for visceral leishmaniasis, and doxycycline for malaria (Begley et al., 2021; Farha and Brown, 2019). Similarly, azithromycin and clarithromycin were used to treat COVID-19 patients during the COVID-19 pandemic. Azithromycin was found to be effective when given in combination with hydroxychloroquine (Yacouba et al., 2021). Anticancer drugs possess promising antibacterial activity and are therefore used for treating several bacterial infections. For example, 5-fluorouracil (5-FU), an anticancer drug, showed excellent antibacterial activity against drug-resistant *E. coli*, *P. aeruginosa*, and *S. aureus*. Mitomycin C, another anticancer drug, showed notable antibacterial properties against *S. aureus*, *A. baumannii*, *E. coli*, and *P. aeruginosa*(Cruz-Muñiz et al., 2017).

The existing drugs that exhibit in vitro antibacterial activity against several bacterial pathogens are considered as 'nonantibiotics' (Boyd et al., 2021). Combining antibiotics with repurposed nonantibiotic compounds has been reported to inhibit the bacterial resistance determinants or enhance the antibiotic activity, which offers a sustainable and strategic approach to confront MDR bacteria. These drug–drug interactions may be synergistic, show no interaction, or may be antagonistic (Liu et al., 2018). Anthracyclines are known to have bacteriostatic and bactericidal effects against different MDR bacteria. In a recent study, anthracyclines, particularly mitoxantrone (MX) and colistin under tissue medium RPMI, were found to exhibit strong synergistic effects against *mcr*-positive *E. coli*, methicillin-resistant *Staphylococcus aureus*, and *Salmonella enteritidis* by interrupting iron homoeostasis (She et al., 2023b). Drug repurposing with nonantibiotic agents, for example, a combination of tetracycline and erythromycin with ivermectin, and amoxicillin/clavulanic acid with flunixin, showed remarkable potential against *S. aureus* inhibition (Shafique et al., 2025). Curcumin, an anticancer drug, combined with ciprofloxacin, was found to exhibit potential antibacterial activity against *S. aureus* and *P. aeruginosa* (Kamurai et al., 2020). Infections caused by drug-resistant *Neisseria gonorrhoeae* are a critical health concern. Only a few antibiotics, such as azithromycin and expanded-spectrum cephalosporins, remain effective against this pathogen. According to reports, dichlorophenamide, acetazolamide, and ethoxzolamide have been repurposed for gonorrhoea treatment, whereas zoliflodacin and gepotidacin, novel bacterial gyrase inhibitors, are under clinical trial (Pisano et al., 2024).

Another FDA-approved drug, niclosamide, which was earlier used to treat tapeworm infections, has recently been repurposed against Gram-positive *Staphylococcus* and *Streptococcus* due to its ability to inhibit α-haemolysin enzyme (Zhang et al., 2022). Target-based phenotypic screening methods also identified this drug as a potent inhibitor of *P. aeruginosa*, inhibiting the production of acyl-homoserine lactone, quorum-sensing signalling molecules, and regulating virulence factors (Farha and Brown, 2019). Several FDA-approved drugs that have been repurposed against AMR bacteria are listed in Table 21.3.

21.4 Possibilities and limitations of target-centric drug repurposing

Target-centric drug repurposing offers a promising and systematic approach to address the urgent need for alternative therapeutics to deal with emerging AMR. Drug repurposing identifies several nondrug compounds that have intrinsic inhibitory activity against both common and rare diseases (Pushpakom et al., 2018). Target-centric drug repurposing facilitates the development of combinatorial treatment therapies by precise targeting of specific disease-driven pathways. Metabolomic profiling of molecular pathways associated with disease provides an initial framework for target screening. Thus this enables subsequent evaluation of the repurposed drug in a more reduced timeline and efficient manner. It also simplifies discrepancies in interacting compounds and their biological effects (Tanoli et al., 2018).

TABLE 21.3 Some FDA-approved drugs that have been repurposed for treating different antimicrobial resistance (*AMR*) bacterial infections.

Drug name	Approved use	Target pathogens	Mechanisms of action	Works in combination or alone	Evidence type	References
Tyrothricin	Antibiotic for Gram-positive bacteria	*Acinetobacter baumannii*	-		In vitro	Chopra et al. (2010)
Ciclopirox	Antifungal	*Escherichia coli*	Affects galactose metabolism and lipopolysaccharide (LPS) biosynthesis	Alone	In vitro	Carlson-Banning et al. (2013)
DP7	Antimicrobial	*Staphylococcus aureus*	-	Azithromycin and vancomycin	In vitro	Wu et al. (2017)
Fluspirilene, Bay 11–7082	Anticancer	*baumannii*	-	Azithromycin and colistin	In vitro	Cheng et al. (2019)
Metformin	Antidiabetic drug	*Bacillus pumilus, P. aeruginosa and S. aureus*	Binds to β-lactamases and penicillin-binding proteins (PBPs)		In silico and in vitro	Kapoor et al. (2020)
Amitriptyline	Antidepressant					
Ellipticine hydrochloride (EH)	Anticancer	*E. coli*	Binds to *E. coli* topoisomerase IV, reduces the level of pro-inflammatory factors, such as TNF-a and IL-6	Colistin	In vitro and in vivo	Lu et al. (2020)
Docetaxel	Anticancer drug	*E. coli*	Inhibits biofilm formation, reduces bacterial load in urine, kidney, and bladder, and blocks histdinol phosphate aminotransferase (HisC) of uropathogenic *E. coli*	Alone	In silico, in vitro, and in vivo	Kaur et al. (2022)
Auranofin	Antiprotozoal	*K. pneumoniae*	Pentamidine disrupts the bacterial membrane, enhances permeability to auranofin, and thus improves the antibacterial activity of auranofin	Pentamidine	In vitro	Yu et al. (2022)
Tamoxifen	Breast cancer therapeutic	*Mycobacterium tuberculosis* (*Mtb*)	Increases lysosomal activation, and hence empowers the host to control intracellular infections by *Mtb*	Alone	In vitro and in vivo	Boland et al. (2023)
9-Aminoacidine (9AA)	Anticancer	*K. pneumoniae*	Interacts with specific bacterial DNA and disrupts the proton motive force in *K. pneumoniae*	Rifampin	In silico	She et al. (2023a)
Tafenoquine (TAF)	Antimalarial	*S. aureus*	Disrupts bacterial cell membrane	Alone	In silico, in vitro, and in vivo	
Teniposide	Antitumor drug	*S. epidermis*	Affects biofilm viability	Alone	In vitro	Dell'Annunziata et al. (2024)

Despite these advantages, target-centric drug repurposing has several pitfalls that complicate the translation of in silico findings to clinical practices. There might be some mechanistic uncertainties that persist as a critical barrier in developing novel therapeutics. Several drug candidates lack well-defined molecular targets or fail to penetrate effectively against biofilm and MDR Gram-negative bacteria due to outer membrane impermeability and active efflux pump systems (Pushpakom et al., 2018). Moreover, interspecies variations in pharmacokinetics and immune response to microbial pathogens appear as translational obstacles, which weaken the ability of preclinical models to accurately forecast the efficacy of humans or targeted organisms (Theuretzbacher et al., 2023; Li et al., 2025). Some repurposed drugs, initially designed for specific indications, may not perfectly satisfy the requirements of target product profiles (TPP) in a new therapeutic context. Such misalignments can lead to challenges in pharmacology, safety, formulation, and cost to prioritise the existing molecule over the TPP (Hamid et al., 2024). Although many researchers are starting to rely on drug repurposing for its cost-effectiveness, another group of scientists considers it a high-risk investment. An already approved drug has an established safety profile, which may allow phase I and II trials to be skipped; however, earlier data might not be efficient when repurposing the same drug for new indications (Sharma et al., 2021). Thus rigorous experimental validations and translational frameworks are necessary to overcome these challenges and to ensure the effective clinical use of the repurposed drug candidates.

References

Ahmed, S.K., Hussein, S., Qurbani, K., Hussein Ibrahim, R., Fareeq, A., Mahmood, K.A., et al., 2024. Antimicrobial resistance: Impacts, challenges, and future prospects. Journal of Medicine, Surgery, and Public Health 2, 100081. Available at https://doi.org/10.1016/j.glmedi.2024.100081.

Alam, S., Khan, F., 2018. Virtual screening, docking, ADMET and system pharmacology studies on *Garcinia* caged xanthone derivatives for anticancer activity. Scientific Reports 8 (1) Available from. Available at https://doi.org/10.1038/s41598-018-23768-7, www.nature.com/srep/index.html.

Anderson, A.C., 2003. The process of structure-based drug design. Chemistry & Biology 10 (9), 787–797. Available at https://doi.org/10.1016/j.chembiol.2003.09.002.

Aparoy, P., Kumar Reddy, K., Reddanna, P., 2012. Structure and Ligand based drug design strategies in the development of novel 5- LOX inhibitors. Current Medicinal Chemistry 19 (22), 3763–3778. Available at https://doi.org/10.2174/092986712801661112.

Barh, D., Tiwari, S., Jain, N., Ali, A., Santos, A.R., Misra, A.N., et al., 2011. In silico subtractive genomics for target identification in human bacterial pathogens. Drug Development Research 72 (2), 162–177. Available at https://doi.org/10.1002/ddr.20413.

Batool, M., Ahmad, B., Choi, S., 2019. A structure-based drug discovery paradigm. International Journal of Molecular Sciences 20 (11), 2783. Available at https://doi.org/10.3390/ijms20112783.

Baxevanis, A.D., Francis Ouellette, B.F., 2002. Information retrieval from biological databases. Wiley, Bioinformatics: A Practical Guide to the Analysis of Genes and Proteins 43, 155–185. Available at https://doi.org/10.1002/0471223921.ch7.

Begley, C.G., Ashton, M., Baell, J., Bettess, M., Brown, M.P., Carter, B., et al., 2021. Drug repurposing: misconceptions, challenges, and opportunities for academic researchers. Science Translational Medicine 13 (612) Available from. Available at https://doi.org/10.1126/scitranslmed.abd5524, https://www.science.org/doi/10.1126/scitranslmed.abd5524.

Berdigaliyev, N., Aljofan, M., 2020. An overview of drug discovery and development. Future Medicinal Chemistry 12 (10), 939–947. Available from. Available at https://doi.org/10.4155/fmc-2019-0307, http://www.future-science.com/loi/fmc.

Boland, R., Heemskerk, M.T., Forn-Cuní, G., Korbee, C.J., Walburg, K.V., Esselink, J.J., et al., 2023. Repurposing tamoxifen as potential host-directed therapeutic for tuberculosis. American Society for Microbiology, Netherlands mBio 14 (1) Available from. Available at https://doi.org/10.1128/mbio.03024-22, https://journals.asm.org/doi/10.1128/mbio.03024-22.

Boucher, H.W., Talbot, G.H., Bradley, J.S., Edwards, J.E., Gilbert, D., Rice, L.B., et al., 2009. Bad bugs, no drugs: No ESKAPE! An update from the infectious diseases society of America. Clinical Infectious Diseases 48 (1), 1–12. Available at https://doi.org/10.1086/595011.

Boyd, N.K., Teng, C., Frei, C.R., 2021. Brief overview of approaches and challenges in new antibiotic development: a focus on drug repurposing. Frontiers in Cellular and Infection Microbiology 11 Available from. Available at https://doi.org/10.3389/fcimb.2021.684515, http://www.frontiersin.org/Cellular_and_Infection_Microbiology/archive.

Bruch, E.M., Petrella, S., Bellinzoni, M., 2020. Structure-based drug design for tuberculosis: challenges still ahead. Applied Sciences 10 (12), 4248. Available at https://doi.org/10.3390/app10124248.

Cao, C., Moult, J., 2014. GWAS and drug targets. BMC Genomics 15 (S4). Available at https://doi.org/10.1186/1471-2164-15-s4-s5.

Carlson-Banning, K.M., Chou, A., Liu, Z., Hamill, R.J., Song, Y., Zechiedrich, L., 2013. Toward repurposing ciclopirox as an antibiotic against drug-resistant *Acinetobacter baumannii*, *Escherichia coli* and *Klebsiella pneumoniae*. PLoS One 8 (7) Available from. Available at https://doi.org/10.1371/journal.pone.0069646, http://www.plosone.org/article/fetchObjectAttachment.action;jsessionid=DFA8DC717C194F025C5709F30C8A5C09?uri=info%3Adoi%2F10.1371%2Fjournal.pone.00669646&representation=PDF.

Cheng, Y.S., Sun, W., Xu, M., Shen, M., Khraiwesh, M., Sciotti, R.J., et al., 2019. Repurposing screen identifies unconventional drugs with activity against multidrug resistant *Acinetobacter baumannii*. Frontiers in Cellular and Infection Microbiology 9 Available from. Available at https://doi.org/10.3389/fcimb.2018.00438, http://www.frontiersin.org/Cellular_and_Infection_Microbiology/archive.

Chinemerem Nwobodo, D., Ugwu, M.C., Oliseloke Anie, C., Al-Ouqaili, M.T.S., Chinedu Ikem, J., Victor Chigozie, U., et al., 2022. Antibiotic resistance: the challenges and some emerging strategies for tackling a global menace. Journal of Clinical Laboratory Analysis 36 (9) Available from. Available at https://doi.org/10.1002/jcla.24655, http://onlinelibrary.wiley.com/journal/10.1002/(ISSN)1098-2825.

Chopra, S., Torres-Ortiz, M., Hokama, L., Madrid, P., Tanga, M., Mortelmans, K., et al., 2010. Repurposing FDA-approved drugs to combat drug-resistant *Acinetobacter baumannii*. Journal of Antimicrobial Chemotherapy 65 (12), 2598–2601. Available at https://doi.org/10.1093/jac/dkq353.

Cruz-Muñiz, M.Y., López-Jacome, L.E., Hernández-Durán, M., Franco-Cendejas, R., Licona-Limón, P., Ramos-Balderas, J.L., et al., 2017. Repurposing the anticancer drug mitomycin C for the treatment of persistent *Acinetobacter baumannii* infections. International Journal of Antimicrobial Agents 49 (1), 88–92. Available from. Available at https://doi.org/10.1016/j.ijantimicag.2016.08.022, www.elsevier.com/locate/ijantimicag.

Davies, J., 1996. Origins and evolution of antibiotic resistance. Microbiología (Madrid, Spain) 12 (1), 9–16.

de Kraker, Marlieke E.A., Stewardson, Andrew J., Harbarth, Stephan, 2016. Will 10 Million People Die a Year due to Antimicrobial Resistance by 2050? PLOS Medicine 13, e1002184. Available at 10.1371/journal.pmed.1002184.

Dell'Annunziata, F., Folliero, V., Marca, R.D., Palma, F., Sanna, G., De Filippis, A., et al., 2024. Repurposing the antibacterial activity of the drug teniposide against gram-positive bacteria. Cellular Microbiology 2024 Available from. Available at https://doi.org/10.1155/2024/9389729, https://www.hindawi.com/journals/cmi/.

Dickson, M., Gagnon, J.P., 2004. Key factors in the rising cost of new drug discovery and development. Nature Reviews. Drug Discovery 3 (5), 417–429. Available from. Available at https://doi.org/10.1038/nrd1382, http://www.nature.com/nrd/index.html.

Dogiparthi, L.K., Bukke, S.P.N., Thalluri, C., Thalamanchi, B., Vidya, K.P., Sree, G.N., et al., 2025. The role of genomics and proteomics in drug discovery and its application in pharmacy. Discover Applied Sciences 7 (6) Available from. Available at https://doi.org/10.1007/s42452-025-07155-2, https://link.springer.com/journal/42452.

Drawz, S.M., Bonomo, R.A., 2010. Three decades of β-lactamase inhibitors. Clinical Microbiology Reviews 23 (1), 160–201. Available from. Available at https://doi.org/10.1128/CMR.00037-09, http://cmr.asm.org/cgi/reprint/23/1/160.

Dubey, K.K., Indu, Sharma, M., 2020. Reprogramming of antibiotics to combat antimicrobial resistance. Archiv der Pharmazie 353 (11) Available from. Available at https://doi.org/10.1002/ardp.202000168, http://onlinelibrary.wiley.com/journal/10.1002/(ISSN)1521-4184.

Farha, M.A., Brown, E.D., 2019. Drug repurposing for antimicrobial discovery. Nature Microbiology 4 (4), 565–577. Available from. Available at https://doi.org/10.1038/s41564-019-0357-1, www.nature.com/nmicrobiol/.

Fatoba, A.J., Okpeku, M., Adeleke, M.A., 2021. Subtractive genomics approach for identification of novel therapeutic drug targets in *Mycoplasma genitalium*. Pathogens 10 (8) Available from. Available at https://doi.org/10.3390/pathogens10080921, https://www.mdpi.com/2076-0817/10/8/921/pdf.

Ferreira, F.J.N., Carneiro, A.S., 2025. AI-driven drug discovery: a comprehensive review. ACS Omega 10 (23), 23889–23903. Available at https://doi.org/10.1021/acsomega.5c00549 Available from: pubs.acs.org/journal/acsodf.

Ferreira, R.L., Da Silva, B.C.M., Rezende, G.S., Nakamura-Silva, R., Pitondo-Silva, A., Campanini, E.B., et al., 2019. High prevalence of multidrug-resistant klebsiella pneumoniae harboring several virulence and β-lactamase encoding genes in a brazilian intensive care unit. Frontiers in Microbiology 10 Available from. Available at https://doi.org/10.3389/fmicb.2018.03198, https://www.frontiersin.org/journals/microbiology#.

Fortman, J.L., Mukhopadhyay, A., 2016. The future of antibiotics: emerging technologies and stewardship. Trends in Microbiology 24 (7), 515–517. Available from. Available at https://doi.org/10.1016/j.tim.2016.04.003, www.elsevier.com/locate/tim.

Gupta, R., Srivastava, D., Sahu, M., Tiwari, S., Ambasta, R.K., Kumar, P., 2021. Artificial intelligence to deep learning: machine intelligence approach for drug discovery. Molecular Diversity 25 (3), 1315–1360. Available from. Available at https://doi.org/10.1007/s11030-021-10217-3, www.wkap.nl/journalhome.htm/1381-1991.

Hamid, A., Mäser, P., Mahmoud, A.B., 2024. Drug repurposing in the chemotherapy of infectious diseases. Molecules (Basel, Switzerland) 29 (3), 635. Available at https://doi.org/10.3390/molecules29030635.

Hosen, M.I., Tanmoy, A.M., Mahbuba, D.A., Salma, U., Nazim, M., Islam, M.T., et al., 2014. Application of a subtractive genomics approach for in silico identification and characterization of novel drug targets in *Mycobacterium tuberculosis* F11. Association of Scientists in the International Association of Scientists in the, Bangladesh Interdisciplinary Sciences – Computational Life Sciences 6 (1), 48–56. Available from. Available at https://doi.org/10.1007/s12539-014-0188-y, http://www.springer.com/life+sciences/systems+biology+and+bioinformatics/journal/12539.

Huang, Y., Dong, D., Zhang, W., Wang, R., Lin, Y.C.D., Zuo, H., et al., 2024. DrugRepoBank: a comprehensive database and discovery platform for accelerating drug repositioning. Database 2024 Available from. Available at https://doi.org/10.1093/database/baae051, http://database.oxfordjournals.org/.

Hukerikar, N., Hingorani, A.D., Asselbergs, F.W., Finan, C., Schmidt, A.F., 2024. Prioritising genetic findings for drug target identification and validation. Atherosclerosis 390 Available from. Available at https://doi.org/10.1016/j.atherosclerosis.2024.117462, https://www.sciencedirect.com/science/journal/00219150.

Indrajith, Chowdhury, Meghanathan, Gopal, Muthupandian, 2021. Molecular insights of Carbapenem resistance *Klebsiella pneumoniae* isolates with focus on multidrug resistance from clinical samples. Journal of Infection and Public Health 14, 131–138.

Iredell, J., Brown, J., Tagg, K., 2016. Antibiotic resistance in Enterobacteriaceae: mechanisms and clinical implications. BMJ (Online) 352 Available from. Available at https://doi.org/10.1136/bmj.h6420, http://www.bmj.com/content/bmj/352/bmj.h6420.full.pdf.

Jakubec, D., Skoda, P., Krivak, R., Novotny, M., Hoksza, D., 2022. PrankWeb 3: accelerated ligand-binding site predictions for experimental and modelled protein structures. Nucleic Acids Research 50 (W1), W593. Available at https://doi.org/10.1093/nar/gkac389.

Kamurai, B., Mombeshora, M., Mukanganyama, S., 2020. Repurposing of drugs for antibacterial activities on selected ESKAPE bacteria *Staphylococcus aureus* and *Pseudomonas aeruginosa*. International Journal of Microbiology 2020 Available from. Available at https://doi.org/10.1155/2020/8885338, http://www.hindawi.com/journals/ijmb/.

Kapoor, Y., Sharma, R., Kumar, A., 2020. Repurposing of existing drugs for the bacterial infections: an in silico and in vitro study. Infectious Disorders - Drug Targets 20 (2), 182–197. Available from. Available at https://doi.org/10.2174/1871526519666181126094244, http://www.eurekaselect.com/node/167716.

Kaul, R., Risinger, A.L., Mooberry, S.L., 2019. Microtubule-targeting drugs: more than antimitotics. Journal of Natural Products 82 (3), 680–685. Available from. Available at https://doi.org/10.1021/acs.jnatprod.9b00105, http://pubs.acs.org/journal/jnprdf.

Kaur, H., Kalia, M., Chaudhary, N., Singh, V., Yadav, V.K., Modgil, V., et al., 2022. Repurposing of FDA approved drugs against uropathogenic *Escherichia coli*: In silico, in vitro, and in vivo analysis. Microbial Pathogenesis 169, 105665. Available at https://doi.org/10.1016/j.micpath.2022.105665.

Khan, K., Alhar, M.S.O., Abbas, M.N., Abbas, S.Q., Kazi, M., Khan, S.A., et al., 2022. Integrated bioinformatics-based subtractive genomics approach to decipher the therapeutic drug target and its possible intervention against brucellosis. Bioengineering 9 (11) Available from. Available at https://doi.org/10.3390/bioengineering9110633, www.mdpi.com/journal/bioengineering.

Khan, M.T., Mahmud, A., Iqbal, A., Hoque, S.F., Hasan, M., 2020. Subtractive genomics approach towards the identification of novel therapeutic targets against human Bartonella bacilliformis. Informatics in Medicine Unlocked 20 Available from. Available at https://doi.org/10.1016/j.imu.2020.100385, https://www.sciencedirect.com/science/journal/23529148.

Krishnamurthy, N., Grimshaw, A.A., Axson, S.A , Choe, S.H., Miller, J.E., 2022. Drug repurposing: a systematic review on root causes, barriers and facilitators. BMC Health Services Research 22 (1) Available from. Available at https://doi.org/10.1186/s12913-022-08272-z, http://www.biomedcentral.com/bmchealthservres/.

Latif, K., Ullah, A., Shkodina, A.D., Boiko, D.I., Rafique, Z., Alghamdi, B.S., et al., 2022. Drug reprofiling history and potential therapies against Parkinson's disease. Frontiers in Pharmacology 13 Available from. Available at https://doi.org/10.3389/fphar.2022.1028356, http://www.frontiersin.org/Pharmacology.

Leung, E.L., Cao, Z.-W., Jiang, Z.-H., Zhou, H., Liu, L., 2013. Network-based drug discovery by integrating systems biology and computational technologies. Briefings in Bioinformatics 14 (4), 491–505. Available at https://doi.org/10.1093/bib/bbs043.

Li, Y., She, P., Liu, Y., Guo, S., Huang, G., Xiao, D., et al., 2025. Drug repurposing: an emerging strategy in *Staphylococcus aureus* infections. Journal of Microbiology and Biotechnology 35 (10). Available at https://doi.org/10.4014/jmb.2507.07048.

Liu, P., Li, X., Luo, M., Xu, X., Su, K., Chen, S., et al., 2018. Risk factors for carbapenem-resistant *Klebsiella pneumoniae* infection: a meta-analysis. Microbial Drug Resistance 24 (2), 190–198. Available at https://doi.org/10.1089/mdr.2017.0061.

Lu, H., Liu, M., Lu, W., Wang, C., Wang, G., Dong, W., et al., 2020. Repurposing ellipticine hydrochloride to combat colistin-resistant extraintestinal pathogenic *E. coli* (ExPEC). Frontiers in Microbiology 11. Available at https://doi.org/10.3389/fmicb.2020.00806.

Lutgring, J.D., Limbago, B.M., 2016. The problem of carbapenemase-producing-carbapenem-resistant-enterobacteriaceae detection. Journal of Clinical Microbiology 54 (3), 529–534. Available from. Available at https://doi.org/10.1128/JCM.02771-15, http://jcm.asm.org/content/54/3/529.full.pdf+html.

MacGowan, A., Macnaughton, E., 2017. Antibiotic resistance. Medicine 45 (10), 622–628. Available at https://doi.org/10.1016/j.mpmed.2017.07.006.

Maia, E.H.B., Assis, L.C., de Oliveira, T.A., da Silva, A.M., Taranto, A.G., 2020. Structure-based virtual screening: from classical to artificial intelligence. Frontiers in Chemistry 8 Available from. Available at https://doi.org/10.3389/fchem.2020.00343, http://journal.frontiersin.org/journal/chemistry.

Majumder, S., Panigrahi, G.K., 2025. Advancements in contemporary pharmacological innovation: mechanistic insights and emerging trends in drug discovery and development. Intelligent Pharmacy 3 (2), 118–126. Available from. Available at https://doi.org/10.1016/j.ipha.2024.10.001, https://www.keaipublishing.com/en/journals/intelligent-pharmacy/.

Mayorga-Ramos, A., Zúñiga-Miranda, J., Carrera-Pacheco, S.E., Barba-Ostria, C., Guamán, L.P., 2023. CRISPR-cas-based antimicrobials: design, challenges, and bacterial mechanisms of resistance. ACS Infectious Diseases 9 (7), 1283–1302. Available from. Available at https://doi.org/10.1021/acsinfecdis.2c00649, http://pubs.acs.org/journal/aidcbc.

Meng, X.Y., Zhang, H.X., Mezei, M., Cui, M., 2011. Molecular docking: a powerful approach for structure-based drug discovery. Current Computer-Aided Drug Design 7 (2), 146–157. Available from. Available at https://doi.org/10.2174/157340911795677602, www.eurekaselect.com/582/journal/current-computer-aided-drug-design.

Mudgal, B., Verma, D., Venugopal, D., Atram, S.V., Mitra, D., Gupta, S., 2025. Subtractive genomics approach: a guide to unveiling therapeutic targets across pathogens. Journal of Microbiological Methods 232-234, 107127. Available at https://doi.org/10.1016/j.mimet.2025.107127.

Murthy, D., Attri, K.S., Gokhale, R.S., 2013. Network, nodes and nexus: systems approach to multitarget therapeutics. Current Opinion in Biotechnology 24 (6), 1129–1136. Available at https://doi.org/10.1016/j.copbio.2013.02.009.

Naghavi, M., Vollset, S.E., Ikuta, K.S., Swetschinski, L.R., Gray, A.P., Wool, E.E., et al., 2024. Global burden of bacterial antimicrobial resistance 1990–2021: a systematic analysis with forecasts to 2050. The Lancet 404 (10459), 1199–1226. Available from. Available at https://doi.org/10.1016/S0140-6736(24)01867-1, https://www.sciencedirect.com/science/journal/01406736.

Navon-Venezia, S., Kondratyeva, K., Carattoli, A., 2017. Klebsiella pneumoniae: a major worldwide source and shuttle for antibiotic resistance. FEMS Microbiology Reviews 41 (3), 252–275. Available at https://doi.org/10.1093/femsre/fux013.

Niederman, M.S., 2005. Principles of appropriate antibiotic use. International Journal of Antimicrobial Agents 26 (3), S170. Available from. Available at https://doi.org/10.1016/S0924-8579(05)80324-3, www.elsevier.com/locate/ijantimicag.

Olatunji, I., Bardaji, D.K.R., Miranda, R.R., Savka, M.A., Hudson, A.O., 2024. Artificial intelligence tools for the identification of antibiotic resistance genes. Frontiers in Microbiology 15 Available from. Available at https://doi.org/10.3389/fmicb.2024.1437602, https://www.frontiersin.org/journals/microbiology#.

Oselusi, S.O., Dube, P., Odugbemi, A.I., Akinyede, K.A., Ilori, T.L., Egieyeh, E., et al., 2024. The role and potential of computer-aided drug discovery strategies in the discovery of novel antimicrobials. Computers in Biology and Medicine 169 Available from. Available at https://doi.org/10.1016/j.compbiomed.2024.107927, https://www.sciencedirect.com/science/journal/00104825.

Paananen, J., Fortino, V., 2020. An omics perspective on drug target discovery platforms. Briefings in Bioinformatics 21 (6), 1937–1953. Available from. Available at https://doi.org/10.1093/bib/bbz122, http://bib.oxfordjournals.org.

Parvathaneni, V., Kulkarni, N.S., Muth, A., Gupta, V., 2019. Drug repurposing: a promising tool to accelerate the drug discovery process. Drug Discovery Today 24 (10), 2076–2085. Available from. Available at https://doi.org/10.1016/j.drudis.2019.06.014, www.elsevier.com/locate/drugdiscov.

Pascucci, M., Royer, G., Adamek, J., Asmar, M.A., Aristizabal, D., Blanche, L., et al., 2021. AI-based mobile application to fight antibiotic resistance. Nature Communications 12 (1) Available from. Available at https://doi.org/10.1038/s41467-021-21187-3, http://www.nature.com/ncomms/index.html.

Patel, J., Sridhar, D., 2022. The pandemic legacy of antimicrobial resistance in the USA. The Lancet Microbe 3 (10), e726. Available at https://doi.org/10.1016/s2666-5247(22)00227-0.

Paterson, D.L., Bonomo, R.A., 2005. Extended-spectrum β-lactamases: a clinical update. Clinical Microbiology Reviews 18 (4), 657–686. Available at https://doi.org/10.1128/CMR.18.4.657-686.2005.

Piddock, L.J.V., 2006. Clinically relevant chromosomally encoded multidrug resistance efflux pumps in bacteria. Clinical Microbiology Reviews 19 (2), 382–402. Available at https://doi.org/10.1128/CMR.19.2.382-402.2006.

Pisano, L., Giovannuzzi, S., Supuran, C.T., 2024. Management of *Neisseria gonorrhoeae* infection: from drug resistance to drug repurposing. Expert Opinion on Therapeutic Patents 34 (6), 511–524. Available from. Available at https://doi.org/10.1080/13543776.2024.2367005, http://www.tandfonline.com/loi/ietp20.

Projan, S.J., Shlaes, D.M., 2004. Antibacterial drug discovery: is it all downhill from here? Clinical Microbiology and Infection 10 (4), 18–22. Available at https://doi.org/10.1111/j.1465-0691.2004.1006.x.

Pun, F.W., Ozerov, I.V., Zhavoronkov, A., 2023. AI-powered therapeutic target discovery. Trends in Pharmacological Sciences 44 (9), 561–572. Available from. Available at https://doi.org/10.1016/j.tips.2023.06.010, www.elsevier.com/locate/tips.

Pushpakom, S., Iorio, F., Eyers, P.A., Escott, K.J., Hopper, S., Wells, A., et al., 2018. Drug repurposing: progress, challenges and recommendations. Nature Reviews. Drug Discovery 18 (1), 41–58. Available from. Available at https://doi.org/10.1038/nrd.2018.168, http://www.nature.com/nrd/index.html.

Raman, K., Yeturu, K., Chandra, N., 2008. targetTB: a target identification pipeline for *Mycobacterium tuberculosis* through an interactome, reactome and genome-scale structural analysis. BMC Systems Biology 2. Available at https://doi.org/10.1186/1752-0509-2-109.

Rathored, J., Budhbaware, T., 2025. Integrative strategies against multidrug-resistant bacteria: synthesizing novel antimicrobial frontiers for global health. Microbial Pathogenesis 208, 108018. Available at https://doi.org/10.1016/j.micpath.2025.108018.

Raval, K., Ganatra, T., 2022. Basics, types and applications of molecular docking: a review. IP International Journal of Comprehensive and Advanced Pharmacology 7 (1), 12–16. Available at https://doi.org/10.18231/j.ijcaap.2022.003.

Rossolini, G.M., Arena, F., Pecile, P., Pollini, S., 2014. Update on the antibiotic resistance crisis. Current Opinion in Pharmacology 18, 56–60. Available at https://doi.org/10.1016/j.coph.2014.09.006.

Sahoo, B.M., Ravi Kumar, B.V.V., Sruti, J., Mahapatra, M.K., Banik, B.K., Borah, P., 2021. Drug repurposing strategy (DRS): emerging approach to identify potential therapeutics for treatment of novel coronavirus infection. Frontiers in Molecular Biosciences 8. Available at https://doi.org/10.3389/fmolb.2021.628144 Available from: journal.frontiersin.org/journal/molecular-biosciences.

Salvato, F., Hettich, R.L., Kleiner, M., 2021. Five key aspects of metaproteomics as a tool to understand functional interactions. in Host Associated Microbiomes Advances in Clinical Immunology, Medical Microbiology, COVID-19, and Big Data. Jenny Stanford Publishing, United States, pp. 647–660. Available from. Available at https://www.jennystanford.com/9789814877848/advances-in-clinical-immunology-medical-microbiology-covid-19-and-big-data/.

Serral, F., Castello, F.A., Sosa, E.J., Pardo, A.M., Palumbo, M.C., Modenutti, C., et al., 2021. From genome to drugs: new approaches in antimicrobial discovery. Frontiers in Pharmacology 12 Available from. Available at https://doi.org/10.3389/fphar.2021.647060, http://www.frontiersin.org/Pharmacology.

Shafique, L., Khan, M.F., Parveen, S., Xu, Y., Zhu, P., 2025. Overcoming multidrug resistance in *E. coli* and *Salmonella* Isolates from Nile tilapia: synergistic effects of novel antibiotic combinations. Molecular Biotechnology Available from. Available at https://doi.org/10.1007/s12033-025-01462-0, https://www.springer.com/journal/12033.

Shahid, F., Shehroz, M., Zaheer, T., Ali, A., 2019. Subtractive genomics approaches: towards anti-bacterial drug discovery. Frontiers in Anti-infective Drug Discovery 8, 144–158. Available from. Available at https://doi.org/10.2174/9789811412387120080007, https://www.eurekaselect.com/173730/volume/8.

Shao, Y., He, X., Harrison, E.M., Tai, C., Ou, H.Y., Rajakumar, K., et al., 2010. mGenomeSubtractor: a web-based tool for parallel in silico subtractive hybridization analysis of multiple bacterial genomes. Nucleic Acids Research 38 (2), W194. Available at https://doi.org/10.1093/nar/gkq326.

Sharma, P.P., Bansal, M., Sethi, A., Poonam, Pena, L., Goel, V.K., et al., 2021. Computational methods directed towards drug repurposing for COVID-19: advantages and limitations. RSC Advances 11 (57), 36181–36198. Available from. Available at https://doi.org/10.1039/d1ra05320e, http://pubs.rsc.org/en/journals/journal/ra.

Sharma, S., Singh, R., Kant, S., Mishra, M.K., 2025. Integrating AI/ML and multi-omics approaches to investigate the role of TNFRSF10A/TRAILR1 and its potential targets in pancreatic cancer. Computers in Biology and Medicine 193, 110432. Available at https://doi.org/10.1016/j.compbiomed.2025.110432.

She, P., Li, Y., Li, Z., Liu, S., Yang, Y., Li, L., et al., 2023a. Repurposing 9-aminoacridine as an adjuvant enhances the antimicrobial effects of rifampin against multidrug-resistant *Klebsiella pneumoniae*. Microbiology Spectrum 11 (3) Available from. Available at https://doi.org/10.1128/spectrum.04474-22, https://journals.asm.org/doi/pdf/10.1128/spectrum.04474-22.

She, P., Yang, Y., Li, L., Li, Y., Liu, S., Li, Z. et al., 2023b. Repurposing of the antimalarial agent tafenoquine to combat MRSA. mSystems 8 (6) Available from. Available at https://doi.org/10.1128/msystems.01026-23, https://journals.asm.org/doi/10.1128/msystems.01026-23.

Shulga, D.A., Ivanov, N.N., Palyulin, V.A., 2022 In silico structure-based approach for group efficiency estimation in fragment-based drug design using evaluation of fragment contributions. Molecules (Basel, Switzerland) 27 (6) Available from. Available at https://doi.org/10.3390/molecules27061985, https://www.mdpi.com/1420-3049/27/6/1985/pdf.

Szklarczyk, D., Gable, A.L., Lyon, D., Junge, A , Wyder, S., Huerta-Cepas, J., et al., 2019. STRING v11: protein-protein association networks with increased coverage, supporting functional discovery in genome-wide experimental datasets. Nucleic Acids Research 47 (1), D607. Available from. Available at https://doi.org/10.1093/nar/gky1131, https://academic.oup.com/nar/issue.

Talat, A., Bashir, Y., Khan, A.U., 2022. Repurposing of antibiotics: sense or non-sense. Frontiers in Pharmacology 13 Available from. Available at https://doi.org/10.3389/fphar.2022.833005, http://www.frontiersin.org/Pharmacology.

Talat, A., Khan, A.U., 2023. Artificial intelligence as a smart approach to develop antimicrobial drug molecules: a paradigm to combat drug-resistant infections. Drug Discovery Today 28 (4), 103491. Available at https://doi.org/10.1016/j.drudis.2023.103491.

Tanoli, Z., Alam, Z., Ianevski, A., Wennerberg, K., Vähä-Koskela, M., Aittokallio, T., 2018. Interactive visual analysis of drug-target interaction networks using drug target profiler, with applications to precision medicine and drug repurposing. Briefings in Bioinformatics 21 (1), 211–220. Available from. Available at https://doi.org/10.1093/bib/bby119, http://bib.oxfordjournals.org.

Theuretzbacher, U., Baraldi, E., Ciabuschi, F., Callegari, S., 2023. Challenges and shortcomings of antibacterial discovery projects. Clinical Microbiology and Infection 29 (5), 610–615. Available at https://doi.org/10.1016/j.cmi.2022.11.027.

van Montfort, R.L.M., Workman, P., van Montfort, R.L.M., Workman, P., 2017. Structure-based drug design: aiming for a perfect fit. Essays in Biochemistry 61 (5), 431–437. Available at https://doi.org/10.1042/ebc20170052.

Vaara, M., 2019. Polymyxins and their potential next generation as therapeutic antibiotics. Frontiers in Microbiology 10 (JULY). Available at https://doi.org/10.3389/fmicb.2019.01689.

Verma, A.K., Aggarwal, R., 2021. Repurposing potential of FDA-approved and investigational drugs for COVID-19 targeting SARS-CoV-2 spike and main protease and validation by machine learning algorithm. Chemical Biology and Drug Design 97 (4), 836–853. Available from. Available at https://doi.org/10.1111/cbdd.13812, http://onlinelibrary.wiley.com/journal/10.1111/(ISSN)1747-0285.

Vora, N., Shah, S., Patel, P., Shah, M., 2025. Artificial intelligence and multi-omics in drug discovery: a deep learning-powered revolution. Cure & Care 1 (1), 100011. Available at https://doi.org/10.1016/j.ccwv.2025.100011.

Wang, J.H., Byun, J., Pennathur, S., 2010. Analytical approaches to metabolomics and applications to systems biology. Seminars in Nephrology 30 (5), 500–511. Available at https://doi.org/10.1016/j.semnephrol.2010.07.007.

Wang, T., Wu, M.B., Zhang, R.H., Chen, Z.J., Hua, C., Lin, J.P., et al., 2016. Advances in computational structure-based drug design and application in drug discovery. Current Topics in Medicinal Chemistry 16 (9), 901–916. Available from. Available at https://doi.org/10.2174/1568026615666150825142002, http://www.eurekaselect.com/615/journal/current-topics-medicinal-chemistry.

Wang, X., Wei, X., Thijssen, B., Das, J., Lipkin, S.M., Yu, H., 2012. Three-dimensional reconstruction of protein networks provides insight into human genetic disease. Nature Biotechnology 30 (2), 159–164. Available at https://doi.org/10.1038/nbt.2106.

Wu, X., Li, Z., Li, X., Tian, Y., Fan, Y., Yu, C., et al., 2017. Synergistic effects of antimicrobial peptide DP7 combined with antibiotics against multidrug-resistant bacteria. Drug Design, Development and Therapy 11, 939–946. Available from. Available at https://doi.org/10.2147/DDDT.S107195, https://www.dovepress.com/getfile.php?fileID=35608.

Xing, H., Gardner, T.S., 2007. The mode-of-action by network identification (MNI) algorithm: a network biology approach for molecular target identification. Nature Protocols 1 (6), 2551–2554. Available at https://doi.org/10.1038/nprot.2006.300.

Yacouba, A., Olowo-okere, A., Yunusa, I., 2021. Repurposing of antibiotics for clinical management of COVID-19: a narrative review. Annals of Clinical Microbiology and Antimicrobials 20 (1) Available from. Available at https://doi.org/10.1186/s12941-021-00444-9, https://ann-clinmicrob.biomedcentral.com/.

Yadav, B.S., Tripathi, V., 2018. Recent advances in the system biology-based target identification and drug discovery. Current Topics in Medicinal Chemistry 18 (20), 1737–1744. Available from. Available at https://doi.org/10.2174/1568026618666181025112344, http://www.eurekaselect.com/166616/article.

You, J., McLeod, R.D., Hu, P., 2019. Predicting drug-target interaction network using deep learning model. Computational Biology and Chemistry 80, 90–101. Available at https://doi.org/10.1016/j.compbiolchem.2019.03.016.

Yu, Y., Zhao, H., Lin, J., Li, Z., Tian, G., Yang, Y.Y., et al., 2022. Repurposing non-antibiotic drugs auranofin and pentamidine in combination to combat multidrug-resistant gram-negative bacteria. International Journal of Antimicrobial Agents 59 (5), 106582. Available at https://doi.org/10.1016/j.ijantimicag.2022.106582.

Zhang, W., Ran, J., Shang, L., Zhang, L., Wang, M., Fei, C., et al., 2022. Niclosamide as a repurposing drug against Gram-positive bacterial infections. Journal of Antimicrobial Chemotherapy 77 (12), 3312–3320. Available from. Available at https://doi.org/10.1093/jac/dkac319, http://jac.oxfordjournals.org/.

Zhou, Y., Zhang, Y., Zhao, D., Yu, X., Shen, X., Zhou, Y., et al., 2024. TTD: therapeutic target database describing target druggability information. Nucleic Acids Research 52 (D1), D1465. Available at https://doi.org/10.1093/nar/gkad751.

Summary: a concluding overview

The introduction of antibiotics in the mid-20th century marked one of medicine's most spectacular successes, converting once lethal infections into manageable diseases and paving the way for modern surgery, cancer chemotherapy and organ transplantation. But within decades of the discovery of penicillin by Alexander Fleming in 1928, these wonder drugs faced increasingly sophisticated opponents, as microorganisms showed an astonishing capacity to adapt and evolve mechanisms to evade, neutralise and survive antibiotic attack. Nowadays, antimicrobial resistance (AMR) is recognised as one of the most urgent global health challenges of the 21st century, and it could undermine a century of medical progress by ushering in a postantibiotic era in which common infections could once again kill mankind. In this book, contributors take an interdisciplinary examination of AMR, scientific, clinical, environmental, economic and social-political contexts, across the spectrum, using general principles to provide an integrated understanding of these multifaceted problematic microbial and humane challenges and the strategies to mitigate their impact, action by action solutions.

The response of microorganisms, which have evolved over billions of years to survive in the most hostile environments, is the main issue with AMR. Chapter 1 presents an introduction to the common mechanisms by which bacteria gain resistance and includes enzyme-mediated degradation, modification of drug targets, alteration of membrane permeability and activation of efflux pumps to extrude antimicrobial agents from within bacterial cells. Mechanisms of resistance are not just academic curiosities but have important clinical implications, directly impacting treatment failures, increased length of stay, higher mortality rates and greater healthcare costs that negatively impact health systems globally. These foundational concepts are necessary for context about the complexity and heterogeneity of the resistance crisis and potential multifaceted solutions. AMR is in reality a constellation of interrelated problems that spans molecular biology, clinical medicine, public health, agriculture, environmental science, economics and policy. In Chapter 2, this is expanded, first with a discussion of the genetic and biochemical basis of resistance, then turning attention to how bacteria can acquire and transfer resistance genes between themselves by horizontal gene transfer mechanisms such as conjugation, transformation and transduction. This molecular machinery for resistance, ranging from beta-lactamases that hydrolyse penicillin-based antibiotics to aminoglycoside-modifying enzymes and even altered penicillin-binding proteins, represents eons of microbial evolution, compressed into human timeframes by the selective pressure of antibiotic use. New studies employing genomic surveillance approaches have provided information about the complex genetic architecture of resistance determinants, as demonstrated by the involvement of mobile genetic elements (plasmids, transposons and integrons) in mediating rapid dissemination of antibiotic resistance traits among disparate species and genera. This chapter discusses the powerful genes for resistance to one or multiple classes of antimicrobials, which have moved from environmental bacteria into important human pathogens, and these organisms are capable of becoming multidrug resistant, thus challenging even our most highly active antimicrobial agents. In Chapter 3, resistance is approached from the perspective of evolutionary biology. Antibiotic use exerts harsh selective pressures on bacteria, which encourages wily resistant forms. The phrase 'survival of the fittest' seems to gain immediate contemporary urgency as we observe adaptive evolution of antibiotic-resistant bacteria in the clinical, agricultural and environmental environments in real time, with the emergence and spread of resistant populations across locations. Resistance reduces susceptibility to antibiotics but is complicated by the trade-off between resistance and fitness, compensatory mutations that can restore bacteria's fitness without restoring sensitivity and survival of phase variants called persister cells, which survive antibiotic treatment in the absence of genetic or phenotypic resistance. Recognising these evolutionary dynamics is fundamental to antimicrobial stewardship and the design of new therapeutic strategies that capture bacterial adaptation. This chapter reviews mathematical modelling approaches and experimental evolution studies that are beginning to inform our understanding of resistant trajectories towards emerging strains.

Chapter 4 explores the human causes of the resistance crisis, focusing on antibiotic misuse and overuse. Inappropriate prescriptions, such as using antibiotics for viral infections, improper dosing regimens and early treatment discontinuation, lead to suboptimal drug concentrations, which are favourable regions for resistance selection. The patient's noncompliance with antibiotic prescriptions, which results in the survival and expansion of partially resistant bacterial populations by virtue of not carrying out the complete antibiotic regimen. This chapter explores the many psychological, social and systems determinants that compel inappropriate antibiotic use, encompassing patient expectations of healthcare providers to prescribe antibiotics. Physician prescribing patterns and norms, marketing of antimicrobials by pharmaceutical manufacturers to prescribers and patients, pay-for-performance incentives that reward treatment over prevention, and public health failures to adequately describe the consequences of AMR as a global crisis across sectors.

The environment is not only a major reservoir of resistant bacteria and resistance genes but also an important transmission route for them, as discussed in Chapter 5. Wastewater treatment plants, agricultural runoff, effluents of pharmaceutical manufacturing and contaminated soil and water bodies act as silent hot spots where resistant bacteria from human, animal and environmental sources are combined by gene exchange to be released back into the environment. Until recently, environmental aspects of AMR had been underappreciated, but growing evidence suggests that environmental bacteria act both as reservoirs for resistance genes and as drivers of resistance evolution and dissemination. Concentration of residual antibiotics in the environment puts selective pressure that is responsible for persisting and enhancing resistance, while environmental factors favour horizontal transfer among highly diversified microbial communities. This chapter uses a One Health perspective since the crisis of resistance cannot be understood without taking into account human, animal and environmental health.

As discussed in Chapter 6, contrary to their mission of healing, healthcare facilities have become hotbeds for the emergence and spread of resistance. Multidrug-resistant organisms acquired in the hospital are associated with morbidity and mortality in these vulnerable populations, such as methicillin-resistant *Staphylococcus aureus*, vancomycin-resistant *Enterococcus* and carbapenem-resistant Enterobacteriaceae. The presence of antibiotics, immunocompromised patients, the use of invasive medical devices and opportunities for cross-transmission make hospital settings particularly ripe for resistance selection and dissemination. This chapter addresses the complex aetiology of healthcare-associated AMR nurtured by the various contributory drivers of inadequate infection control practices, shortcomings in antibiotic stewardship and the costly challenges associated with instilling evidence-based interventions to combat AMR within resource-constrained settings. Approaches to overcome hospital-based resistance, including improved surveillance and isolation measures, antimicrobial stewardship programmes and decontamination of the environment, are thoroughly discussed. As described in Chapter 7, the challenge of resistance goes well beyond hospitals and healthcare settings to community environments where increasingly resistant infections are affecting otherwise healthy individuals. Although infections caused by antibiotic-resistant bacteria are less common than community-acquired infections, they have reached a concerning threshold in several parts of the world, particularly among children with ear infections and adults with urinary tract, skin, and soft tissue infections. Antibiotic use in the community (any or outpatients), person-to-person spread and environmental exposure (contamination of resistant imports abroad and incoming persons by immigration and travel) contribute to a reservoir of resistance. This chapter examines community-based resistance and some of its unique challenges, focusing on the barriers to application of stewardship interventions in outpatient settings (which often are geographically dispersed), how antibiotic availability without prescription has affected natural selection processes for more resistant strains and how public education is important to instilling appropriate behaviours regarding antibiotic use. Chapter 8 expands this view further still, to consider resistance within the community setting: schools and daycare facilities; retail food establishments; public transport; swimming pools. Although not recognised as a high-risk setting for AMR, these settings are typically characterised by the aggregation of people, potential transmission of pathogens and links to contaminated environmental sources, which then contribute to a collective burden of resistance. In this chapter, the author points out that fighting resistance cannot be limited to clinical and agricultural hot spots. It must include the many other environments where humans contact microorganisms and where resistant bacteria can establish themselves in communities. While it is crucial to diagnose resistant infections in a timely manner for both adequate treatment and infection control, the potential diagnostic hurdles are considerable (Chapter 9). Methodology based on traditional culture methods needs days to detect pathogens and assess antimicrobial susceptibility; thus patients still receive empirical therapy, which is adequate only in a reduced percentage of cases and promotes poor outcomes and further resistant selection processes. This chapter reviews new technologies for rapid diagnosis, particularly molecular techniques that identify resistance genes and molecular techniques for pathogen identification, as well as point-of-care tests that can guide therapeutic choices within hours instead of days. Besides this, the coupling of genomic surveillance with bioinformatics tools is transforming our capacity to monitor resistance spread, predict resistance phenotypes from genetic data and develop precision antimicrobial treatment strategies for individual infections.

New antibiotics are a crucial part of the solution to AMR. The pipeline for new antibiotics has been starkly empty in recent decades (Chapter 10). Pharmaceutical investment has been stunted on the back of scientific, regulatory and economic challenges in antibiotic discovery and development, producing a market failure that threatens both our ability to keep pace with changing resistance. Proposed new directions for antibiotic discovery include studies of the vast diversity of previously unculturable environmental bacteria, the use of artificial intelligence and machine learning to discover novel microbial compounds with activity against antimicrobial-resistant pathogens, and rational design of antibiotics that target essential bacterial processes. New ways to run clinical trials, provide regulatory headroom for new antibiotics and deliver public–private partnerships are heading to revive the dwindling antibiotic pipeline while ensuring that new drugs arrive where they will do most good, in the treatment of patients. Chapter 11, aware that new antibiotics will not be enough to solve the resistance crisis, looks at alternative routes to therapeutic targets that do not directly use conventional antibiotic mechanisms. After decades tucked away in a drawer, phage therapy, the use of bacterial viruses to specifically eliminate disease-causing bacteria, is setting off as an alternate route. Antimicrobial peptides from the innate immune systems of diverse organisms provide unique mechanisms of action with diminished potential for resistance. New paradigms to prevent and cure infectious diseases have emerged with immunotherapies enhancing host defences, antivirulence strategies disarming pathogens without killing them, and microbiome-based interventions restoring protective microbial communities. This chapter examines the potential value and limitations of these new therapies from scientific validation through regulatory approval to clinical implementation.

As always, prevention is better than cure; the key role of vaccines and other preventative measures in decreasing the burden of infections and hence the requirements for antibiotics is discussed further in Chapter 12. Data from vaccines against such bacterial pathogens as *Streptococcus pneumoniae* and *Haemophilus influenzae* indicate that effective vaccination significantly reduces antibiotic use and resistance rates in the vaccinated populations. Vaccines against other resistant pathogens and vaccines against major pathogens could have a great influence on resistance. Outside of vaccination, preventing transmission through better sanitation, access to clean drinking water, practice guidelines for infection control and even public health interventions that prevent disease transmission all play a role in lessening consumption of antibiotics as well as the selection pressure for resistance.

Antimicrobial stewardship, which is the systematic effort to optimise antibiotic use, as discussed in Chapter 13, is one of the approaches to the mitigation of resistance. Stewardship programmes aimed at optimising antibiotic use are comprehensive and implement evidence-based interventions to evaluate the need for an antibiotic in the first place and ensure the most appropriate agent is selected and optimal dosing and duration. This chapter considers stewardship strategies across the continuum of care from dedicated infectious disease pharmacist-focused programmes in a hospital setting through primary care interventions and community-wide campaigns. There is strong evidence on the effectiveness of stewardship, with substantial decreases seen for antibiotic consumption and resistance rates in both humans and livestock, as well as associated healthcare costs and adverse drug events. Implementation remains challenging, especially in low-resource settings where stewardship infrastructure and expertise may be limited.

Chapter 14 examines how effective governance and policy frameworks are vital for coordinating responses to resistance across sectors and regions. AMR knows no borders, increasing the need for global cooperation. WHO is developing a Global Action Plan on Antimicrobial Resistance, which will provide a guide to developing national action plans, surveillance networks and coordinated research. Policy challenges that are reviewed include ensuring appropriate regulation of antibiotic use in human and animal health, enforcement of prescription-only access, incentivising antibiotic innovation and equitable distribution of antimicrobials. The challenge of balancing access to life-saving antibiotics with conservation for retention of effect has been a core policy tension, particularly in low- and-middle income countries where the burden from infectious disease remains high. In Chapter 15, the landscape of poor access to effective antimicrobials and resistance is discussed as a perfect storm. In certain markets, poor quality and falsified antibiotics result in treatment failures and selection for resistance. Inadequate laboratory facilities restrict the surveillance and necessary treatment. This chapter focuses on solutions for low- and middle-income countries that are appropriate in context, including system-level improvements, rapid access to diagnostic capacity and implementation of sustainable stewardship programmes with equitable access to both existing and novel antimicrobials.

In Chapter 16, the pharmaceutical industry is central to both the resistance problem and potential solutions. The high costs of research and development, the long pathway for regulatory approval and a limited market opportunity due to stewardship practices that limit use have left development prospects unsatisfyingly slim, with many pharmaceutical companies abandoning their research efforts into antibiotics. This chapter describes how the practice of an industry promoting its products from this marketing to releasing manufacturing waste containing active pharmaceutical ingredients into the environment has led to resistance.

As detailed in Chapter 17, agriculture is a large global user of antibiotics, where over half of the total antimicrobial production used is directed to livestock and aquaculture. Antimicrobial use in food animals for growth promotion and disease prevention generates selective pressures favouring the development of resistance in bacterial pathogens, commensal bacteria and environmental microorganisms. Antibiotic-resistant pathogens and resistance genes spread to humans primarily by direct contact, through the consumption of contaminated food products (transfer along the food chain) and through environmental pathways such as manure application onto agricultural land and runoff movement into water systems. This chapter studies the evidence for a link between agricultural use of antibiotics and human health impacts. It describes various regulatory approaches taken in different countries to restrict such agricultural use of antibiotics, as well as alternatives to support animal health and productivity without antimicrobial growth promoters.

Attitudes, cultural beliefs, social norms and habitual behaviours have a significant impact on antibiotic use and resistance, as addressed in Chapter 18. In some cultures, antibiotics are seen as a miracle cure for any disease, whereas in other places, traditional medicine customs may take precedence and delay or avoid taking proper antibiotics. Antibiotic consumption patterns are shaped by a multitude of factors, including patient expectations for receiving antibiotic prescriptions, physician prescribing behaviours driven by time pressures and their assessment of uncertainty in diagnosing bacterial infection, and community-level norms around sharing medications prescribed to others or self-medicating. This chapter analyses the way cultural contexts shape antibiotic use and efforts to promote appropriate use through culturally tailored interventions. A better understanding of these behavioural and cultural dimensions will improve stewardship as well as efforts to engage the public.

As outlined in Chapter 19, infection prevention and control measures are among the most basic approaches to the reduction of antibiotic usage through preventing infections from occurring in the first place. Hygiene practice improvements, including hand hygiene, environmental cleaning and healthcare facility design to reduce transmission, are already evidence-based strategies that have shown efficacy in infection control and antibiotic use. Infection prevention is much more than a technical challenge but rather a behavioural and organisational one that requires commitment over time, training and culture change, as this chapter notes. Chapter 20 analyses the economic and social implications of antimicrobial resistance, which go well beyond health care costs. The resistance is associated with loss of productivity, economic output, increased poverty and deepening health inequities. According to estimates from the World Bank, AMR could inflict damage to the economy that would rival the financial crisis, with its worst fallout expected in the poorest parts of our world. The social implications of resistance include the moral toll suffered by patients with resistant infections, ethical questions of prioritising limited but effective antibiotics and potential tensions inherent in utilising antibiotic expansion as a means to reduce health inequities. This chapter explores economic evaluations of the consequences of resistance and ways to counteract it, showing that controlling infections by investing in prevention, stewardship and innovation is cost-saving compared with a scenario without any control measures.

Finally, target-centric drug repurposing based on the identification of existing approved drugs that can be considered as new potential antimicrobials is described in Chapter 21 with an emphasis on overcoming resistance mechanisms directed against essential bacterial functions. Such a strategy for the identification of potential indications has various benefits such as shorter timelines, known safety profiles and lower costs compared to de novo discovery. Researchers are discovering surprising antibacterial activities in drugs designed for other therapeutic purposes by systematically screening existing libraries of small molecules against bacterial targets and resistance mechanisms. This pragmatic next-generation approach utilises currently available pharma-era compounds to rapidly expand our arsenal of antimicrobials.

Together, these 21 chapters provide an overview of the problem, spanning from mechanistic science all the way to global policy, from the hospital ward to the agricultural field, from individual behaviour to international governance. Understanding the complexity of AMR, this book displays that AMR cannot be solved with piecemeal interventions in any particular sector or discipline. Integrated and multisectoral approaches are required to address the complex web of drivers of resistance emergence and spread. The One Health framework, which embraces the interconnections of human health with animal and environmental health, offers a conceptual basis for dealing with resistance in a comprehensive way. This book will discuss tensions of access and excess, the need for progress and conservancy, global solidarity equity and urgent action across sectors and geographies.

The resistance crisis we face today is, by large, the result of our own actions with decades of misuse and overuse of antibiotics in human medicine, agriculture and aquaculture, nested with limited infection prevention, investment in new antimicrobials and environmental dispersal of antibiotics and resistant bacteria. However, this crisis is also an opportunity to reimagine our relationship with the microorganisms that share our planet and our bodies, to develop more sustainable approaches to prevention and treatment of infectious disease and to create health systems better suited to the emerging threats. We know the solutions: better stewardship, better prevention, new therapeutics and vaccines, surveillance and evidence-based policy. What is now needed is a combined desire to realise that these solutions at scale are funded. So that they can be implemented across the board, and an ongoing dedication to maintaining antimicrobial functionality for generations to come.

Index

Note: Page numbers followed by "*f*" and "*t*" refer to figures and tables, respectively.

B

C

D

E

F

G

S

T

U

V

W

Z

9780443447488